Springer Series in Information Sciences

3

Editor: Manfred R. Schroeder

Springer Series in Information Sciences

Editors: King-sun Fu   Thomas S. Huang   Manfred R. Schroeder

Wolfgang Hess

Pitch Determination of Speech Signals

Algorithms and Devices

With 254 Figures

Springer-Verlag
Berlin Heidelberg New York Tokyo 1983

Dr.-Ing.-habil. Wolfgang Hess

Lehrstuhl für Datenverarbeitung im Institut für Informationstechnik,
Technische Universität München, Arcisstraße 21,
D-8000 München 2, Fed. Rep. of Germany

Series Editors:

Professor King-sun Fu

School of Electrical Engineering, Purdue University,
West Lafayette, IN 47907, USA

Professor Thomas S. Huang

Department of Electrical Engineering and Coordinated Science Laboratory,
University of Illinois, Urbana, IL 61801, USA

Professor Dr. Manfred R. Schroeder

Drittes Physikalisches Institut, Universität Göttingen, Bürgerstraße 42–44,
D-3400 Göttingen, Fed. Rep. of Germany

ISBN 3-540-11933-7 Springer-Verlag Berlin Heidelberg New York Tokyo
ISBN 0-387-11933-7 Springer-Verlag New York Heidelberg Berlin Tokyo

Library of Congress Cataloging in Publication Data. Hess, W. (Wolfgang), 1940-. Pitch determination of speech signals. (Springer series in information sciences ; v. 3). Bibliography: p. Includes index. 1. Speech processing systems. I. Title. II. Series. TK7882.S65H47 1983 621.389'598 83-450

Offset printing: Beltz Offsetdruck, 6944 Hemsbach/Bergstr. Bookbinding: J. Schäffer OHG, 6718 Grünstadt
2153/3130-543210

To my wife Helga
and to my children Patrizia and Eurydice
for their love, encouragement, and support

Preface

Pitch (i.e., fundamental frequency F_0 and fundamental period T_0) occupies a key position in the acoustic speech signal. The prosodic information of an utterance is predominantly determined by this parameter. The ear is more sensitive to changes of fundamental frequency than to changes of other speech signal parameters by an order of magnitude. The quality of vocoded speech is essentially influenced by the quality and faultlessness of the pitch measurement. Hence the importance of this parameter necessitates using good and reliable measurement methods.

At first glance the task looks simple: one just has to detect the fundamental frequency or period of a quasi-periodic signal. For a number of reasons, however, the task of pitch determination has to be counted among the most difficult problems in speech analysis.

1) In principle, speech is a nonstationary process; the momentary position of the vocal tract may change abruptly at any time. This leads to drastic variations in the temporal structure of the signal, even between subsequent pitch periods, and assuming a quasi-periodic signal is often far from realistic.

2) Due to the flexibility of the human vocal tract and the wide variety of voices, there exist a multitude of possible temporal structures. Narrow-band formants at low harmonics (especially at the second or third harmonic) are an additional source of difficulty.

3) For an arbitrary speech signal uttered by an unknown speaker, the fundamental frequency can vary over a range of almost four octaves (50 to 800 Hz).

4) The glottal excitation signal itself is not always regular. Even under normal conditions, i.e., when the voice is neither hoarse nor pathologic, the glottal waveform exhibits occasional irregularities. In addition, the voice may temporarily fall into vocal fry, which is a nonpathologic mode of voice excitation with irregular intervals between subsequent glottal pulses.

5) Additional problems arise in speech communication systems, where the signal is often distorted or band limited.

Literally hundreds of pitch-determination methods and algorithms have been developed. None of them work perfectly for every voice, application, and environmental condition. A great number of these methods, however, operate rather well when the external conditions are specified more precisely.

This book presents a detailed survey of the principal methods of pitch determination as well as a discussion of individual solutions. The problem of pitch determination is looked at in three different ways.

1) The problem is discussed from the point of view of the *process of speech communication.* Speech is a means of human communication; it is an acoustic signal generated by a human speaker in order to be perceived by a human listener. We can thus look at the speech signal from the point of view of speech production as well as of speech perception, or we can regard it as an ordinary acoustic signal without taking into account its origin or destination. In doing so, we arrive at different representations of the parameter *pitch,* such as *rate of vocal-fold vibration, elapsed time between adjacent glottal pulses, fundamental frequency* and *period,* or *perceived* (spectral or virtual) *pitch.* For a stationary periodic signal all these definitions are uniquely related; the speech signal, however, which is time variant, causes them to diverge. Temporary failure of a particular method is thus often due to a mismatch between the underlying definition of pitch and the signal. The question of defining the parameter *pitch* constitutes the main criterion for categorizing pitch determination methods; the primary categorization (Sect.2.6) separates *time–domain* pitch determination methods (Chap.6), which process the signal period by period, and *short–term analysis* methods (Chap.8), which deal with the signal in a more global way. The various definitions of the parameter *pitch* are explicitly formulated in Section 9.1.1. To better understand how pitch is produced and how it is perceived, Chaps.3 and 4 provide a short introduction to the mechanisms of the voice source and to pitch perception.

2) The problem is discussed from the point of view of the respective *application.* There are four main areas of application for pitch determination algorithms: a) *speech communication;* b) *phonetics, linguistics,* and *musicology;* c) *education;* and d) *phoniatrics* and *medicine.* With respect to the problem of pitch determination this means that the requirements on the respective algorithms are rather diverse, if not contradictory. The problem is briefly discussed in Sect.1.2; a more detailed discussion follows in Section 9.4. Prospective readers of the present book, if they come from all these areas of application, thus have rather different scientific backgrounds. On the other hand, describing all the methods of pitch determination in a unified way can only be done in terms of *digital signal processing.* A great number of algorithms cannot run other than digitally; however, any method realized in analog technology can be implemented in a digital way as well. Hence this book contains a short tutorial introduction to digital signal processing (Chap.2); more details on short-term analysis and spectral transformations are added later (Sect.8.1).

3) The problem is discussed from a *functional* point of view. Each pitch determination algorithm is subdivided into the three functional blocks *preprocessor, basic extractor,* and *postprocessor* (Sect.2.6). For each of these blocks there exist a number of specific methods which may improve the performance when they are combined, but which may also be mutually exclusive. Arranging the pitch determination methods according to this criterion forms an alternative to categorization according to 1) and is discussed in Section 9.1.

Voicing determination is a necessary prerequisite for pitch determination, i.e., it is necessary to locate the segments in the signal where pitch is present. Although the tasks of voicing and pitch determination can be combined in one algorithm, they are two separate tasks. Voicing determination is not the main topic of this book; it is reviewed in Section 5.3.

A survey of *instrumental* methods of pitch determination, which derive the information on pitch from the glottal excitation signal without analyzing the speech signal, completes the text (Sect.5.2). These methods have found wide application in phoniatrics; yet they can be used in other areas of application as well.

This book is the outgrowth of a thesis written in order to obtain the *venia legendi* in digital signal processing at the Technical University of Munich. I am indebted to the thesis supervisors, Professor T. Einsele and Professor E. Terhardt, for their careful review and constructive criticism of this work. My special thanks go to Professor Terhardt for his valuable suggestions with respect to the survey of frequency-domain PDAs (Sect.8.5) and the review of pitch perception theories (Chap.4). Other colleagues who have been involved directly with the preparation of this book include Mr. H. Dettweiler who compiled a part of the bibliography. Dr. M. Cartier and Dr. G. Mercier of CNET (Lannion, France) as well as Dr. A. Knipper (Moscow) provided me with relevant Soviet literature on speech processing. Mr. A. Bladon of Oxford University and Dr. J. Holmes of JSRU (Cheltenham, England) reviewed several chapters and gave numerous valuable comments especially on Chapter 2. The help and support of all these colleagues is gratefully acknowledged.

The layout of this book was prepared by means of a text processing system and a graphic software package developed at the author's laboratory. I wish to thank Mrs. I. Kunert who edited most of the text. I am also indebted to Dr. H. Lotsch, Mr. R. Michels, and Dr. M. Wilson of Springer-Verlag; without their help and patience many of my intentions would not have been realized.

Munich, March 1983   Wolfgang Hess

Contents

1. Introduction

1.1 Voice Source Parameter Measurement and the Speech Signal

Speech sounds are produced by modulating the air flow in the speech tract in order to create some kind of acoustic source. One distinguishes between two categories: *voiceless* sounds which originate from turbulent noise created at a constriction somewhere in the vocal tract, and *voiced* sounds, excited in the larynx by periodic vibrations of the true vocal cords. These two sources of sound can act fairly independently of each other; in a first approximation one might even treat them as mutually exclusive.

The voice source (voiced as well as voiceless) supplies the "carrier" which is necessary to transmit the higher-level (phonetic, syntactic, semantic, emotional, etc.) information (Dudley, 1940a). The speaker modulates this carrier with the information he wants transmitted, thus generating and uttering an acoustic sound pattern, the speech signal (Fig.1.1):

The illustration chosen for the carrier wave of speech is a talker's sustained tone such as the sound *ah.* In the idealized case there is no variation of intensity, spectrum or frequency. This carrier then is audible but contains no information, for information is dynamic, ever changing. The carrier provides the connecting link to the listener's ear over which information can be carried. Thus the talker may pass information over this link by starting and stopping in a prearranged code the vocal tone as in imitating a telegraph buzzer. For transmitting information it is necessary to modulate this carrier with the message to be transmitted.

For the second illustration, message waves are produced as muscular motions in the vocal tract of a "silent talker" as he goes through all the vocal effort of talking except that he holds his breath. The message is inaudible because the motions are at slow syllabic rates limited by the relatively sluggish muscular actions in the vocal tract. Nevertheless, these motions contain the dynamic speech information, as is proved by their interpretation by lip readers to the extent that visibility permits. ... The need of an audible "carrier" to transmit this inaudible "message" is obvious.

The final example, to illustrate the modulating mechanism in speech production, is from a person talking in a normal fashion. In this example are present the message and carrier waves of the previous examples, for both are needed if the former is to modulate the latter. However, the mere presence of the carrier and message waves will not make speech, for if they are supplied separately, one by a silent talker and the other by an intoner, no speech is heard but only the audible intoned carrier.

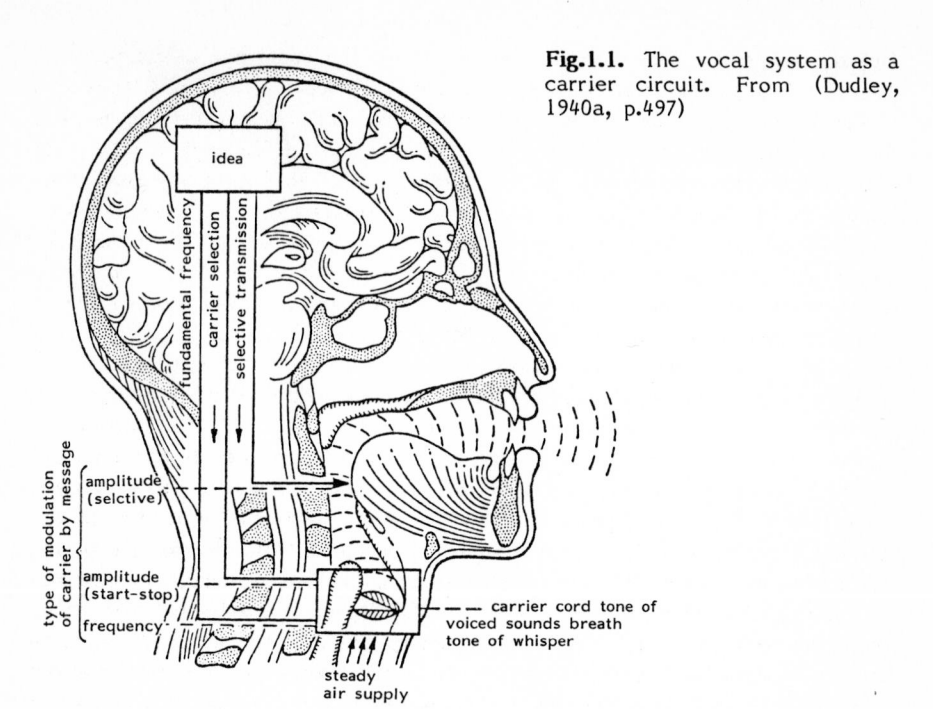

Fig.1.1. The vocal system as a carrier circuit. From (Dudley, 1940a, p.497)

Ordinary speech results from a single person producing the message waves and the carrier waves simultaneously in his vocal tract, for then the carrier of speech receives an imprint of the message by modulation. (Dudley, 1940a, p.497)

There is of course much that separates the "carrier" provided by the voice source from the carrier in a technical system, such as a broadcast transmitter. Firstly, the excitation function, according to the higher-level information, is switched on and off as well as to and fro between voiced and voiceless excitation, and sometimes both excitations act simultaneously. Secondly, as an amplitude modulation it permanently changes its intensity. Thirdly, the voiced excitation function can vary its "carrier frequency," i.e., the fundamental frequency F_0 of the speech signal within a wide range (more than two octaves for a normal voice). These features are deliberately used by the speaker to support the process of phonation in the vocal tract; they yield a complexity to the voice source signal which makes the estimation of the source parameters difficult.

The difficulty estimating and measuring voice source parameters is further increased by the fact that the only signal easily accessible is the speech signal, i.e., the final product of the modulation (and convolution) process. The separation of voice-source and vocal-tract properties presents an additional problem in estimating source parameters from the speech signal. Hence voice source parameter measurement is a subset of the measurement techniques of speech analysis. To estimate the parameters of the vocal tract, especially its transfer function, several theories have been developed (Chiba and Kajiyama, 1941; Ungeheuer, 1962; Fant, 1960, 1970), and measurement techniques have been designed, such as formant analysis (cf. Flanagan, 1972a, for a survey) or linear-prediction analysis (cf. Markel and Gray, 1976). Applying these techniques to the speech signal, the vocal tract parameters and the voice source parameters can be separated to a great extent. However, it is usually not necessary to completely remove the influence of the vocal tract in order to measure the basic voice source parameters, such as fundamental frequency or the presence and absence of voicing. These parameters can be directly estimated from the speech signal, although the varying temporal structure of the speech signal tends to degrade the quality and accuracy of the measurement.

To measure the voice source parameters a multitude of methods and techniques have been developed. We have to distinguish between those techniques which only investigate the speech signal, and those techniques which attempt to determine the source parameters by direct or indirect measurement of the vocal cord vibrations during the process of phonation. The former techniques are nowadays more or less a matter of electroacoustics, electronics, and signal processing; the older mechanical methods (kymograph, tuned reeds) are of no more than historic or didactic interest. Among the direct or indirect measurement techniques for vocal cord vibration we find methods such as high-speed photography, stroboscopy, ultrasonic measurement, radiography, and electric measurement of glottal vibrations (laryngography).

What are the source parameters to be measured? The most elementary and at the same time the most important acoustic parameters are 1) the *type of phonation* (voiced, voiceless, silence), and 2) the *periodicity* (i.e., fundamental frequency, fundamental period duration, or individual fundamental periods) of the vocal cord wave if the signal is voiced. Much work has been carried out in order to develop measurement techniques for the determination of these voice source parameters. This report intends to give a detailed survey of methods, devices, and algorithms dealing with pitch determination from the speech signal. A summary will be presented of the measurement techniques applied to determine more complex parameters like the glottal spectrum or the glottal waveform. The book will also contain an abridged summary of those techniques applied to directly observe the movements of the glottis. More detailed reports on those subjects, however, are left to another author or another time.

1.2 A Short Look at the Areas of Application

Possible applications of devices measuring voice source parameters from the speech signal may be grouped into the following four main groups.

1) **Application in speech communication:** analysis, transmission, synthesis of speech; speech and speaker recognition. In this domain the estimation of

source parameters presents itself as a subset of the measuring techniques of speech analysis. Detailed surveys of the problems of speech analysis have been given by Flanagan (1972a), by Pirogov (1974), or by Rabiner and Schafer (1978), the latter with special respect to digital processing of speech signals. A great deal of the recent articles in this field have appeared in the IEEE Transactions on Acoustics, Speech, and Signal Processing, to which the reader is referred. Shorter surveys have been presented by Schroeder (1966) on vocoders, by Schroeder (1970d) on speech parameters, by Flanagan (1971, 1972b, 1976) on speech analysis and synthesis, or by Gold (1977) on speech transmission problems.

2) **Application in phonetics and linguistics:** investigation of prosodic and phonetic features. Among the speech parameters the fundamental frequency F_0 is probably the one which is influenced and controlled in the most complex way. At least five more-or-less independent sources of influence act on the same parameter:

a) *unintended articulation constraints,* such as *intrinsic pitch,* or *micro-prosody* at fast transitions. The origin of these effects is not yet fully understood. According to recent results they are supposed to result from interactions between the voice source and the vocal tract although this hypothesis does not yet seem fully verified (Ohala, 1973; Guérin et al., 1977). For more details see publications such as Peterson and Barney (1952), Fairbanks (1940), Lehiste and Peterson (1961), and Boë et al. (1975).

b) *phonemic tone.* In some tone languages like Thai or Chinese the phonemic information is transmitted not only by a sequence of phonemes (or other discrete units), but in addition by the *word* and *sentence melody.* For details see references like Pike (1948), Lehiste (1970), or Léon and Martin (1969). Other languages, such as the Indo-European language family, do not contain this feature.

c) *word stress (word accent).* This feature exists in languages that are not tone languages, like the Indo-European ones. It is an alternative way to distinguish between homonyms, i.e., words with identical phoneme sequence but different meaning. This feature is, for instance, extensively used in Russian, whereas in other languages, like French, it serves more or less only as an anchor point to identify the beginning or the end of a word (Vaissière, 1976a,b). Usually one distinguishes between three prosodic (or *suprasegmental*) parameters or manifestations in speech: *quantity* (duration), *intensity,* and *fundamental frequency* (Lehiste, 1970). Word accent is shown to influence at least intensity and fundamental frequency. In most languages it also influences duration (this is valid, e.g., for Russian and Italian, whereas in German, for instance, and even more distinctly in Hungarian, stress and vowel duration are controlled by independent pronunciation rules). The manifestation of the influence upon F_0 at the word level is called *tone* by some researchers (Lehiste, 1970).

d) *sentence melody ("intonation").* In all languages there exist variations of F_0 dictated by syntax and semantics. The manifestation of the influences upon F_0 at the sentence level is called *intonation* (Lehiste, 1970). The intonation rules are extremely different from language to language, and a vast literature has been published in this field. We must distinguish between at least two levels of intonation: the *syntactic* level which indicates the type of sentence (e.g., enunciative phrase) as well as its syntactic structure, and the *semantic* level related to the content of a sentence and its relative importance in the context.

e) *emotional aspects*: the general mood of the speaker as well as the emotional and psychological aspects of the message inherent in the speech signal. There are relatively few quantitative measurements dealing with this aspect (for instance, Fonagy and Magdics, 1963; Scherer, 1981a,b).

f) *individual aspects*: range of F_0 of an individual speaker; condition of the voice (normal, temporarily or permanently pathologic), ability and willingness of the speaker to make use of his articulatory and phonatory repertoire.

Not all of these aspects (which are grouped here in a sequence extending from short-term influences to long-term or even speaker-invariant influences) are the subject of phonetic and linguistic research. In this domain, however, the analysis and determination of fundamental frequency is a necessary presumption to obtain linguistic and phonetic knowledge. It is of great importance to the researchers in this field to have devices at hand that automatically determine F_0 and yield a great amount of data within a short time. A general introduction to the problems of intonation and tone is given by Léon and Martin (1969) and by Léon (1972); the reader is also referred to the collection of articles edited by Bolinger (1972). (None of these surveys, however, treats the problem of voicing detection.) A bibliography covering much of this work is given by Léon and Martin (1969) and Léon (1972). The largest bibliography on prosody which contains more than 4000 items has been compiled by Di Cristo (1975).

3) **Application in education.** This task is subdivided into two subtasks: a) aids for the deaf to support their efforts in learning how to speak correctly (Dolansky et al., 1971), and b) aids to efficiently control and exercise correct intonation of a foreign language. A deaf person (and a deaf child in particular) has enormous difficulties to acquire correct intonation, even to acquire an adequate range of pitch. In this area even very crude visual or tactile indicators of F_0 can help very much. The difficulties with the prosody of a foreign language, on the other hand, are familiar to anyone who has ever had to learn one.

4) **Application in medicine, pathology, and psychology.** Determination of source parameters from the signal can serve as a quick and easily accessible help for voice diagnostics and for examining the progress of voice therapy. In phoniatric practice direct measurement and investigation of the speech organs is usual and natural, and pitch determination instruments (cf. Sect.5.2) are a most valuable aid; deriving source parameters from the signal, however, is a hopeful alternative, in particular for early detection of developing voice diseases and for diagnostic evaluation of slight pathologies. A recent survey of this application has been presented by Davis (1978).

Outside the immediate speech and voice domain *musicology* is another domain of application. In this area the device has to fulfil tasks as an automatic or interactive aid for music transcription. This task, however, is so related to those in phonetics and linguistics that we will be able to treat these application areas in the same way.

So far, the wide variety of applications imply a great variety in the design of processing devices to extract voice source parameters. The difficulty of this task which seems so simple when regarded superficially, greatly increases the variety of proposals made and approaches realized. Literally hundreds of devices and algorithms for fundamental frequency and voicing detection are in existence. None of them, however, has solved the problems for all signals and all circumstances. Partial solutions, on the other hand, perform well for a restricted variety of applications or for a restricted number of conditions, such as signal distortions and degradations, and range of F_0.

1.3 Organization of the Book

The first and principal aim of this book is to present a unified view of the multitude of approaches, methods, devices, and algorithms for the one task of pitch determination of speech signals. The second aim is a necessary consequence of the first one: to bring together the various application areas. Hence it is necessary to make this book understandable both to the speech communication engineer and to the phonetician, linguist, or speech pathologist. The point of view from which the book is written will be that of its author: speech communication engineering with special emphasis on digital signal processing and computer application. This point of view has two advantages: 1) any method can be simulated on the computer, and thus any method can be described in algorithmic form; and 2) the most powerful methods in all application areas use the computer and cannot be run other than digitally. A unified description of all the methods for pitch determination therefore is only possible if it is done in terms of computer algorithms.

The extensive use of digital signal processing and computer simulation as well as the claim of this book to serve as a reference for both engineers and phoneticians (to name two groups of prospective readers) make a comparatively broad scope of general knowledge necessary in order to understand all these methods. Speech analysis can be performed originating from one of three premises (Zwicker et al., 1967):

1) the *production* point of view. The signal is analyzed using knowledge about the way it is generated in the human speech tract, and the parameters extracted by such methods are related to the control parameters and signals of speech production.

2) the *perception* point of view. The signal is processed in a way similar to the processing in the human ear, and parameters extracted by such methods are related to speech perception.

3) the *signal processing* point of view. In the extreme case, the analysis looks for parameters that describe the signal as it is in some optimal way whatsoever, regardless of its origin or the manner of perception.

In a less restricted sense, however, and with appropriate modeling of speech production and perception mechanisms, the signal processing point of view can include all these premises. The book will thus first define a basic terminology and then present the most basic knowledge on digital signal processing. This is done in Chap.2 together with a first categorization of the methods of pitch determination into time-domain and short-term analysis methods. Chapter 3 deals with the aspect of speech production: after a short look at the anatomy and the main theory of phonation, emphasis is switched to the question of voice registers and the influence of the vocal tract on the waveform of the glottal excitation. It is obvious that this book shall not serve as a reference on digital signal processing or on speech production. Any reader who is familiar with one of these aspects may skip the corresponding chapter since nothing will be new to him. For the reader who is not familiar with digital signal processing, the introduction may be too concise; on the other hand, reference is given to textbooks where this topic is treated in a much broader way. Chapter 4 deals with the perception point of view. The main question in that chapter, however, is not how pitch is perceived, but what consequences arise from the perceptional facts and hypotheses for the range, reliability, and accuracy of the measurement.

The main review of the methods of pitch determination starts with Chapter 5. This chapter presents the methods around the central problem: manual pitch determination, "instrumental" pitch determination (i.e., pitch determination other than from the speech signal alone), and voicing detection. Chapters 6 through 8 deal with the individual methods for pitch determination from the speech signal: Chap.6 presents the time-domain methods, whereas the short-term analysis methods are dealt with in Chapter 8. In Chap.7 an individual time-domain method is discussed in more detail. Chapter 9 finally draws the conclusions from the survey of all the methods: it focuses on topics like performance evaluation or the requirements on methods and devices for a particular application area.

2. Basic Terminology. A Short Introduction to Digital Signal Processing

The aim of this chapter is twofold: 1) to create and define the terminology needed to survey the literature, and 2) to provide the reader with the necessary basic knowledge of digital signal processing.

Since the terminology in the literature is not uniform and sometimes even contradictory, it is necessary to define the basic terminology related to the problem of pitch determination. This is done in Sect.2.1 with reference to the proposal by McKinney (1965). Sections 2.2-4 form an introduction to digital signal processing (DSP). They present the basic principles of DSP - as far as they are needed throughout the book - in concentrated form. This part is intended to serve as a quick reference for those readers who are not (too) familiar with DSP and the digital way of representing signals and systems. Sections 2.5,6 define the task from the signal processing point of view as well as present a preliminary categorization of the methods of pitch determination into the two general classes: 1) time-domain methods, and 2) short-term analysis methods.

2.1 The Simplified Model of Speech Excitation

A simplified model of the excitation of voiced speech sounds is used to set up the basic terminology (McKinney, 1965; Fig.2.1). According to this model,

a volume velocity *glottal excitation function* $u_g(t)$ excites a linear passive system with a *supraglottal transfer function* which represents the transmission and radiation characteristics of the entire structure above the glottis (Fig.2.1a[1]). The resulting sound pressure will be called the *acoustic speech wave*. Although the glottal excitation of the vocal tract frequently is modeled as a pulse train, a nonnegative function of time which takes on the value zero at some time between each consecutive pair of pulses, it is often not true that the volume velocity at the glottis goes to zero during each vibration cycle of the vocal cords. In order to allow for this possibility, $u_g(t)$ will be considered to be the sum of the pulse

---

[1] Figures mentioned in quotations are shown in this volume; their numbering has been adapted to conform to that used here. The same applies to the numbering of equations.

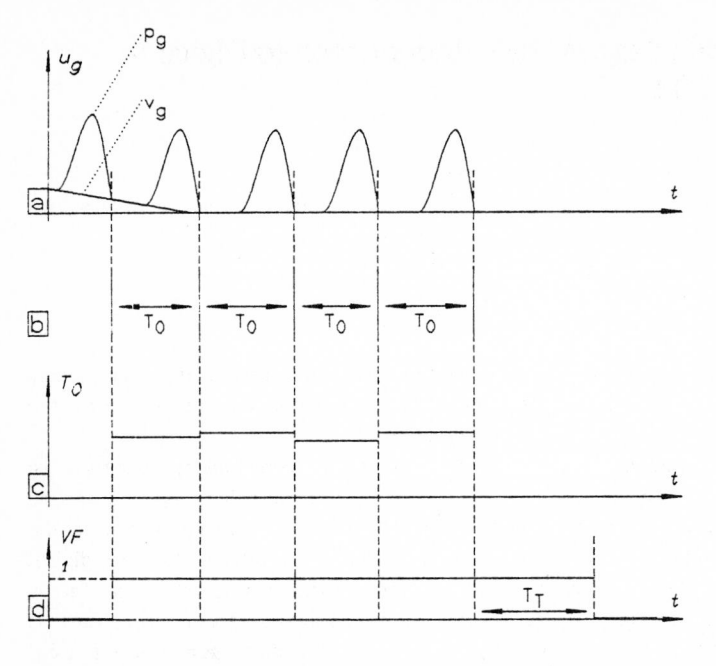

Fig.2.1a-d. Basic definitions dealing with voice source parameters. After (McKinney, 1965). **(a)** Excitation signal, **(b)** excitation pulse time and period durations, **(c)** course of fundamental period T_0, **(d)** voicing function. (T_T) Maximum duration of pitch period. The voicing function is defined later in this section. For more details, see text

train $p_g(t)$ and a slowly varying function $v_g(t)$. $p_g(t)$ will be called the *excitation pulse function* (Fig.2.1a), and each of its constituent pulses is an *excitation pulse*. To each excitation pulse is assigned a time of occurrence, its *excitation pulse time*. In order to make it coincide as nearly as possible with the principal excitation of the formant resonances, the excitation pulse time is defined to be the time at which the excitation pulse function reaches zero value at the end of an excitation pulse.[2] In case the glottis is closed during each cycle, this time equals the instant of glottal closure (McKinney, 1965, pp.13-14).

Throughout this book, most signals, time functions, and parameters are referred to as *digital* signals. A brief introduction to digital signal processing follows in Sects.2.2-4. For the moment it is sufficient to know that any time function $f(t)$ is converted into the corresponding digital signal $x(n)$ by sampling the analog signal $f(t)$ at - usually constant - time intervals $t = nT$. If the bandwidth of $f(t)$ is finite and the sampling interval T has been appropri-

---

[2] Details of the underlying theory as well as quantitative modeling of the excitation signal will be supplied in Chap.3, and some experimental data verifying this will be given in Section 6.3. For the moment we can take these statements as established facts; they are mainly used here to build up the terminology.

ately chosen, x(n) can theoretically be reconverted into the analog representation without any difference in waveform between the original and the reconverted analog signal. Thus both time and frequency, which are usually measured in seconds and hertz in the analog domain, may be represented as a number or index of samples where the sampling interval or the associated sampling frequency must be known in order to reconvert these measures into real values in the time or frequency domain.[3]

In the following the acoustic speech wave as well as the glottal excitation function will be referred to as *signals* regardless of whether they are represented as sound pressure or as the electronic or numeric equivalent thereof. The acoustic speech wave will thus simply be the *speech signal,* and the glottal excitation function will be referred to as the *excitation signal.* The segment of the signal placed between two successive excitation pulses is a *fundamental period* or a *pitch period* or simply a *period;*[4] usually a period is thought to be located *in correct phase* (in its proper position), that is, the beginning and end of the period coincide with the definition of excitation pulse time given further above. An individual period can theoretically be extracted from the speech signal by weighting the latter by a rectangular window whose length equals T_0, which is the *period duration* T_0 or *period length* or, simply, the *period.* T_0 is equal to the elapsed time between two successive excitation pulses.

The frequency associated with period duration is the *fundamental frequency* F_0; the (sinusoidal) waveform which contains only the frequency F_0 is the *first harmonic* or *first partial* of the signal. (The term *first partial* is somewhat more general since it also comprises the case when the signal is not exactly periodic. Thoughout this book, however, we will not strictly distinguish

---

[3] All signals are regarded as *digital* signals when algorithms are quantitatively described since this is the only representation in which all the algorithms and methods can be treated in a unified manner. The *graphic representation* of signals, however, corresponds to plots of analog signals unless it is necessary to display the individual samples (such as in Fig.2.2).

[4] The meaning of the term *period* is ambiguous. On the one hand, it can refer to an individual segment x(n), n = q, q+1, ... q+K-1. According to this definition the signal is delimited by the boundaries at n = q and n = q+K-1, respectively. This signal is defined as extracted from the periodic signal x, where K is the reciprocal of the fundamental frequency F_0. On the other hand, it can also refer to the duration of such a segment regardless of its actual location. Since the speech signal is not exactly periodic, the use of different definitions may lead to different results. If, in spite of this, we continue using the term *period* for both meanings, only because it will usually be possible to distinguish between them according to the respective context. If this is not easily possible, however, then we will distinguish - at the expense of some tautology - between a *period* as the individual segment x(n), n = q, q+1, ... q+K-1 of the signal x, and the pertinent *period duration* or *period length* T_0 = K. A third meaning of the term *period,* by the way, will follow in Section 2.5. The whole problem will be taken up again in Sect.9.1.1 where a formal definition of the term *period* in its different meanings will be given.

between exactly harmonic and nearly harmonic signals so that there will be no difference in use of these two terms.) These terms will be used for both the time-domain waveform and the spectral representation in amplitude and power spectrum (ambiguities can be easily avoided by the context). The term *pitch* is used when the question of the domain of operation is intentionally left open, e.g., when both time and frequency domain are dealt with in the same statement. Thus a measuring device of any kind which gives an estimate of pitch will be called a device for pitch determination or - since the digital representation is preferred - a *pitch determination algorithm* (PDA) whenever pitch is derived from the speech signal alone, and a *pitch determination instrument* (PDI) otherwise. Parts of an algorithm, device, or instrument will be referred to as *routines* (or *subroutines*) and *circuits*. In case a PDA is realized as a piece of hardware (analog or digital), we shall also speak of a *pitch determination device* (PDD).

The author is aware of the fact that the term *pitch* has originally referred to a percept and not a matter of production, though admittedly (McKinney, 1965) the two are closely correlated in this case.[5] In the technical literature the term *pitch* is widely used as an expression for both fundamental frequency F_0 and period T_0, which is incorrect in the sense of linguistic or psychoacoustic terminology (Hammarström, 1975). On the other hand, a general statement on pitch determination sometimes covers both fundamental frequency and period. In this case it would be clumsy to speak of "fundamental frequency and/or period determination" instead of "pitch determination" and incorrect to speak of one of the two alone. To speak of "fundamental frequency" in connection with a time-domain circuit, for instance, may lead to inaccurate statements such as that "fundamental frequency is measured as the distance between two consecutive zero crossings of equal polarity in the time-domain signal." For this reason the term *pitch* will be employed in those cases where a distinction as to the domain of operation is impossible (because the statement is too general) or explicitly not intended. The use of this term, however, will be restricted to such cases and to the composite expressions like *PDA* or

---

[5] From the viewpoint of a phonetician and editor of a journal on phonetics, Kohler (1981, unpublished letter) writes in regard to this question: "I am fully aware of the established use of the term *pitch* - at least in linguistic circles - with exclusive reference to perception, as against the physical measures of *fundamental frequency* and *rate of vocal fold vibration*. But I feel that we need a general name to refer to all the different manifestations of that aspect of speech in which *vocal fold vibration, fundamental frequency, pitch perception,* and *phonological, linguistic and communicative functions* are intimately related to each other. The obvious name for this whole field is *pitch* because it has already been used in this wider sense in the literature on signal processing and it refers to a clearly defined physical correlate, whereas *prosody* or *intonation* are either vague or at least multi-parametric in their physical connection."

pitch period. If a percept is really meant, we will speak of *perceived pitch*[6], a term that is clearly defined per se, although again at the expense of a certain tautology.

The excitation pulse function as well as the fundamental frequency F_0 and period T_0 are defined only in those segments where the signal is *voiced*. To indicate a voiced signal, the *voicing function*[7] (Fig.2.1d), a binary function of time, is defined to take on the value 1 and thus to indicate voicing as long as the elapsed time between the current instant and the previous excitation pulse does not exceed the *maximum period duration* T_T (Fig.2.1d). Besides the voiced segments there are segments of *voiceless* signals and segments of *silence*. Both of these will be called *unvoiced*. For the operation of a PDA, only the distinction *voiced-unvoiced* is needed; it is not necessary to further classify the unvoiced segments into *voiceless speech* and *silence* since they will be disregarded by the PDA in any case. As a first approximation this distinction is a binary one although a more elaborate subdivision may be necessary for other purposes (Dreyfus-Graf, 1977). The *voicing detection algorithm* (VDA) - which will be the common label for routines or devices doing this job - is normally not identical with the PDA itself. Thus these algorithms will be treated separately from the proper pitch detection problem unless the procedures are identical in a particular approach.

It is usual but not necessary to set the value of F_0, or T_0, to an arbitrary value outside the measuring range, frequently to zero outside voiced intervals. Doing so, one can waive the separate voicing function and define the signal to be unvoiced as soon as F_0 or T_0 take on a value of zero. This may be dangerous, however, when the course of F_0 (which is referred to as the *pitch contour, fundamental frequency contour,* or *fundamental period contour*) is to be smoothed or low-pass filtered, since it now exhibits great discontinuities at voiced-unvoiced transitions.

---

[6] Psychoacoustic research avoids the problem with the term *pitch* in its own way. Psychoacoustics rigorously distinguishes between a sensation quantity (that can only be measured or verified by subjective evaluation and testing), and the same quantity when it is quantitatively formed and measured in a functional model. For the case of pitch perception, several theories and models have been developed which derive a quantitative measure for the sensation quantity *perceived pitch*. To distinguish the model-oriented measurements from the subjective evaluation, the quantitative estimates of pitch at the output of these models have been assigned the special names *virtual pitch* (Terhardt, 1974), *residue pitch* (Ritsma, 1962), and *periodicity pitch* (Goldstein, 1973).

[7] This definition of the voicing function is a very crude simplification which may prove sufficient for pitch determination as an indication of segments where a reasonable measure of pitch can be expected, as opposed to those segments where there is definitely no voicing, so that a PDA need not operate there. In reality the voicing function is much more complex; mixed excitation, i.e., voiced and voiceless excitation simultaneously, is possible, and the transitions between voiced and voiceless segments are much more gradual than this binary function would suggest. More details on the voicing problem are found later in this chapter and in Section 3.5.

2.2 Digital Signal Processing 1: Signal Representation

Acoustic signals belong to a class of functions depending on one independent variable. This variable is *time*. The dependent variable is *sound pressure* or some electrical equivalent. In the digital representation a signal can only be described as a sequence of numbers. Thus the analog signal a(t) has to be converted into a number sequence. This is done in two steps.

1) **Sampling** (Figs.2.2,3). This means discretization of the independent variable (in case of the acoustic signal, time). Sampling is usually carried out at a constant *sampling interval* T which corresponds to the *sampling frequency* (or *sampling rate*) 1/T:

$$a(t) \longrightarrow a(nT) = a(n) \; . \tag{2.1}$$

It is usual to omit the sampling interval T from the argument of the discrete function a(n). Of course, the sampling interval T is then assumed to be *constant*. The function or - as it will henceforth be called - the *signal* a(n) then is a *sampled signal*, and the individual value a(n) for a particular n is simply a *sample*. Every sample a(n) is characterized by its amplitude a and by its *address* n, i.e., the time (relative to some given zero, usually the beginning of the data) of its occurrence. The choice of the sampling rate is limited downward by the sampling theorem (for details see, for instance, Rabiner and Schafer, 1978; Marko, 1980; Kunt, 1980; or Shannon, 1948) which requires the analog input signal a(t) to be band limited in the frequency domain to a maximum frequency less than 1/2T. The reason for this is, qualitatively speaking, that sampling of the input signal causes its Fourier spectrum to become *periodic* with the period 2/T (or 2_π/T when radian frequencies are regarded) in the spectral domain (Fig.2.2). The analog signal will thus be

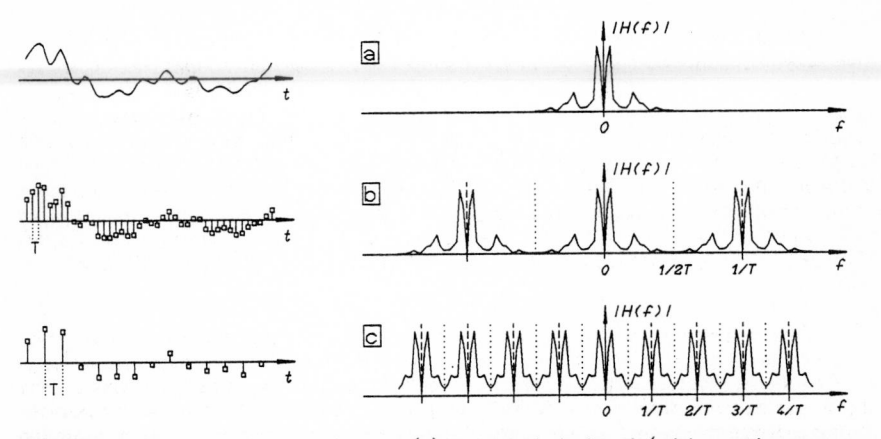

Fig.2.2a-c. Illustration of sampling. (a) Unsampled signal (with pertinent spectrum); (b) signal sampled at a suitable sampling rate; (c) signal sampled at a low sampling rate so that the sampling theorem is violated. (Left-hand side) Signals; (right-hand side) Fourier spectra

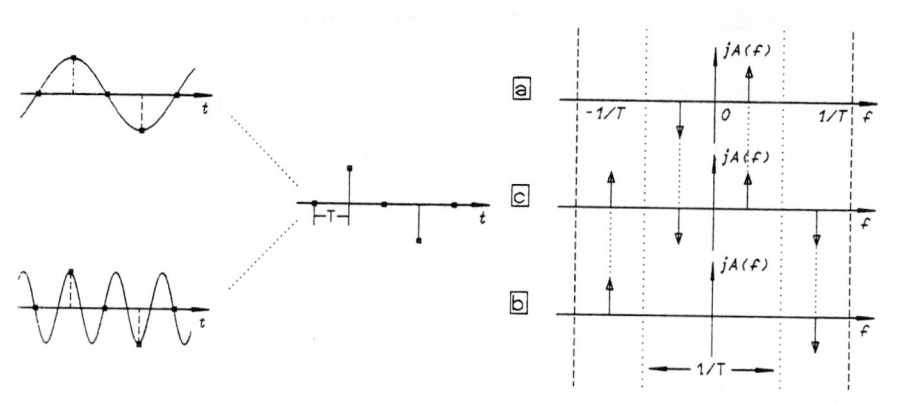

Fig.2.3a–c. Choice of the sampling interval. (**a**) Sinusoid which will be sampled correctly (1/4 of sampling frequency), together- with the pertinent spectrum; (**b**) sinusoid which will not be sampled correctly (3/4 of sampling frequency), (**c**) result of sampling for both (a,b). (Left-hand side) Signals; (right-hand side) spectra

completely represented by the sampled version if the nonzero part of its spectrum falls within one period of the spectrum of the sampled signal for a given sampling rate 1/T. Violation of the sampling theorem will result in typical distortions (*aliasing,*[8] see Figs.2.2,3).

2) **Quantizing** (Figs.2.4,5). This means discretization of the dependent variable,

$$x(n) = [a(n)]_Q = a_Q(n) k_Q .$$ (2.2)

Given the representation and its accuracy, it is usual to express the amplitude of individual samples as a nondimensional entity with the value a_Q.

Quantization results in a limitation of the possible range of the signal. The upward limitation of the overall amplitude (overload limit) is the minor problem; analog amplifiers have similar characteristics. The major quantization problem is that the increments of the dependent variable are not allowed to become arbitrarily small (Fig.2.4). This causes the *quantization noise* which depends on the *word length* of the digital representation and on the input signal itself: quantization noise is *signal correlated.* If the signal is digitally stored in binary form, the possible number of discrete steps available will be

---

[8] Aliasing distortions are linear distortions; when aliasing occurs, the signals still obey the superposition theorem. However, this is not a linearity in the restricted sense, as in linear passive systems where a sinusoid with an arbitrary frequency f at the input leaves the network as a sinusoid with a possibly different amplitude and phase but with the same frequency f. A sinusoid that is distorted by aliasing remains a sinusoid, but its frequency is changed in the same way as in a linear modulation or demodulation network.

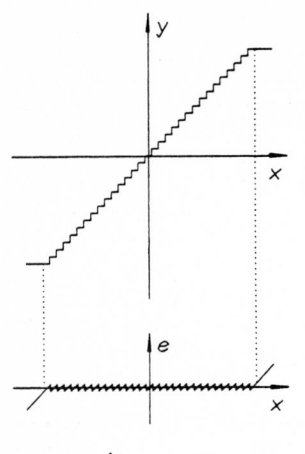

Fig.2.4. Characteristics of a sample quantizer (linear); (x) input signal, (y) output signal, (e) quantization error

$$K = 2^k \qquad\qquad (2.3)$$

where k is the word length (in bits) and K the number of steps. For k considerably greater than one, the maximum signal-to-noise ratio with respect to quantization noise will be approximately

$$S/N \approx k \cdot 6 \text{ dB} . \qquad\qquad (2.4)$$

Equation (2.4) is a crudely simplified measure. It holds approximately for a stationary signal with maximum amplitude and for the case that *linear quantizing* [as in (2.2) and Fig.2.4] is applied. Speech signals, however, have time-variant amplitudes so that the local signal-to-noise ratio can be much worse. For sophisticated speech transmission or storage systems a nonlinear quantization is often preferred which gives a better match of the quantization characteristic and the amplitude statistics of the signal [for details see, for example, Flanagan et al. (1979)]. As soon as processing (other than mere quantization, transmission, and storage) is performed on the signal in the digital domain, linear quantization is necessary since otherwise the whole system would be nonlinear even if all the components of the digital processing system except the quantizer were linear. For the remainder of this book we will thus assume *linear quantization* to be given. Since we deal with speech analysis and measurement methods, we can even assume that the word length of the signal representation used is so large that we can simply ignore the effects of quantization.

The chain between digital and analog representation is the *analog-to-digital converter* (ADC) and the *digital-to-analog converter* (DAC). For a long time, the ADC was the weakest and the most expensive component[9] of a digital

---

[9] Large-scale integration and modern circuit technology have nowadays made it possible to design and realize high-quality DACs and ADCs at relatively low cost. On the other hand, a word length of 12 bits already corresponds to a signal-to-noise ratio of about 72 dB.

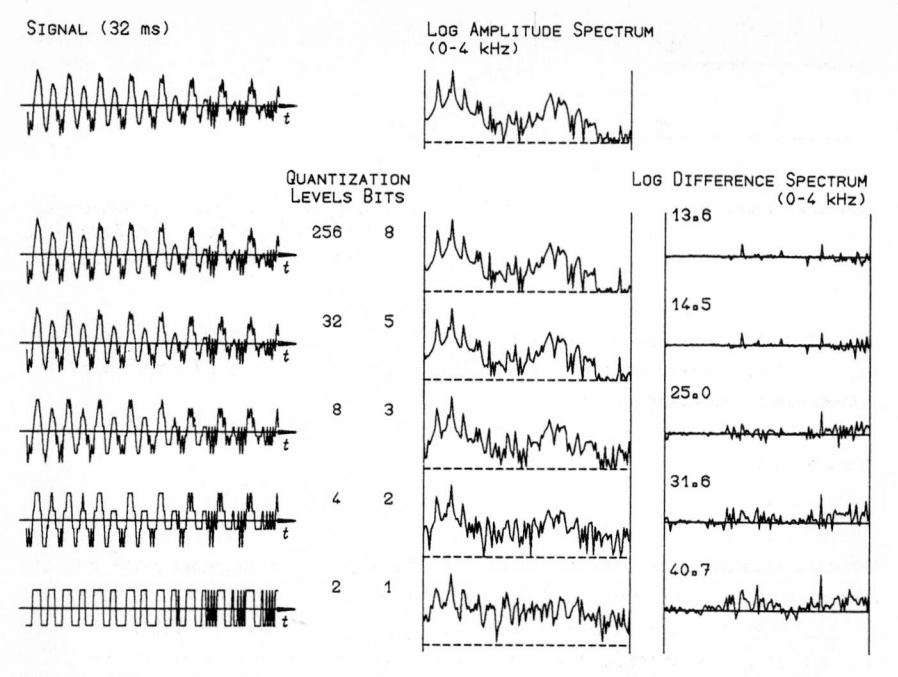

SIGNAL (32 ms)   LOG AMPLITUDE SPECTRUM
                                  (0-4 kHz)

                                  QUANTIZATION
                                  LEVELS BITS   LOG DIFFERENCE SPECTRUM
                                                                   (0-4 kHz)

                                      256   8

                                      32   5

                                      8   3

                                      4   2

                                      2   1

Fig.2.5. Illustration of the effect of quantizing. Signal: part of the utterance "Hello operator," male speaker. Range of the log amplitude spectra: 40 dB. The spectra were normalized before plotting. The numbers at the left-hand side of the difference spectra indicate the maximum difference (in dB)

signal processing system. For speech signals, its word length is usually limited to 12 bits; more sophisticated ADCs (for high fidelity purposes) can process up to an accuracy of as much as 19 bits. On the other hand, within the DSP system word length is more or less a matter of cost and effort, unless special time requirements make a short word length necessary; an internal word length between 16 and 24 bits for fixed-point arithmetic is usual.

As shown before, the digital signal has a periodic spectrum; the period (in the frequency domain) equals $2\pi/T$ where T is the sampling interval. This spectrum can be computed by the *discrete Fourier transform* (on the imaginary axis of the s plane) or by the *discrete Laplace transform*. The periodic spectrum, however, is difficult to handle. Thus a different form of spectral representation is commonly used, the *z-transform*. For a given signal a(n) this transform is defined by

$$A(z) = \sum_{n=0}^{\infty} a(n) \, z^{-n} \tag{2.5}$$

In the z-transform the sampling interval T is eliminated; also the (frequency-domain) periodicity disappears.

The z-transform is an infinite power series in the variable z^{-1} where the sequence values a(n) play the roles of coefficients in the power series. If a(n)

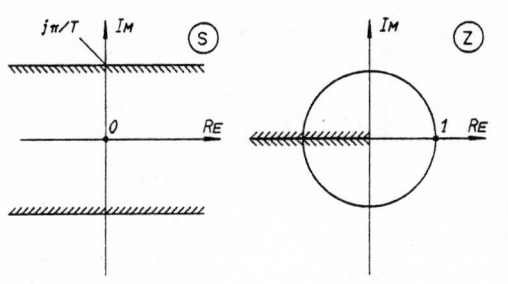

Fig.2.6. Relation between the s plane (Laplace transform) and the z plane (z-transform)

is forced to zero for n less than a given constant,[10] A(z) converges in a region of the complex z plane outside the circle $|z| = R$, where R is the *convergence radius* which depends on the values of a(n).

Between the z plane and the s plane of the Laplace transform there exists the relation (Fig.2.6)

$$z = e^{sT} .$$ (2.6)

By this relation the imaginary axis of the s plane is mapped into the unit circle of the z plane so that $s = 0$ (zero frequency) is converted into $z = 1$, and $s = 2\pi j/2T$ (half the sampling frequency; $j = \sqrt{-1}$) into $z = -1$. This means that one period of the periodic spectrum A(s) is mapped on the whole z plane. The next frequency domain (s plane) period again occupies the whole z plane so that the point $s = 2\pi j/T$ (which represents the sampling frequency) is also converted into $z = 1$. This is valid for all possible periods of A(s) so that any point

$$s = 2\pi jk/T , \quad k = -\infty \ (1) \ -1, \ 0, \ 1 \ (1) \ \infty$$ (2.7)[11]

is mapped into $z = 1$. For the following considerations, however, this fact is only of theoretical interest.[12] Applying the z-transform, the band limitation in the frequency domain is automatically taken account of. For more details, see textbooks such as Jury (1973), Doetsch (1982), Kunt (1980), Rabiner and Schafer (1978), or Rabiner and Gold (1975). Some details on the use of the z-transform in digital filtering will be explained in the next section.

---

[10] In the (one-sided) z-transform, as in the Laplace transform, it is presupposed that any signal be zero for time values less than a given limit. This limit is conveniently set at $t = 0$, or $n = 0$, so that the signal disappears for negative values of the independent variable.

[11] The notation $i = n (k) K$ stands for the statement "for i = n step k until K do ....". This notation will be used throughout the book.

[12] It must not be concealed, however, that a number of powerful digital signal processing techniques deliberately make use of this feature. Examples occur several times in this book; see, for instance, Sect.7.5.1 (modulation), Sect.8.2.5 (decimation and interpolation, i.e., down- and upsampling), or Sect.8.5.4 (spectral rotation).

2.3 Digital Signal Processing 2: Filters

In the analog domain a *filter* is a device that converts an input signal x(t) into an output signal y(t) in such a way that there exists a definite relation (mostly an analytical one) between the spectrum $X(j\omega)$ of the input signal and the spectrum $Y(j\omega)$ of the output signal. The term "filter" is often confined to a *linear* filtering system. A linear filter satisfies the superposition theorem

$$k_1x_1(t) + k_2x_2(t) \rightarrow k_1y_1(t) + k_2y_2(t) \qquad (2.8)$$

when

$$x_1(t) \rightarrow y_1(t) \text{ and } x_2(t) \rightarrow y_2(t)$$

hold for arbitrary input signals x_1 and x_2, as far as their samples take on finite values; k_1 and k_2 are arbitrary finite constants.

In the digital domain any device which converts an input signal x(n) into an output signal y(n) is a *digital filter* (DF). In this book the term *digital filter* is confined to *linear* digital filters; the superposition theorem is identically formulated for both digital and analog signals. (Of course it must be presupposed that the sampling rate T be the same for all signals subject to superposition). Digital filters are well described in the literature; for more information see textbooks such as those by Cappellini et al. (1978), Rabiner and Gold (1975), Peled and Liu (1976), Schüssler (1973), Lacroix (1980), or Kunt (1980).

The (linear) DF has a transfer function H(z) which defines the analytical relation between the z-transforms X(z) and Y(z) of the input and output signals,

$$H(z) = \frac{Y(z)}{X(z)} \, . \qquad (2.9)$$

For the linear filter, H(z) can be formulated as the quotient of two polynomials P(z) and Q(z),

$$H(z) = \frac{P(z)}{Q(z)} \, . \qquad (2.10)$$

In the practical realization the linear digital filter can be built from the following elements: 1) addition or subtraction of signals, 2) multiplication of signals by constant coefficients, and 3) delay (storage) of signal samples for one sampling interval T or a multiple thereof. According to the delay theorem of the z-transform it is usual to characterize the delay elements by z^{-1} since the delay of a signal x(n) by one sampling interval causes a multiplication of its z-transform by z^{-1}. With these elements the linear DF is built up as shown in Figure 2.7. For the following considerations the filter is idealized so that the effect of quantization can be neglected. The momentary state of the filter is then described by the following difference equation:

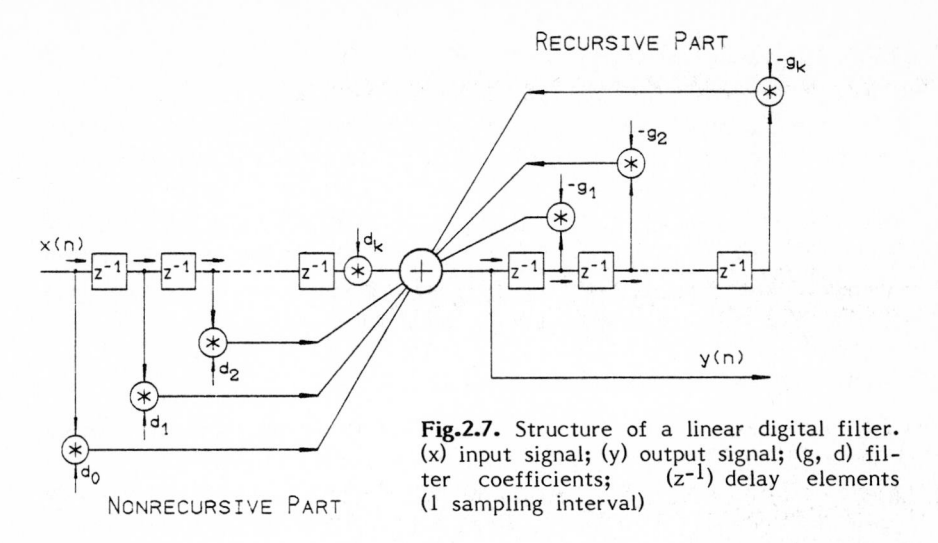

Fig.2.7. Structure of a linear digital filter. (x) input signal; (y) output signal; (g, d) filter coefficients; (z^{-1}) delay elements (1 sampling interval)

$$d_0 x(n) + d_1 x(n-1) + \ldots + d_k x(n-k) \tag{2.11a}$$

$$= y(n) + g_1 y(n-1) + g_2 y(n-2) + \ldots + g_k y(n-k) \, ,$$

where $x(n)$ is the input signal, and $y(n)$ the output signal. If this equation is resolved according to the output signal $y(n)$, we obtain

$$y(n) = d_0 x(n) + \ldots + d_k x(n-k) - g_1 y(n-1) - \ldots - g_k y(n-k) \, . \tag{2.11b}$$

This form corresponds to the filter structure displayed in Figure 2.7. Note that 1) this equation and this structure already represent the most general case of a linear DF, and that 2) any linear DF can be realized by this structure.

With $g_0 = 1$, (2.11a) is equivalent to

$$\sum_{i=0}^{k} d_i x(n-i) = \sum_{i=0}^{k} g_i y(n-i) \; ; \quad g_0 = 1 \, , \tag{2.11c}$$

where k is the order of the filter. A value of zero can be assigned to all the coefficients g_i except g_0 and all but one of the coefficients d_i. If we apply the z-transform to (2.11a-c), we obtain the transfer function[13]

---

[13] For those readers who are not familiar with the properties of the z-transform, a qualitative derivation of the transfer function (2.12) from the state equation (2.11a) reads as follows. The delay theorem of the z-transform says that the delay of a signal $x(n)$ by q samples - which converts the signal $x(n)$ into the delayed signal $x(n-q)$ - corresponds to a multiplication of the z-transform $X(z)$ by z^{-q}, i.e., converts $X(z)$ into $X(z)z^{-q}$. If we now multiply the state equation (2.11a) by z^{-n} and sum up over all values of n from zero to infinity, we obtain the z-transform on both sides of the equation. The delay theorem will then be applied to the delayed terms $x(n-q)$ and $y(n-q)$, $q = 1\,(1)\,k$, and $H(z)$ is calculated according to (2.9) as the quotient of $Y(z)$ and $X(z)$.

$$H(z) = \frac{\sum\limits_{i=0}^{k} d_i z^{-i}}{\sum\limits_{i=0}^{k} g_i z^{-i}} = \frac{\sum\limits_{m=0}^{k} d_{k-m} z^m}{\sum\limits_{m=0}^{k} g_{k-m} z^m} \quad ; \; g_0 = 1 \tag{2.12}$$

which fulfils condition (2.10). From this equation it is obvious that the *recursive* part of the filter [the part which provides the feedback of the delayed output signal y(n-i), see Fig.2.7] is responsible for the poles of H(z), whereas the *nonrecursive* part is responsible for the zeros. The absolute value of H(z) on the unit circle is the *frequency response* and is defined as a function of the angle Ω (between 0 and π),

$$H_f(\Omega) := |\, H(z=e^{j\Omega}) \,| \; . \tag{2.13}$$

In the same way the *phase response* is defined

$$\phi\,(\Omega) := \phi(z=e^{j\Omega}) = \arctan \frac{\mathrm{Im}\{H(z)\}}{\mathrm{Re}\{H(z)\}} \Big/ \; z=e^{j\Omega} \tag{2.14}$$

where $\mathrm{Im}\{H(z)\}$ and $\mathrm{Re}\{H(z)\}$ are the imaginary and the real parts of the complex transfer function, respectively.

In order to guarantee *stability* of the filter, H(z) must exist and be regular on the unit circle. This means, however, that the z-transform of the output signal y(n) must converge on the unit circle [provided that the z-transform of the input signal x(n) also converges on the unit circle, of course]. Hence H(z) must not have poles on or outside the unit circle; the roots of the denominator polynomial must lie within the unit circle, i.e., their magnitude must be less than unity.

The *direct* structure given in Fig.2.7 is not optimized in any way.[14] On the other hand, *any* linear digital filter can be realized by this structure. The multitude of existing DF structures represent variations with respect to practical realization and may be optimal in the sense of a special application. For more details on this issue see a textbook on digital filters, such as Schüssler (1973), Rabiner and Gold (1975), or Peled and Liu (1976).

If the input signal x(n) is an impulse, i.e., if

$$x(n) = \mathrm{imp}(n) = \begin{cases} 1 & n=0 \\ 0 & \text{otherwise ,} \end{cases} \tag{2.15}$$

---

[14] The algorithmic realization of higher-order digital filters in the direct structure often involves small differences of large quantities and may thus be unfavorable due to quantization effects, i.e., when the accuracy of the signal representation or the computation is limited. These effects may even temporarily move poles of the transfer function outside the unit circle and render the filter temporarily unstable (Holmes, 1982, private communication). The arithmetic unit and the accuracy of the computation often decide which structure a digital filter is finally implemented in.

the corresponding output signal represents the *impulse response* h(n) of the filter. The z-transform of h(n) is the transfer function H(z) since the z-transform of imp(n) equals unity.

One more structure will be mentioned here because it is important for understanding the process of speech production. It results from a different representation of the transfer function. Once the roots of the polynomials in (2.12) have been found, this equation can be written in product form,

$$H(z) = \frac{\displaystyle\prod_{i=1}^{k} (z-z_{0i})}{\displaystyle\prod_{i=1}^{k} (z-z_{pi})} \,, \tag{2.16}$$

where z_{0i}, $i = 1\,(1)\,k$ are the zeros, and z_{pi}, $i = 1\,(1)\,k$ the poles of the transfer function. Since the filter coefficients are real (and not complex), the poles and zeros will be either real or conjugate complex. Hence we can directly realize this filter as a *cascade* of first- and second-order blocks,

$$H(z) = \frac{\displaystyle\prod_{i=1}^{k_{20}} (z-z_{0i})(z-z_{0i}^{*}) \cdot \prod_{i=1}^{k_{10}} (z-z_{0i})}{\displaystyle\prod_{i=1}^{k_{2p}} (z-z_{pi})(z-z_{pi}^{*}) \cdot \prod_{i=1}^{k_{1p}} (z-z_{pi})} \,, \tag{2.17}$$

where $k_{10}+k_{20} = k_{1p}+k_{2p} = k$. The corresponding structure is referred to as the *cascade structure* (Fig.2.8). Again any linear DF can be realized using this structure.

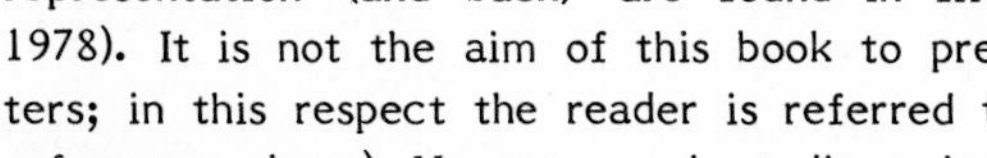

Fig.2.8. Digital filter in cascade structure

The standard solutions for (recursive) frequency selective filters, such as low-pass, high-pass, band-pass or band-stop filters, are well known; their properties as well as their transformation into the reference low-pass filter representation (and back) are found in filter catalogs (for instance, Saal, 1978). It is not the aim of this book to present detailed data on digital filters; in this respect the reader is referred to a textbook on this matter (see references above). However, a short discussion follows. It deals with two types of digital filters: a) linear-phase filters in general, and b) comb filters in particular.

Linear-phase digital filters are of great importance in signal processing. Their phase response must have the form

$$\phi(\Omega) = k \cdot \Omega \tag{2.18}$$

where k is a constant. This condition can also be formulated as

$$\tau_{G} = d\phi/d\Omega = k = const \,, \tag{2.19}$$

i.e., linear-phase filters have a constant group delay time. This condition is fulfilled if and only if the poles and zeros of the filter transfer function are either situated in $z = 0$ or symmetrically with respect to the unit circle. Since poles are not permitted outside the unit circle for reasons of stability, it is clear that all the poles must be located in $z = 0$, and that the linear-phase filter is necessarily *nonrecursive*. If the filter coefficients take on real values (which is always assumed to be the case), complex zeros appear in quadruples; given a complex zero z_{01}, the other three zeros are

$$z_{02} = z_{01}^*; \quad z_{03} = 1/z_{01}; \quad z_{04} = 1/z_{01}^* . \tag{2.20}$$

If there are additional zeros on the unit circle, the phase response of the filter is piecewise linear; it contains a jump of $m \cdot \pi$ whenever a zero of order m on the unit circle is crossed.

Any nonrecursive filter has a *finite impulse response*; from the state equation (2.11a), with all recursive coefficients g_i, $i = 1 (1) k$ equal to zero, it follows that

$$y(n) = \sum_{i=0}^{k} d_i x(n-i) \tag{2.21}$$

and the impulse response becomes

$$h(n) = \begin{cases} d_{k-n} & n = 0 \ (1) \ k \\ 0 & \text{otherwise} . \end{cases} \tag{2.22}$$

In addition, the coefficients of linear-phase filters are symmetrical,

$$d_i = \pm d_{k-i} . \tag{2.23}$$

According to the sign of (2.23) in an actual implementation, the coefficients (and, in consequence, also the impulse response) show an even or odd symmetry with respect to $K = k/2$.

To make the design easier, a negative delay of K samples is usually introduced in linear-phase filter theory. Doing this, the phase function is forced to zero or $\pm\pi/2$. The filter becomes noncausal; adding a positive delay of K sampl s, however, heals this ff ct in a practical implementation.

One group of filters which must be mentioned since they play a major role in pitch determination is formed by the *comb filters*. The overall frequency response of these filters is neutral; they are neither high-pass nor low-pass filters, but the frequency response consists of a regular periodic sequence of passbands and stopbands. The comb filters that are piecewise linear phase are the important ones for pitch determination. All the zeros of these filters are situated on the unit circle, and they form regular, equally spaced patterns. Figure 2.9 gives some examples. In the following the term *comb filter* will be used in this restricted sense; it will be reserved for those comb filters which are linear phase and have their zeros on the unit circle.

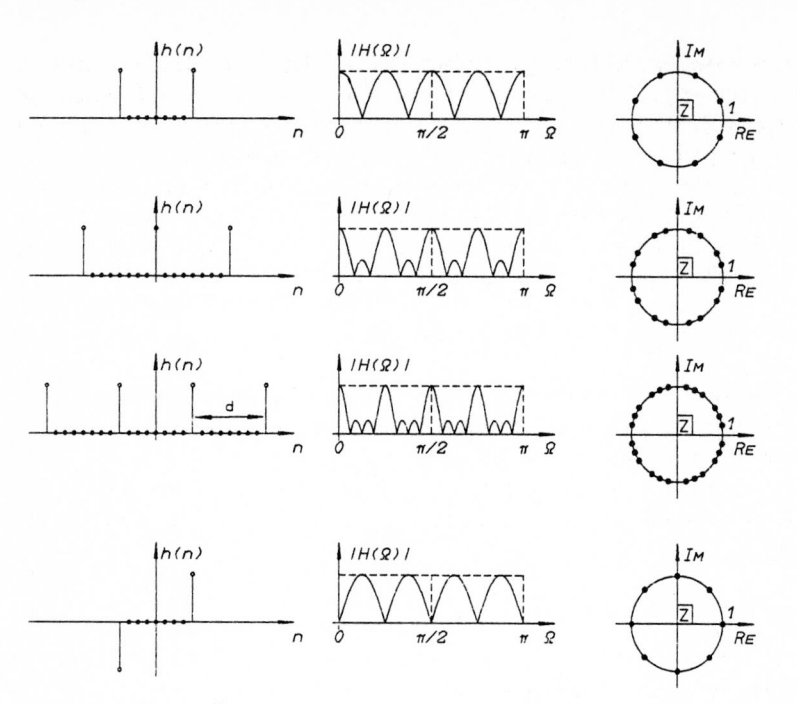

Fig.2.9. Examples of comb filters: impulse response, frequency response, and location of zeros in the z plane. Value of the delay parameter d: 8 samples

The computational burden of comb filters is extremely low: they do not need any multiplications. The simplest comb filters are the *two-pulse filters*: their impulse response simply consists of two pulses which are separated from each other by a delay d,

$$y(n) = x(n-d/2) \pm x(n+d/2) \qquad (2.24)$$

which gives the transfer functions (in the noncausal representation)

$$H_+(\Omega) = 2 \cos \Omega d/2$$
$$H_-(\Omega) = 2j \sin \Omega d/2 \qquad (2.25)$$

for equal signs or opposite signs of the two pulses. According to whether the pulses have equal signs or not, the filter is best suited for enhancement or suppression of a periodic signal whose period T_0 matches the delay parameter d of the filter.

If the impulse response contains more than two pulses (let us assume equal signs), then the filters become more selective with respect to the periodic signal. In this generalized case the impulse response would read

$$y(n) = \sum_{i=0}^{K-1} \left\{ x \left[n + \frac{(2i+1) \cdot d}{2} \right] + x \left[n - \frac{(2i+1) \cdot d}{2} \right] \right\} \qquad (2.26a)$$

for an even number, or

$$y(n) = x(n) + \sum_{i=1}^{K} [\ x(n+d) + x(n-d)\ ] \qquad\qquad (2.26b)$$

for an odd number of pulses. Filters with opposite signs of the pulses are of minor importance in speech processing if the number of pulses exceeds two.

Comb filters are widely used in the determination of the (unknown) period of periodic signals. In Sects.7.1, 8.3, and 8.6 pitch determination algorithms based on this filter principle will be discussed.

2.4 Time-Variant Systems. The Principle of Short-Term Analysis[15]

The speech signal is *time variant*, i.e., the essential control parameters, among them pitch and voicing, change with time. Figure 2.10 shows an example. It is therefore of primary importance to determine their momentary values rather than their overall (long-term) average. For this purpose a *short-term (short-time) analysis*[16] is carried out. The *short-term representation* of a signal x(n) is obtained by *windowing*, i.e., weighting the signal x(n) by the *windowing function* (or *weighting function* or *window*) w(k),

$$x_S(n,q) = x(n)\ w(q-n)\ . \qquad\qquad (2.27)$$

The window w(k) is an arbitrary time function which is significantly different from zero only within a small interval, conveniently situated around or starting at $k = 0$, and zero outside that interval. An example is the rectangular window,

$$w_r(k) = \begin{cases} 1 & k = -K/2\ (1)\ K/2-1 \\ 0 & \text{otherwise} \ . \end{cases} \qquad\qquad (2.28)$$

Within the interval $k = -K/2\,(1)\,K/2-1$, which corresponds to the interval $n = q-K/2\,(1)\,q+K/2-1$ (the *window interval*) of the original signal, the parameters to be extracted from the signal $x_S(n,q)$ are not normally expected to change significantly so that one can treat them as constant without causing significant errors. Parameter estimation algorithms are repetitively applied to the signals

---

[15] This section discusses the general aspects of short-term analysis methods in the domain of speech processing. The implications of short-term analysis on pitch determination in particular are dealt with in Section 8.1.

[16] In connection with time functions the term *short-time analysis* is often used instead of *short-term analysis*. In this book *short-term analysis* is preferred since the meaning is somewhat more general; it permits labelling this analysis principle in the same way even if the independent variable of the signal involved is different from time.

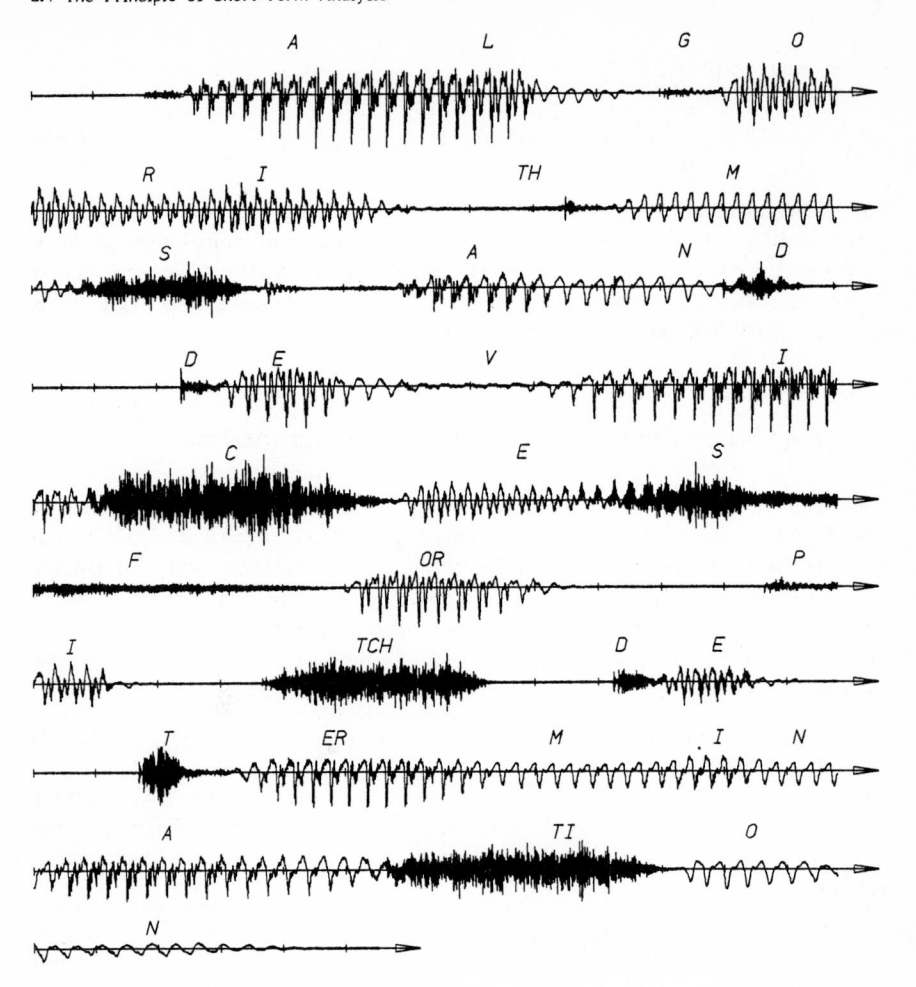

Fig.2.10. Example of a speech signal. Spoken sentence: "Algorithms and devices for pitch determination." Speaker: MIW (male). Scale: 320 ms per line

$x_S(n,q)$ for all those values of q for which the parameters are desired. A single parameter set extracted for an individual point q is called a *frame*[17] (Fig.2.11). The length K of the window is the *frame length.* This value is chosen in such a way that, on the one hand, the parameters to be measured remain constant or at least quasi-constant and, on the other hand, that there are enough samples of x(n) within the frame to guarantee reliable parameter determination. Obviously the choice of the windowing function influences the values of the short-term parameters; the shorter the window, the greater

---

[17] It is usual to assign the term *frame* to the short-term signal $x_S(n,q)$ for an individual value of q as well.

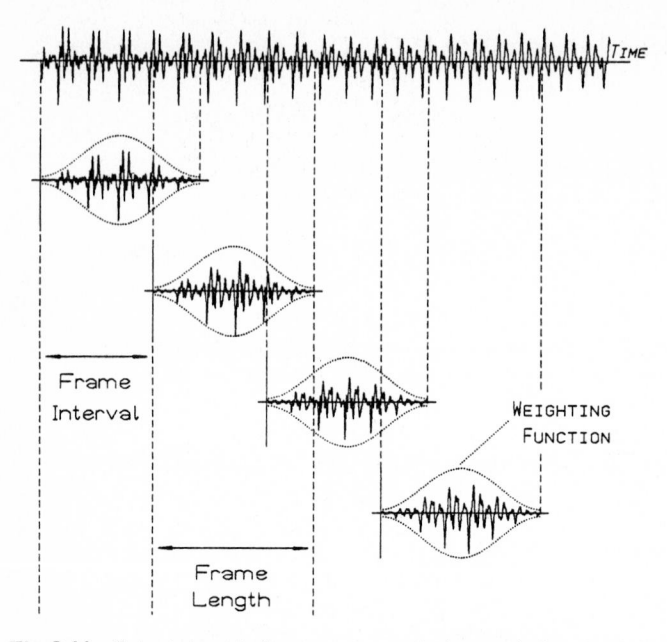

Fig.2.11. Principle of short-term analysis - window and frame. (Upper figure) Original signal; (lower figures) signal after windowing (solid line) and window (dotted line). A Hamming window was applied as the windowing function in the example displayed

this influence. Further details can be found in the literature, especially in a paper by Harris (1978) who presented a comparative review of windows and their performance in filtering and spectral analysis, or in a paper by Geçkinli (1979). As an example, Fig.2.12 shows four common windows and their Fourier spectra.[18]

To obtain a higher degree of freedom for the analysis (and thus a wider choice of analysis methods), it is convenient to declare the short-term signal $x_S(n,q)$ *undefined* outside the window interval, i.e., to define $x_S(n,q)$ as follows,

---

[18] The multiplication of the signal $x(n)$ by the window $w(k)$ in the time domain causes a convolution of the pertinent spectral functions $X(z)$ and $W(z)$ in the frequency domain. This convolution introduces spectral distortions: 1) smearing of the peaks of the original spectral function due to the bandwidth of the main lobe of $W(z)$, with the consequence of measurement incertainties, and 2) spurious peaks which result from the side lobes of $W(z)$. The distortions will be a minimum when $W(z)$ has a narrow and distinct main lobe and negligibly small side lobes. These two requirements, however, are conflicting so that any window represents a compromise. For this reason a large number of window proposals have been made in the literature.

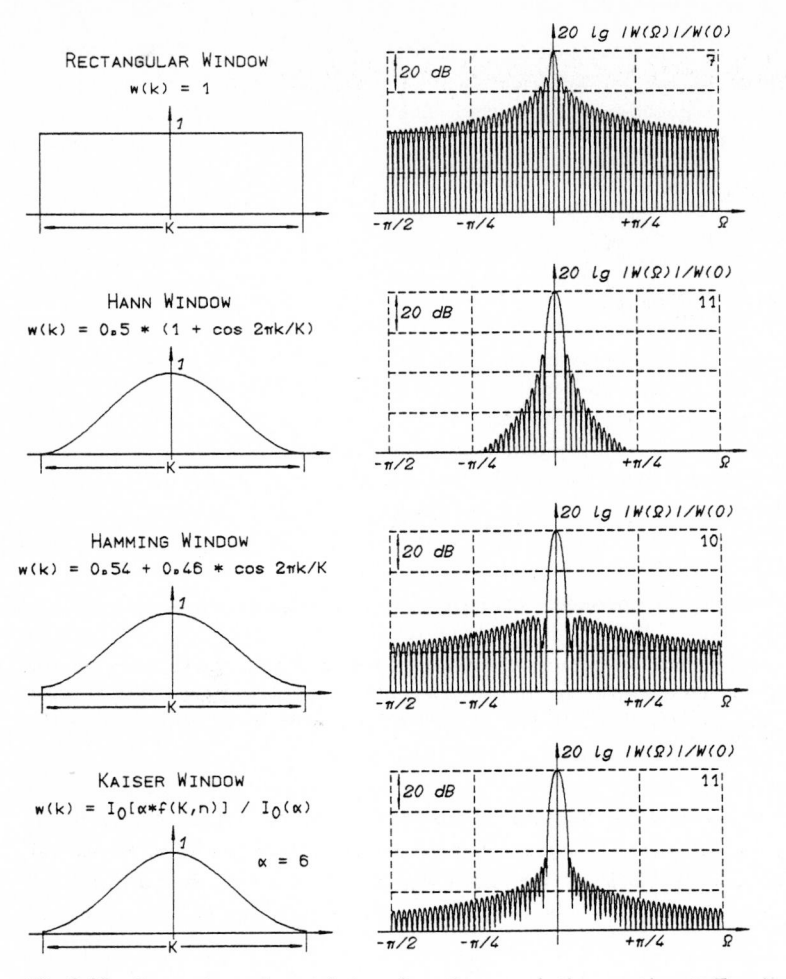

Fig.2.12. Examples of weighting functions and the pertinent Fourier spectra. The numbers in the upper right-hand corner of the spectra denote the 3-dB bandwidths (in spectral samples) of the main lobes. In the example displayed the transformation interval N has a length of 1024 samples; K equals 128 samples

$$x_S(n,q) := \begin{cases} x(n)\, w(q-n) & n = q-K/2 \ (1) \ q+K/2-1 \\ \text{undefined} & \text{otherwise}. \end{cases} \qquad (2.29)$$

Using definition (2.29), arbitrary values can be assigned to $x_S(n,q)$ outside the window interval. These values are chosen in such a way that the subsequent analysis is optimally performed. There are three usual ways of defining $x_S(n,q)$ outside the window interval for speech.

1) **Stationary analysis.** The short-term signal $x_S(n,q)$ is forced to zero outside the window interval so that (2.27), for a given value of q, is valid for

all values of n. The signal $x_S(n,q)$ thus turns into a signal of finite duration and finite energy, and all the methods used to analyze infinite time-invariant signals become applicable to this signal. Examples are autocorrelation analysis (see Sect.8.2) and the autocorrelation method of linear prediction (Markel and Gray, 1976). The stationary approach is suitable when the frame is permitted to be relatively long. For pitch determination this means that the frame is to contain several periods.

2) **Nonstationary analysis.** The analysis and parameter estimation is confined to the finite interval $n = q-K/2\,(1)\,q+K/2-1$. If samples outside this interval are needed, they are taken from the original signal (with or without previous windowing). With this approach only analysis methods that assume the signal to be time variant can be applied. In case other methods are applied, they have to be redefined in a suitable way. Examples of nonstationary analysis methods are the covariance method of linear prediction in its classical formulation (Atal and Hanauer, 1971; Markel and Gray, 1976), or - in pitch determination - the average magnitude difference function (see Sect.8.3). This approach is most suitable when the frame, for any reason whatsoever, must be comparatively short.

3) **Synchronous** or **periodic analysis.** The window interval of $x_S(n,q)$, $n = q-K/2\,(1)\,q+K/2-1$, is regarded as one period of a periodic signal $x_{SP}(n,q)$ with the period K. Harmonic analysis can then be performed on $x_{SP}(n,q)$. Some measurement methods implicitly require this analysis to be applied. The most prominent example is the discrete Fourier transform (DFT) which is derived from harmonic series analysis (see the discussion further below). If this analysis is to be performed on a nonperiodic signal or on a signal whose period T_0 does not match the frame length K, then the signal must first be forced to zero outside the frame, and care must be taken that the transitions between adjacent "periods" of the signal $x_{SP}(n,q)$ do not have a disadvantageous effect on the results of the analysis.

There is no difference between these three methods of short-term analysis as long as the frame length K can take on arbitrary values, and as long as the measurement can be kept strictly within the frame. This is possible for measurements such as zero crossings rate or intensity. Some short-term measurement methods, as we have seen, implicitly require a certain principle to be applied. For pitch determination, stationary and nonstationary short-term analysis are mainly used; synchronous analysis is not suitable since the frame length K usually does not match the period T_0; it is therefore not used unless one is forced by the algorithm to do so. Once the period T_0 is known, however, the principle of synchronous short-term analysis can be successfully used to reduce the distortions introduced by the windowing process to a minimum by choosing the frame length K equal to the period T_0 or equal to a multiple thereof. In speech processing this analysis method is known as *pitch-synchronous* analysis. The problem of windowing in short-term analysis will again be taken up in Chapter 8.

Parameter extraction is a method of data reduction. The parameters are thus postulated to vary much more slowly with time than the signal itself. Parameters in time-variant systems, however, can be regarded as time functions like the signals from which they have been extracted. They are thus subject to the sampling theorem in the same way.[19] From the signal processing point of view, the main difference is that the sampling rate of the parameters, which is called the *frame rate*, can be much lower than the sampling rate of the signal. The *frame interval*, as the reciprocal of the frame rate, thus defines the positions q_i of the individual frames. Often the structure of the speech signal supports the application of an adaptive frame interval rather than of a constant one. The size of the frame interval often corresponds to that of the frame length, but the two need not be equal (Fig.2.11). For some measurements a certain overlap in processing is required, i.e., a frame interval which is shorter than the frame length. Usual values in speech analysis are

-- 20 to 50 ms for the frame length, and
-- 5 to 25 ms for the frame interval.

The most widespread short-term measurement is spectral analysis. In the digital domain this is done by means of the *discrete Fourier transform* (DFT),

$$X(m) = \sum_{n=0}^{N-1} x(n) \exp(-2\pi jnm/N) \qquad (2.30)$$

This transform converts the N samples $n = 0\,(1)\,N-1$ of the input signal $x(n)$ into the Fourier spectrum $X(m)$. In this transform the Fourier spectrum $X(m)$ is also sampled. The accuracy in the frequency domain, i.e., the frequency

---

[19] With regard to this question, Holmes (1982, private communication) made the following comment: "There seems ... to be an inherent difference in the way one should look at the sampling theorem for speech signals and parameter signals. Because of the frequency selective operation of physical systems, and in particular of the peripheral auditory system, it makes sense to define a bandwidth of a speech signal, and provide proper band limiting so that the sampling theorem applies. At the much slower rates of variation of parameter signals it is obviously acceptable to use a much lower sampling rate. However, the physical processes underlying the parameters are not obviously band limited to anything like the frequency range implied by such a rate. Obvious examples, where a major change occurs in a few milliseconds, are 1) amplitude change at stop consonant release, and 2) formant frequency change at the boundary between vowels and nasal or lateral consonants. Because of these factors the sampled-data representation of parameters at a low sampling rate may actually be a 'better' representation (in some sense) of the underlying parameters if proper anti-aliasing filters are *not* used." - A good example how dangerous the application of anti-aliasing or other low-pass filters on parameter signals can be, is given in Sect.6.6.2 in connection with smoothing and low-pass filtering of pitch contours. Low-pass filters in parameter signals should thus be used with extreme caution, and it may be better, as Holmes stated, to accept certain aliasing effects in the parameter representation where such major changes occur.

interval between two adjacent spectral samples, directly depends on the length N of the time function,

$$f = 1/NT ,\qquad\qquad (2.31)$$

where T is the sampling interval of the signal. Since sampling in the frequency domain induces periodicity in the time domain, this DFT is nothing but the harmonic analysis of a periodic, sampled time function. The DFT and its fast algorithmic implementation, the *fast Fourier transform* (FFT), have been extensively discussed in the literature (see, for instance, G-AE Subcommittee, 1967; Nussbaumer, 1980). It is often convenient to write (2.30) in matrix form,

$$\mathbf{X} = \mathbf{W}\,\mathbf{x}\qquad\qquad (2.32)$$

where $\mathbf{X}^T = [X(0), X(1), X(2) \ldots X(N-1)]$ is the spectral vector, and $\mathbf{x}^T = [x(0), x(1), x(2) \ldots x(N-1)]$ is the signal vector; T means that the column vector ($\mathbf{x}$ or $\mathbf{X}$) has been transposed into a row vector (which is done for ease of writing). Hence the signal vector $\mathbf{x}$ is transformed into the spectral vector $\mathbf{X}$ by matrix multiplication of the transformation matrix $\mathbf{W}$ and the signal vector $\mathbf{x}$. For the DFT the matrix $\mathbf{W}$ is written as follows:

$$\mathbf{W} = \begin{pmatrix} w^0 & w^0 & w^0 & \cdots & w^0 \\ w^0 & w^1 & w^2 & \cdots & w^{N-1} \\ w^0 & w^2 & w^4 & \cdots & w^{2(N-1)} \\ \vdots & \vdots & \vdots & & \vdots \\ w^0 & w^{N-1} & w^{2(N-1)} & \cdots & w^{(N-1)(N-1)} \end{pmatrix} \quad ; \; w = \exp(-2\pi j/N) \qquad (2.33)$$

This formulation is very suitable since any other discrete transform, provided it is built up as a sum of products, can be written in the form of (2.32); it is only necessary to replace the transformation matrix $\mathbf{W}$ since all the properties of the transform are concentrated there.

Frequency is the independent variable in the spectral domain when the Fourier transform has been applied to the signal. This need not be the case for an arbitrary short-term transformation. There are a number of short-term transformations that do not leave the time domain although they can also computed via (2.32), with a different matrix $\mathbf{W}$, of course. In order to avoid confusion with running time (index n or q), the target domain of any short-term transform will be labeled *spectral domain* regardless of its independent variable. In case the independent variable of the spectral domain is time, it will be labeled *delay time* or *lag* (index d) in order to distinguish it from running time, which is the independent variable of the original signal.

The length N of the DFT (the *transform interval*) is not necessarily equal to the frame length K. Often the desired spectral resolution or the implementation of a special fast algorithm demand that the transform interval be longer. In this case the defined samples $n = q-K/2\,(1)\,q+K/2-1$ of an individual frame have to be extended by zeros or in some other suitable way in order to meet the length requirement of the applied short-term transformation.

2.5 Definition of the Task. The Linear Model of Speech Production

Three parameters determine the voice source: 1) the fundamental period T_0, 2) the amplitude A of the oscillation, and 3) the voice source spectrum. To a first (but quite good) approximation we can model the voice source as a linear system; in addition the backward influence of the supraglottal structure is usually neglected since the acoustic impedance of the vocal tract is assumed to be small compared to that of the glottis. (This is not always true; in particular, it does not hold very well for the lower formants.) In this case we can separate the three parameters and arrange them in an arbitrary sequence (Fig.2.13). The voice source is then defined to consist of the pulse generator P(z), the source filter S(z), and the amplitude component A. The pulse generator is defined to generate a train p(n) of ideal pulses of constant power, as defined in (2.15).[20] The interval between two successive pulses equals a fundamental period. The source filter contributes the voice source spectrum and converts the excitation pulse sequence into the excitation signal as it exists at the human glottis. The amplitude of the excitation signal is controlled by the amplitude component A. In order to complete the model for voiced speech, we must switch in series the transfer function H(z) of the vocal tract (see Sect.3.4 for a more detailed discussion) and the radiation R(z). At the output of the model we then obtain the speech signal x(n). This model helps to define the task of pitch determination from a signal processing point of view which is in this case closely related to speech production. The manifestation of pitch is given by P(z) or - more exactly - by the intervals between successive pulses of the signal p(n). Each interval between two consecutive pulses of p(n) was defined to be a *period* (cf. Sect.2.1). This definition also holds when the signal is irregular or not strictly periodic. Hence the task of pitch determination in terms of the model is the estimation of p(n) or P(z) from the speech signal x(n). The results, however, will be somewhat different according to the definition of fundamental period T_0 or fundamental frequency F_0, respectively. We can define T_0 locally as the elapsed time between the actual pulse of p(n) and the preceding one, that is, as the duration (or the exact location) of the individual fundamental period which pertains to a given value of the time variable *n* (this leads to the two "*instantaneous*" definitions of the term *period*, as introduced in Sect.2.1). On the other hand we can define T_0 in a more global way, for instance as the average elapsed time between successive pulses, averaged (not necessarily as the arithmetic average) over a given short-term interval ("*short-term*" definition of the term *period*). In this sense T_0 will also be called the *local*

---

[20] Quantization of the interval between consecutive pulses of p(n) due to the sampling of the signal (at usual sampling rates) may cause an audible roughness of the pitch contour in synthetic speech; the effect becomes more distinct for high fundamental frequencies as well as for low sampling rates. This issue is further discussed in Chap.4; for the formation of the present model it has been ignored.

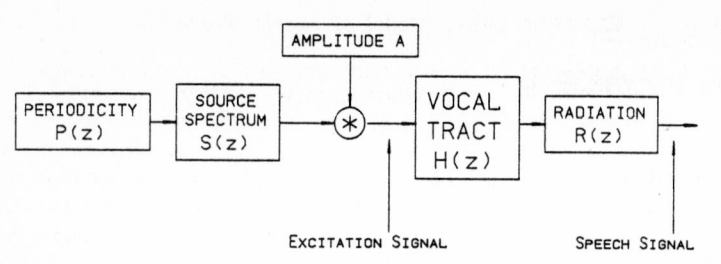

Fig.2.13. The linear model for production of voiced speech

average period if it is not clear from the context which definition is meant. All these definitions are valid, and there are PDAs avaliable for any of them.

From the above definition we find that it is the fundamental period that primarily varies with time, not the fundamental frequency. This agrees with the process of speech excitation; the excitation signal is set up as a (properly filtered) train of excitation pulses that are individually generated in the larynx; in principle, each excitation pulse can be regarded independently from its neighbors (although there is interaction between successive excitation pulses due to the production mechanism; see Chap.3). It is therefore suitable to attach the definition of fundamental frequency to that of period duration,

$$F_0 := 1/T_0 \, , \tag{2.34}$$

regardless of the actual definition of T_0 for a particular application.[21,22]

There remains the question of voicing. Besides pitch, voicing and the pertinent voice source features are the parameters controlling the behavior of the voice source. Solving the voicing determination problem is a premise for solving the pitch determination problem: it must be known *where the periods are* before it is possible to determine their duration. A partial solution is sufficient, however; it is only necessary to find those segments of speech where the voiced source is active, regardless of what the rest of the segments are like.

In order to define voicing in terms of the linear model, let us consider a voiced-unvoiced transition (Fig.2.14). In this model, voicing can be forced to

---

[21] The sign ":=" is taken from programming and means *is defined to be.*

[22] It is obvious that in (2.34) T_0 must be measured in *seconds,* and F_0 in *hertz.* For reason of convenience, however, both T_0 and F_0, in connection with the notation used throughout this volume, will be expressed in *samples* unless otherwise specified. To avoid confusion, the reader is therefore requested, before applying (2.34), to verify how T_0 and F_0 are represented in the particular context, and, if necessary, to replace the "1" in (2.34) by the appropriate constant which takes account of the current sampling rate and/or interval.

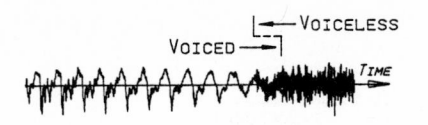

Fig.2.14. Example of a voiced-unvoiced transition. Speaker: MIW (male)

stop in two different ways: 1) by forcing the interval between successive excitation pulses to go to infinity, i.e., to exceed a given maximum period duration T_T; or 2) by forcing the amplitude of the oscillation to zero. In speech production both are done simultaneously and certainly by a nonlinear process. In the linear time-variant model, however, it appears more suitable to force the amplitude to zero than to increase the period. In doing so, we are able to completely separate the problems of voicing detection and pitch determination, and to assign an arbitrary value (according to what will be optimal for further processing of the obtained pitch contour) to T_0 outside the voiced interval where a fundamental period does not exist and is therefore not defined.

In order to further consider the problem of voicing, the linear model of speech production has to be completed. The source for voiceless sounds is set up in a way similar to that of voiced sounds. The noise generator $N(z)$ supplies the source filter $S_L(z)$ whose output signal is controlled by the voiceless amplitude component A_L. Since the amplitudes are allowed to go to zero we can simply add the output signals of the two sources before they pass the vocal tract. The control of voicing is then performed by the two amplitude components[23] (Fig.2.15).

As stated before, the voicing problem has to be partially solved as a premise for solving the pitch determination problem. In terms of the model the step necessary for pitch determination reduces to estimating whether the amplitude in the voiced path is zero or not. This leads to a redefinition of the voicing function which is then defined to take on a value of zero if the amplitude in the voiced path of the model is zero, and a value of one if not. In this respect, it does not matter whether a simultaneous voiceless excitation is present or not. Voiced fricatives and other sounds with mixed excitation

---

[23] This model is not satisfactory if one intends to deal with simultaneous periodic and noise excitation since the vocal tract transfer function is different for the voiced and the voiceless components due to the different place of excitation. The only way to handle this fact in the present model is to group the difference of the two transfer functions into the source spectrum of the voiceless path. This may be a clumsy approach if phenomena of mixed-excitation mode are to be handled. In the actual case, however, the model is only used to determine a crude estimate of voicing for pitch determination, and to quantitatively define the task of pitch determination itself. For such a purpose this model seems justified. On the other hand, we must not conceal that speech analysis methods which handle the signal from the *signal processing* point of view (see p.6 in Chap.1), such as linear prediction, implicitly make use of a model like this. The question of voicing will be further discussed in Sects.3.5 and 5.3.

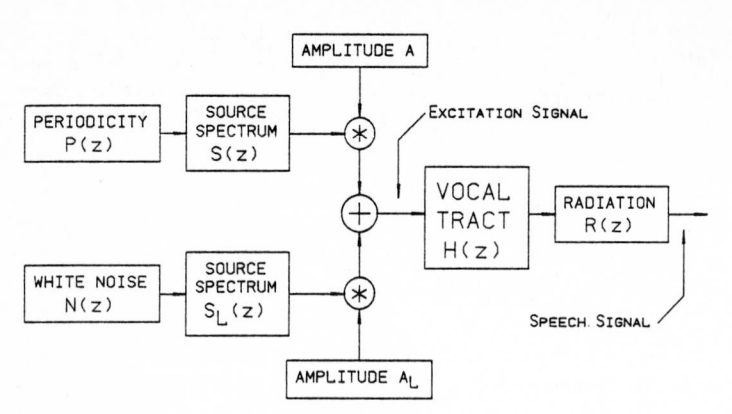

Fig.2.15. The simplified linear model of speech production

must be counted among the *voiced* sounds as long as voicing is really present in the actual utterance. In practical PDAs the determination of the voicing function is frequently performed simultaneously with pitch determination. Some realizations even set the voicing function to zero at segments where a periodicity cannot be found. This is dangerous, however, since one must allow the voiced signal to become irregular for short intervals (Lieberman, 1963; Fujimura, 1968); gross voicing determination errors can be the consequence of such an invalid premise.

In this book voicing determination and pitch determination are both assumed and treated as separate problems. The solution of the voicing problem is a premise for the solution of the pitch determination problem, but only to such an extent that the course of the voicing function must be known in order to correctly interpret the results supplied by the PDA. Knowledge of the voicing function is usually not a prerequisite for the operation of a PDA (although it is very suitable for reasons of implementation and effort to stop the PDA when the signal is definitely unvoiced). That means that the PDA and the VDA can (and should) be realized separately. Since there are a multitude of solutions to the voicing detection problem it would go beyond the limits of this book to give a detailed survey of them. An abridged survey will be presented in Section 5.3. For the survey of the PDAs the voicing problem is regarded as if it had been solved, and is not further pursued unless it directly influences the realization or the design of a particular PDA.

2.6 A First Categorization of PDAs

According to McKinney (1965) the usual realization of a PDA is subdivided into three main blocks which are passed through successively (Fig.2.16): 1) the *preprocessor,* 2) the *basic extractor,* and 3) the *postprocessor.* This subdivision

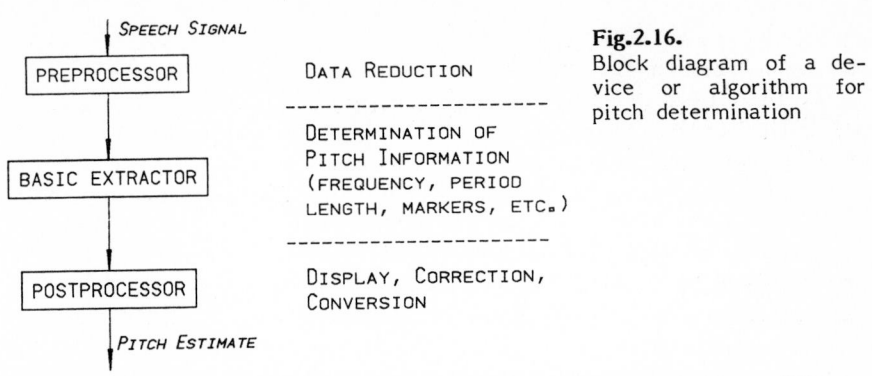

SPEECH SIGNAL

PREPROCESSOR → DATA REDUCTION

DETERMINATION OF
PITCH INFORMATION
(FREQUENCY, PERIOD
LENGTH, MARKERS, ETC.)

BASIC EXTRACTOR

DISPLAY, CORRECTION,
CONVERSION

POSTPROCESSOR

PITCH ESTIMATE

Fig.2.16.
Block diagram of a device or algorithm for pitch determination

is maintained here although the individual blocks are interpreted in a different way. The basic extractor performs the main task of measurement: it converts the input signal into a sequence of pitch estimates. The main task of the preprocessor is data reduction in order to increase the ease of pitch extraction. The postprocessor is a block which performs more diverse tasks, such as correction, error detection, smoothing of an obtained contour, time-to-frequency conversion and vice versa, and display of the parameter(s).

A first categorization of PDAs can be made according to their domain of operation (Fig.2.17), which is defined as the domain of the input signal of the basic extractor. If there is a time-domain signal at this point that has the same time base as the input signal, the PDA works in the *time domain*. The alternative is a PDA working in domains other than the time domain - the lag domain if the input is a correlation function, or the frequency domain if the input is a Fourier spectrum or some function derived therefrom. For reason of uniformity, all these domains (other than the time domain itself) will be labeled *spectral domains*. The common feature of all these PDAs is that a short-term transformation (such as the discrete Fourier transform or the computation of the short-term autocorrelation function) is included in the preprocessor. Hence these methods will be referred to as *short-term analysis* (STA) PDAs.

In the case of a time-domain PDA the output signal of the basic extractor is a train of laryngeal pulse estimates which are called *pitch markers* or simply *markers*. These markers form the *period boundaries* of the *measured periods*. They may or may not be in correct phase (see the above definitions as well as Chaps.3 and 4 and Sect.9.1 for more details). The preprocessor consists of a linear or nonlinear filter which performs a certain degree of data reduction. The time-domain PDA thus presumes the local definition of T_0 and permits the signal to be processed period by period. In the case of a

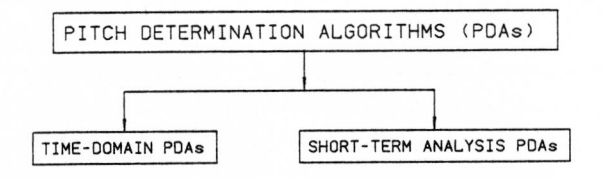

PITCH DETERMINATION ALGORITHMS (PDAs)

TIME-DOMAIN PDAs   SHORT-TERM ANALYSIS PDAs

Fig.2.17.
Preliminary categorization of pitch determination algorithms (PDAs)

short-term analysis PDA the output of the basic extractor is an estimate of the average pitch period duration or fundamental frequency within a given frame. This "average" is a short-term average which is derived from a small number of individual pitch periods. The essential step in the preprocessor is the short-term transform by which the time domain is left. The short-term analysis PDA follows the more global definition of T_0 as an average estimate, averaged in some defined way over a given frame so that finally one estimate per frame is supplied. Short-term analysis PDAs will be discussed in Chapter 8. Together with formal definitions of T_0 and F_0 in their respective meanings, the general issue of categorizing the PDAs will be taken up again in Sect.9.1 where the review of the individual PDAs will have been completed.

The simplest time-domain basic extractor [which is quite consistently called *frequency meter* in the literature (McKinney, 1965; Léon and Martin, 1969; Baronin, 1974)] is the *threshold analysis basic extractor* (TABE). This is defined to be a circuit (or routine) whose output signals the occurrence of threshold crossings (positive and/or negative) of the input signal. In this book, the name "frequency meter" for this component is avoided since it is more than that. Apart from the fact that it measures a period length and not a frequency, it locates the exact instant of each threshold crossing and is thus able to indicate the individual period boundaries by markers. The simplest case is given when the threshold is zero; the basic extractor is then called a *zero crossings analysis basic extractor* (ZXABE). The more complicated time-domain basic extractors investigate the temporal structure of the input signal in detail. Such an extractor is referred to as a *structural-analysis basic extractor*. All these basic extractors will be discussed in Chapter 6. The extractors of short-term analysis PDAs are usually simple *peak detectors*. They will not have a separate label in this book.

Two expressions associated with computer processing are *on-line* and *real-time* processing. They are often used as synonyms. In this book, however, they will be used with slightly different meanings. A PDA implemented on a computer or represented by a digital piece of hardware will run in *real time* when all the operations necessary for determining pitch from any given utterance take a shorter time than the utterance itself. On the other hand, a PDA will run *on-line* if it is able to determine pitch in a single step without requiring a substantial lookahead which would cause a significant processing delay. Hence a PDA is *instantaneous* only when it works both on-line and in real time. Otherwise it will cause a substantial delay in the total course of processing (e.g., in a vocoder or a speech transmission system). A nonsignificant delay of up to 50 ms, however, can be tolerated when speaking of instantaneous performance.

Finally a word on the errors which occur during operation of a PDA. The most common error is the detection of a higher harmonic instead of the first one; this is known as *higher-harmonic detection* (if continued for a longer time, *higher-harmonic tracking*). The opposite would be *subharmonic detection* (or tracking, respectively). Errors of this kind are also referred to as *octave errors* if the second harmonic or the subharmonic at $F_0/2$ is involved. Another error associated with time-domain techniques is *jitter*, i.e., some uncertainty of the position of individual markers. The estimate of T_0 then becomes noisy. If an individual marker is missed, this will cause a *hole*; in contrast to this there are *extra markers* (or *chirps*) which are the time-domain equivalent to higher-harmonic detection. A *hop* occurs when an individual marker is misplaced (Gold, 1962; Reddy, 1966); it is a peak of jitter confined to a single marker.

3. The Human Voice Source

This chapter presents a brief qualitative outline of the mechanism of the human voice source. It intends to give the reader an idea why the speech signal is as it is, and a feeling for what problems arise in estimating source parameters from the temporal structure of the signal. Like parts of Chap.2, the reader may skip this chapter if he is familiar with the basic aspects of voice production.

3.1 Mechanism of Sound Generation at the Larynx

The periodic source signal of voiced sounds is generated at the larynx by periodic vibrations of the (true) *vocal cords*[1]. The vibrations are caused as a consequence of forces that are exerted on the laryngeal walls when air flows through the *glottis* (the gap between the vocal cords).

The human vocal folds consist of two fleshy cords attached to the inside walls of the larynx. In a very general sense they serve as a constriction, to govern the amount of air flow into and out of the lungs. During phonation they appear as a thick membrane stretched (horizontally from anterior to posterior, i.e., from front to back) across a nearly circular tube (Titze, 1973a, 1974). Figure 3.1 shows a schematic diagram of the larynx in different views. The tension of the vocal cords is mainly governed by the vocalis muscle which forms the body of the vocal cords (Hirano, 1976; Fig.3.2). Above the vocalis muscle the cover consists of the mucosa and the lamina propria (vocal ligament). Cover and body have different degrees of stiffness and can move independently of each other to a certain extent. The transition layer (see Fig.3.2) was found to behave in about the same way as the cover. For more details the reader is referred to the original publication (Hirano, 1976 as well as previous and later publications) or to a tutorial handbook on larynx anatomy. The theory of larynx vibration, as it is understood today, has to a great extent been developed by van den Berg (1955b, 1958). The classic paper "On the Air Resistance and the Bernoulli Effect of the Human Larynx" (van

---

[1] The terms *vocal folds* and *vocal cords* (or *chords*) are more or less used as synonyms in the literature.

den Berg et al., 1957) provides details about the pressure losses in the glottis which play a vital role in today's understanding of larynx mechanics. For a more simplified case, as given in falsetto voice, comparable results had been derived earlier by Wegel (1930a,b). Flanagan (1958a) computed the glottal area and the volume velocity through the glottis from high-speed cinegraphic data (Farnsworth, 1940; Fletcher, 1950). Both approaches, that of van den Berg et al. (1957) as well as that of Flanagan (1958a), make use of a model of the glottal orifice as a rectangular constriction of about 18 mm length, 3 mm thickness, and variable width, the latter being the independent variable which causes the modulation of the air flow.

In simple terms, the course of one vibratory cycle (for a normal voice) consists of two parts: the *open-glottis* cycle and the *closed-glottis* cycle. Consider for the moment the glottis to be closed. In this case no air can pass it, and the *subglottal* (or *subglottic*) *pressure* P_s, which is usually defined as the pressure difference between the pressure in the lungs and the trachea on the one hand and the atmospheric pressure outside the human body on the other hand, is built up below the larynx. It exerts a force on the vocal cords which causes them to move outward. Now the glottis opens, and the air passes it. Due to the glottal constriction the air develops a considerable particle velocity. The Bernoulli force associated with this velocity builds up a negative pressure in the glottis which forces the vocal cords to move toward each other again until the orifice of the glottis is closed, and the air stream ceases. This cycle is repeated over and over. From this model assumption, Flanagan and Landgraf (1968a) have developed a one-mass model of the voice source for speech synthesis systems[2].

Evidence has been given, however, that the assumption of one uniform section of width w and thickness L is too great a simplification (Ishizaka and Matsudaira, 1968; Ishizaka and Flanagan, 1972; Baer, 1975). Figure 3.3 shows the course of one vibratory cycle according to a more sophisticated model of the vocal cords by Hirano (1976) and Lecluse (1977) which in essence follows the anatomical structure displayed in Hirano's paper (see Fig.3.2). Of course, simplifications have been made in this model; the transition layer and the cover have been merged together, and average (i.e., constant) values of stiffness have been assumed for cover and body. Stevens (1977) has carried out computations on the glottal pressure according to the proposal by Ishizaka and Matsudaira (1968) and according to the two-mass model (Ishizaka and Flanagan, 1972) derived therefrom. He assumed the constriction of the glottis

---

[2] This theory of voice generation is called the *myoelastic/aerodynamic* theory of phonation. It explains the oscillations of the vocal cords as more-or-less forced oscillations of elastic material. In opposition to this theory Husson (1962 and previous publications) postulated that every vibratory cycle of the vocal cords is individually controlled by a neurophysiological process. The contradiction between these two theories has challenged researchers' activities for a long time until so much experimental evidence was found against Husson's theory that is has practically been disproved.

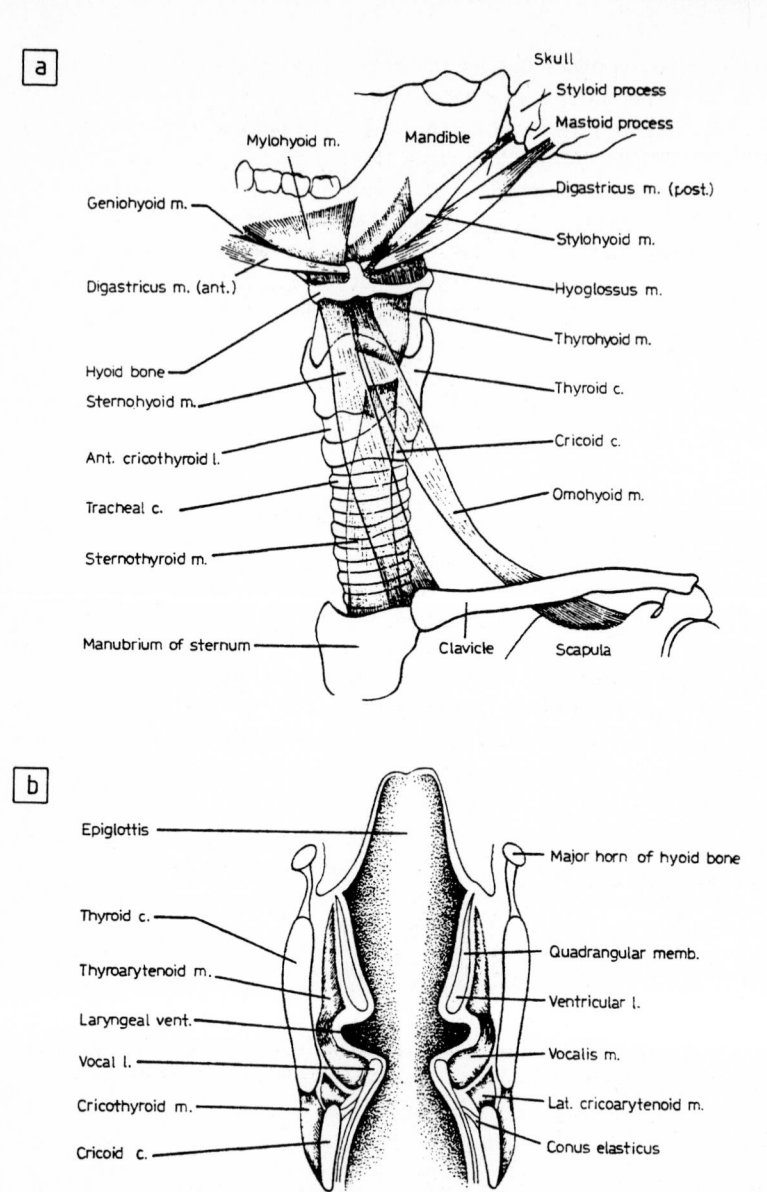

Fig.3.1a,b. Diagrammatic views of the larynx. (**a**) Extrinsic muscles (lateral view), (**b**) coronal view. (m.) Muscle, (l.) ligament, (c.) cartilage. From (Fant and Scully, 1977, p.250)

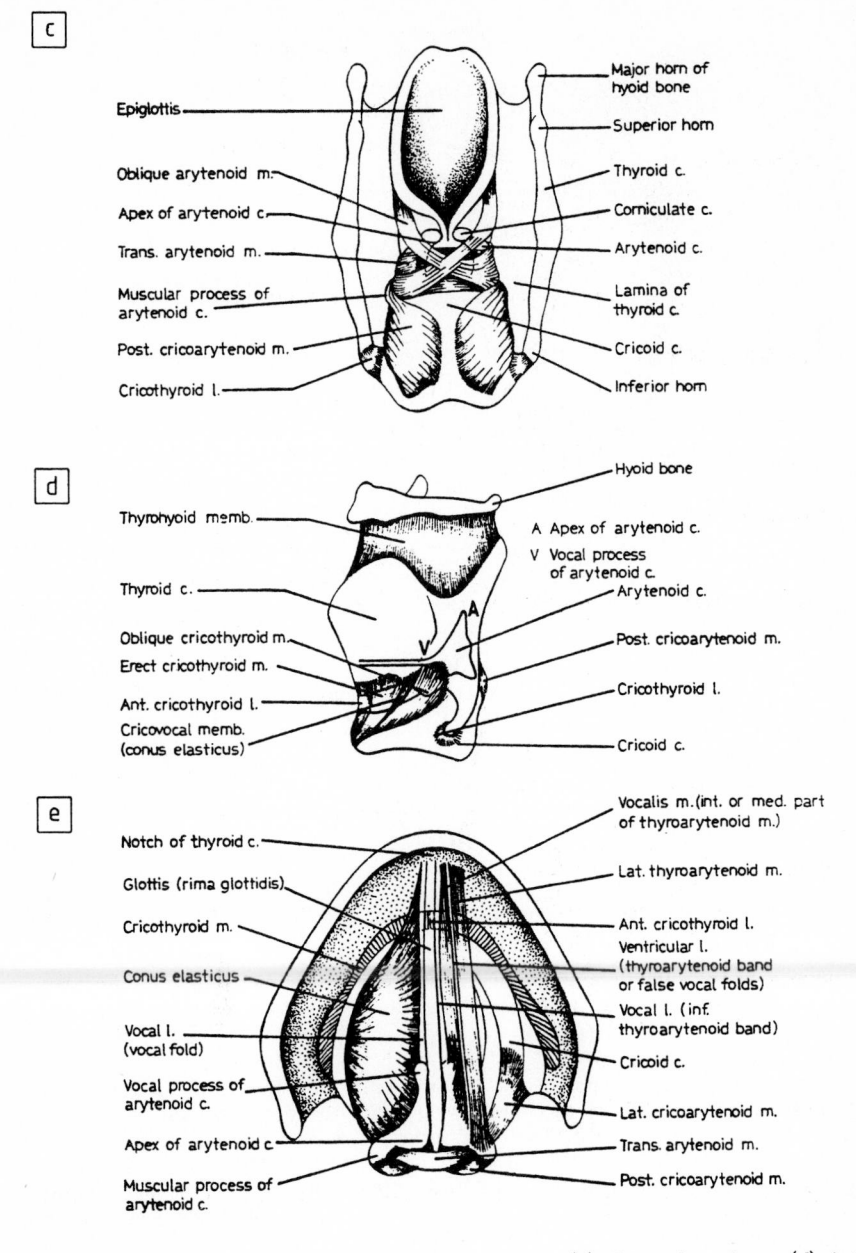

Fig.3.1c–e. Diagrammatic views of the larynx. (**c**) Posterior view, (**d**) lateral view, (**e**) superior view. (m.) Muscle, (l.) ligament, (c.) cartilage. From (Fant and Scully, 1977, p.251)

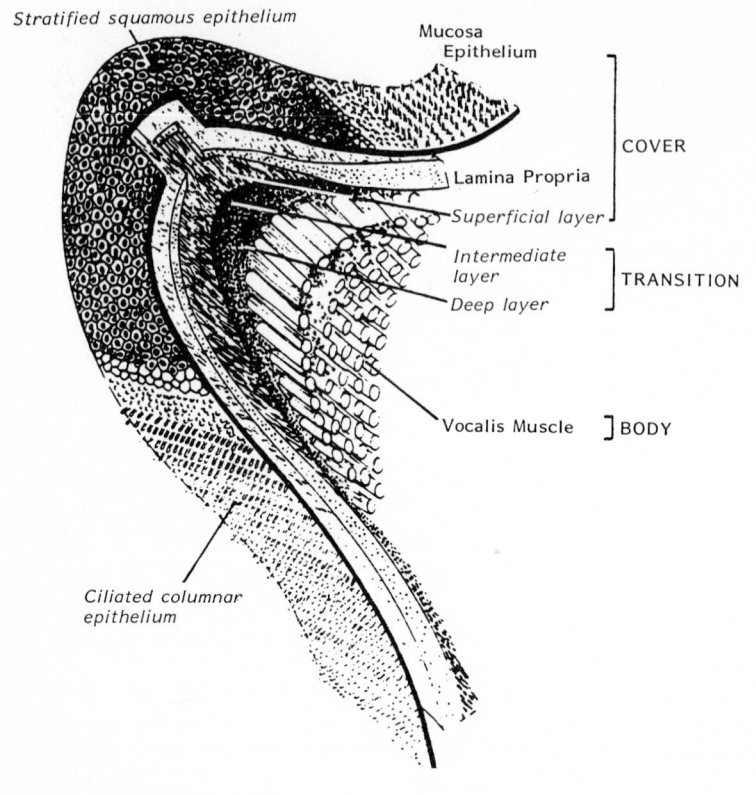

Fig.3.2. Schematic presentation of the structure of the human vocal cord. From (Hirano, 1976, p.20)

to consist of two uniform sections of width w_1 and w_2 and thickness d_1 and d_2 (Fig.3.4b). From the theory of fluid flow he calculated the pressure P_1 in the inferior section (w_1) of the glottis, depending on the widths w_1 and w_2 for a given value of the subglottal pressure P_s. He then explained the mechanism of phonation as follows.

The mechanical movements of the glottal walls take the form of a surface wave that propagates in an upward direction over a vertical length of 5-10 mm and with a velocity of the order of 1 m/s (Baer, 1975). Hirano (1974) has noted that the thickness of the surface 'cover' that participates in this wave propagation (Fig.3.4a) can be controlled in part by contraction of the vocalis muscle that lies beneath the surface. Ishizaka and Matsudaira (1968), Ishizaka and Flanagan (1972), and others, have approximated this wave phenomenon by representing each vocal cord by two coupled masses, one (m_1) for the inferior section of the cord and the other (m_2) for the superior section including the vocal ligament, as shown in Figure 3.4b. Each of these masses has a stiffness of its own (k_1 and k_2), which tends to return the mass to a rest position if it is displaced away from that position by some external force. There is also an interaction between the two masses, represented by a coupling stiffness k_c. The coupling

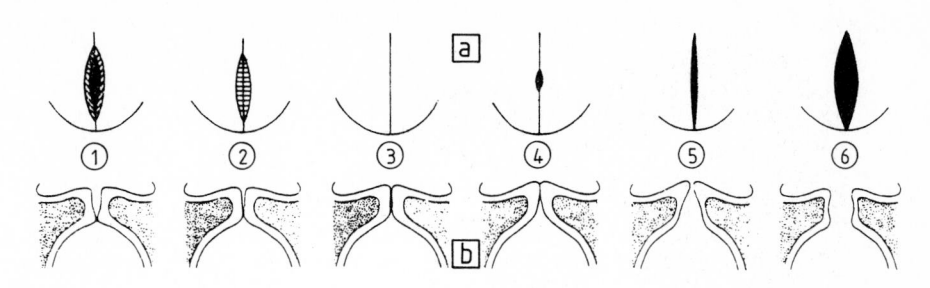

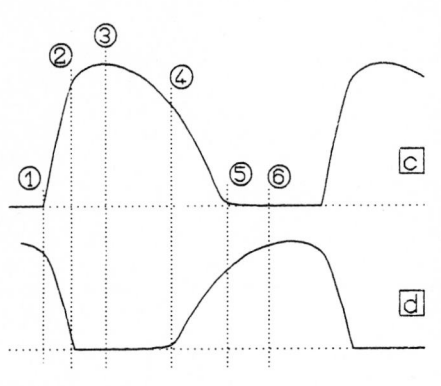

Fig.3.3a–d. Schematic display of the vocal-cord movement (male speaker, phonation in modal register - chest voice). The corresponding display for falsetto register is shown in Figure 3.7. (a) Superior view of the vocal cords (anterior at bottom), (b) cross-sectional view of the vocal cords, (c) pertinent glottogram (laryngo-gram); (d) excitation signal (for comparison). For more details on the glottogram see Sect.5.2.3 of this book or Lecluse's original work (1977), from which displays (a,b) are taken (p.71)

stiffness represents the fact that as one of the masses is displaced relative to the other, there is a force tending to restore the masses to their equilibrium position relative to one another, and this force is proportional to the difference between the displacements of the two masses. (Losses are omitted in this simplified model.) This simple representation of each vocal fold as two coupled masses is inadequate in several ways, but has some validity for explaining in a qualitative way some phenomena associated with laryngeal behavior. ...

In order to explain how vocal-fold vibration occurs, let us examine the shape of the vocal folds at various instants of time throughout a cycle of vibration. Figure 3.5b shows a cross section through the glottis at each of eight points in time throughout the cycle, together with approximations of the glottal shape in terms of two uniform sections. These sections were prepared by Baer (1975), and are based on his detailed measurements of the vibration of an excised larynx. The two-section approximations are identified with points on the chart that have been shown previously (Fig.3.5a), and these points are connected by a contour that represents the sequence of conditions that occur throughout one cycle of vibration. In our analysis of the mechanism of vibration, we start with the vocal folds in the configuration shown in panel 0 of Figure 3.5b. The superior section (mass m_2) is closed, and hence there is no airflow. The subglottal pressure applies an outward force to mass m_1, causing it to move laterally, as indicated in panels 1 and 2. In fact, the subglottal pressure causes a 'peeling apart' of the vocal folds, so that in panel 2 only a very thin section remains closed. As a consequence of the coupling between lower and upper masses, the abducted lower portion causes the superior section to open, as shown in panel 3. The outward movement of mass m_2 is rapid, and the configuration in panel 4 is reached one eighth of a period later. When the vocal folds are in a configuration represented by panels 3 and 4,

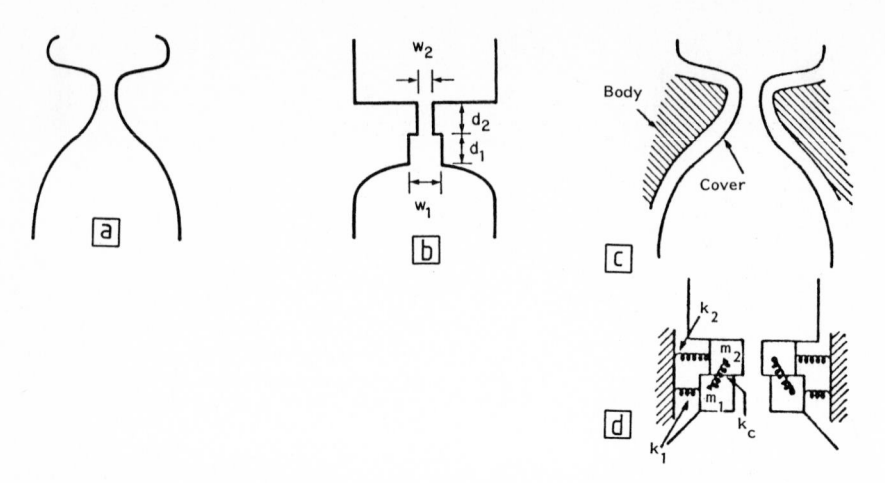

Fig.3.4a–d. Two conceptions of the vocal-cord mechanism. (**a**) Cross section through the vocal cords (after van den Berg et al., 1957); (**b**) representation of the glottis in form of two uniform sections; (**c**) simplified view of the vocal cords (Hirano, 1976) as consisting of a body (ligament plus muscle) and a cover (mucosa and parts of the transition layer) both of which can move independently to a certain extent; (**d**) two-mass mechanical model. From (Stevens, 1977, pp.265,267)

the force on mass m_1 drops abruptly, since, when there is airflow, the pressure within the inferior section is no longer equal to the subglottal pressure. Thus mass m_1 begins an inward movement since there is a restoring force due to the stiffness of this section of the vocal folds. Mass m_2 also begins an inward movement since it is coupled to m_1. When the configuration represented by panel 5 is reached, an inward force is generated on mass m_1 due to airflow through the constricted passage formed by m_1. Mass m_2 continues its inward motion, pulled by m_1, and eventually the superior section closes in panel 7. The lower section does not close completely for this mode of vibration. The entire cycle then repeats, beginning again with panel 0.

The fact that the trajectory in Fig.3.5 executes clockwise movements indicates that energy is being fed from the aerodynamic system into the mechanical system - a necessary requirement for the maintenance of vibration. Thus, for example, as the width w_1 increases (panels 7, 0, 1, 2, 3), the pressure P_1 is positive, and tends to assist the outward movement of the glottal walls; as w_1 decreases (panel 5), P_1 is negative, and tends to assist the inward movement of the walls in the inferior section of the glottis. An analogy can be made to a pendulum or a swing, in which oscillation is maintained by giving a push in the right direction at the proper point in the cycle. The amount of energy that the aerodynamic system transfers to the mechanical system (to make up for energy losses) is proportional to the area of the trajectory in Figure 3.5b. The position and shape of the trajectory depend upon the static opening of the glottis and on the physical properties of the vocal folds. When the area of the trajectory becomes too small, vocal-fold vibration is no longer maintained since the energy supplied from the aerodynamic system cannot make up for the losses in the mechanical system.

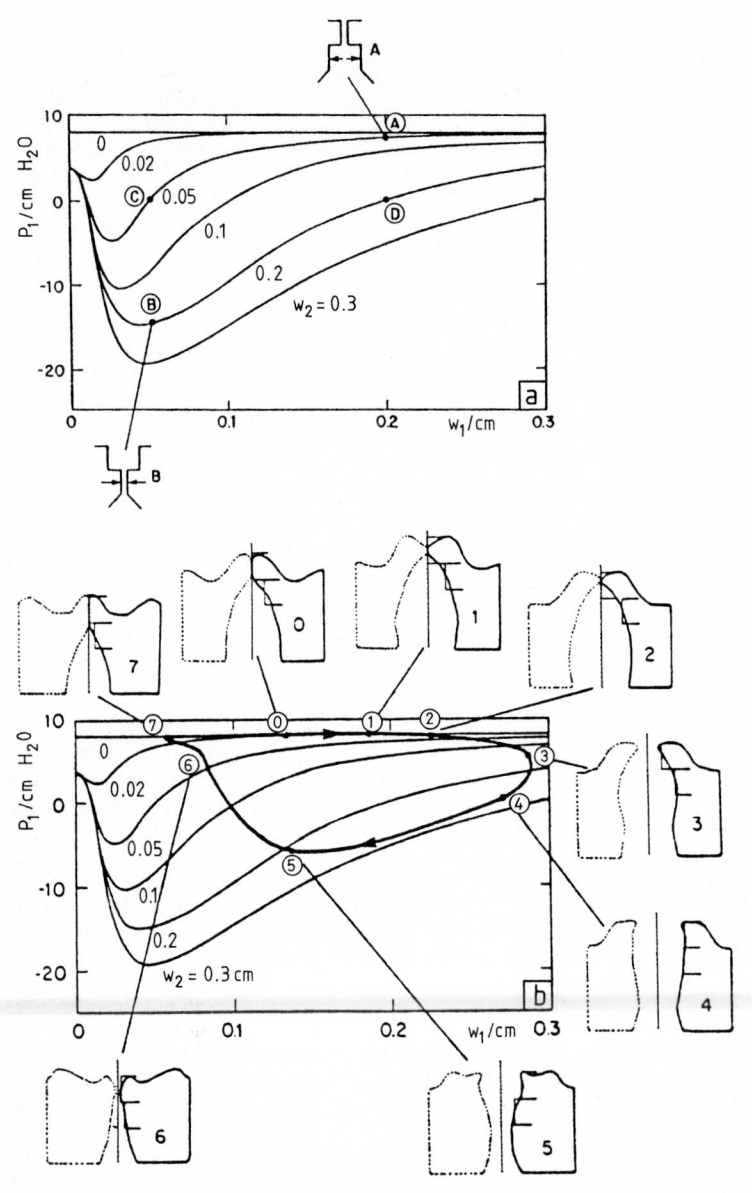

Fig.3.5a,b. Pressure in the lower section of the glottis and vocal-cord vibration. (a) Pressure in the lower section of the glottis for various widths of the upper constriction; (b) pressure and vocal-cord shape for 8 instants of time equally spaced throughout one cycle of vibration. Data after Baer (1975) and Ishizaka and Matsudaira (1968). Subglottal pressure: 8 cm H_2O; (w_1) width of the lower part of the glottal constriction (mass m_1); (P_1) pressure (relative to supraglottal pressure) at the lower part of the glottal constriction; (w_2) parameter of the pressure curves: upper constriction width (mass m_2). From (Stevens, 1977, pp.266,268)

The trajectory shown in Fig.3.5b is assumed to be representative of vocal-fold vibration for a normal configuration and stiffness of the folds. If the vocal folds are positioned farther apart, the trajectory shifts to the right and, for some degree of spreading, closure does not occur, so that the trajectory does not touch the upper line ($w_2 = 0$). On the other hand, adduction of the folds causes the trajectory to shift to the left in Figure 3.5b. When the coupling stiffness k_c between the upper and lower masses in the model becomes larger, there is less movement of one mass relative to the other, and the trajectory becomes flatter and smaller in area, since w_1 tends to remain about the same as w_2. If k_c becomes too large, oscillation will cease. Likewise, there is less total vocal-fold displacement and a smaller area if the stiffnesses k_1 and k_2 are increased.

This qualitative description of vocal-fold vibration suggests that there are a number of parameters that can be manipulated in the larynx to change the properties of the sound generated at the larynx. These parameters control the manner of vibration and the frequency of vibration, and determine whether or not vibration will occur. The parameters include the pressure across the glottis, the spacing between the vocal folds, the stiffness and mass of each fold, and the coupling stiffness between the inferior and superior sections of the fold. (Stevens, 1977, pp.265-270)

Figure 3.6 shows photographic pictures from one vibratory cycle of a human larynx, together with the pertinent glottogram. Note that the waveform of the glottogram displays the conductance of the larynx. It is thus some "negative" of the glottal signal and by no means the excitation signal itself (for more details, see Sect.5.2.3).

There are a variety of degrees of freedom in this system. The main factors of control are 1) the stiffness of the vocal cord which is mainly influenced by the tension of the vocalis muscle and the strain of the vocal cords, i.e., the elongation per unit of length (Hirano, 1976), and 2) the spreading of the cords which is mainly due to the distance of the arytenoid cartilage parts which delimit the vocal cords at the posterior end. The average voice source signal, however, is subject to major variations: interspeaker variations as well as intraspeaker variations due to the mode of phonation (register), to intensity, to fundamental frequency, all of which are controlled by the human speaker in a complicated way which is only partially understood.

It might be necessary to add some numeric data to the above considerations (Stevens, 1975; Flanagan, 1972a). The subglottal pressure usually takes on values ranging from 2 cm H_2O (very low voice) to 20 cm H_2O (very loud voice). The glottal air flow is found around 1 cm^3 per glottal cycle, with major variations according to intensity, fundamental frequency, and how well trained the speaker is (e.g., at singing). Catford (1964) gave values between 50 and 250 cm^3/s for vowels, while Lehiste (1970) found higher values (500 cm^3/s). As one might expect, the airflow (at constant intensity) is lowest for professional singers who thus provide a greater overall effectiveness of the pressure-to-sound conversion (Sundberg and Gauffin, 1978, 1979).

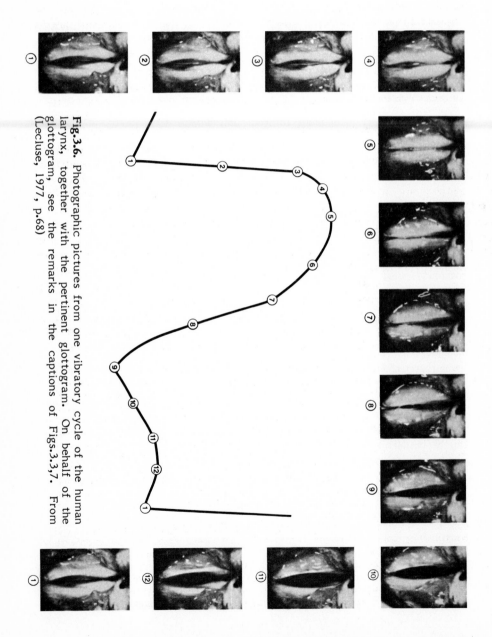

Fig.3.6. Photographic pictures from one vibratory cycle of the human larynx, together with the pertinent glottogram. On behalf of the glottogram, see the remarks in the captions of Figs.3.3,7. From (Lecluse, 1977, p.68)

3.2 Operational Modes of the Larynx. Registers

According to Catford (1964, 1968), man uses only a very small part of his acoustic repertoire for phonation. Apart from the facility to control intensity and fundamental frequency there are a number of vibration modes, i.e., different shapes of vibration of the larynx. Catford counted more than ten as being linguistically significant (for at least some languages) which cover different ranges of F_0 and show a different quality of voice. These modes - or *registers*[3] - represent different waveforms of laryngeal oscillation. In the literature there is no agreement as to the number of different registers or as to the terminology. Van den Berg (1960, 1968b) has significantly contributed to the explanation of register phenomena. According to the proposal by Hollien (1972), three major registers are to be distinguished for speech: 1) the *modal* register, 2) the *falsetto* register, and 3) *vocal fry*. Hollien has given some data on the frequency range of these registers. There is normally no overlap in F_0 between normal (modal register) phonation and vocal fry whose fundamental frequency is situated one octave or more below the corresponding values for the modal register. Between the modal register and the falsetto ("head") register, there is usually some overlap, i.e., some values of F_0 can be reached with both registers, though normally not with equal energy. Figure 3.7 displays a schematic view of the vocal-cord movements for the falsetto register, same as Fig.3.3 for the modal register [both figures from (Lecluse, 1977)]. Vocal fry apparently corresponds to what is called *creak* (Catford, 1964), *Strohbass* (van den Berg, 1968; Köster, 1972), or *laryngealization* (Lehiste, 1970). It is characterized by an extremely low F_0 and by a relatively small mean rate of airflow. Figure 3.8 shows an example of speech signals in modal, falsetto, and vocal-fry registers. Lehiste (1970) characterized some of these features as follows:

> Hollien et al. (1966) and Hollien and Wendahl (1968) define vocal fry as a phonational register of frequencies lying below the frequencies of the modal system. They measured a large number of samples of intentionally produced "vocal fry", finding that most of them involved repetition rates below any level that might be expected for a normal mode of phonation. The ranges were from approximately 28 to 73 pulses per second. One of the samples involved biphasic phonation, which had been previously described by Timcke, von Leden, and Moore (1959a,b). In that study, ultraslow motion pictures revealed that as the "fry" was produced, the vocal folds vibrated in a biphasic pattern: each vibratory cycle consisted of a rapid sequence of two movements, followed by a prolonged period of approximation. The transition from normal phonation to biphasic phonation did not change the

---

[3] The term *register* has been used for centuries to characterize vocal phenomena, especially in singing. It was defined by Garcia (1855), the inventor of the laryngoscope, in the following way: "By the word *register* we mean a series of succeeding sounds of equal quality on a scale from low to high, produced by the application of the same mechanical principle, the nature of which differs basically from another series of succeeding sounds of equal quality, produced by another mechanical principle".

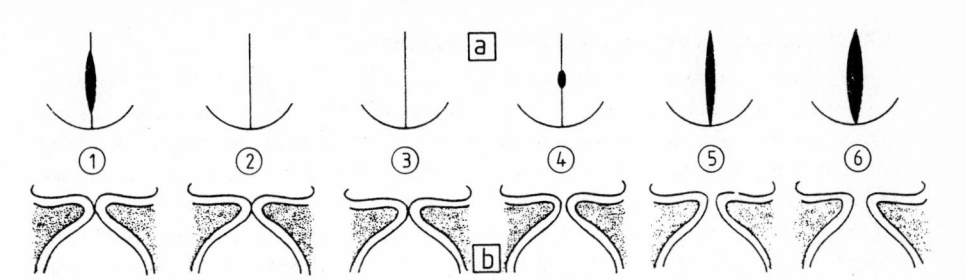

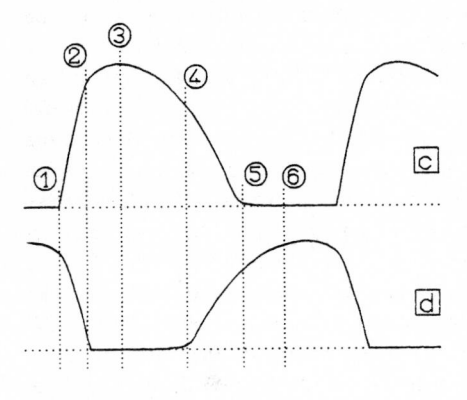

Fig.3.7a-d. Schematic display of the vocal-cord movement (male speaker, phonation in falsetto register). The corresponding display for modal register is shown in Figure 3.3. (**a**) Superior view of the vocal cords (anterior at bottom), (**b**) cross-sectional view of the vocal cords, (**c**) pertinent glottogram (laryngogram); (**d**) excitation signal (for comparison). For more details on the glottogram see Sect.5.2.3 of this book or Lecluse's original work (1977), from which displays (a,b) are taken (p.72)

number of individual pulses per second, but did change their arrangement. When a sound that had a frequency of approximately 150 Hz changed to "fry", the length of the complete (biphasic) vibratory cycle corresponded to a frequency of 75 Hz, and a "subharmonic" appeared at this frequency, which was audible in conjunction with the primary tone at 150 Hz. The two movements differed in amplitude: the amplitude of the first excursion of the biphasic cycle was 20 to 25% smaller than the amplitude of the second excursion. While evidence for this doubling activity is available from acoustical sources, the more common pattern present in laryngealization appears to be one of a single pulse followed by a period of no excitation. Another characteristic of vocal fry is nearly complete damping of the vocal tract between successive excitations ... When these modes of vibration are used in a linguistically significant way, I prefer to use the term laryngealization, referring to irregular, biphasic, or unusually slow vibration of the vocal folds (Lehiste, 1970, p.60).

The main reason why a section of this book has been devoted to the discussion of these aspects is the fact that things like vocal fry or abrupt register changes may occur anywhere in normal speech (perhaps except for stressed vowels). In the event that they do occur, they cause severe perturbations and irregularities in the speech signal. In trying to cope with this sort of signal any PDA may be led into systematic failure.

It must be noted that vocal fry, though irregular, is perceptually well distinguished from a hoarse voice which also generates irregular fundamental periods, but apparently in a perceptually (and optically) different way (Hollien, 1972).

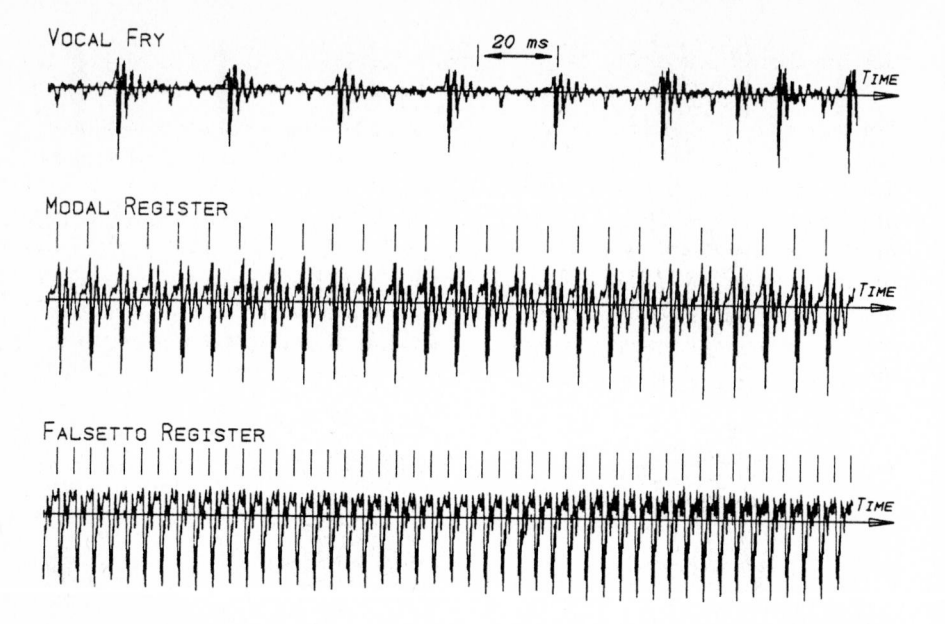

Fig.3.8. Examples of different registers. Spoken: sustained vowel /ε/. Speaker: WGH (male). Scale: 200 ms per line. The vertical lines indicate the results of pitch determination (using the PDA described in Chap.7). For vocal fry the PDA failed since the fundamental frequency of the utterance (around 30 Hz!) fell below the measuring range (lower limit: 50 Hz)

The usual register of phonation, i.e., the modal register (which by some researchers is subdivided into subregisters like *chest* or *mid* register) as well as the higher falsetto register (head register) exhibit regular F_0 patterns. Of course there are always slight perturbations which actually contribute to the "naturalness" of the speech sound[4]. Further details will be presented in Chapter 4. Although both are regular, the modal and falsetto registers sound very different for a normal speaker (assumed not to be a trained singer). Professional singers, however, are able to equalize these registers to such an extent that they sound like one (Large, 1973).

---

[4] In some special cases alternate laryngeal pulses are more similar to each other than successive ones. If this fact becomes perceptually significant, one might be obliged to redefine the fundamental period to occur between alternate laryngeal pulse times. This phenomenon is called *paired pulsing* by McKinney (1965). It occurs occasionally in normal speech and frequently in certain pathological voices. See, for instance, (Rabiner et al., 1976), or (van den Berg, 1954a).

3.3 The Glottal Source (Excitation) Signal

Let us return to the considerations made from a point of view more close to the source signal. What is the source signal like in terms of a waveform? The source signal itself is not directly accessible. The speech signal is generated from the source signal in the vocal tract and thus essentially the result of the convolution by some time-variant linear filter (Fant, 1960, 1970). To reconstruct the source signal from the speech signal one must thus carry out a deconvolution (see Chap.8) or must neutralize the influence of the vocal tract in some way. Another possibility for estimating the source signal is by direct or indirect instrumental measurement of the movement of the vocal cords.

Flanagan (1958a) computed the glottal area and volume velocity as a function of time from high-speed motion pictures supplied by Fletcher. Figure

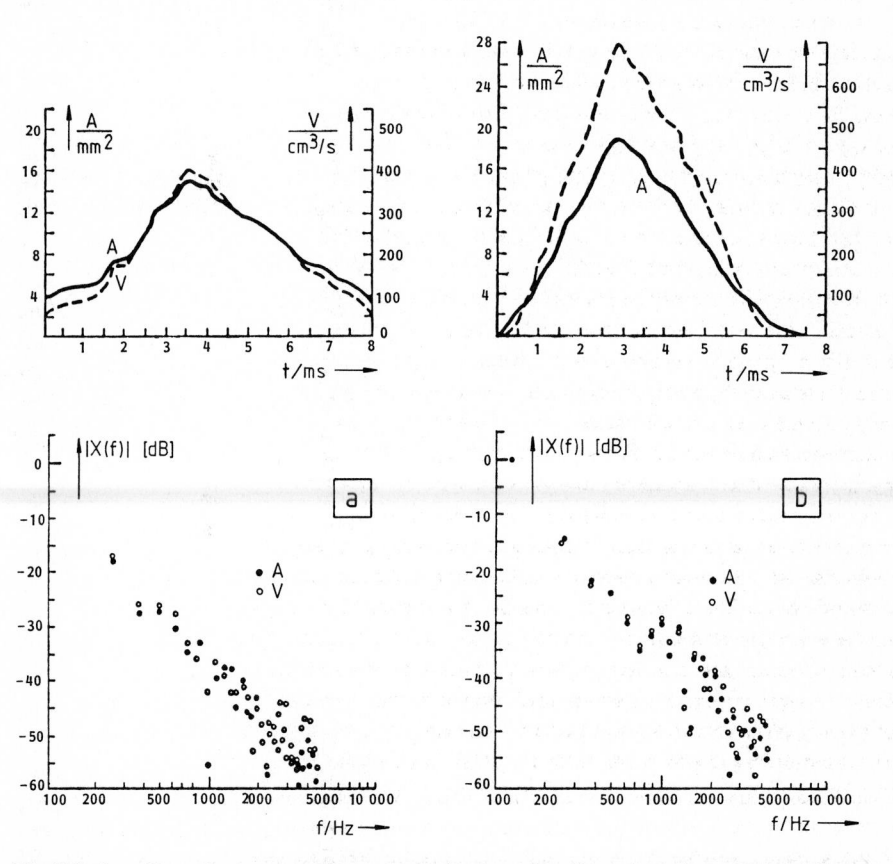

Fig.3.9a,b. Glottal area, volume velocity, and voice source spectrum for a male speaker. (a) Soft voice; subglottal pressure 4 cm H_2O; (b) loud voice, subglottal pressure 8 cm H_2O. (A) Area; (V) velocity. From (Flanagan, 1958a, p.106)

3.9 shows the results (waveform and pertinent harmonic Fourier spectrum) for a male speaker. The voice source waveform and its spectrum have been the object of extensive research. The main measurement methods are 1) determination of area and velocity functions from high-speed motion pictures (Flanagan, 1958a); 2) determination of the glottal area function from electric, i.e., laryngographic (Fourcin and Abberton, 1971), or ultrasonic (Hamlet and Reid, 1972) measurements; 3) determination of the acoustic source signal by mechanically neutralizing the vocal-tract influence (Sondhi, 1975a,b); 4) inverse filtering, i.e., inverting the transfer function of the vocal tract (Miller, 1959; Holmes, 1962b,c; Lindqvist, 1964, 1965, 1970; and many others); 5) results of model simulations and experiments (van den Berg et al., 1957; Ishizaka and Flanagan, 1972; Titze, 1973a, 1974); and 6) measurements on excised larynges (van den Berg et al., 1957; Baer, 1975). The individual measurement techniques are not of interest here; they will be discussed in Sects.5.2 and 6.5. The results obtained (which depend somewhat on the particular method applied) indicate the following.

1) The voice source signal (glottal excitation signal) is an asymmetric waveform of roughly triangular shape.

2) For normal speaking mode the glottis closes completely during a certain part of the fundamental period. The duty factor (Flanagan, 1958a, 1972a), i.e., the ratio of open-glottis time to period duration usually varies between 0.3 and 0.7. This depends on the intensity, the applied register, and the voice quality.

3) Due to the approximately triangular shape of the excitation signal the "usual" source spectrum exhibits a higher harmonic attenuation of about 12 dB/octave. This is only a basis of valuation, however. There are great deviations in the single harmonics from individual to individual and even from period to period. According to R.L.Miller (1959), *"perhaps the most important result which emerges from this work"* (i.e., an extensive inverse-filter study) *"is the knowledge that a uniform harmonic distribution in the glottal wave is a rarity"*.

Using the mechanical inverse-filter tube by Sondhi (1975a,b), Monsen and Engebretson (1977) investigated a number of excitation signals in different modes of voicing, including vocal fry. Some waveforms and source spectra of normal phonation in modal register are displayed in Figure 3.10.

According to Stevens (1975), there are different voice qualities (within the same register) which lead to different overall slopes of source spectra (Fig.3.11). Stevens distinguishes between the normal voice, "breathy" voice (the glottis is always open, the air flow is relatively high, the intensity is low) and "creaky" voice or screaming (the duty factor is small, the air flow is low, at least relative to the high subglottal pressure, and the high frequencies in the spectrum are enhanced which gives the voice a shrill character). Thus the beginning of the -12 dB decay of the higher harmonics can be shifted along the frequency axis by laryngeal control. The degree of discontinuity in the derivative of the glottal signal at the closure point (indicated by

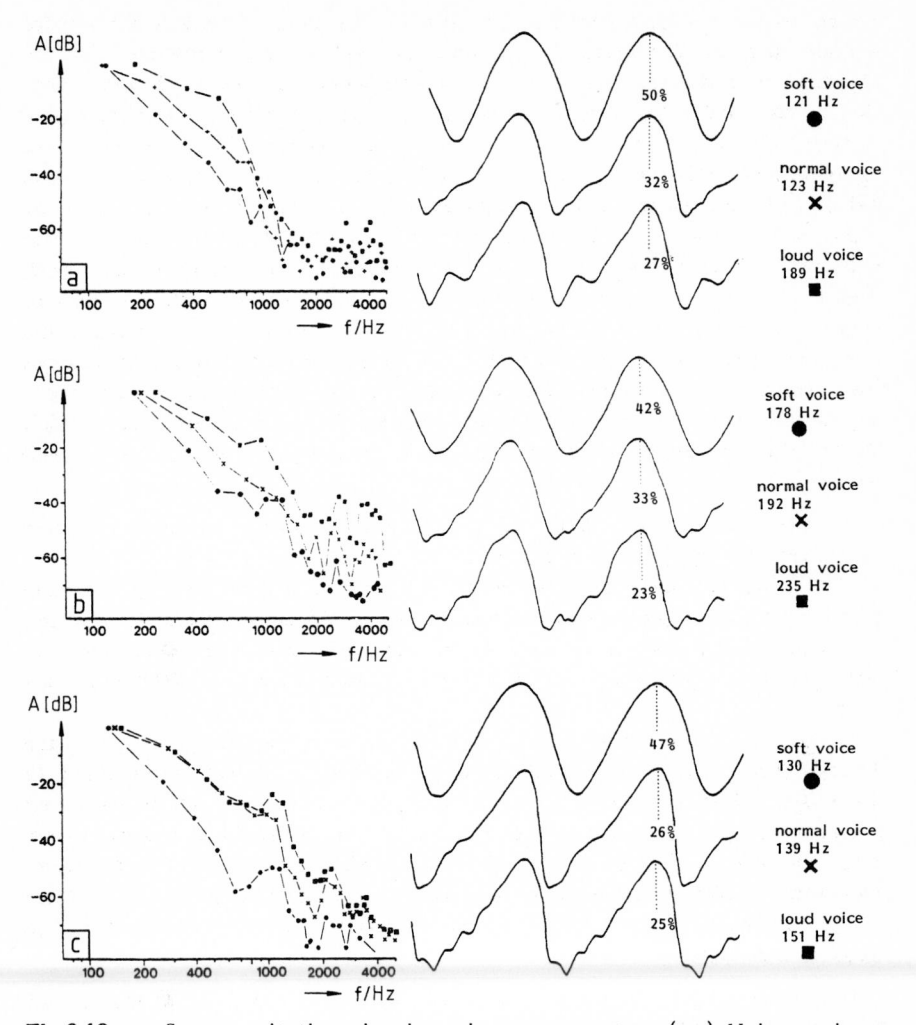

Fig.3.10a–c. Some excitation signals and source spectra. **(a,c)** Male speakers; **(b)** female speaker. The closing portion of the waveform is indicated as a percentage of the total period. The data were obtained using the mechanic inverse-filter tube by Sondhi (1975a,b). (For more details on this mechanic PDI, see Section 5.2.2.) From (Monsen and Engebretson, 1977, p.988)

an arrow in Fig.3.11) is mainly responsible for the high-frequency behavior of the signal.

Based on earlier investigations by Cederlund et al. (1960), by Sundberg and Gauffin (1978), and by Wakita and Fant (1978), Fant (1979b) proposed a model of the excitation signal. In contrast to the models that simulate the mechanics of the vocal cords (Flanagan and Landgraf, 1968a; Ishizaka and Flanagan, 1972; Titze, 1973a, 1974; and others), this approach models an idealized time-domain

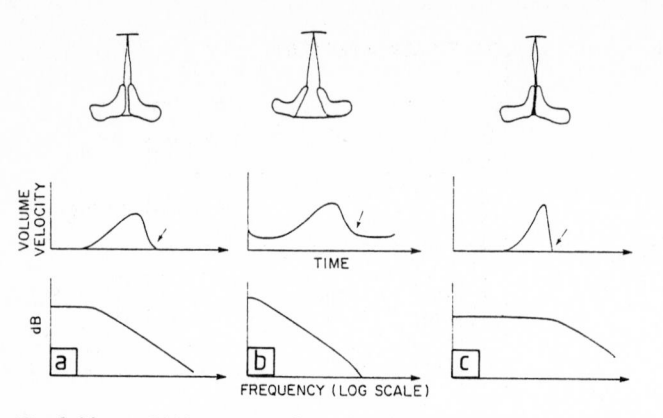

Fig.3.11a–c. Different modes of voice quality in the same register. (a) Normal, (b) breathy, (c) creaky. (1) Schematic view of the glottis, (2) glottal area function (volume velocity), and (3) schematic source spectrum envelope. From (Stevens, 1977, p.273)

contour of the excitation signal without reference to a mechanical system. The glottal cycle consists of three phases:

1) the phase of *glottal opening* (pulse rise time) beginning from zero with a smooth onset,

$$s_1(n) := \hat{s} \cdot 0.5\,(1 - \cos \omega_R n) \quad \text{for } n = 0\,(1)\,n_2; \quad n_2 = \pi/\omega_R \ . \qquad (3.1)$$

The frequency ω_R is defined as the pulse rise frequency; the signal $s_1(n)$ extends over the first quarter of the cycle of a sinusoid, and the phase ends when the positive peak is reached.

2) the *falling branch*, given by

$$s_2(n) := \hat{s}\,[\,K \cos(\omega_R n - \pi) - K + 1\,]$$
$$= \hat{s}\,\{\,1 + K \cdot [\cos \omega_R(n-n_2) - 1\,]\,\} \quad \text{for } n = n_2\,(1)\,n_3 \ . \qquad (3.2)$$

K is defined as the steepness factor. The falling branch ends at $n=n_3$ when $s_2(n)$ crosses the zero axis,

$$n_3 = \frac{1}{\omega_R}\,\text{arc cos}\,\frac{K-1}{K} \qquad (3.3)$$

3) the *closed-glottis interval*,

$$s_3(n) = 0 \quad \text{for } n = n_3\,(1)\,T_0 \ . \qquad (3.4)$$

The steepness factor K and the pulse rise frequency ω_R (which is independent of T_0!) determine the duty factor of the glottal cycle for a given value of T_0. The model simulates the source behavior for modal register when glottal closure during the excitation cycle is guaranteed (thus the model works only for $K > 0.5$ and $T_0 > 2\pi/\omega_R$).

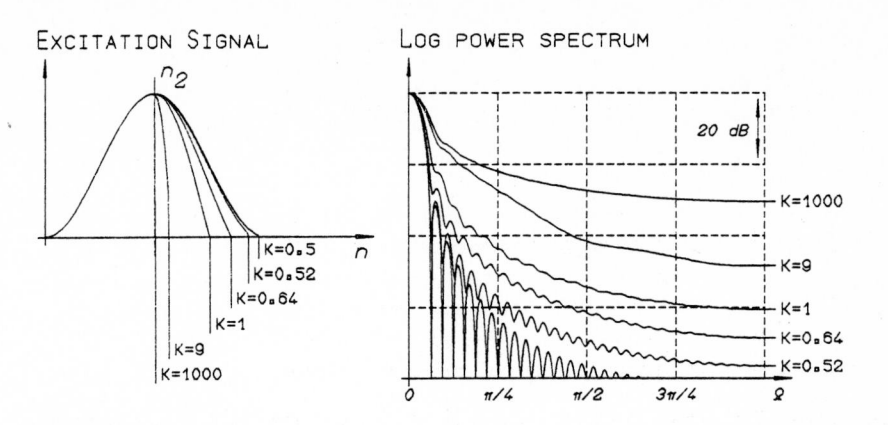

Fig.3.12. Waveforms and spectra from Fant's model of the voice excitation signal (Fant, 1979b). The model (with the parameter K) is quantitatively explained in (3.1-5)

Figure 3.12 shows the waveforms and the corresponding source spectra for different values of K. The model permits estimating the z-transform of the voiced source spectrum in analytic form,

$$S(z) = \sum_{n=0}^{n_2} s_1(n)z^{-n} + \sum_{n=n_2}^{n_3} s_2(n)z^{-n} . \tag{3.5}$$

It is the main advantage of this model that it permits representing the abrupt closure of the glottis in simple mathematical terms. It represents a good approximation of real voice source spectra, especially for values of K slightly above 1.

3.4 The Influence of the Vocal Tract Upon Voice Source Parameters

The stationary behavior of the vocal tract is well described by the acoustic theories of vowel production (Fant, 1960, 1970; Ungeheuer, 1962). In first approximation these theories describe the vocal tract as a nonuniform tube with a high-impedance excitation by the glottal source and a low-impedance load by mouth and lip radiation. The vocal tract is modeled as a linear passive transmission system which is defined to have the transfer function H(z). After introducing an additional transfer function R(z) which takes into account the radiation, the output impedance of the vocal tract can approximately be set to zero. In the neutral position where the vocal tract can be regarded as a uniform tube the resonances of the vocal tract occur at sound wavelengths of

$$\lambda = 4L \ / \ (2k-1)c \ ; \quad k = 1 \ (1) \ ... \ .$$ (3.6)

With $L = 17$ cm (average value of vocal-tract length) and a sound propagation speed of $c = 340$ m/s, the frequencies of these resonances will be

$$Fk = (2k-1) \cdot 500 \ Hz \ ; \quad k = 1 \ (1) \ ... \ .$$ (3.7)

The frequencies Fk, $k = 1 \ (1) ...$ are called *formant frequencies*. For speech transmission the first four or five formants (up to 5 kHz) are of interest. For speech recognition and vowel production the first two formants (F1 and F2) are of essential importance. When the vocal tract utters a vowel other than the neutral *schwa*, it leaves the neutral position, and the tube model becomes nonuniform. This results in a shift of the lower formants (especially F1 and F2) towards lower or higher frequencies which are characteristic for the respective sounds. Formant frequencies are documented in the well-known formant charts (e.g., Peterson and Barney, 1952) which are established for most languages. In addition to the theory, the measurement techniques for formant analysis are well developed; for a survey see (Flanagan, 1972a) or (Markel and Gray, 1976). The resonances of the vocal tract, the formants, are the poles of the transfer function H(z). In digital terms one can write the transfer function H(z) of the vocal tract as

$$H(z) = \frac{X(z)}{S(z)} = \frac{C \cdot z^{2K}}{\displaystyle\prod_{i=1}^{K} (z - z_{pi})(z - z_{pi}^{*})} \ .$$ (3.8)

In this equation X(z) is the spectrum of the speech signal, S(z) the voice source spectrum, and C a constant. The influence of the radiation is disregarded. The locations of the formants in the z plane are given by z_{pi}; the asterisk denotes conjugate complex. Comparing (3.8) to (2.16), it is easily realized that this transfer function can be modeled by a purely recursive digital filter, since it is an all-pole transfer function. K is the number of formants taken into account in the model transfer function. In the neutral position of the vocal tract the formants are regularly spaced every 1000 Hz. Since the higher formants tend to keep their frequencies even when the lower formants are changed, the order of the digital filter necessary to model the vocal tract at a given sampling rate $1/T$ is given by

$$2K = 1 \ / \ (T \cdot 1000 \ Hz) \ .$$ (3.9)

The representation (3.8) of the transfer function permits immediate realization as a digital filter in cascade structure. In linear prediction analysis (see Markel and Gray, 1976, or Sect.6.3), where such an all-pole transfer function is modeled in direct structure, K is usually increased by 2 or 3 to allow for a contribution of the radiation and the voice source spectrum.

The transfer function H(z) of the vocal tract has a profound influence on the shape and the spectrum of the signal. The bandwidths of the formants, at least of the low formants, are in the order of 50 Hz (Dunn, 1961; Flanagan, 1972a; Fig.3.13). This is normally much less than the fundamental

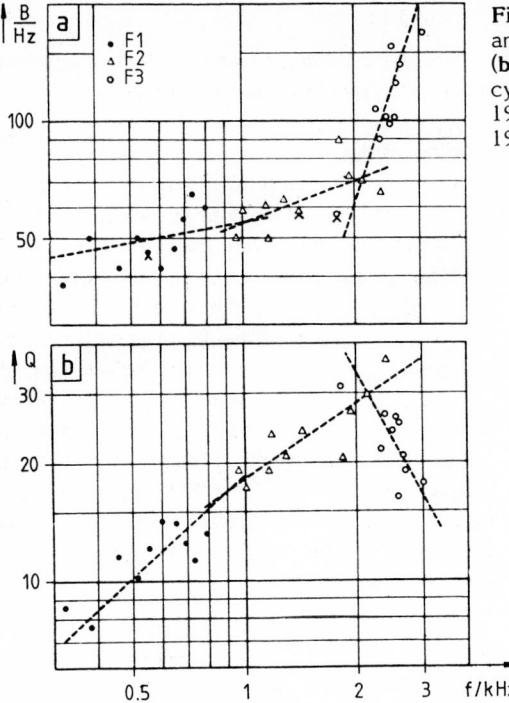

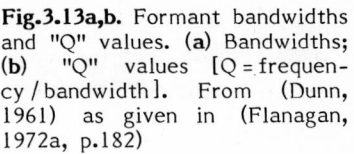

Fig.3.13a,b. Formant bandwidths and "Q" values. (a) Bandwidths; (b) "Q" values [Q = frequency / bandwidth]. From (Dunn, 1961) as given in (Flanagan, 1972a, p.182)

frequency F_0. Formants, especially the first formant, thus can and do enhance single partials of the excitation signal to such an extent that they dominate the waveform and the spectrum, while damping the other partials. In particular the first partial is greatly attenuated by the vocal tract since it is usually situated well below the first formant. Hence the influence of the vocal tract on the ease of pitch determination can and will be detrimental, especially when the formant F1 coincides with the second partial.

To illustrate this, Fig.3.14 shows the spectra of synthetic vowels (Fant, 1968). The discrepancy can be seen by comparing these spectra to the source spectra from Figure 3.10.

If the nasal tract is coupled to the vocal tract to produce a nasal or a nasal vowel, this acts like a shunt to the vocal tract. Nasal formants are added; the main effect of the shunt, however, is the introduction of zeros to the transfer function of the vocal tract. These zeros change the spectrum, but they do not immediately influence parameters like the amplitude of the first harmonic. Since the first nasal formant usually has a low frequency and comparatively large bandwidth, the extraction of source parameters like F_0 is somewhat easier for nasals than for vowels.

Besides the steady-state influence of the vocal tract on the source spectrum, there is also a mutual influence of source, vocal tract, and radiation which is partly steady state, partly dynamic. The vocal-tract model discussed above assumes a zero radiation impedance and an infinite glottal impedance.

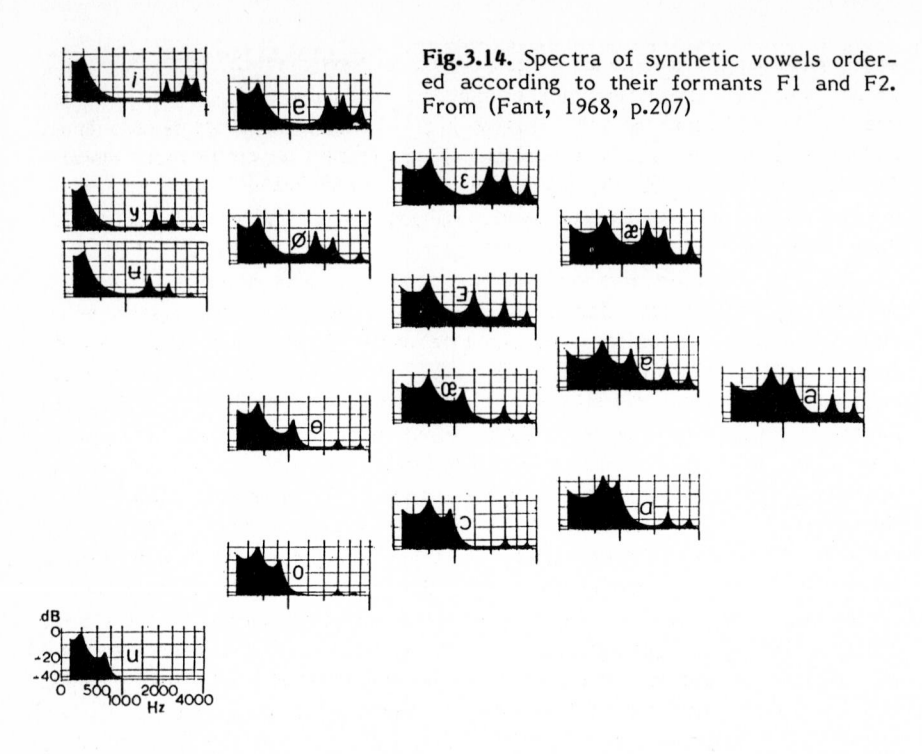

Fig.3.14. Spectra of synthetic vowels ordered according to their formants F1 and F2. From (Fant, 1968, p.207)

The assumption of a finite radiation impedance leads to slight shifts of the formant frequencies and to increased bandwidths primarily of the higher formants (Rabiner and Schafer, 1978). Thus the radiation impedance can be disregarded in the investigation of source parameters. If there is no interaction between glottal excitation and vocal tract, the spectrum of the speech signal can be modeled as a cascade of second-order digital filters, and we arrive at the simplified linear model of speech production as established in Sect.2.5 for voiced speech,

$$X(z) = P(z)\ S(z)\ H(z)\ R(z)\ A = P(z)\ G(z)\ H(z)\ A\ . \tag{3.10}$$

In this equation $H(z)$ represents the vocal-tract transfer function, and $G(z)$ combines both the long-term characteristics of the voice source and the characteristics of the radiation. The notation used in (3.10) is very convenient since the normal task of pitch determination is only interested in $P(z)$, not in $G(z)$, and in some methods, such as linear prediction, it is easier to separate $P(z)$ from the combined transfer function $G(z)\ H(z)$ than to separate the total source and radiation transfer function $S'(z) = P(z)\ G(z)$ from the vocal-tract transfer function $H(z)$.

There is, however, mutual influence between the glottal excitation and the vocal tract since the acoustic impedance of the glottis is finite. The acoustic impedance of the glottis can be modeled by a time-variant resistance and a time-variant inductance in series (van den Berg et al., 1957; Flanagan, 1972a). For this reason the glottal impedance increases with increasing frequency. The influence of the resistive part on the vocal tract results in an additional loss and thus in an additional attenuation which is most distinct for the lower formants. The bandwidth of the first formant, for instance, is mainly determined by this glottal loss (Fant, 1970; Guérin and Boe, 1977; Fant, 1979a,c).

What has just been discussed was the static effect of interaction between source and vocal tract. The interaction also has a dynamic aspect which divides the pitch period into two separate parts. As the voice source waveform indicates, the whole system is nonlinear so that we must distinguish between the *closed-glottis interval* and the *open-glottis interval,* at least for normal phonation in the modal register. When the glottis is closed, the glottal resistance can really be assumed to be infinite since the vocal tract is practically decoupled from the subglottal system. The spectrum $X(z)$ then is the product of $G(z)$ and $H(z)$. If the glottis is open, however, things are different. Not only the finite glottal impedance, but even the low-impedance subglottal system can and does influence the vocal-tract transfer function. A considerable change in the overall spectrum is the consequence of the glottal coupling. During the open-glottis cycle the formant bandwidths increase, and the formant frequencies rise; F1 is most affected. During each pitch period the formants, especially F1, are thus wobbled by a small but perceptible amount of frequency and bandwidth. Therefore accurate measurements of acoustic vocal-tract parameters are preferably carried out over the closed-glottis interval, when the vocal tract is undisturbed. The determination of the closed-glottis interval, however, is a refinement of the more general problem of pitch determination. It will be discussed in more detail in Section 6.5. For this accurate task only time-domain PDAs are suitable, and the signal to be processed must not be distorted, either in amplitude or in phase.

3.5 The Voiceless and the Transient Sources

Compared to the voiced source, the voiceless source does not raise this variety of questions. Sound is excited by a turbulent flow at a constriction in the vocal tract. The constriction can be situated at almost any point, starting at the glottis, and ending at the lips. Figure 3.15 shows a schematic diagram of both voiced and voiceless source spectrum (Stevens, 1975). The voiceless spectrum occupies the higher frequencies and is comparatively lower in energy. This figure as well as data by Flanagan (1972a) and others indicate that the overall effectiveness (i.e., the ratio of signal power to airflow) is less for the voiceless source than for the voiced source.

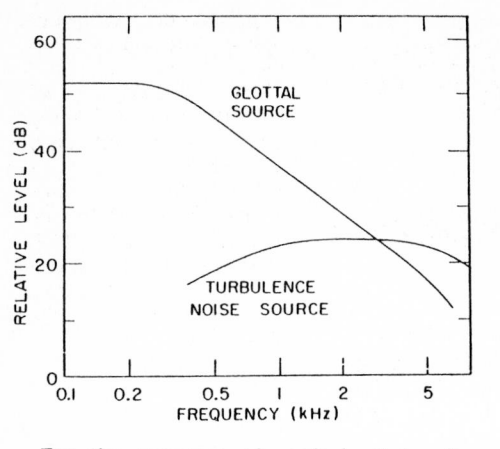

Fig.3.15. Schematic diagram of voice source spectra: glottal excitation and turbulent noise source. From (Stevens, 1975)

For the purposes of analysis it is often assumed that the voiced and the voiceless source are mutually exclusive, i.e., that no significant turbulent noise is present when there is a glottal excitation signal and vice versa. This abstraction is definitely not more than a first approximation, and it has been questioned by many researchers (e.g., Dreyfus-Graf, 1977; Gold, 1977). From the phonetician's point of view there are - in almost every language - voiced fricatives that demand by definition that both the voiced and the voiceless source to be present simultaneously. The speech signal itself, however, justifies the approximation of mutual exclusiveness to a much greater extent. From the above discussion we know that the voiced source is more "effective" than the voiceless source. In addition, the voiced fricatives are comparatively weak, and the feature *fricative* is expected to be more important for correct perception of the spoken phoneme than the feature *voiced*. All this together may explain the widely observed tendency towards devoicing of voiced fricatives. In these cases the voiced fricative is in fact pronounced voiceless, and the perceptual voiced/voiceless discrimination must be carried by other cues, such as duration of the preceding vowel[5,6].

---

[5] Results obtained by Anthony (1979, personal communication) as well as the investigations by Stevens (1977) suggest that this is due to a mismatch between the air flow necessary to generate voiced excitation and turbulent noise. The lowest level where turbulent noise at the desired place of articulation can exist is usually higher than the lowest level to maintain voiced excitation. For many speakers it is thus more convenient to devoice the voiced fricatives. This holds especially for languages like German, where the phonetic inventory contains only very few voiced fricatives.

[6] In a spoken-digit recognition experiment carried out by the present author (Zwicker et al., 1967) involving 50 speakers, the initial /z/ in the German words "sechs" (6) and "sieben" (7) was devoiced by 49 speakers, and the machine promptly committed an error when trying to recognize these two words spoken by the fiftieth one.

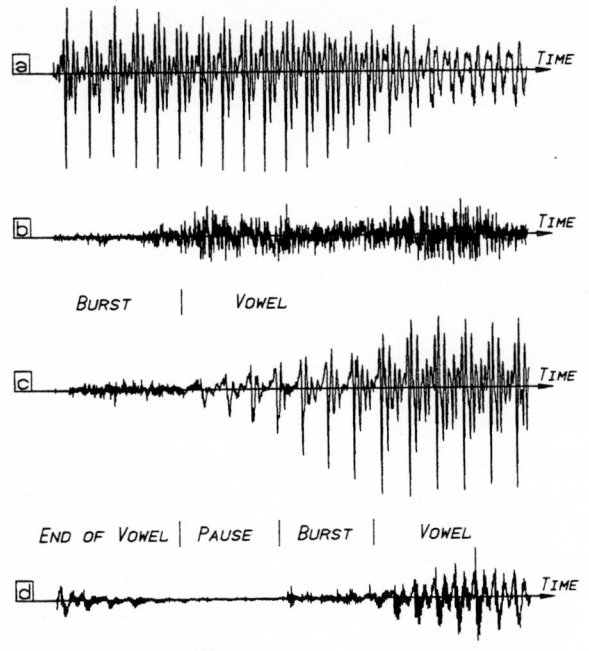

Fig.3.16a–d. Examples of speech signals with different modes of excitation. Speaker: WGH (male); scale: 100 ms per line. (**a**) Voiced source: transition /a/-/l/; (**b**) voiceless source: phone /ʃ/; (**c**) transient source: transition /t/-/a/, starting at the beginning of the burst; (**d**) transient source: transition /i/-/g/-/i/

Finally a third source of speech signals has to be regarded, the *transient source*. In stop consonants the vocal tract is temporarily closed at a defined place. At the same time the glottis opens, and the subglottal pressure is built up behind the closure in the vocal tract. When the pressure is released by reopening the closure, a transient turbulence is created before voicing starts again. If the closure is maintained for a time long enough, there will be a gap of some ten milliseconds in the speech signal during which all voicing ceases. To pronounce voiced stops, however, additional space can be created in the vocal tract by larynx lowering. Thus the larynx is able to maintain voicing for several additional cycles even when the vocal tract has already been closed. The transients are especially difficult to handle in speech analysis and voice source parameter estimation since the general assumption of short-term analysis that the signal parameters be constant during any frame under consideration is no longer valid for these sounds. Figure 3.16 shows some examples.

To complete this brief discourse on the basics of speech production, Fig.3.17 displays a speech signal and the corresponding spectral and formant data (Hess, 1976b). The pitch periods were determined by the PDA to be discussed in Chapter 7.

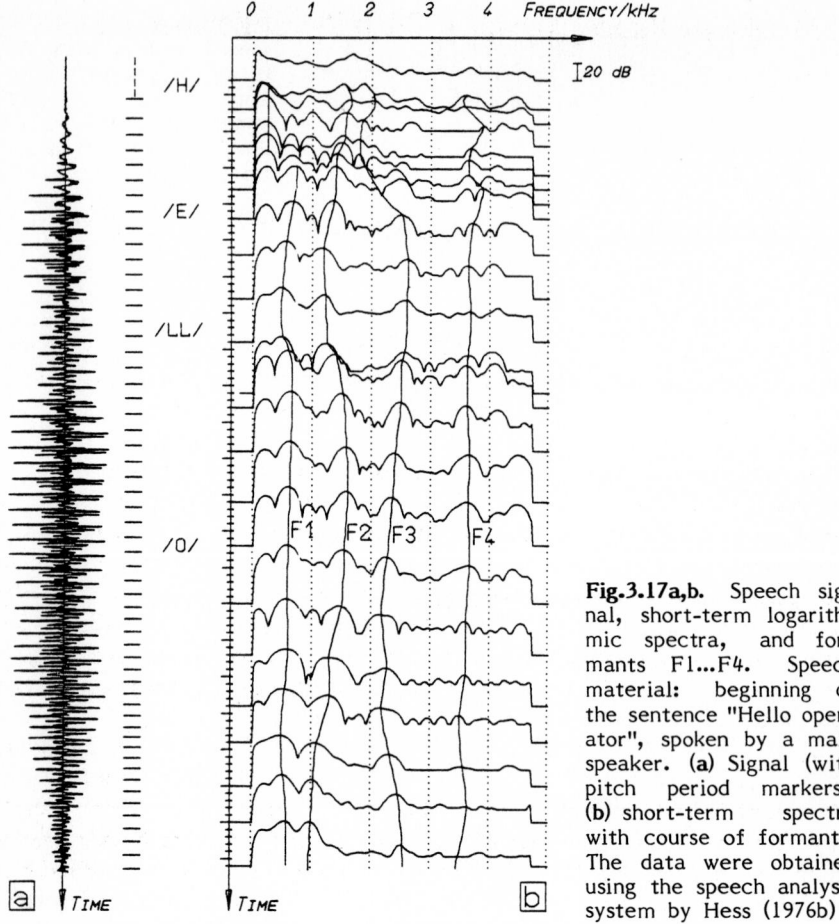

Fig.3.17a,b. Speech signal, short-term logarithmic spectra, and formants F1...F4. Speech material: beginning of the sentence "Hello operator", spoken by a male speaker. **(a)** Signal (with pitch period markers); **(b)** short-term spectra with course of formants. The data were obtained using the speech analysis system by Hess (1976b)

4. Measuring Range, Accuracy, Pitch Perception

The requirements to be met by a PDA, such as frequency range, accuracy, and rate of fundamental frequency change, are dictated by the characteristics of speech production and perception as well as by the special purpose for which the PDA is designed.

In a transmission system it is only possible to transmit signals that can be generated by the source; it is useless to transmit signals that cannot be further processed due to limitations of the receiver. In our case of spoken communication between humans, which can be viewed as a transmission system, the speech tract is the signal source, and the ear is the receiver. It is easy to show that, with respect to both measuring range and accuracy of the system, the ear outperforms the speech tract by far; the main constraints on the range of F_0 and the accuracy of the realization of a periodic signal in speech are due to the process of speech production.

Hence we start this chapter with a collection of some data on the range of F_0 in the human voice and compare these data with the measuring range of a number of implemented PDAs (Sect.4.1). The following section deals with pitch perception and outlines the basic ideas of the prevailing theories on this matter. In Sect.4.3 the issue of measurement accuracy is discussed. Section 4.4 presents an outlook on the features from which pitch information can be derived, and Sect.4.5 mentions some problems associated with performance evaluation.

The point of view from which this chapter has been conceived is similar to that of Chaps.1 and 9; the level of knowledge available, however, is very different in these three chapters. In Chap.1 we have just formulated the problem in an introductory way without presuming anything. In the preceding two chapters, basic data on voice excitation (Chap.3) as well as on signal processing (Chap.2) have been presented, a unified terminology has been created, and a first categorization of the different methods has been carried out. Together with the discussion on pitch perception in Sect.4.2, these discussions provide the knowledge necessary to understand the individual methods presented in Chapters 5-8. The present chapter thus serves to set up the framework within which the PDAs, PDDs, and PDIs are intended to operate. Since the individual methods have not yet been presented, many points will merely be outlined. The detailed discussion is delayed until Chap.9, i.e., until the main methods of pitch determination will have been reviewed.

4.1 The Range of Fundamental Frequency

The fundamental frequency of human utterances can take on a wide range. However, only a small part thereof is used for conversational speech. Mörner, Fransson, and Fant (1964) found that, for arbitrary human voice utterances, the fundamental frequency can vary between 33 and 3100 Hz whereas speech is confined to the lower part of this range; their data indicate that for each kind of voice the range of F_0 in conversational speech is situated somewhere in the lower register of that individual voice.

Catford (1964) confined the "voice" type of phonation to a F_0 range from 70 to 1100 Hz (excluding vocal fry register). Music, in this case singing, requires a somewhat wider range of fundamental frequency. In the operas by Mozart, for example, who poses most extreme demands with respect to the frequency range of the singers, tones range from C_2 (bass) to F_6 (soprano), which means frequencies of 66 and 1400 Hz, respectively. If one regards that a professional singer must even reach the musical interval a third beyond that range in order to have the required scale available at any time, then this range may extend from 50 to 1800 Hz. Speech, however, is not vocal music, and the fundamental frequency range observed in human speech is narrower. Some data taken from the literature are listed in Table 4.1.

For an individual speaker the distribution depends on the experimental conditions, especially on whether the distribution is taken from conversational speech or from read text. F_0 distributions taken from read text show quite consistently that the interval of an octave is rarely exceeded; displayed with F_0 in terms of musical intervals, the fundamental frequency distribution

Table 4.1. Range of the fundamental frequency observed in human speech and voice utterances. The "phonation" entries include "creak" (vocal fry) and falsetto register. The study by Keating and Buhr (1978) refers to children and infants only

Speech	Fairbanks (1940)	65 - 450 Hz
	Risberg (1961b)	50 - 310 Hz
	Hadding-Koch (1961)	50 - 500 Hz
	Shaffer (1964)	110 - 500 Hz
	Hollien (1972)	80 - 300 Hz
	Rabiner et al. (1976)	50 - 500 Hz
	Monsen and Engebretson (1977)	110 - 250 Hz
Phonation	Catford (1964)	28 - 1100 Hz
	Mörner et al. (1964)	33 - 3100 Hz
	Hollien (1972)	27 - 1200 Hz
	Keating and Buhr (1978)	30 - 2500 Hz
	Monsen and Engebretson (1977)	35 - 1200 Hz
Singing	Classical music (Mozart)	50 - 1800 Hz
	Mörner et al. (1964)	65 - 1350 Hz

comes close to a normal distribution (Risberg, 1961b; Schultz-Coulon, 1975b), whereas for other scales (for instance, a linear scale in F_0 or T_0) the deviation from the normal distribution is significant (Mikheev, 1971). Numerous F_0 distributions for individual speakers or for certain groups of speakers (frequently for selected social groups) have been published (Metfessel, 1926; Cowan, 1936; Dempsey, 1955; Risberg, 1960a,b; McGlone and Hollien, 1963; Winckel and Krause, 1965; Takefuta et al., 1972; Boe and Rakotoforinga, 1975; Schultz-Coulon, 1975b; and others). A bibliography is found in the paper by Boe and Rakotoforinga (1975); a number of these references has been included in the bibliography of this book.

Table 4.2. Range of the fundamental frequency implemented in existing PDAs. The number in the column "subranges" indicates that the PDA has a manual restriction which can be a preselection of the measuring range or a manual preset of the expected value (as in the PDA by Reddy, 1967); a dash in this column means that the PDA covers the whole range indicated without manual preset. The labels are explained in Chaps.6 and 8 and in Table 9.1. COMP stands for the comparative evaluation of PDAs

Reference	Label	Section	Implemented Range	Sub-ranges
Dudley (1937)	PiTN	6.1	80 - 400 Hz	
Grützmacher and Lottermoser (1937)	PiTN	6.6	80 - 500 Hz	2
Halsey and Swaffield (1948)	PiTN	6.1	90 - 300 Hz	
Dempsey et al. (1950)	PiTL	6.1	50 - 1000 Hz	3
Gruenz and Schott (1949)	PiTE	6.2	100 - 600 Hz	
Dolansky (1955)	PiTE	6.2	66 - 500 Hz	4
Risberg (1960a,b) and Fant (1962)	PiTN	6.1	60 - 300 Hz	-
Yaggi (1962)	PiTA	6.3	70 - 330 Hz	-
Carré et al. (1963)	PiTN	6.1	80 - 400 Hz	4
McKinney (1965)	PiTN	6.4	50 - 500 Hz	
Noll (1967)	PiFM	8.4	66 - 1000 Hz	-
Reddy (1967)	PiTS	6.2	75 - 400 Hz	1
Gold and Rabiner (1969)	PiTS	6.2	50 - 500 Hz	-
Léon and Martin (1969)	PiTLM	6.4	70 - 500 Hz	-
Teston (1971)	PiTL	6.1	50 - 1000 Hz	2
Baronin and Kushtuev (1974)	Survey		70 - 500 Hz	
Erb (1974)	PiTL	6.1	75 - 500 Hz	2
Moorer (1974)	PiCD	8.3	75 - 225 Hz	-
Ross et al. (1974)	PiCD	8.3	70 - 300 Hz	-
Fedders and Schultz-Coulon (1975)	PiTLM	6.4	50 - 4800 Hz	-
Rabiner et al. (1976)	COMP	9.3	50 - 500 Hz	
Un and Yang (1977)	PiCDL	8.3	66 - 400 Hz	-
Friedman (1978)	PiE	8.6	75 - 500 Hz	-
Tucker and Bates (1978)	PiTS	6.2	40 - 2400 Hz	-
Hess (1979a,c)	PiTN	7.5	50 - 830 Hz	-
Martin (1981)	PiFH	8.5	50 - 1000 Hz	-

The range of F_0 observed in human speech is not necessarily identical to the range implemented in an actual PDA, PDD, or PDI. *"Numerical estimates about laryngeal frequency parameters,"* so McKinney (1965), *"in connection with speech bandwidth compression systems design ... usually represent a compromise between fully adequate and inexpensive instrumentation."* This statement has certainly been valid at least for speech bandwidth compression systems for which the majority of PDAs were designed. It is strongly correlated with the necessity of an individual PDA to use a preset, i.e. an a priori estimate of expected pitch. In addition, many of these systems, at least most of the earlier ones, were confined to processing speech from male speakers. Some data indicating the implemented measuring ranges, taken from the literature, are listed in Table 4.2. The more recent and more elaborate algorithms tend to process a wider range of F_0. A measuring range that exceeds the range of F_0 for voice utterances is necessary when the respective PDA is to be applied to signals generated by music instruments.

4.2 Pitch Perception. Toward a Redefinition of the Task

There are three prospective "users" of the results produced by PDAs: 1) the human ear for the task of speech communication, 2) the human eye for most other tasks which are directed towards persons (for instance, visual inspection of pitch contours in phonetics, linguistics, and education), and 3) the computer (in tasks such as speaker recognition). If the computer is the final user, one can usually confine oneself to the performance of the PDA as it is and leave the matter of correction to higher level processing steps. The human visual system, on the other hand, has a remarkable ability to correct erroneous contours and substitute outliers by the correct values. Gross pitch determination errors can thus easily be corrected, and the accuracy of the eye with respect to fine errors is rarely better than a few percent.

   The discussion on accuracy and reliability of a PDA can thus concentrate on the case where the human ear is the final user of the results, i.e., on the tasks of speech communication. In order to get an idea what the requirements are for a PDA which serves as a component in the channel between the speaker at the source and the human ear at the receiver's side, we must briefly review the mechanism of auditory pitch perception. This will then lead us to a redefinition of the task by which the term *pitch determination* (cf. the discussion in Sect.2.1) will be finally justified.

4.2.1 Pitch Perception: Spectral and Virtual Pitch

From the perception point of view, the speech signal is looked at in a much more global way than from the production or the signal processing point of view. The ear is an instrument designed for a multitude of input signals among which speech is an important case, but by far not the only one. Regarding the fact that speech is a time-variant waveform with a complicated structural pattern, it is understandable that work on psychoacoustic problems such as the perception of pitch as a quantity was first done for simpler sounds, such as pure tones, and that it later proceeded on to stationary complex harmonic sounds. The main theories of pitch perception were conceived for stationary complex sounds, and yet little is known about the perception of time-variant quasi-periodic signals with changing fundamental frequencies, although today's pitch perception theories can easily be extended to such signals.

Because of its fundamental importance in speech research and musicology, the problem how pitch is perceived has attracted the attention of researchers for a long time. Much of the early work on this question has been summarized by Plomp (1976) in his book *Aspects of Tone Sensation.*

> The publication of a paper by Seebeck in 1841 was the beginning of the controversy as to what physical property underlies the low pitch of a complex tone. Seebeck experimented with siren discs, so it is not surprising that he considered the repetition rate of air puffs to be the physical correlate of pitch. Ohm's "definition of tone"[1] was the alternative interpretation of these siren experiments (Ohm, 1843). He considered the presence of a sinusoidal component to be a necessary condition for hearing a corresponding pitch. Since the pitch of a complex tone is equal to the pitch of its fundamental, this component is, apparently, the most important harmonic from which pitch is derived. In accepting Ohm's view, von Helmholtz (1863) assumed that the fundamental's role originates from its dominant amplitude in common sounds. (Plomp, 1976, p.111)

In the early research, the fundamental was regarded as playing the dominant role in pitch perception. In later years, however, the neglect of the higher harmonics in the theory of pitch perception was severely criticized. In the late 1930s, finally, counterevidence was found against the dominance of the fundamental. To follow again Plomp's description,

> ... the first successful attack on the significance of the fundamental in low-pitch sensation was made by Schouten in some pioneer investigations carried out from 1938 to 1940. Schouten (1938) studied the pitch of periodic sound waves produced by an optical siren in which the fundamental of 200 Hz was cancelled completely. This was verified by the absence of any audible beats when a probe tone of 206 Hz was introduced. The pitch of the complex tone, however, was the same as that prior to the elimination of the fundamental. (Plomp, 1976, p.112)

---

[1] About Ohm's "definition of tone", Plomp (1976, p.2) wrote: "The concept that the ear can be compared with a frequency analyzer was accurately formulated for the first time by the German physicist Ohm in 1843. His 'definition of tone,' nowadays known as 'Ohm's acoustical law,' states that a tone with frequency f is only heard if the complex sound contains $\sin(2\pi ft+\phi)$ as a component (Ohm, 1843). In von Helmholtz's phrasing, this definition reads: 'Every motion of the air which corresponds to a composite mass of musical tones, is, according to Ohm's law, capable of being analyzed into a sum of simple pendular vibrations, and to each such single simple vibration corresponds a simple tone, sensible to the ear, and having a pitch determined by the periodic time of the corresponding motion of the air' (von Helmholtz, 1870; English translation, p.33)."

Schouten's results, as well as that of others, showed that the phenomenon of pitch perception is not evoked by the fundamental harmonic, at least not for the range of F_0 covered by normal speech. It was found that a pitch of a harmonic complex sound is perceived which equals that of its fundamental even when the first harmonic is totally absent or masked by noise.

Looking for a mechanism that explains this phenomenon, De Boer (1956) found that it is *not* due to combination tones, i.e., to a reconstruction of the fundamental harmonic by the intrinsic nonlinearity of the ear.

> De Boer (1956) ... shifted a carrier of 2000 Hz, modulated in amplitude by a signal composed of tones of 200, 400, and 600 Hz, over Δf Hz. This means that the harmonic stimulus consisting of components of 1400, 1600, 1800, 2000, 2400, and 2600 Hz is transformed into an inharmonic complex of 1400+Δf Hz, 1600+Δf Hz, etc. De Boer found that the low pitch of this stimulus changed slightly as a function of Δf. This effect excludes the possibility that low pitch is determined by a difference tone of 200 Hz because the frequency of such a tone does not change with the frequency shift. It excludes also the possibility that the pitch is derived from the *envelope* of the amplitude-modulated signal because the periodicity of this envelope does not change either. (Plomp, 1976, p.113)

For more details of this work, the reader is referred to Plomp's survey where a number of additional references are found (some of which are included in the bibliography of this book).

More recent theories and models, according to experimental evidence, postulate that pitch perception is performed by harmonic pattern matching. Each harmonic which can be individually resolved by the ear evokes a *spectral pitch* which corresponds to the frequency of that harmonic (and for these spectral pitches Ohm's "acoustical law" as well as Helmholtz's formulation do apply). All the spectral pitches together contribute to the overall pitch percept. Since this percept does not require the first harmonic to be physically present, it is called *residue pitch* (Schouten, 1940; Ritsma, 1962a), *periodicity pitch* (Goldstein, 1973), or *virtual pitch* (Terhardt, 1974).

Let us again quote Plomp (1976) who outlined the essential ideas of today's most important pitch perception models as follows.

> In recent years some authors have proposed models of the way in which low pitch may be derived from the resolved lower harmonics of complex tones. Three models, representing quite different approaches, are mentioned here.
>
> a) *Wightman's (1973b) pattern-transformation model.* This model contains three successive stages: 1) a limited-resolution power-spectrum analyzer, to be identified with the peripheral auditory system, 2) a Fourier transformer, assumed to be a specially wired network of neural interconnections, and 3) a pitch extractor operating upon the positions of maximal activity in the pattern coming from the former stage. Essentially, the model can also be described in terms of autocorrelation. Though they are mathematically equivalent, Wightman preferred a Fourier transform of the power spectrum to a real autocorrelation in the time domain. This is because the nervous system cannot possibly compute a "true" autocorrelation function, which would be required to account for the finding that the low pitch is insensitive to phase. The model predicts, qualitatively at

least, how the value and the distinctness of the low pitch of a complex sound depend upon its main physical parameters.

b) *Goldstein's (1973) optimum-processor model.* The processor is considered to make an optimal estimate of the fundamental frequency on the basis of "noisy" representations of the frequencies of the resolved harmonics. Assuming periodic stimuli with adjacent harmonics, the processor computes the harmonic numbers of the components present and from these data the frequency of the fundamental. Goldstein developed algorithms to compare estimates from his model with the experimental results on low pitch from various investigators. Generally, good agreement with the data was found.

c) *Terhardt's (1974) learning-matrix model.* The heart of this model (for an earlier version see Terhardt, 1972a,b) is a learning matrix with the spectral-pitch cues and the lowest spectral-pitch cue as (horizontal and vertical) input signals. In the learning phase, assumed to be a part of the childhood learning process for acquiring the ability to recognize speech sounds, the correlations between the two input signals for complex tones (vowels) impress their "traces" on the learning matrix. These traces provide the "virtual" low pitch, to be evoked every time that the matrix is activated by the spectral pitches of a group of harmonics. Just as in the two previous models, any stimulus initiates a number of low-pitch cues the strongest of which determines the final solution.

The models mentioned have many features in common. All three include a peripheral stage characterized by a limited frequency analysis and a central stage in which low pitch is derived. The models are different descriptions of how a pattern-recognition process may be organized. (Plomp, 1976, pp.140-141)

Since Plomp's book came out (1976), these models have been further developed and refined (Patterson and Wightman, 1976; Goldstein et al., 1978; Gerson and Goldstein, 1978; Terhardt, 1979b; Terhardt et al., 1982a,b). The following discussion, which supplies some more details about pitch perception, will widely follow Terhardt's theory as it has been outlined in his publications (Terhardt, 1970, 1972a,b, 1974, 1979a,b, 1980; Terhardt et al., 1982b).

A single, isolated pure (sinusoidal) tone evokes a *spectral pitch* in the ear. The speech signal, however, is not a sinusoid, but a complex sound; let us assume it to be strictly periodic for the following considerations. As a complex periodic sound this signal has many harmonics each of which (being a pure sinusoid) may produce a spectral pitch of its own. The spectral pitches are (centrally) combined to a common sensation, the *virtual pitch*, which is the perceptual counterpart to fundamental frequency. Terhardt (1972a) defined spectral and virtual pitch as follows.

1) A sinusoidal (pure) tone evokes a sensation of pitch which is directly related to the place of greatest excitation in the Organ of Corti; this pitch is known as *spectral pitch*.

2) The spectral pitches belonging to the partials of a sound can be individually perceived by the subject when his attention is suitably directed, and as long as the difference in frequency of the partials does not fall below a certain level. In the case of a *harmonic* sound the spectral pitches of the first 6-8 harmonics can each be perceived individually.

3) Apart from the pitches mentioned in 1) and 2) a sound generally evokes a dominant "global pitch", which in the case of harmonic sounds corresponds to the fundamental frequency. ... This sensation of pitch represents a completely different perceptual phenomenon to spectral pitch and is known as *virtual pitch*.

4) The partials which evoke individual spectral pitches under the conditions given in 2) have proved to be those characteristics of a sound
which are decisive for its virtual pitch. (Terhardt, 1972a, p.63; original in
German, cf. Appendix B)

In a later publication Terhardt (1979b) established a functional model of
virtual pitch formation by means of which a quantitative estimate can be
derived.

A relatively concise and yet precise specification of the transformation
of sound parameters into virtual pitch is enabled by the established fact
that virtual pitch can be deduced exclusively from the spectral pitches
which are evoked by the complex stimulus (Terhardt, 1972a, 1974). Hence,
for the present purpose [of virtual pitch calculation] the physical and
physiological mechanisms which link the physical stimulus parameters to
virtual pitch, are of secondary interest. ... The problem is reduced to two
questions:

a) Which is the functional relation between physical stimulus parameters
and spectral pitch?
b) Which is the functional relation between simultaneous spectral pitches
and virtual pitch?

The answer to the first question is simple: To a particular spectral pitch
there corresponds one particular spectral component (partial), and the spectral pitch magnitude is essentially dependent on the component's frequency.
A preliminary answer to the second question is: Virtual pitch is obtained
by a specific matching process of subharmonics pertinent to certain determinant spectral pitches. This process is dependent on a previous perceptual
mode of learning, in which the knowledge of harmonic pitch intervals has
been acquired. (Terhardt, 1979b, p.156)

A flow chart of the basic model is shown in Figure 4.1. The learning phase
is assumed to be performed at a very early age, and the speech signals
children hear from their environment are supposed to contribute significantly
towards establishing the perceptual knowledge, so that at last only the recognition path of the model will be further needed.

Figure 4.2 shows the block diagram of the model as it was used by
Terhardt et al. (1982) to build a multipurpose PDA which is not restricted
to speech signals. The formation of virtual pitch is performed as follows. A
partial (as defined above) may contribute in the form of a sample of *spectral
pitch* if and only if 1) it can be resolved by the ear as an individual component, and 2) it is not masked by other simultaneously present components of
the same sound. In this case its level (*sound pressure level* SPL) exceeds the
masking level at that particular frequency. Since the remainder of the spectrum under consideration (except the partial for which the masking level is
calculated) represents a complex sound, the total masking level is determined
by the masking effect of all the other spectral components of the signal. As
the result of the masking effect, only those spectral components contribute
individual samples of spectral pitch whose *SPL excess* (sound pressure level
exceeding the overall masking level at the frequency of the partial under
consideration; cf. the discussion further below) is greater than 0 dB. There is
considerable discrepancy between the spectrum of a vowel sound and the

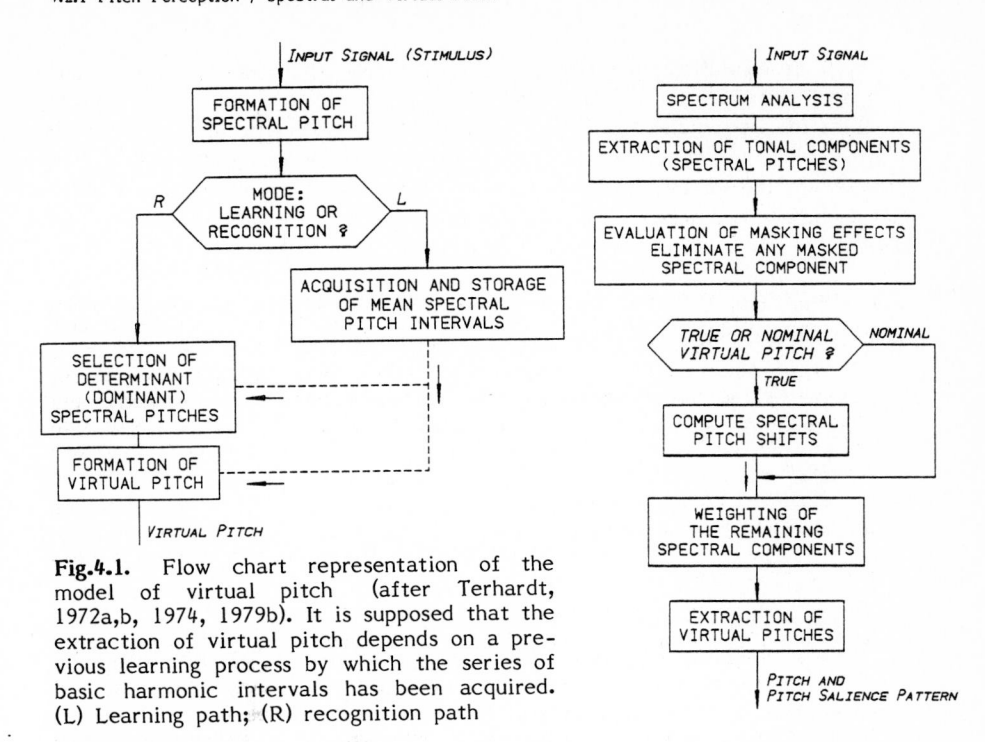

Fig.4.1. Flow chart representation of the model of virtual pitch (after Terhardt, 1972a,b, 1974, 1979b). It is supposed that the extraction of virtual pitch depends on a previous learning process by which the series of basic harmonic intervals has been acquired. (L) Learning path; (R) recognition path

Fig.4.2. Outline of the model of virtual-pitch formation, as it was used by Terhardt et al. (1982b) to design a multipurpose PDA

histogram of the SPL excess at the individual harmonics, indicating that even some strong higher formants might be prevented from contributing individual spectral pitches (although their spectral energy of course contributes to the timbre of the complex sound), whereas seemingly unimportant pitch information in the valleys between the formants may remain audible. As an example, Fig.4.3 shows synthetic vowel spectra calculated by Terhardt (1979b) according to data by Fant (1960, 1970) and their auditory counterparts consisting of those harmonics whose SPL excess is greater than 0 dB. The figure also shows that for any of these sounds (and more or less for any speech signal) the higher harmonics above 1000 Hz significantly contribute to the virtual pitch sensation. This fact confirms the results obtained by experiments with realized PDAs, such as the one by Friedman (1978; Sect.8.6) or Hess (1979a,c; Sect.7.5) which work better when high frequencies are not excluded from processing.

The effect of mutual masking of the individual harmonics of a complex sound can be derived from the masking patterns observed with masking by pure tones or critical band noise. When displayed in terms of SPL (ordinate) over critical band rate (abscissa), the masking pattern produced by a pure tone (the *masker*) may be regarded as roughly triangular in shape and independent of the frequency of the masker. The slope of the masking patterns towards lower values of the critical-band rate (i.e., towards frequencies lower than that of the masker) is approximately given by

$$S_L = 27 \text{ dB/Bark} , \tag{4.1a}$$

whereas for frequencies higher than that of the masker there is a dependence on frequency (which decreases rapidly for frequencies above 100 Hz) and on the SPL of the masker,

$$S_H = \{ 24 + 0.23 \, [f(m)/kHz]^{-1} - 0.2 \, L(m)/dB \} \text{ dB/Bark} \tag{4.1b}$$

where $f(m)$ and $L(m)$ are the frequency and the SPL of the masker. This calculated masking level is in good agreement with experimental data (Terhardt, 1979a,b; Zwicker and Feldtkeller, 1967). The masking level at the frequency $f(k)$ of the kth partial, caused by the masker with frequency $f(m)$, is then obtained as

$$L_M(m,k) = L(m) - S_H \, [b(k) - b(m)] \qquad \text{for } b(k) > b(m) , \tag{4.2a}$$

$$L_M(m,k) = L(m) - S_L \, [b(m) - b(k)] \qquad \text{for } b(k) < b(m) . \tag{4.2b}$$

In these equations, $b(i)$ is the equivalent for the frequency $f(i)$ after transformation into the bark scale.

The sound pressure level excess (SPL excess) ΔL is now defined as the level difference between the level of the kth partial with the frequency $f(k)$ and the masking level at $f(k)$ due to the masker at $f(m)$,

$$\Delta L(k) = L(k) - L_M(m,k) . \tag{4.3}$$

When there are several masking components, the composite masking level is found by adding the *amplitudes* which correspond to the individual masking levels. For the masking amplitude $A_M(k)$ at $f(k)$ we obtain

$$A_M(k) = \sum_{m=1}^{k-1} A_M(m,k) + \sum_{m=k+1}^{N} A_M(m,k) \tag{4.4}$$

if the complex cound contains N partials. The amplitude $A_M(m,k)$ reads as

$$A_M(m,k) = 10^{L_M(m,k)/20dB} . \tag{4.5}$$

The final value of $\Delta L(k)$, which has to take into account the absolute hearing threshold $T(k)$ at that frequency [in order to assure that any level $L(k)$ which falls below the absolute hearing threshold will be regarded as masked and thus inaudible] and a correction factor I for nonperiodic, noiselike components in the spectrum, is calculated as follows,

$$\Delta L(k) = L(k) - 10 \, \lg \left[A_M^2(k) + 10^{T(k)/10dB} + I \right] \tag{4.6}$$

The correction factor I is computed as the sum of the spectral amplitudes in the critical band around $f(k)$. This term vanishes if the signal is periodic.

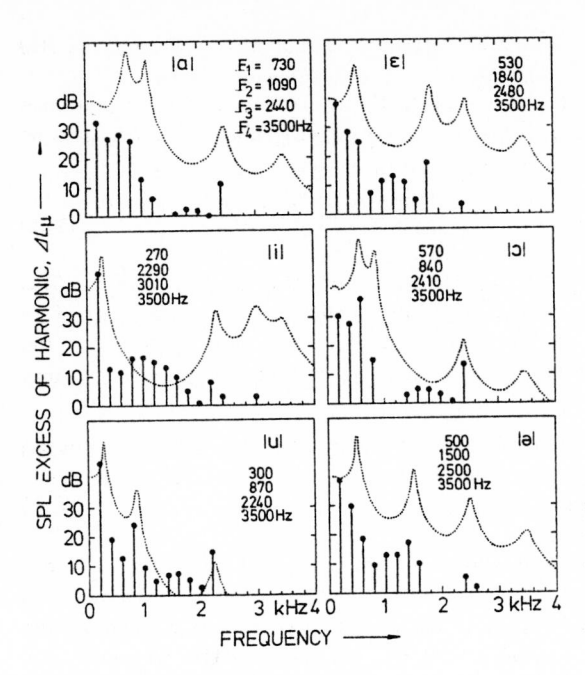

Fig. 4.3. SPL excess of the harmonics of synthetic vowels. (Dotted line) Shape of amplitude spectra. Assumed fundamental frequency: 200 Hz. From (Terhardt, 1979b, p.160)

From the individual spectral pitches (partials) which have survived the masking process, the components which really contribute to *virtual pitch* are selected by the existence of a *dominant region*.

> The existence of that region may be considered as being caused by the tendency of the "central processor" (i.e., the subsequent formation of virtual pitch) to optimize its own performance, i.e., to operate in a frequency region where relevant spectral parameters such as threshold of hearing and frequency resolution are optimal. This process of optimization seems to be rather complex ... so that a precise and general description as yet appears to be hardly possible. (Terhardt, 1979b, p.168)

However, Terhardt then described a simplified model:

> a) Spectral pitches below 300 pu [*pu*, pitch unit, is a unit for pitch sensation which at first approximation can be replaced by the unit of frequency, the hertz] and beyond 4 kpu are considered to be not relevant;
> b) from the remaining spectral pitches the lowest one is most determinant while the following contribute with descending weight. (Terhardt, 1979b, p.168)

In the most recent publication, Terhardt, Stoll, and Seewann (1982b) gave a quantitative description of the dominant region by introducing a weighting factor for the individual spectral pitches.

From the selected spectral components, virtual pitch is calculated by subharmonic matching. The model is able to take account of established psychoacoustic phenomena such as (level- and frequency-dependent) shifts of spectral pitches, or harmonic interval stretching. In the simplified version, however, only the true subharmonics with integer frequency ratios (1:2, 1:3, etc.) are regarded, thus computing the *nominal virtual pitch*[2] (instead of the *true virtual pitch*) which is equivalent to the fundamental frequency F_0 if the spectrum is harmonic.

Consider a given short-term (power, amplitude or logarithmic) spectrum $X(f)$ to contain peaks (that are not masked) at frequencies f_i, $i = 1\,(1)\,K$ within the dominant region. K represents the number of peaks actually present there. Then F_0 will be a subharmonic of the lowest partial f_1, say f_1/m. The task of the algorithm is now to find the one subharmonic f_1/m, $m = 1\,(1)\,M$ which coincides with the subharmonics of the other relevant spectral components, i.e., f_i/n, $i = 2\,(1)\,K$, $n = 2\,(1)\,M$. M delimits the range of subharmonics to be scanned; this value indicates the number of the highest harmonic which can contribute to the virtual pitch at the fundamental. The algorithm permits detecting the coincidence of any subharmonic f_i/n, $n = 2\,(1)\,M$ of a given spectral component f_i (i greater 1) with one single arithmetic check. Around any F_0 candidate $p = f_1/m$, a tolerance interval is established such that

$$p(1-d) < f < p(1+d) \, . \tag{4.7}$$

Any frequency $f = f_i/n$ within this interval will be regarded as coincident (or *near-coincident*, to use Terhardt's term). The width of the tolerance interval is typically set to a few percent. To check whether any subharmonic f_i/n coincides with p, it is only necessary to check whether the condition

$$\text{floor} \; [(1+d) \; m \; f_i/f_1] \; > \; (1-d) \; m \; f_i/f_1 \tag{4.8}$$

is fulfilled. *Floor(x)* (taken from PL/I notation) denotes the greatest integer less than or equal to the argument x. If the condition is fulfilled by a spectral peak f_i, the integer subharmonic f_i/n (n unspecified) falls into the tolerance interval and leads to an indication of coincidence. The value for

---

[2] There is experimental evidence that a spectral pitch depends not only on frequency, but also on the amplitude of the pertinent sinusoid as well as on the presence or absence of competitive spectral pitches at other frequencies. Shifts of spectral pitches due to these effects rarely exceed a few percent, but they are perceptually significant. For details the reader is referred to the literature (Terhardt et al., 1982a,b; Terhardt, 1979b; Plomp, 1976; Walliser, 1969), where additional references can be found. There is one reason, however, not to take into account this psychoacoustic fact in a practical PDA. Since the ear already performs these spectral pitch shifts, they must not be performed elsewhere in the processing system (unless some compensation is provided) if the output of this system again is an acoustic signal. This holds for any application in speech transmission and synthesis. If the output is a graphic display, things may be different. Taking into account the pitch shifts, however, requires the individual partials to be separated, which is difficult unless the PDA operates in the frequency domain. For this reason, and because there is no PDA whatsoever (except the model by Terhardt et al., 1982b) that makes use of this phenomenon, the question of spectral pitch shifts in pitch perception is not further pursued in this book. Hence the term *virtual pitch*, unless specified otherwise, will always mean *nominal virtual pitch.*

which all the peaks f_i, $i = 2(1)K$ signal coincidence, is then taken as the (true or nominal) virtual-pitch estimate,

$$\hat{p} = f_1/m \ . \tag{4.9}$$

In sounds that are more complex or not exactly periodic, one cannot expect that there is a frequency f_1/m for which all the partials contribute in form of a single (near-)coincidence. One might even find situations in which several competitive virtual pitches exist. (These situations may easily occur in speech, e.g., when the voice of the speaker tends towards paired pulsing). The great advantage of the most recent model (Terhardt et al., 1982b) is that it permits statements on pitch perception even in such cases. Thus the decision as to which component represents the pitch of the signal is not a binary one, but several virtual pitches are permitted to exist simultaneously. To make this possible a weight is assigned to each spectral component which depends on its SPL excess and its frequency with respect to the dominant region. Since each coincidence of subharmonics of any two spectral pitches is a possible candidate for virtual pitch, weights of importance are also assigned to the individual virtual-pitch estimates. Last but not least, the coincidence event itself is weighted; the better the coincidence (i.e., the smaller the distance between the two subharmonics within the tolerance interval), the larger the pertinent weight.

> Following the principles of the virtual-pitch theory, the significance of a virtual pitch is represented by a numerical weight which depends on the following criteria:
> 1) The number of relevant spectral components which provide the same, or nearly the same, virtual pitch; the weight should increase with the number of components.
> 2) The spectral-pitch weight of the relevant components; virtual-pitch weight should increase with spectral-pitch weight of involved components.
> 3) The subharmonic numbers involved; the weight should decrease with increasing subharmonic number.
> 4) The accuracy of the near-coincidences mentioned in 1); the weight should increase with accuracy, attaining a maximum with perfect coincidence. (Terhardt et al., 1982b, p.680)

Instead of making a binary decision, this model gives a measure for *pitch salience* or *pitch strength*, thus indicating which value in the virtual pitch pattern is dominant when the overall situation is ambiguous.

Further details on the PDA derived from these principal assumptions are supplied in Section 8.5.2.

In the model by Goldstein (1973), further developed by Goldstein et al. (1978), virtual pitch (labeled *periodicity pitch* by these authors) is obtained as the optimal match between successive harmonics of a complex sound. The criterion for "best match" is the minimum squared error,

$$E = \sum_{k=1}^{N} \left[\frac{\hat{n}(k)\hat{p} - f(k)}{\sigma(k)} \right]^2 \ . \tag{4.10}$$

In this equation, E is the squared error, N the number of partials involved, $\hat{n}(k)$ the optimal harmonic number for the partial at the frequency f(k), $\hat{p}$ the optimal pitch estimate, and $\sigma^2(k)$ the expected variance of the spectral pitch estimate at f(k). [The parameter k is only a count variable and does not mean that the partial at f(k) is the kth harmonic.] The variance $\sigma^2(k)$ roughly corresponds to the auditory difference limen for pure sinusoids. Goldstein (1973) derived qualitative results from a former experiment (Houtsma and Goldstein, 1972) on the perception of complex tones comprising two successive harmonics.[3]

The simultaneous optimization of $\hat{p}$ and $\hat{n}(k)$ in the presence of noise is done by a maximum-likelihood approach. The final estimate for F_0 reads as

$$\hat{p} = \frac{\sum_{k=1}^{N} f(k)\,\hat{n}(k)\,/\,\sigma^2(k)}{\sum_{k=1}^{N} \hat{n}(k)\,/\,\sigma^2(k)} . \qquad (4.11)$$

This formula can also be used to improve the accuracy of frequency-domain PDAs in case the harmonic numbers of the partials are known but the frequency estimates f(k) are noisy (including quantization noise). In the latter case σ^2 can be assumed constant (and thus neglected). Further details on the PDA designed according to this formula (Duifhuis et al., 1978, 1979a-c, 1982) are added in Section 8.5.2.

4.2.2 Toward a Redefinition of the Task

For speech signals we can widely assume that virtual pitch and fundamental frequency correspond to each other. This is valid as long as the fundamental frequency[4] F_0 is understood as the reciprocal of the fundamental period[4] T_0 [according to (2.34)]. Fundamental frequency therefore must not be understood as the greatest common divisor of the individual partials ("harmonics") in a strictly mathematical sense. This latter definition holds only as long as the signal is strictly periodic. It might happen, for example, that a slight additive frequency shift is imposed on the speech signal which may be sufficiently great for the signal to become inharmonic (the temporal structure of

---

[3] In his basic paper, Goldstein (1973) postulated the presence of a number of *successive* harmonics in order to determine virtual pitch. From the beginning, this assumption looked like a kind of "implementation restriction" for the model rather than a psychoacoustically justified fact. Indeed, Gerson and Goldstein (1978) found experimental evidence against this restriction from new pitch shift experiments. They proposed a modification of the original theory which led to an identical mathematical formulation no longer requiring the constraint of successive harmonics.

[4] We have already seen from the discussion in Sect.2.1 that some caution is necessary with respect to the terms *fundamental frequency* and *fundamental*

the signal is more robust against such a shift). In this case the fundamental frequency (in the strict definition as the largest common divisor of the frequencies of the partials) would change drastically, whereas the virtual pitch - the same as the temporal structure, and with it the period duration T_0 - is only slightly influenced. Thus the definition of the pitch determination of speech signals given by Terhardt (1979b) appears as a good way to combine the temporal structure of the signal and the perception point of view.

> *Virtual pitch* is the type of pitch which is produced by complex signals, in contrast to *spectral pitch*, which is the type of pitch evoked by pure tones. While spectral pitch may be considered as a product of relatively peripheral auditory analysis, the perception of virtual pitch probably is dependent on higher (i.e. more central) stages of auditory processing. In the perception of complex tones, e.g., voiced speech sounds, both types of pitch play a role. While some lower harmonics (including the fundamental) may produce simultaneous, corresponding spectral pitches, virtual pitch is the prevalent pitch mode of those signals and corresponds to the "fundamental frequency". "Extraction of fundamental frequency" is in some respect equivalent to extraction of virtual pitch. In a strict sense, however, the frequency which corresponds to virtual pitch, and the fundamental frequency [defined as the largest common divisor of the partials] are in general not identical. ... Hence in the analysis of auditory signals such as speech and music actually not *extraction of fundamental frequency* is the real aim but rather *extraction of the frequency which corresponds to virtual pitch.* (Terhardt, 1979b, p.156)

By this redefinition of the task the term *pitch determination* is finally justified. Actually there is a much better correspondence of the (quasi)periodicity of the (time-variant) signal to the concept of virtual pitch than to the (strictly harmonic) concept of fundamental frequency. In this respect, the statement by Filip (1969) points in the same direction:

> For measurements based on linear filtering, the concept of fundamental frequency may serve quite well although it has its full meaning only in true periodic functions with the amplitude of the fundamental spectral component not equal to zero. However, when pure nonlinear methods are adopted and inharmonic components are assumed in nearly periodic portions of the input signals and transitions, the concept of fundamental frequency can no longer be treated with full satisfaction. The operation of the proposed [cf. Sect.6.2.1] ... and ... any equivalent detector seems to be

---

period because we deal with *time-variant* signals that have time-variant parameters. Fundamental frequency is thus generally not the frequency of the fundamental (i.e., the frequency of the first partial), and fundamental period is generally not the period duration of the fundamental. The two would only be equal if the signal were time-invariant and exactly periodic. For didactic reasons a complete definition of these terms cannot be given yet; this will be given in Section 9.1.1. In an actual PDA, PDD, or PDI, we will regard fundamental period as the primary quantity if the measurement is carried out in the time domain (or some equivalent domain), and fundamental frequency if the measurement is carried out in the frequency domain. The other quantity - fundamental frequency and fundamental period, respectively - is then defined as the reciprocal of the primary quantity according to (2.34).

more adequately interpreted in the time domain, in terms of periodicity, although it should be admitted that the instantaneous (fundamental) frequency, defined as reciprocal of the period ... might form an equivalent conceptual basis. (Filip, 1969, p.720)

In this book the definition (2.34) of fundamental frequency as the reciprocal of fundamental period T_0 will always be obeyed. Most PDAs implicitly follow this concept. They determine T_0 directly (in the time domain, as correlation lag, or as quefrency), or as its reciprocal $1/T_0$ (in form of spectral distances in the frequency domain). That does not mean, however, that the concept of fundamental frequency as the largest common divisor of the partials is invalid. On the contrary, the pitch perception theories apply the principle of subharmonic matching - which actually represents the search for a common (perceptual) divisor for the individual spectral pitches - for determination of perceived pitch. The small but significant difference to the rigorous mathematical definition is that a small tolerance interval is created around each of these subharmonics retrieved. If some subharmonic of another partial falls into this tolerance interval, a (near)coincidence is signalled which actually contributes to the formation of virtual pitch. Thus the model explains why a small frequency shift (for instance via mistuned SSB modulation) does not produce a significant change in pitch perception.

The relation between the concept of virtual pitch (disregarding the effects of pitch shift) and fundamental frequency in the mathematical sense is, as one could say, a frequency-domain counterpart to the time-domain relation between long-term and short-term signals. In the time domain most long-term measurements can be transferred to short-term signals (cf. Sects.2.4 and 8.1 for a detailed discussion), but the necessary windowing process introduces an intrinsic inaccuracy - a "perceptual inaccuracy" of the model, as one could circumscribe it here. In the case of true perception by human listeners, where it is represented by the tolerance intervals, this perceptual inaccuracy may in part even be the outcome of some built-in time-domain windowing process in the human ear.

4.2.3 Difference Limens for Fundamental-Frequency Change

The surprisingly low perceptive threshold (*difference limen*, DL) for the audibility of fundamental-frequency change in (synthetic) speech signals can also be explained by the theories of pitch perception. This DL is one order of magnitude smaller than the corresponding DL for formant frequencies (Flanagan and Saslow, 1958; Flanagan and Guttman, 1960a,b). According to the measurements by Flanagan and Saslow, the DL for (synthetic) vowels is situated in the range of 0.3% to 0.5% (measured for the F_0 range of a male voice). This is even considerably less than the 3-Hz DL for pure tones in the low frequency region (Zwicker and Feldtkeller, 1967). According to the virtual-pitch theory, the sensation of virtual pitch is "made" by the first determinant spectral pitch(es) in the dominant region (i.e., above 300 to 500 Hz). These spectral pitches are subject to the DL for pure tones above 500 Hz 0.3% to

0.6%, after Zwicker and Feldtkeller (1967)], so that the DL for F_0 would then equal the DL of the determinant spectral pitch(es)[5] and not the DL of the pure tone with frequency F_0, which can be considerably higher, depending on the actual value of F_0. Other investigations suggest that the DL for the audibility of F_0 changes in (natural) speech signals is higher. Isačenko and Schädlich (1970) found 5% (at $F_0 = 150$ Hz), and Rossi and Chafcouloff (1972b) found 4% (at $F_0 = 195$ Hz). [In these experiments, however, short segments of natural speech were offered to the listeners. The higher DL is thus likely to be partly due to perturbations because of the intrinsic jitter and shimmer of the natural speech signal (Chafcouloff, personal communication).]

Results by Cardozo and Ritsma (1968) are likely to resolve this contradiction. Cardozo and Ritsma carried out comprehensive experiments with jittered pulse trains. They used band-pass filtered pulse trains with a repetition rate of 100 and 333 Hz. The center frequencies of the band-pass filters applied were 1000, 1400, and 2800 Hz, and the bandwidth of the filters was about one-third octave. The jitter[6] applied was quasi-Gaussian.

In the first experiment to be reported here, Cardozo and Ritsma measured[7] the DL for the audibility of the jitter. They found values between 0.17% and 0.7% for pulse trains longer than 100 ms. For shorter pulse trains the DL increased rapidly; for a 50-ms pulse train duration the DL was found to be twice as high. The DL depended on the ratio of the center frequency of the filter and the pulse repetition rate; it was found to be lowest (0.17%) for the combination (333 Hz, 1400 Hz; pulse repetition rate named first), and highest (0.7%) for the combination (100 Hz, 1400 Hz). The two combinations (100 Hz, 2800 Hz) and (333 Hz, 1000 Hz) were not used.[8] The general tendency is that

---

[5] In this respect it is irrelevant whether the pitch percept is determined by a single higher partial, as in Terhardt's theory (1979b), or by a weighted interaction of all the significant partials, as postulated by Goldstein (1973). At this point, both theories lead to identical statements.

[6] Jitter (see Sect.2.6) means (stochastic) small frequency changes and modulations of a signal, in contrast to shimmer which refers to small changes in amplitude.

[7] The measurement of these DLs was carried out as a forced-choice experiment. The subjects listened binaurally to a sequence of two pulse trains of equal amplitude and duration. One was jittered, the other perfectly periodic, with the order randomized. The listeners were forced to label one of the two impulse trains as jittered. The difference limen was then found as the value of the jitter for which 72% of the responses were correct. Flanagan and Saslow (1958) applied the same procedure.

[8] The combination (333 Hz, 1000 Hz) was excluded from the experiments since the signal would then contain only a single harmonic. The ratio of the two frequencies in the combination (100 Hz, 2800 Hz) is so high that a virtual pitch cannot be derived from these pulses any more. In all the other combinations applied the formation of virtual pitch from the pulses is possible.

this DL decreases with increasing pulse repetition rate when the relative jitter, i.e., the standard deviation of the pulse distances divided by the average pulse distance, is taken as the reference. This result suggests that jitter is more readily perceived in utterances by female than by male speakers.

In a second experiment Cardozo and Ritsma (1968) investigated the DL for single misplaced impulses in a short pulse train (in the terminology used in this book, one could call this DL a difference limen for hops). Each pulse train contained 8 impulses of equal amplitude and with equal distances except for the one impulse which was delayed by a certain amount of time. Cardozo and Ritsma found the DL for the relative delay (delay time divided by average pulse distance) to be about 2% when the delayed impulse was located in the center of the pulse train, and 10% for the leading impulse. These results are in agreement with findings by Michaels and Lieberman (1962) that voice perturbations are more readily perceived when they occur in the center of a stationary vowel. The results are also in agreement with the first experiment if one takes into account the short duration of these pulse trains (less than 80 ms), and if a "root-mean-square jitter" for the whole pulse train is evaluated from the delay of the single impulse.

This experiment also clearly indicates that virtual-pitch formation is involved in the process of perception of irregularities. The values of 2% to 10% applies to the combinations (pulse repetition rate, center frequency) where virtual-pitch formation is possible. For the combination (100 Hz, 2800 Hz) the same DL was found to be about 9% for impulses located in the center of the pulse train, and 20% for the leading or trailing impulses.

In a third experiment Cardozo and Ritsma (1968) measured the DL for jitter in nonstationary pulse trains. The pulse repetition rate was decreased from 400 Hz at the beginning of the pulse train to 200 Hz at its end. (In another test, the sweep ranged from 500 Hz to 100 Hz.) The duration of the pulse trains was 100, 250, and 500 ms. The results show that the DL increases; for moderate sweep rates,[9] however, the difference is comparatively small. The results suggest "*that jitter is perceived as soon as it is of the same order of magnitude as the systematic change of the pitch period*" (Cardozo and Ritsma, 1968).

The results of all these experiments must be interpreted with some caution in the respect that deviations of more than 100% between individual data need not indicate that the results are contradictory. There are always discrepancies between individual subjects which easily exceed 100% or even 200%; in addition, the choice of the subjective test conditions plays a major role (Rosenblith and Stevens, 1953). A number of reasons, however, support the results by Flanagan and Saslow (1958): 1) the agreement between their data and the DL for frequency changes of pure tones above 500 Hz via the

---

[9] There is little agreement in the literature as to how the rate of change of F_0 or T_0 should be quantitatively indicated. First of all, it depends on how

current theories of virtual-pitch formation, 2) the agreement between the data by Flanagan and Saslow and the data by Cardozo and Ritsma (1968) found for jitter, and 3) the fact that a quantization of pitch in digital speech synthesis systems due to sampling can cause audible effect even if the quantization error is less than 1% (Holmes, 1973). The data by Cardozo and Ritsma (1968) indicate that the virtual-pitch formation mechanism is involved in the perception of irregularies and jitters in quasiperiodic signals. For speech we can state that, due to the general nonstationarity and the short durations of stationary segments, irregularities are less readily perceived, so that the DL as indicated by Flanagan and Saslow (1958) represents the worst case. At the same time, however, this DL represents the ultimate demand with respect to the accuracy of a PDA.

Just because a fundamental-frequency change is audible does not at all mean that this change must be linguistically significant. In fact, there is considerable discrepancy between the DL of *audibility* and a corresponding "DL" for *linguistic significance* of a F_0 change; the latter is at least by an order of magnitude higher. McKinney (1965) summarized experimental results important in this connection:

An abrupt change in laryngeal frequency of as little as 5 Hz appears to be capable of signalling linguistic stress. ... As a cue in perception of stress in synthetic speech, a small abrupt change in F_0 has been found capable of overriding both duration and intensity. Shifts in the location of such an abrupt change make little difference as long as no syllable boundary is crossed. ... This finding for synthetic speech has been partially

---

T_0 or F_0 is changed. Even if the change is linear, it will make a difference whether T_0 or F_0 is linearly changed. In the present experiment by Cardozo and Ritsma, it is F_0 which is linearly changed; that means that the change of T_0 has a hyperbolical shape. The second problem is the measure by which the rate of change shall be indicated. Reasonable frequency-domain measures are *hertz per second* or *octaves per second*; the hertz per second are more easily converted into time-domain measures, and the octaves per second give a better correspondence to psychoacoustics and music. In the time-domain, since pulse trains as well as speech signals consist of a series of discrete events, the *(absolute or relative) change of duration from one period to the next* may be a suitable measure. (Converting these measures into each other requires knowing additional parameters such as the value of the pulse repetition rate, which sometimes do not directly influence the results and are thus omitted from the publications.) In the present experiment, the pulse repetition rate, i.e., a frequency-domain measure, was linearly decreased. If we evaluate this quantitatively for the case of the slowest (400 to 200 Hz in 500 ms) and the fastest sweep (500 to 100 Hz in 100 ms), this would mean that the rate of change took on values of 400 and 4000 Hz per second, respectively. For the given frequencies this would correspond to 2 and about 23 octaves per second. Since the pulse distance changes hyperbolically, the relative change between one period and the next ranged from 0.25% at the beginning to 1% at the end of the pulse train for the slowest, and from 1.6% to 40% for the fastest sweep. These details may look rather trivial; nevertheless, they are important for the case that these psychoacoustic results shall be related to the accuracy demands and the quantitative performance of PDAs.

verified for natural speech. ... The shape of somewhat less abrupt changes in F_0 may also carry linguistic information. Slow changes (1-3 Hz) and absolute levels of frequency, on the other hand, apparently do not function to signal linguistic units, such as stress. In natural speech, the laryngeal frequency, intensity, and duration of speech sounds covary in an apparently complicated manner. For this reason, an analyzer [i.e., a PDA] should make independent measurements of these parameters, and a corresponding task for the acoustic phonetician and linguist is to discover those correlations which are physiologically conditioned or linguistically conditioned. (McKinney, 1965, pp.32-33)

Since most of the experiments summarized by McKinney deal with speech uttered by male speakers, it is to be expected that for arbitrary voices these changes are about equal in percentage, but not about equal in absolute value. If we assume a fundamental frequency of about 120 Hz for McKinney's data, this would give a DL for linguistic significance between 4% and 5%. (As we will see in the next section, the question of the DL for linguistic significance is not a matter of pitch perception alone, but also a matter of speech production.) Earlier data (Cowan, 1936; Black, 1967) as well as more recent investigations indicate that the perceptive threshold for linguistic significance is still much higher. In a fairly complicated experiment, 't Hart (1981) found that the DL for a safe discrimination of two pitch contours as *different* is in the order of three to four semitones, i.e., between 18% and 25%! In his experiment, 't Hart offered his subjects a sequence of two four-syllable Dutch numerals where the third syllable was the stressed one. This syllable (together with the last syllable of the word) was systematically raised or lowered in pitch with respect to the first two (unstressed) syllables. The task of the subjects was to indicate whether the first or the second numeral had the larger step in F_0; the subjects were allowed, however, to mark the two steps as equal if they were unable to make a decision. In order to bring the experiment as close to real speech as possible, 't Hart provided the stimuli themselves as well as the sequences of two stimuli with the overall declining basic intonation contour, as is usual in connected speech. Such an experiment, where nonstationary speech signals are used, is of course much more difficult to solve than a similar experiment with musical instrument notes where a semitone (in oriental music even half a semitone) should be easily discriminable, at least by trained listeners. The control experiment by 't Hart, where the speech signals were replaced by piano tones, showed that the DL for the discrimination of these instrument tones is smaller than for speech.[10]

These results are consistent with the fact that the fundamental-frequency range of connected speech for a particular voice rarely exceeds one octave (unless particular emphasis is intended by the speaker), and with the postulate by a number of intonation researchers (e.g., Pike, 1948) that the syntactic

---

[10] These results suggest that the auditory system performs extremely well in the frequency domain, but quite badly with respect to temporal changes. For this reason quantitative results of auditory pitch determination in speech are practically impossible to obtain.

intonation rules are confined to four *intonation levels*. Four levels within one octave, together with the general decline of the pitch contour within a sentence, would suggest the difference between adjacent levels to be situated between two and three semitones.

Just these few examples show that there are a number of questions still open in this domain. This brief review, however, far from being complete, is only intended to give some hints from the perceptional point of view dealing with the question of accuracy of PDAs which is to be discussed in the next section.

4.3 Measurement Accuracy

The auditory difference limen for F_0 changes - which is by far the most sensitive differential threshold in the auditory system - is certainly the ultimate requirement of accuracy for a PDA. Linguistically significant changes of F_0 must be considerably larger since the requirement number one for a linguistic distinction is that its acoustic representation is - clearly - heard.

In the majority of the publications on actual PDAs no statement is made with respect to the accuracy of the measurement. Most evaluative statements deal with critical reviews of the "gross" errors, such as higher harmonic tracking and holes. Some authors, such as Gold (1962b), check their PDAs against manual analysis or against a voice-excited vocoder. An evaluation as to fine measurement errors was not carried out until 1976 (Rabiner et al.; cf. Sect.9.3 for a detailed discussion). Manual analysis, on the other hand, is generally not accurate enough to meet the auditory difference limen. Thus for analog systems a general statement for the accuracy actually achieved seems impossible. Some remarks, however, can be made for digital systems. The intrinsic accuracy of a time-domain PDA is limited ultimately by the sampling interval due to quantization of the estimation of elapsed time from which the period estimate is derived. Using a sampling frequency of 10 kHz, the achievable accuracy would be 1% at a fundamental frequency of 100 Hz (if T_0 is only permitted taking on integer multiples of the sampling interval). Even this value is above the difference limen, which means that a time-domain PDA would cause quantization noise in the pitch estimate even at this low fundamental frequency; the accuracy decreases with increasing F_0. The same holds for short-term analysis PDAs whose basic extractor operates in the lag domain (cepstrum, autocorrelation, etc.). For frequency-domain analyzers, the basic problem is even more serious.[11] The unrealistically great frame length of 100 ms (which could be achieved, however, by padding a frame of reasonable length with zero samples) causes the spectral samples to

---

[11] Frequency-domain PDAs, on the other hand, have the advantage over any PDA whose basic extractor works in the time or a lag domain, that the

be spaced at intervals of 10 Hz; a direct measurement of F_0 would then result in an intrinsic accuracy of 10% at 100 Hz and 2% at 500 Hz.

One way out of this problem is interpolation. Interpolation is obvious and readily realized in frequency-domain analysis, easily done in any other short-term analysis method when a peak detector is used, and hard for a time-domain PDA where it strongly depends on the applied basic extractor. The accuracy of peaks and zero-crossing locations can be improved by interpolation. For nonzero threshold crossings, however, the results will be rather arbitrary. Another way out is smoothing (Sect.6.6). Smoothing eliminates the quantization noise, but its contribution to the actual accuracy of the measurement of individual pitch periods is sometimes questionable.

On the other hand there is the accuracy of speech production. Except for the applications where the output of a PDA is used to generate synthetic speech (which is then presented to the ear), it does not seem necessary to determine F_0 more accurately than it is produced by the vocal source, provided that the source produces T_0 less accurately than the DL would require. Measurements exist on the accuracy of glottal pulses, especially on stochastic variations. Gill (1962a,b) found (qualitatively) that there were more variations in shape than in period length of the excitation signal. Lieberman (1963) investigated small pitch period perturbations. For this he determined the relative difference in duration of successive periods. According to his results, this difference is greater than 1% in 30% of all pitch periods, and greater than 3% in 10% of the periods investigated. Similar results have been reported by Hollien et al. (1973) and by Horii (1979a). According to Horii's results, the mean value of the jitter (expressed in time-domain terms as absolute value of the deviation of T_0 between successive periods) depends somewhat on the momentary value of F_0; Horii found mean values of 51 μs for $F_0 = 98$ Hz and 24 μs for $F_0 = 298$ Hz. This means that the *relative* jitter slightly increases with increasing F_0. It must be noted, however, that this may be partly due to an artifact caused by the temporal resolution due to sampling. Although relatively high sampling rates (up to 80 kHz) were used in these investigations, both Horii (1979a) and Hollien et al. (1973) reported that part of the increase of the relative jitter for F_0 above 200 Hz could be due to the contribution of quantization noise and thus not reflect the real pertur-

---

individual harmonics are well separated in the spectrum and can be individually identified. F_0 can thus be computed as the appropriate subharmonic of some higher harmonic, by which the quantization error is decreased by a factor equal to the harmonic number of the higher harmonic taken. F_0 can even be computed as the weighted average of several subharmonics of this kind, as it is done by Goldstein (1973) in his pitch perception model. This procedure increases the measurement accuracy of frequency-domain PDAs by as much as an order of magnitude without requiring a better spectral resolution. In spite of the seemingly bad shape of frequency-domain pitch determination due to the basic quantization problem (see Fig.4.4), the effects of quantization can be largely avoided so that, after all, frequency-domain PDAs are least affected by quantization errors. See Sect.8.5 for more details.

bations of the human voice. The distribution of the jitter in Horii's data shows that it exceeds 100 μs for 10% of the periods. Assuming an average value of 200 Hz for F_0 this would give a perturbation of 2% or more in 10% of all pitch periods.

Comparing these results to the difference limen of fundamental frequency, we see that these slight perturbations (which represent the natural "noise" in the pitch signal) are clearly audible.[12] (With respect to accuracy, the ear thus outperforms the speech production system by far.) They were found to contribute to the "naturalness" of the signal (in contrast, e.g., to synthetic speech where the perturbations are zero). Certainly they should not contribute to linguistic information (such as linguistic stress); in order to be clearly discriminated, a linguistic distinction is in fact represented by a F_0 change well above this "noise". If only linguistically relevant properties of pitch contours are to be analyzed, one definitely need not measure pitch as accurately as specified by the DL for audibility of F_0 changes. *"Absolute judgments of laryngeal-frequency changes in real speech are not so acute"* (McKinney, 1965). On the other hand, PDAs must be as accurate as to reliably indicate those changes which are perceptually significant because they signal prosodic boundaries and marks. If one neglects the aspect of naturalness to a certain extent, an accuracy of a few percent seems sufficient from the linguistic point of view. This can be easily achieved by digital analysis, even without interpolation; it corresponds to the accuracy of manual pitch determination using the spectrogram.

As to the rate of change in F_0 to process by a PDA, an upper bound has to be defined rather than a lower one. It is nonsense to measure "frequency changes" in intervals shorter than one pitch period. Unless it commits an error, a time-domain PDA which yields individual markers is never in danger of missing an abrupt F_0 change there. This might rather happen to short-time analysis PDAs when the frame interval is too great. If one proceeds from the point of view that most significant parameters, such as formants or intensity, are influenced by the same production process and thus vary with comparable speed, one might expect that the usual frame interval of 10 to 25 ms in analysis-synthesis systems will not violate the sampling theorem for the parameter F_0.

---

[12] *Audible* does not mean that such perturbations are distinguishable as discrete events; they pass too rapidly to be individually perceived. They rather contribute to the overall percept in form of slight variations of timbre, or in form of a slightly increased roughness (Terhardt, personal communication). Qualitatively these perturbations do contribute to the sensation of voice naturalness (Schroeder and David, 1960), and they are well distinguishable from artificial perturbations in synthetic speech due to quantization noise (Holmes, 1976) or stochastic jitter [which was implemented to improve the naturalness of synthetic speech, but was then found increasing the roughness of the synthetic signal without necessarily contributing to the naturalness (Makhoul et al., 1978, Rozsypal and Millar, 1979)].

An upper bound to the rate of change in F_0 has to be defined simply in order to detect a measurement error (and to be able to correct it). A usual bound for time-domain PDAs is 10 to 15% for subsequent periods (Reddy, 1967). For changes of F_0 which are intended by the speaker, the upper limit will be of that order. Data by Sundberg (1979) indicate that the maximum rate of change of frequency that a voice can normally achieve is in the order of 1%/ms. (This would correspond to a change of somewhat more than 3 octaves per second if the relative change rate is kept constant.) Converted into time-domain terms, this would correspond to a 2% change in T_0 between adjacent periods for $F_0 = 500$ Hz, and a 10% change in T_0 for $F_0 = 100$ Hz. Except for very low fundamental frequencies, T_0 will thus never change more than 10% from one period to the next when the signal is regular. Average change rates due to prosody are lower than that. Results by Black (1967) derived from data by Cowan (1936) and Takefuta (1966) indicate a rate of change in F_0 between 0.2%/ms and 0.65%/ms due to prosody during pitch inflections, corresponding to a change of T_0 by 2% to 6.5% between successive periods for $F_0 = 100$ Hz.

On the other hand, there are changes which are not intended by the speaker and can be much more drastic. Irregularities of individual periods or period sequences occur occasionally in normal speech (Dolansky and Tjernlund, 1968; Fujimura, 1968). In some pathological voices these irregularities occur more frequently. Vocal fry, which can form part of a normal speech signal, is per se irregular. With an upper limit of frequency change rate defined, all those irregularities would be detected and treated as "measurement errors."

Dolansky and Tjernlund (1968) investigated a large number of pitch periods with respect to irregularities in the waveform. In this study they compared the speech signal and the simultaneously recorded output signals of a glotto-graph (cf. Sect.5.2.3) and a contact microphone (cf. Sect.5.2.2). Their main conclusions read as follows.

1) Of about 30 000 pitch signals, 78 were judged irregular.

2) In about 20% of these, the corresponding glottal excitation is not irregular.

3) Irregularities in the beginning of the utterances are usually of a single-period type, while at the end, trains of irregulaties are en-countered. ...

4) Irregular excitation in the ending portions consists of alternating complete and incomplete glottal closures. Usually an incomplete closure is followed almost immediately by a complete closure, but a larger distance is found between the complete closure and the following incomplete clos-ure. [This might indicate a temporary register change of the voice into vocal fry.]

5) Even disregarding the multiplicity of irregularities in the terminating portion of waveforms, the irregularities at the end outnumber the ones at the beginning about four to one.

6) Most of the irregularities occur when the pitch is low. ...

7) Rapid rates of variation of formant frequencies or fundamental fre-quency do not appear to cause any waveform irregularity. (Dolansky and Tjernlund, 1968, pp.55-56)

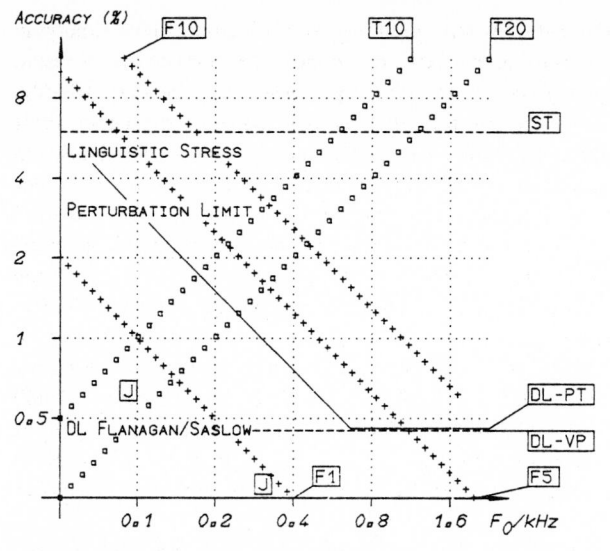

Fig.4.4. Data concerning the accuracy of PDAs. (DL Flanagan-Saslow) Difference limen for F_0 change according to Flanagan and Saslow (1958); (J) difference limen for jitter according to Cardozo and Ritsma (1968); (Linguistic
stress) data by Bolinger and Fry indicating the minimal F_0 change which can
signal stress; (Perturbation limit) 90%-limit of pitch period perturbations (after
Lieberman, 1961, 1963). (T10, T20) Intrinsic accuracy of time-domain PDAs
with a sampling frequency of 10 and 20 kHz, respectively, without interpolation; (F1, F5, F10) intrinsic accuracy of frequency-domain PDAs with a frequency resolution of 1, 5, and 10 Hz, repectively, without interpolation or
higher harmonic matching. [Note that higher harmonic matching increases the
accuracy of the frequency-domain PDA by a factor equalling the harmonic
number of the harmonic from which F_0 is derived.] (ST) Frequency change
corresponding to the musical interval of a semitone. (DL-VP) Hypothetic
difference limen derived from Terhardt's model of virtual pitch (assumed
value: 0.45%); (DL-PT) difference limen of a pure tone (after Zwicker and
Feldtkeller, 1967). Assumed value for high frequencies: 0.45%

The results by Dolansky and Tjernlund suggest that the tendency of PDAs
to introduce an upper bound for the rate of change and to regard every
change exceeding this bound as candidate for a measurement error is reasonable, at least in a first approximation. According to these data, less than 0.5%
of the pitch periods of an utterance are irregular for a healthy voice. The
number of gross errors to be expected from a usual PDA is as much as an
order of magnitude higher unless the PDA is fairly sophisticated and a reliable voiced-unvoiced discrimination is incorporated. Therefore, when a significant deviation between successive pitch estimates is encountered, it is much
more likely that the PDA has temporarily failed than that the voice was
irregular. (Pitch determination during an unvoiced segment, in this respect, is
regarded as temporal failure of the PDA.)

Figure 4.4 collects the data on the accuracy of and around PDAs, as far as discussed in this chapter. To extend the difference limen data by Flanagan and Saslow, a simplified hypothetical difference limen is derived according to the virtual-pitch model (Terhardt, 1979b). In this hypothetical difference limen the first spectral pitch above 500 Hz is assumed to determine the accuracy of the virtual-pitch sensation.

4.4 Representation of the Pitch Information in the Signal

Three definitions of periodicity and period duration T_0 are possible: 1) the long-term definition with T_0 as the period duration of a strictly periodic signal, 2) the short-term definition with T_0 as average period duration, averaged in some way over a finite number of individual period, and 3) the instantaneous definition with T_0 as the elapsed time between the actual and the preceding pulse of a pulse train or between the last two occurrences out of a series of repetitive events. Since the signal varies with time, the long-term definition of T_0 is not applicable. The other two definitions lead to the first categorization of PDAs as carried out in Chap.2: the short-term analysis PDAs follow the short-term definition, and the time-domain PDAs follow the instantaneous definition. The instantaneous definition does not permit processing other than in the time domain; in order to leave the time domain a transformation is necessary which requires at least a short-term representation of the signal. In this case it is unimportant whether the independent variable of the spectral domain is again time (in form of a cepstrum quefrency or an autocorrelation lag) or frequency; the short-term transformation focuses the information on pitch that is distributed over a whole frame into one single peak in the spectral domain and indicates pitch by averaging and a nonlinear decision process. In Sect.9.1, when all the principles of PDAs will have been discussed, the issue of defining pitch is taken up again and discussed in more detail; also a number of formal definitions will then be given.

The categorization into time-domain and short-term analysis PDAs forms the basis of the organization of this book. It is not the only possibility for categorizing PDAs, however. Pitch is represented in the signal in many ways. To appropriately arrange these representations, let us go back to the discussion at the end of Chapter 1. As stated there, the analysis of speech can start from one of the three points of view (Zwicker et al., 1967): 1) the production point of view, 2) the perception point of view, and 3) the signal processing point of view, the latter in principle without reference to production or to perception. From the signal processing point of view, we adopt the features that indicate periodicity in the signal: in the time domain this is the presence of a repetitive structure. If present in the signal, the first partial also indicates T_0. The repetitive temporal pattern corresponds to the harmonic spectral structure in the frequency domain with its equidistant peaks. The temporal structure as well as the harmonic pattern can be focused into one peak by a suitable short-term transform.

Our knowledge about speech production is mainly time-domain oriented. Speech production tells us what the repetitive temporal structure looks like. The speech tract is modeled as a linear passive system excited by a pulse generator. Hence the pitch period is the sum of a finite number of exponentially damped sinusoids whose amplitude take on a maximum at the beginning of the period and decay toward the end. The temporal structure of the excitation signal is also known: a roughly triangular shape with a discontinuity in the first derivative at the point of glottal closure. This structure can also be exploited provided that one succeeds in reconstructing the excitation signal.

The contribution of speech perception is more oriented towards the frequency domain. It is known that the sensation and formation of perceived (virtual or residue) pitch of a complex sound does not need the presence of the first partial; even if it is present in a speech signal, it is of subordinate significance due to the particular range of F_0 in speech which is situated below the frequency range that dominates the sensation of spectral pitches (Terhardt, 1972a, 1979a,b). It is important to know that the sensation of virtual pitch is performed in the ear using "near coincidences," i.e., matching subharmonics that coincide only approximately. Besides the two realizable definitions of T_0 the perceptual model of virtual pitch is a third, independent approach. It leads to the definition of pitch in a frequency-domain unit to which a frequency value F_0 can be straightforwardly assigned. Yet no PDA is found which strictly adheres to this category since there are severe problems

Table 4.3. Features used for pitch determination

| Representation | Domain | | |
(Dominant Feature)	Time	Frequency	Other
Periodicity	Repetitive temporal structure	Harmonic structure	Correlation: peak at T_0 (max or min)
Fundamental Harmonic	Waveform of first partial	Peak at F_0	
Production	Pitch period as sum of decaying exponentials		
Excitation signal	Triangular shape; discontinuity at glottal closure		
Perception		Virtual (or residue) pitch as matching of subharmonics	

with the realization. According to Terhardt (1979b), virtual pitch is set up as a matching process of the subharmonics of spectral pitches. There are PDAs which perform the task this way (Terhardt, 1979a,b; Terhardt et al., 1982b; Duifhuis et al., 1978, 1982; Schroeder, 1968), but they obtain the values of spectral pitch from an ordinary short-term Fourier analysis.

The process of virtual-pitch formation suggests that some form of short-term analysis takes place in the ear, but neither the shape of the pertinent window is sufficiently known, nor do we know whether this process, if performed in the human ear, can be modeled by a linear system at all. For these reasons perception-oriented frequency-domain PDAs are not treated as a separate category, but in the same way as ordinary short-term analysis PDAs. Table 4.3 gives a list of the features that can be used for the purpose of pitch determination.

4.5 Calibration and Performance Evaluation of a PDA

Calibration of a measuring instrument requires a standard or another instrument of at least the same accuracy as the one being tested. In pitch determination, however, the problem is that both a standard and a high-accuracy instrument are missing to some extent, at least under normal speaking conditions. The standard for a perceptually perfect pitch determination was evaluated in Sect.4.2: an accuracy of 0.3% to 0.5% according to the auditory difference limen for fundamental frequency. For a linguistically perfect determination (that means, an analysis which meets all the requirements imposed by phonetical and linguistic constraints), a lower accuracy may be accepted (2% to 4%). At least the perceptual standard is hardly reached by existing PDAs. From the production point of view, there is a standard: the laryngeal excitation function. This function and with it the definition of fundamental period

> ... is relatively difficult to instrument directly because of the comparative inaccessibility of the larynx and because of the sensitivity of the larynx and much of its anatomical environment. This has led to the use of various indirect means of measuring the frequency, even to provide a standard for comparison for the performance of laryngeal-frequency analyzers. (McKinney, 1965, p.36).

The laryngeal excitation function, in the meantime, can be easily measured by PDIs such as the laryngograph. However, this facility of calibrating a PDA by comparing its output to that of a laryngograph or a similar instrument has practically not been used up to now. The most widespread method of evaluating the performance of a new PDA is to check it against another PDA whose characteristics are known to be good, the latter preferably in connection with a speech communication system. Gold (1962b), for example, used a voice-excited vocoder to make comparisons

... between excitations produced by the "voice," the "eye," and TX8 [Gold's time-domain PDA]. ... All listeners agreed that each of the excitations resulted in synthetic speech which was not too different from ordinary telephone speech. ... It is our opinion that "eye" detected pitch is sufficiently comparable in quality and intelligibility to voice-excited pitch to justify our original assumption that the eye could serve as the "ideal" for our experiments. (Gold, 1962b, p.921)

Since manual determination of pitch is extremely time consuming (cf. Sect.5.1), most researchers have confined themselves to check their PDAs against some PDA(s) already in existence. Cepstrum pitch determination (cf. Sect.8.4) has served as such a standard PDA for a long time. It emerges partially from the principle of operation of these "standard" PDAs that most evaluations of PDA performance are confined to a check against gross errors, such as higher harmonic tracking, or to a formulation of the requirements towards the recording conditions, such as band limitation or surrounding noise. Hence the performance evaluation is mostly the weakest part of publications dealing with the implementation of a new PDA. Rabiner et al. (1976), for instance, complain that

... in spite of the proliferation of pitch detectors, very little formal evaluation and comparison among the different types of pitch detectors has been attempted. (Rabiner et al., 1976, p.393)

Consequently Rabiner et al. (1976) carried out a comparative evaluation of several existing PDAs according to objective as well as to subjective criteria of measurement. Their standard for the objective measurement is an interactive PDA (McGonegal et al., 1975) which combines the ease of pitch determination achieved by automatic analysis with the faultlessness of the manual analysis. For the subjective analysis a rank ordering procedure is applied relative to original speech and speech resynthesized using the original pitch information. In Sects.9.2,3 these evaluations, together with the whole issue of calibration, will be discussed in more detail.

It may be partially due to such a relative lack of standards that the performance of PDAs with respect to irregular signals has only been partially explored. Most PDAs give priority to the detection of a "regular," i.e., periodic pitch pattern. The majority of the short-term analysis PDAs even operate with the presumption of periodicity and fail systematically when there are drastic deviations from periodicity in the signal. Such PDAs will thus assume a measurement error when a nonperiodic pattern is encountered; frequently they will even try to "correct" it. The inability of most existing PDAs to distinguish between measurement errors and real signal irregularities (which are numerous, as we have learned from the discussion in this as well as in the preceding chapter) is one of the big problems still unsolved in pitch determination.

5. Manual and Instrumental Pitch Determination, Voicing Determination

This chapter summarizes a number of problems and tasks in the neighborhood of the task of pitch determination.

The oldest techniques of pitch determination are *auditory* and *manual* pitch determination (Sect.5.1). Auditory pitch determination is hardly applicable to speech signals. The ear performs extremely well with respect to frequency discrimination, but the human observer is unable to quantitatively reproduce the rapid variations during speech. Manual pitch determination from visual displays of speech has played a significant role, especially in phonetics and linguistics; it is subdivided into *time-domain* manual pitch determination and *frequency-domain* (spectrographic) manual pitch determination, according to the type of visualization used.

Section 5.2 deals with *pitch determination instruments* (PDIs). There are a variety of instruments available, most of them in clinical use, but also in phonetics, linguistics, and education. The section also contains a short survey of the clinical methods of larynx inspection and some examples of comparative evaluations of the performance of PDIs which, in contrast to the situation with PDAs, are rather numerous.

Section 5.3 deals with *voicing determination algorithms* (VDAs). Pitch determination and voicing determination are the two subproblems of voice source analysis. Voicing determination is a prerequisite to pitch determination since it calculates those segments "where the periods are." But it is not necessary to know this beforehand. The PDA can in principle process voiceless or silent segments as well; of course it will waste a lot of computer time if operating on such segments. The combination of a VDA and a PDA switched in series is thus a matter of implementation rather than of principle; the two problems can be treated independently. Rather than presenting a comprehensive survey, Sect.5.3 gives some typical examples for the three categories of VDAs: simple VDAs using a small number of parameters, combined VDA-PDA systems, and pattern recognition VDAs which treat the problem as a task of classification.

5.1 Manual Pitch Determination

From the instrumentation point of view, the manual method is the simplest solution to the problem of pitch determination. In order to obtain information on pitch it is merely necessary to visually display the signal in the time domain or in some other suitable representation; the rest is done by the (human) researcher.

Pitch determination started with the manual methods. It started under conditions we hardly can imagine in our modern, well-equipped laboratories. For a long time manual pitch determination methods were the only ones available. The first successful attack on their predominance came with the development of a PDD capable to present a graphic display (Grützmacher and Lottermoser, 1937). The invention of the sound spectrograph (Koenig et al., 1946) again signified a great impact for the manual method since the spectrograph enabled the researchers not only to look at the global shape of a sequence of short-term spectra, but also at the course of individual harmonics from which the fundamental frequency is readily derived. Even in the age of computer-based PDAs, manual methods have been able to survive since computer algorithms, superficially regarded, have been found not universally applicable, not reliable enough, too complicated to be easily reimplemented according to the description given in a publication, and not normally provided with a suitable display (cf. Sect.9.4 for further discussions of this issue). Interactive PDAs (cf. Sect.9.1) may be a viable alternative in tasks where up to now manual methods were applied; they may combine the speed, accuracy, and storage facility of the automatic procedure with the reliability of the manual method. There is no doubt that manual analysis outperforms every automatic algorithm with respect to reliability: once a researcher has gained some practice with signals and spectrograms, he can hardly do anything wrong again.

Manual analysis methods are simple in instrumentation but extremely tedious with respect to human working time. No on-line experiments can be carried out using these methods. For this reason, manual pitch determination has more or less been confined to applications in phonetics, linguistics, and pathology.

Manual pitch determination methods can be subdivided into two categories: *time-domain* and *frequency-domain* manual pitch determination. Time-domain manual pitch determination only requires a visual display of the signal. Thus time-domain manual pitch determination is the oldest means of quantitative pitch determination. Frequency-domain manual pitch determination came into existence with the development of the sound spectrograph. Compared to time-domain manual pitch determination, this method allows a more global view of the temporal behavior of the signal and permits F_0 to be determined with much less human effort.

In this connection, let us dwell for a moment on the oldest "method" of pitch determination: the judgment by the human ear. *Auditory pitch determination,* as we have seen in the previous chapter (in contrast to auditory pitch *perception*) has its limitations in the central processor located in the brain rather than in the human ear which performs this task very reliably and accurately. The limitations deal with time-domain behavior rather than with frequency-domain performance, which may be quite accurate for a minority of individuals.[1] The human mind is unable to follow all the short-term variations of pitch in the voice, and completely unable to quantitatively reproduce them. Hence auditory pitch determination is confined to tasks where a stationary behavior of the voice can be presupposed. Such tasks occur in musicology (for instance, transcription of exotic folklore) or in pathology (e.g., scanning the frequency range of a voice).

5.1.1 Time-Domain Manual Pitch Determination

Time-domain manual pitch determination means determination of the fundamental period from a visual display of the speech signal. Each period, at least each period which is found significant for the overall pitch contour, has to be investigated, and its duration has to be measured. This is an extremely fatiguing and time-consuming job. The accuracy of the measurement strongly depends on the time scale used in the display.

A look at the devices used for visual display of speech signals is necessarily a look into history. In this short survey we will widely follow the description given by Léon and Martin (1969).

> One of the first measuring devices in instrumental phonetics was the kymograph described by Rousselot (1897-1908) in his *Principles of Experimental Phonetics.*
>
> The principle of the apparatus (Fig.5.1) is simple. The sound waves of the word are transmitted by a rubber tube to a drum, which is caused to vibrate. A recording stylus mounted on the drum inscribes the vibrations on a sheet coated with lampblack attached to a cylinder revolving at a constant speed.
>
> The drum, acting as a low-pass filter, fails to pick up the higher harmonics of the sounds of the word. The resulting curve brings out at most the first and second harmonics. ... A kymographic tracing therefore does not suffice to analyze the distinctive sounds of the word (its phonemes), but it is good enough to study the three prosodic parameters, duration, intensity and pitch. ...
>
> Pitch can be determined by calculating the frequency of each phone. To do this it suffices to note how many vibrations there are during its emission. If the phone is of brief duration, say around fifty milliseconds, it is enough to allow for a single average frequency. But if the phone is

---

[1] The ability of a person to quantitatively determine the (fundamental) frequency of a periodic sound without a known reference frequency is called *absolute pitch*; it is a well-known feature among musicians. A rich literature has been published about this phenomenon. In this book the question of absolute pitch is not further discussed.

Fig.5.1. The kymograph (Rousselot, 1908, as given in Léon and Martin, 1972, p.30)

long (for example a long vowel), it is necessary to average the frequency every fifty milliseconds so as to catch the possible changes in pitch. ...

It is especially the pitch of the vowels that counts in the perception of intonation, and it is often pointless to compute the vibrations of the consonants, as some researchers have done.

One needs a good magnifying glass,[2] good eyesight, and plenty of patience to make out the intonation even of a simple phrase by using this method, which phoneticians regularly employed in the heroic days of instrumental phonetics. Nevertheless, a good many important studies were carried out with the kymograph at the turn of the century - and not a few even of fairly recent date. (Léon and Martin, 1969, as given in Bolinger, 1972, pp.30-33)

At the time of Rousselot, instruments comparable to the kymograph had already been in use - or, at least, been proposed - for half a century.[3] According to Scripture (1902), who gave an excellent survey of this historical development (in those days it represented the most recent history!), the first attempt at recording and visualizing speech was done by Scott in 1856.

---

[2] Since there were no amplifiers in those days, the amplitudes of kymographic recordings were rather small so that their evaluation had to be done under the microscope (Scripture, 1902). Later kymographs, however, were equipped with amplifiers and galvanometers so that larger amplitudes could be achieved in the recordings (Mettas, 1971).

[3] Instruments for graphic registration of signals based on the kymographic principle came in use in the 18th century first in meteorology, then in other areas of physics. In 1847, Karl Ludwig was the first to apply such an instrument in physiology for measuring blood pressure curves. The name *kymographion* (which means "curve-writing device" in Greek), also originates from Ludwig (Rousselot, 1908; Mettas, 1971).

Scott was a proof reader; noticing a picture of the ear in the proof-sheet of a textbook on physics, he believed that he could get a record of speech by imitating the structure of the ear. In his *phonautograph* a large parabolic receiving trumpet carried at its end a thin membrane whose movement caused a small recording lever to write upon the smoked surface of a cylindrical drum. The sounds of the voice passing down the receiver agitated the membrane and caused the lever to draw the speech curve on the drum. (Scripture, 1902, p.17)

According to Scripture, this device, published in 1861, represented

... the prototype of the later machines that make speech records by registering the vibrations of a diaphragm on a moving surface by means of a lever. (Scripture, 1902, p.17)

This device was used and modified by a number of researchers in the late 19th century (Barlow, 1874; Schneebeli, 1878; Hensen, 1887; Hermann, 1889). A contemporary report (Jenkin and Ewing, 1878) indicates that Edison's phonograph (1877), although aiming at acoustic reproduction and not at the visualization of sounds, was used for phonetic investigation as well. For more details on this early work, the reader is referred to Scripture's survey (1902).

A remarkable aid in time-domain manual pitch determination was presented by Meyer (1911). One can attribute the development of the first postprocessor to Meyer; it applies a mechanic time-to-frequency conversion (cf. Sect.6.6.1 for more details on this issue). Due to his experience in the experimental investigation of intonation, Meyer was well aware of the necessity of displaying the intonation contour in terms of frequency on a logarithmic scale, i.e., in terms of musical intervals. He was also well aware that this time-to-frequency conversion constituted a major part of the whole work in manual pitch determination.

The total conversion of recorded voice signals into melody curves ... consists of the following steps: 1) microscopic measurement of the individual period lengths; 2) calculation of the logarithm of the period lengths; 3) drawing of the period lengths along the abscissa; 4) drawing of the logarithmic values of the period lengths along the ordinate at the pertinent abscissa points; 5) connection of the obtained points to form a curve; 6) calculation of the ordinates which correspond to the tones of the musical scale, according to the speed of the cylindrical surface of the drum at the time of recording (Meyer, 1911, pp.7-8; original in German, cf. Appendix B).

To make the work easier, Meyer developed a mechanic apparatus by which steps 2 to 4 of the above list could be mechanized. The apparatus converts a horizontal displacement (as which the period length is represented in the graphic recording) into a vertical straight line. Starting at a fixed horizontal line which marks the time axis of the display, the vertical straight line ends in the desired ordinate point. In the mechanic realization this point is marked by the intersection of a vertical ruler fixed at the beginning of the period and the curved edge of a metal templet which can be moved horizontally and is positioned to the beginning of the next period (Fig.5.2). The relation between the horizontal displacement and the vertical ordinate value is defined by the shape of the curved edge of the templet. Meyer claimed that this time-to-frequency converter reduced the human work necessary to manually evaluate a pitch contour by at least 80%. In manual pitch determination this device found numerous applications (Meyer, 1937; Fant, 1957, 1958; Mettas, 1971).

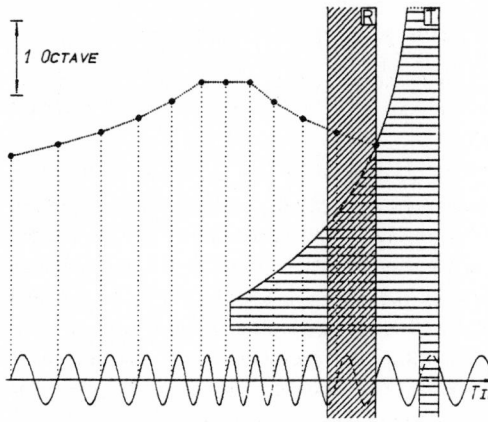

Fig.5.2. Principle of the mechanic time-to-frequency converter by Meyer (1911). The vertical ruler R is placed at the beginning of the period; the templet T which can be moved horizontally is positioned to the end of the period. The intersection between the templet T and the ruler R marks the desired point of the pitch contour. In the figure the signal is a sinusoid with varying frequency, and the shape of the curved edge of the templet gives a logarithmic display of the period durations. The original device is capable of compressing the time axis in an arbitrary scale. This technical detail, although quite important for the practical use of the device, has been omitted in the figure for reasons of simplicity

When electronic instruments came into use, the oscilloscope and oscillograph became the successors of the kymograph (although mechanic oscillographs had already been in use long before that time). In contrast to the kymograph, these instruments are not band limited within the frequency range of the speech signal and are thus able to give an exact visual display. These instruments are so well known that no description is necessary. The technique of pitch extraction, however, using the oscillograph and its time-domain display, is essentially the same as using the kymograph.

A disadvantage of the oscillograph (unless a storage oscilloscope is used) is that it needs a high amount of special paper (film or similar material) in order to give a permanent visual display, whereas the smoked surface of the kymograph could be used over and over. As an additional advantage (especially for phoneticians), sophisticated kymographs are able to record several signals and parameters simultaneously (Mettas, 1971). For cathode ray oscilloscopes the realization of this feature is expensive, and only few of them have more than two input channels; ink oscillographs and similar instruments can display up to eight signals simultaneously.

A significant step forward, even in manual pitch determination, was marked by the use of computers in speech processing. Nowadays the oscillograph is being replaced more and more by graphic terminals which display the waveform on a screen (and produce a paper copy if desired) and allow this analysis to be performed without excessive consumption of material. Lightpens and other graphic input facilities are an additional help. Of course, once a computer is used in research, the researchers will usually not hesitate to implement an automatic or at least an interactive algorithm (or they will connect a PDI to the computer) in order to reduce the work load associated with this analysis. Hence, with computing facilities available in many laboratories, and with graphic terminals and operating systems that permit interactive man-machine communication, one can say that at least time-domain manual pitch determination on the basis of the signal display has become a matter of the past and is of no more than historical interest today.

5.1.2 Frequency-Domain Manual Pitch Determination

Frequency-domain manual pitch determination became possible when the sound spectrograph was developed (Koenig, Dunn, and Lacy, 1946). The sound spectrograph enables speech to be displayed in a pseudo-three-dimensional visualization: with time and frequency as the abscissa and ordinate, respectively, and the short-term spectral amplitude indicated in terms of density, i.e., darkness. This graphic representation allows a global view of the essential parameters of the speech signal and their course during a whole utterance (2.5 s maximum).

The basic principle of the sound spectrograph is described in the original publication as follows.

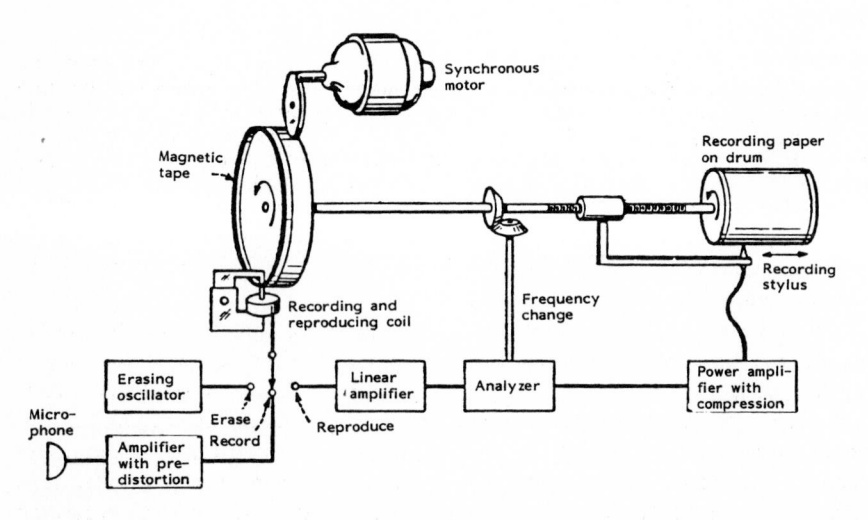

Fig. 5.3. Schematic representation of the basic method of the sound spectrograph. The sound is recorded on the loop of magnetic tape, and is analyzed while reproduced repeatedly. The fluctuating analyzer output builds up a pattern of light and dark areas on the electrically sensitive paper. From (Koenig et al., 1946, p.23)

Figure 5.3 shows in highly schematic fashion the basic method of the
sound spectrograph, as originally proposed by Mr.R.K.Potter. It is necessary, first, to have a means of recording the sound in such a form that it
can be reproduced over and over. The means shown here is a magnetic
tape, mounted on a rotating disk. In recording, some predistortion of the
signal may be desirable and is therefore indicated in connection with the
recording amplifier. With speech, for example, it has been found advantageous to raise the amplitude of the higher frequencies by about 6 dB per
octave in order to equalize the representation of the different energy
regions.
    Second, a means of analyzing must be provided. Most convenient is the
heterodyne type of analyzer employing a fixed band pass filter, with a
variable oscillator and modulator system by which any portion of the
sound spectrum can be brought within the frequency range of the filter.
    Finally, the output of the analyzer must be recorded in synchronism
with the reproduced sound. The simplest method is by means of a drum,
on the same shaft with the magnetic tape, carrying a recording medium
which should be capable of showing gradations of density depending on the
intensity of the analyzer output. Each time the drum revolves, the stylus
which marks the paper is moved laterally a small distance, and the oscillator frequency is changed slightly. Thus a picture is built up which has
time as one coordinate and frequency as the other, with intensity shown
by the density or darkness of the record. It may be necessary or desirable
to distort the amplitudes in the analyzer output, depending on the recording medium used and the use to be made of the spectrograms. This
function is indicated in the figure by the compression in the last amplifier.
(Koenig et al., 1946, p.21)

How can the sound spectrograph be used for pitch determination? There are
two possibilities, both of which have already been mentioned in the original
publication.

Pitch inflection ... may be brought out by two methods. If the analyzing
band is made narrow enough, the separate voice harmonics appear and
their rise and fall with pitch can be seen. The other method is to permit
the beats between harmonics in a wide band to register, producing characteristic striations vertically across the pattern. (Koenig et al., 1946, p.23)

The shape of the spectrogram crucially depends on the frequency resolution,
i.e., on the bandwidth of the band-pass filter applied in the analyzing circuit.
In the commercially available *Sonagraph* this bandwidth can take on two fixed
values: 45 and 300 Hz.

For male speakers, the wide-band spectrogram (bandwidth 300 Hz) cannot
resolve the individual harmonics, but it resolves the individual periods whose
beginnings can be seen as faint black lines parallel to the frequency axis.
The reason for this is quite obvious: in any period all the harmonics are
excited simultaneously at the instant of glottal closure (cf. the discussion in
Chap.3) and decay exponentially during the remainder of the period. The
fundamental period T_0 (as well as individual periods) can thus be measured
from the distances of these lines in a wide-band spectrogram. This also holds
when the signal is irregular. The accuracy of these measurements, however,
is fairly low since the time axis (which cannot be stretched in the original
device) has a scale of about 50 ms/cm; i.e., the display of a single period

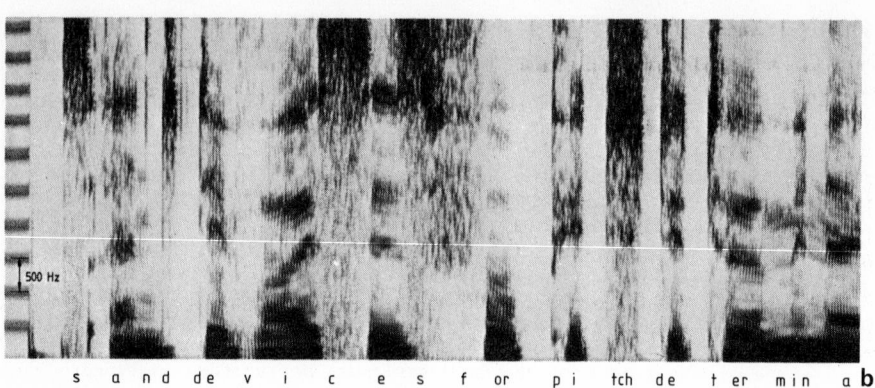

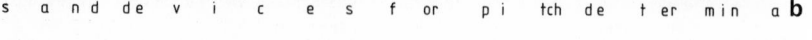

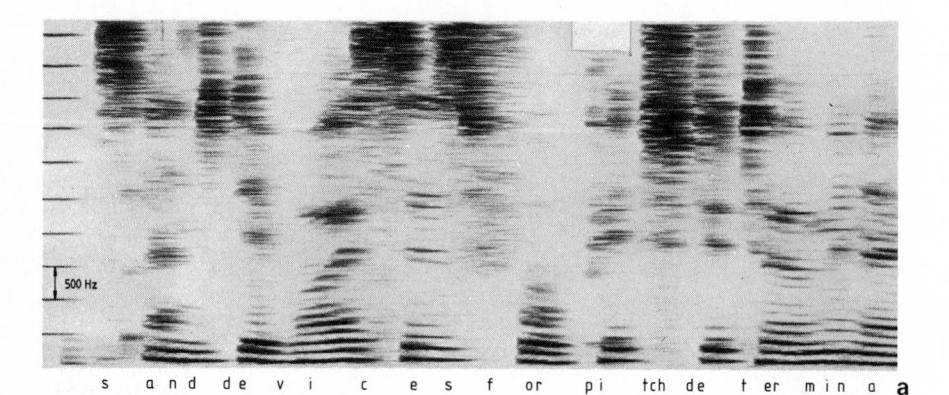

Fig.5.4a,b. Wide-band **(a)** and narrow-band **(b)** spectrogram. Speech signal: part of the utterance "Algorithms and devices for pitch determination," speaker MIW (male). Frequency scale: 0 to 5 kHz

thus rarely exceeds a few millimeters. As an example, Fig.5.4a shows a wide-band spectrogram.

The narrow-band filter (45 Hz) resolves the individual harmonics of the signal almost for the whole measuring range of F_0 (cf. the example in Fig.5.4b). It enables the researcher to read the shape of the F_0 contour directly from the spectrogram. Narrow-band spectrograms have thus found wide application in pitch determination, whereas the wide-band spectrogram has proved more suitable for the determination of vocal tract parameters, such as the formants. In addition to these basic facilities, the *Sonagraph* provides a number of helpful features: the investigation of a reduced (and thus stretched) frequency band (Fig.5.5) improves the accuracy of the measurement; frequency and time marks are useful calibration points.

The technique of pitch determination from a narrow-band spectrogram is simple. It is only necessary to measure the frequency F_0 directly or derive it

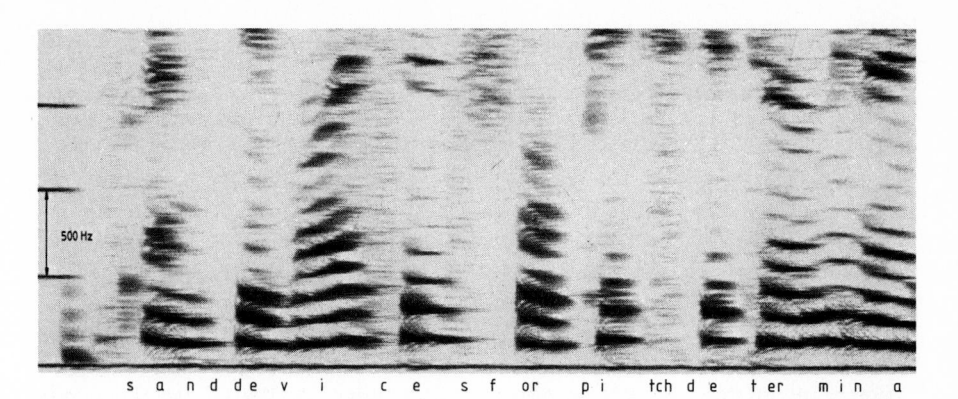

s a n d de v   i   c e   s f or   pi   tch d e   t er   m in   a

Fig.5.5. Narrow-band spectrogram with stretching of the frequency axis. Same utterance as in Fig.5.4. Frequency scale: 0 to 2 kHz

from the frequency of a higher harmonic, the latter improving the accuracy of the method.

> The spectrogram reveals the arrangement of harmonics with their varia-tions in frequency shown in the undulating movements of the horizontal lines. With a spectrum of this type the changes in intonation can be observed by following the curve of the fundamental. However, since the bottom line, that of the fundamental, has a comparatively low elevation and errors of measurement are liable to be serious, it is better to take some higher harmonic as the basis for measurement. ...
> If for any reason whatever (noise, filtering, etc.) the fundamental is absent from the spectrum of a vowel, one can always deduce its pitch from that of any two consecutive harmonics. By the same token it is advisable to calculate the frequency from the tenth harmonic, for example: the chances of error in measuring the fundamental are divided by ten. ...
> Most modern phonetics laboratories have a spectrograph and the number of studies using it are quite numerous. Improvements in technique have greatly reduced the time required to make a spectrographic analysis. Nevertheless the procedure is slow and melodic analysis using the apparatus will be truly interesting only when it can be carried out in real time - as is the case with melodic analyzers. If the spectrograph continues to be widely used for melodic analysis, it is mainly because of its reliability. With a little practice it is almost impossible to go wrong in calculating harmonic frequencies. (Léon and Martin, 1969, as given in Bolinger, 1972, p.38)

Like the other manual pitch determination methods, spectrographic manual pitch determination cannot be performed on-line and in real time. Hence the application areas are again phonetics, linguistics, and pathology. Although the method is quite straightforward, some problems may arise for unexperienced researchers, so that, for instance, McKinney (1965) complained that

> ... another difficulty [except that it does not work in real time] with spectrographic analysis is that reading single-valued acoustic parameter

values from sound spectrograms requires pattern recognition of a higher order on the part of a human observer than does reading them from an oscillographic trace of the parameter i.e., a visual display of the pitch contour . Both of these are shortcomings of spectrographic analysis which have helped to spur investigators in the development of more direct parameter analysis and analysis systems. (McKinney, 1965, pp.6-7)

Although there are a number of open questions with respect to the display (as we have seen), the readability and plausibility of the display are one of the strengths of the spectrogram. Nevertheless there have been a number of efforts to improve the display. Pisani (1975) connected the spectrograph to a PDA (Pisani, 1971, cf. Sect.6.1) in order to have the pitch contour and the (wide-band) spectrogram displayed simultaneously on the same sheet of paper. Approaches of this kind will be discussed in Section 9.4. Digital signal processing paved the way for digital sound spectrographs which allow for a number of additional features, such as continuous adjustment of the analyzer filter bandwidth within a wide range, stretching of the time axis, simultaneous display of the time-domain waveform, and so on. Such systems were proposed, for instance, by Strong and Palmer (1975) or by Holmes et al. (1978). For linguistic research in general, the question of visual display is crucial. In manual pitch determination, however, the basic method is not affected much by these special features; the researcher may only gain an improved insight into the fine structure of the signal. Figure 5.6 shows an example of an expanded spectrogram made on the spectrograph developed by Holmes et al. (1978).

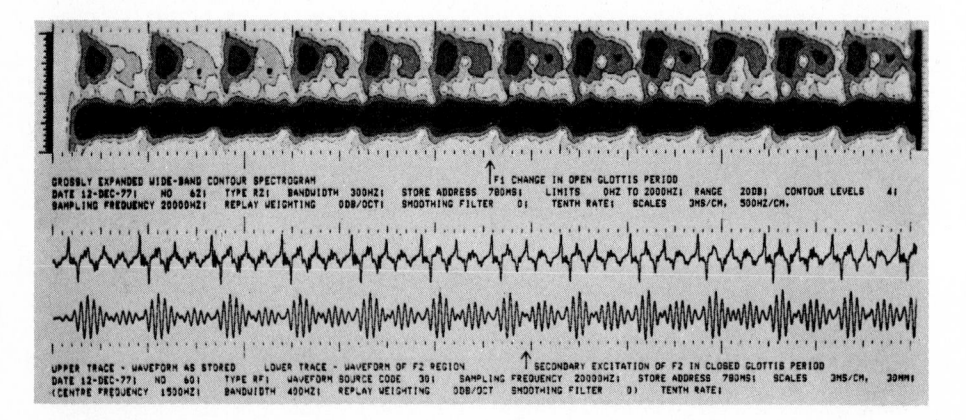

Fig.5.6. Example of a digital spectrogram with waveform display and stretched time scale. Signal: vowel part of the word "bird" from the utterance "A bird in the hand is worth two in the bush"; male speaker. From (Holmes et al., 1978, p.46). [British Crown Copyright; reproduced with the permission of the Controller of Her Brittanic Majesty's Stationery Office]

5.2 Pitch Determination Instruments

Pitch determination instruments (PDIs) are a useful and valuable alternative to pitch determination algorithms and devices. In this book we define any measuring device as a PDI which derives its information on pitch from sources other than the (unprocessed) speech signal. In a wider sense we can label any instrument a PDI that serves for (clinical) larynx inspection, voice source analysis, or pitch determination; in the more restricted sense the PDI is an instrument which is used to determine pitch from continuous speech.

The section presents only a short survey of instrumental techniques that serve for clinical larynx investigation and for voice source analysis. The feature common to all these instruments is that they inhibit normal phonation and articulation and, hence, cannot be used for analyzing normal speech. The PDIs in the more restricted sense are presented in more detail; they hardly affect the speaker's articulatory ability and are thus well suited for the normal tasks of pitch determination in speech.

If we start from the block diagram of a PDA with its three components - preprocessor, basic extractor, and postprocessor - and if we apply this scheme to the PDI, then the PDI completely replaces the preprocessor, greatly simplifies the basic extractor, and has little influence on the postprocessor except that error correction may no longer be necessary. The PDI usually obtains its information on pitch directly from the vibrations of the vocal cords. The main problem of PDAs, i.e., the influence of the supraglottal system (vocal tract plus radiation) does not exist with a PDI. The temporal structure of the output signal of the PDI is thus much more regular than in the case of the preprocessor output signal of a PDA. The basic extractor which is to follow the PDI can thus be rather simple; a zero crossings analysis basic extractor (see Sect.6.1.1 for discussion), a simple threshold analyzer, or an event detector which marks a significant feature in the waveform of the PDI will prove sufficient in most cases. In some PDIs the basic extractor and the postprocessor together with a display are incorporated into the instrument (Fourcin and Abberton, 1971). Since the structure of the output signal of the PDI is so regular, the PDI can also be applied in such situations where a PDA systematically gets into trouble, for instance when the glottal excitation signal becomes irregular. This makes the PDI a real alternative to the PDA in phonetics, linguistics, and education. In the field of voice pathology and phoniatrics the PDI has always played a more important role than the PDA, which is only slowly beginning to gain access to this area of application. In speech communication, however, the PDI cannot be applied. If a PDI were applied to a task like vocoder telephony or in a voice input command system, the speaker would always be obliged to carry the instrument and to attach the electrodes, transducers, or contact microphone to his neck when communicating with the system. This is simply asking too much of both system and speaker, and hence speech communication clearly remains the domain of the PDA which can do the task analyzing the signal alone, i.e., without requiring the physical presence of the speaker at the place where the analysis is done.

Section 5.2.1 gives a brief survey of some methods of optical larynx inspection: *stroboscopy, high-speed larynx photography*, and the *fiberscope*. These instruments are not really PDIs since they permit investigating laryngeal behavior rather than pitch. Most of these instruments, in addition, inhibit the speaker from articulating normal and intelligible speech.

Section 5.2.2 deals with *mechanic* PDIs; it is divided into two parts. First, there are mechanical means which are typical for voice source investigation; they represent useful aids in inverse filtering. The *pneumotachographic mask* (Rothenberg, 1972) compensates the characteristics of sound radiation at the mouth and the nostrils. The *vocal-tract extension tube* (Sondhi, 1975b) represents a reflectionless termination of the vocal tract and completely cancels its transfer function. With both instruments it is obviously impossible to articulate as in normal speech. On the other hand, there are mechanic instruments that pick up the vibrations of the throat walls; here *throat microphones, contact microphones,* and piezo-electric transducers (which act as *accelerometers*) are applied. These instruments are very good for pitch determination because their output signal is very much like a sinusoid with the frequency F_0. Since it is difficult to draw conclusions about events like glottal closure from this signal, these PDIs are only of moderate value for detailed voice source investigations.

In Sect.5.2.3 *electric* PDIs are discussed. The different realizations, which range under the names of *laryngograph, glottograph,* or *electroglottograph,* all apply the principle of measuring the variations of the laryngeal impedance due to the vocal-cord movement. These instruments have found numerous areas of application; they are well suited for both pitch determination and voice source investigation since one can easily draw conclusions about the significant events of the glottal cycle from the output signal of the PDI, at least during the closed-glottis phase. The technical problems encountered in their realizations have led to some artifacts which, however, can easily be taken into account (and compensated for) if they are known.

Ultrasonic PDIs are treated in Section 5.2.4. In these PDIs ultrasound (usually continuous-wave ultrasound) is transmitted through the vocal cords. When the glottis is closed, the ultrasound passes it; otherwise it is reflected. The output signal of this PDI is similar to that of the electric PDI; the position of the ultrasound transducers with respect to the larynx, however, can be a source of error.

Section 5.2.5 deals with the *photoelectric* PDI. In this instrument a strong light source is attached to the exterior of the neck somewhat below the larynx. The light source transilluminates the neck tissue; if the glottis is open; the light can also pass into the pharynx. A photoelectric transducer (which is inserted into the pharynx by a thin tube via the nose) picks up the light in the pharynx. This device gives good information about the open-glottis phase; its use, however, may cause some discomfort to the speaker.

In contrast to PDAs where *comparative evaluations* are rather rare, the performance of PDIs has been the subject of various studies and critical reviews, some of which are discussed in Section 5.2.6.

5.2.1 Clinical Methods for Larynx Inspection

The instruments for visual larynx inspection are not really PDIs since they do not serve to detect pitch, but to investigate the behavior and the state of the larynx. They cannot be used for pitch determination since either they inhibit the speaker from uttering normal speech, or their use is too expensive to justify the purpose of mere pitch determination. In clinical research, where the actual state of the larynx has to be evaluated, and in basic phonetic research, where the movement and operation of the larynx are investigated from a functional point of view, the value of these instruments cannot be overestimated. Our knowledge about the operational mode of the larynx would be much less if these instruments were not available to verify theories and to find new results. For this reason, and for sake of completeness, a short review of these instruments is presented here although they are not used for pitch determination in the narrower sense. This review will consist of a short historical survey together with some examples of current work in progress.

Historical surveys about larynx investigation were published, among others, by Hast (1975) and by van den Berg (1962, 1968a), to which the reader is referred for details. The history of research on the experimental physiology of the larynx, having started with Galen of Pergamon (130-200 A.D.), was resumed by Ferrein (1741). In his experiments on excised dog larynxes, Ferrein showed that a dead vocal organ could still produce sound. It was also Ferrein who first used the term *cordes vocales.* In the 19th century, Johannes Müller (1837) carried out fundamental experiments on laryngeal physiology which formed the basis of the myoelastic-aerodynamic theory of the vibration of the vocal cords (Van den Berg, 1958, 1962). Müller also validated Ferrein's theory that the voice originates from vibration of the vocal cords.

Up to that time, all investigations on larynx physiology had to be made on excised cadaveric larynxes; the living larynx was inaccessible to medical and physiological investigation. This changed when Garcia (1855) invented the laryngeal mirror. With this instrument it was possible for the first time to observe the living larynx during phonation. Although the method has been much refined since that time, the basic technique of the laryngoscope has not changed much in over 100 years.

Not much later, Czermak (1861) published his findings on the first attempt of laryngeal photography. His results were only moderate, and some twenty years later the method was improved by French (1883).

Another landmark in visual larynx investigation was the application of stroboscopy to larynx observation and photography (Oertel, 1878). Although the stroboscopic principle had already been discovered in 1829 by the physicist Plateau of Brussels and - independently - in 1832 by Stampfer in Vienna, and although it had been applied to investigation of an excised larynx by Harless in 1852, i.e., even before the invention of the laryngeal mirror, it was not until 1878 that this technique was applied to the investigation of the living larynx (Schönhärl, 1960; Wendler, 1975). Today the laryngostroboscope is a standard tool in phoniatric practice (Schönhärl, 1960).

Stroboscopy is perhaps the most straightforward example for the validity of the sampling theorem. According to the sampling theorem, the speed of the vocal-cord vibrations can seemingly be brought down to a near-zero value by sampling the observation with a frequency a little below the fundamental frequency of the signal. If F_0 is the fundamental frequency, and if the observation of the larynx is sampled with a frequency $f_S = F_0 - \Delta f$, then for the observer it looks as if the larynx vibrates with a frequency Δf (Fig.5.7). Of course one must assume that neither the laryngeal frequency nor the shape of the laryngeal movement change substantially during the observation, nor that the laryngeal excitation function becomes irregular. Modern laryngostroboscopes can even approximately cope with irregular excitation, as frequently occurs with pathologic voices; the sampling is synchronized with the voice by means of a small mechanic PDI (see Sect.5.2.2) using an accelerometer or a throat microphone which is attached to the subject's neck; and the special synchroniz-ation makes it possible to arbitrarily shift the phase angle between the sampling instant and some reference point in the glottal waveform, for instance the point of glottal closure (Wendler, 1975). For a detailed descrip-tion the reader is referred to a textbook on this subject, such as Schönhärl (1960) or Luchsinger and Arnold (1970). A good introduction to the principle has been given in the booklet by Wendler (1975).

Further significant progress was made with high-speed cinematography of the vocal cords. The first publication in this field came from Farnsworth (1940). He filmed the vibrating larynx using a high-speed camera developed at Bell Laboratories operating at a speed of 4000 frames/s. With this device avaliable, it was possible to study the moving larynx in much greater detail. By viewing the developed film at a standard speed of 16 or 24 frames/s and with frame-by-frame analysis the speech scientist could now gain much infor-mation on both normal and abnormal laryngeal function. In the meantime a

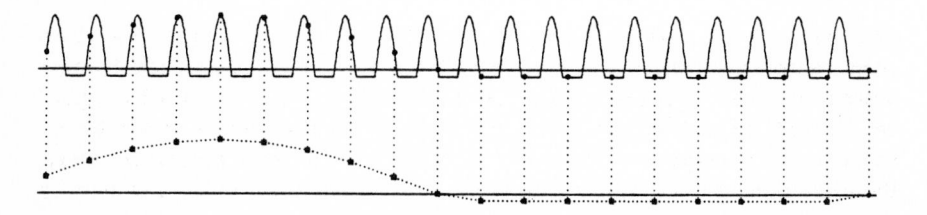

Fig.5.7. Example for the principle of stroboscopy. A periodic waveform (in this figure a clipped sinusoid) with the period T_0 (corresponding to the funda-mental frequency F_0) is sampled with the sampling rate $f_S = F_0 - \Delta f$, which corresponds to the sampling interval $T_0 + \Delta t$. The sampled waveform is then periodic with the period $1/\Delta f$. In the figure the ratio $\Delta t/T_0$ equals 0.05

number of high-speed films on the laryngeal function have been made (Moore and von Leden, 1958; Soron, 1967; Hayden and Koike, 1972; Koike and Hirano, 1973; Gould, 1976; Hirano, 1976; Childers et al., 1972, 1976, 1980). Frame-by-frame analysis of these films, which is necessary to evaluate the dynamic shape of the glottis, is very cumbersome and requires an extreme amount of human work. First efforts have been made toward developing at least interactive analysis procedures using the computer. Methods of image processing and visual pattern recognition in order to automatically determine the cross-sectional area of the glottis have been introduced to make this job easier (Childers et al., 1980).

All these methods have the disadvantage that they inhibit the speaker from normal phonation so that only some sustained vowels can be investigated. A further gap towards optical observation of the laryngeal function in normal environment was closed when fiberoptic photography was invented (Sawashima and Hirose, 1968; Sawashima et al., 1970; Sawashima and Miyazaki, 1974; Niimi et al. 1976; Fujimura, 1976, Sawashima, 1976). Although fiberoptics does not permit the use of such high frame rates as high-speed photography, valuable information about laryngeal behavior in normal speech is obtained. Recent investigations apply a stereofiberscope (Sawashima and Hirose, 1968; Fujimura, 1976; Sawashima, 1976), and thus enable the investigator to cope with the vertical movements of the larynx which make observation difficult otherwise.

Electromyography is a completely different means for the investigation of laryngeal physiology. Electromyography is an electric method to measure muscle activity by neural abduction. In larynx physiology it is used to investigate the activities of the laryngeal muscles. It can give insight into the coordination of the intrinsic and extrinsic muscles of the larynx, especially on the control of the larynx position, and on the combination of articulatory gestures in connection with vertical laryngeal displacement (Hirose, 1977). Electromyography, however, does not show individual vibrations of the vocal cords (which is another support for the theory that vocal-cord vibration is caused by myoelastic and aerodynamic effects).

5.2.2 Mechanic Pitch Determination Instruments

The speech signal is the outcome of a convolution process between the three components *excitation signal, vocal-tract transfer function,* and *radiation characteristics.* These components (except the periodicity in the excitation signal) can be modeled by linear passive networks whose transfer functions are represented by their respective poles and zeros. In order to draw conclusions about the excitation signal from the output signal, it is necessary to reverse part of this convolution process, i.e., to perform a (partial) deconvolution.

Deconvolution of the speech signal in order to obtain the glottal waveform is called *inverse filtering* since a filter is applied whose transfer function is the reciprocal of the transfer function of the vocal tract (usually including the radiation transfer function). The main difficulty in determining the inverse filter characteristics is not so much the measurement of the poles and zeros but the decision as to which poles (and/or zeros) are due to the vocal tract and radiation, and which ones pertain to the excitation component.

> There being no unique decomposition of a speech signal into excitation and vocal-tract influence, the success or failure of the inverse filtering technique depends entirely upon the constraints and criteria imposed by the experimenter. These constraints and criteria are based upon *a priori* information about the nature of the speech production apparatus and may take such forms as restrictions in the locations and number of poles, degree of smoothness of the output of the inverse filter, etc. While, in general, the *a priori* information is reliable, it is important to keep this limitation in mind. Quite frequently it is not possible to decide whether a feature in the recovered waveform is to be attributed to the glottal source or to an improper adjustment of the inverse filter. Thus, for instance, demanding the smoothest possible output from the inverse filter may obliterate genuine oscillations in the glottal wave. (Sondhi, 1975b, p.229)

Excitation signal determination is more than pitch determination. In pitch determination it is only necessary to determine the periodicity component P(z) of the whole process, whereas in excitation signal determination the voice source spectrum S(z) or its time-domain equivalent has to be estimated as well. This is possible with carefully selected high-quality signals (without any distortions whatsoever), preferably from sustained vowels. Even under these restricted conditions excitation signal determination requires a lot of manual interaction so that only few data are available since researchers' time is limited.

The whole complex is discussed in Sect.6.5.1 as one of the tasks of high-accuracy PDAs. In the present section the question is whether there are instrumental aids to render this task easier. For inverse filtering this means that part of the deconvolution must be performed by the instrument: the vocal-tract transfer function (if possible); at least, however, the radiation characteristics must be cancelled. This may imply that the speech signal becomes unintelligible. Since excitation signal determination is almost exclusively performed to reveal characteristics of individual voices, this price can be paid.

Rothenberg (1972, 1973b) developed a mask which cancels the radiation component at the mouth.

> One of the methods used previously is illustrated at the top of Figure 5.8. In this technique, the radiated pressure waveform from the microphone is processed by a filter having a transfer characteristic the inverse of that of the vocal tract. This *inverse filter* has a zero (antiresonance) for every pole (resonance) of the vocal tract in the frequency range of interest, including a pole at zero frequency. The pole at zero frequency, which is an integrator in the time domain, is a fundamental source of inaccuracy in this method. Since the integrator increases the low-frequency response of the filter, this method is very sensitive to low-frequency noise. Secondly, since it is impossible to use a true integrator, the dc level of the glottal waveform is lost in the process. And lastly, the amplitude calibration depends on factors such as the distance of the microphone from the

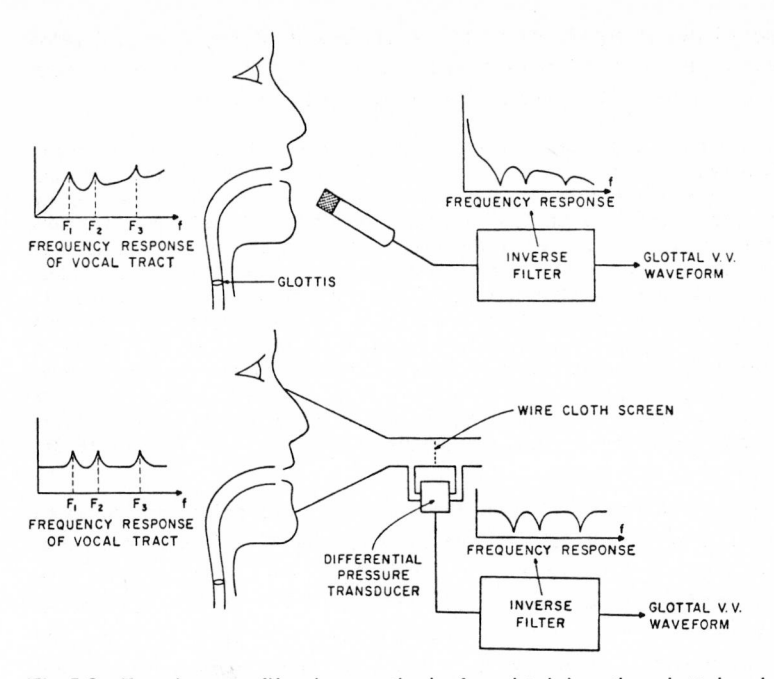

Fig.5.8. Two inverse-filtering methods for obtaining the glottal volume velocity waveform. (Upper figure) inverse filtering of the radiated pressure waveform; (lower figure) inverse filtering of the volume velocity waveform at the
mouth. From (Rothenberg, 1972, p.380)

lips and cannot be made accurately (Miller, 1959; Holmes, 1962b,c; Miller
and Mathews, 1963).

In our work, we use the *volume velocity* at the mouth, instead of the
pressure, as the input to an inverse filter, as shown at the bottom of
Figure 5.8. The inverse filter for volume velocity does not require an
integrator and, therefore, provides an accurate zero level and is not sensitive to low-frequency noise. There is also no problem with amplitude
calibration. The volume velocity at the mouth is shown sensed by a
standard wire screen pneumotachograph mask, of the type used in respiratory measurements.

A pneumotachograph mask more suitable for inverse filtering is shown
in Figure 5.9. To reduce speech distortion caused by acoustic loading of
the vocal tract, and improve the frequency response of the pneumotachograph, the wire screen is distributed around the circumference of the
mask, as close to the face as possible. To keep the pressure difference
across the screen linearly related to volume velocity, the transducer port
sensing internal pressure is shown protected from the direct impact of the
breath stream. We use a disk of polyurethane foam for this purpose.

If the inverse filter is adjusted accurately, the fidelity of the inverse-
filtered glottal waveform is limited by the pneumotachograph mask. The
response time of our mask was about 1/2 ms, limiting the system to
glottal waveforms with a fundamental frequency no more than 150 to
200 Hz.

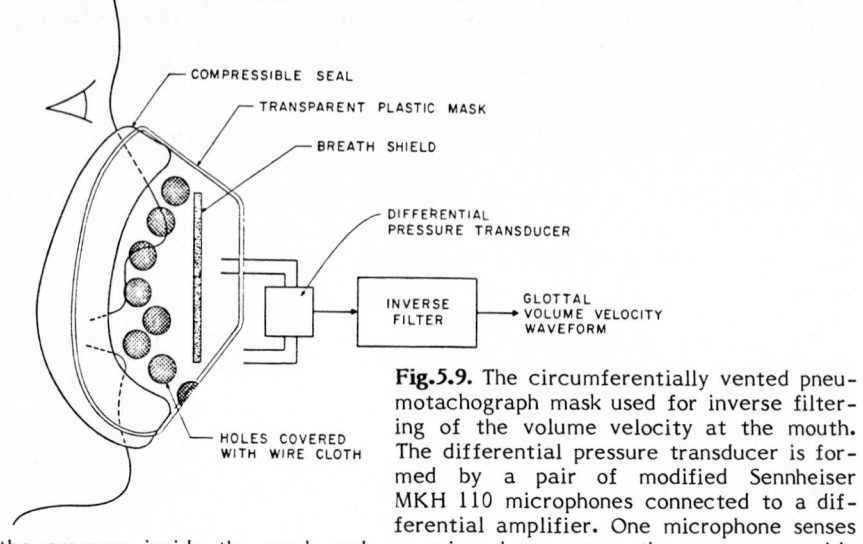

Fig.5.9. The circumferentially vented pneumotachograph mask used for inverse filtering of the volume velocity at the mouth. The differential pressure transducer is formed by a pair of modified Sennheiser MKH 110 microphones connected to a differential amplifier. One microphone senses the pressure inside the mask and one microphone senses the pressure outside the mask. From (Rothenberg, 1972, p.381)

For a reasonably good approximation to the glottal waveform, formant frequency and damping settings are not critical if measurements are made during vowels having a first formant at least three or four times higher than the fundamental frequency. For this reason we favored vowels such as /a/ and /ɛ/. The inverse-filter was normally adjusted from spectrogram measurements of the first two formants, and the settings retouched, if necessary, by observing the inverse-filtered waveform (Holmes, 1962b,c). The third and higher formants were removed by low-pass filtering. (Rothenberg, 1972, pp.381-382)

A later version (Rothenberg, 1977) of the mask was used by Sundberg and Gauffin (1978) for a study of the voice source. To realize the pressure transducer, they developed a special microphone.

The mask is a standard respiratory plastic mask in which holes have been drilled. The holes are covered with a fine mesh screen. The acoustic resistance of the screen is chosen so as to give a pressure drop which is sufficient for the microphone transducer at normal air flow values and at the same time low enough, so as not to distort the speech.

The microphone fits into the hole for the mask fitting. It is a pressure differential transducer with a metalized mylar membrane having a diameter of 15 mm. One side of the membrane is open to the inside of the mask while the other side is open to the outside. The deflection of the membrane is sensed by an optical system consisting of an integrated light-emitting diode and a photo transistor. The photo transistor measures the amount of light reflected by the membrane. The temperature drift of this system is inherently low and slow. It would be possible to reduce it, if the light emitting diode voltage is used as a temperature sensor inside the transducer. The membrane has a resonance at 1.3 kHz, approximately, with a Q value of 5. This resonance is compensated for in the preamplifier

> which also has means for calibration of the output. The preamplifier is
> contained in a small box permanently connected to the connection wire of
> the microphone. Microphone and preamplifier together are designed to
> give a flat frequency response up to 1.5 kHz. (Sundberg and Gauffin,
> 1978, pp.36-37)

Due to the high-frequency band limitation of the mask, the inverse filter was
realized only for the formants F1 and F2, thus permitting the study of the
glottal waveform up to 1.5 kHz. The filter must still be tuned manually; the
task, however, has been greatly simplified, and the investigator is free from
constraints imposed by the recording conditions and able to do the experiment
on-line.

> The voice source was studied in five males. During the experiments the
> subject was seated in an ordinary room and phonated sustained vowels
> holding the mask firmly to his face. The vowels /a/ and /ε/ were used
> because of their concentration of sound energy in the low-frequency region.
> The experimenter was seated next to the subject and adjusted the inverse
> filters. He recorded the inverse filter output as soon as the filter settings
> appeared to be appropriate. At the same time he determined the SPL
> reading. (Sundberg and Gauffin, 1978, p.38)

A device which claims to completely replace the inverse filter in voice
source studies was proposed by Sondhi (1975b). The underlying idea is that the
formants which "disturb" the shape of the excitation signal, and which are
due to the impedance mismatch at the end of the vocal tract, can be
canceled by mechanical means. The low-impedance termination at the mouth,
together with the high-impedance excitation at the glottis evokes reflections
in the vocal tract which cause the resonances (Fant, 1960). All these resonances are canceled when the vocal tract is terminated without reflections at
its end. To achieve this, Sondhi developed an acoustic tube which is used as
an extension of the vocal tract (Fig.5.10).

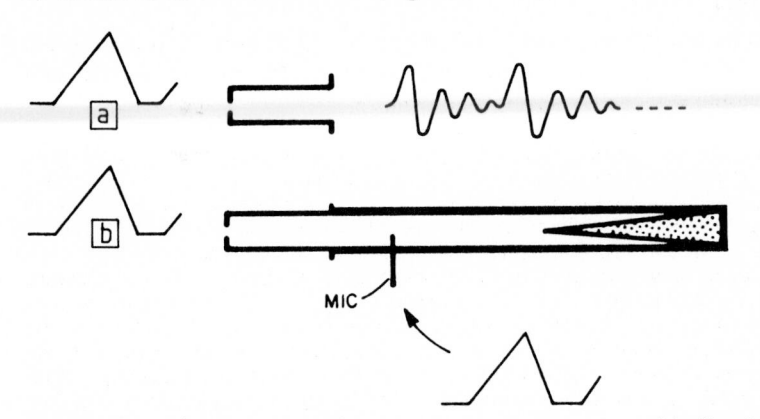

Fig.5.10a,b. (a) Waveform of the velocity output of an idealized hard-walled
uniform vocal tract; (b) waveform of the velocity output when the tract is
extended by a reflectionless acoustic tube. From (Sondhi, 1975b, p.229)

Consider the lower part of Fig.5.10, where we illustrate what happens when the idealized uniform vocal tract opens into a hard-walled uniform tube (of cross section identical to that of the tract) with a matched termination at the far end. If the termination is perfect (i.e., reflectionless) then the glottal source feeds into a uniform lossless tube which is effectively infinite in length. It is evident without any analysis that a probe microphone placed anywhere in this tube will pick up a waveform which, except for a delay, is identical to the input waveform. We thus conclude that in order to measure the vocal-cord wave we need only ask the speaker to phonate the non-nazalized neutral vowel into a reflectionless tube. ...

The extension tube is of brass, about 6 ft [1.8 m] long with an inner diameter of 1 in [2.5 mm]. The reflectionless acoustic wedge is of fiberglass, about 3 ft [90 cm] long and approximately conical. The probe is a 1/4-in-diameter [6.25 mm] electret microphone placed about 1 ft [30 cm] from the open end. In order to prevent a buildup of pressure, which would alter and eventually extinguish the vocal-cord oscillation, a small hole has been etched in the periphery of the acoustic wedge. (Sondhi, 1975b, pp.229-230)

Sondhi further showed that articulatory deviations from the neutral vowel only cause slight distortions of the waveform at the microphone. Resonances are introduced to the system when the cross-sectional area of the acoustic tube is not uniform (in case of the vocal tract this applies for any sound except the neutral vowel). Due to the reflectionless termination of the acoustic tube, however, these resonances are highly damped.

Rothenberg's mask and Sondhi's tube are for use in voice source investigation exclusively. Rothenberg's mask cancels the radiation characteristics of the mouth. It is thus of great help for the adjustment of an inverse filter. One cannot expect, however, that it renders the task of pitch determination in normal speech any easier. Sondhi's tube cancels the transfer function of the whole vocal tract. It may thus be the easiest "inverse filter" to be realized whatsoever, but it completely disables the speaker from articulating intelligible speech. Hence these instruments are extremely well suited for voice source investigations, but cannot be applied to normal speech (and thus, in a strict sense, are not really *PDIs*).

In contrast to this there are a number of mechanic instruments that can be used for pitch determination of arbitrary speech signals. All these instruments exploit the fact that the larynx causes the body wall of the throat to vibrate. The instruments pick up this vibration whose temporal structure is closely related to that of the excitation signal. A subsequent simplified PDA (which consists of a simple basic extractor and - optionally - of a postprocessor) extracts the pitch information from this signal instead of taking it from the speech signal.

A throat microphone[4] (also called larynx microphone) was used for this purpose by Sugimoto and Hiki (1962a,b), Porter (1963), and Tjernlund (1964b).

---

[4] The basic idea of the mechanic PDI is much older. Prototypes of throat "microphones" have already been reported by Rousselot (1908), Krüger and

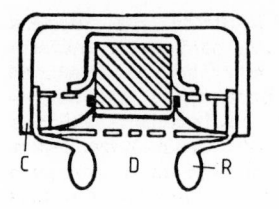

Fig.5.11. Larynx microphone. (C) Case; (R) rubber cushion; (D) dynamic system. From (Tjernlund, 1964b, p.32)

Tjernlund, who developed his PDI for use in educating deaf people, gave a detailed description of the microphone (Fig.5.11):

> The main difficulty of the larynx microphone is to find a suitable micro-
> phone which besides being cheap and mechanically robust has a good low
> frequency response, down to about 80 Hz. ... The microphone has been
> mounted in an aluminium case (see Fig.5.11) and is coupled to the throat
> via an air cavity. The problem is to get an airtight connection between
> microphone and throat, in order not to lose the low-frequency response.
> This was solved by using a rubber ring which gave a reasonably satisfac-
> tory fit.
>   The most suitable microphone position was found to lie on the center
> line of the throat below the larynx. This is about the same position as
> found by Sugimoto, Hiki and Porter (Tjernlund, 1964b, pp.32-33)

The total setup consists of the microphone and the pertinent amplifier, a time-domain analog PDA (type PiTE, see Sect.6.2.1), and the display facilities for the special purpose (see above). The output signals depicted in the original publication look very similar to the excitation signals obtained by inverse filtering (except that, obviously, the cartilages and the tissue between the larynx and the skin act like a low-pass filter). This shape of the waveform makes the simple peak-detection circuit (see Sect.6.2.1) particularly well suit-ed for this task.

Using this special microphone Tjernlund apparently has overcome problems which may arise otherwise, i.e., when ordinary throat or neck microphones are used. These microphones are much more sensitive to background noise; in particular, the signal directly picked up from the larynx interferes with the ordinary speech signal from the same speaker. Temporary changes of the position of the microphone with respect to the larynx may cause additional problems. These difficulties have led McKinney (1965) to the statement that throat microphones do not greatly facilitate the job of a PDA and thus are not of a great value in pitch determination. Peckels and Riso (1965) also complained about throat microphones:

Wirth (cited by Meyer, 1911), and Meyer (1911), in connection with the kymograph (cf. Sect.5.1.1). In these devices, called *Kehltonschreiber* (Meyer, 1911), the sound receiving capsule was attached to the throat, and the sound was transferred to the recording lever by a rubber tube, as usual with the kymograph.

> When using a microphone, it is difficult to find a small type having a good low-frequency response, and the result is a bad S/N ratio and frequency distortion at the output. In any case a very tight coupling and extremely good isolation at the throat is needed to avoid interference with environmental noises or speech of the talker itself, also considered as "noise". To assure this coupling, the talker must wear a neck-band holder, which results in discomfort. (Peckels and Riso, 1965, p.2)

Peckels and Riso made a different proposal which results in a different waveform but avoids these difficulties (see the discussion later in this section).

The contradiction between McKinney and the results of Tjernlund (1964b) may be partly due to differences in the practical realization of the microphone. Nevertheless, this principle seemed not to be very successful in speech research since it was rarely taken up again.

A different mechanic PDI with direct pickup of the vibrations at the larynx was reported by Peckels and Riso (1965), and later by Askenfelt, Gauffin, Kitzing, and Sundberg (1977, 1980). In both cases an *accelerometer* is used for the conversion of the body wall vibrations into an electric signal. The accelerometer measures the acceleration rather than the displacement of mechanic bodies; it is well known in acoustics for the measurement of mechanic shock and vibration in buildings. Due to the different measuring technique the output signal is no longer similar to the glottal waveform, but rather to its second time derivative. In normal voice (modal register) the second derivative of the excitation signal shows a sharp peak at the instant of glottal closure. The instant of glottal opening is more difficult to find, since it is not as well represented in the signal. The excitation signal can be obtained by filtering the accelerometer output signal with a double integrator filter which attenuates the higher frequencies by 12 dB/octave. The fact that the double integrator filter can be realized only approximately (due to the double pole at zero frequency which would render the filter unstable if implemented exactly) and the resulting sensitivity to low-frequency distortions (cf. the discussion in Sect.6.1.1 on the use of low-pass filters in connection with PDAs) make this PDI not well suited for excitation signal determination. For pitch determination, however, the shape of the signal (which is similar to that of the epoch detector PDA, see Sect.6.3.2) is extremely well suited; a simple PDA which serves as the basic extractor can do the task quite reliably. In both of the publications cited, a simplified analog PDA of the peak-detector type (PiTE, cf. Sect.6.2.1) is applied.

Peckels and Riso (1965) did not further specify the accelerometer they used. The accelerometer employed by Askenfelt et al. (1980) contains a piezo-electric ceramic disk as the pickup unit. This disk is enclosed in a metal container of 15 mm in diameter and 5 mm in thickness and weighs about 20 g. A small preamplifier with high input impedance is placed close to the microphone along the connection wire in order to minimize hum.

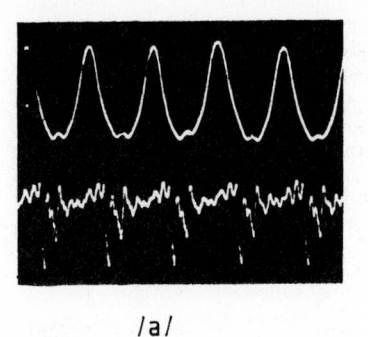 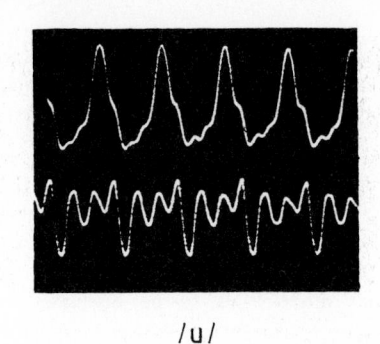

/a/   /u/

Fig.5.12. Output signal of a mechanic PDI (upper trace) and speech signals (lower trace) for two sustained vowels produced by a male speaker. Fundamental frequency is 118 Hz. From (Stevens et al., 1975, p.596)

In their comparative evaluation, Askenfelt et al. (1977, 1980) found the performance of the mechanic PDI (which they called *contact microphone*) somewhat inferior to the electric PDI (see the next section) but nevertheless good. The mechanic PDI can be a viable alternative for certain subjects for whom the electric PDI does not work. A problem is the relatively heavy microphone arrangement which can cause some discomfort to the speaker when worn for a longer period of time. A solution to this problem has been presented by Stevens, Kalikow, and Willemain (1975) who applied a lightweight accelerometer housed in a cylindrical case only 8 mm both in diameter and in height. The accelerometer weighs 1.8 grams. The casing even contains a pre-amplifier circuit.

> The accelerometer may be easily attached to the neck or nose of a speaker by means of double-sided adhesive tape. Once in place, the accelerometer is barely perceived by subjects. They can wear it for hours without discomfort. (Stevens et al., 1975, p.595)

Figure 5.12 shows some speech signals and the pertinent output signals of the mechanic PDI (Stevens et al., 1975). It seems difficult, if not impossible, to determine the instant of glottal closure. Since the tissue between the larynx and the neck acts like a low-pass filter, the output signal of the mechanic PDI is nearly a sinusoid. The small superimposed ripple, according to Stevens et al. (1975), is presumably the result of acoustic excitation of a natural resonance in the subglottal system in the vicinity of 500 Hz. Pitch determination from such a signal, however, is extremely simple. Moreover, the signals in Fig.5.12 give evidence that the problem of interference between the mechanic signal and the acoustic speech signal has really been solved; the PDI output signal in Fig.5.12 is certainly free from any spurious contribution due to the speech signal.

Stevens et al. (1975) also investigated the sensitivity of this PDI to changes in the position of the accelerometer with respect to the larynx.

Experiments were performed to determine the amplitude of the accelerometer output and the amount of harmonic content as a function of the mounting position on the neck. The results indicate that a point midway between the thyroid cartilage and the sternal notch and in the midline [i.e., just in front of the larynx] gives waveforms with maximal amplitude and minimal harmonic content. With the accelerometer in this location, there is little variation in the waveform from one vowel to another. The exact positioning of the accelerometer in the vicinity of this point is not critical. (Stevens et al., 1975, p.597)

Like most other PDIs to be discussed in the following sections (and in contrast to the majority of PDAs), mechanic PDIs have the great advantage that they are able to detect voicing almost without error. Voiceless sounds usually do not make the throat walls vibrate (perhaps except for strong pharyngeal fricatives, such as /x/). A PDI cannot distinguish between voiceless sounds and silence, whereas any voiced sound gives a distinct signal. This means that the voicing decision, i.e., the discrimination voiced-unvoiced (the latter containing both voiceless excitation and silence) is performed very reliably.

The publications proposing mechanic PDIs are rather rare. Apart from the difficulties reported by some researchers (McKinney, 1965), those investigators who have mastered the technical realization of the microphone or the transducer have obtained good results although there are occasional errors which render a correction or nonlinear smoothing routine in the postprocessor (see Sect.6.6) advisable (Askenfelt et al., 1978).

The application of these instruments is mainly in education (i.e., teaching the intonation of foreign languages and using the device in the education of the deaf). The mechanic PDI yields an excellent signal for pitch detection; for glottal waveform investigation, however, its output signal is not equally well suited since the detection of the instant of glottal opening and closure is difficult, if not impossible. This may be a reason that mechanic PDIs which operate on the basis of throat microphones, contact microphones, and accelerometers have been relatively rare although they can be realized and operated with comparatively little cost and effort.

5.2.3 Electric PDIs

Direct measurement of the laryngeal vibrations by electrical means uses the change of the electric impedance of the larynx due to opening and closure of the glottis. Since its invention (Fabre, 1957), this PDI principle has found wide range of applications, and a rich literature has been published on it (for instance, Fabre, 1958, 1959, 1961; Fabre et al., 1959; Husson, 1962; Fant et al., 1966; Ondráčková et al., 1967; Vallancien and Faulhaber, 1967; Fourcin and West, 1968; Loebell, 1968; Lhote et al., 1969; Lhote, 1970, 1980; Fourcin and Abberton, 1971; Smith, 1974, 1977, 1981, 1982; Fourcin, 1974; André et al., 1976; Lecluse, 1977; Kitzing, 1977; Boves, 1979a,b; Boves and Cranen, 1982a,b). Fabre's original device has been further developed and modified in a number of laboratories. It is commercially available under the names *glotto-*

graph[5] (Frøkjaer-Jensen, 1968) and *laryngograph* (Fourcin and Abberton, 1971). In the literature the term *electroglottograph* can also be found (Lecluse, 1977). In this book all the instruments that operate according to this principle will simply be called *electric PDIs*.

The basic idea of the electric PDI is simple. The impedance of the larynx varies according to whether the glottis is open or closed. When the glottis is open, the impedance reaches a maximum; it decreases rapidly when the vocal cords clap together. Two electrodes are placed on the neck at the two sides of the thyroid cartilage, i.e., at the level of the vocal cords. The electrodes are fed by a weak high-frequency current originating from an electric current generator. The voltage between the two electrodes is a measure for the laryngeal impedance.

> An alternating current of very high frequency - in the region of 200 kHz - but of very low intensity is made to flow through the larynx via the skin of the neck. Two longitudinal metal electrodes are moistened with a kaolin paste and applied to opposite sides of the prominence of the thyroid cartilage. Some of the current flows through the glottis so that it flows the more easily the more closed the glottis is. The amplitude of the current will thus be modulated by the adduction of the vocal folds. The amplitude of the modulated current will be proportional to the degree of glottal closure since the resistance to the current decreases as the surface increases where the vocal folds are in contact. After detection, the recording of the modulated current will thus give a faithful picture of glottal activity. (Fabre, 1957, pp.67-68; original in French, cf. Appendix B)

Figure 5.13 shows the block diagram of the electric PDI, as originally developed by Fabre (1957) and modified by Hlaváč (unpublished; cited in Fant et al., 1966). The high-frequency current source is formed by a HF oscillator in connection with the resistance R which is much larger than the impedance of the larynx. The electrodes have an effective area of about 2 cm^2 each and have a narrow elliptical shape. This shape minimizes the effect of gross up and down movements of the larynx. The changing impedance of the larynx modulates the envelope of the voltage between the two electrodes, one of which is connected to ground. The demodulator extracts the envelope from the HF voltage at the electrodes, and the subsequent high-pass filter removes unwanted low-frequency components which come from impedance changes by vertical displacement of the larynx or by changing muscle tension.

Figure 5.14 shows the schematic diagram of the output signal[6] of the electric PDI. The individual instants, as explained in the figure, were defined

---

[5] The *glottograph* should not be confused with the *photoglottograph*, which is the photoelectric PDI to be discussed in Section 5.2.5.

[6] The output signal of the electric PDI (displayed on a sheet of paper) is frequently called *laryngogram* or *glottogram*. In a similar way the ordinary time-domain display of the speech signal is labeled *phonogram*, and the display of the short-term spectrum is called *spectrogram*, as we have seen in Section 5.1.2.

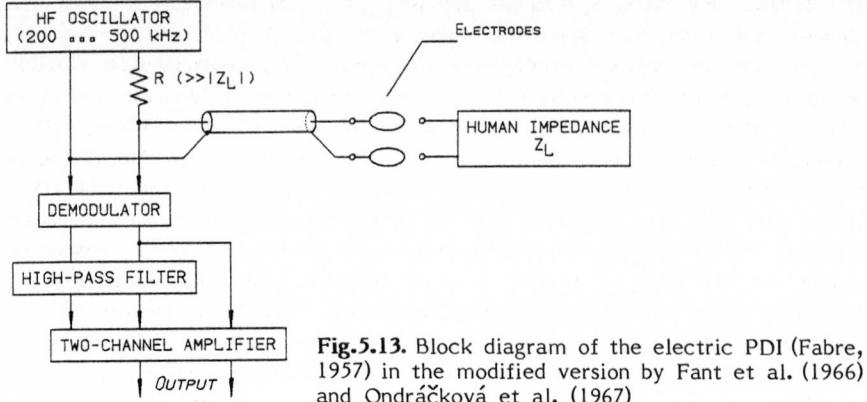

Fig.5.13. Block diagram of the electric PDI (Fabre, 1957) in the modified version by Fant et al. (1966) and Ondráčková et al. (1967)

by Lecluse (1977). During the opening phase of the glottis the resistance is nearly constant so that no statement can be made about the cross-section-al area of the glottis once the glottis is completely open. The force exerted by the Bernoulli effect is quite large so that the vocal cords close abruptly. Once they have come to contact, this contact proceeds in vertical direction, and the resistance diminishes rapidly (instants 2 and 3 in Fig.5.14). The reopening of the glottis, forced by the subglottal pressure, is more gradual (instants 3 to 5 in Fig.5.14). The relation between the state of the vocal cords and the output signal of the electric PDI has been a cause of contro-versy in the literature for a long time; it was verified as explained above by Fourcin (1974) and Lecluse (1977) via simultaneous observation using the PDI and a laryngostroboscope (see Sects.5.2.1,6). Results of this investigation were already displayed in Chap.3 (Figs.3.3,6,7).

The principle of the electric PDI is simple and obvious. The technical realization, however, faces some severe problems which has led to a number of controversial discussions in the literature. First of all, the impedance change due to the vocal-cord movement only makes up a small percentage of the total impedance between the electrodes - 0.1%, according to André et

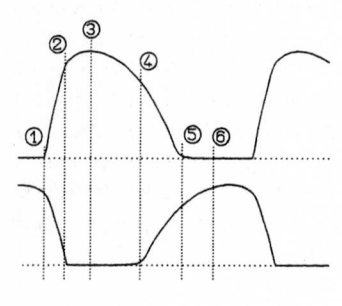

Fig.5.14a,b. Schematic diagram (a) of the out-put signal of the electric PDI (Lecluse, 1977), and (b) of the glottal airflow. Time instants: (1) the instant of initial closure at a single point; (2) the instant at which the closure is complete over the whole length of the glottis, but not in the vertical plane; (3) the instant of maximum closure; (4) the instant where open-ing begins; (5) the first moment at which the whole length of the glottis is open

al., (1976), but 1% according to Lecluse (1977). Low-frequency impedance changes due to vertical larynx movement or to changing contact between the electrode and the skin can exceed the amplitude of the impedance change due to vocal-cord movement by a factor of more than twenty (Fourcin, 1974). In addition, the device itself may be a source of trouble. The high-pass filter necessary to separate the variations of the laryngeal impedance due to vocal-cord movement and those due to low-frequency effects (see the above discussion) induces phase distortions which affect the shape of the output signal (Lecluse, 1977). Finally, the high-frequency current source and the demodulating circuit must be designed with relatively high impedance, thus being sensitive to picking up hum (and even signals of local AM broadcasting stations).

A number of investigators complained about microphonic effects which tend to become nonnegligible when there is fat tissue between the larynx and the skin:

> The equivalent circuit of the human subject as sensed from the electrodes contains an electrode-skin contact resistance with a capacitance in parallel followed by the transfer resistance which is modulated by the movements of the vocal folds and all laryngeal structures. The contact impedance can be made small compared with the true object of measurements by introducing a sufficiently high carrier frequency. Thin male subjects are ideal in the sense that the relative modulating effect of the vocal cords is large. With more heavily built subjects there is a certain loss of sensitivity due to the shunting effect of fat tissues covering the front of the thyroid cartilage. In such cases a microphonic effect may be noted in the form of a weak formant ripple superimposed on the waveform. The microphonic effect probably derives from a vibratory modulation in the electrode-to-skin impedance which is then no longer small compared with the useful glottal modulation. (Fant et al., 1966, p.16)

Smith (1977, 1981, 1982) assigned the microphony effect to pressure changes in the vocal cords as well as in the tissue around the larynx. These pressure changes cause impedance changes which influence the output signal of the PDI. According to Smith, this effect questions the success of the whole method since it induces a position-dependent phase shift of the first partial of the output signal (Smith, 1977). Lecluse (1977), who tested a number of different realizations of the electric PDI, found that the microphonic effect is measurable but decreases when the carrier frequency is increased. One of the PDIs tested in his investigation operated with a comparatively low carrier frequency (between 20 and 50 kHz). This device showed strong microphonic effects, whereas the other devices which used much higher carrier frequencies [Fabre (1957): 200 kHz; Frøkjaer-Jensen (1968): 300 kHz; one of the PDIs involved in Lecluse's investigation even used 10 MHz as carrier frequency] were not sensitive in this respect. These results suggest that spurious signals due to microphonic effects, regardless whether they come from variations of the electrode-skin impedance or pressure variations in the tissue, represent a capacitive component in the total impedance whose influence can be decreased by increasing the carrier frequency. For the electrode-skin impedance this was clearly formulated already by Fant et al. (1966).

Lecluse (1977) made quantitative measurements of the larynx impedance in humans. The equivalent circuit diagram is shown in Figure 5.15.

In an experiment with an impedance meter and the glottograph electrodes attached to the neck (electrode paste was used to improve the conductivity), it was shown that with increasing oscillator frequency the magnitude of the impedance of the subject's neck decreases, and the phase of the impedance goes to zero. For the model in Fig.5.15a this means that the overall impedance behaves like a capacitor in series with a resistor. If the glottis is open, C_E and R_S are dominant elements; we can assume that the capacity between the electrodes and the neck tissue is large compared to the capacity of the neck tissue itself, the glottal resistance being very large. The capacities C_T and C_S are small compared to R_T and R_S and can be neglected. Since the phase of the complex impedance formed by C_E and R_S is small we can also neglect C_E when using an oscillator frequency of 300 kHz or more. The glottal capacity C_G is cut short when the glottis closes. Just before the glottis closes, however, this capacity takes on its maximal value. If this capacity were of any significance, it would have affected the relation between the glottogram and the stroboscopic picture. We were not able to find such an influence in our experiments. That means, we can also neglect C_G. By this reduction we obtain the circuit depicted in Fig. 5.15b; a model that only consists of resistors switched in parallel. The total larynx resistance R_L is then calculated as follows.

$$1/R_L = 1/(2R_T+R_G) + 1/R_S \tag{5.1}$$

We can now estimate R_S when the glottis is open. ... In this case R_S equals R_L. For the larynx preparation [Lecluse made his experiments using excised human cadaveric larynxes as well as working with living subjects] R_S equals 130 Ω, for the subject we obtain $R_S = 30$ Ω. In the human subject, R_S is smaller than in the larynx preparation. This results from the fact that in the human subject there is much more subcutaneous tissue, the resistance of which is inversely proportional to the cross section

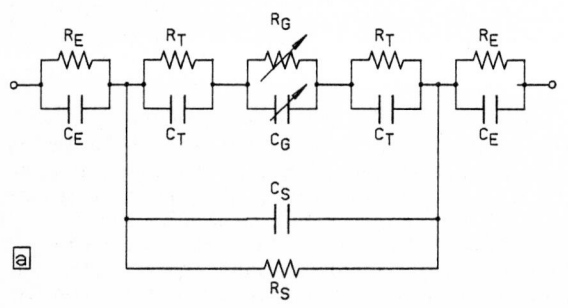

Fig.5.15a,b. Circuit diagram of the electric model of the larynx (Lecluse, 1977): (a) complete diagram; (b) simplified diagram. (R_E, C_E) electrode impedance; (R_T, C_T) tissue impedance in the larynx; (R_G, C_G) glottis impedance; (R_S, C_S) shunt impedance

area. The excised larynx which is almost naked has much less tissue that can serve as a shunt. In our model the resistance R_G is infinite when the glottis is open and zero when it is completely closed. The vocal cords thus act like a switch. We now know that the total resistance of the excised larynx changes by 10% giving $117\,\Omega$, and the total resistance of the subject's larynx by 1% giving $29.7\,\Omega$, respectively. If from this we compute R_T, we obtain $2R_T = 1170\,\Omega$ for the preparation and $2970\,\Omega$ for the subject. (Lecluse, 1977, pp.151-152; original in Dutch, cf. Appendix B)

Lecluse was only able to give quantitative estimates of his model because he assumed the glottal resistance R_G to vary between infinity for a completely open glottis and zero for a completely closed glottis. This assumption meets reality as soon as every bypass of the glottal surface is packed into the shunt impedance R_S. When the glottis is only partly closed, R_G takes on a finite nonzero value. The shunt resistance R_S is responsible for the fact that the variations of the total resistance R_L due to vocal-cord movement comprise only about 1% of the actual value of R_L. For women and children with smaller larynxes and a softer subcutaneous tissue this ratio can even be worse, so that in extreme cases the PDI cannot be used for individual subjects (Askenfelt et al., 1977; Fourcin and Abberton, 1976).

The same principle but a different technical realization is pursued in the laryngograph by Fourcin and Abberton (1971). Due to an improved design many of the technical shortcomings were avoided. Firstly, Fourcin and Abberton's laryngograph employs an oscillator frequency of 1 MHz and thus safely avoids any microphonic effect. Secondly, it uses a voltage source rather than a constant-current source at the electrodes, lowering the impedance of the demodulator circuit. The whole circuit is thus much more robust with respect to hum and spurious signals. In addition, the low-impedance design in connection with the constant-voltage source made the use of the laryngograph independent of an individual adjustment of the resistor R_1 (see Fig.5.13) according to the larynx impedance of the individual speaker. This enabled Fourcin and Abberton to provide the electrodes with guard rings (at earth potential) and thus to greatly reduce conduction of the current along the skin. This results in a better S-N ratio of the output signal.

Our own work (Fourcin and West, 1972) on this approach involved first the study of the electric-impedance variations which were likely to be encountered in practical situations, and the subsequent development of an arrangement that was electrically capable of compensating at least in part for the large impedance variations which can occur between one subject and another and for the slow impedance changes that occur in a particular subject during the course of speaking, and which result primarily from the continuous laryngeal adjustment that takes place as the speaker monitors and controls his voice output. The intersubject variation has been reduced by the use of a constant-voltage source rather than a constant-current or bridge output as the input to the subject's neck, and by detecting the current output from the neck in a low-impedance circuit. Both input and output are now at a low impedance level and the electrodes can be given guard rings, at earth potential, to reduce surface conduction across the skin. (Fourcin, 1974, p. 316)

According to Fourcin and Abberton (1971), a S-N ratio of more than 40 dB can be reached with male speakers within a frequency range below 5 kHz, and even with a newborn baby usable results were obtained. Fourcin and Abberton's laryngograph thus solved a number of the technical problems associated with previous versions of this PDI.

The substantial discussion in the literature on the temporal relation between the waveform of the laryngogram and the state of the glottis - this point was finally cleared by the comparative investigations of Fourcin (1974) and Lecluse (1977) - had its origin in a misinterpretation of Fabre's results by the author himself (1957). Fabre correctly stated that the resistance of the larynx goes down when the glottis is closed (see the above quotation), but then, in the same paper, he presented a laryngogram and assigned the wrong polarities of the signal to the open-glottis and closed-glottis phase. This misinterpretation was taken over by some other researchers, until Fant et al. (1966) gave the correct interpretation. They calibrated the instrument switching on and off a large resistance in parallel to the electrodes.

> At the first glance it would be tempting to correlate the rather level upper part of the curve with vocal-cord closure and the more symmetrical downwards deflected peak as the maximum opening. This was the interpretation given by Fabre (1957) in this original work and quoted by Husson (1960). The resistance calibration ... on the other hand, indicates increasing resistance, i.e., increasing glottal opening upwards. This is the true interpretation as also noted by Ohala (1966) and implicit in the later presentation of Husson (1962). (Fant et al., 1966, p.17)

To complete the confusion, it was possibly overlooked that the realization of the PDI according to Fabre's principle (1957) and according to Fourcin's design (Fourcin and Abberton, 1971) necessarily yields output signals of opposite polarity. This is easily seen from Figure 5.16.

In Fabre's PDI the larynx acts as a load resistor for the constant-current source. The envelope of the voltage at the load resistor yields the output signal. If the resistance of the larynx goes down, the voltage between the electrodes will also go down, thus yielding a negative peak at the instant of maximum glottal closure. In contrast to this, in Fourcin's PDA the larynx is switched between the constant-voltage generator and a constant load resistor which is supplied by the demodulating network. Seen from there, the larynx acts like a time-variant internal generator resistance. If the resistance of the larynx, i.e., the internal resistance of the generator, goes down, the voltage at the load goes up, thus yielding a positive peak at the instant of glottal closure. Lecluse (1977) has devoted considerable detail to this question, and he postulated that any laryngogram - regardless of the PDI on which it has been made - should be recorded in such a way that it shows a positive peak at the instant of maximum glottal closure. To achieve this, any electric PDI employing the constant-current source should use an inverting amplifier in the demodulation circuit.

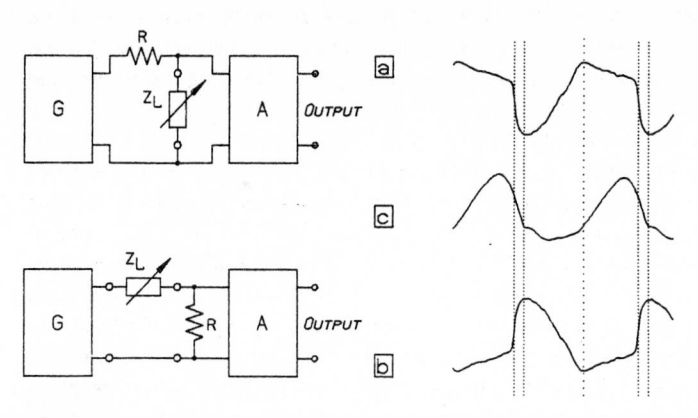

Fig.5.16a-c. Block diagram of the electric PDI: (**a**) schematic diagram of the realization by Fabre (1957) with an example of the output signal (upper display on the right-hand side); (**b**) schematic diagram of the realization by Fourcin and Abberton (1971) with an example of the output signal; (**c**) excitation signal (obtained by LPC inverse filtering). (·····) Instant of glottal closure; (·····) instant of glottal opening; (G) HF generator; (R) internal generator resistance (a), load resistance (b); (A) amplifier; (Z_L) laryngeal impedance

One final word about phase distortions. As we have seen, the high-pass filter in the amplifier is absolutely necessary to separate the low-frequency components due to gross position changes of the larynx from the useful signal due to the vocal-cord vibration. This high-pass filter with a standard time constant of 25 ms (corresponding to a cutoff frequency of about 7 Hz) causes considerable phase distortion in the laryngogram. Some researchers thus prefer the high-pass filter to have a much lower cutoff frequency at the expense of having difficulty in finding a suitable baseline of the laryngogram. The respective application should determine what one should do. In pitch determination, for instance, the missing baseline cannot be tolerated; phase distortions, however, can easily be coped with since they have little influence on the characteristic feature of the glottogram, i.e., the discontinuity of the signal at the instant of glottal closure (see Sect.9.2.1 for a detailed discussion of this issue). In phonetic research it may sometimes be useful to get data on the overall laryngeal behavior, i.e., vocal-cord vibrations *plus* vertical movements, so that the lower cutoff frequency makes sense. The problem is more difficult for voice source analysis. Here we need both the baseline and the fine structure of the signal without phase distortions. There is one way out of the problem when digital signal processing methods are available. Voice source investigation involves relatively short signals and does not really need on-line processing. In this case one can reapply the same high-pass filter while processing the signal in backward direction. This action cancels all the phase distortions. A corresponding procedure can be applied in an analog realization. The signal is recorded on an analog tape recorder and is rerecorded in back-

ward direction with the same high-pass filter in both the PDI and the rerecording channel. In this manner all phase distortions - including those by the
tape recording (in case the tape recorders are identical) - are canceled, and
the phase response of the whole system becomes zero.

Pitch determination from the output signal of an electric PDI is simple.
The most prominent feature is the large jump of the signal at the instant
of glottal closure. By computing the first derivative of the signal the instant
of glottal closure will thus be marked by a pulse which can be easily detected by a peak detection (cf. Sect.6.2.1) or a threshold analysis (cf. Sect.6.1.1)
basic extractor.

In summary, the electric PDI (laryngograph or glottograph) is an ideal
instrument for pitch determination in phonetics, linguistics, education, and
pathology. It is robust, easy to handle, and does not cause any discomfort to
the speaker. In some rare cases it might not work for an individual speaker;
when it does work, however, one can hardly do anything wrong with it.

5.2.4 Ultrasonic PDIs

Medical ultrasound has been used to study the vocal cords since the early
1960's. The principle of this method is based on the fact that the acoustic
impedance of air is extremely different from that of flesh, cartilages, and
tissue. If a focused ultrasound beam is transmitted through the vibrating vocal
cords, it will only be able to pass if the glottis is closed. If the glottis is
open, the ultrasound wave is almost totally reflected due to the impedance
mismatch between the tissue and the air in the glottis.

To investigate the vocal cords with this method, two separate procedures
appear possible. In the first case, the *pulse-echo method*, a train of short
ultrasound pulses is transmitted into the larynx from one side, and the reflected wave is picked up at the same side of the neck. In the second case, the
continuous-wave method, a continuous-wave ultrasound is transmitted through
the larynx at the level of the vocal cords. The vibrating vocal cords modulate
the wave, and the envelope of the received ultrasound signal is a measure of
this vibration.

Besides these two principal methods a number of researchers applied a
combination of pulse echo and pulse through transmission with the transmitter
and the receiver on different sides of the neck (Van den Berg, 1962; Kitamura
et al., 1964; Kaneko et al., 1974).

To design a PDI, the continuous-wave method appears most promising. The
transmitter and the receiver are constructed as a pair of frequency-matched
ultrasound piezoelectric transducers, i.e., barium titanate crystals. Transmitter
and receiver are positioned on either side of the neck at the level of the
vocal cords. Between these transducers a continuous ultrasonic wave propagates through the larynx.

When the vocal folds are open, the intensity of the ultrasound passing
through the air-filled glottis is low. This depends on the fact that the
acoustic impedance of air is much smaller than that of the surrounding

tissue and, consequently, strong reflections from the surface of the vocal folds are to be expected. However, when the vocal folds are in contact with each other within an unbroken ultrasound beam, the ultrasound passes through the contact surface. The intensity transferred changes with the variations of the contact area of the vocal folds. This variation of the received ultrasound intensity is refolded and is proportional to the fundamental frequency of the vocal folds. (Holmer and Rundqvist, 1975, p. 1073)

This principle was first applied by Bordone-Sacerdote and Sacerdote (1965), later by Hamlet (1971, 1972, 1978, 1980), Hamlet and Reid (1972), and Holmer and Rundqvist (1975). The block structure of this PDI is very similar to that of the electric PDI (see Figs.5.13 and 5.16b). An ultrasound generator replaces the HF generator, and the ultrasonic transducers replace the electrodes. Extensive high-pass filtering of the signal is unnecessary since the modulation depth of the output signal is much higher for the ultrasonic than for the electric PDI. The frequency of the ultrasound takes on values in the megahertz region [Bordone-Sacerdote (1965): 1 MHz; Hamlet (1980): 1.8 MHz; Holmer and Rundqvist (1975): 2.2 MHz; Hamlet and Reid (1972): 3.2 MHz; Hamlet (1972): 5 MHz]. The transducers are small; they have a diameter of 6 mm (Hamlet and Reid, 1972) or a rectangular shape [6 mm x 2 mm (Hamlet, 1980) or 16 mm x 12 mm (Holmer and Rundqvist, 1975)]. A good contact between the transducers and the skin must be provided, and the application of a special contact paste is required (Bordone-Sacerdote and Sacerdote, 1965). Using an acoustic power of less than 10^{-3} W/cm^2 (Hamlet and Reid, 1972) this method is absolutely harmless to the subject under investigation.

In contrast to the electric PDI (the current flows through the whole tissue) the ultrasonic beam is well focused and quite narrow. At the vocal cords its main intensity is confined to an area about 4 mm in diameter when circular transducers are used (Hamlet and Reid, 1972). This is an advantage and a drawback of the method at the same time. If the ultrasonic wave passes the vocal cords, the modulation depth equals nearly 100% (in contrast to the electric PDI where the impedance change is in the order of 1%). At the same time, the method is sensitive to the positioning of the transducers.

When the transducers are moved vertically away from the position described above, the signal quickly disappears and no transmission is received. This is to be expected from the anatomy of the larynx, since there are air spaces immediately above and below the vocal cords. On the other hand, a change in anteroposterior transducer positioning alters the form of the signal, since the anterior and posterior boundaries of the vocal cords are continuous with the inner walls of the larynx. When the ultrasound beam is directed at the extreme anterior or posterior parts of the vocal cords, some transmission may be received continuously through the walls of the larynx. The signal then appears as an amplitude-modulated wave with a continuous carrier. (Hamlet and Reid, 1972, p.35)

Due to temporary vertical larynx displacement in running speech, the amplitude of the output signal may change drastically because of deviations in the position of the transducers with respect to the vocal cords. The larynx can easily move up and down by more than 2 cm; the maximum displacement is obtained for swallowing and for major F_0 changes with untrained speakers (Hamlet, 1980; Shipp, 1975). Holmer and Rundqvist (1975) found that the 16 mm x 12 mm transducers were not sufficient to give a good signal for any position of the larynx. So they designed a 30 mm x 5 mm transducer for use in a PDI for linguistic research; however, they did not give quantitative data detailing how good this design really is. Since in all cases the vocal cords are located within a distance of less than 3 cm from the transducers, they are still in the acoustic near field so that the shape of the ultrasonic beam is still very similar to the shape of the transmitting transducer. In this respect the shape of the receiving transducer is less critical (Hamlet, 1980); of course both transmitter and receiver must have the same resonance frequency.

Hamlet (1980) carried out a detailed investigation on behalf of the transducer design. She found that a setup with a 6 mm x 2 mm unfocused transmitter and a 6 mm x 6 mm receiver could cope with a vertical larynx movement of 20 mm when a 20-dB amplitude variation of the output signal due to vertical larynx movement was considered as tolerable.

In summary, ultrasonic PDIs using continuous-wave ultrasound show an output signal similar to that of electric PDIs. The ultrasound passes the neck when the glottis is closed; otherwise it is reflected. Compared to the electric PDI, we almost have a 100% amplitude modulation when the beam passes through the vocal cords. On the other hand, the ultrasonic PDI is much more sensitive to vertical positioning of the transducers. As with the electric PDI, the active elements are attached to the neck from outside and can be worn for a long time without discomfort. Figure 5.17 shows an example of the output signal of this PDI.

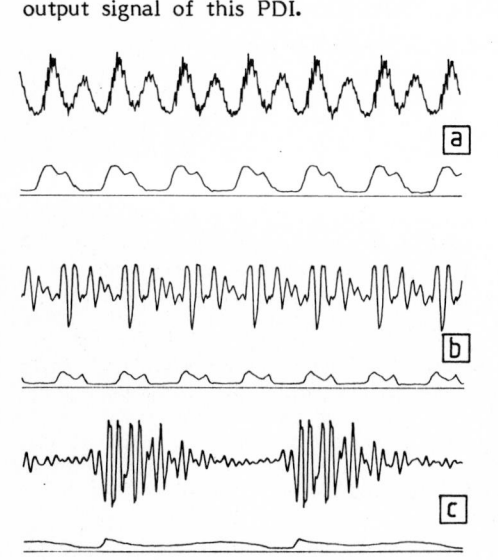

Fig.5.17a-c. Simultaneously recorded speech and ultrasonic signals. **(a)** Vowel /i/; **(b)** vowel /ɔ/; **(c)** vowel /a/, vocal fry. (Upper curve) Speech signal, (lower curve) ultrasonic signal. Time scale: 55 ms per line. From (Hamlet, 1971, p.1561)

In contrast to continuous-wave ultrasonic PDIs, the pulse-echo method (as well as the combination of pulse-echo and pulse-through methods) is preferred for diagnostic use; in phonetic and linguistic research it is used for instance to investigate whispered and voiceless speech (Hamlet, 1972). In the pulse-echo method a train of ultrasonic pulses is transmitted into the neck tissue. The reflected pulses are received by the same transducer which transmitted the original pulses. From the elapsed time it is possible to determine position of the vocal cords. Ultrasound velocity in tissue is about 1500 m/s; so the elapsed time between the transmission of a pulse and the reception of the reflected one will be 30 to 40 μs if the pulse is reflected at the glottis; the same time must be allowed for a pulse to travel through the larynx if the pulse-through method is applied. Of course this limits the repetition rate of the pulse train to about 10 kHz. The advantage of the method is that each vocal cord can be watched individually; no other method can achieve this without surgical interaction or at least the insertion of a receptor probe into the pharynx (via mouth or nostrils).

The single-transducer pulse-echo method was the first ultrasonic method to be applied in larynx investigation (Mensch, 1964; Asano, 1968; Hertz et al., 1970; Hamlet, 1972; Holmer et al., 1973). Hamlet (1972) used the method (with a pulse duration of 2 μs, a pulse repetition rate of 1000 Hz and an ultrasound frequency of 5 MHz) to obtain data on the vocal-cord position in whispered sibilants. For a PDI, however, this method has never been used; probably because the positioning of the transducer is too critical, and because the pulse repetition rate cannot be made high enough so as to avoid major quantization noise.

5.2.5 Photoelectric PDIs (Transillumination of the Glottis)

A single PDI has to be reviewed in this section: the photoelectric PDI with transillumination of the glottis. The principle is due to Sonesson (1960); the PDI is commercially available (Frøkjaer-Jensen, 1967).

A strong light source is placed at the neck below the glottis. Part of the light passes through the skin and the tissue into the trachea. If the glottis is open, the light can also pass into the pharynx; if the glottis is closed, the light is absorbed by the vocal cords, and the pharynx remains dark. A phototransistor in the pharynx, which works as a light transducer, picks up the temporal variations of light in the pharynx due to glottal opening and closure. In contrast to the electric and the ultrasonic PDI, the photoelectric PDI gives a measure of the cross-sectional area of the glottis, not of the degree of glottal closure.

A problem is presented by the photoelectric receiver. It has to be inserted into the pharynx in some way. In the first development by Sonesson (1960) a plastic rod carrying the photoelectric element had to be inserted through the mouth and was placed in the pharynx. Normal speech was thus hardly possible, only sustained vowels such as /ɛː/ or /oeː/. In a development by Abramson et al. (1965) the light source was inserted via the mouth, and the photo transducer was placed in contact with the skin just below the larynx. However, this device proved sensitive to vertical movements of the cartilages of the larynx. Malécot and Peebles (1965) attached the light source to the neck below the glottis and inserted the phototransducer via the nose into the pharynx. This setup was the first to enable the investigator to carry out measurements during normal speech. Similar PDIs were reported by Ohala (1966) as well as Slis and Damsté (1967). Frøkjaer-Jensen (1967), finally, made the PDI commercially available. His papers contain a detailed discussion of the device, especially of the problems with the lamp and the transducer.

> The light source consists of a 50-W projector lamp with blower. The light is led to the skin through a 32-cm long acryl rod with an outer diameter of 10 mm. This acryl rod makes it more comfortable for the subject because of the greater distance between the subject's face and the hot light house of the projector. Only a few per cent of the light is lost in the transmission rod which is mounted in the lense holder instead of the front lenses (objectives) of the projector.
>
> In order to avoid the 50-Hz ripple of the light, which will interfere very strongly with the oscillations from the vocal cords, the current for the projector lamp is rectified and smoothed. In the first experiments ... the light sensitive transducer in the pharynx was a photo transistor ... . This photo transistor is very small (diameter 2 mm) and has a shape which makes it very well suited for mounting in a polyethylene tube with an internal diameter of 2 mm and an external diameter of 3 mm. Thus mounted in the end of the small tube it goes down quite easily into the pharynx. If the subject has very sensitive mucous membranes and strong reflexes, it can be necessary to give a little amount of local anaesthesia, but this is not normally needed. ... The distance from the nostrils to the lens of the photo transistor is for a normal subject 180 cm tall between 13 and 16 cm. Therefore, we have placed a little mark on the polyethylene tube at a distance of 13 cm from the photo transistor. The photo transistor must be fixed by means of glue in the tube in order to avoid that saliva percolates into the tube, thereby causing partial shortcut between the terminals. ... The photo transistor is connected through a shielded cable to an amplifier, which is mounted in the housing of the projector. (Frøkjaer-Jensen, 1967, p.6-7)

The big problem with this PDI is how to fix the transducer in the pharynx. It cannot be fixed mechanically so that its main position is somewhat undefined. In addition, it is moved around by articulatory gestures.

> It is, however, difficult to fix this position of the transducer for most subjects. The articulatory movements disturb it, especially movements of the soft palate and movements of the epiglottis. We have tried to fix the transducer by mounting it midway in the polyethylene tube, which is then swallowed down into the oesophagus. Although the photo transistor is placed far back in the pharynx, experiments with different subjects have shown that this mounting is the best one, at least if the lens of the photo transistor is grinded.
>
> I have tested different types of photo transistors and photo diodes. The type which was used originally had a light sensitive angle of only 15°, which was probably too little. ... The convex lens which forms the front part of the photo transistor house, can easily be grinded away. As a result of this, it is much easier to pick up the light, and the position of the

photo transistor in the pharynx is not so critical as before. (Frøkjaer-Jensen, 1967, p.10)

The output signal depends on the degree of glottal opening. In a first approximation the signal is proportional to the cross-sectional area of the glottis although there are a number of uncertainties which make quantitative measurements of the glottal area difficult with this PDI. First of all, there is the unsolved problem of the transducer position. Second, since the light is diffuse, it cannot be guaranteed that the glottis is always illuminated at its whole length although control experiments show that this is usually the case (Frøkjaer-Jensen, 1967). Third, the relation between the light intensity and the signal amplitude at the transducer may be nonlinear (Hutters, 1976). Except for the last effect which may be compensated if known, these problems influence the long-term behavior of the signal amplitude and make quantitative statements on the glottal area questionable (Hutters, 1976). The short-term performance of this instrument, however, is excellent. There is an exact synchronization both with the point of glottal opening and glottal closure. Measurements are also possible when the glottis does not close completely (due to a voice disease or a breathy voice in normal speech) - a situation where both the electric and the ultrasonic PDIs fail. This PDI is thus ideal for voice source investigation in basic phonetic and linguistic research or for research in logopedics and phoniatrics. Its use causes some discomfort to the speaker - not everyone tolerates the transducer in pharynx and nose cavity, and the speaker must remain immobile with the light-conducting acryl rod pressed against his neck. This is certainly one reason why the instrument is confined to the areas mentioned and does not find application in the field of education.

5.2.6 Comparative Evaluation of PDIs

In contrast to PDAs (cf. Sect.9.3) the number of comparative evaluations of PDIs is quite large. This may have several reasons. First, evaluation of the performance of PDIs is simpler than that of PDAs. Failures are rather rare; the evaluation is thus a performance investigation while the device is properly working rather than an error analysis which is much more difficult. Second, and perhaps more important, the widespread use of these instruments in medical and clinical practice made it necessary to provide a good calibration and to perform extensive tests. In addition, it has become quite usual, if PDIs are applied in basic speech and voice source research, to apply several of them simultaneously. This situation makes comparative evaluations of their performance very likely. The majority of these evaluations have been done in the domains of medicine and pathology; some relevant work was also carried out in phonetic laboratories. Much of this work served to test the electric PDI (glottograph, laryngograph) against other instruments whose behavior was thought to be more straightforwardly known. The motivation to perform these investigations may have arisen from the controversy about the physiological relevance of the output signal of the electric PDI (see the discussion in Sect.5.2.3).

Such an evaluation was carried out, for instance, by Vallancien, Gautheron, Pasternak, Guisez, and Paley (1971). In this evaluation four signals were measured and recorded simultaneously: 1) the ordinary speech signal; 2) the output signal of a laryngograph (Fourcin and Abberton, 1971); 3) the output signal of a photoelectric PDI [in the version after Lisker and Abramson (1969) with the light source in the pharynx and the photo element at the surface of the neck]; and 4) the output signal of a subglottal pressure transducer (which is a special application of the mechanic PDI). An example of the simultaneous recording of the photoelectric PDI and the electric PDI is shown in Fig.5.18. Vallancien et al. commented the results of their study as follows:

> During the first phase ("separation of the vocal cords") the amount of light increases slowly[7] while the impedance increases rapidly. These variations reflect the slow increase in a beam of light and also the rapid separation of the vocal cords.
>
> In the second phase ("opening") the amount of light increases rapidly and the impedance reaches saturation point.
>
> During the third phase ("closure") the amount of light decreases rapidly and the impedance decreases slowly indicating the beginning of contact between the vocal cords. Finally, during the fourth phase ("contact") the amount of light drops to zero, the impedance decreases abruptly, and an acoustic impulse of large amplitude is produced by the rapid closure of the vocal cords. (Vallancien et al., 1971, pp.375-76; original in French, cf. Appendix B)

From this investigation Vallancien et al. drew the conclusion that the electric PDI is an exact instrument whose output signal has physiologic significance.

The investigation of the electric PDI by Lecluse (1977), who tested several PDIs against each other and then verified the temporal relation between the electric PDI and the glottal waveform by means of stroboscopy, was discussed in Section 5.2.3. The findings by Lecluse and by Vallancien et al. are more or less identical. A similar investigation was published by Pedersen (1977).

In the investigation by Kitzing (1977) the photoelectric PDI and the electric PDI were compared. Kitzing presented data on voices with irregular excitation and showed that the performance of the two instruments is in some way complementary. The electric PDI yields an excellent indication for the instant of glottal closure, whereas the photoelectric PDI indicates the instant of glottal reopening very well.

------

[7] The slope of the output signal of the photoelectric PDI during this phase (phase A in Fig.5.18) need not be due to increasing transillumination of the glottis, but can be an artifact caused by the phase distortions in the analog tape recorder. Vallancien et al. recorded all the output signals simultaneously (and additionally) on a mingograph - without low-frequency phase distortions. The original publication contains such a mingograph display where this part of the waveform is nearly flat. However, the authors obviously drew this conclusion using the phase-distorted analog tape recorder display.

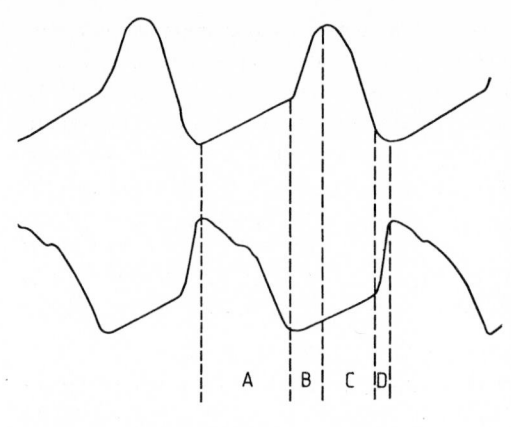

Fig.5.18. Comparison of the output signals of the photoelectric (upper trace) and the electric PDI (lower trace). (A) Closed-glottis phase; (B) open glottis phase, the vocal cords move apart; (C) open glottis phase, the vocal cords move together; (D) instant of glottal closure. The waveforms were taken from an analog recording so that there might be substantial phase distortions which cause the rising slope in the flat parts of the signals. From (Vallancien et al., 1971, p.375)

Dejonckere (1981) compared the electric PDI to the two types of photo-electric PDIs (depending on whether the light source is at the neck surface and the photosensor is located in the pharynx or vice versa). According to his findings, the two photoglottographic methods give very similar results.

Askenfelt, Gauffin, Kitzing, and Sundberg (1977, 1980) carried out a different evaluation. They investigated the performance of an electric PDI in comparison with a mechanic PDI (using an accelerometer) for measurement of pitch coutours. They made extensive measurements for a number of speakers (13) and found good agreement between the results of the two instruments. Fundamental frequency was measured using each of the PDIs together with a simple time-domain PDA (Larsson, 1977), and F_0 histograms were computed for the subjects examined. The performance of the electric PDI was found superior for normal voices in chest register, whereas for falsetto and especially for breathy voice, where the glottis never closes completely, difficulties were found with the electric PDI because the amplitude of the signal fell drastically. In addition, the electric PDI failed for three of the subjects. Some of these difficulties might be due to the rather conservative design of the electric PDI applied in these experiments, or due to the subjects who were selected in such a way that difficulties with the electric PDI were likely to occur.

> Both techniques examined ... have strong and weak points. The electro-glottograph [i.e., the electric PDI] rarely fails on those subjects for whom it works at all: those voices that can be measured are generally measured accurately. It records the glottal closures rather than the sound generated by the vocal-fold vibrations. It is often difficult to use on subjects having a thick layer of subcutaneous tissue on the neck, and on subjects in which the glottis never closes such as in the case of vocal-fold paralysis. The contact microphone [i.e., the mechanic PDI using an accelerometer], on the other hand, can probably be used in all subjects, even in cases where the electroglottograph fails. Its output signal mirrors the sound generated by the vocal-fold vibrations. There is no exact agreement between the fundamental frequency data obtained by the two methods. Mainly, this is a consequence of the fact that variations in the contact microphone output signal waveform cause errors in the fundamental frequency measurement

[for the mechanic PDI local errors were much more numerous than for the electric PDI]. ...

For the practical use in the phoniatric clinic, it seems that the electro-glottograph should be preferred as long as it works well, and it should be replaced by the contact microphone for remaining patients. (Askenfelt et al., 1980, p.272)

The investigations by Boves (1979a,b) and Boves and Cranen (1982a,b) are remarkable since they are not confined to PDIs, but use the PDI to evaluate the performance of glottal inverse filtering procedures (see Sect.6.5.1 for a detailed discussion). Boves and Cranen measured 1) the speech signal, 2) the output signal of a laryngograph (Fourcin and Abberton, 1971), 3) the output signal of a photoelectric PDI, and 4) the subglottal pressure variations. Measurement 4 is carried out by means of a small transducer which is inserted via the nose, the pharynx, and the glottis into the trachea. According to the authors, some local anaesthesia is necessary for this. From the speech signal, they compute 5) the LPC residual and 6) the (reconstructed) glottal waveform using the inverse filtering technique to be tested. Digital recording and linear-phase digital filters were applied to remove spurious low-frequency noise from the signals without introducing phase distortions. Figure 5.19 shows some typical results. The main conclusions read as follows:

If we take the moment of glottal closure as evidenced from the electro-glottogram as a reference and combine all evidence available for determin-ing the instant of opening of the glottis, it is clear that on the average the reconstructed flow waveform matches these instances quite well. The different duty cycles in the vowels are reproduced correctly, even in those cases, like the vowel /o/, where the closed glottis interval is extremely short.

The subglottal pressure signals very consistently show a pronounced peak at the moment of glottal closure. In all registrations a clear, but highly damped resonance structure can be seen. This high pressure peak at the moment when the vocal cords meet to close the glottis suggests that the familiar description of the mechanics of vocal-cord vibration (in which the opening of the glottis is ascribed to a recovery of the subglottal pressure after the cessation of the flow that pushes the folds apart) is in need of revision.

The normalized prediction error appears to be an excellent indicator of the moments of opening and closing of the vocal folds in some cases, but in other cases it clearly is not. (Boves and Cranen, 1982a, p.1990)

This investigation by Boves and Cranen (1982a,b) might be the only one where a PDA, in this case a high-accuracy PDA for glottal waveform estimation, is evaluated with the help of a PDI. Similar investigations by the present author using ordinary time-domain and short-term analysis PDAs are in progress. For the evaluation of PDAs the PDI is an ideal calibration instrument since it works almost free of error, and since it can give a reliable estimate, at least for normal voices, whether the glottal source is active or not.

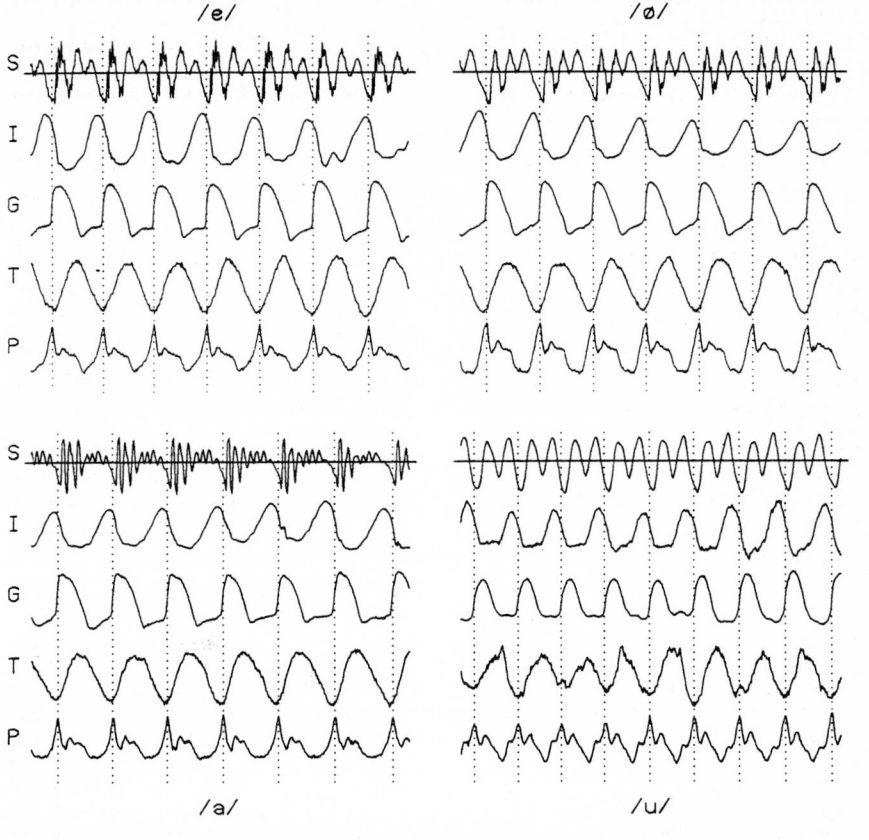

Fig.5.19. Simultaneous recordings of various PDI output signals for four sustained vowels (as indicated in the figure). (S) Speech signal; (I) excitation signal (reconstructed by LPC inverse filtering; see Sect.6.5.2); (G) glottogram (electric PDI); (T) photoelectic glottogram (transillumination of the larynx); (P) subglottic pressure. From (Boves, personal communication)

5.3 Voicing Determination - Selected Examples

Strictly speaking, voicing determination is the determination of the operational mode of the voice source, i.e., 1) detecting the presence or absence of a *voiced excitation*, and 2) detecting the presence or absence of a *voiceless excitation*. In case neither of the two sources is active we speak of a *pause* or a segment of *silence*; segments where both sources act simultaneously are called *mixed-excitation* segments. In this context the terms *voiced-voiceless-silence discrimination* or *voiced-unvoiced* (V-UV) *discrimination* are frequently used in the literature. According to the terminology established in Sect.2.1, any algorithm which deals with the operational mode of the voice source will be called a *voicing determination algorithm* (VDA), or a *voicing determination device* (VDD) if it is conceived as a piece of hardware.

One more word about terminology. If the vocal cords vibrate during a segment of speech, the segment is *voiced*. Correspondingly, if there is a turbulence in the speech tract causing a noiselike signal during a segment (and no vocal cord vibration) the segment is *voiceless*. If we label a segment *unvoiced* we only mean that it is not voiced; it can thus be voiceless or silent. This distinction between voiceless and unvoiced can be important for pitch determination. From the same point of view, the voiced-unvoiced discrimination will group segments with mixed excitation together with the voiced segments.

In this section selected examples of VDAs are presented. The aim is not to give a general survey of the problems of voicing determination; this would exceed the scope of this book. It is rather intended to present some typical examples, to elucidate the interactions between voicing determination and pitch determination, and to point out some pitfalls which might lead to incorrect conclusions.

To start with the latter, it is a rather trivial (but fundamental and correct) statement that *pitch can exist only where voicing exists*. Unvoiced segments do not have a pitch. On the other hand, there is the widespread misconception that this statement is reversible, i.e., that *voicing exists only where a pitch can be found*. As we have seen from the discussions in Chaps.3 and 4, this is not true. A good deal of voiced signals have an irregular excitation. The causes for this are manifold; they range from permanent voice pathology over (temporary) hoarseness to excitation in the vocal fry register, which may last only for isolated vowels and other limited segments. In addition, there are the breakdowns of the PDA which are possible for any signal. It would lead to grossly erroneous results to label such segments unvoiced although the particular PDA which analyzes them cannot find a periodic excitation.

Voicing determination should thus be kept apart from pitch determination although the two are closely interrelated. Voicing determination can be regarded as a procedure prior to pitch determination. In order to determine the quantitative value of pitch, one must first know *where the periods are*. However, is it really necessary to know this beforehand? If the answer is yes, then for reasons of implementation, but not for principal reasons. In principle any PDA can be run on unvoiced segments. It will yield some "pitch estimate" for the unvoiced segments as well. In this case the VDA is incorporated in the postprocessor of the PDA; from the above consideration, however, it follows that regularity of the pitch estimate should not be the only criterion for the (subsequent) voiced-unvoiced discrimination. From the point of view of the PDA, the only task of the VDA is to separate voiced segments from unvoiced ones in order to establish those segments of the signal where a meaningful output of the PDA can be expected. In most systems, however, the VDA is intended to be able to do more: it is supposed to at least give an estimate of voiced, voiceless, and silent segments. Accurate VDAs also provide a means to determine mixed excitation. This latter task is the most difficult one for the VDA (Siegel and Bessey, 1982). It will not be treated in much detail in this survey since it is not too important for pitch determination.

In principle the task of voicing determination comprises two binary decisions: 1) *voiced* versus *unvoiced*, and 2) *voiceless* versus *not voiceless*. For the case of mixed excitation, when the excitation is both voiced and voiceless, and only for this case, it is adequate to define a *degree of voicing*, that is, some kind of measure of how the total energy of the signal is distributed between the voiceless and the voiced sources (Dreyfus-Graf, 1977; Makhoul et al., 1978). In many speech analysis systems VDAs serve as the first step of a *segmentation algorithm*, the so-called primary segmentation (Hess, 1976); this routine yields the so-called *acoustic segments* (Olson et al., 1967).

If the VDA incorporates a speech-silence discrimination routine, there must always exist an assumption about the *background noise*. Such an assumption is quite obvious when signal energy is the criterion for the decision. If other parameters are also taken into account, detailed measurements of the "silent" intervals may prove necessary, and the performance of a VDA can critically depend on how well the actual environmental conditions meet the underlying model of the background noise (Atal and Rabiner, 1976).

The VDAs can be subdivided into three main categories (Fig.5.20): 1) simple threshold analyzers using a small number of parameters, 2) VDAs that apply a pattern recognition approach or use a classifier, and 3) VDAs that are incorporated into a PDA. Like the PDA, the typical VDA consists of three processing steps. The threshold analyzer or classifier of the VDA makes the decisions and determines the voicing parameters; it corresponds to the basic extractor of the PDA and will be referred to by the same name here. The preprocessor calculates the speech parameters necessary for the basic extractor. The main tasks of the postprocessor are local error correction and exact positioning of the boundaries.

In a number of VDAs the voicing parameters are determined in several steps. It is thus possible that they have several preprocessors and several basic extractors switched in series. If the VDA is run independently of the PDA, its operational mode necessarily corresponds to that of short-term analysis PDAs. This means that the signal is first subdivided into frames. Then the parameters are measured within each frame, and the VDA gives a

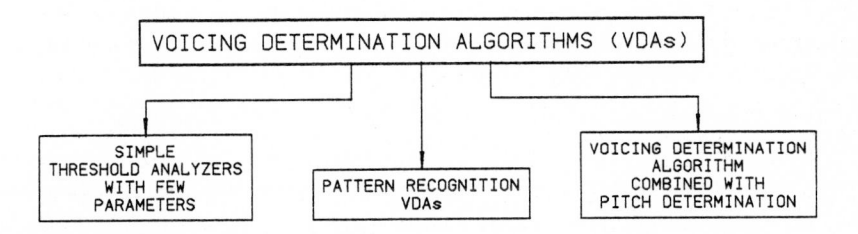

Fig.5.20. Categorization of VDAs into three main categories

single estimate for the whole frame. For tasks which apply a short-term analysis PDA or no PDA at all this may be sufficient; in connection with a time-domain PDA which analyzes the signal period by period, the instant of voiced-unvoiced transition (which is the only point of uncertainty with respect to time) can be exactly determined in such a way that the PDA is switched between the basic extractor and the postprocessor of the VDA, and that the local corrections at the voiced-unvoiced boundaries are performed according to the results of the PDA.

The remainder of this section will be split up in four parts. In Sect.5.3.1 we will examine the speech parameters that are used to run a VDA. In Sect. 5.3.2 simple VDAs using a small number of parameters as well as combined VDA-PDA systems will be discussed. In Sect.5.3.3 sophisticated pattern recognition VDAs will be dealt with, and Sect.5.3.4, finally, will present some conclusions.

5.3.1 Voicing Determination: Parameters

The question of selecting adequate parameters for voicing determination is discussed in detail by Baronin (1974b), Abry et al. (1975), Atal and Rabiner (1976), Rabiner et al. (1977), Boulogne (1979), Siegel (1979b), Siegel and Bessey (1980, 1982), and others. From these references the following list of parameters has been compiled which does not claim, however, to be complete.

Energy (Level) of the Signal. The energy is defined as the short-term root mean square (rms) value of the signal,

$$E_S = \sqrt{\sum_n x^2(n)} .$$

(5.2)

For reasons of computational effort it can be suitable to take the average magnitude

$$|\bar{x}| = \sum_n |x(n)|$$

(5.3)

or the peak-to-peak amplitude

$$\hat{x} = \max [x(n)] - \min [x(n)]$$

(5.4)

instead of the rms value. Frequently the logarithm of E_S is taken rather than E_S itself. In all cases the summation (or peak determination) is performed over the length of the frame (see Sect.2.4). This parameter is mostly used for speech-silence discrimination (Reddy, 1966; Atal and Rabiner, 1976; Hess, 1976b; Siegel and Bessey, 1982).

Normalized Short-term Autocorrelation Coefficient at Unit Sample Delay. This parameter is given as

$$r(1) = R(1) / R(0) = \sum_n [x(n)x(n+1)] / \sum_n x^2(n)$$

(5.5)

where R(d) are the short-term AC coefficients of the signal for the delay d (see Sect.8.2.2 for a detailed discussion). Since voiced speech has a quasi-periodic excitation with high correlation between subsequent samples, and since voiceless speech is noiselike, this parameter is expected to provide a good measure of the voiced-unvoiced discrimination.

A related measure is the *ratio of the energy of the differenced signal and the energy of the ordinary signal* (Hess, 1972b, 1976b),

$$Q_S = P[x(n+1)-x(n)] \; / \; 2P[x(n)] \; , \tag{5.6}$$

where P(x) means one of the energy measures given in (5.2-4). In the definition of (5.6) this parameter takes on a low value for voiced signals and a value near unity for voiceless signals.

Another related parameter is the *first spectral moment* which one could also call the *center of the spectral energy distribution*,

$$f_E = \sum_f [\, f \cdot |A(f)| \,] \; / \; \sum_f |A(f)| \; . \tag{5.7}$$

Instead of the spectral magnitude $|A(f)|$, the samples $|A(f)|^2$ of the power spectrum can also be taken to compute this parameter.

Average Number of Zero Crossings. This very simple parameter also gives a good measure of the predominant frequency in the signal and is deliberately taken for voiced-voiceless discrimination (Reddy, 1966; Ito and Donaldson, 1971; Siegel and Bessey, 1982).

Normalized Minimum Error of Linear Prediction. This parameter is calculated as the ratio of the energy [according to (5.2)] of the LPC residual signal and the energy of the original signal. For more details on linear prediction, see Section 6.3.1. The value is a measure for the quality of linear prediction. In voiced speech LPC works well, and the unpredictable part of the signal is low. With noiselike excitation, the prediction error increases by as much as an order of magnitude.

First Predictor Coefficient. The transfer function of the vocal tract is estimated by the LPC coefficients (see Sect.6.3.1) which refer to a digital filter in the direct structure (see Sect.2.3). In simple words, the first predictor coefficient equals the first sample of the impulse response of the inverse filter (see Sect.6.3.1 for more details); a large difference between voiced and voiceless sounds has to be expected.

Ratio of Energy in a High-Frequency and a Low-Frequency Subband. For voiced sounds the energy concentrates in the low-frequency band below 1 kHz; for voiceless sounds the energy is expected to concentrate above 2 kHz.

Quality of Periodicity Detection. A number of short-term analysis PDAs, among them the autocorrelation PDA with nonlinear preprocessing (center clipping, Sect.8.2.4), the AMDF PDA (Sect.8.3), the cepstrum PDA (Sect.8.4.2), and the maximum likelihood PDA (Sect.8.6.1) provide a reference from which the salience of the pitch estimate can be derived. Low salience of the pitch measurement suggests an unvoiced frame; we have already seen, however, that it is dangerous to rely on this feature alone since irregular excitation or even a drastic change within the actual frame can have the same effect (Sreenivas, 1981). The feature is extremely useful in cases where other parameters indicate that voicing may be questionable.

Other parameters are presence or absence of narrow-bandwidth formants in the frequency band below 1200 Hz and the distribution of the instantaneous values x(n) of the signal which is approximately normal for voiceless sounds, but strongly different for voiced segments (Baronin, 1974b).

Besides the usefulness and effectiveness, which is of course the most essential selection criterion for any parameter, the computing effort for its calculation is a second important matter. The effectiveness of the parameters will be discussed in the next sections. For the following reason, the question of computing effort strongly depends on whether the VDA has to operate on-line or not. Human speech consists of many pauses; even in continuous reading there are silent intervals about 30% of the time. Once a segment has been safely labeled as silent, no further processing whatsoever is necessary. If the speech analysis system (the VDA being a part of it) does not operate on-line (which implies real-time processing), a substantial amount of processing time is saved when these silent segments are found as early as possible and by means of readily calculable parameters. In such systems, parameters like short-term energy, absolute average, or zero crossing rate will be preferred to more "costly" parameters, such as the first correlation coefficient, the first LPC coefficient, or the signal energy ratio of subbands. If the system is a real-time vocoder, which must operate instantaneously frame by frame, its computing capacity has to be designed for the worst case, i.e., according to the requirements during speech segments. In this case it does not matter whether the processor runs idle or performs the full analysis during the pauses as well. If the system performs the full analysis regardless whether the signal is speech or silence, then the VDA can make use of all the parameters needed anywhere in the course of the analysis. In principle, on-line speech recognition systems must complete their job in real time, but they operate in an asynchronous mode and can thus be designed like off-line speech analysis systems. They must recognize the utterance as quickly as possible without the need for keeping on-line during each frame. Such a system is of course considerably speeded up when the pauses are recognized very early and with low effort.

5.3.2 Voicing Determination - Simple VDAs; Combined VDA-PDA Systems

A simple two-step VDA as part of a speech recognition system was presented by Reddy (1966). The first step discriminates between silence and speech, and the second step separates voiced from voiceless intervals. Mixed excitation is not taken into account. The speech-silence discrimination is performed via threshold analysis with the signal level [computed according to (5.4) for segments of 10 ms each] as the only parameter. Since the whole utterance is normalized with respect to amplitude before processing, a fixed threshold (about 20 dB below maximum amplitude) can be applied. The voiced-voiceless discrimination is performed using the zero crossings count as the decision parameter. Again a threshold analysis is carried out. Two thresholds, TH_1 and TH_2, are defined, TH_2 being greater than TH_1. In order to be labeled voiceless, an independent frame must have a zero crossings count above the upper threshold TH_2. If this is the case, all the surrounding frames whose zero crossings counts exceed the lower threshold TH_1 are also labeled voiceless.

For good-quality signals with almost time-invariant environmental conditions such an ad hoc nonadaptive VDA can prove sufficient. Similar VDAs have been applied in many other systems, with the same or other parameters (Flanagan, 1972; Baronin, 1974b). For example, Boulogne's VDA (1979) counts the zero crossings of the autocorrelation function for delays between zero and 200 ms (which causes a huge computing effort). Systems that apply a filterbank (for instance, channel vocoders, but also speech recognition systems), especially if they are built up in analog technology, preferably apply parameters such as the energy ratio in low- and high-frequency subbands (Gold, 1977; Abry et al., 1975; Ruske and Schotola, 1978).

The VDA by Hess (1972b, 1976b), implemented as a subroutine of an automatic speech recognition system, applies an adaptive threshold analysis for speech-silence discrimination and a fixed two-threshold analysis for voiced-voiceless separation. The underlying assumption is that the background noise is slowly varying. So, if the levels (computed as the average magnitude of the signal) of successive frames are taken to create a histogram, a distinct maximum is expected at the background noise level whereas the contribution of speech is more diffuse. The speech-silence discrimination threshold is set 1-2 dB above the level where the significant maximum has been found. The algorithm needs a learning phase of about 300 ms of silence in order to detect the maximum. In this primitive version the algorithm was found to label weak fricatives as silence when the recording had some low-frequency noise. In order to avoid this error, the same histogram was computed for the level of the differenced speech signal. This histogram, however, if used alone, tends to suppress weak nasals and voice-bars before voiced plosives. Hence the two histograms are used simultaneously, and a frame is labeled silent only when both levels are found below their respective thresholds.

For the voiced-voiceless discrimination a simple two-threshold analysis is carried out using a single parameter, the level ratio Q_S, as defined in (5.6). The two thresholds TH_1 and TH_2 ($TH_2 > TH_1$) are used to separate definitely voiced signals ($Q_S < TH_1$) and definitely voiceless signals ($Q_S > TH_2$) from an in-between area where the decision is ambiguous. These ambiguous frames are resolved according to the results of the subsequent PDA (see Chap.7). In the ambiguous segments some periodicity must be found, otherwise the frame is labeled voiceless. This procedure makes sure that the PDA operates on any signals where voicing may exist, and takes account of the possibility that voiced frames may be irregular. The two thresholds TH_1 and TH_2 are fixed; they were found to depend somewhat on the environmental conditions, especially on the sampling frequency and on low-frequency band limitation; this dependence, however, is not too critical. It is possible to build up a histogram of Q_S which has two significant maxima, one at the value for sustained voiceless sounds and the second one somewhere in the voiced region. In between there is a minimum since the transitional frames between voiced and voiceless excitation occur less frequently. Yet the learning phase before this minimum can be exploited is rather long; about half a minute of speech is needed to establish it.

A number of short-term PDAs combine their VDA with the PDA. A good example for this is the cepstrum PDA by Noll (1967). The PDA itself is discussed elsewhere in this book (Sect.8.4.2). As we will see (in Sect.8.4.1) the cepstrum, being the inverse Fourier transform of the logarithm of the power spectrum, performs a deconvolution of the signal and at the same time throws off all phase relations. The periodic component P(z) of the vocal tract model (Chap.3) appears in the cepstrum as a pulse train p(d), starting at quefrency $d = 0$, and showing the other pulses at $d = T_0$, $2T_0$, and so on. For an infinite periodic signal which is the impulse response of a linear system to a pulse train with the repetition rate T_0, the amplitude of the pulses $p(kT_0)$ at integer multiples of T_0 in the cepstrum domain will be equal to the amplitude at $d = 0$, i.e., at zero quefrency. For finite signals the cepstrum looks similar although the amplitude of the pitch peaks tend to decrease with increasing quefrencies (Noll, 1967). Due to the effect of deconvolution, the amplitude of the pitch peak at $d = T_0$ shows only a slight dependence on the amplitude of the signal. Thus the amplitude of this peak can be taken as a measure of the salience of the pitch measurement. Noll's PDA (1967) defines a threshold THR (whose value remains unspecified in the publication) and applies a three-frame window, consisting of the frames k-1 through k+1, in order to determine whether frame k is voiced. The performance of this method is best shown in form of the truth table of a three-variable Boolean function (Table 5.1).

The input variables are the voicing function V(k-1) of the previous frame and the threshold functions T(k) and T(k+1) of the actual and the successive frame (definition of the functions in the caption of Table 5.1). In general the decision is performed in a local mode using T(k) alone except when both V(k-1) and T(k+1) are different from T(k). The influence of the PDA is given by manipulation of the threshold THR which is lowered around $d = T_0$ when a regular periodicity has been detected (see Sect.8.4.2).

The algorithm is insensitive to failures of the PDA which result from the presence of two or more large peaks. Voicing determination errors, however, may result from irregular excitation or rapid changes of F_0 within the frame. According to an investigation by Timme et al. (1973) who built a real-time cepstrum VDD (using a real-time spectrum analyzer for computation of the logarithmic spectrum and a third-octave filterbank for the formation of the cepstrum), this principle works very reliably; it fails occasionally when the

Table 5.1. Voicing decision in the VDA-PDA system by Noll (1967). [V(i)] Voicing function of the ith frame; [T(i)] threshold function of the ith frame. A (binary) one means for the voicing function that the frame is labeled voiced, and for the threshold function that the amplitude of the significant peak exceeds the threshold THR. For more details, see text

V(k-1)	T(k)	T(k+1)	V(k)	Remark
0	0	0	0	Continuously unvoiced
0	0	1	0	
0	1	0	0	Isolated pitch peak; to be corrected
0	1	1	1	Voicing begins
1	0	0	0	Voicing has ended
1	0	1	1	Isolated failure of PDA; to be corrected
1	1	0	1	Still voiced
1	1	1	1	Continuously voiced

background noise has a harmonic structure or when the speech signal is distorted by another speech signal in the background. Timme et al. (1973) used the cepstrum only for the voiced-unvoiced discrimination so that the crude realization of the transformation into the cepstrum domain is justified.

Combined VDA-PDA systems where the voicing indication is based on the salience of the significant peak of a short-term function were published, among others, by Seneff (1978, Sect.8.5.3) and by Nguyen and Imai (1977, Sect.8.3). It must again be emphasized that all these VDAs do not really perform a voiced-unvoiced discrimination; rather they provide a sufficient (but not necessary) criterion for the presence of voicing. Hence the performance of these VDAs is clearly biased: they will commit a very small number of unvoiced-to-voiced errors (i.e., labeling frames voiced that, in reality, are unvoiced), but under adverse conditions the number of voiced-to-unvoiced errors can be substantial.

An interesting VDA of a similar type, the so-called *phase plane method*, was published by Lobanov (1970). The underlying idea is very similar to that of the vector PDA (Rader, 1974, see Sect. 6.3.2).

The structural properties of signals can be exhibited by analyzing the characteristics of the relationship between the instantaneous signal levels and its linear transformations ... For the discrimination of noisy and quasiperiodic signals one can make use of the fact that for noisy signals with a normal distribution there are several linear transformations by which the original signal x(t) and the transformed signal Lx(t) become independent at coincident times. The most important of these linear transformation are differentiation and the Hilbert transform. With these transformations the path of the image point on the phase plane [x(t), Lx(t)] for a noise signal represents a random curve, which sweeps out a certain area about the origin, while for a periodic signal it traces out a time-stable closed curve. (Lobanov, 1970, p.353)

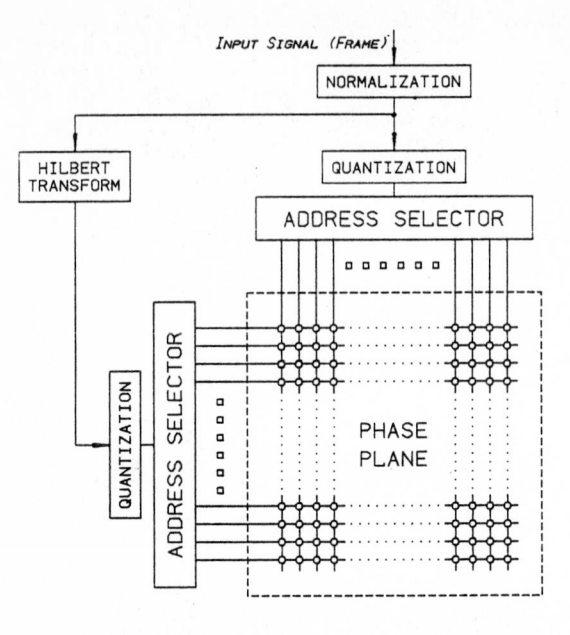

Fig.5.21. VDA by Lobanov (1970): block diagram of the preprocessor

The block diagram of the VDA is shown in Fig.5.21, and some typical examples are presented in Figure 5.22. Lobanov's original device was implemented in analog technology.[8] In digital terms, the implementation is rather simple. What is needed is a filter for the Hilbert transform [which can be done in good approximation by a linear-phase digital filter (Schüssler, 1973; Parks and McClellan, 1972)], an amplitude normalization, a quantization routine, and a certain amount of one-bit memory with the pertinent address computation facility in order to realize the phase plane. This memory is arranged in the form of a matrix. The current sample x(n) of the original signal selects the row of the matrix according to its value x; in the same way the column is selected by the current sample y(n) of the Hilbert transform. The memory cell at the intersection of the selected row and column is set to 1 in order to indicate that the point (x, y) of the phase plane has been passed. It is shown that for noise signals the (x, y) curve will gradually cover all the points of the phase plane, whereas for voiced signals, due to the closed-curve

---

[8] The performance of this VDA in a digital version may be degraded due to sampling. This can be clearly seen from Figure 5.22. In the center column the trace of the signal has not been quantized; in addition, the trace has been interpolated between the individual samples. It thus represents a closed curve. In the displays of the quantized phase plane on the right-hand side, however, the trace has only been marked at the sampling instants. Especially for voiceless signals the trace is thus strongly interrupted, which results in a degradation of the performance. The algorithm therefore requires an analog realization or - at least - relatively high sampling frequencies.

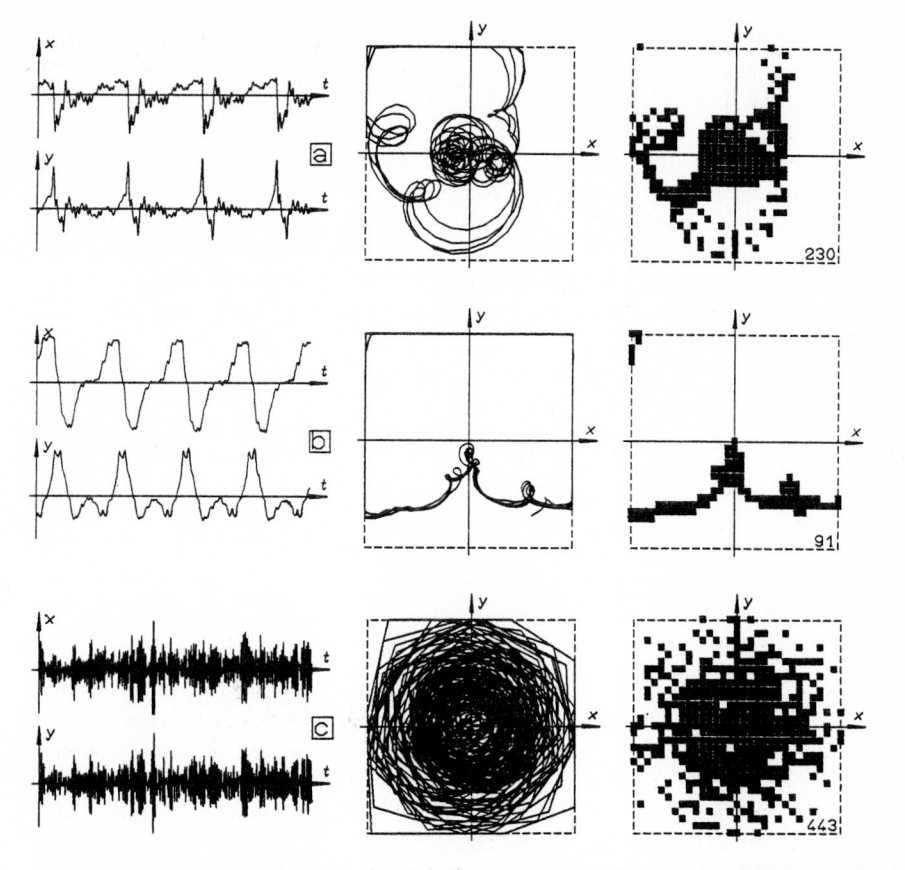

Fig.5.22a-c. VDA by Lobanov (1970/71): some typical examples of the performance. (a) Vowel /ɛ/; (b) nasal /m/; (c) fricative /s/. (Left column) Speech
signal (x) and signal after Hilbert transform (y); (center column) x-y plane
(phase plane) with continuous trace of the signal; (right column) phase plane
after sampling and quantization (31 quantization steps for each signal). The
numbers in the bottom right-hand corner of the quantized phase plane indicate
the number of coincidences for a given frame. Sampling frequency: 40 kHz
(see footnote 8). The signal was normalized in such a way that the maximum
amplitude exceeded the range of the phase plane by 50%. According to the
underlying theory this is necessary in order to reach the most distant point of
the phase plane within a given amount of time

behavior, many points in the phase plane will never be selected. This statement rigorously holds only for infinite signals and infinite measuring time. For
finite duration of measurement, where the voiced-unvoiced discrimination is
performed counting the number M of selected points in the phase plane and
performing a simple threshold analysis on the obtained count, a measure of
discrimination salience can be derived by just dividing the count M_N for

voiceless sounds by the count M_P for voiced sounds. For each pair of voiceless and voiced sounds this measure depends on a number of parameters, in particular the center frequency f_N of the noise, the duration T_m of the measurement, the center frequency f_P of the voiced sound, and the fundamental period T_0. The minimum measurement duration is influenced by the size of the quantization step of the phase plane in relation to the variance of the noise signal; it is the mean time required for the normalized random process to pass even the most distant point of the phase plane. The measure of discrimination salience is then given by

$$S = \text{const} \cdot \frac{f_N T_m}{f_P T_0} .$$
(5.8)

When the measurement duration is long enough, it seems always possible to make S greater than unity. Lobanov reported that for the worst case in Russian, i.e., the discrimination between /a/ and /x/, where $f_N \approx f_P$, for a low-pitched male voice, the value S equal to 2 was obtained for a measurement duration of 15 ms and a 100×100 memory matrix.

This VDA presents an extension to the principle of voiced-unvoiced discrimination with the presence or absence of a quasiperiodic excitation as the decision criterion. It is not the principle of this method to discriminate between a periodic and a nonperiodic process, but rather between a random process and a deterministic process which repeats for each glottal cycle. This distinction, however, is valid as long as voiced excitation is pulselike, even if the pulses are irregularly spaced.

5.3.3 Multiparameter VDAs. Voicing Determination by Means of Pattern Recognition Methods

The main routine of a pattern recognition algorithm is the classifier which assigns a certain label (the *class*) to the input data (which is usually a list of parameter values, the *pattern* or *template*, and is represented in the form of a matrix or a column vector) according to certain decision and classification criteria. The main reason that classifiers are powerful instruments is the fact that the knowledge concentrated in the decision and classification criteria does not have to be known beforehand, but is aquired and updated by the classifier during a *learning phase*; it is then applied to unknown patterns in the *test* or *recognition phase*.

For the learning phase there are two general strategies: 1) *supervised* learning, and 2) *unsupervised* learning. In a supervised learning phase a representative set of prelabeled patterns is made available to the classifier. From these the classifier knows which class they pertain to, and it builds up a set of *reference patterns* from which the membership of an unknown pattern to one of the classes is derived. In the unsupervised learning phase a representative set of patterns is made avaliable to the classifier without previous labeling. The classifier does not know how many classes there are, nor where to establish the decision boundaries between them. The only thing the classifier can do is just to analyze the patterns and to see whether they concentrate in several parts of the multidimensional parameter space established by the input parameters. If the patterns do so, i.e., if they form *clusters*, the classifier can offer suggestions as to how many classes are readily separable and where to put the decision boundaries in the multidimensional space.

It is obvious that for supervised learning there is much more manual work to be done than for unsupervised learning; however, for supervised learning the classifier is immediately adapted to the correct formulation of the problem when the recognition phase begins.

This brief qualitative introduction into pattern recognition has to be sufficient for this context; for more details, the reader is referred to a textbook on this subject (for instance, Nilsson, 1965; Meisel, 1972; Duda and Hart, 1973; Tou and Gonzalez, 1974; Niemann, 1974, 1981).

Returning to our problem of voicing determination we see that, in contrast to pitch determination, the present problem is well suited for a pattern recognition approach. We have a limited number of classes (two to four according to the particular task to be carried out), and we usually have a limited number of parameters from which we can easily establish suitable pattern vectors. We even have the possibility of easily performing manual labeling of a large number of patterns. Hence the preconditions for applying a pattern recognition algorithm are ideal. The only real problems in this respect may be that we have to establish reference patterns for a multitude of background noise conditions or to cope with the fact that the algorithm may fail systematically when environmental conditions arise for which the classifier has not been trained. A pattern recognition VDA as a component of a speech recognition system has been proposed by Atal and Rabiner (1976). It is based on the measurement of five parameters: 1) energy of the signal; 2) zero crossings rate; 3) autocorrelation coefficient $r(1)$; 4) first predictor coefficient of a 12-pole LPC analysis; and 5) energy of the normalized prediction error. The algorithm segments the signal into the three classes voiced, voiceless, and silence.

Figure 5.23 shows the block diagram of the algorithm in the recognition phase. The current values of the five parameters mentioned form the input pattern. The classifier applies a classical minimum-error-probability decision rule. For details the reader is referred to the original publication; a brief qualitative outline reads as follows. The five input parameters establish a five-dimensional parameter space within which a point is assigned to each pattern according to the actual values of the parameters. The classifier now assumes that there is a normal (Gaussian) distribution for each output class and each parameter so that the five parameters together yield a five-dimensional normal distribution for each output class. The better the approximation of the actual distribution(s) of the parameters to this normal distribution, the better the results will be. The decision rule which minimizes the probability of misclassifications states that an unknown parameter vector $\mathbf{x}$ should be assigned to class i if

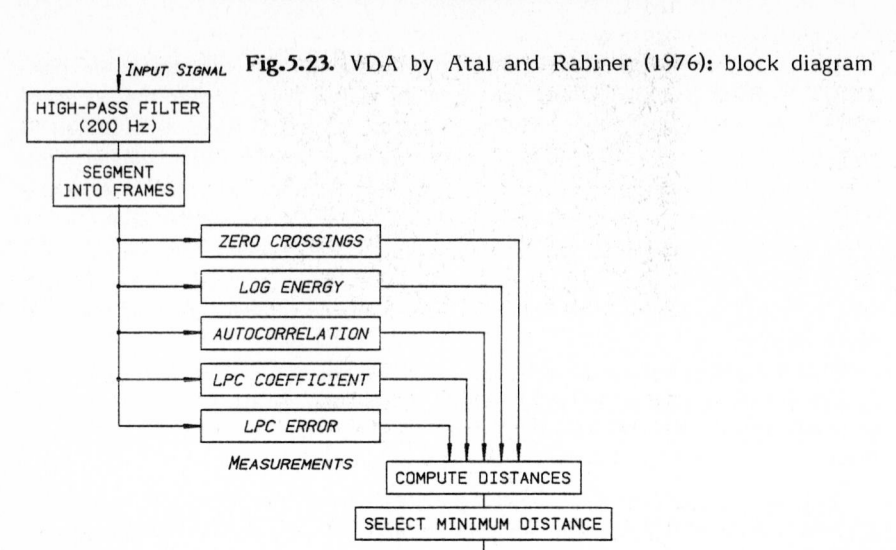

Fig.5.23. VDA by Atal and Rabiner (1976): block diagram

$$p_i g_i(x) > p_j g_j(x); \quad i \neq j \tag{5.9}$$

where p_i is the a priori probability that **x** pertains to class i (this is important in the actual case since we may have 50% silence, 40% voiced intervals.
and only 10% of unvoiced signals in an actual recording); $g_i(x)$ is the value of
the multidimensional distribution for class i and the actual values of **x**. Since
the distribution is assumed to be Gaussian, one can evaluate the logarithm of
this decision rule and thus simplify the expression. The actual decision is
based on a distance measure which gives a weighted distance d_i of the actual
parameter vector **x** from the three reference vectors x_i, i = 1, 2, 3, for each
of the three classes (which form the mean of the normal distribution and are
established during the learning phase); the class i is assigned to **x** for which
d_i becomes a minimum.

The learning phase is supervised; it is carried out using hand-labeled
frames of about 6 s of speech for several speakers with good-quality signals.

> For a training set to be representative of the entire population, it must
> include a variety of speech utterances spoken by a number of speakers.
> Second, manual segmentation of speech into the three classes is not always
> perfect. The influence of segmentation errors can be minimized by using a
> large training set. It is also necessary for the satisfactory operation of the
> algorithm that the recording conditions - frequency response of the recor-
> ding system, background noise conditions, etc. - remain reasonably stable.
> A new set of training data is usually needed if there are drastic changes
> in the recording conditions. In practice, a good training set for voiced or
> unvoiced speech can be obtained without difficulty. The greatest difficulty
> is encountered in the characterization of the "silence" class which is
> strongly dependent on the speaking environment. (Atal and Rabiner, 1976,
> p.205)

The performance of the algorithm was found to be very good for high-quality speech. Occasional problems arose when the signal-to-noise ratio changed from speaker to speaker. The long-term amplitude normalization applied in the system (which is permissible for this special task of a speech recognition system) tends to leave the parameter distribution for the two speech classes (voiced and voiceless) almost unchanged, but to strongly influence the parameter distribution of the "silence" class. A number of misclassifications of silent intervals was the consequence (which would be tolerable in a vocoder due to the low signal amplitudes during silence, but can be fatal in a speech recognition system). For this reason Atal and Rabiner added a (nonlinear) smoothing routine to the algorithm which corrects local errors.

One of the motivations to apply a PR algorithm to the voicing determination problem was the wish to get away from the conjunction of voicing determination and pitch determination.

> Methods for voiced-unvoiced (V-UV) decision usually work in conjunction with pitch analysis. For example, in the well known cepstral pitch detector (Noll, 1967), the V-UV decision is made on the basis of the amplitude of the largest peak in the cepstrum. There are two disadvantages in this approach to V-UV decision. First, the decision is based on a single feature - the degree of voice periodicity. Voiced speech is only approximately periodic; sudden changes in articulation and the idiosyncracies of vocal cord vibration can produce speech waveforms which are not periodic. In such cases, a feature such as the amplitude of the largest cepstral peak will fail to distinguish voiced speech from unvoiced. In practice, additional features, such as the rate of zero crossings of the speech waveform, the ratio of low to high-frequency energy, etc. must be included in the decision procedure. Second, the V-UV decision is tied to pitch detection which may be acceptable for speech synthesis applications. But, for other applications, such as speech segmentation or speech recognition, the linking of V-UV decision to pitch detection can result in unnecessary complexity as well as in poorer performance, particularly at the boundaries between voiced and unvoiced speech. For pitch detection, a large speech segment, 30-40 ms long, is necessary, which can result in unwarranted mixing of voiced and unvoiced speech. By separating the V-UV decision from pitch detection, it is possible to perform the V-UV decision on a much shorter speech segment, thereby enabling one to track fast changes in speech from one class to another. (Atal and Rabiner, 1976, p.201)

The algorithm was included in the comparative evaluation of PDAs by Rabiner et al. (1976) where it was incorporated into the PDA by Atal (unpublished, see Sect.8.3.3). It may seem ironic that just in this system where the VDA and the PDA are completely separated, the overall subjective performance of the PDA was underestimated because of a systematic failure of the VDA[9] (McGonegal et al., 1977). If there is any example to underline the necessity for separating voicing determination and pitch determination even in subjective evaluations, then it is given here.

_____

[9] During the training phase the VDA had been adapted to high-quality speech and was then applied to telephone speech with the reference data unchanged. The result was a large number of unvoiced-to-voiced errors.

The insight gained into this problem was used to modify and improve the algorithms based on the principle of pattern recognition. Rabiner, Schmidt, and Atal (1977) adapted the original algorithm to operate on telephone speech. They took a large number of parameters (up to 70) including LPC coefficients, cepstrum coefficients, correlation coefficients for delays greater than 1 samp-le, and the energy of the differenced signal. In a statistical evaluation of the relevance of the individual parameters for the voicing determination problem they selected only a small number of them for the final algorithm (the selected parameters were not identical to those used by Atal and Rabiner in the original algorithm). They found that in telephone speech the voiced-un-voiced discrimination - which is the important one for the following PDA - can be done almost as well as for good-quality speech (with a different set of reference patterns, of course), whereas the problem is to reliably separate between speech and silence, in particular between voiceless speech and silence.

Rabiner and Sambur (1977a,b) tried to improve the performance of the algorithm by changing the classification routine. The parameters of their algorithm were 1) the log energy of the signal, and 2) an LPC distance measure based on a proposal by Itakura (1975) which involves all the LPC coefficients and the correlation (or covariance) matrix for the current frame. This measure is thus representative for the spectral distance of the current frame to the reference data for each class.

Figure 5.24 shows the block diagram of the decision routine. It is interes-ting to note that, most of the time, the decision between speech and silence is based on the energy of the signal alone. For the rest of the routine the figure is self-explanatory. The algorithm was found to have an error rate of about 5%, the majority of errors occurring at the transitions between different excitation modes.

The pattern recognition approach to voicing determination on a purely parametric statistical basis has three disadvantages: 1) the dependence on the statistical assumptions on the reference data, in particular on the (normal) distribution; 2) the concentration of errors around the transitions; and 3) the complexity of the learning phase which needs a good deal of manual work. Feature 2 may be common to all pattern recognition problems in that the safety of the decision decreases when the pattern is located near an inter-class decision boundary, and may be common to all VDA in that the decision is safer for stationary segments than for transitions. As the algorithm by Rabiner and Sambur (1977b) shows, the first disadvantage becomes less relevant when the classifier is somewhat adapted to the particular problem and not purely based on statistical assumptions which might be far from reality. The problem of the learning phase remains; it requires a lot of hand labeling of data and does not readily provide a measure whether the collected data is already representative enough to terminate the learning phase.

Since the number of classes in the voicing determination problem is low, especially when the problem is reduced to the two-class task of voiced-un-

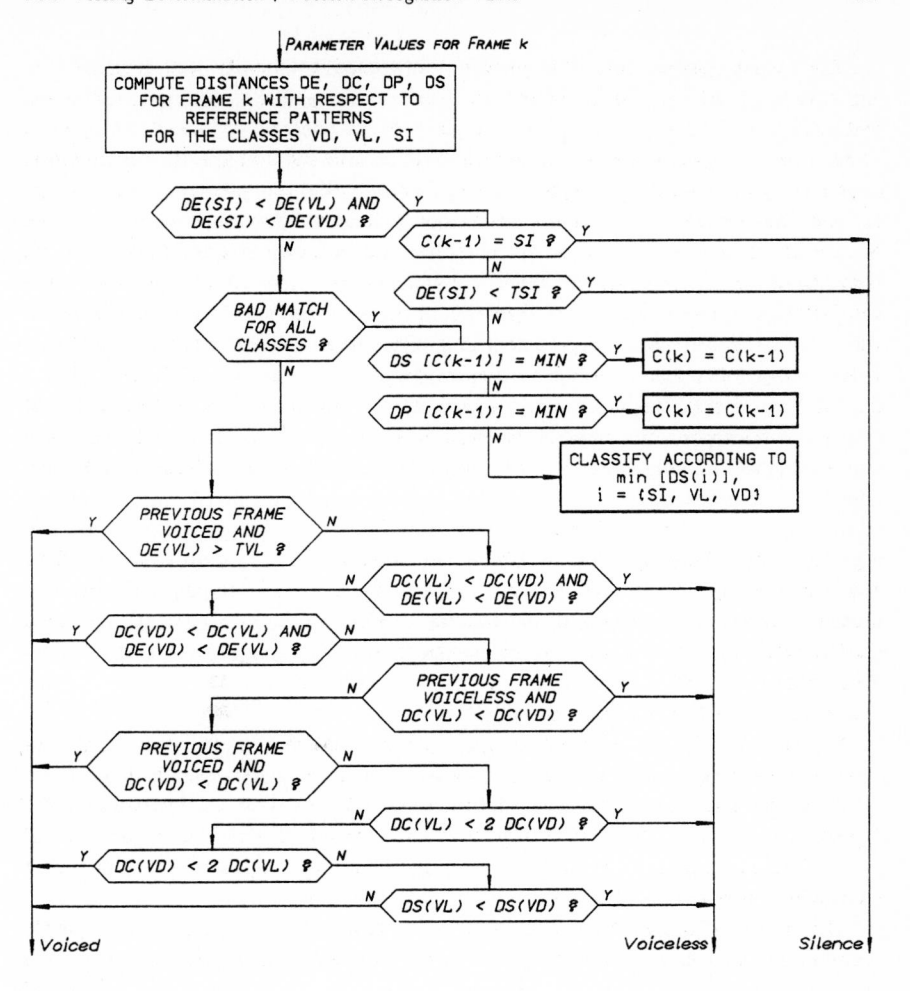

Fig.5.24. VDA by Rabiner and Sambur (1977b): block diagram of the decision routine. (VD) Voiced, (VL) voiceless, (SI) silence. (DE) Distance measure for log energy; (DC) distance measure for the LPC coefficients; (DS) sum of DC and DE; (DP) product of DC and DE; [C(k-1)] classification of previous frame (for k = 1 this value is set to "silence"). The energy distance DE has been normalized to the standard deviation of the reference distribution for each class. Hence, fixed thresholds can be used for a number of decisions (bad-match threshold, threshold TSI for silence, threshold TVL for the decision "definitely not voiceless")

voiced discrimination, a different learning strategy may prove simpler and may even give better results. The usual way learning is performed in pattern recognition is to define a set of input patterns which are as typical for the respective class as possible. In the recognition phase a distance measure of the unknown pattern to the reference template of each class is computed in order to find the minimum distance or the optimal match. The interclass boundary, whatever its shape in the multidimensional parameter space, is estimated or calculated according to the statistical characteristics of the classifier applied.

In contrast to this, it might be possible to know the shape and location of the interclass boundary as well as possible without the necessity of knowing to much about the distribution within the individual classes. The membership of the unknown pattern to one of the classes is then determined according to the side of the interclass boundary on which the pattern is found. The shape of the hypercurve which represents the interclass boundary is best estimated when patterns are used for the learning phase that are typical for transitions between adjacent classes, but still undoubtedly pertain to one of the classes (and not to both). This approach has the advantage that no assumption has to made with respect to the intraclass distribution of the parameters. The computational complexity of this approach increases exponentially with the number of classes, but is appropriate for this problem where the number of classes is low.

A VDA of this type was developed by Siegel (1977, 1979b). Starting with one speaker and a low number of parameters, the classifier is built up switching to and fro between the learning and the test mode and gradually increasing the number of speakers and parameters. In Siegel's algorithm

> ... the decision making process is viewed as a pattern classification problem. Training of the classifier is accomplished by defining a set of speakers from which frames of speech may be selected to characterize voiced and unvoiced segments, and a set of features which may be used in making the classification. In order to use as few frames as possible in the training and as few features as possible to make the V-UV decision, a procedure is developed in which the contributions of these two sources of information are interleaved. The notions of covering and satisfaction are defined in order to determine what information is needed to improve the performance of the classifier. The failure of a classifier to cover a set of speakers indicates that more training information from those speakers [i.e., addition-al frames] in necessary to define the classifier. The failure of a classifier to satisfy a set of speakers indicates that the performance of the classi-fier could be improved by the use of more features in making the V-UV decision. By achieving covering and satisfaction on successively larger sets of speakers, a classifier was obtained which performed with a misclassifi-cation rate of less than 0.5% on tests containing 8600 classifications over eleven speakers, including seven not used in the training. Of the 37 misclassifications made on the training and test sentences, only one audible misclassification was detected. (Siegel, 1979b, p.83)

The final classifier was established from a set of only 179 patterns from four speakers, with 2600 test classifications made during the training procedure in order to check covering and satisfaction. A similar algorithm was developed by Siegel and Bessey (1980, 1982) to develop a voiced-unvoiced-mixed excita-tion classification algorithm.

A pattern recognition VDA which was built without a substantial learning phase has been reported by Cox and Timothy (1979, 1980). It applies a rank ordering procedure. Rank ordering enables the investigator to compare the variances of two stochastic or nonstochastic processes without having to know the law of distribution which they follow. In the actual case such a procedure is applied to the speech signal which is subdivided into four subbands between 0.2 and 3.2 kHz. The classification is made according to the subband in which the largest deviation from the background noise is found.

5.3.4 Summary and Conclusions

As stated in the introduction to Sect.5.3, this review was intended to represent a collection of typical examples of VDAs rather than an exhaustive survey of the state of the art. The main conclusions from these examples can be drawn as follows.

1) The voicing determination problem can be formulated as the task of estimating two binary parameters: a) presence versus absence of a voiced excitation, and b) presence versus absence of a voiceless excitation. This leads to four modes of excitation: voiced excitation, voiceless excitation, mixed excitation, and silence. A VDA may investigate all four modes or confine its task to discrimination between three (voiced-voiceless-silence) or even two modes (voiced-unvoiced or speech-silence).

2) The problem of voicing determination is about as difficult as the pitch determination problem itself (Gold, 1977; Rabinar and Sambur, 1977). For both tasks the reason is the same: there is no alternative to these parameters once the particular application requires them to be determined.

3) Without making statement 2) invalid, the problem can have remarkably simple solutions for good-quality signals and known environmental conditions. In many of these algorithms an ad hoc decision routine based on a small number of parameters proves sufficient. For more adverse conditions, such as telephone speech, however, more complex VDAs, like the pattern recognition VDAs, are required.

4) Voicing determination is often seen and implemented in conjunction with pitch determination. Well understood and well done, such a conjunction can greatly reduce the computing effort of the speech analysis system as a whole (Reddy, 1966; N.Miller, 1974; Hess, 1976b). A crude VDA used as a prerequisite for the PDA can determine those segments of the signal which are definitely unvoiced and thus need not be processed by the PDA. A VDA based on periodicity or regularity criteria, on the other hand, does not really indicate voicing; it rather establishes a sufficient (but not necessary) criterion to segment the signal into parts that are definitely voiced, and the remainder which is not definitely voiced. This remainder can then consist of unvoiced segments, voiced segments with irregular excitation, and other voiced segments where the PDA has failed for some reason.

Voicing determination and pitch determination are the two components of the more comprehensive problem of voice source analysis. The two problems are strongly interwoven; nevertheless, they should be kept apart from each other and treated as two separate tasks.

6. Time-Domain Pitch Determination

Time-domain pitch determination is the oldest means of automatic pitch measurement. Figure 6.1 shows the block diagram of a typical time-domain PDA with the three "classical" blocks preprocessor, basic extractor, and postprocessor, as described by McKinney (1965). The main property of the basic extractor is the ability to process every pitch period individually. This is an advantage in accuracy since the PDA can handle rapid changes in F_0 or even slightly irregular signals; it is a disadvantage on the other hand since the PDA may have difficulties when individual periods are disturbed by noise or spurious signals.

The output of the basic extractor is usually a sequence of pulses (*markers*) which serve to mark a defined significant point of every individual period. This can be, for instance, the largest maximum or minimum, or the zero crossing between the principal extremes. Since at the input of the basic extractor there is a signal which is directly derived from the speech signal (the input signal of the PDA), the task of the preprocessor is that of data reduction by signal filtering, in order to facilitate the job of the basic extractor. The tasks of the postprocessor are more oriented towards the respective application(s). In older devices this was mainly a conversion from time (marker distance, i.e., elapsed time between successive periods) to frequency, i.e. to a voltage proportional to F_0. At that time considerable effort was put into analog postprocessors which nowadays are merely of historical interest. The development of computer algorithms and - in more recent time - of low-cost complex digital hardware has mostly affected ‹the postprocessor, has degraded the time-to-frequency conversion to a simple

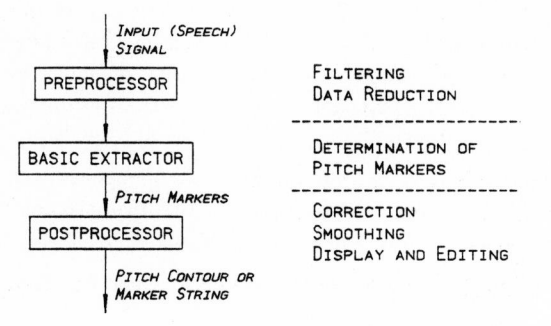

Fig.6.1. Block diagram of a time-domain PDA

numerical division, and has opened the way for more complex tasks such as detection and correction of errors, indication of regularity, or smoothing and editing of the obtained contour.

What are the significant features from which periodicity can be derived in the time domain? For any periodic or nearly periodic signal, we may observe

1) the presence of a *fundamental harmonic*, i.e., a first partial whose period equals the fundamental period T_0; and

2) a *structural pattern* which repeats exactly or at least approximately from period to period.

From the linear model of speech production we know that the voiced excitation signal can be represented by a pulse generator followed by an approximately time-invariant linear filter. This assumption yields two additional features from which T_0 can be derived:

3) Since the vocal tract can be assumed to be a linear passive system (Fant, 1960), its impulse response consists of the sum of exponentially decaying sinusoids. Hence we will find *high amplitudes* at the *beginning* of each period and low amplitudes toward the end.

4) If a linear system is excited by a pulse train, the resulting signal may exhibit *discontinuities* at those instants where the individual pulses occur. If this does not hold for the signal itself, it will at least hold for the first- or

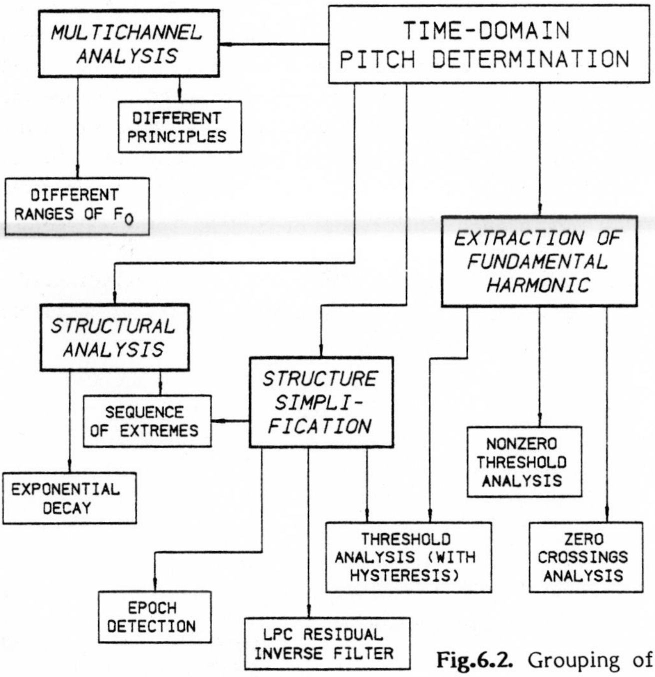

Fig.6.2. Grouping of time-domain PDAs

a higher-order time derivative of the signal. For speech, it has been shown to apply to the second derivative (Ananthapadmanabha and Yegnanarayana, 1975).

Regarding these aspects, the time-domain PDAs have been grouped accor- ding to the representation in Figure 6.2. *Fundamental-harmonic extraction* and *structural analysis* are antagonistic principles. In the former case, the whole burden of data reduction is placed on the preprocessor, and the basic extractor is comparatively simple. In the latter principle, the basic extractor operates directly on the speech signal. If a preprocessor is necessary at all, it will be reduced to a simple low-pass filter. The two principles are also antagonistic with respect to linearity: fundamental-harmonic extraction may be performed by a linear system, whereas structural analysis is essentially a nonlinear decision and discrimination process.

The third way, i.e., *simplification* and *transformation* of the *temporal structure*, takes on an intermediate position. The burden of data reduction is distributed between the preprocessor and the basic extractor.

In this chapter we will first deal with those devices which apply funda- mental-harmonic extraction. The structural-analysis algorithms will be treated next. Section 6.3 will cover the algorithms which apply temporal-structure simplification and transformation. The multichannel algorithms will follow in Section 6.4. The review of the algorithms is completed in Sect.6.5 by discus- sing those methods which perform special tasks, such as determining the point of glottal closure or reconstructing the excitation signal. Section 6.6 deals with the particular tasks of the postprocessor.

6.1 Pitch Determination by Fundamental-Harmonic Extraction

Fundamental-harmonic detection is the classical technique for time-domain pitch determination using analog technology. An elaborate linear or nonlinear filter, which forms the preprocessor, is followed by a comparatively simple basic extractor. There are three main types of simple basic extractors (Fig.6.3): the zero crossings analysis basic extractor (ZXABE), the nonzero threshold analysis basic extractor (TABE), and finally the TABE with hyster- esis (two-threshold basic extractor). It is obvious that the ZXABE, as the simplest device, needs the highest filtering effort.

A different categorization can be carried out when the preprocessor is regarded. Here we must distinguish between those PDAs which apply nonlinear distortion in the preprocessor and those PDAs which operate on a linear basis alone. This leads to the grouping listed in Table 6.1.

The problems of fundamental-harmonic extraction PDAs have been discussed in great detail in the surveys by McKinney (1965) and by Baronin (1974a). When McKinney's book came out in 1965, its appearance just coincided with the advance of the computer in the domain of signal processing. Most analog PDDs, especially those which apply fundamental-harmonic extraction, had

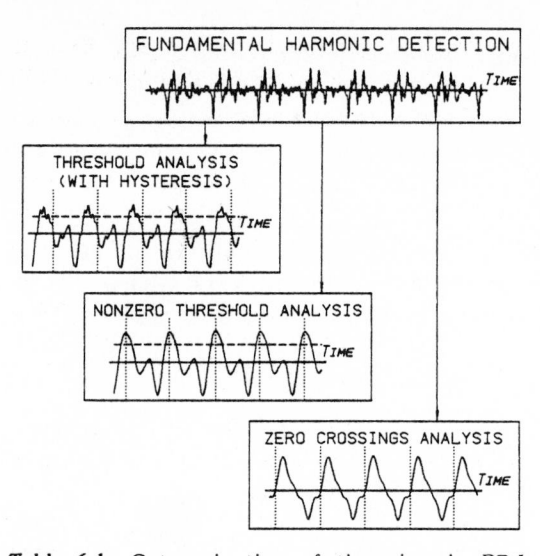

Fig.6.3. Grouping of time-domain algorithms that apply the principle of fundamental-harmonic determination, according to their basic extractor

Table 6.1. Categorization of time-domain PDAs which apply the principle of fundamental-harmonic determination

Label[a]	Description
PiTL	Linear device: detection of the fundamental frequency or waveform in the signal. The fundamental is necessarily enhanced by a linear filter, such as a low-pass filter or a chain of integrators
PiTN	Nonlinear device: detection of the fundamental frequency or waveform in the signal after its enhancement by a nonlinear procedure or function

[a] Throughout this book, labels are introduced to briefly characterize each type of PDA or PDD. In these labels, Pi stands for "pitch determination algorithm" and T for "time-domain." The consecutive letter stands for the particular method; in the case of Table 6.1 we find L for linear preprocessing, and N for nonlinear processing. Further labels will be explained in later sections.

already been developed at that time. McKinney's survey of these PDAs is thus practically complete; few analog systems with principally new ideas have been developed thereafter. Baronin's report adds important details on the design of such PDDs and, in particular, on their performance in a vocoder.

The technique of fundamental-harmonic extraction in the time domain requires the first partial to be present in the waveform. This restricts the application of these PDAs to those cases where the signal is not band limited unless nonlinear preprocessing reconstructs the first partial. In addition to this drawback, the PDAs using this principle are in general sensitive to low-frequency signal distortions. In a system where the environmental situation is unpredictable, these PDAs have been replaced by more robust (but at the same time more complex) devices. For applications where highly accurate

results are not required, and where the environmental conditions are predictable (as they occur especially in education, but also in phonetics and pathology), this principle is highly attractive due to its simplicity.

6.1.1 The Basic Extractor

For fundamental-harmonic extraction three main types of basic extractors are
used: 1) the *zero-crossings analysis basic extractor* (ZXABE), 2) the (nonzero)
threshold analysis basic extractor (TABE), and 3) the *TABE with hysteresis*.
The principle of operation of these basic extractors is shown in Figure 6.4.
Their output is usually a train of pulses (markers) of constant shape and
amplitude. The ZXABE generates an individual impulse as soon as the zero
axis is crossed with a definite polarity. For the TABE, the same is valid
when its nonzero threshold is crossed. A variation is the TABE with hysteresis;
it generates an impulse after the signal has crossed two different thresholds
(one of which might be zero) in a defined sequence. This basic extractor has
the advantage that the upper threshold as well as the lower threshold can be
crossed individually without generating a marker. An impulse is generated if
and only if both lines have been crossed successively.

What behavior is required from the signal and the preprocessor in order to
drive these basic extractors correctly? For the ZXABE, the simplest realization
of a basic extractor, the requirements must be rather hard to meet since
many investigators report that they have difficulties with this device. To
answer this question a lower bound for the relative amplitude of the first
harmonic has to be found which guarantees that the corresponding waveform
has two and only two zero crossings per period.

Detailed discussion of this problem is found in McKinney (1965) and Erb
(1972a,b). After McKinney, a necessary and sufficient condition to guarantee

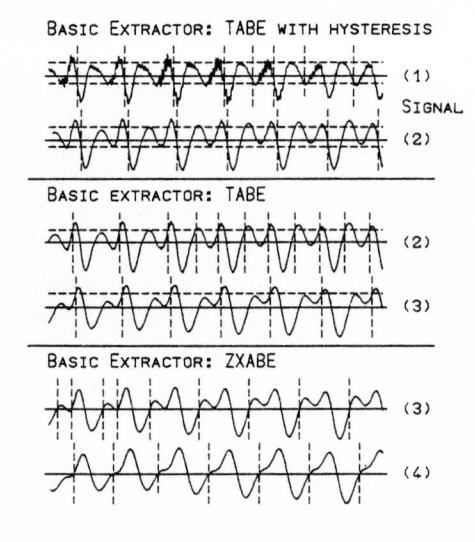

Fig.6.4. Performance of simple basic extractors. Sample signal: /eː/ spoken by a female speaker. For each extractor, normal operation (upper line) and failure (lower line) are displayed. Signals: (1) original; (2) after a single integrator filter, i.e., attenuated by 6 dB/octave; (3) attenuated by 12 dB/octave; (4) attenuated by 18 dB/octave

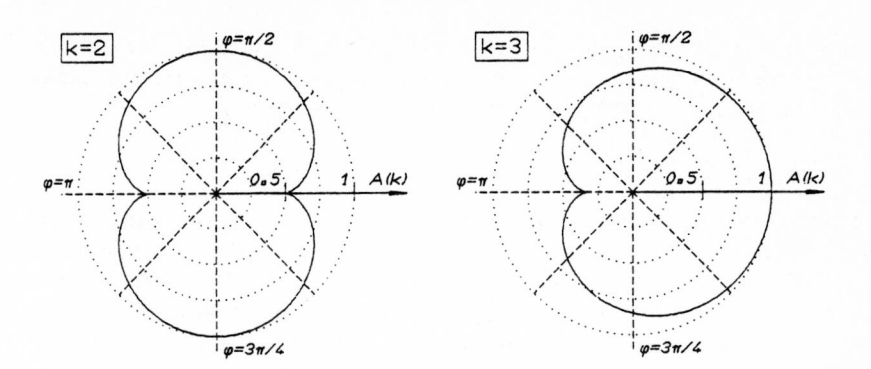

Fig.6.5. Operating range of a zero crossings basic extractor (ZXABE) for two sample signals that consist of a first and its second and third harmonics, respectively, depending on the phase of the higher harmonic. The amplitude of the lower harmonic is assumed to one, its phase to zero. The figure shows the number of zero crossings of the function $x(n) = \sin bn + a_k \sin (kbn+\phi_k)$, depending on the amplitude a_k and the phase ϕ_k of the higher harmonic. The case of the second and third harmonic

two and only two zero crossings per period regardless of the phase of the individual harmonics is given by the following condition:

$$\text{RS1} := 20 \log \frac{A(1)}{\displaystyle\sum_{k=2}^{M} k \, A(k)} > 0 \ (\text{dB}) \ . \tag{6.1}$$

In this formula, RS1 is the ratio (in dB) of the amplitude $A(1)$ of the first harmonic and the arithmetic sum of the amplitudes $A(m)$ of the relevant higher harmonics, weighted by their respective harmonic numbers. In this definition, the amplitude of the component at zero frequency is supposed to be zero. If condition (6.1) is violated, there may be still two zero crossings per period for certain phase relations between the first harmonic and the higher ones. Examples for the second and third harmonic are given in Figure 6.5. It can easily be seen that condition (6.1) really represents the worst-case condition. Assuming that all harmonics are given in sinewave representation,

$$a_k(n) = A(k) \sin (kn+\phi_k) \ ,$$

the worst case is given when all the higher harmonics take on a phase opposite to that of the first harmonic. In this case, the derivative of the signal at the zero crossing of the first harmonic (from negative to positive) is given by

$$a_k'(n=0) = A(1) - \sum_{k=2}^{M} k\, A(k) \qquad\qquad (6.2)$$

with a phase of zero assumed for the first harmonic. This value remains positive as long as the sum at the right-hand side of the equation is less than the amplitude $A(1)$ of the first harmonic, i.e., as long as condition (6.1) is fulfilled. If the derivative $a'(0)$ takes on a value less than zero, however, the single zero crossing at this point is replaced by at least three zero crossings, and the ZXABE cannot operate any more (see Fig.6.6).

For the TABE which applies a nonzero threshold a condition similar to (6.1) for the presence of two and only two threshold crossings per period is not easy to formulate since the amplitude of the threshold (or the position of the threshold relative to the maximum amplitude of the signal) represents an additional degree of freedom. Experiments show that the low-pass preprocessor filter can attenuate the higher harmonics by about 6 dB/octave less than necessary to drive a ZXABE (see Fig.6.4). On the other hand, there is the problem of normalizing the threshold with respect to the signal amplitude (or vice versa). This can be solved, however, using a dynamic compressor in the preprocessor or adapting the threshold to the short-term signal amplitude.

Due to the second threshold, the TABE with hysteresis offers an additional degree of safety with respect to the influence of higher harmonics. The simplest device of this kind is the Schmitt trigger which is frequently applied [for instance, by Martin (Léon and Martin, 1969), or by Fedders and Schultz-Coulon (1975)]. The application of such a basic extractor, if the two thresholds are well apart, may enable the designer to omit another 6 dB/octave high-

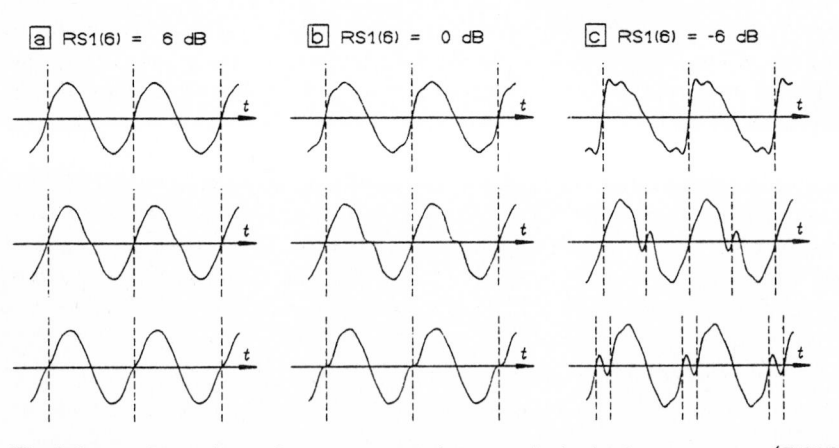

Fig.6.6a–c. Operation of a zero crossings analysis basic extractor (ZXABE). **(a)** Condition (6.1) satisfied; **(b)** condition (6.1) just satisfied; **(c)** condition (6.1) violated. Signal: $x(n) = \sin(an) + \Sigma\, f/k \sin(kn+\phi)$, $k = 2\,(1)\,6$. The factor f is used to control the higher harmonics as indicated in the figure. (Upper line) All partials in phase ($ = 0$); (center line) the odd harmonics are shifted by 180°; (lower line) phase shift by 180° for all the higher harmonics

frequency attenuation in the preprocessor filter. The question of amplitude normalization is the same for both the TABE and the TABE with hysteresis.

The three basic extractors discussed so far can be characterized as *event detectors*. When they detect the significant event they have been designed for, they generate a marker. Whatever the actual representation of the marker is (the occurrence of an impulse in an analog realization, changing the contents of a data register, or setting a binary variable in a digital or algorithmic implementation), the most important feature is the time of its occurrence. If the basic extractor operates correctly, the elapsed time between two consecutive markers represents a pitch period, and the position of each marker has a defined temporal relation to the pertinent glottal pulse.

Up to this point, the discussion in this section around condition (6.1) for the ZXABE and equivalent conditions for the TABEs has only dealt with the question of gross errors, i.e., whether or not the basic extractor is able to generate one and only one marker per period. There is also the problem of fine errors, or, positively speaking, the problem of exactly where to place the marker once the significant event has occurred. For the ZXABE it is quite straightforward to place the marker at the point of the zero crossing. For the TABE the question cannot be answered as straightforwardly since the threshold is asymmetric with respect to the signal, and one may obtain different results depending on whether the marker is set at the instant where the threshold is exceeded or at the instant where the signal returns below the threshold. The same applies to the TABE with hysteresis.

The signal at the input of the basic extractor has been derived directly from the speech signal. If we regard the marker sequence at the output as a signal as well, and if the marker sequence has been found by a (zero or nonzero) threshold analysis event detector, then this process is clearly non-linear.[1] The basic extractor thus generally consists of a nonlinear circuit or routine. Its operation is always connected with data reduction, in this special case with the loss of all information other than that on periodicity so that the nonlinear process is irreversible. This means that systematic errors may remain undiscovered if they are not detectable in the postprocessor.

Besides the gross errors which we can regard as being under control (provided the preprocessor fulfills the necessary requirements), there are some systematic fine errors which may arise in the basic extractor due to the fact that the input signal is time variant. Insufficient preprocessing in connection with a ZXABE, even if (6.1) is fulfilled, can cause the position of the zero crossings to be slightly shifted due to the influence of the higher harmonics. This shift, which depends on the mutual phase relations between the harmo-

---

[1] The only possibility of obtaining the sequence of markers by purely linear means would be exact inverse filtering, i.e., applying a filter whose transfer function is the reciprocal of the vocal-tract transfer function and the radiation and voice source transfer functions. For various reasons such a PDA can hardly be realized.

nics, can bring about measurement inaccuracies which render the obtained pitch contour noisy. The worst case would apply if the phase of the first and the higher harmonics were uncorrelated (Niedźwiecki and Mikiel, 1976). To force these inaccuracies below a reasonable limit, say 1%, even more severe low-pass filtering would be necessary than is specified in (6.1) (Gubrynowicz, 1972, as cited by Niedźwiecki and Mikiel, 1976). Fortunately, in speech the first and higher harmonics are phase correlated so that the problem may be not too severe. Nevertheless, these effects should not be neglected; as the display in Fig.6.4 (lowest line) shows, this error is still visible, even after a preprocessor filter with a slope of -18 dB/octave.

Another systematic error, this time for the TABE, arises from the amplitude sensitivity of the nonzero threshold analysis. If the amplitude of the input signal changes rapidly, the instant of the threshold crossing varies with respect to the location of the subsequent peak or the previous zero crossing. If no information on the amplitude change is preserved after the basic extractor, it looks as if the TABE has converted the amplitude modulation into a frequency modulation at the output (see Fig.6.7). A comparison of the data in the figure with the difference limen specified in Sect.4.2.3 shows that the error is clearly audible. To compensate for this, one could take the average of the time values for the upward and downward threshold crossings. This works well with a TABE; using a TABE with hysteresis, i.e., with two thresholds, both thresholds are influenced in the same direction if they have opposite polarity. If therefore a Schmitt trigger, for instance, is applied as a basic extractor and a signal-silence threshold discriminator at the same time, this error will be intrinsic to the pertinent PDA.

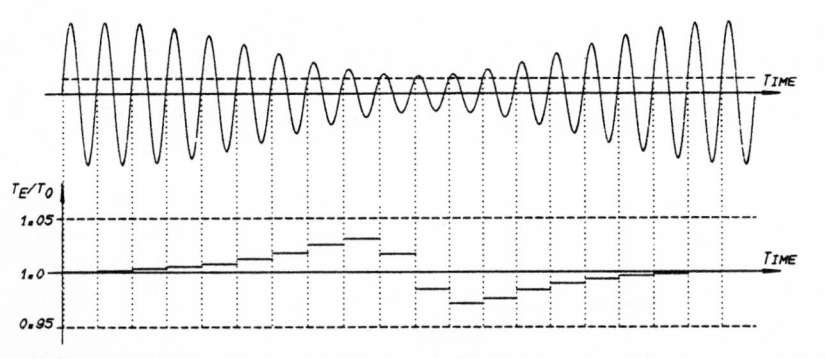

Fig.6.7. Performance of a TABE for signals with varying amplitude. The signal (———) in this example is a sinusoid with the period T_0 whose amplitude varies by a factor of 4 (i.e., 12 dB). The scale used corresponds to a 12-dB amplitude change within 30 ms when a fundamental frequency of 200 Hz is assumed. In speech the amplitude changes can be even more drastic. The threshold of the TABE (- - - - -) is assumed to be situated at 20% of the maximum amplitude (i.e., at 80% of the minimum amplitude). A marker (·····) is assumed to be set whenever the threshold is crossed in positive direction. (T_E) Estimate of the PDD

This simple example shows that the question of the correct marker location is a nontrivial one. In various experiments in connection with a TABE
with hysteresis, the present author (Hess, 1972b, 1976a,b) found that the first
zero crossing of the signal (the axis is crossed toward negative values if the
original polarity of the speech signal has been maintained) after the occurrence
of the glottal pulse represents a good anchor point where the markers can be
positioned, although this point is definitely not identical to the instant of
glottal closure (see Sect.7.1 for more details).

Another error intrinsic to all PDAs using fundamental-harmonic extraction
arises from the time-variant phase response of the vocal-tract transfer function at the frequency F_0. For a male voice this effect is small since the first
harmonic is usually well below the first formant. For a female voice, however,
interaction between the first harmonic and the formant F1 can result in phase
shifts of the first harmonic of as much as 90°. During a formant transition
the time-variant phase distortion of the first harmonic is converted into a
frequency modulation of the corresponding marker sequence (see the example
in Fig.6.8).

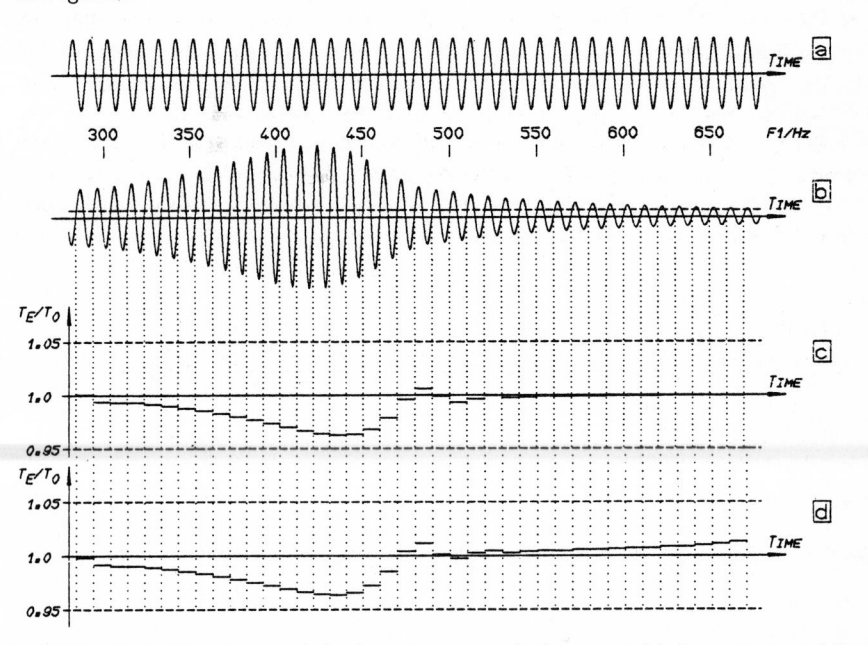

Fig.6.8a–d. Performance of basic extractors during a rapid formant transition.
(a) Input signal; the figure shows a sinusoid with $T_0 = 2.5$ ms ($F_0 = 400$ Hz).
(b) Same signal at the output of a second-order time-variant filter which
simulates a formant transition from 300 to 650 Hz in 90 ms (corresponding to
a glide-vowel transition or a diphthong /ia/ in real speech); (c) T_0 estimate
for a ZXABE; (d) T_0 estimate for a TABE whose threshold is set to 10% of
the maximum amplitude of the signal. (T_E) Estimate of the PDD. The bandwidth of the formant is assumed to be 50 Hz

Whether these small systematic errors can be tolerated, depends on the respective application. In principle, they are audible; it is, however, unlikely that they have much influence on the naturalness of vocoded speech. Admittedly, these effects are small, and filter and quantization noise due to sampling in digital systems may more strongly affect the results. Nevertheless, for high-accuracy measurements in phonetics and linguistics (for instance in investigations of microprosody), these PDAs should be used with some caution.

As in any time-domain PDA, the basic extractors discussed so far are able to track the signal period by period. In many applications this feature is not required; instead, a (continuous) indication of the momentary fundamental frequency or period (corresponding to the output of a short-term analysis PDA) is wanted. This is obtained by an interval-to-voltage converter (McKinney, 1965). The output voltage of this circuit, which is discussed in more detail in Sect.6.6.1, can be proportional to T_0 or F_0 or $\log F_0$. If the voltage is proportional to F_0, the combination of basic extractor and interval-to-voltage converter is often called a *frequency meter* in the literature (McKinney, 1965; Léon and Martin, 1969; Baronin, 1974a). To avoid confusion, however, we will not use the term *frequency meter* in this book in connection with a basic extractor which works in the time domain. In a number of PDDs, mostly in early realizations, the basic extractor has been incorporated into the interval-to-voltage converter so that the interval-to-voltage converter is directly driven by the input signal. These PDDs (labeled PDDs *with basic extractor omitted*) will be discussed in Section 6.6.2.

In summary, the main types of basic extractors used for fundamental-harmonic detection are threshold analyzers with one or two zero or nonzero thresholds. They act as unipolar event detectors; i.e., they generate a marker whenever the signal crosses the given threshold in a given direction. Figuring as event detectors, these basic extractors are able to operate in a period-synchronous way.

6.1.2 The Simplest PDD - Low-Pass Filter and Zero (or Threshold) Crossings Analysis Basic Extractor

Most of the early circuits use a simple basic extractor, i.e., a ZXABE or a TABE. This circuit requires a preprocessor which has to perform a very elaborate prefiltering. The preprocessor of this type of PDD

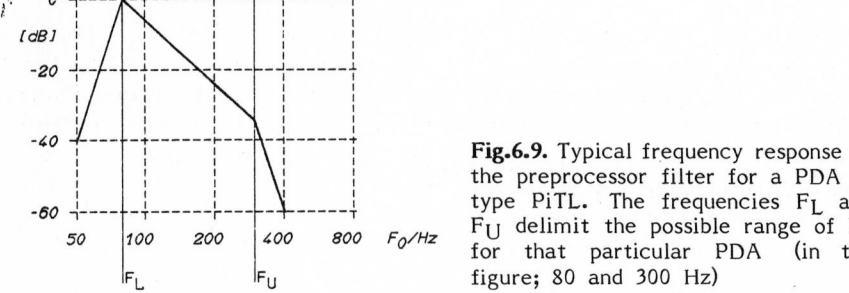

Fig.6.9. Typical frequency response of the preprocessor filter for a PDA of type PiTL. The frequencies F_L and F_U delimit the possible range of F_0 for that particular PDA (in the figure; 80 and 300 Hz)

... includes a spectrum weighting filter whose purpose is to attenuate higher-harmonic components of the signal more than the fundamental component. This is to ensure that the assumption of two zero crossings per laryngeal period may be fulfilled much of the time. A nonlinear transformation is often an essential part of the preprocessing.

The way in which the spectrum weighting filter functions in the circuit may be described in terms of ... three component operations (Fig.6.9): 1) sharply increasing attenuation above some upper cutoff frequency (e.g., 300 Hz) to attenuate harmonic components with a frequency higher than the cutoff frequency; 2) slowly increasing attenuation over the range ... to be measured, and 3) rapidly decreasing attenuation up to some lower cutoff frequency (e.g., 80 Hz) to reduce the effects of power line frequency noise, room noise, aspiration puffs, etc. (McKinney, 1965, pp.46-47)

Many PDDs simply consist of a low-pass filter followed by a ZXABE. Some of these PDDs, as already mentioned by McKinney, are additionally supplied with a high-pass filter with a cutoff frequency well below the lower limit of the measuring range in order to remove unwanted low-frequency components. They depend on the presence of the first harmonic in the speech signal at an adequate level, and upon being able to filter it out and to measure it without a nonlinear transformation in the preprocessor (Obata and Kobayashi, 1937, 1938; Dempsey et al., 1950, 1953; Munson and Montgomery, 1950; Riesz, 1952; Acoustics Lab. MIT, 1955, 1957; David, 1956; Plant, 1959; Glass, 1960; Stead and Jones, 1960; Righini, 1963; Dukiewicz, 1965; Pisani, 1971; Erb, 1972a,b, 1974a,b; Lacroix, 1976; Lacroix and Hoptner, 1977; Simon, 1977; and others). However, there exists considerable motivation for enhancing the relative level of the fundamental component by nonlinear means: difficulties due to hum and low-frequency rumble (Stead and Jones, 1960), attenuation of the fundamental harmonic by a recording equipment with poor low-frequency response (R.L.Miller, 1953), and so on. The most severe restrictions of these PDDs are that 1) the first harmonic must be present in the signal, and 2) a considerable degree of low-pass filtering has to be carried out in order to guarantee the occurrence of two and only two zero crossings per period. This holds for the ZXABE. For the TABE with a nonzero threshold, the problem is not as serious, especially when the circuit operates with a hysteresis (Fig.6.4). However, most nonzero TABEs require an adaptive threshold or some kind of automatic dynamic range compression, which again complicates the task.

As stated in the previous section, the relatively hard condition (6.1) for the ZXABE can only be met by extensive low-pass filtering. This is fairly easy when the PDD only covers a narrow measuring range for F_0, such as one or two octaves. In this case, one has to cope with a rather low number of harmonics, and a low-pass filter with an attenuation of 6 to 12 dB/octave within the measuring range of the PDD will be sufficient. (The attenuation outside this range may be arbitrarily high.) Of course, the attenuation of the filter above this range must be much higher in order to guarantee that no harmonic above the frequency range under consideration is included in the signal. Flanagan (1955), for instance, mentioned 300 Hz for the cutoff frequency of the low-pass filter (which would mean that the range of pitch to be

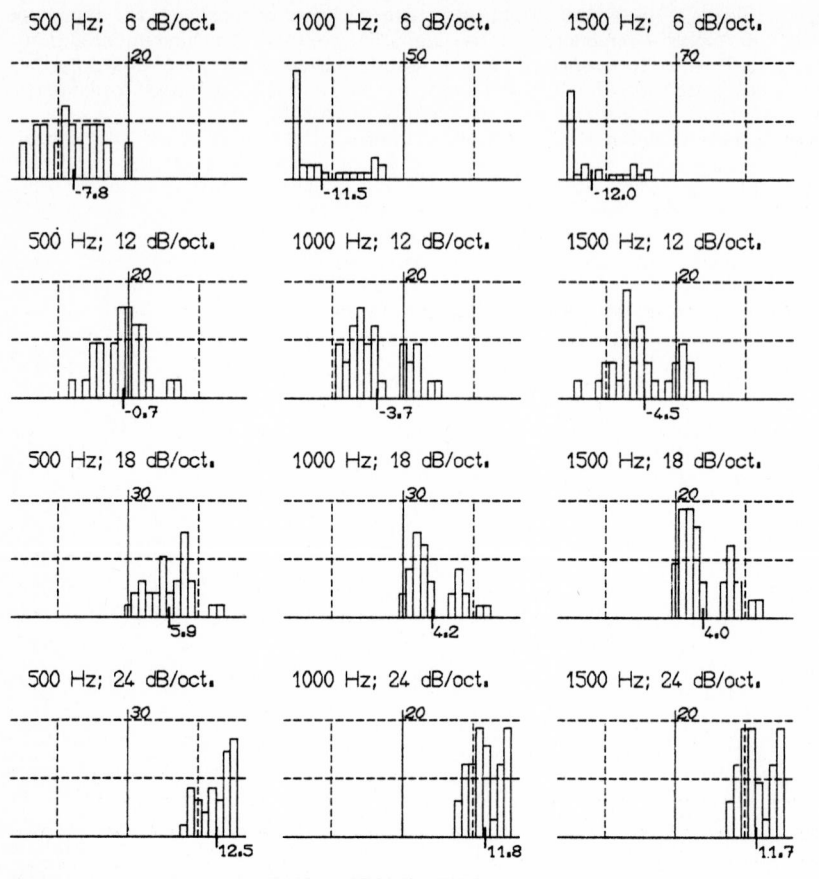

Fig.6.10. Relative amplitude RS1(6K) of first harmonic, according to definitions (6.3,4). The cutoff frequency of the low-pass filter and the attenuation within the measuring range of F_0 are given at each figure. Abscissa: value of RS1 (-15 to +15 dB); ordinate: percentage of occurrence; number below each figure: average of RS1 (averaged over the dB scale). The ZXABE will always operate correctly (i.e., without gross errors) when RS1 takes on a value above 0 dB

measured ends up at about 250 Hz) and used a filter with an attenuation of 12 dB/octave within the range of 100 to 300 Hz. Flanagan himself characterized this device as "considerably limited in accuracy." Plant (1959) used a filter with a -18 dB/octave asymptote.

If the PDD has to cover a wider range, more harmonics are involved, and condition (6.1) is more difficult to fulfil. Low-pass filtering by a single integrator (-6 dB/octave) would change the condition to

$$RSI(6) = 20 \log \frac{A(1)}{\displaystyle\sum_{k=2}^{M} A(k)} > 0 \text{ (dB)} \qquad\qquad (6.3)$$

Consequently, a cascade of K integrators would give

$$RSI(6K) = 20 \log \frac{A(1)}{\displaystyle\sum_{k=2}^{M} k^{-(K-1)} A(k)} > 0 \text{ (dB)} . \qquad\qquad (6.4)$$

Figure 6.10 shows the value $RSI(6K)$, $K = 1 (1) 4$, corresponding to integrator filters with an asymptote of -6, -12, -18, and -24 dB/octave, and for different cutoff frequencies (1200, 800, and 400 Hz, respectively). The values of RSI were computed from the sample pitch periods used for the experiment to be reported in Chap.7 (2 male speakers with F_0 around 150 Hz; the female speaker was omitted from this consideration). The figure displays that, for the given samples of speech, a frequency response with an attenuation of at least 18 dB/octave is necessary to guarantee successful performance of the PDD over a measuring range from 80 to 350 Hz. (Note that the upper cutoff frequency of 1000 Hz in the figure in connection with the given samples corresponds to the mentioned range with a speech sample having its fundamental frequency at the lower cutoff frequency.) A large high-frequency attenuation (say, 18 dB/octave), necessary to cover a wider measuring range, on the other hand, limits the measuring range again, this time because of the dynamic-range problem. Even a very elaborate preamplifier can only process a limited dynamic range. An attenuation of 18 dB/octave, for instance, means that over a range of 3 octaves (i.e., 50 to 400 Hz) a dynamic range of 54 dB has to be covered due to fundamental-frequency variations alone, in addition to the intrinsic dynamic variations of the speech signal which may exceed 30 dB. Apart from the fact that such a PDD would be extremely sensitive to hum and low-frequency spurious signals, this range is beyond the scope of a simple realization. Hence, this principle does not seem straightforwardly applicable to a device which has to process wide variations of F_0. If the respective application permits a smaller range of F_0 or the subdivision of the covered range into narrower subranges, however, this principle can be successfully used. Baronin (1974), with reference to earlier investigations, wrote on behalf of this problem:

> The ratio $H_L(F_0)/H_L(2F_0)$ [$H_L(f)$ is the transfer function of the low-pass filter in the preprocessor] must be sufficiently high over the whole measuring range of F_0. The higher this ratio, the lower the probability that the result is influenced by the second harmonic. If the ratio is very high, however, the amplitude of the first harmonic at the filter output for high values of F_0 could become comparable to the level of the low-frequency noise components which are amplified by the filter since under these conditions the speech signal is strongly attenuated. The PDA starts committing errors when dealing with high-pitched voices. On the other hand, if the ratio $H_L(F_0)/H_L(2F_0)$ is decreased, the error rate increases for low-

pitched voices because of the increasing influence of the higher harmonics. (Baronin, 1974a, pp.116-117; original in Russian, cf. Appendix B)

The question of the slope of the preprocessor filter transfer function within the measuring range of the PDD was experimentally investigated by Baronin (1964). In one of those experiments Baronin used low-pass filters with an attenuation of 3, 6, 9, and 12 dB/octave and a sharp cutoff above the measuring range (80-480 Hz) in connection with a ZXABE. For 6 male and 5 female speakers, the overall optimum was obtained with a slope of 9 dB/octave; in this case the error rate was 2.4% for the male voices (mostly higher-harmonic tracking) and 5% for the female voices (mostly due to the dynamic-range problem). For the male voices alone (but with the measuring range unchanged) the optimal slope was 12, for the female voices 6 dB/octave, with error rates of 1.2% and 2.6%, respectively.

Compared to condition (6.1) and the results displayed in Fig.6.10, these results suggest that the phase relations between the first and the higher harmonics are such that they usually do not represent the worst case.

Another way out of the problem with the dynamic range is the use of time-variant elements in the low-pass filter. A number of researchers applied dynamic compressors (McKinney, 1965). Winckel (1964) realized the low-pass filter in his PDD in three stages. The first filter stage removes all frequencies above 400 Hz. The second and third stage are low-pass filters with a 6-dB-per-octave attenuation within the measuring range. A dynamic compressor which reduces the dynamic range by 20 dB is switched between the second and third stage. More recent PDDs which apply this principle (Teston, 1972; Niedźwiecki and Mikiel, 1976) use filters with variable cutoff frequencies. These PDDs are discussed in Section 6.1.4. Instead of one filter with variable cutoff frequency, a bank of low-pass filters with fixed or variable cutoff frequencies is frequently used (Peterson and Peterson, 1968; Léon and Martin, 1969; Dibbern, 1972; Kubzdela, 1976, and others). In this book, such approaches are counted among the multichannel PDDs and will be discussed in Section 6.4.1.

6.1.3 Enhancement of the First Harmonic by Nonlinear Means

In a periodic signal which contains many harmonics, like the speech signal, the first harmonic can be enhanced by nonlinear distortion. The first harmonic is emphasized as the sum of the difference tones of adjacent higher harmonics. Moreover, any nonlinear distortion flattens the spectrum of the signal so that the formant frequencies lose part of their predominance. Even one of the earliest PDDs, the one by Hunt (1935), applied nonlinear distortion. A preprocessor circuit consisting of a nonlinear amplifier (further specifications are missing) was followed by a low-pass filter:

> ... a filter cutoff frequency of 270 Hz is satisfactory for recording the intonation of almost all speech sounds for both male and female voices. With this compromise selection of the filter cutoff it is found that the 'oo' sound (as in loose) gives some difficulty frequently recording at double

frequency, since the strong second harmonic usually falls within the pass-band of the filter with male voices. (Hunt, 1935, as given in McKinney, 1965; p.50)

Since that time, various forms of nonlinear transformation have been proposed and tried in preprocessors for time-domain PDDs of this type. McKinney (1965) settled upon the full-wave rectifier:

The full-wave rectifier, whose output time function is the absolute value of the input function, offers an easy way to realize a simple nonlinear transfer function. This factor, together with its apparent beneficial effect on the signal, has made the use of full-wave rectifiers in preprocessors popular from the beginning. (McKinney, 1965, pp.56-57).

Full-wave rectification was applied, for example, by Riesz (1939), Dudley (1941), Stockhoff (1959), Risberg (1960a,b, 1961a), Carré et al. (1963), or by David (1956) who found that this nonlinear distortion avoids, to some degree, the difficulty of a predominant second harmonic on low frequency voices. Lawrence et al. (1956) also used full-wave rectification:

Distortion by full-wave rectification was akin to concentrating the instantaneous energy at the instants of high larynx excitation into pulses at the larynx repetition frequency. (Lawrence et al., 1956, as given in McKinney, 1965; p.57)

Full-wave rectification belongs to the group of even nonlinear functions. The other important representative of this group applied (not only) in the earlier systems is squaring, as implemented, for instance, by Grützmacher and Lottermoser (1937, 1938a,b, 1940), by Obata and Kobayashi (1940; in an improved version of their original circuit), or by Kühlwetter[2] (1981).

There remains the necessity of attenuating the higher harmonics, even after nonlinear distortion. Moreover, the distortion itself requires some more linear circuitry in the preprocessor. Two linear components definitely have to follow the nonlinear element. The first one is the inevitable low-pass filter, this time, however, with only a moderate degree of attenuation over the measurement range of T_0 (since the amplitude of the first harmonic is expected to be higher now). The other component is a high-pass filter to remove the component at zero frequency (Fig.6.11). This high-pass filter, however, has to be laid out in such a way that it does not attenuate the first harmonic.

There is some discussion in the literature as to the necessity of a filter before the nonlinear circuit. Probably in order to avoid interference of the reconstructed first harmonic with the one originally present in the signal, some approaches use a narrow band-pass filter with the lower cutoff frequency above the range of F_0 to be measured (Dudley, 1937; Grützmacher and

---

[2] This reference contains several proposals of (digital) PDDs, which are quite different from each other. However, most of them follow the principle PiTN, and one of them applies squaring.

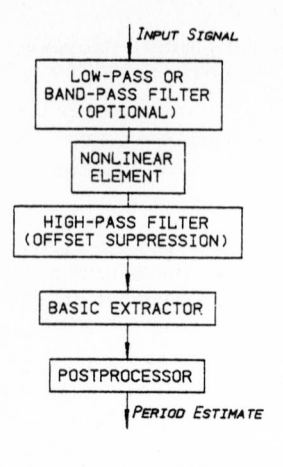

INPUT SIGNAL

LOW-PASS OR
BAND-PASS FILTER
(OPTIONAL)

NONLINEAR
ELEMENT

HIGH-PASS FILTER
(OFFSET SUPPRESSION)

BASIC EXTRACTOR

POSTPROCESSOR

PERIOD ESTIMATE

Fig.6.11. Block diagram of a typical PDA (PiTN) with nonlinear preprocessing

Lottermoser, 1937; Kriger, 1959, Kühlwetter, 1981). This band-pass filter, with a passband of about 1 octave (e.g., 250 to 500 Hz) could be designed in order to remove all harmonics except two adjacent ones (generally the second and third harmonic). Examples given by Kriger, however, show that a fixed passband may cause a great loss in signal level when the harmonics in the passband do not contain a significant amount of the total signal energy, i.e., the first formant. Thus Kriger settled on a moderate low-pass filter (1000 Hz) and mentioned that a tunable filter might be better for some voices. Others, like Riesz (1939), David (1956), Risberg (1961a) or Fant (1962b) did not apply a filter before the nonlinear circuit (or only a low-pass filter with a cutoff frequency above 2 kHz). The majority of researchers at that time (up to the early sixties), however, preferred the presence of few harmonics at the nonlinear element to avoid strong difference tones of nonadjacent harmonics (Peterson, 1952a,b).

Besides the two even nonlinear functions mentioned, there were other nonlinear characteristics under consideration, such as peak clipping (Toffler and Wade, 1964; Kryukova, 1968; Kühlwetter, 1981), one-way rectification (Kallenbach, 1951), or rectified single-sideband modulation (SSB), as proposed by Risberg et al. It was not until 1959, however, that the effect of nonlinear functions was systematically explored. Kriger (1959) carried out a theoretical investigation concerning the amplitude of the difference tone after squaring and cubing a function consisting of three adjacent harmonics (fundamental missing), and he found squaring definitely preferable to cubing. Moreover, he implicitly argued in favor of even nonlinear functions. For reasons of implementation and dynamic range, he then settled upon full-wave rectification. In this context, it must not be forgotten that the earlier devices had to struggle with severe technological problems dealing with the realization of nonlinear distortions. For instance, Kriger argued that "... *the full-wave rectifier must be carefully balanced in order to prevent cubic distortion and its less desirable effect upon the resulting signal* ...". Thus, some of the shortcomings and inadequacies of these circuits might be due to a lack of accurate realization, whereas for some signals an individual device may have worked only because its circuitry was not ideal.

Risberg, Möller, and Fujisaki (1960) reported having tried various nonlinear systems for regenerating or enhancing the fundamental component of the speech wave:

A speech material of 10 seconds each from 4 male and 4 female speakers was processed by the various methods. Narrow band spectrograms of the results were studied and evaluated with regard to the relative intensity of the voice fundamental. The percentage of pitch periods which were judged to require only a moderate filtering in the following stages were counted. The results obtained are listed in Table 6.2.
These results do not pertain to the overall performance of a complete F_0 meter. The half-wave rectification is phase sensitive. The direct unprocessed speech provides the best material for female voices which is due to the natural prominance of their fundamental. Full-wave rectification tends to produce a frequency multiplication. In these instances the second harmonic is highly boosted which accounts for the low figure of merit, 22% for female voices.
Rectified single-sideband modulation provides the best results. The shortcomings of single-sideband processing are mostly due to instances in which the speech wave either was of low intensity or was dominated by the voice fundamental. These findings conform with results from a theoretical analysis made by Fujisaki (1960). At low signal levels the rectifier characteristics approximated a square-law function which accounts for a second power dependency of the amplitude of modulation products on the input signal amplitude. In case the input signal has a flat spectrum envelope and the harmonics are linearly related in phase it may be shown that the ratio of the fundamental to the second harmonic after the SSB-rectification, becomes $(M-1)/(N-2)$, where N is the number of harmonics present in the input. [M is a constant > 1.] An input band consisting of two harmonics is thus optimal and it has been shown that the presence of a formant structure will favorably influence the ratio of fundamental to second harmonic. (Risberg et al., 1960, as given in McKinney, 1965, pp.63-64)

Apparently in this report the SSB signal was full-wave rectified (using a nonideal rectifier). The main advantage of SSBM in connection with nonlinear distortion function is that the frequency band of the SSB-modulated signal at

Table 6.2. Results obtained by Risberg et al. in a comparative performance evaluation of nonlinear functions. The numbers indicate the percentage of pitch periods which needed "only a moderate degree of filtering" after the nonlinear process. (Risberg et al., 1960, as given in McKinney, 1965, p.64)

Method	Speaker	
	Male	Female
Direct (Linear)	49	85
Half-wave rectification (positive)	35	67
Half-wave rectification (negative)	50	87
Full-wave rectification	47	22
Rectified SSB modulation	83	82
Voice channel: 50...3000 Hz input		

the input of the rectifier and the frequency band of the output signal do not overlap so that there is no interference between existing harmonics and those reconstructed by difference tone. On the other hand, it was pointed out that this nonlinear function, just as the others, had its shortcomings so that, at least from Risberg's report, no "ideal" single nonlinear function could be proposed.

McKinney (1965) himself further pursued this issue. He investigated seven nonlinear functions of different characteristics; the speech material consisted of 27 utterances (single words) spoken by several speakers of different languages.

Utterances and portions of utterances were sought which seemed the most likely to result in incorrect measurement of the laryngeal frequency when preprocessed by linear filtering alone. Thus, a large percentage of the phonetic sound types in the analysis set are high vowels, fricatives, and stop consonants. The vowel /u/ provides a particularly troublesome speech wave … . (McKinney, 1965; p.188).

The types of transformations[3] studied fall rather naturally into three groups, both according to their mathematical type and according to their effects upon the signals. The squaring, full-wave rectification, and Gaussian transformations are even transformations, and the identity and logarithmic transformations are odd. Positive and negative half-wave rectification are neither, and the resulting spectra were, roughly speaking, intermediate between those resulting from the even and the odd transformations. …

The even transformations usually produced the most helpful effects upon signals where help was needed, and the most detrimental effects upon signals which originally needed only mild filtering. These effects are perhaps more than incidentally analogous to the effect of even function transformation on a sine wave. In most practical cases, an even transformation of a sine wave produces another periodic wave with twice the fundamental frequency of the first.

If the level of the first harmonic component of the speech wave is high relative to the levels of the neighboring harmonics, then little preprocessing of the signal is needed before it is transmitted to the frequency meter. In such a case, mild low-pass filtering will usually suffice. An even transformation of such a speech signal, however, was found in almost every such case to depress the level of the first harmonic component relative to the levels of its neighbors, and of course this is undesirable.

The outcome was rather different where the second harmonic component had somewhat the highest level of all and the first harmonic component level was not very low. Of the first four harmonics of the transformed wave, the first was usually higher in level in this case than any other except perhaps the fourth. More generally, if one harmonic component level was somewhat higher than the rest, then in almost every observed case the even transformations depressed the level of that component and raised the level of the component of twice that frequency. …

On the whole, both polarities of half-wave rectification were much less effective in increasing the level of the first harmonic component relative

---

[3] McKinney uses the term *transformation* instead of *nonlinear function*. The "identity transformation" is nothing but processing without any nonlinear distortions. The "logarithmic transformation" takes the logarithm of the signal magnitude (plus 1) multiplied by the sign of the signal.

to the levels of neighboring harmonic components than were any of three even transformations studied. In all of the data examined there were almost no instances in which half-wave rectification appeared to be more effective than the even transformations or the identity transformation. Thus, there seems little to be gained from half-wave rectification under the conditions which the analysis set typifies. ...

The logarithmic transformation did not have as notable an effect upon the signal spectra as did the other transformations. This suggests that the function chosen was not sufficiently different from a straight line to give a clear indication of the advantages and disadvantages of a function with this general shape. On the other hand, if two transformations, one odd and the other even, are identical for one polarity of the input signal, then it seems reasonable that in most cases the effects of the odd transformation upon the spectrum of the input signal will be much less pronounced than the effects of the even transformation.

In general, odd (or nonrectifying) transformations resulted in very little dc level in the output signal, while all the other (rectifying) transformations produced a resulting dc component whose level was comparable to the highest of the harmonic component levels. (McKinney, 1965; pp.215-218).

The similarity between the effects of the three even transformations investigated was much greater than was the similarity between any of them and any of the other transformations investigated. In fact, there were striking similarities between ... the different even transformations of any particular speech wave. ... There were, however, some consistent differences between them which are worthy of note. The most readily predicted one is that the overall signal level variations in the output function were less for full-wave rectification than for the square-law transformation. Variations in the level of the squared signal were roughly twice ... the corresponding variations in the level of the full-wave rectified signal. ... This effect of squaring might be called level expansion, to contrast with the level compression produced by the logarithmic transformation. A compensating virtue of squaring is that the level of the dc or average component of the resulting wave, relative to the level of the first harmonic component, is quite often from one to six decibels lower for squaring than for full-wave rectification. (McKinney, 1965; p.220)

In spite of the evident shortcomings of the even nonlinear functions, McKinney settled upon them in the final conclusion of his report:

An even function nonlinear transformation serves a useful purpose in pre-processing the speech wave. ... A practical even function nonlinear transformation is full-wave rectification. (McKinney, 1965, p.254)

This statement directly contradicts the findings of Risberg et al. (1960) where full-wave rectification provided the least adequate results. As we can conclude so far from the literature, no nonlinear function has been propagated as being especially advantageous. Odd NLFs, including linear processing, are of little help when the relative amplitude of the first harmonic is low. Even NLFs produce detrimental effects when the first harmonic has a high amplitude. "Mixed" NLFs, such as half-wave rectification, are a compromise, but not optimal. More complicated NLFs, such as rectified SSBM, certainly deserve further investigation. From a theoretical point of view, however, it is to be suspected that (full-wave) rectified SSBM, if performed in an exact way (e.g., by digital simulation) will behave similarly to any other of the even NLFs. One significant advantage of rectified SSBM, however, is the following. If the signal consists of a sinusoid with the (fundamental) frequency F_0, this waveform will be converted into a waveform with the fundamental frequency $2F_0$ when nonlinearly distorted by an ordinary even function. Usually a measurement error will result from this constellation since the second harmonic falls into the measuring range of the PDA. By the previous SSB modulation, however, the sinusoid with the frequency F_0 is converted into a sinusoid with the frequency F_0+F_M, if F_M is the modulation frequency. As already stated, the frequency bands of the original and the modulated signals do not overlap if the carrier frequency is chosen high enough. The subsequent even nonlinear function converts this signal into a signal with the fundamental frequency $2(F_0+F_M)$, which is far outside the measuring range since F_m is usually much greater than F_0. So in this case no signal will remain within the measuring range, and the PDA can easily realize that something has gone wrong.

At the end of this discussion there remains nothing but to conclude that the application of a single nonlinear function in the preprocessor apparently does not prove appropriate for all signals and all recording conditions. What way out can be gone? Four implementation principles may yield possible solutions.

1) Realize a compromise, like half-wave rectification, provided the results are good enough.

2) Realize a PDA with several NLFs and provide a manual mode switch between them.

3) Realize a PDA with several parallel channels, each using a different NLF, and combine them in an appropriate way (see Sect.6.4).

4) Abandon the principles PiTL and PiTN and settle upon structural analysis, or even abandon time-domain PDAs altogether.

As to point 1, half-wave rectification or a comparable NLF may be the only solution if a single channel PDA (PiTN) is mandatory. For a universal application, however, this PDA is too phase sensitive. Principle 2 may be adequate for phonetic and linguistic research when the type of signal and the approximate range of F_0 are known. It is definitely not adequate for speech communication systems. Moreover, the question whether a single NLF is adequate for every signal uttered by one speaker is still open. Principle 3 is promising, but it poses the problem of automatic selection of the optimal channel. More about that will follow in Sect.6.4 and Chapter 7. The most pessimistic view given as "principle" 4 has been pursued for a long time. Nevertheless, a fairly large number of PDAs in category PiTN and PiTl have been developed after the appearance of McKinney's report (Léon and Martin, 1969; Teston, 1971; Dibbern, 1972; Hess, 1972b; Erb, 1974a,b; Kubzdela, 1976; Niedźwiecki and Mikiel, 1976; and others). Until the contrary has been proved, however, one should not assume that the principle of fundamental-harmonic extraction is inferior to other principles with respect to its performance. The question of what can be achieved by a PDA pertaining to these groups will be taken up again in Chap.7 when the survey of existing time-domain PDAs has been completed.

6.1.4 Manual Preset and Tunable (Adaptive) Filters

The requirement that the low-pass filter in a PDD of type PiTL (and also PiTN, at least partially) must have an attenuation of 18 dB/octave or more within the permitted range of F_0 limits this range of F_0 due to dynamic constraints. A PDD which covers the full range, for instance from 70 to 500 Hz, is not possible without major effort. A viable compromise appears, however, when the range of F_0 can be divided into subbands.

A very early example of this technique is the Purdue Pitch Meter (Dempsey et al., 1950; type PiTL), with three manual range settings: 50 to 250 Hz, 100 to 500 Hz, and 200 to 1000 Hz. The low-pass filter of its preprocessor consists of a cascade of three RC integrators whose 3-dB frequency can be manually set to a value of 26, 53, or 106 Hz, respectively. Other PDAs of this type have two range settings: roughly speaking, one for "male" and one for "female" speech (Riesz and Schott, 1946; Dawe and Deutsch, 1955; Murav'ev, 1959; Erb, 1972a). Such a manual preset seems justified for phonetic research or an application in education, but not for use in a speech communication system where the speaker is completely unknown.

Instead of manually switching between various subranges, the measuring range can be adaptively restricted to the range of the voice being analyzed. For this the PDA requires a tunable element (which is the preprocessor filter in the case of fundamental-harmonic extraction) and some prerequisite information on the momentary range of F_0.

Two principal arrangements are possible (Baronin and Kushtuev, 1974). In both cases the tunable preprocessor filter is one of the elements of a control system (Fig.6.12) which can consist of an open-loop circuit or a closed-loop circuit.

The PDD with an open-loop control element combines the principle of fundamental-harmonic extraction with the multichannel principle. PDDs of this type have been reported by Peterson (1952a, b), R.L.Miller (1953), Barney (1958), Martynov (1958), and Yaggi (1962).[4] The main channel of this PDD consists of the tunable preprocessor filter (which can be low-pass or band-pass) and the basic extractor. It operates using a restricted measuring range (1 octave or even less) and therefore requires a crude estimate of the actual measuring range in order to be able to start the measurement. This value is provided by the auxiliary channel which consists of a crude PDD with an interval-to-voltage converter (cf. Sect.6.6.1). This PDD generates a voltage proportional to F_0 (or T_0) which is then used to tune the preprocessor filter in the main channel. The accuracy of such a PDD is much better than that

---

[4] This reference contains 3 different PDDs, one of which applies a tunable band-pass filter.

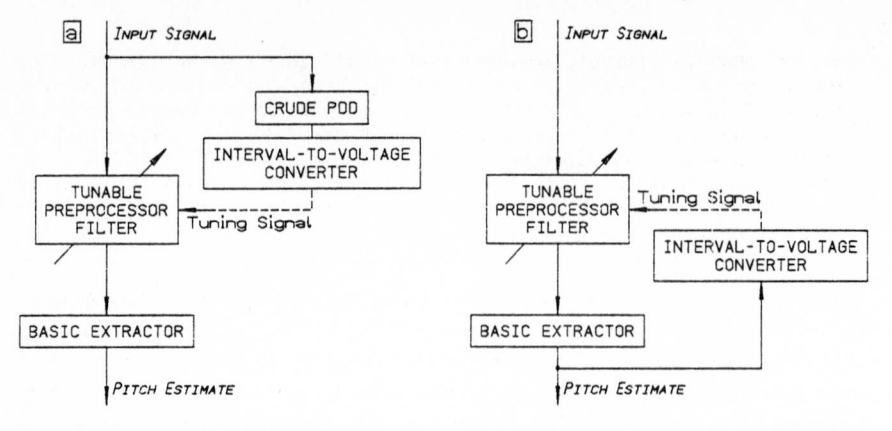

Fig.6.12a,b. PDD with tunable preprocessor filter: block diagrams. **(a)** Open-loop control system; **(b)** closed-loop control system

of a PDD with a time-invariant preprocessor filter. For instance, if F_0 is approximately known, one can use a tunable band-pass filter with a bandwidth of one octave so that the output signal does no longer contain any harmonic other than the first. In connection with a ZXABE inaccuracies resulting from insufficiently damped higher harmonics (see Fig.6.4 and the discussion in Sect.6.1.1) are thus completely avoided. On the other hand, the reliability of this PDD, i.e., its performance with respect to gross errors, is not better than that of the crude PDD in the auxiliary channel; the main channel immediately fails when the auxiliary channel gives a grossly wrong estimate. This was clearly seen by Baronin and Kushtuev (1974) in connection with the PDD by Martynov (1958). This PDD uses a tunable band-pass filter which is tuned to the first harmonic. If the condition is to be satisfied that the passband of this filter may never contain more than one harmonic (i.e., the first), then the filter must have a bandwidth slightly less than the value of F_0 at the low-frequency end of the measuring range, and the error of the crude PDA must not exceed half this bandwidth. In Martynov's PDA the measuring range starts at 70 Hz so that an error of less than 35 Hz may be tolerated for the crude PDD.

When one reads that the average of F_0 is about 200 Hz, then the possible measurement error of the crude PDD is about 17.5%. It seems unlikely that the design and implementation of a PDD with such a low requirement with respect to the performance will be very cumbersome. But one should keep in mind that this value of 17.5% refers to the maximum of the permissible error. If the error were normally distributed, then the permitted average error rate would be about 6%-8%. As the experiments show, however, the distribution of the error is largely different from a normal distribution.

Let the crude PDA, for instance, have an average error rate of 6%, and let the errors be such that they occur in clusters and that there are no errors except second-harmonic tracking. The PDD then gives results that are practically correct 94% of the time. During those 6% of the time where the crude PDA is in error, the tunable band-pass filter will be tuned to the second harmonic, and the whole PDD gives wrong results. This means that the overall performance of the PDD crucially depends on the performance of the crude PDA. By this, even if the average performance of the crude PDA is relatively good, major errors (for instance, higher-harmonic tracking if the PDD is of the filter type; wrong reaction in the case of an amplitude-selection PDD [cf. Sect.6.2.1]), particularly if they occur in clusters, significantly decrease the performance of such a system. (Baronin and Kushtuev, 1974, pp.150-151; original in Russian, cf. Appendix B)

The PDDs with tunable filters that apply a closed-loop control system do not need an additional PDD to obtain the control signal. They extract it, rather, from the output of the PDD itself (Fig.6.12b). The output signal of the basic extractor is fed into an interval-to-voltage converter which directly generates the control signal for the adaptive component. Various PDDs of this type have been proposed (Riesz, 1952; Feldman and Norwine, 1958; Sapozhkov, 1963; Yasuo, 1962; Pirogov, 1963). The most straightforward motivation for the design of such a PDD has been the wish to have a PDD that fulfills condition (6.1) but avoids the dynamic-range problem. Such a PDD simply applies a low-pass filter with a slope of 18 or 24 dB/octave but with a variable cutoff frequency. Such a PDD was realized, for example, by Riesz (1952).

> The signal wave may be passed through a variable selecting arrangement that is so constructed and arranged that its upper cutoff frequency ... progressively lowers as the frequency of the wave components transmitted therethrough is lowered. (Riesz, 1952, as given in McKinney, 1965, p.76)

The PDD by Niedźwiecki and Mikiel (1976; type PiTL) applies a low-pass filter with an attenuation of 24 dB/octave (within the measuring range), followed by a ZXABE. The cutoff frequency of the low-pass filter is externally controlled by the amplitude of the output signal. If the amplitude is zero, the cutoff frequency equals the upper limit of the measuring range (500 Hz). If there is a signal at the output, however, the cutoff frequency is lowered until the amplitude of the signal goes down. Thus the cutoff frequency is always situated at or slightly above F_0. The performance of this PDD has been reported to be somewhat dependent on the level of the input signal, and some 20 ms are needed to adjust the cutoff frequency after the onset of the signal.

Other researchers, like Feldman and Norwine (1958), use tunable band-pass filters instead of low-pass filters. To Edson and Feldman (1959), however, this use of band-pass filters appeared dangerous:

> ... In a sense the root of the difficulty lies in the fact that prior art systems have imposed two incompatible requirements on the pitch tracking filter. The first requirement is that its passband shall be sufficiently narrow to discriminate on the low frequency side against noise and on the high frequency side against harmonics. The second requirement is that the filter shall recognize the fundamental pitch component whenever it is present. When the filter is designed to meet the first requirement, and when it happens momentarily to track the second or third harmonic of the

speech, it is quite insensitive to the return of the fundamental pitch component, which now lies outside of its field of view: below its passband. Hence it fails to meet the second requirement. (Edson and Feldman, 1959, as given in McKinney, 1965, p.80).

In consequence, Edson and Feldman built a "fault-tolerant" PDD with two channels (Fig.6.11). Each of these channels consists of a tunable band-pass filter and a simple basic extractor (ZXABE or TABE). Channel one is designed to be tuned to the fundamental frequency F_0; the passband of channel two is always centered around half the center frequency of the passband of channel one. In addition, channel two has a "voicing indicator." If channel one is tuned to F_0, which corresponds to the "normal" mode of operation, no speech signal will pass through channel two. If channel one tracks a higher harmonic, however, channel two will be activated, and the final selector will switch to this channel. With the data therefrom, the PDD is able to return to the normal mode. A further development of this PDD even realizes a third channel which allows for recovering from third-harmonic tracking. A similar PDD has previously been reported by R.L.Miller (1953).

Another possibility realized in several PDDs is a tunable band-pass filter in connection with nonlinear distortion. From Kriger (1959) we know that a fixed band-pass filter preceding the nonlinear element may often cause a great loss in intensity and thus a greater sensitivity to spurious signals. Similar results were reported by Yuan (1960). To avoid this effect, Peterson (1952), for instance, used tunable band-pass filters whose passbands will contain two adjacent harmonics. This signal is nonlinearly distorted by a bridge modulator with the aim of reconstructing the first harmonic from the two adjacent harmonics in the filters. Even with tunable band-pass filters,

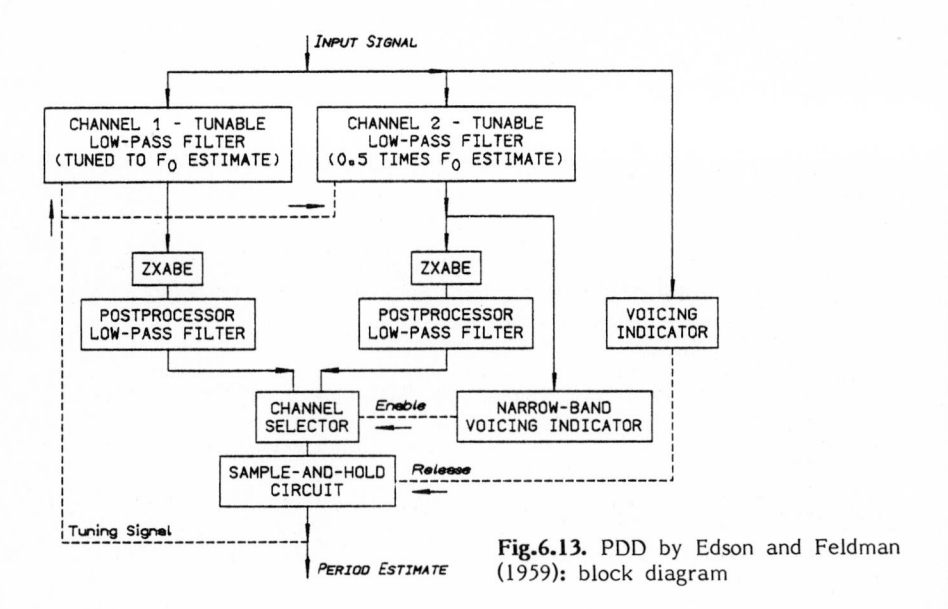

Fig.6.13. PDD by Edson and Feldman (1959): block diagram

however, this principle does not work satisfactorily since frequently a pair of adjacent harmonics with sufficiently high energy does not exist in the signal (this applies especially to signals where a narrow-band first formant directly coincides with a harmonic). The algorithm by Hess (1972b), following the principle PiTL, uses a tunable digital band-stop filter which cancels the first formant. It thus removes the strongest harmonic from the signal.

The PDD by Baronin and Kushtuev (1971) is a good example how a PDA with a tunable band-pass filter in a closed-loop control circuit can be success-fully realized. Rather than tuning the band-pass filter, this PDD tunes a voltage-controlled oscillator (VCO) which is used to drive a modulator and demodulator circuit. The PDD works as follows (Fig.6.14).

The essential component of the PDD is the voltage-controlled oscillator which is used to provide the carrier for the two modulators M1 and M2. (Except for the bandwidth of the filter BP3, the oscillator is the only tunable element in this PDD.) The signal is first band limited to the measuring range (70-740 Hz); then it passes a dynamic compressor. The amplitude modulator M1 transposes the signal into a higher frequency band (the carrier frequency must be at least twice as high as the highest possible value of F_0). The band-pass filter BP2 selects the lower sideband of the modulated waveform and thus converts it into a single-sideband modulated signal, the fundamental of the input signal (with the frequency F_0) being transposed to the frequency F_G-F_0, where F_G is the frequency of the tunable oscillator. At the same

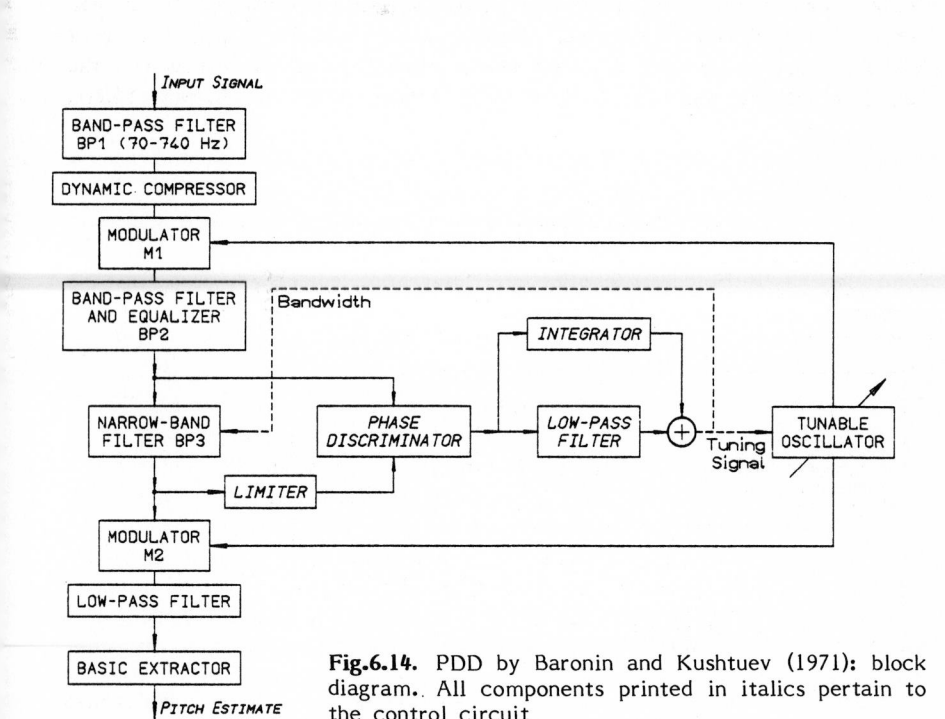

Fig.6.14. PDD by Baronin and Kushtuev (1971): block diagram. All components printed in italics pertain to the control circuit

time the filter BP2 acts as an equalizer filter which emphasizes the frequencies around F_G-F_0. The band-pass filter BP3 is a narrow-band filter with an initial bandwidth of 70 Hz. Only the first harmonic is intended to fall into the passband of this filter. The subsequent modulator M2 transposes the signal back to the original frequency band. Since the two modulators are driven by the same carrier, unwanted frequency shifts due to mistuned modulators cannot occur. The low-pass filter, which terminates the preprocessor, removes unwanted high-frequency modulation products.

The control voltage which is used to tune the oscillator is gained from a phase discrimination of the signal before and after the narrow-band filter BP3. The limiter in the control loop serves for reducing the influence of amplitude variations. The phase discriminator measures the phase difference between its two input signals; it is thus able to determine whether the frequency of the waveform in the passband of the filter BP3 is greater or less than the center frequency of the filter. The output signal of the phase discriminator is fed into a low-pass filter with an integrator in parallel which give the control element a proportional-plus-integrating (PI) characteristic. The output signals of these two components are added and form the tuning signal of the VCO which is then tuned in such a way that the first harmonic of the input signal takes on a frequency as close to the center frequency of the filter BP3 as possible. In addition, the tuning signal is used to control the bandwidth of the filter BP3. Since the bandwidth of 70 Hz is only appropriate at the low-frequency end of the measuring range, it is increased as soon as the tuning signal indicates that F_0 is higher.

With zero tuning signal the PDD is tuned to the low-frequency end of the measuring range. This ensures that during the adaptation the fundamental is always passed first (i.e., before the system can erroneously lock onto the second harmonic). Baronin and Kushtuev, however, reported some problems with higher-harmonic tracking after rapid downward inflections of F_0.

The characteristic of the control circuit is intended to provide a long-term adaptation of the preprocessor to the voice analyzed. For this reason the controller has a PI characteristic; the integrating component preserves the most recent estimate of F_0 for a certain amount of time so that the PDD remains correctly tuned during brief pauses and voiceless segments. On the other hand, the PDD needs a few periods to reach correct tuning if the current frequency of the VCO does not match the actual value of F_0.

It is not easy to categorize this PDD into one of the groups specified in Figure 6.12. On the one hand, the control circuit definitely represents a closed loop; on the other hand, the control signal is not derived from the basic extractor. It is thus possible to regard the control circuit as a crude PDD; the tuning signal then represents the equivalent of a crude T_0 estimate. Although this scheme looks rather complicated, it is more readily realized than a tunable-filter PDD because it is much easier to design a VCO than a filter with time-variant elements. Since the dynamic-range problem does not exist in this PDD, the device works more reliably than a conventional PDD of type PiTL. On the other hand, considerable phase distortions of the output signal (caused by the band-pass filter BP3) may accompany major changes of the carrier frequency during the action of tuning the VCO; these distortions will appear at the basic extractor as small pitch "inflections" and will affect the final estimate in much the same way as a rapid formant change in the signal (see the example in Fig.6.8).

Instead of a tunable filter (or the combination of a fixed filter and a tunable modulator) a phase-locked loop (PLL) is occasionally applied in the control circuit (Righini, 1968; Baronin and Kushtuev, 1974). A quantitative derivation of the PLL would exceed the scope of this book since the use of a PLL never gained great significance in pitch determination. For details, the reader is referred to a textbook on communication engineering, such as the one by Lindsey and Simon (1973). Qualitatively speaking, the PLL consists of a closed-loop control system, the main components of which are a voltage-controlled oscillator (VCO), a phase discriminator, and an amplifier with a low-pass filter. The phase discriminator compares the instantaneous phases of the input signal (in this case the preprocessed speech signal) and the waveform generated by the VCO. The output signal of the phase discriminator consists of a low-frequency component, which is a function of the phase difference of the two signals compared, and some unwanted high-frequency components emerging from the measurement procedure[5] which are suppressed in the subsequent low-pass filter. The low-frequency component is amplified and fed back to control the oscillator frequency of the VCO. The circuit has locked onto the input signal if the phase difference between the input signal and the waveform of the VCO becomes constant. Since instantaneous frequency is the time derivative of momentary (instantaneous) phase, the output signal of the VCO then has the same instantaneous frequency as the input signal. For technical reasons this works only when the instantaneous frequency varies slowly, i.e., when the input signal is nearly a sinusoid,[6] and when the frequen-

---

[5] A simple way of comparing the instantaneous phase of two signals is as follows. Let $x(t) = X(t) \sin \phi_X(t)$ be the input signal with the instantaneous amplitude $X(t)$ and the instantaneous phase $\phi_X(t)$, and let $v(t) = V \sin \phi_V(t)$ be the VCO output signal with the constant amplitude V and the instantaneous phase $\phi_V(t)$. The phase difference to be measured then is $\Delta\phi(t) = \phi_X(t) - \phi_V(t)$. If we perform the measurement for the instant when $\phi_X(t)$ is zero, i.e., simply in the moment when a zero crossing of $x(t)$ occurs, then the momentary value of $v(t)$ represents $V \sin \Delta\phi(t)$ and can be taken as an estimate for the phase difference. In this case, the estimate of $\Delta\phi(t)$ will be sampled; $\Delta\phi(t)$ must change so slowly that it obeys the sampling theorem, the sampling interval being equal to the (instantaneous) period of the input signal $x(t)$. The low-pass filter following the phase discriminator is then simply an anti-aliasing filter applied in order to remove the high-frequency components in the output signal, especially the components in the vicinity of the frequency of the input signal $x(t)$.

[6] For a sinusoid, instantaneous frequency, defined as the time derivative of (instantaneous) phase, and (long-term) frequency, given in Hz, i.e., as the

cy deviation between the oscillator frequency and the frequency of the input signal is small. Usually a PLL has a capturing range of a few percent. Hence, what a PLL demands is an uninterrupted input signal with a frequency that varies slowly and within a narrow band. PLLs are thus ideal in information transmission and in frequency demodulation. What the PLL gets when used for pitch detemination, however, is a signal which is interrupted from time to time and which changes its (fundamental) frequency rapidly and over a wide range. This drawback can be partly overcome by transferring the input signal into a higher frequency band via single-sideband modulation (Righini, 1968; Baronin and Kushtuev, 1974) or by applying PLLs in multichannel PDDs which use narrow-band filters in the preprocessor (Baronin and Kushtuev, 1974). However, it seems as if such PDDs have never been very reliable so that, apart from these occasional investigations, it has been rather quiet concerning PLLs in the area of pitch determination.

When a closed-loop control system is used, the basic problem is that, due to the properties of the input signal, the control system has several stable states. This applies to all the realizations of this principle, regardless to whether tunable low-pass filters, tunable band-pass filters, or phase-locked loops are used. Only one of these states is correct, i.e., the state where the tunable band-pass filter traces F_0, or, if a tunable low-pass filter is applied, the cutoff frequency of the filter is located between F_0 and $2F_0$. Other stable states can and do exist for any higher harmonic of sufficiently high energy that falls into the measuring range. If the control system is in the correct state, it will work perfectly. The problem is how to force the system back into the correct state once it is incorrectly tuned. Here we must distinguish between the two cases that the system is incorrectly tuned 1) because there is no F_0 estimate available (onset of voicing), or 2) because the estimate available is wrong. To handle the first case, the default tuning of the system is such that the correct state is always the first to be reached from the starting point. The second case is more difficult. If the system has completely lost the right track and has locked onto a higher harmonic, the error may continue during a whole voiced segment. This happens most easily for a tunable band-pass filter and a controller with P characteristic, as realized by Feldman and Norwine (1958). Using a controller with increased inertia which only follows the long-term variations of F_0 helps partly (Baronin and Kushtuev, 1974). During a rapid downward inflection of the voice, however, all these PDDs may fail systematically.

---

number of cycles per second, are identical since the instantaneous phase of a sinusoid is a linear function of time. If the measurement of the phase difference $\Delta\phi(t)$ is carried out as described in footnote 5, the input signal may deviate from a sinusoid as long as it behaves like a sinusoid in the vicinity of the zero crossings. For instance, the input signal $x(t)$ may have been processed by an amplitude limiter which clips the high-amplitude parts of the signals but leaves the position of the zero crossings unchanged.

In conclusion, tunable or tracking filters mark one possible way out of the problem when a wide range of F_0 has to be covered without subdivision. The investigations carried out with such PDAs show that the problem with the dynamic range of the amplifiers is really avoided. On the other hand, the main problem, as formulated succinctly by Edson and Feldman (1959; see the quotation further above in this section), only seems transferred, not really solved. The linear devices (PiTL), regardless which filters they apply in the preprocessor, still need the presence of the first harmonic in the signal. For nonlinear devices, a tunable band-pass filter before the nonlinear element is at least of questionable necessity; the main justification for this filter - according to Peterson - has been the fear that the basic extractor might otherwise pick up difference tones resulting from nonadjacent harmonics. It has never been verified theoretically or experimentally, however, that such a difference tone really has a spurious effect. (In Chap.7 this issue will be picked up again.) Tunable filters create a new problem: the one of "cranking" the PDA. This applies especially to tunable band-pass filters in a closed-loop circuit (Edson and Feldman, 1959); low-pass filters and band-stop filters are less critical since they are usually not in danger of filtering away the first harmonic. Tunable preprocessor filters in open-loop control systems avoid these problems. However, they require a second PDD in the auxiliary channel which gives a crude estimate of T_0. Such an arrangement increases the accuracy of the PDD but not its reliability. This means that, if the PDD works, it will provide a good estimate; the probability of failures and gross errors, on the other hand, depends solely on the performance of the PDD in the auxiliary channel and is thus not better than for the PDD in the auxiliary channel alone.

6.2 The Other Extreme - Temporal Structure Analysis

The speech signal is the time-variant response of the vocal tract to the pulse train provided by the voice source. Every fundamental period of the speech signal represents the first part of the (momentary) impulse response of the vocal tract. It consists of the sum of exponentially damped sinusoids; the dominant waveform is usually that of the first formant.

Accordingly, one should be able to "read" the periodicity of the signal out of its temporal structure. In this respect the computer program is intended to model the process which takes place in the human eye and brain when pitch is visually determined from an oscillogram of the speech signal: a complex pattern matching process which is only partially understood and in no way easily brought into algorithmic form.

A possible subdivision of the algorithms is shown in Table 6.3 and Figure 6.15. It appears worthy to note that practically all PDAs of type PiTE are realized in analog technology whereas all PDAs of type PiTS go digitally. Doesn't this subdivision therefore force together two independent principles

Table 6.3. Subdivision of PDAs that apply the principle of temporal structure analysis

Label	Description
PiTE	Modeling of signal envelope and search for discontinuities which mark the beginning of individual periods
PiTS	Direct investigation of the temporal structure by algorithm; search and extraction of anchor points from which periodicity is derived

rather than present a real subdivision? From history, this may seem so. Nevertheless, it appears justified to group these two principles together since they share a number of common features.

1) They start from the pitch period as the representation of the vocal tract impulse response.

2) They seldom use one single detector: extraction of periodicity is done by repetitive application of similar or identical subroutines and/or circuits, or by appropriate matching of the output of parallel processors.

3) The principal mode of operation is nonlinear; the algorithmic decision statement, for instance

if a <u>less</u> b <u>then</u> c; <u>else</u> d;

where c and d stand for algorithmic statements whatsoever, corresponds to the commonly used diode circuit in analog technology.

4) They are unbiased with respect to the errors that will occur. The principles PiTL and PiTN were clearly biased towards higher harmonic tracking, at least at the output of the basic extractor. In structural analysis we can find any type of error: holes, hops, chirps, and so on.

How do these algorithms work? The principle PiTE can be explained in just a few words: the *envelope* within the pitch period is modeled by a *decaying exponential*. A new pitch period is assumed whenever the signal

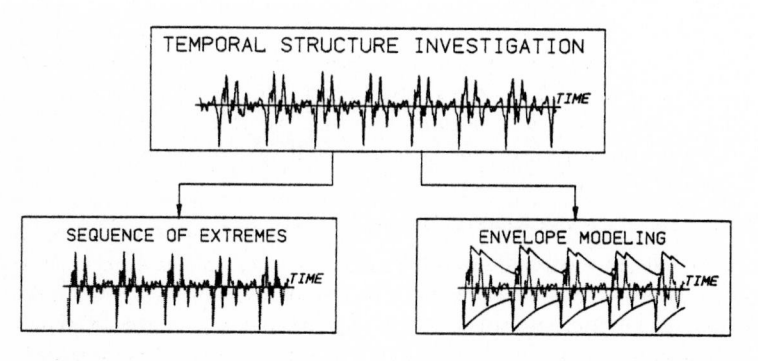

Fig.6.15. Subdivision of PDAs that apply the principle of temporal structure analysis

exceeds this model function. Matching the decay time constant of the exponential to the decay time constant(s) of the signal, i.e., of the formants, is the great problem of this PDA. The principle PiTS, on the other hand, is hard to explain in brief: all these algorithms apply *heuristic models* of the *temporal signal structure*. Hence, this principle grants a maximum of individual freedom to the designer, and the respective solutions are correspondingly different. For didactic reasons, it appears impossible to explain them other than each one on its own. The main idea of all these PDAs, however, is that of algorithmic data reduction within the basic extractor. This is performed by the appropriate choice of anchor points from which the individual pitch period limitations (*markers*) are derived by an iterative process of selection and elimination. A sample algorithm could look as follows.

1) Choose an appropriate feature and select all those samples of the signal to which this feature applies. Eliminate the rest of the samples (first step of data reduction).

2) Select those of the remaining samples which are likely to represent a period delimiter, and reject those which are unlikely (second step of data reduction). Repeat this step if necessary.

3) Check all the selected delimiters as to whether they form a "regular" string of markers. Correct obvious errors.

Maxima and *minima* as well as *zero* or *nonzero threshold crossings*, for example, serve as primary features for these algorithms. Since practically all of the processing is done within the basic extractor, the preprocessor is totally omitted or reduced to a (moderate) low-pass filter. The programming effort can be relatively high. Nevertheless, these algorithms run very fast on a computer since they perform decisions and branchings rather than time-consuming arithmetic operations. Moreover, the primary feature selection which forms the first step greatly reduces the number of samples involved in the further process. The great number of errors at the output of the basic extractor may require a global correction routine in the postprocessor. If such a step is needed, care has to be taken that it does not induce a substantial processing delay. This can easily happen if the correction step has to collect a series of markers from a voiced segment, for instance, before it corrects them all at once. Such a delay would make the PDA unsuitable for application in a speech communication system which needs on-line performance.

In this section we will first deal with the analog solutions of type PiTE. After that, some of the algorithms of type PiTS will be reviewed.

6.2.1 Envelope Modeling - the Analog Approach

In terms of a simplified model of speech production, the voiced speech signal is the response of the vocal tract to the pulse train generated by the voice source. The individual pitch period, at first approximation, is the truncated version of the impulse response of the vocal tract. Since the vocal tract is equivalent to a lossy, approximately linear passive circuit, the pitch period is represented as the sum of several simultaneously excited exponentially damped sinusoids. The algorithms model the envelope of the period by a decaying exponential with properly adjusted time constants. The starting point of the exponential equals the principal peak of the period, and the exponential is expected to be situated above the signal for the rest of the period. If there is an intersection between the signal and the exponential, the beginning of a new period is assumed.

There exist many proposals (mostly older than the digital algorithms) to realize this principle in an analog way: Vermeulen and Six (1949), Gruenz and Schott (1949), Dolansky (1954, 1955), Acoustics Lab. MIT (1955, 1957), Bado-yannis (1957), Adolph (1957), Martynov (1958), Anderson (1960), Licklider (1960), Jenkins (1961), Winckel (1963), Filip (1962, 1967, 1969), Ungeheuer, Rupprath, and Friedrich (1965), Rokhtla (1969), Admiraal (1970), Schunk (1971), Larsson (1977), and Zurcher (1972, 1977). Although the basic idea is similar, the digital and the analog devices differ completely. Firstly, an analog device has practically no possibility of storing signal data such as needed for repeated scans of a total voiced segment. Thus the analog devices are "instantaneous"; they must be built up more reliably since there is no correction step. If this is not possible, restrictions have to be imposed on these PDDs, usually resulting in a limitation of the measurement range of F_0 (to about two octaves or even less). Secondly, the number of decision elements in analog circuitry is limited. The only one applied almost universally is the diode circuit which is equivalent to the (digital) statement

$$\underline{if} \ x(n) \ \underline{greater} \ y(n) \ \underline{then} \ y(n+1) = x(n); \ \underline{else} \ y(n+1) = y(n); \ . \qquad (6.5)$$

For the connection of the diode and a resistor-capacitor (RC) circuit, this statement would read

$$\underline{if} \ x(n) \ \underline{greater} \ y(n) \ \underline{then} \ y(n+1) = x(n);$$
$$\underline{else} \ y(n+1) = y(n) * \exp \ (-p); \ , \qquad (6.6)$$

where p is a decrement which corresponds to the decay time constant of the RC circuit. The asterisk denotes multiplication.

Let us start with those PDDs that apply a unipolar peak detection circuit. The most "extreme" PDD, with respect to the degree of nonlinearity, is undoubtedly that by Dolanský (1955). It emerged from an earlier device by Gruenz and Schott (1949) and was later modified by Anderson (1960). In this review, we will mostly follow Anderson's concept.

The isolation of principal peaks is performed in two ways: 1) enhancement of principal peaks; 2) suppression of secondary peaks. Both principles are realized alternatively in this PDD:

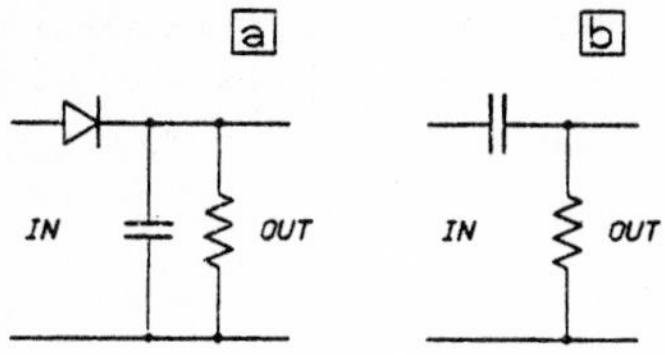

Fig.6.16a,b. Pitch determination by envelope modeling: principle of operation. (a) Peak detector, (b) RC differentiator

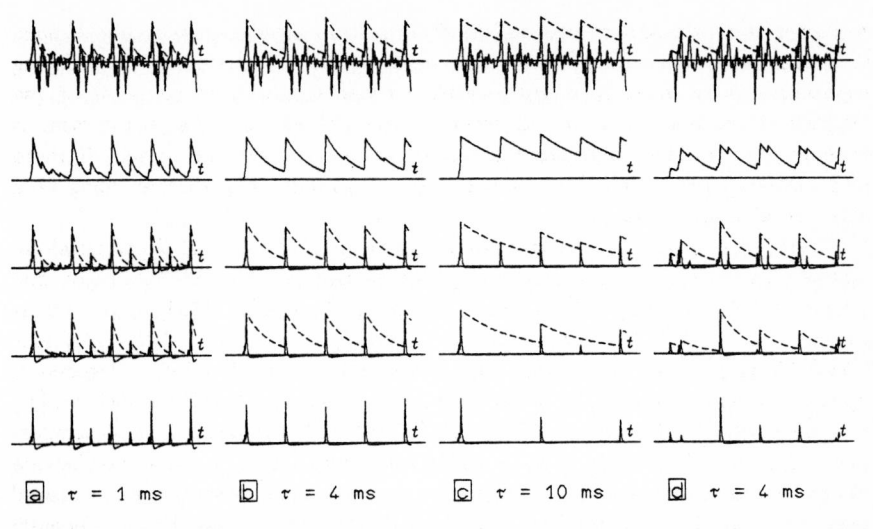

|a| τ = 1 ms |b| τ = 4 ms |c| τ = 10 ms |d| τ = 4 ms

Fig.6.17a–d. Pitch determination by envelope modeling: example of performance. Signals (from top to bottom): original signal (30 ms); signal after the first peak detector; signal after the first differentiator; signal after two stages (each consisting of peak detector and differentiator); signal after three stages. Dashed lines: envelope as modeled by the circuit. Displayed examples: **(a)** peak detector time constant too small; **(b)** time constant appropriate; **(c)** time constant too large; **(d)** signal with wrong polarity, time constant as in **(b)**. In all cases the signal amplitude is assumed to be zero at the beginning of the displayed signal frame

Each pulse produces basically a damped oscillation in the vocal tract, and it becomes clear that the commencement of each fundamental cycle is characterized by a) maximum slope and b) maximum amplitude.

A circuit is sought which will recognize these two properties and produce one output pulse per fundamental pitch period. A suitable solution is shown in Figure 6.16a. The speech waveform is fed to the diode D, which charges C_1 as long as the input voltage exceeds that present across C_1. Beyond the peak P1 of the first swing conduction in D ceases and C_1 discharges exponentially through R_1 as shown by the dashed curve in Fig.6.17, upper line. During this decay there may be further points on the waveform ... where the input voltage momentarily exceeds that on C_1, when D will again conduct and C_1 will receive further small charges. The final waveshape appearing at the output then resembles the signal shown in the second line of Figure 6.17. This is then partly differentiated in the circuit shown in Fig.6.16b to produce the waveshape shown in Fig.6.14, third line. ... It is at once evident that the initial peak has been noticeably enhanced. The two basic properties of the first peak are responsible for the operation of this circuit, maximum amplitude being recognized by the diode and C_1 and maximum slope by the differentiating network. This process can be repeated as many times as is required to remove all but the initial peaks of the original waveshape. In each stage the ratio of the amplitudes of initial to subsequent peaks is increased. (Anderson, 1960, p.1067)

The peak detection circuit is thus responsible for secondary peak suppression, whereas the differentiation circuit enhances the principal peak. In all these PDDs several of these "primary extractors" are switched in cascade in order to provide complete suppression of all secondary peaks (Gruenz and Schott: two circuits; Anderson: three circuits; Schunk: four circuits; Ungeheuer et al.: four circuits with additive superposition of the original signal after the second stage; Dolanský: as many as six circuits). At the output of the last primary extractor, the number of peaks per period has been reduced to one. Where the primary extractor is realized more than three times, the time constants of the peak detectors tend to decrease with increasing distance from the input.

Figure 6.18 shows the block diagram of the PDD by Anderson (1960). In the preprocessor (blocks A and B) we find an automatic gain control in order to reduce the dynamic range of the signal. Blocks C through H represent the basic extractor. The gate I is provided to inhibit the basic extractor during silence or unvoiced sounds. Contrary to the principles PiTL and PiTN which usually have zero output at the basic extractor when the signal is not voiced, this PDD yields irregular pulses during unvoiced segments. Thus a voiced/unvoiced discrimination is incorporated (blocks M through T). It stops the output of the pulse limiter during unvoiced sounds. The remainder of this PDD (blocks J, K, L, U, and V) makes up the preprocessor.

The whole principle has one major problem, namely, the time constants applied in the primary extractors. This problem is discussed in almost every

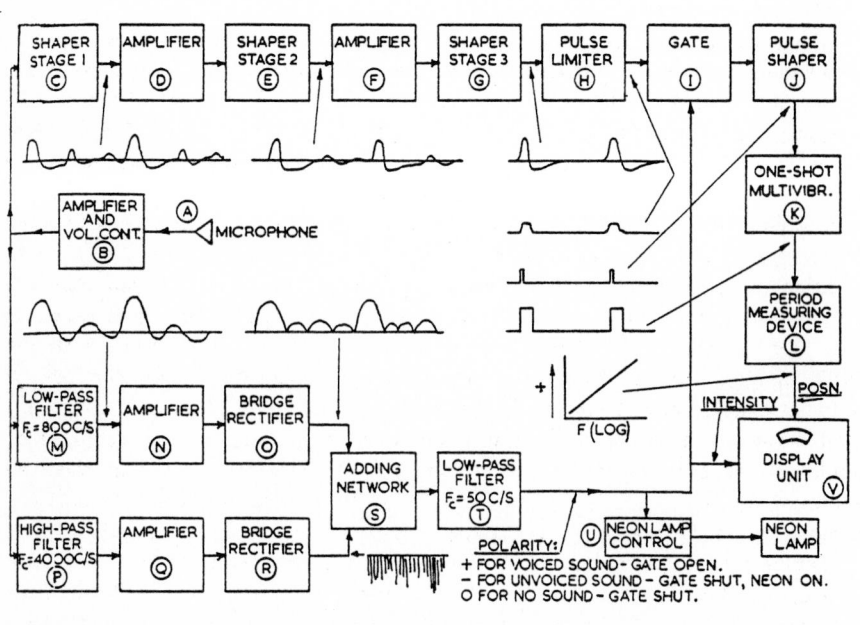

Fig.6.18. PDD by Anderson: block diagram. From (Anderson, 1960, p.1071)

of the papers mentioned. Gruenz and Schott as well as Anderson preferred a single constant as a compromise over the whole range of F_0 to be covered, whereas Dolansky used four different time constants for four subranges of F_0. The compromise time constants given by Anderson and by Gruenz and Schott (same as, later, by Filip) are nearly identical (5 ms) and correspond to the time constants for the lowest subrange in the first stage of Dolansky's PDD.

It is of importance to choose a suitable time constant for the product R_1C_1 [see Fig.6.16a]. Making this small will allow subsequent peaks to become large, thus requiring more stages of detection and differentiation (Fig.6.17a); making it large causes the output to become very small. This is shown in Figure 6.17c. The final value of 5 ms is a compromise which in practice is satisfactory for the frequency range concerned. For the same reason R_2C_2 was made 0.5 ms. (Anderson, 1960; p.1067)

The fact that the number of stages must be greater when the time constants are smaller is clearly demonstrated when the three circuits are compared; Dolansky's PDD, for instance, has twice the number of stages compared to Anderson's; at the same time, the time constants in Dolansky's PDD are on the average by a factor of two smaller than those implemented in the PDD by Anderson. When realized with too small a time constant, the PDD tends to insufficiently suppress secondary peaks, thus leading to higher harmonic tracking. Presumably for this reason, Dolansky's PDD (unlike the others) has an additional blocking circuit which adaptively blanks out spurious extra markers as soon as an estimate of period duration has been established. Gruenz and Schott counteracted higher harmonic tracking by a preprocessor equalizer filter which enhances frequencies around 150 Hz. In Anderson's paper, this point remains unspecified.

Choosing a time constant which is too long, on the other hand, is also dangerous. Anderson mentioned that a time constant which is too large leads to an attenuation of the principal peaks of the periods after a sharp rise in intensity. This fact is true, but it alone is not the essential issue. The problem is twofold. Firstly, a principal peak that has already been "damaged" by a large time constant in the first primary extractor risks being totally suppressed when reexposed to a time constant of similar value in the successive extractor(s) (see Fig.6.17c). Local amplitude variations in the input signal additionally contribute to this problem. Secondly, the time constant must be so small that the device can still work when the overall amplitude is decreasing. This is indeed a hard problem, and presumably for this reason, both the PDDs by Anderson and by Gruenz and Schott (as opposed to Dolansky) use an automatic gain control circuit in the preprocessor. Does automatic gain control, however, really help? At the output of the basic extractor, constant amplitude is certainly needed for the pulse signal. But this can also be achieved by some limiting circuit following the primary extractors or by a monostable multivibrator in this place. The ideal realization of a gain control circuit is the multiplication of the signal by the reciprocal of its short-term envelope. This envelope, however, is usually measured by a diode peak detector circuit with a neglectably small rise time constant and a fairly large time constant for recovery.[7] That means this circuit is realized in a way similar to the primary extractors. If now the overall amplitude of the signal is decreasing, the amplification of the gain control circuit rises exponentially, and the circuit directly counteracts the basic extractor.[8] If gain control is to be applied, a more elaborate circuit is needed. This point of discussion will again be taken up further below.

A closer look on the conditions for assigning the right value to the time constant in the basic extractor shows that the problem is really not easy to solve. The RC circuit is intended to model the envelope of the signal within the pitch period. The temporal structure of the signal, and with it the envelope, is primarily determined by the waveform of the first formant. Consequently, the time constant of the RC circuit should correspond to the bandwidth of the formant F1. For low values of F1, the bandwidth ΔF1 takes on a value around 50 Hz (Dunn, 1961). The pertinent decay time constant then becomes

$$t_D = 1 / (\pi \cdot \Delta F1) \approx 6.25 \text{ ms} \tag{6.7}$$

From visual inspection of the signal, on the other hand, it is easily seen that the overall amplitude of the signal can change drastically even between subsequent pitch periods. It appears safer to assign a value to the time constant which is a bit too small compared to what is desirable with respect to the bandwidth of F1, and to repeat the circuit if necessary. Assigning different time constants for different subranges of F_0, on the other hand, appears unjustified; the temporal structure, in particular the amplitude of the largest secondary peak in the period, depends on frequency bandwidth of F1 and rather than on the momentary value of F_0.

In any case, one must put up with the fact that this PDD has no general bias towards a certain category of errors. It will produce both holes and chirps. Errors are introduced due to insufficient separation of intraperiod temporal structure and interperiod amplitude changes. However, the PDD will tend towards producing chirps when the short-term amplitude of the signal is rising, and towards producing holes when the amplitude is falling. With respect to the time constant of the primary extractor circuit, it will preferably produce chirps when this time constant is too small, and holes when it is too

---

[7] "Short-term," in this respect, has to be taken relatively. The time constant for recovery of the circuit must be small enough to follow the global intensity changes but large compared to the length of the pitch period - an often violated condition!

[8] In an evaluation of the performance of PDAs on musical melodies (sung by male and female singers as well as played by various instruments), Larsson (1977) examined PDAs of type PiTE both with and without dynamic compressor. According to his results the compressor did not improve the performance of the PDA. There may be differences from device to device, however (cf. the discussion on the dynamic compressor later in this section).

large. With respect to the global amplitude of the signal, the PDA will tend to produce chirps when the amplitude is rising, and holes when it is falling.

Baronin and Kushtuev (1974) reported having carried out a statistical performance evaluation of this type of PDD. For both correct marker sequences and false alarms they computed the distributions of the amplitude ratio versus the temporal distance, taken for two successive peaks. Since only two speakers were involved, the two distributions showed distinct maxima. As one could expect, the distribution for correct sequences had its maximum at an amplitude ratio equal unity and at a temporal distance of $d = 6$ ms (i.e., all the utterances had an average F_0 of about 170 Hz). The false-alarm distribution clearly showed a maximum for the second harmonic ($d = 3$ ms) with a preferred amplitude ratio in the vicinity of 0.4. From these distributions Baronin and Kushtuev tested the separability of true markers and false alarms by means of a minimum-distance classifier, and they found that even for this nearly ideal case (only 2 speakers with similar fundamental frequencies, only stationary sounds) the classifier yielded an error rate of 6%.

These PDDs use only one polarity of the signal (corresponding to one-way rectification). Dolansky as well as Gruenz and Schott confine themselves to the positive polarity, whereas Anderson provided a manual switch:

> The speech wave is in general not symmetrical about the zero axis. The microphone is originally poled in such a way that the direction in which the highest peak occurs is that which causes current flow through the diode. As the polarity of this peak is determined by the direction in which the breath flows, and as speech is invariably produced upon exhaling, there is no need to change the polarity for different speakers. There is, however, no telling what a deaf child will do! (Anderson, 1960, footnote p.1067)

All these PDDs take for granted that the speaker speaks directly into the microphone. Thus the signal is expected to be free from phase distortions which might drastically influence the temporal structure (Holmes, 1975). Since the frequently observed asymmetry of the waveform in voiced speech is to a great extent due to the presence of the first harmonic (Endres, 1967; Loo, 1972; Wolf and Brehm, 1973), low-frequency distortions such as occur in tape recorders are likely to inpredictably influence the polarity where the principal peak is most prominent. In case phase distortions have to be taken into account, an automatic discrimination between the two polarities thus seems desirable. Even this can lead to a multichannel solution. For instance, the proposal by Badoyannis (1957) pointed in that direction:

> ... to avoid locking of the pitch frequency extractor to the second harmonic, as sometimes happens with the vowel sounds /u/ and /i/, a possible arrangement could be to compare and select the detected pitch frequencies as obtained after positive and negative halfwave rectification and stretching. It was observed that by the outlined method of pitch frequency detection second harmonic locking occurs only with positive or negative rectification at any one time. In other words, the pitch frequency could be monitored at the two outputs shown and then selection of the lowest of the two by appropriate electronic circuits could result in a continuously true pitch frequency determination free of second harmonic jumps. (Badoyannis, 1957; as given in McKinney, 1965; p.101).

Badoyannis's proposal would result in the application of the minimum-frequency selection principle applied to discriminate between the two polarities which are processed in two separate channels. To apply this "minimum-frequency selection principle" (see Sect.6.4 for details), however, one must be sure that the PDD, if it is in error, produces a chirp but not a hole. For the PDAs of group PiTE this is absolutely not guaranteed although their performance can be biased in that way by choosing the time constants relatively low. Dolansky mentioned the same problem when he found that full-wave rectification is inadequate for his device.

A few years later, Filip (1969) presented a PDD that operates according to the same principle. This PDD, which had already been developed by 1962, was designed for a variety of input signals especially from the musical domain. Filip's paper is remarkable not only for its theoretical background but also for the simplicity of the proposed circuit (Fig.6.19).

Filip showed that differentiation and secondary-peak suppression (integration) can be unified in one single circuit. The basic circuit is given in Figure 6.19a. With an ideal diode, the output voltage v_o equals zero if the voltage v_r at the resistor R_1 is negative, and equals the voltage v_r otherwise. Thus the total resistance r seen from the capacitor C equals R_1 when the diode does not conduct, and otherwise equals the resistance R_0 resulting from the shunt connection of R_1 and R_2 (Fig.6.19b). The circuit can be considered linear between two successive zero crossings of the voltage v_r. It is easy to see that the voltage v_r is negative as long as the voltage v_c is larger than

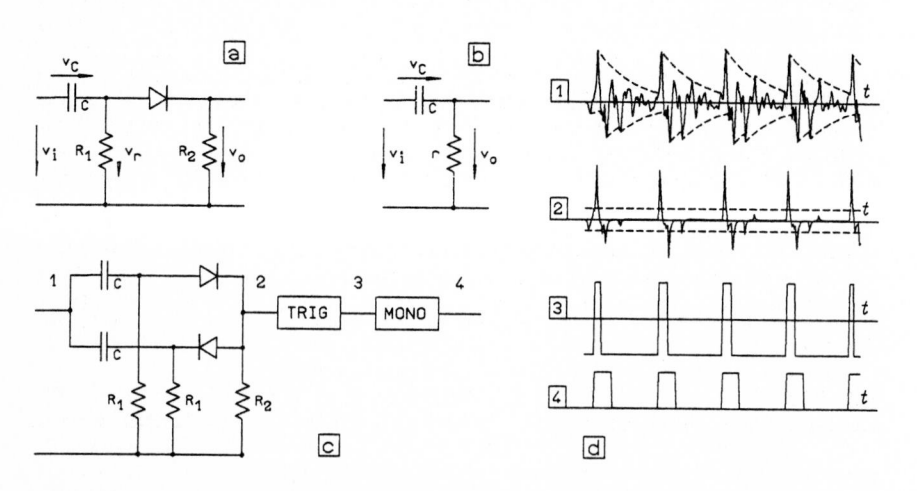

Fig.6.19a–d. PDD by Filip (1969): basic circuit and principle of operation. **(a)** Basic circuit (unipolar); **(b)** equivalent circuit, valid for time intervals between two successive zero crossings of v_r; **(c)** basic circuit (bipolar) and extractor; **(d)** example of operation (bipolar). (TRIG) Bistable circuit (Schmitt trigger); (MONO) monostable circuit; (1–4) points of measurement (corresponding to the signals displayed in the right hand part of the figure). Input signal: /ε/ from "Hello...", male speaker

the momentary voltage v_i of the input signal. It is also easy to see that this condition is going to be fulfilled when the input signal has just passed a peak. If the time constant R_1C, which is valid for this state of the circuit, is large compared to the period length of the first formant, the capacitor will slowly discharge thus keeping the voltage v_r at the diode below zero until the next major peak is reached. At the time the voltage v_r crosses the zero axis towards positive values, the diode starts conducting and thus changes the time constant of the circuit to the much smaller value R_0C. Now the circuit behaves like a RC differentiator, and the voltage v_r which, for this moment, equals the output voltage vo, reaches zero again when the input signal has arrived at its principal peak. Thus the whole circuit acts as a peak detector, enhancing the principal peak and suppressing the spurious peaks in the period.

Filip also took up the point of polarity:

It is well known that vowel waveforms most often have major peaks of the same sign ... However, the major-peak sign is not necessarily constant if a more general inventory of input signals is to be processed. To quote only one example, the excitation function produced by bowing in stringed instruments exhibits alternate phase reversals at each reverse of bowing. A bipolar detector is needed, the sign of the processed half-wave no longer being predetermined by the diode connection. (Filip, 1969, p.724).

He then gave a remarkably simple proposal for the bipolar detector:

Since a positive and a negative peak can never occur simultaneously, the output resistor may be common to both basic unipolar circuits. Amplified output is fed to a bistable circuit (Schmitt trigger) with the dc bias set at the midpoint of the hysteresis interval. ... In each period, there is at least one positive and one negative pulse to trigger the bistable circuit. Should the number of pulses of one sign increase arbitrarily, the bistable circuit will continue producing a fundamental periodicity square wave if the number of opposite sign pulses remains one per period. Thus the bistable circuit provides selection of the "more favorable half-wave" in the sense of the major-peak condition, and delivers an output signal at fundamental repetition rate as long as the input signal has a major peak of arbitrary sign. ... Thus, the major peak is required only to exist, regardless of its sign and, consequently, ... no other restriction ... is imposed ... upon the variety of input functions. In fact, even this need be not always fulfilled; it is sufficient if all pulses of one sign form a group that is not interrupted (within a period) by alternating pulses opposite in sign. (Filip, 1969, p.724).

Together with the following Schmitt trigger[9], the bipolar basic detector realizes Badoyannis' proposal of two channels, each for one polarity, with a

---

[9] An equivalent proposal is found in the PDD by Ungeheuer et al. (1965). The four-stage primary extractor is realized separately for the two polarities of the signal. The outputs of the extractors are then combined in a flipflop which is set whenever a positive peak has passed the pertinent extractor, and reset whenever a negative peak has passed the extractor operating on the inverse polarity.

decision element thereafter which is not a minimum frequency selection element (although it might look that way) but behaves equivalently to Reddy's definition of the significant peak (see Sect.6.2.2 for details). The monostable circuit triggered by the differentiated output of the Schmitt trigger has a pulse length which is slightly shorter than the minimum period duration to be measured. If the range of F_0 to be determined is not too large, this offers a good means to prevent second harmonic tracking.

The problems with the time constants are the same as in the previous circuits of this type. As long as the signal is strictly periodic, the time constant R_1C responsible for secondary peak suppression can be chosen large. For the time-variant speech signals, however, a smaller value is necessary:

> Owing to statistical fluctuations inevitable in "natural" signals, the major peak value is no longer a constant over a large number of periods and the same is concerned with transitions. A large time constant would then cause spurious response of the detector which may omit smaller major peaks subsequent to higher ones, thus delivering occasionally a subharmonic output. (Filip, 1969, p.724)

Filip evaluated the performance for several artificial signals rather than for natural speech. Since music, not speech, is the main field of application for this PDD, no practical evaluation with speech signals was presented in his paper, except the following more general comments:

> It is perhaps worth mentioning that, in general, the human voice is somewhat easier to process correctly than some of the orchestral instruments ... investigated. ... The common disadvantage of all envelope periodicity detectors is their sensitivity to amplitude fluctuations and noise. ... Amplitude fluctuations and transitions involved in "natural" signals often cause spurious responses between portions where the fundamental is detected correctly. The sensitivity to noise ... is inherent in this class of devices since no input reduction of the signal bandwidth takes place. It may be much improved by appropriate prefiltering. Some components of input signals, such as introduced by reverberation, are very hard or impossible to eliminate just because the time-domain mode of operation implies phase sensitivity. (Filip, 1969, pp.730,732)

Why is a bipolar circuit necessary at all? We have stated before that the circuit models the intraperiod temporal structure of the signal which is mainly determined by the bandwidth of the first formant. This is true as long as the first harmonic is weak, and as long as the interference between the formants F1 and F2 is negligible, i.e., as long as the frequencies of F1 and F2 are situated well apart from each other. Hence, we have the strange case that this PDD might work better for a band-limited signal than for an undistorted signal! In case the first harmonic is strong (this holds especially when the speaker is close to the microphone), it creates an asymmetry in the waveform which will favor one of the two polarities. Which polarity - positive or negative - this will be depends on the external recording conditions. For a given recording the favored polarity will remain the same if the signal is not phase distorted. In the case of phase distortions and in the presence of a strong but not predominant first harmonic, a bipolar circuit is indispensable. The case of interaction between Fl and F2, disregarded in this consideration, further complicates the issue.

There is no uniform opinion whether a low-pass filter should be applied in the preprocessor or not. A number of PDDs analyze the signal as it is (Dolansky, 1955; Anderson, 1960; Filip, 1969; etc.); others perform a (mostly moderate) low-pass filtering at the beginning (for instance, Vermeulen and Six, 1946; Winckel, 1963; Carré et al, 1963; Ungeheuer et al., 1965; Zurcher, 1972; Larsson, 1977) in order to reduce uncertainties of the peak locations due to interactions between Fl and higher formants. Schunk (1971) applied a high-pass filter. Eliminating the first partial he obtained a waveform whose amplitude distribution is symmetric with respect to the zero axis; he thus avoided the need for a bipolar detection circuit.

An interesting variation of this principle was proposed by Zurcher (Zurcher, 1972, 1977; Zurcher, Cartier, and Boe, 1975; also available on a commercial base). The overall buildup of the PDD resembles those by Unge-heuer et al. (1965) and Filip (1969). Two separated peak detection circuits (one for each polarity), are combined at the output by a circuit which realizes the condition that at least one peak of each polarity has to be detected in order to make the circuit work properly. The peak detection unit is the most interesting detail since it does not need a differentiating circuit for peak enhancement; this job is performed by nonlinear feedback. The circuit works so well that it need not be applied several times.

Figure 6.20 shows a diagram of the peak detection circuit. The diode usual in the other PDDs has been replaced by a comparator CMP that controls the two switches S_1 and S_2. The input signal (which has been one-way rectified before the circuit in order to enforce the right polarity) first passes the differential amplifier (depicted as an adder in the figure). The peak memory circuit operates as follows. If its output voltage v_m is greater than or equal to the output voltage v_a at the output of the amplifier, the compara-tor output signal takes on a value of zero, and the two switches by which the capacitors C_1 and C_2 can be charged are open. If the voltage v_a rises towards a major peak, it becomes greater than v_m, and the comparator changes its state, signalling a major peak and closing the switches S_1 and S_2 at the same time. Now the capacitors C_1 and C_2 are charged, and the charge resistors R_C ensure that the comparator remains in this state as long as the voltage v_a is rising. The voltage v_a having reached the peak, the comparator changes back to zero; the switches are reopened, and the capacitors C_1 and C_2 start discharging via the resistors R_1 and R_2. Since this is a double RC circuit, the voltage v_m does not simply decay exponentially but according to the following relation:

$$v_m(t) = \hat{v}_m \left[\frac{k_2}{k_2 - k_1} \exp(k_1 t) - \frac{k_1}{k_2 - k_1} \exp(k_2 t) \right] \qquad (6.8)$$

with the initial condition

$$dv_m/dt = 0 \qquad (6.8a)$$

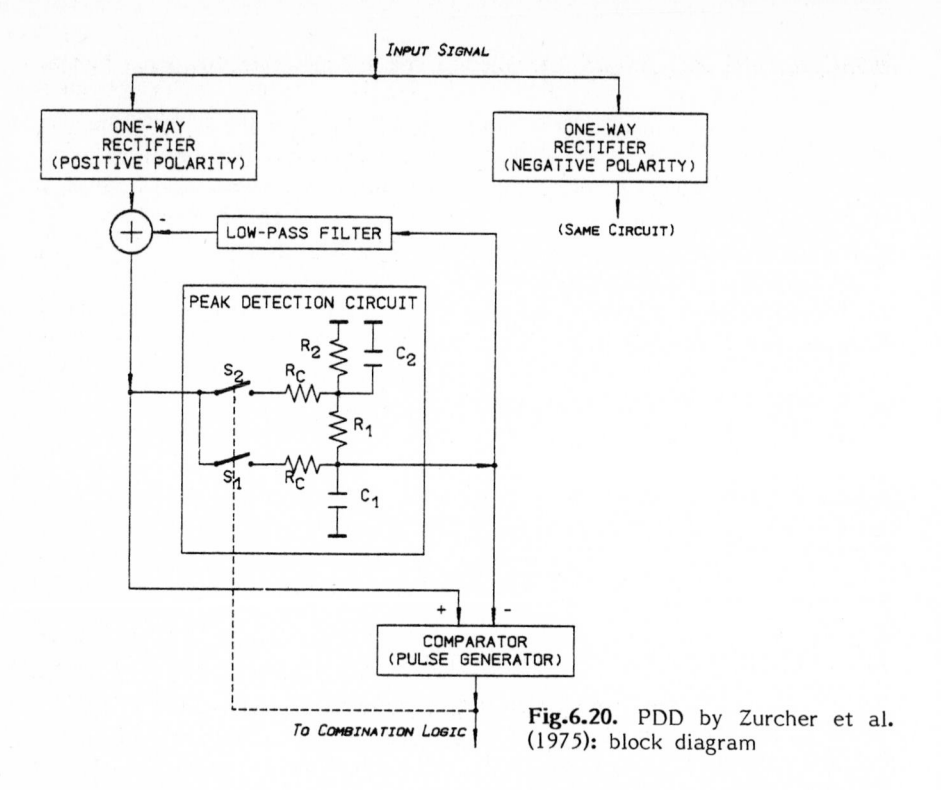

Fig.6.20. PDD by Zurcher et al. (1975): block diagram

which is fulfilled since the switch is opened when the voltage reaches a peak. The values k_1 and k_2 are linear combinations of time constants which depend on the values of C_1, C_2, R_1, and R_2; and $\hat{v}_m$ is the peak value of the voltage v_m. Due to the horizontal slope at the beginning of discharge, the circuit shows a good enhancement of the primary peak and good suppression of the following secondary ones. This fact is enhanced by the nonlinear feedback circuit (which is nonlinear since it contains the switches). The low-pass filter in the feedback loop delays the relatively abrupt rise of the voltage v_m by about 1 ms. After that time v_m counteracts the input signal at the differential amplifier and thus further suppresses the secondary peaks. The amplifier recovers to the same extent as the voltage v_m decreases. Since the input signal is low-pass filtered (with a cutoff frequency of 400 Hz and a slope of 12 dB per octave), there is no interference from higher formants so that the output signal of the comparator can directly be used to drive the flipflop at the end of the basic extractor. The PDD by Zurcher is the only one of category PiTE which has a rudimentary memory (3 periods) and performs some final correction. Details about this are added in Section 6.6. Figure 6.21 gives an example of the performance of this PDD.

The second interesting feature of Zurcher's PDD is the way the dynamic compression is realized. In order to work properly, the dynamic compressor is indispensable for Zurcher's PDD. The performance of this component is desired

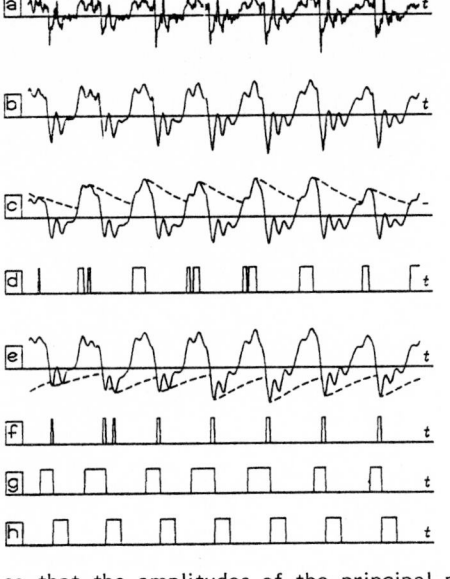

Fig.6.21a–h. PDD by Zurcher et al. (1975): example of performance. Signal: beginning of the utterance "Algorithms ..." (51.2 ms), speaker cessed by the dynamic compressor shown in Figs.6.22,23.   (a) Input signal; (b) signal after low-pass filter; (c) signal at the input of the peak detection circuit (positive polarity); (d) output signal of comparator (positive polarity); (e,f) same as (c,d), but for negative polarity; (g) matched comparator output; (h) final marker sequence

so that the amplitudes of the principal peak are kept as constant as possible. As already stated, a single-loop dynamic compressor is insufficient; during recovery, its temporal envelope would attenuate the input signal and thus directly counteract the performance of the peak detector. Zurcher applied a dynamic compressor which was developed by Courbon (1972) for this purpose. Figure 6.22 shows its block diagram.[10]

The crucial feature of such a compressor if designed for application in pitch determination is its short-term behavior. In this special device, the level determination is performed by two peak detection circuits, one for each polarity. The two partial signals (positive and negative half-wave) can then be processed separately. The rise time of the peak detection circuits is neglegibly small; the decay time constant itself is time-variant. In normal position, when the input level is approximately constant, the decay time constant is fairly large (2 s) and does not affect the temporal structure within the periods. The level of the output signal is additionally determined by full-wave rectification and peak detection with a much shorter time constant. After a certain amount of smoothing (by a low-pass filter), the derivative of the output level is used to influence the time constants of the peak detectors. Let now the level of the input signal decrease rapidly. In a conventional circuit with a

---

[10] Figures 6.22,23 display the results of a digital simulation of Courbon's dynamic compressor. This compressor, however, is only well suited for analog systems. In a digital system, the delay by one sample which is necessary in the closed-loop version may cause isolated peaks to occur in the signal which will not occur in an analog system. On the other hand, the fact that there is no drift or parameter change over time in a digital system would favor an open-loop solution if implemented digitally.

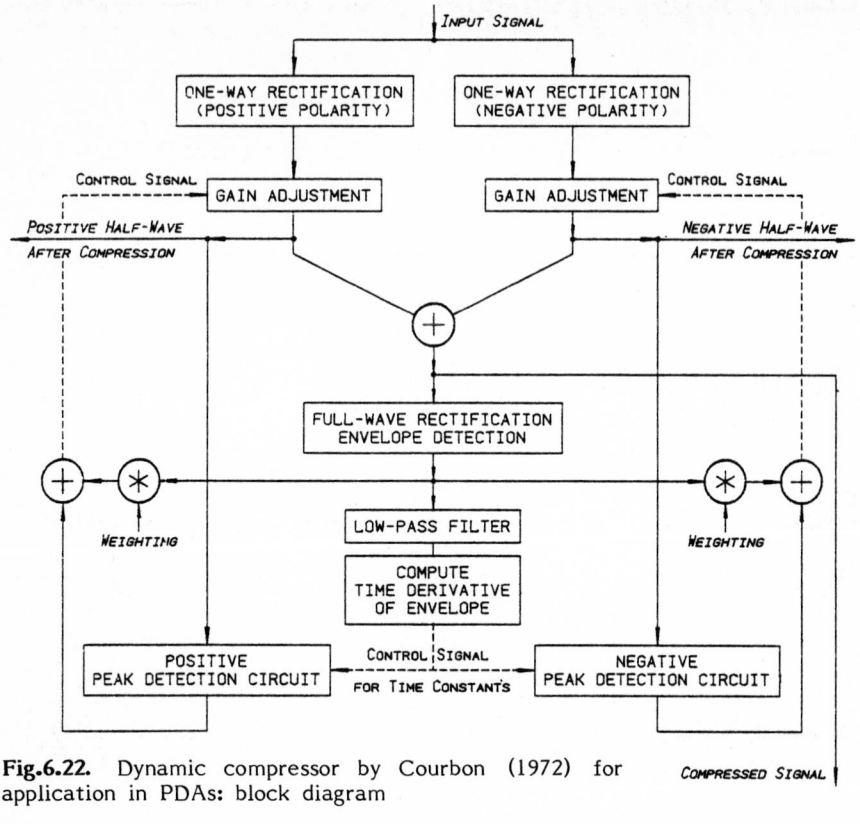

Fig.6.22. Dynamic compressor by Courbon (1972) for application in PDAs: block diagram

long decay time constant, the compressor cannot follow the envelope fast enough; in consequence, the output level begins to decrease too. In the present circuit, however, this situation (presence of a negative derivative of the output level) is used to momentarily diminish the time constants of the peak detectors in the compressor (down to 22 ms). The recovery of the compressor is thus accelerated, and the circuit is able to follow a rapidly decaying signal envelope.

In the most recent analog version of his PDD, Zurcher (1977) incorporates such a compressor separately for each polarity after low-pass filtering. With this configuration he achieved an even better degree of amplitude compression for the principal peaks. A digital version was built by El Mallawany and Zurcher (1977). It was found, however, that the digital realization of this principle is comparatively time-consuming since it needs a lot of multiplications, and that the design of the dynamic compressor had to be changed in the digital system (El Mallawany and Zurcher, 1977).

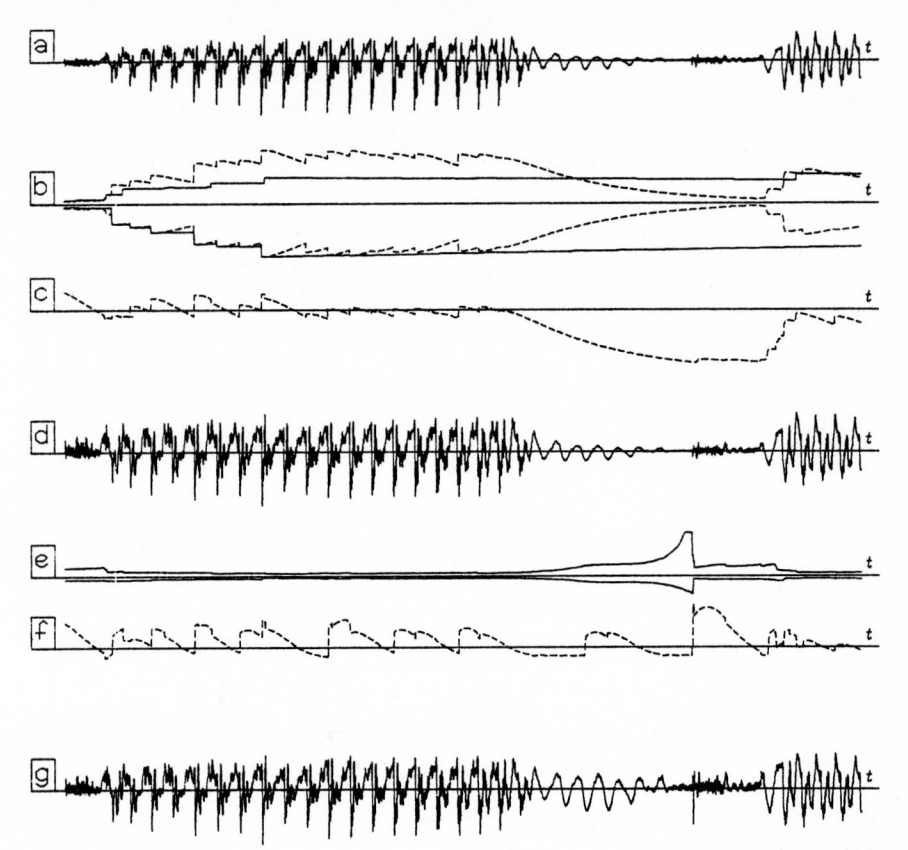

Fig.6.23a-g. Dynamic compressor by Courbon (1972): example of performance. Signal: beginning of the utterance "Algorithms..." (512 ms), speaker MIW (male), undistorted signal. (a) Original signal, (b) solid line - long-term envelope of the input signal (time constant 2 s), separate for each polarity; dashed line - short-term envelope (time constant 30 ms) of the input signal; (c) time derivative of the short-term envelope of the input signal, in this example computed as the difference between the short-term envelope and the average of the long-term envelopes; (d) signal after compression where both polarities are separately compressed, but without changing the time constants in the compressor; (e) gain of the compressor; (f) time derivative of the short-term envelope in the final version of the compressor where the time constants of the compressor are controlled; (g) compressed signal

El Mallawany (1977c) carried out a quantitative performance evaluation of this PDD in several versions. As a base of comparison he selected Markel's SIFT algorithm (1972; see Sect.8.2); the evaluation showed that the errors committed by the algorithms were quite different although the overall performance was comparable. Some more details will be added in Section 9.3.1.

6.2.2 Simple Peak Detector and Global Correction

One of the best known algorithms in the area is that by Reddy (1966, 1967). This algorithm is not the first one existing in the area, but it is perhaps the one most easily understood. It shall thus form the first point of discussion. The algorithm uses sequences of significant maxima and minima as anchor points for pitch markers. It starts with the determination of all local maxima and minima in a given segment (about 25 ms) of the signal (Fig.6.24a). After that, it proceeds towards the determination of "significant" maxima and/or minima (Fig.6.24b):

> Of the above local maxima and minima, we are interested only in those maxima and minima which are likely candidates as pitch markers. These significant maximum and minimum peaks may be defined as follows.
> An *absolute maximum* is defined to be the maximum of all the local maxima in the given segment of speech.
> A *significant max peak* is defined to be one that
> 1) is positive,
> 2) does not occur within 2.5 ms[11] from the previous significant maximum, and
> 3) is a) either greater than 0.9 times the absolute maximum, b) greater than the linearly extrapolated value from the previous two significant maximum peaks, or c) if neither condition a) nor condition b) is satisfied within 13.5 ms of speech from the previous significant maximum, then the maximum of all the local maxima in that 13.5 ms of speech.
> A *significant min peak* is defined similarly.
> Figure 6.24b illustrates the significant max and min peaks. ... It can be seen that the markers indicated by the minimum peaks are already good indicators of the pitch periods. This need not always be the case ...
> By examining Fig.6.24, one can see that whenever there exists a significant max peak that would be a good pitch marker there also exists a corresponding significant min peak within a small neighborhood. This phenomenon is to be expected since the greatest perturbations in the speech wave occur soon after the vocal tract is excited by the release of the pressure built up behind the vocal cords. ... Since, at this stage, there will be a significant max peak corresponding to every significant min peak, we only need to keep track of one of these peaks, which shall henceforth be called a *significant peak*. ... A significant peak is defined to be a significant max peak which has a significant min peak within 3.5 ms of its occurrence (Reddy, 1966; pp.56-59).

For many periods, these significant peaks can already serve as pitch markers. However, there remain numerous errors which have to be corrected in the subsequent correction subroutine.

Reddy's algorithm was originally designed for a speech recognition system. Therefore, it could tolerate the fact that

---

[11] The limitations given in Reddy's paper (2.5 and 13.5 ms, respectively) for the range of the pitch period length would indicate a range of F_0 to extend from 75 to 400 Hz. The formant F1, which is responsible for the distance between sig max and sig min peak, would be permitted to go down till 150 Hz.

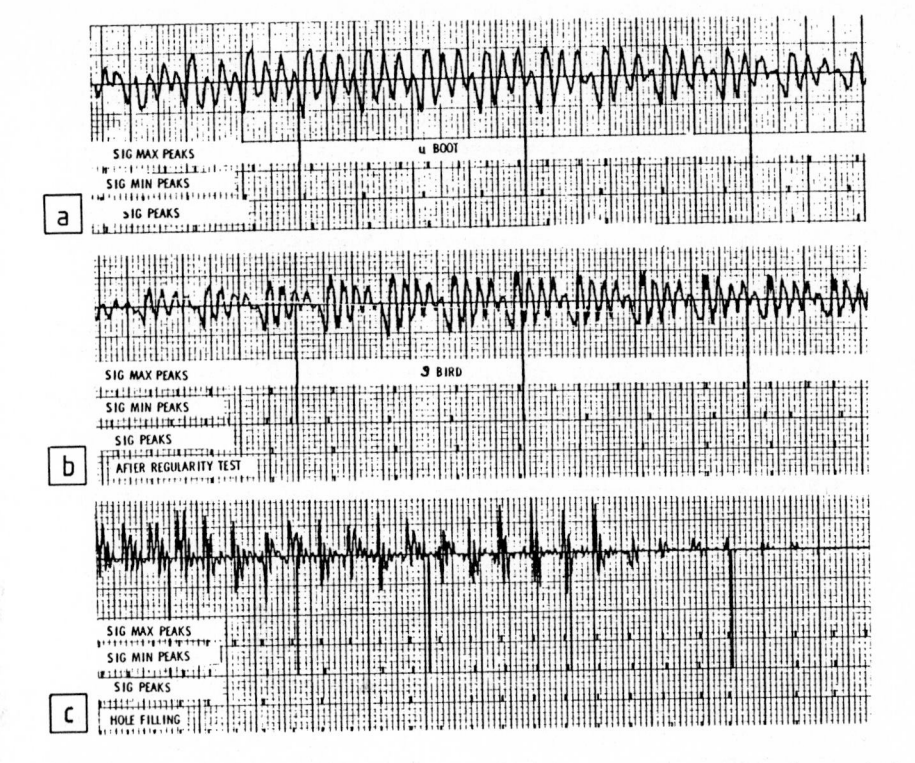

Fig.6.24a–c. PDA by Reddy (1966, 1967): principle of operation (examples). (a) Extraction of significant maxima and minima and determination of significant peaks; (b) determination of significant peaks and regularity test; (c) determination of significant peaks and first correction step (hole filling). From (Reddy, 1966, p.58)

> pitch periods may be specified after examining the whole utterance rather than almost 'instantaneously' as in the case of vocoder applications ... Original data will be available for repeated scans, if necessary, during error detection and correction ... (Reddy, 1966, p.53).

The extraction of the significant peaks is performed locally for small blocks of speech (about 25 ms). The correction routine,[12] however, cannot start until all the significant peaks of a whole utterance have been determined. The flow chart of this subroutine is displayed in Figure 6.25. Most of it is self-explanatory, but some further comments are needed. To judge the regularity of an individual pitch period, a preset for the expected period duration is necessary. If the length of an individual period (the "indicated pitch") falls within this

---

[12] The correction routine performs a *list correction*. A detailed discussion of such routines in general follows in Section 6.6.3.

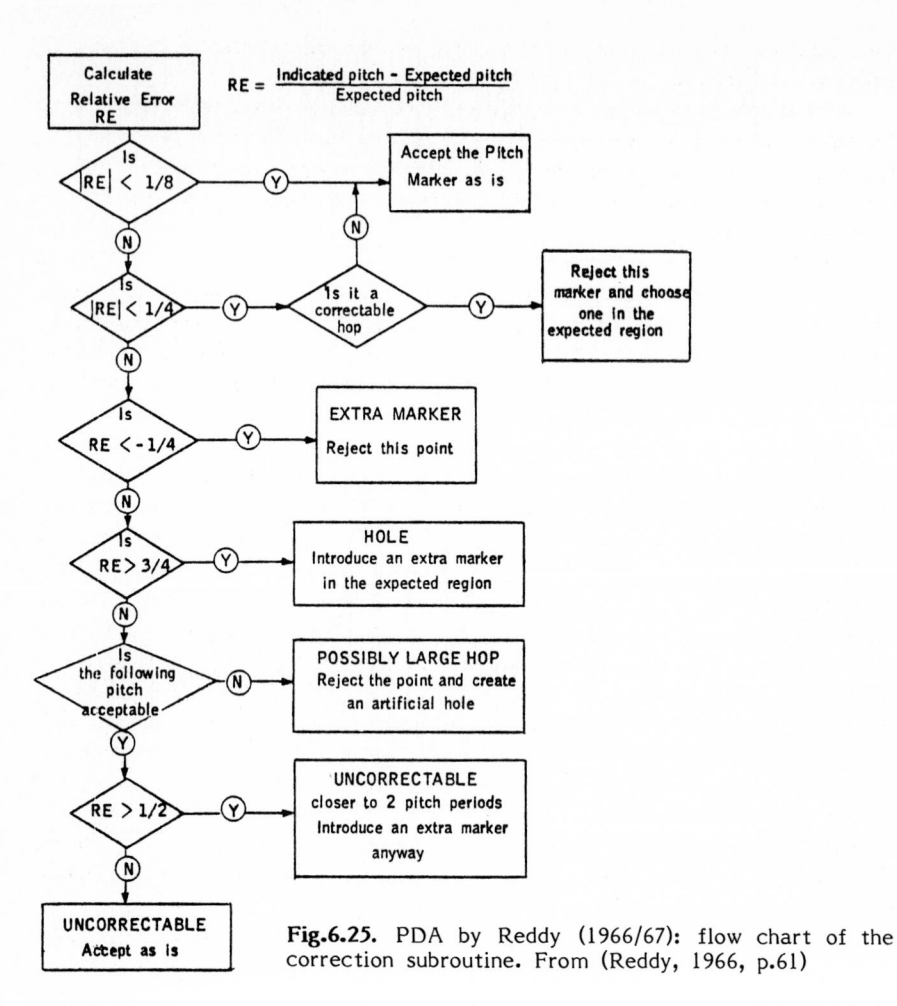

Fig.6.25. PDA by Reddy (1966/67): flow chart of the correction subroutine. From (Reddy, 1966, p.61)

interval, the corresponding marker is labeled "regular." This "expected" pitch is easy to obtain when the regularity of several periods is already evident to the PDA. To obtain an initial estimate, however, the distances of all consecutive significant peaks in the whole utterance are needed to form a distribution out of which the "most likely" period length is computed. A certain amount of signal is needed to build up that distribution; this is the main reason why the local processing of the signal has to be followed by a more global processing mode. In the correction subroutine, the local value of the "expected pitch" is the duration of the previous period if that period has been regular; otherwise it is the most likely period as defined above. The relative error is calculated as indicated in Figure 6.25. If its absolute is below 12.5%, the marker is regular. A "correctable hop" is given when the marker changes peaks so that errors in adjacent pitch periods are of a cancelling nature. In this case the largest peak of the signal within the tolerance interval, if existing, is chosen as the new pitch marker.

"Correctable hop," "hole," and "extra marker" (chirp) are corrected straightforwardly. The rest of the corrections, however, need further iterations. The "following pitch" is acceptable if the marker under consideration can be embedded in a "regular" future, for instance, if the marker is the first one in a voiced segment.

The algorithm was tested with 30 sentences (by one speaker). It "... *results in correct pitch indication in most cases but not all the cases ... In present form the algorithm is not suitable for those applications ... where the data has been band-compressed and phase-distorted.*"[13] The algorithm ran within one or two times real time on an IBM 7090 computer (for 1966 this was extremely fast).

Together with the PDA by Gold (1962a,b; to be reviewed in Sect.6.2.4), this algorithm is the first to systematically exploit the facilities offered by the computer and only by the computer, namely the ability of branching and decision making on the one hand, and the ability of storing repeated aand backward scanning on the other hand. The greatest advantage of the algorithm undoubtedly is its simplicity, not so much because of the performance of the algorithm itself but rather for didactic reasons. This PDA shows (to a much greater extent than the one by Gold) that computer-aided pitch determination according to the principle of structural analysis does work, and that it works very fast and for a great number of cases with a comparatively simple, well understandable algorithm. This PDA has thus opened the way for a number of simple and fast special purpose PDAs. Some of them will be discussed later in this section.

Besides the advantages given by the simplicity of this PDA, one should not overlook its shortcomings. The basic extractor is probably still too simple to be suitable for a universal application. Even for the restricted application in the speech recognition system, for which the PDA was designed, the basic extractor produces such a large number of errors that a global correction step appears necessary. This global correction subroutine, however, forms the main point of criticism. Referring to a more detailed discussion of this problem In Sect.6.6.3, the main contradictions are summarized as follows:

1) the global correction subroutine introduces a substantial delay in processing;

2) the routine may become unstable when the errors in the basic extractor become too numerous, and when the markers are scanned several times.

---

[13] In the opinion of the present author, Reddy's PDA does not require the presence of the first harmonic in the signal. It may be sensitive, however, to low frequency phase distortions which are likely to increase the probability of hops.

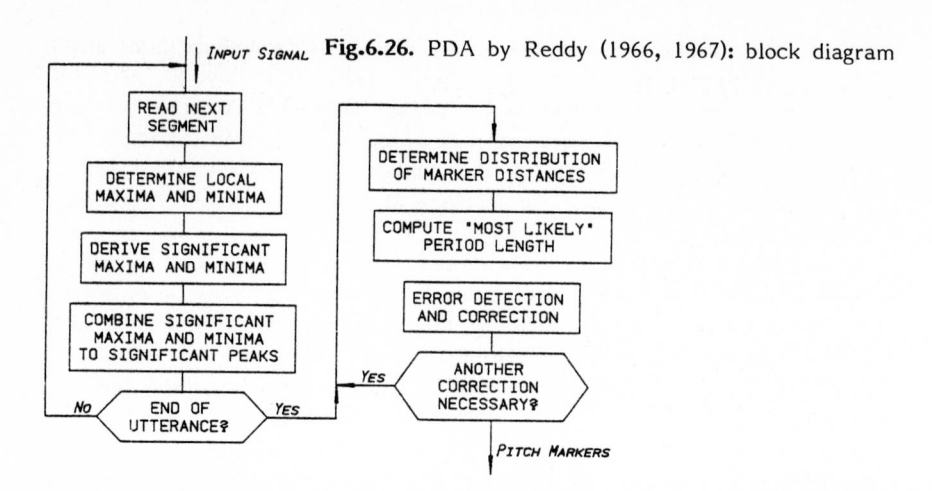

INPUT SIGNAL **Fig.6.26.** PDA by Reddy (1966, 1967): block diagram

The latter statement was verified by Liedtke (1971) who investigated Reddy's algorithm in detail and found that this PDA tends to yield a great number of incorrect periods in the vicinity of the lower limit of the permitted period duration range (2.5 ms).

6.2.3 Zero Crossings and Excursion Cycles

Peaks are popular as primary cues for time-domain structural fundamental period determination. They are not the only cues, however. In this section, PDAs are presented whose primary cues are different: zero crossings.

Figure 6.27 shows the block diagram of the PDA by N.Miller (1974, 1975). The first two steps deal with what is called excursion cycles. These steps perform a very efficient data reduction. The further steps - basic extraction and correction - are realized as usual, i.e., in a manner similar to Reddy's algorithm. The essentially novel feature of this PDA is how data reduction is carried out. The *excursion cycle* (EC) is defined to consist of all the samples between two consecutive zero crossings of the signal. In voiced speech signals, the length of an EC usually equals half the period length of the first formant. Hence, a significant data reduction is achieved when the signal is represented by its ECs, since the number of ECs, for voiced sounds, is at least one order of magnitude less than the number of samples. A "data structure" is compiled from the ECs as follows.[14]

> Each pitch period consists of a small number of excursion cycles. The first excursion cycle that occurs in a pitch period can be called its significant excursion cycle. The beginning of a pitch period is defined to be the first nonzero sample of its significant excursion cycle. Using this definition,

---

[14] In the following and all other quotations of the articles by N.Miller, his original term *principle cycle* has been replaced by the term *significant excursion cycle*.

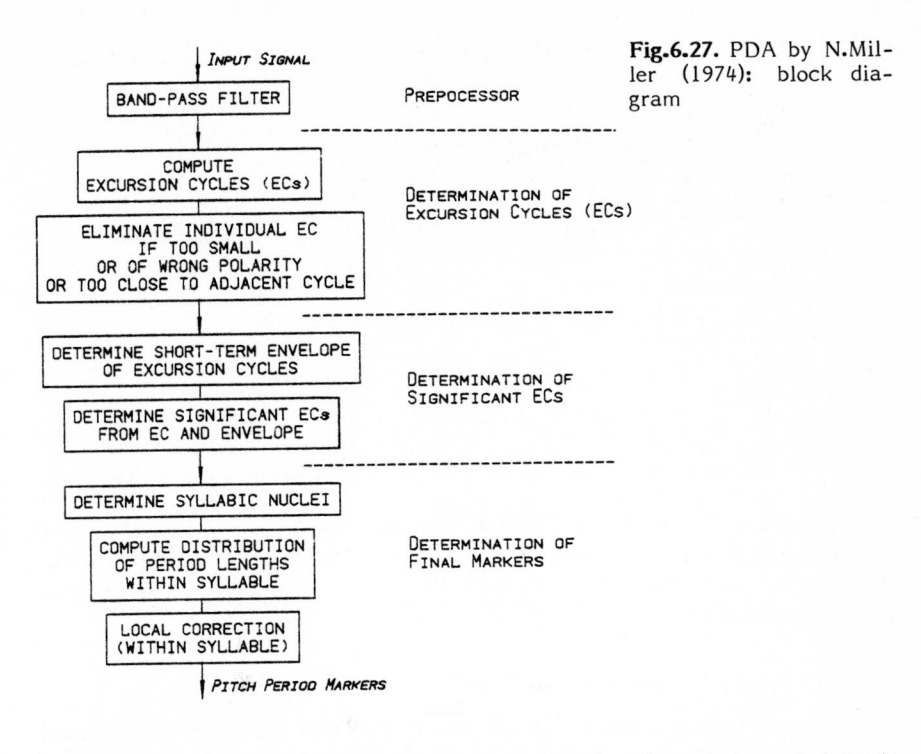

Fig.6.27. PDA by N.Miller (1974): block diagram

identification of the significant excursion cycle of a pitch period leads directly to identification of its beginning. Therefore, the excursion cycles which are compiled into the data structure can be limited to those which are possible significant excursion cycles.

Each excursion cycle is numerically analyzed as a potential significant excursion cycle. This analysis is based on two properties of speech signals. First, the speech waveform is generally characterized by greater amplitudes at the beginning of a pitch period than near its end. The significant excursion cycle is the first excursion cycle in a pitch period, and therefore often contains the greatest amplitude of the pitch period. Accordingly, each excursion cycle can be characterized by its amplitude. Secondly, unvoiced portions of a speech signal have shorter time intervals between zero crossings than voiced portions. Thus excursion cycles of unvoiced portions have short durations. Significant excursion cycles tend to have large amplitudes and long durations. Consequently, they have considerable energy. The excursion cycles can therefore be analyzed as to their energy. This process sums the samples within each excursion cycle, computing a coarse energy approximation. The magnitudes of the sums from significant excursion cycles generally exceed any sums computed from unvoiced sections. A threshold value can be estimated which exceeds the sums computed from unvoiced sections. Computed sums for each excursion cycle are compared to this threshold value. Only those excursion cycles whose sum exceeds the threshold value are entered in the data structure. This selective elimination of excursion cycles from unvoiced portions further reduces the data to be processed. The sampled waveform of a typical unvoiced and voiced signal is shown in Fig.6.28a. Figure 6.28b displays the magnitudes of sums for corresponding excursion cycles. (Miller, 1975, p.73).

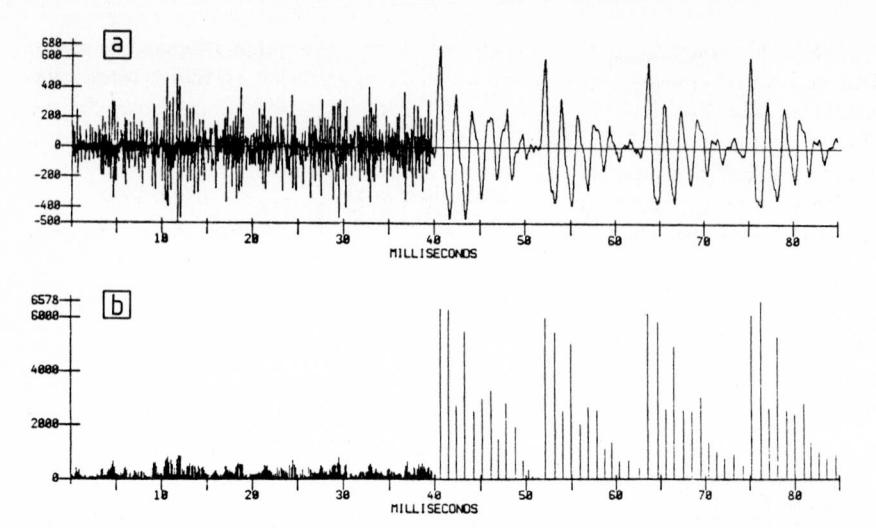

Fig.6.28a,b. PDA by N.Miller: Extraction of excursion cycles. **(a)** Sample signal; **(b)** sample sums of ECs. The left-hand side of the figure shows a voiceless signal; on the right-hand side the signal is voiced. The value of the sum $s(q) = \Sigma \, x(n)$ of an EC (which is assumed to start at the instant $n = q$) depends on the amplitude of the signal within the EC *and* the duration of the EC. Since the ECs of voiceless sounds are much shorter than the ECs of voiced sounds, there is a considerable amplitude difference between voiced and voiceless signals if the sum function $s(q)$ is regarded (even if the amplitude of the original signal is comparable for voiced and voiceless sounds). This feature permits the VDA to be incorporated into the PDA in this special case. From (Miller, 1974, p.123)

In this algorithm the voiced-unvoiced decision is so closely associated with period determination that it must be discussed in connection with the discussion of the PDA. The first reduction of the number of ECs is performed by threshold analysis. Since the sum of amplitudes within a single EC, which is simply the sum over all the samples in that particular cycle, is proportional to the signal amplitude and, in addition, to the length of the EC, those ECs compiled from voiceless signals are excluded since they are too short, whereas voiced ECs of low amplitude but relatively long duration may give sums well above the threshold.

The second and third reduction of the number of ECs is done by two different signal constraints. Firstly, the significant EC of the signal maintains its polarity during most of the time, and this polarity can easily be determined by amplitude discrimination. Once this polarity is known, all the ECs of opposite sign can be omitted. Secondly, the range of fundamental frequency is limited. This PDA has a value of 500 Hz as an upper limit. Hence, each EC is checked as to whether there exists another EC with a greater maximum amplitude within an interval of 2 ms on either side. If yes, the EC under consideration can be dropped.

The ECs remaining in the data structure are then characterized by three data entries: 1) the starting address which is the address of the first nonzero sample in the EC, 2) the amplitude of the peak of the EC, and 3) the address of this peak. The author reported that the compilation of this data structure requires about half the total computation time.

The rest of the algorithm represents the basic extractor which determines the markers in the typical manner applied to PDAs of type PiTS. Like the

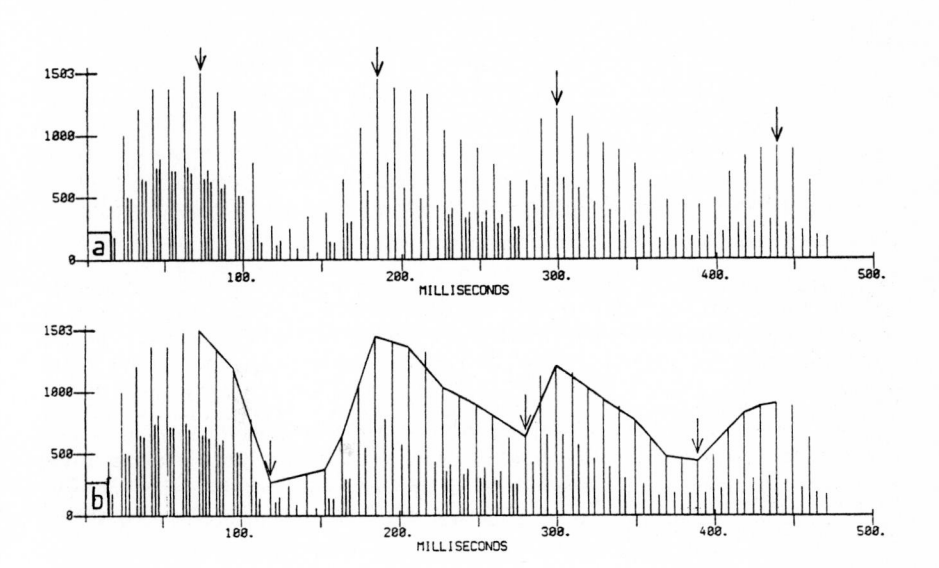

Fig.6.29a,b. Determination of syllabic nuclei, syllable boundaries, and envelope in the PDA by N.Miller. Arrows indicate (**a**) syllabic nuclei; and (**b**) syllabic boundaries. From (Miller, 1974, p.126)

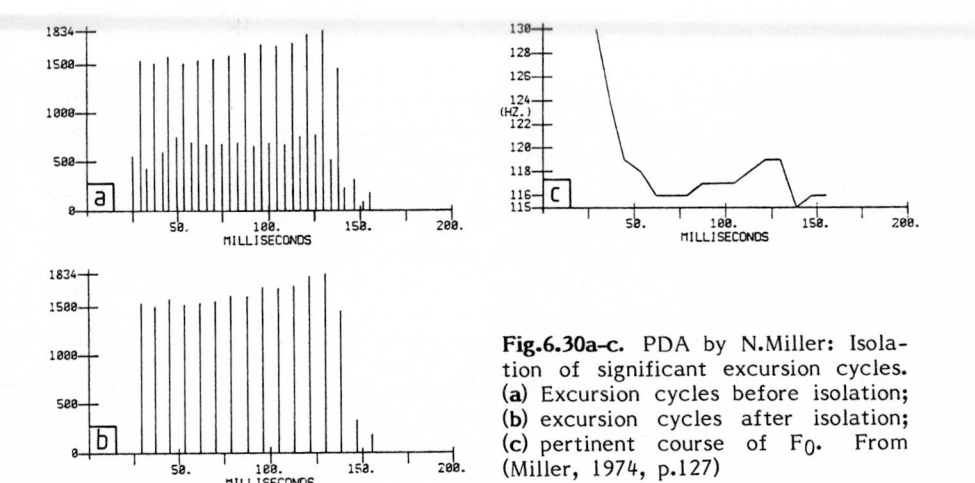

Fig.6.30a–c. PDA by N.Miller: Isolation of significant excursion cycles. (**a**) Excursion cycles before isolation; (**b**) excursion cycles after isolation; (**c**) pertinent course of F_0. From (Miller, 1974, p.127)

PDA by Reddy, this PDA is a single channel algorithm. The individual steps are realized in a way similar to Reddy's implementation. Accordingly, this PDA also has a global correction step which prevents it from on-line operation unless a delay of several hundred milliseconds is tolerated. Moreover, also the basic extractor is not quite on-line either. It needs information about syllabic nuclei and boundaries and processes the signal syllable by syllable.

Miller set the lower limit of the range of F_0 to 50 Hz. That means, at least one EC which is a candidate for a significant excursion cycle must be present every 20 ms. An interval greater than 20 ms between consecutive ECs indicates a voicing discontinuity; single ECs with no neighbors within 20 ms are regarded as unvoiced and excluded from further processing[15].

Each voiced segment of speech contains at least one syllabic nucleus (usually the vowel). Here the syllabic nucleus is defined to be an EC whose maximum amplitude exceeds that of all neighbors in an environment of 60 ms. Once the first syllabic nucleus in a voiced section has been found, the consecutive one is looked for in the same way; if present at all, it must be located at least 80 ms apart from the nucleus previously found. These duration thresholds were experimentally verified. (Fig.6.29).

The syllabic boundary which separates two consecutive nuclei is located by constructing an amplitude envelope between them. First, the time interval is subdivided into segments of 20 ms duration. Each segment is then represented by the EC with the largest amplitude. The envelope is constructed by straight line connection of the amplitudes of the ECs that represent the individual segments; then the minimum of the envelope between two consecutive nuclei is marked as syllabic boundary. Once a syllabic boundary has been found, the delimited syllable is further processed individually. A certain overlap between consecutive syllables prevents concatenation errors.

In the next step, which is carried out at the level of syllables, the significant ECs are further isolated. Any EC whose amplitude reaches the envelope by more than 90% is labeled a preliminary significant EC. Using these ECs, a preliminary period duration distribution is taken within the syllable. The maximum of this coarsely quantized distribution is then defined to represent the expected period duration within the syllable, i.e., the preset for the final step in the basic extractor. Since the period duration is not expected to vary more than one third of an octave within an individual syllable, an EC is eliminated from further processing if another EC with greater amplitude is found within a distance less than 70% of the expected period duration. By now, most of the markers have already been correctly determined [Fig.6.30, step from (a) to (b)].

---

[15] It is always dangerous to regard a segment as "unvoiced" only because no pitch estimate can be found and the segment is obviously irregular. In this particular case, however, the examination at the level of significant ECs is identical to that at the level of ordinary ECs; it forms a part of the simultaneous determination of pitch and voicing which is typical for this individual algorithm.

The final correction step (not included in Fig.6.27) operates on the basis of a whole voiced segment. This step corresponds to the global list correction subroutine described in Sect.6.6.1; a discussion of further details is unnecessary.

This algorithm avoids some of the possible errors associated with Reddy's PDA. In particular, the application of the distribution of period durations on the level of the syllable appears advantageous. On the other hand, the PDA may be sensitive to second harmonic tracking, especially when the signal is band limited. Another problematic point is the beginning of the period. Due to ambiguous and time-varying phase relations between the excitation function and the speech signal (Anathapadmanabha and Yegnanarayana, 1975), the zero crossing before the first EC in the period is unreliable. If the ratio of F1 and F_0 is odd, the last cycle of the previous period and the first EC of the new period may be combined in such a way that there is only a local minimum between them instead of the expected zero crossing. In that case the significant EC of the new period is strictly connected to the last cycle of the previous period, and serious hops in the course of the markers may be the consequence. For an example, see Figure 6.31.

Another PDA using zero crossings intervals as primary cues was proposed by Geçkinli and Yavuz (1977). This PDA claims to follow "the visual pitch extraction part" of the interactive PDA by McGonegal et al. (1975; see Sect.9.2.2). In fact it is a pure time-domain PDA which is entirely based on the sequence of zero crossing intervals, i.e., the distances between adjacent zero crossings. Hence it can work with infinitely clipped speech and is about as fast as the PDA by N.Miller.

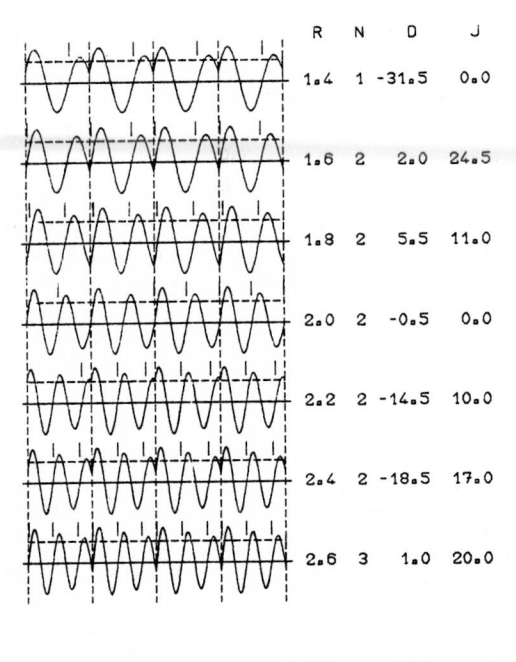

R	N	D	J
1.4	1	-31.5	0.0
1.6	2	2.0	24.5
1.8	2	5.5	11.0
2.0	2	-0.5	0.0
2.2	2	-14.5	10.0
2.4	2	-18.5	17.0
2.6	3	1.0	20.0

Fig.6.31. Influence of the ratio F1/F0 on the position of markers. To generate the signal displayed, a weakly damped second order digital filter was excited by a pulse train. (R) Ratio F1/F0; (N) number of "markers" per period; (D) deviation in position (percentage of real period length); (J) jitter (% of average marker distance). Dashed vertical lines indicate the locations of the excitation pulses. The markers are defined to be situated in the last positive zero crossing before the threshold of the TABE (dashed horizontal line) is crossed. This corresponds to the starting point of the excursion cycle as defined by N.Miller (1974, 1975)

Only a brief outline of the principle will be given here. Whenever the first formant F1 does not directly coincide with a multiple of F_0, i.e., for an odd ratio $F1/F_0$, there will be at least one zero crossing (ZX) interval in every pitch period which deviates in length from the others. If we assume that the vocal tract is excited by a pulselike waveform, the resultant impulse response will be the sum of complex exponentials with frequencies equal to the formant frequencies F1, F2, and so on. If the signal is low-pass filtered to 900 Hz, as done in this PDA, all the formants except F1 (and sometimes F2) are removed. The ZX intervals of the filtered signal will thus equal half the period length of F1, at least immediately after the beginning of the period where the waveform of F1 dominates the signal. This fact has been frequently exploited to determine F1 in time domain; it is also used by the present author in his PDA (see Sect.7.1). The ZX interval remains fairly constant until the beginning of the next period where the damped oscillation is disturbed by the new laryngeal pulse[16]. To demonstrate this, Fig.6.31 shows the synthetic example of a "formant" which moves from 1.4 to 2.6 times the fundamental frequency. The most unreliable ZX of the whole period is the last (or first) positive zero crossing before the principal maximum.

The use of this ZX as starting point of the pitch period formed one point of criticism of Miller's PDA; for Geçkinli and Yavuz it is the anchor point and the reason that their method does work at all. Geçkinli and Yavuz subdivide the ZX intervals of subsequent pitch periods into "stable" and "unstable" intervals. The stable, i.e., approximately constant length ZX intervals correspond to the F1 waveform after the beginning of the period (in the closed glottis interval); they are expected to occur again in the same order in the consecutive period(s). The unstable ZX intervals towards the end of the period may or may not occur again in subsequent periods. If the period is correctly detected, the sum of a single sequence of stable and unstable ZX intervals equals its length. The PDA thus works as follows.

1) A voiced frame of the utterance (about 25 ms) is scanned in order to find out the most stable ZX interval length. This is done setting up the distribution of the lengths of all the ZX intervals in that frame.

2) Starting from this most stable ZX interval, stable and unstable intervals are separated. A ZX interval is labeled "stable" if it does not depart from the "most stable" interval length by more than a given threshold value.

3) The most likely sequence of stable and unstable ZX intervals to form a sequence of regular pitch periods is determined from the ZX interval sequence.

This PDA has a severe limitation which requires a constraint on the range of F_0. In case F1 coincides with a low harmonic of F_0 (second or third harmonic), all ZX intervals will be equal in length. Thus, the PDA will fail system-

---

[16] In real speech the oscillation of F1 is even disturbed when the glottis opens at the beginning of a new laryngeal pulse (Holmes et al., 1978).

atically unless the measurement range of F_0 is confined to less than one octave. With this restriction, second harmonic tracking becomes impossible. The very narrow preset of one octave is achieved by a speaker-adaptive training phase. Four to five seconds of speech are reported to be necessary in order to adapt the algorithm to a new speaker. In the opinion of the present author, even this restriction of one octave may not be enough when the signal is band limited to the telephone channel. Second harmonic tracking will undoubtedly be impossible, and for undistorted signals the fundamental harmonic is usually strong enough to produce at least one deviating ZX interval when F1 coincides with the third harmonic. For band-limited signals, however, the coincidence of F1 with the third harmonic can produce six ZX intervals of equal length in one pitch period. Since all the information on amplitude has been removed, the algorithm can fail by tracking 2/3 or 4/3 of the real period duration instead of the true pitch. One must thus be cautious when applying this PDA to band-limited signals. When the algorithm works, however, it works very fast, mostly less than 0.7 times real time on a PDP-8 computer with a 1.5-μs memory cycle. To exactly determine the ZX intervals, relatively high sampling frequencies (24 to 30 kHz) are required.

The PDA by Czekajewski and Wojczynski (1968) combines the investigation of distances between adjacent zero crossings and the principle PiTN. The signal is first low-pass filtered; from the distance of consecutive zero crossings a new signal is then derived as follows,

$$y(n) = n_k - n_{k-1} \; ; \quad n = n_k \ (1) \ n_{k+1} - 1 \ , \tag{6.9a}$$

where n_i ($i = k-1$, k, $k+1$) are the instants where the zero axis is crossed (from negative to positive values). This means that the value of $y(n)$ always equals

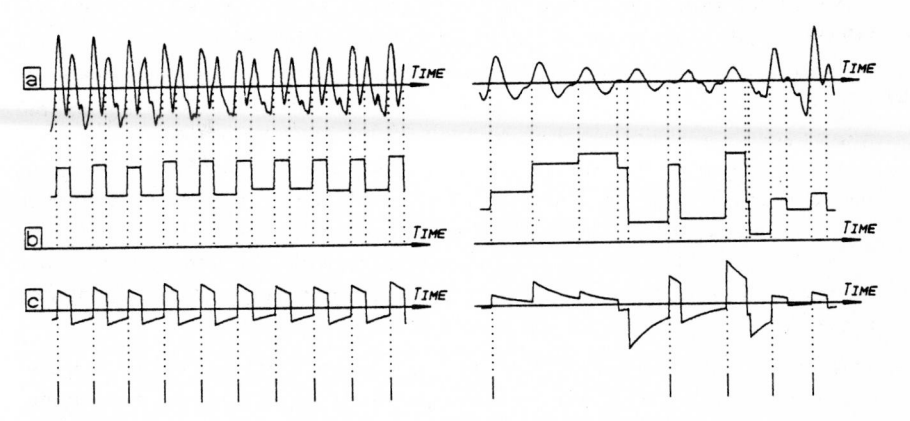

Fig.6.32a–c. PDA by Czekajewski and Woyczynski (1968): examples of performance. (a) Speech signal after some low-pass filtering; (b) zero crossings distance function $y(n)$; (c) zero crossings distance function $x_1(n)$. Signals: stationary vowel /o/, female speaker (left-hand side); transition, male speaker (right-hand side). The short-term average $\tilde{y}(n)$ was computed via a first-order digital filter (coefficient $g_1 = 0.95$)

the distance between the two most recent zero crossings; y(n) thus has the shape of a steplike nonnegative function. If the (short-term) average $\bar{y}(n)$ is subtracted from y(n),

$$x_1(n) = y(n) - \bar{y}(n) ,\tag{6.9b}$$

the signal $x_1(n)$ is identical to y(n) except that $x_1(n)$ varies around zero. Since y(n) is constant between adjacent zero crossings of x(n), the signal $x_1(n)$ has fewer zero crossings than x(n). The same procedure is now repeated for $x_1(n)$; it generates a signal $x_2(n)$ which again has fewer zero crossings. The algorithm always terminates with a signal $x_{i+1}(n)$ identical to zero. If the original signal x(n) is approximately periodic, the signal $x_i(n)$ will have two zero crossings per period so that it can be analyzed by a ZXABE.

The algorithm is simple and fast. However, it has considerable problems. Figure 6.32 shows an example. Like the PDA by Geçkinli and Yavuz, it fails systematically when the formant F1 exactly coincides with a higher harmonic. In addition, if the zero crossings pattern changes during nonstationary segments, this may result in a drastic change of the short-term average of y(n) so that the signal $x_1(n)$ becomes irregular.

6.2.4 Mixed-Feature Algorithms

The algorithms discussed in this section are the most convenient and powerful ones in group PiTS without necessarily being the most complicated ones. They combine the complexity and the individual freedom of algorithmic structure investigation with features adopted from other areas such as temporal structure transformation by preprocessing (Tucker and Bates, 1978) or parallel processing (Gold and Rabiner, 1969).

The best known algorithm in this section is probably that by Gold and Rabiner (1969). This PDA performs parallel processing in the basic extractor. Six identical "primary extractors" work on six different signals derived from the maxima and minima of the speech signal. The results of these six extractors are then combined in a final pitch period computation block (Fig.6.33).

The basic idea of this algorithm had previously been expressed in an earlier algorithm by Gold (1962a,b, 1964c). It is a combination of "regularity" scan and derivation of "peakedness." From a study of the literature Gold (1962b) concluded that

> ... the pitch detectors based solely on regularity [cited are time-domain PDAs, such as (Gruenz and Schott, 1949); or autocorrelation techniques, such as (Gill, 1959); or (Fujisaki, 1959)] fail when either the pitch or the spectrum of the speech change too rapidly; the pitch detectors based only on peakedness [cited are time-domain PDAs, such as those by Dolansky 1955); or Lerner (1959)] fail when the wave is insufficiently peaked (as often happens), even though the wave may be perfectly regular at the time. Thus, one type of pitch extractor may fail where the other succeeds, and vice versa. (Gold, 1962b, p.917)

Gold claimed in his program to have exploited both peakedness and regularity. He further concluded:

> Designers of pitch detectors have, of course, tried to make their circuits simple, and, to that end, have usually tried to find *the* one operation which will give a good pitch indication. There is a serious doubt, however, as to whether any one rule will suffice to weed out the pitch from as complicated a waveform as speech. (Gold, 1962b, p.917)

In consequence, he applied three "instantaneous" rules to the maxima of the signal. Each maximum in the signal is tagged according to the particular rule which is applicable. In the following regularity test, a window consisting of five consecutive peaks is moved throughout the signal. This regularity test operates in various modes. Each mode investigates the peaks in a certain manner according to the associated tags; the test proceeds from the most to the least restrictive constraints until it succeeds in finding a regular pattern, or until it has unsuccessfully run through all the implemented modes. After that, a final regularity test (global correction subroutine) is carried out over the whole utterance in order to fill up the remaining holes and reject the remaining chirps.

Like the algorithm by Reddy, this PDA is not "instantaneous," i.e., it needs a whole utterance or at least a whole voiced segment to make its final decision. Another shortcoming is the fact that the implementation is based solely on the positive peaks, i.e., on one polarity of the waveform. Like a PDA (PiTN) using half-wave rectification, the algorithm thus becomes phase sensitive.

The further development of this older algorithm, carried out first by Gold alone (1962b, 1964c), later together with Rabiner (1969), simplifies the original algorithm and removes these two shortcomings. Figure 6.33 shows the block

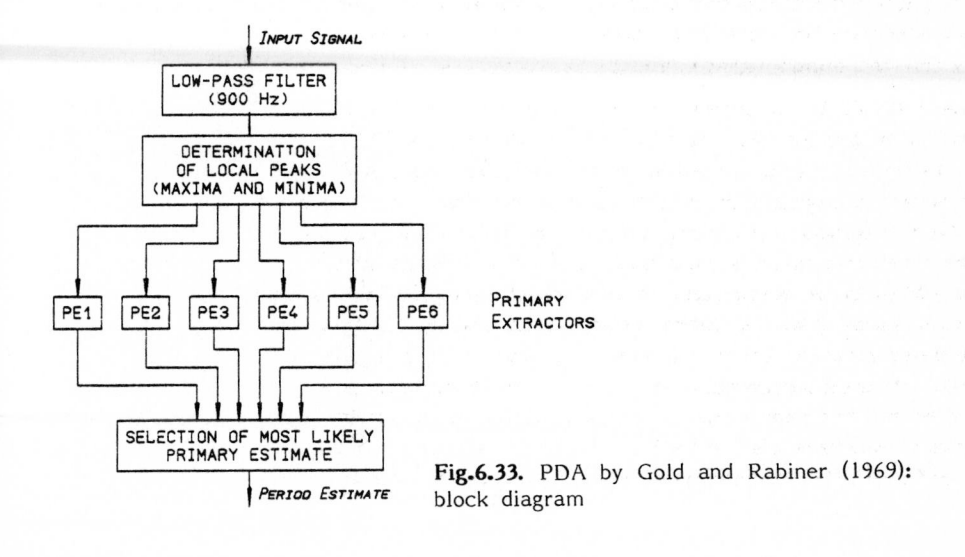

Fig.6.33. PDA by Gold and Rabiner (1969): block diagram

diagram of the PDA by Gold and Rabiner. Unlike Gold's original algorithm (1962a,b), this one has a preprocessor, a low-pass filter with a cutoff frequency of 900 Hz which is well above the measuring range (50 to 600 Hz). The cutoff frequency as well as the specifications of this filter are uncritical. The authors claimed that if the first harmonic is present in the signal the cutoff frequency of the preprocessor filter can even be lower. They further found that

> ... if the cutoff frequency is reduced to about 250 Hz, a noticeable change appears in the estimated pitch periods and this has been shown to produce vocoded speech with a rough quality (Goldberg, 1967). Also, in the case when no fundamental is present and nonlinear elements are used to re-generate the fundamental, these same effects occur; i.e., the vocoder speech has a noticeable rough quality. (Gold and Rabiner, 1969, p.444).

This last statement may be interpreted in such a way that PDAs of type PiTS which derive the markers from the peaks of the signal do not work well after application of a nonlinear function. These algorithms are nonlinear by themselves since any procedure which assigns a tag to selected individual samples and no tag to the others is nonlinear, and this nonlinearity and that of the preprocessor might influence each other in an unpredictable way. In fact there is no algorithm of this extreme type PiTS that applies a nonlinear function in the preprocessor.

To continue with the description of the PDA by Gold and Rabiner, six "functions of peakedness" M1 through M6 are derived at the local maxima and minima of the signal. (It is left open whether these functions are derived only for positive maxima and negative minima; but one can assume this since positive minima and negative maxima should not occur in the low-pass filtered speech). These functions are explained in Fig.6.34a.

> Pulses of height M1, M2, and M3 are generated at every positive peak of the filtered speech while pulses of height M4, M5, and M6 are generated at each negative peak. Measurements M1 and M4 are simple peak (positive and negative) measurements, whereas measurements M2, M3, M5, and M6 depend on previous peaks of the speech. Measurements M2 and M5 are peak-to-valley and valley-to-peak measurements whereas M3 and M6 are peak-to-previous-peak and valley-to-previous-valley measurements. All the M's are converted into positive pulse trains. Measurements M3 and M6 are not permitted to become negative. Hence, if a current peak (or valley) is not as large as the previous peak (or valley) measurement M3 (or M6) is set to zero (Fig.6.34b). For the case when only the fundamental is present (Fig.6.34b at the left-hand side), measurements M3 and M6 fail but meas-urements M1, M2, M4, and M5 provide strong indications of the period. For the case when a very strong second harmonic (same figure, at right), and some fundamental, are present, measurements M3 and M6 will probably be correct, whereas M1, M2, M4, and M5 will incorrectly provide indica-tions of half the period ...
> The six sets of pulse trains are applied to the six individual period extractors PE1 through PE6. The operation of the extractor is illustrated in Fig.6.34c. In essence, each simple pitch period estimator is a peak-detec-

ting rundown circuit. Following each detected pulse there is a blanking interval (during which no pulses can be detected) followed by a simple exponential decay. Whenever a pulse exceeds the level of the rundown circuit (during the decay), it is detected and the rundown circuit is reset. It should be noted that both the rundown time constant and the blanking time of each detector are functions of the smoothed estimate of pitch period of that detector. (Gold and Rabiner, 1969, pp.444-445).

In its structure, the individual extractor corresponds to the analog realizations discussed in Sect.6.2.1 except that the analog PDDs do not apply the blanking interval.

If all the algorithms had their individual labels, this one would be the "fault-tolerant PDA." There is no PDA in which the possibility of committing errors in marker determination is taken into account to such an extent as here. It even tolerates that, like in the example in Fig.6.34b (at the right-hand side), the majority of the primary extractors may be wrong. The algorithm can yield correct results as long as at least one of the period extractors gives the right estimate. The problem of selecting the correct estimate is solved in the final period computation subroutine in a very elegant way. Provided that the blanking time of a primary extractor does not exceed the pitch period duration, one can assume that at least the primary extractors

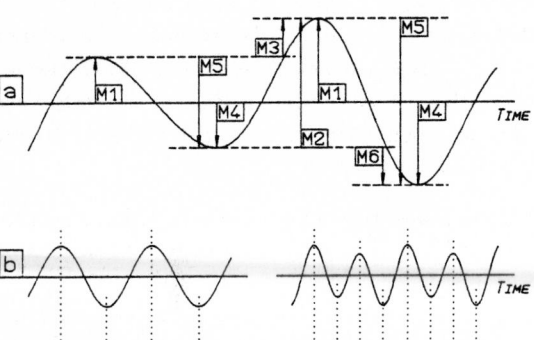

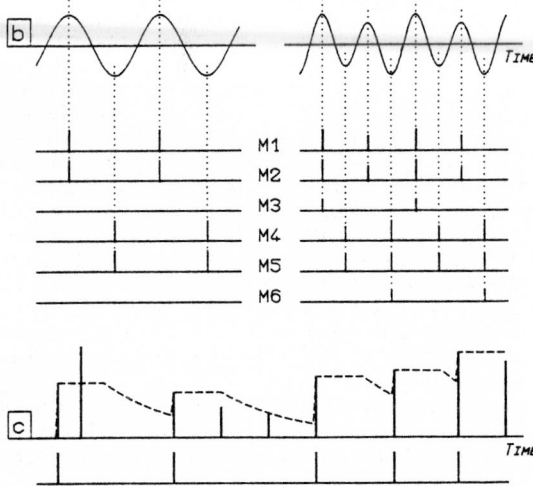

Fig.6.34a–c. PDA by Gold and Rabiner (1969): principle of operation. **(a)** Definition of the six individual peak functions M1 through M6 which serve as input to the six simple period extractors (PE1... PE6); **(b)** example of operation, sample signals: (left) pure sinusoid, (right) signal with a strong second harmonic; **(c)** operation of a simple "period extractor" PE_i, $i = 1\,(1)\,6$. The peak functions M1 through M6 are defined to occur at the same instant as the peaks they are associated with, as shown in (b), unlike (a)

generating the pulses M1, M2, M4, and M5 will never produce a hole since every peak of correct polarity will give an impulse to them. These extractors, on the other hand, can perform higher harmonic tracking very well since the blanking time is adaptively adjusted; it depends solely on the uncorrected period estimate of the primary extractor. Thus, the errors of the primary extractors are forced to be biased towards higher harmonic tracking so that a minimum-frequency selection principle (see also Sect.6.4) can be applied. The results of the six extractors are grouped into a 6×6 matrix whose columns represent the individual detectors and whose rows represent the period length estimates (Fig.6.35). In the first row we find the most recent estimates of the individual extractors which are the candidates for final estimation. To give a correct performance of the PDA, at least one of these values must be correct. In rows 2 and 3 are the estimates of the two previous periods. Rows 4 and 5 contain the sum of rows 1 and 2 and rows 2 and 3, respectively; and row 6 contains the sum of rows 1 to 3. If one primary extractor incidentally traces the second harmonic, the values in rows 4 and 5 will give the correct period; the same applies for row 6 when the extractor traces the third harmonic. The algorithm then compares each of the candidates in the first row to the other 35 entries and computes the degree of coincidence as the sum of individual events of coincidence. An event of coincidence is given when the absolute difference between the candidate under consideration and another entry in the matrix is less than a given threshold. This coincidence is tested four times with increasing thresholds (i.e., with decreasing accuracy requirements); an increasing bias associated with each threshold is subtracted from the number of coincidence events; this bias compensates the fact that the number of coincidences grows with decreasing accuracy. The "winner," i.e., the candidate with the greatest number of coincidences, is chosen as the most likely pitch estimate.

The procedure works very well when the coincidence windows and the biases are carefully balanced. These values are fixed; only the blanking interval and the rundown time constant are changed adaptively. The algorithm does not require the presence of the fundamental and runs very fast.

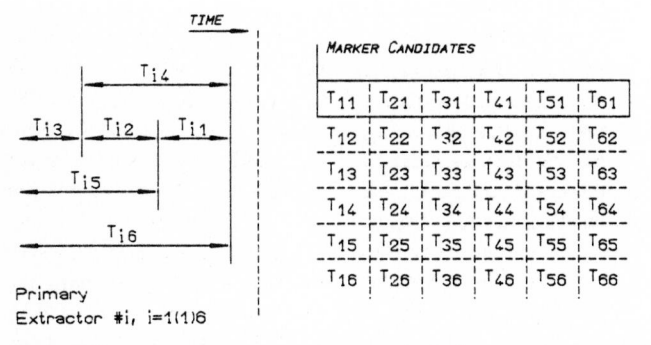

Fig.6.35. PDA by Gold and Rabiner (1969): matrix for channel selection

The authors presented two simplifications to their algorithm for a limited range of F_0 (below 220 Hz, cf. Sect.6.2.5). Both of them realize a subset of the original algorithm and show a considerable gain in computing speed.

Unlike the older algorithm (Gold, 1962a,b), this algorithm is quasi-instantaneous. This feature, together with its robustness and simplicity, makes the algorithm attractive. No global correction step whatsoever is needed, and the length of the local window does not exceed three periods.

For these advantages, however, a certain price has to be paid: the phase relation between the signal and the result is lost. The primary extractors still preserve the beginnings of what they estimate to be a period; but this information is discarded in the subroutine which computes T_0 from the candidate matrix. Since the winner can change permanently, and since the test is repeated whenever one of the primary extractors presents a new pulse (and with it a new estimate), the algorithm will yield several estimates per pitch period. These are estimates of period duration, but not an indication for the starting point of individual periods. In this respect, the PDA by Gold and Rabiner does not quite match the previous definition of a time-domain PDA (see Chap.2). A resynchronization of markers to detect the beginnings of individual periods has proved difficult.

This PDA has been applied in several analysis-synthesis systems (Gold, 1977; Allen, 1976). It also participated in the comparative evaluation by Rabiner et al. (1976).

A more recent PDA of similar type has been presented by Tucker and Bates (1978). The essentially novel feature of this PDA is the application of adaptive center clipping in the preprocessor. The number of peaks to be checked is thus considerably decreased. In short-term analysis PDAs, center clipping is a commonly used technique, especially for autocorrelation PDAs (Sondhi, 1968; Rabiner, 1977; see Sect.8.2 for more details), since it removes the formant structure and retains the periodicity in the signal. In the time domain representation center clipping emphasizes the principal peaks whilst suppressing minor maxima and minima. Therefore it greatly simplifies the job of a PiTS basic extractor.

In the PDA by Tucker and Bates, all the peaks which survive the center clipper are regarded as independent pulses. For each pulse, five features are determined: 1) its amplitude; 2) its width, i.e., the distance between the time at which the clipping level is exceeded and the time at which it is crossed again; 3) its energy (mean square value of the pulse); 4) its polarity (sign); and 5) the ratio of amplitude and (square root of) energy as a shape factor. With these five features (three "primary" and two "secondary" features) which resemble the ones used by Miller (1974) the PDA searches the correct sequence of markers in a way corresponding to the usual PiTS basic extractor already discussed in this section. Like the PDA by Gold and Rabiner, it does not need a global correction step. Since this PDA is mainly applied to musical sounds, it has a remarkably wide range of F_0: 40 to 2400 Hz. The basic extractor is even simpler than that of the PDA by Gold and Rabiner (grace to the center clipping), the performance is comparable[17,18], and the phase relations are preserved (with the principal peaks of the period as anchor points).

6.2.5 Other PDAs that Investigate the Temporal Structure of the Signal

Two types of PDAs shall be discussed in this section: 1) simple PDAs for restricted applications, and 2) algorithms that perform a moderate degree of preprocessing. So far, the algorithms presented in Sects.6.2.2-4 usually do not perform any preprocessing except for an optional moderate low-pass filtering with a cutoff frequency well above the permitted range of F_0. Hence, these algorithms have to deal with the whole variety of temporal structures in the signal. The price which must be paid is either increased complexity or the necessity of a global correction step. There appear to be two ways to avoid paying this price: a) designing special-purpose PDAs for restricted applications, and b) performing some preprocessing. In this section we start with PDAs that perform some preprocessing. Simple PDAs with restricted applications will be discussed in the second part.

Just by looking at the PDA by Tucker and Bates (1978), one can imagine how much can be gained even by a moderate degree of preprocessing. This algorithm was already discussed in the preceding section. It performs some center clipping which eliminates a large number of insignificant peaks before the basic extractor is entered at all.

Ambikairajah, Carey, and Tattersall (1980) presented a PDA whose preprocessor applies a comb filter. Comb filters are well suited to enhance or suppress a periodic signal in an unknown wavelet when their delay parameter d matches the (unknown) period T_0. This principle resembles the average magnitude difference function (AMDF, see Sect.8.3). It can also be regarded as a simplification of the least squares principle (Sect.8.6). Contrary to the AMDF, the present PDA applies comb filters where the individual pulses have equal signs; these filters thus enhance the periodic signal when the period T_0 equals the pulse delay d. The PDA applies comb filters with two to four pulses:

$$y(n) = \sum_{k=0}^{K-1} x(n+kd) ; \quad K = 2, 3, 4 .$$ (6.10)

---

[17] Tucker and Bates, by the way, showed that the PDA by Gold and Rabiner is also able to cope with that wide range of F_0.

[18] Evaluating their PDA with the autocorrelation PDA as a reference, Tucker and Bates (1978) found an exceedingly high number of gross errors for speech signals. Reading the original publication, however, one gets the impression that the criterion for a "gross" error (10% apart from autocorrelation estimate) is too narrow so that this PDA is in reality better than it is ranked in this evaluation.

To measure the amplitude of the output signal $y(n)$ depending on the trial period d, the algorithm simply detects the largest maximum $(n = n_{max})$ and the deepest minimum $(n = n_{min})$ of the signal $y(n)$ within a frame $n = q(1) q + N-1$ and computes their difference

$$D(d) = y(n_{max}, d) - y(n_{min}, d) . \qquad (6.11)$$

From the histogram $D(d)$, the period is derived as the location of the first significant maximum. The PDA can be used in a time-domain and a short-term analysis mode. If only $D(d)$ is regarded, the phase relations between the signal and the period estimate are lost; it is easy, however, to resynchronize the signal and the period estimate if the locations n_{max} and n_{min} of the signifi- cant peaks of $y(n,d)$ are stored. This PDA works well in a noisy environment (an exception is pulselike noise). It does not require the presence of the fundamental harmonic; it works extremely fast since it does not need any multiplication. As a by-product one gets a good measure of the global signal amplitude. As to the number K of pulses in the filter, $K = 2$ proves insuffi- cient; for $K = 3$, rapid changes of T_0 or of the temporal structure of the signal are easily traced, whereas $K = 4$ proves most resistive with respect to noise. For higher values of K, rapid changes of T_0 or of the temporal structure increasingly affect the quality of the measurement. More detail on this attractive principle is added in Section 8.6.3.

In the PDA by Maissis (1972), the period boundaries are determined by a short-term energy measurement. For a given sample n the algorithm computes the difference between the energy of the K following and the K preceding samples:

$$Y(n) = \sum_{i=n+1}^{n+K} x^2(i) - \sum_{i=n-K}^{n-1} x^2(i) \qquad (6.12)$$

This function is expected to show a distinct maximum at the beginning of a new period where the signal energy rises momentarily due to the new excit- ation pulse. (Thus the method is also related to the event detection PDAs discussed in Section 6.3.2.) The basic extractor investigates the temporal struc- ture of the function $Y(n)$ such that it generates a pulse of height $Y(n)$ for each n where $Y(n)$ has a distinct maximum. Spurious pulses are easily detec- ted by their lower amplitude. Maissis found that the method worked even better when the amplitude of the pulses was normalized with respect to the short-term energy of the signal, measured over a whole period. Moderate low-pass filtering before the computation of $Y(n)$ was also found advan- tageous.

If one regards (6.12), it is easily seen that this PDA implicitly follows principle PiTN. If (6.12) is decomposed into two simultaneous equations, it reads as follows

$$u(n) = x^2(n) , \quad Y(n) = \sum_{i=n+1}^{n+K} u(i) - \sum_{i=n-K}^{n-1} u(i) , \qquad (6.13)$$

and shows that this preprocessor consists of an instantaneous nonlinear function (squaring in the present case), a gap-function low-pass filter, and a comb filter in tandem. Since the NLF is even, the PDA will fail systematically when the input signal becomes approximately sinusoidal. This is implicitly admitted by Dours, Facca, and Perennou (1977) who have developed a very similar PDA (with full-wave rectification instead of squaring and a global correction routine in the postprocessor):

> Generally speaking, peak detection is satisfactory. Correction of faulty peaks is only undertaken in the following cases: weak, voiced fricative, at the end of a nasal vowel, and above all at the onset of a vowel preceded by the plosive /p/ ... (Dours et al., 1977, p.69; original in French, cf. Appendix B)

All the cases explicitly described in the above quotation show extremely low values of the formant F1. Both PDAs, the one by Maissis (1972) and that by Dours et al. (1977) seem not to have been tested with material originating from a female speaker.

A third algorithm developed in France by Le Roux (1975) applies a second-order adaptive filter. Its performance being comparable, it must be grouped into the category of temporal structure simplification and will be discussed in Section 6.3.

Let us move on to the simple algorithms with restricted application areas. There are two main areas of restrictions: 1) restrictions with respect to the fundamental frequency range, and 2) restrictions dealing with the environmental conditions. These restrictions may be temporary in such a way that a manual preset or a learning phase is needed before operation, or they may be permanent. A good example for temporary restrictions is the PDA by Geçkinli and Yavuz (1977; see Sect.6.2.3) which confines the range of F_0 to an octave, and this interval has actually been delimited in a previous training phase. Permanently restricted, for instance, are all the PDAs of type PiTL where the presence of the first harmonic is required. Further restrictions may be imposed with respect to phase. If phase distortions are not permitted, this means that the respective PDA can only be used on-line, i.e., in direct conjunction with a microphone (and not even in connection with a good-quality analog tape recorder).

Among the simplest PDAs ever proposed we find the one by Tillmann (1978). The signal is low-pass filtered by a first-order integrator filter,

$$y(n) = \sum_{k=q}^{n} x(k) = y(n-1) + x(n) \tag{6.14}$$

The output signal $y(n)$ shows a distinct minimum at the beginning of each period. This minimum is detected and marked; at the same time the output signal $y(n)$ is forced to zero; i.e., a new integration cycle is started whenever a marker has been set ($n = q$). By this, the integrator filter is forced to remain stable. If the starting point N coincides with the minimum at the beginning of the period, the output signal $y(n)$ will again come close to zero toward the end. If the analysis starts anywhere else, the minimum at the end of the period takes on a negative value. Thus the algorithm easily synchronizes with the proper beginning of the period. The price paid for this simplicity is relatively high; only undistorted signals with a certain predominance of the first harmonic are permitted (female voices, male voices only when they talk at close distance from the microphone). It is easily seen that this implementation closely corresponds to the realization PiTL with a ZXABE and extensive low-pass filtering in the preprocessor. In Tillmann's PDA the condition is that there is only one significant minimum per period. In Sect.6.1 we have shown that the ZXABE condition (6.1) is fulfilled when the signal is low-pass filtered with a slope of -18 dB per octave within the measuring range. Tillmann's PDA implicitly realizes these 18 dB per octave: 6 dB by the filter, another 6 dB by the restriction that the speaker must keep close to the microphone (and thus implicitly cancel the 6 dB per octave rise due to the radiation characteristics of his speech tract), and the remainder by investigating the significant minima instead of the zero crossings. The advantage of Tillmann's PDA, however, is that the problem of excessive dynamic range due to variations in F_0 is avoided (contrary to those PDAs where the 18 dB per octave are achieved by filtering alone). The application area (direct evaluation of utterances for phonetic and educational research) does not pose this problem anyhow since the level as well as the distance from the microphone can be manually adjusted by the speaker.

The PDA by Baudry and Dupeyrat (1976, 1978) is a simplification of Reddy's algorithm (see Sect.6.2.2). Like Tillmann (1978), Baudry and Dupeyrat used their algorithm for an on-line application, this time for a speech recognition system. The PDA examines the peaks of one polarity (in any example given by the authors the minima are checked). Each period is characterized by a significant peak. Let us assume that an estimate $\hat{p}$ of the period duration has been found, and that a marker has been placed at the beginning n_k of the actual period. This corresponds to the steady-state condition of the PDA; the algorithm establishes a tolerance interval around $(n_k+\hat{p})$ and looks for a significant peak within that interval. A significant peak has been found if a) its amplitude exceeds the amplitudes of all the other peaks in the interval, or if b) it exceeds the amplitude of the largest peak by a factor of 0.875 and is situated before the largest peak. Since the two conditions may be contradictory, a rule is set up such that condition b) has priority. Once the peak has been found, a marker is assigned to its location, and the estimate $\hat{p}$ of the period duration is updated. The process is initialized (at the beginning of a voiced section or after an obvious error) by finding two significant peaks whose distance falls within the permitted range of T_0. In a more recent version, a special set of rewriting rules was established for this purpose (Baudry and Dupeyrat, 1980). The algorithm operates in real time on a minicomputer. It has no global correction step; even the local correction step becomes unnecessary since the range of F_0 is restricted to one octave (100 to 200 Hz).

Schäfer-Vincent (1982) has also proposed a PDA which closely related to that by Reddy (1966, 1967). Significant maxima and minima are determined by checking the signal within small intervals around local peaks. Let a local peak occur at the instant $n = q$. This peak is regarded as significant if there is no sample of the same polarity within the interval $n = q-K/2 (1) q+K/2$ which exceeds the amplitude of the peak at $n = q$. By varying the interval length K, the procedure is optimized until the number of significant peaks per pitch period is reduced to one.

Gold and Rabiner (1969), finally, have reported two simplifications of their PDA, again for a restricted range of F_0. The original algorithm had no such restriction, while the first simplified version sets an upper limit of 220 Hz to the permitted measuring range. The modification fixes the length of the blanking interval and the rundown time constant of the individual primary extractors and replaces the complicated coincidence check by a simpler one.

In the second modification the input signal is split up into two frequency bands (80 to 240 and 200 to 600 Hz). In each of these subbands the positive and negative peaks are detected and processed as in the original algorithm (again with fixed blanking interval and rundown time constant). Since only the two previous periods are stored, the new candidate matrix contains rows 1, 2, and 4 of the matrix shown in Figure 6.35. The coincidence check is performed only once with a fixed tolerance interval of 12.5% of the actual estimate. This PDA was implemented in a channel vocoder (Bially and Anderson, 1970); later, it was implemented as a hybrid PDD with switched-capacitor filters in MOS technology and a microprocessor (Dalrymple et al., 1979).

Both these simplifications were implemented in order to speed up the algorithm. The second modification, however, suggests that, other than for the original algorithm, the first harmonic must be present in the signal.

6.3 The Intermediate Device: Temporal Structure Transformation and Simplification

The two extremes of time-domain pitch determination are *extraction of the fundamental harmonic* (as discussed in Sect.6.1) and *direct analysis of the temporal structure* (as discussed in Sect.6.2). Both of them have their shortcomings. The fundamental is often weak or absent. Requiring its presence therefore is a severe drawback for possible application areas of the PDA. Enhancing the fundamental harmonic by nonlinear means often fails, since no NLF enhances this harmonic equally well for an arbitrary input signal. Structural analysis, on the other hand, has to cope with a variety of signal structures which tend to require extensive correction in the postprocessor or - in the case of the analog systems - require the manual preset of a subrange within the possible range of F_0. In addition, these algorithms often are phase sensitive. Of course these restrictions do not apply to any algorithm discussed up to this point, but they signify some general tendency which

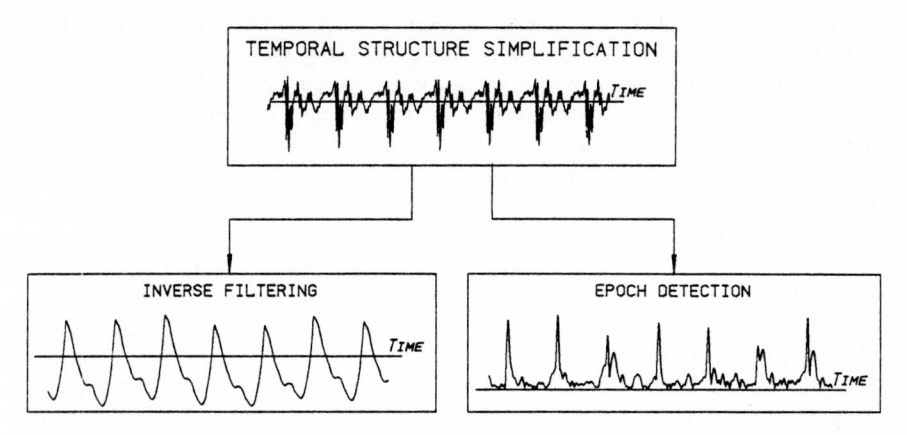

Fig.6.36. Time-domain pitch determination by temporal structure simplification

makes it worthwhile to think about an intermediate solution that combines the advantages of both an elaborate preprocessing and an elaborate basic extractor in such a way that they increase the overall efficiency of the system. In other words, this means that a preprocessor filter of type PiTN or PiTL should be used to reduce the number of possible temporal structures so that a simplified basic extractor of type PiTS could cope with them more easily.

This category of PDAs (henceforth labeled PiTA)[19] is oriented towards two features of periodicity that are typical for the speech signal (Fig.6.36). These two features are closely related. The first one which will be discussed in Sect.6.3.1 is the temporal structure of the excitation signal which is known from the discussion in Chapter 3. To reconstruct the excitation waveform from the speech signal, an *inverse filter* is applied which cancels the transfer function of the vocal tract. The second feature exploited is the abrupt closure of the glottis during normal phonation. At the instant of glottal closure there is a *discontinuity* in the time derivative of the excitation signal which is mainly responsible for its high-frequency components (Fant, 1979a-c). The second derivative of the excitation signal then exhibits an impulse at this instant. Routines that detect and mark this impulse ("epoch") are discussed in the second part of this section.

6.3.1 Temporal Structure Simplification by Inverse Filtering

The basic principle of the inverse filter reads as follows (Fig.6.37): The voice excitation signal (the glottal signal) is reconstructed from the speech signal by a filter whose transfer function is the reciprocal of the vocal tract transfer function H(z). The temporal structure of the glottal waveform is much simpler

---

[19] Since all these PDAs apply some kind of adaptive filtering, they will all be labeled *PiTA.*

and much more regular than that of the original speech signal. Since the transfer function of the vocal tract changes according to the imposed phonetic information, the filter must be time variant, and its parameters must be determined by a suitable short-term analysis method. According to Fant (1960, 1970), vowels and nonnasal consonants can approximately be modeled by an all-pole transfer function. The inverse filter is thus an all-zero filter. In digital terms this is a nonrecursive filter whose state equation, after (2.11), is given by

$$y(n) = d_0 x(n) + d_1 x(n-1) + \ldots + d_k x(n-k) \tag{6.15}$$

and the transfer function, after (2.12), by

$$H(z) = \sum_{i=0}^{k} d_i z^{-i} = \frac{1}{z^k} \sum_{i=0}^{k} d_{k-i} z^i \; . \tag{6.16}$$

With an order of k, $K = k/2$ pairs of complex zeros can be realized [$K = (k-1)/2$ if k is odd]. In this case, H(z) will read as

$$H(z) = \frac{1}{z^k} \; (z-z_{01})(z-z_{01}^*) \ldots (z-z_{0K})(z-z_{0K}^*) \tag{6.17}$$

where z_{0i}^* denotes the conjugate complex value of the zero z_{0i}. The question is now 1) how to design and realize the filter, i.e., how to determine its coefficients or its zeros, respectively, and 2) to determine how accurate an approximation must be (or how inaccurate it is permitted to be) so that it can be taken for pitch determination.[20] If one starts from (6.17), the inverse filter would require a complete formant analysis. Applying the conventional methods of formant analysis, like peak picking or analysis by synthesis (Flanagan, 1972a), the effort is definitely too great. It can be considerably reduced, however, if the formant determination is confined to F1, removing the higher formants by nonadaptive low-pass filtering. To measure F1 from the low-pass filtered signal, very simple methods, such as zero crossings analysis, can be applied. A PDA of this kind was implemented by Hess (1972b). It will be discussed in Chap.7 since it forms the basis for a further development together with the results of the experiments to be reported there.

The other possibility starting from (6.15) is the well-known *linear prediction* (linear predictive coding; LPC). Linear prediction postulates that a signal sample x(n) be predictable from previous samples (Fig.6.38a,b) except for an additive error signal e(n):

$$x(n) = a_1 x(n-1) + a_2 x(n-2) + \ldots + a_k x(n-k) + e(n) \; . \tag{6.18}$$

In this equation, x(n) represents the speech signal, a_i the filter coefficients,

---

[20] Tasks for which an exact inverse filtering is required will be discussed in Section 6.5.1.

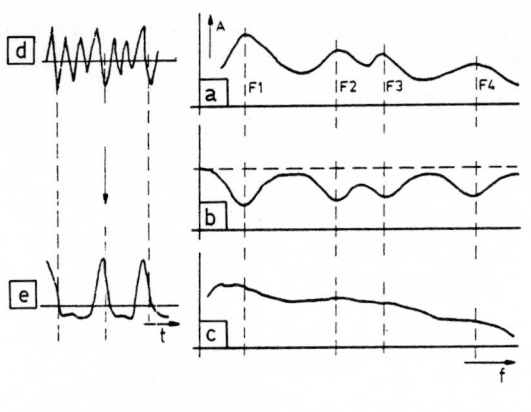

Fig.6.37a–e. Principle of inverse filtering. **(a)** Spectrum of speech signal with formants (smoothed); **(b)** frequency response of inverse filter; **(c)** spectrum of voice source waveform (reconstructed and smoothed); **(d)** speech signal; **(e)** voice source waveform (reconstructed)

and e(n) the error signal[21]. Comparing this equation to the state equation (2.11) of the digital filter, it is easily seen that (6.18) represents a purely recursive digital filter and thus an all-pole model of the vocal tract transfer function. [Except for the sign, the coefficients a_i in (6.18) and g_i in (2.11) are identical.] Including the error signal e(n), (6.18) is exactly valid for all signals and all times. This does not yet imply, however, that the coefficients really represent a useful model of the speech process. To guarantee this, the error signal must be kept small, i.e., it must be minimized in some global sense. Moreover, since the filter coefficients must be time variant, the process of minimization has to be achieved using short-term analysis. The main advantage of this method is the way the filter coefficients are determined. In the classical formulation (Itakura and Saito, 1968; Atal and Hanauer, 1971; Makhoul and Wolf, 1972; Makhoul, 1975b; Markel and Gray, 1976) the optimization of the predictor coefficients is performed by minimizing the energy of the error signal within a given frame. The energy, i.e., the mean square of the error signal, is formulated as a function of the predictor coefficients; the global minimization process then involves straightforwardly solving a set of linear equations for the filter coefficients after calculating easily obtainable coefficients of autocorrelation or covariance functions. Details of this procedure may be obtained from the literature (cf. the references given above). In connection with glottal inverse filtering, some more details will be added in Section 6.5.1.

---

[21] In the literature the error signal e(n) is frequently called *LPC residual signal* or, simply, *residual*. The (nonrecursive) LPC filter is a variation of the inverse filter and is often labeled *LPC inverse filter* (Markel, 1972a, 1973; Markel and Gray, 1976). The residual, however, is not the excitation signal. To distinguish between the (genuine) inverse filter which reconstructs the glottal excitation signal and the (LPC) inverse filter which yields the residual, the LPC inverse filter is sometimes called *residual inverse filter*, as opposed to the *glottal inverse filter*. More details will be added later in this section and in Section 6.5.1.

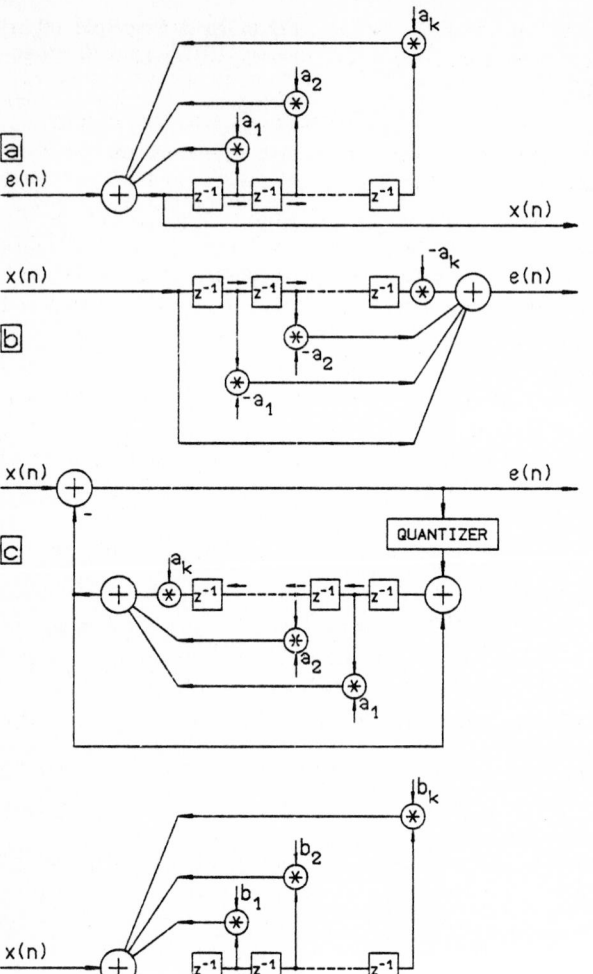

Fig.6.38a–d. Digital filter structures for linear predictive coding (LPC). (a) All-pole LPC model of vocal tract transfer function; (b) inverse filter; (c) first proposal by Maksym (1972b, 1973, see the discussion later in this section); identical to part a except for the quantizer; (d) recursive approach by Maksym. The coefficients a_i in (a,b) and (c) are generally not the same. [(x(n)] Speech signal, [e(n)] prediction error (residual signal)

If one solves (6.18) for e(n), it is easily seen that e(n) is realized by a nonrecursive digital filter, as given in (6.15),

$$e(n) = x(n) - a_1 x(n-1) - a_2 x(n-2) - \dots - a_k x(n-k) = x(n) - x_p(n) , \qquad (6.19)$$

where $x_p(n) = a_1 x(n-1) + \dots + a_k x(n-k)$ represents the fully predicted sample.

If the signal were ideally predictable, the error signal e(n) would be zero. In this case the input signal x(n) would be the sum of a finite number of exponentially decaying sinusoids, according to filter theory. It is easily seen that this assumption can be valid only within the boundaries of a single pitch period. At a pitch period boundary thus a distinct peak is expected for the error signal e(n). That would mean that the periodicity of the speech signal is about the only information on the speech process which is definitely maintained in the error signal.

There are a number of objections to this point, however. The first one is the fact that the filter is linear. Hence, from the preprocessor, a PDA applying this principle would be nothing but a variation of the type PiTL (except that the filter is time-variant) and thus require the presence of the first harmonic in the signal in case a simple basic extractor, like a ZXABE, is used. Of course, such a filter may be used in a PDA of type PiTL with all the restrictions imposed upon this category. Yet, as already pointed out at the beginning of this section, the purpose of this filter is to reduce the number of possible temporal structures in order to simplify the basic extractor of type PiTS to be applied thereafter. The error signal definitely has less variation in its temporal structure than the speech signal itself, so that a simplified PiTS basic extractor might be applied. So this point is no real objection to the use of an adaptive LPC filter; it is only an issue to be known when designing the basic extractor.

The second point, however, is much more serious. The determination of LPC coefficients is nothing but a special method to design a nonrecursive digital filter. To understand what LPC really does, let us for a moment go back to the simplified linear model of speech production (Fig.2.15). The speech signal is there modeled as the output signal of the linear system,

$$X(z) = P(z)\ S(z)\ H(z)\ R(z)\ A = P(z)\ G(z)\ H(z)\ A\ , \qquad (3.10)$$

which corresponds to a time-domain convolution of the individual components,

$$x(n) = A\ \{\ [\ p(n)\ *\ g(n)\ ]\ *\ h(n)\ \}\ . \qquad (6.20)$$

In terms of this model the task of pitch determination is equal to the estimation of the periodicity $P(z)$ from $X(z)$. We also know the problem of determination of the vocal tract transfer function $H(z)$, commonly referred to as formant analysis. In both these problems the investigator knows what to look for. Linear prodiction, however, models the whole speech signal in one digital filter,

$$X(z) = A(z)\ E(z)\ . \qquad (6.21)[22]$$

$A(z)$ is the transfer function of the predictor filter, and $E(z)$ is the z-transform of the residual. The ideal solution to both pitch determination and formant analysis would be obtained if $A(z)$ were equal to $H(z)$ or $H(z)G(z)$, and if $E(z)$ were equal to $P(z)G(z)$ or $P(z)$, respectively. The LPC algorithm,

[22] In this and the following equations, long-term notation is used for ease of writing although linear prediction is usually performed as a short-term method in speech analysis. For the theoretical considerations in this paragraph, however, the difference between long-term and short-term notation is irrelevant.

however, does not necessarily perform the separation in this desired way. Instead, the boundary conditions of LPC demand that 1) A(z) is all-pole, i.e., the transfer function of a purely recursive digital filter, and 2) the energy of the residual becomes a minimum in order to yield an optimal prediction of the forthcoming samples. In addition, it is necessary to remember that it is not the recursive filter A(z), but its reciprocal 1/A(z) that is modeled and optimized by the LPC algorithm. Since 1/A(z) represents a nonrecursive filter which is unconditionally stable, the algorithm does not check the location of the zeros of the modeled transfer function; zeros of 1/A(z) can thus be situated on or even outside the unit circle. That means, the decision which components of X(z) are modeled by A(z), and which ones are not modeled and thus left to E(z), is absolutely unpredictable. It depends on marginal parameters like the order of A(z) or the frame length, or even on the signal itself. Thus it can easily happen that the pole of P(z) on the unit circle[23] at the point that pertains to F_0 is contained in A(z) and not available in E(z). Gold (1977) formulated this problem succinctly:

> The error signal from an LPC analysis has, to a first approximation, removed the vocal tract information via inverse filtering. This reasoning leads to the suggestion that pitch measurement be based on this error signal which presumably has a relatively stronger component of excitation information than existed in the original speech signal. A potential difficulty with this method is that the LPC analysis "doesn't know" that it is supposed to preserve only excitation information; observation of such processing for nasal sounds, for example, [and for female speech as well, especially when F1 coincides with F_0], indicates that much of the excitation information has also been inverse filtered. (Gold, 1977, p.1644)

The problem is further complicated by the common weakness of all inverse filter approaches when used for pitch determination: the tendency to cancel the first harmonic when it coincides with the frequency of the formant F1. This case happens quite frequently with nasal consonants as well as with female voices. The coincidence of F1 and F_0 is a unique case in pitch determination. Since the main pole of P(z) nearly coincides with that of H(z), more than 95% of the total signal energy can be concentrated in the first partial. Hence it is absolutely sure that any adaptive inverse filter will adjust one or even several of its zeros to that frequency. The effect on the temporal structure of the signal is detrimental[24] (see the examples in

---

[23] P(z) must have poles on the unit circle since this function models a pulse generator which is an active circuit, and whose output is a pulse train with constant amplitude. In contrast to this, the passive filter transfer functions G(z) and H(z) are not permitted to have poles on or outside the unit circle.

[24] This coincidence between F_0 and F1 is also the cause for the failure of the SIFT (simplified inverse filtering, see Sect.8.2.5) algorithm by Markel (1972) for high-pitched voices, as reported by Rabiner et al. (1976). That case shows that not even the short-term analysis PDAs are free from this effect.

Appendix A or in Fig.7.4). This phenomenon restricts the application of inverse filters to male voices unless a special logic for recognition of this coincidence is included.

A simple algorithm of this kind was proposed by Le Roux (1975). Le Roux confined his analysis to the frequency band between 300 and 700 Hz; all formants except the first one are thus suppressed. This waveform, sampled at 2 kHz, is then looked at whether it can be computed as valid output signal of a stable recursive second-order filter. The underlying idea is that no excitation has occurred as long as a second-order linear-prediction filter with zero prediction error (which models the waveform as the course of an undisturbed first formant) can be found. If this is not possible, i.e., if the pertinent filter becomes unstable, there must be an external excitation in the signal. From (6.18) we obtain for the second-order filter

$$x(n) = a_1 x(n-1) + a_2 x(n-2) + e(n) \tag{6.18a}$$

From the filter transfer function, the stability criteria for the two coefficients read as follows:

$$|a_2| < 1 ; \quad |a_1| < 1 + |a_2| . \tag{6.22}$$

If we force the prediction error to zero for a given sample n, we obtain an equation for the coefficients a_1 and a_2 which marks a straight line in the a_1-a_2 plane[25],

$$x(n) - a_1 x(n-1) - a_2 x(n-2) = 0 . \tag{6.23}$$

A stable filter is possible if this straight line passes through the triangle in the a_1-a_2 plane set up by the stability criteria (6.22). With the additional constraints that both the original signal and the LPC residual (this time determined using the least-squares criterion) take on a defined polarity, Le Roux's PDA derives a pulse signal which serves as input for the basic extractor. At a given instant n, the pulse signal p(n) takes on the local value of the residual, i.e., $p(n) = e(n)$, if 1) both the speech signal x(n) and the LPC residual e(n) are negative, and 2) no stable second-order filter can be found which satisfies (6.22); otherwise p(n) is zero. Secondary excitations are ruled out by their lower amplitude.

The algorithm has a special routine for the systematic failure which occurs when F_0 coincides with F1. In this case the filter is switched off, and the original signal is regarded:

---

[25] The postulate that the prediction error e(n) be zero for a given sample n leads to an alternative formulation of the linear-prediction problem, i.e., the exact linear-prediction formulation (Markel and Gray, 1976), as opposed to the minimum squared-error criterion usually applied for vocal tract parameter estimation. With a linear predictor of order k, one can force the error signal to zero for k consecutive samples. Nothing is said, however, about the stability of the filter obtained in this way.

No results can be obtained by this method for a number of phonemes spoken by a high-pitched voice when the fundamental frequency coincides with the frequency of the first formant. In such a case the filtered signal has a sinusoidal shape, and its period length is equal to that of the fundamental.

The algorithm is then unable to detect *any* impulse (in contrast to noise where it detects impulses irregularly).

In this case the period of the fundamental is the time separating two crossings of the signal from positive or zero to negative. (Le Roux, 1975, p.8; original in French, cf. Appendix B)

The LPC residual signal is deliberately used in time-domain systems for high-accuracy measurements in connection with voice source parameters, such as determination of the closed-glottis interval. Some of these applications will be discussed in Section 6.5.

The question of nonlinear processing (one-way and full-wave rectification as well as center clipping) in connection with the LPC residual was investigated by Akinfiev and Sobakin (1974). In principle their findings coincide with those discussed elsewhere in this book, namely, that linear inverse filtering and nonlinear distortion are not compatible; the combination of these two methods does not improve the results. Akinfiev and Sobakin, however, found a slight improvement with the center-clipped residual, probably because some secondary excitation at the instant of glottal opening had been removed which would otherwise have caused spurious peaks in the residual.

The PDD by Maksym (1972b, 1973) was implemented as a digital hardware device (one of the first real-time digital hardware PDDs to come out). The predictor applied in this PDD is a special variation resembling the predictors applied in adaptive pulse code modulation (ADPCM) or adaptive delta modulation (ADM). Firstly, the nonrecursive inverse filter (Fig.6.38b) which converts the speech signal $x(n)$ into the error signal $e(n)$ is approximated by a recursive filter (Fig.6.38d). In principle, the frequency response of any nonrecursive filter can be approximated in an arbitrarily exact way by a recursive filter of infinite order (the same holds vice versa). If the order of the recursive filter is limited to a certain value, this filter can be at least an approximation of the intended nonrecursive filter. It is the advantage of the recursive approach, according to Maksym, that the coefficients b_i of this filter can be computed solving a single regression equation. Secondly, the sampling frequency is chosen relatively high (above 40 kHz). For this reason the system works with low prediction error which can then be crudely quantized. The error signal $e(n)$ is reported to have one significant peak at the beginning of every pitch period, and pitch markers can be derived from $e(n)$ by simple triggering. The coefficients of the filter are updated for every sample of the input signal.

It is left open in the paper whether this PDD is suitable for any speech signal uttered by any voice. From the fact that the predictor approximates the transfer function of the inverse filter in a recursive way (i.e., by poles) we can conclude that the filter is not able to completely extinct an undamped oscillation, since the modeled transfer function, if it is approximated by an all-pole filter, cannot have a zero on the unit circle. Thus the information about periodicity will be preserved in the error signal in any case.

In spite of the difficulties already mentioned, a number of further activities have been devoted to the investigation of the LPC residual. A few of them, in connection with short-term analysis, will be reviewed in Chapter 8.

It has been found (Lee and Morf, 1980) that a number of the shortcomings of the LPC residual for pitch determination emerged from the global error minimization process involving all the samples of a whole frame. This analysis takes little account of the fine structure of the signal. An alternative would be linear-prediction analysis over a short frame using nonstationary short-term analysis (which implies the application of the covariance method) and computation of the LPC coefficients as often as possible, preferably for each sample. This would mean that, for each instant n, the LPC filter would be optimal for the prediction of just that one sample x(n), and the prediction error would be a measure of how well this sample can be predicted. It is to be expected that a LPC residual formed in this way will preserve much more information about the periodic excitation than the LPC residual computed from an LPC filter that has been globally optimized over a long frame. (Note that the minimum number of samples necessary to perform the LPC analysis using the covariance method is 2k+1 when k is the order of the predictor filter). The (time-variant) LPC coefficients themselves can also provide significant information about the periodic excitation; they are expected to change much more rapidly when a new glottal impulse is encountered. We can thus additionally define some measure for the rate of change of the LPC coefficients and take it as a parameter for pitch determination.

There is one problem associated with this approach: the computational effort. Each LPC analysis requires computation of a k x k covariance matrix and solving a linear equation system with k unknowns, if k is the order of the LPC filter. The covariance matrix can be recursively updated, thus greatly reducing the effort; the linear equation system, however, cannot be avoided when the classical method of LPC analysis is carried out. However, there are other solutions. With an alternative (but equivalent) filter structure, the so-called *lattice structure*, which involves the computation of the *partial correlation* (PARCOR) *coefficients* (Markel and Gray, 1976), the algorithm becomes more flexible. Originally available only for the autocorrelation method of linear prediction, this calculus has been generalized and made applicable to the covariance method (Morf et al., 1977; Protonotarios et al., 1980; Gueguen and Scharf, 1980). It allows recursive updating of the filter coefficients sample by sample with much less computational effort. In addition, it permits changing the number of signal samples involved and the order of the filter during the computation.[26]

---

[26] When using such recursive algorithms, care must be taken that they remain *numerically stable*. This means that in any algorithm which is recursively executed with finite accuracy, round-off errors in arithmetic operations can accumulate unfavorably so that the results become wrong although the algorithm is, in principle, correct. Working with such an algorithm therefore implies that the operations start from the beginning from time to time; in actual work this means that the covariance matrix is directly calculated from time to time instead of being recursively updated.

Algorithms of these types have become known as *recursive least-squares* signal analysis (Dove and Oppenheim, 1980) or *exact least-squares processing* (Lee and Morf, 1980) and are related to maximum-likelihood analysis methods (Gueguen and Scharf, 1980; Morf et al., 1977). The mathematics involved is rather voluminous; a deduction of the algorithm within the framework of this book is not possible for reasons of space, so that this brief qualitative outline must be sufficient.

It was shown by Dove and Oppenheim (1980) for synthetic nonspeech signals and by Lee and Morf (1980) as well as by Maitra and Jain (1982) for speech signals that a PDA based on this principle is likely to overcome most of the problems with the LPC residual and may serve as a good method to indicate the instants of glottal closure. In this respect the method pertains already to the event-detection PDAs to be discussed in Section 6.3.2. An application in the high-accuracy task of determining the instant of glottal closure or in analyzing aperiodic voiced signals (Sect.6.5.2) could also be thought of. The few PDAs developed so far (Lee and Morf, 1980; Maitra and Jain, 1982) are not more than first attempts to get the method under control; their limitations are not known, and they have not yet been extensively tested. One limitation, however, was already pointed out by Dove and Oppenheim (1980): the LPC residual is sensitive with respect to additive noise. Since noise is not predictable, it will show up in the residual signal and may completely mask the peak at the instant of the glottal pulse. (According to Dove and Oppenheim, this may happen even at a S/N ratio of 30 dB.) However, the already mentioned measure for the rate of change of the coefficients has proved to be much more noise resistant. Lee and Morf (1980) applied a logarithmic likelihood measure which also shows good performance.

In conclusion, recursive least-squares analysis, a generalization of LPC, appears as a promising method to exploit the temporal structure of the LPC residual for pitch determination. Further work is necessary, however, to evaluate its usefulness and limitations for pitch determination.

6.3.2 The Discontinuity in the Excitation Signal: Event Detection

According to the linear model of speech production, the speech signal is excited by a pulselike waveform which drives the vocal tract, a linear passive system. In most cases the lowest resonance of the vocal tract is situated well above the fundamental frequency F_0 of the excitation signal.

According to Fant (1979a-c) and others, the main excitation of the higher frequencies occurs in a pulselike manner at the point of glottal closure (see Sects.3.1,3,4). The idea of event detection is to enhance the features in the temporal structure evoked by the presence of this excitation instant in order to more easily detect it as an isolated *event* or *epoch* (C.P.Smith, 1954; Ananthapadmanabha and Yegnanarayana, 1975). As long as there is no substantial frequency dependent delay in the signal, the instant of excitation is in principle resistive against phase distortions. There is an unwanted effect in connection with phase conditions, however. It is uncertain whether there is a peak or a zero crossing in the output signal at the instant of excitation. For reasons of energy and continuity, a zero crossing is more likely. In this case, however, the instant of excitation is much more difficult to determine than if it is present as a peak. If there is a frequency-dependent phase response in the recording channel (that is the case for any kind of low-frequency phase distortion), the signal at the instant of excitation may temporarily change from peak to zero crossing or may take on an arbitrary position in between. (This even applies to the speech production system itself which has a time-variant transfer function.) To overcome this source of unreliability, a number of PDAs (not all of them) in this category throw out the phase relations by a phase-splitting network or by a Hilbert transform. A different way to become less dependent on phase is to confine the analysis to higher frequencies (above 1 kHz) or to a narrow frequency band, preferably around a formant. For the higher formants the elapsed time between zero crossing and peak becomes smaller; in addition, the damping of the formants increases above 2 kHz (Dunn, 1961) which makes the peak at the beginning of the period more distinct. As can be seen from just these introductory remarks, there is a close relation between the PDAs discussed here and those of group PiTE (Section 6.2.1). Investigation of the LPC residual (Sect.6.3.1) is also closely related.

There are not too many PDAs in this category to be discussed. It begins with the one by C.P.Smith (1954, 1957) which was further developed by Yaggi (1962, 1963). The so-called vector PDA by Rader (1964) has a similar preprocessor but a different basic extractor (which tends more towards nonstationary short-term analysis). The term *epoch detection* was created by Ananthapad-manabha and Yegnanarayana (1975, 1979). Finally there is the PDA by De Mori et al. (1977) which is a multichannel approach based on this principle.

In the PDA by C.P.Smith (1954, 1957), the signal passes a filter bank that consists of 32 second-order band-pass filters. The output signal of each of these 32 channels is full-wave rectified, and all the rectified signals are added to give a new waveform which serves as input signal for the basic extractor (type PiTS, similar to the analog PDDs described in Sect.6.2.1):

> At the rectifier outputs the dc signals represent a demodulation of the signals in the spectrum channels. Summing these dc voltages generates a new voltage parameter that has voice pitch as a prominent periodicity in amplitude ... The build-up and decay of the envelopes is phase-coherent in all of the channels of the spectrum analyzer, the ac signals all building up and decaying in synchronism at the voice pitch frequency, whereas ... residual ac fluctuations tend to add up out of phase and therefore tend to cancel out. ... The signals in the channels of the spectrum analyzer build up to a single peak and then decay. (Smith, 1957, as given in McKinney, 1965, p.120)

After adding the partial signals, the new waveform

> ... is processed in nonlinear circuitry to extract voltage pulses at the inflection points, the troughs of the waveform, where the amplitude changes from negative-going to positive-going. These instants in time specify the onsets of successive pitch epochs of the voice signal.

This method does not require a "fundamental" in the input voice spectrum. Loss of voice frequencies below 200 Hz does not impair the performance of measurement involving relatively large signals, and the utilization of pitch information contained in the total voice spectrum. (Smith, 1957, as given in McKinney, 1965, p.121)

Yaggi (1962, 1963) implemented this PDD within a channel vocoder. The signal passes the band-pass filters of the spectrum analyzer (18 channels which cover a frequency range from 70 Hz to 4 kHz; the bandwidths range from 130 Hz for the 7 lowest channels to 390 Hz for the highest channel) and the subsequent full-wave rectifiers (Fig.6.39). The output signals of the full-wave rectifiers are then smoothed by a first-order low-pass filter; finally they are all added in the summing amplifier.

This procedure acts like an amplitude demodulation. If the output signal of an individual channel contains a formant (and this is the interesting case), then the full-wave rectifier and the subsequent smoothing filter extract its temporal envelope. Since all the formants are simultaneously excited, all the envelopes add coherently; from the resulting signal at the output of the summing amplifier the period markers are derived by simple threshold analysis (Fig.6.40).

The crucial point in inverse filtering - the coincidence between F1 and F_0 - is a minor problem in this analysis. An undamped sinusoid is transformed into a pure DC voltage which does not help for pitch determination, but does not degrade the performance either. Pitch determination, in this case,

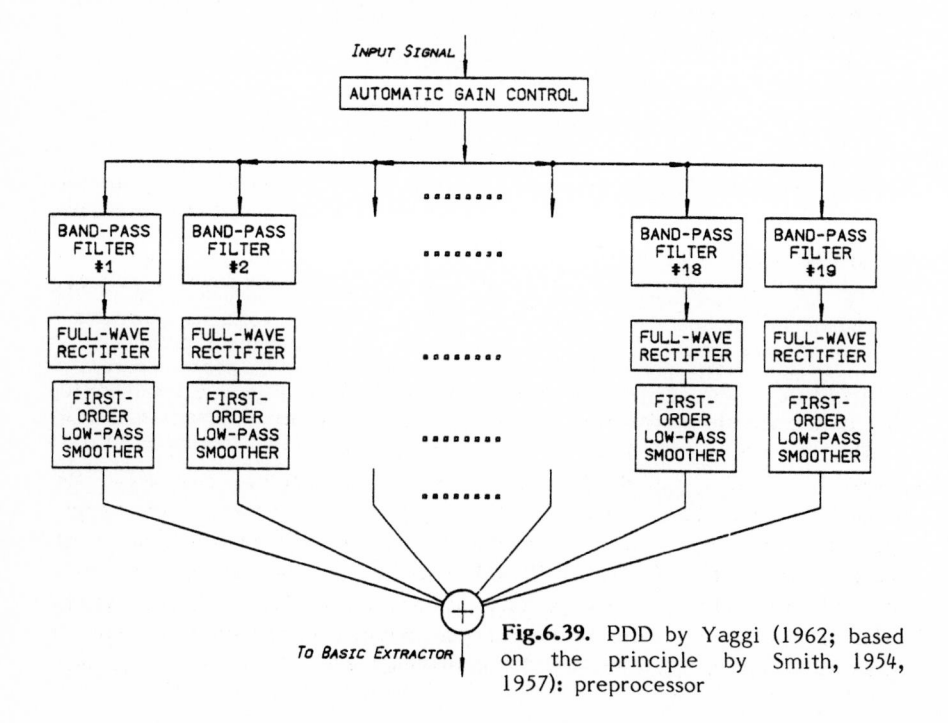

Fig.6.39. PDD by Yaggi (1962; based on the principle by Smith, 1954, 1957): preprocessor

INPUT SIGNAL

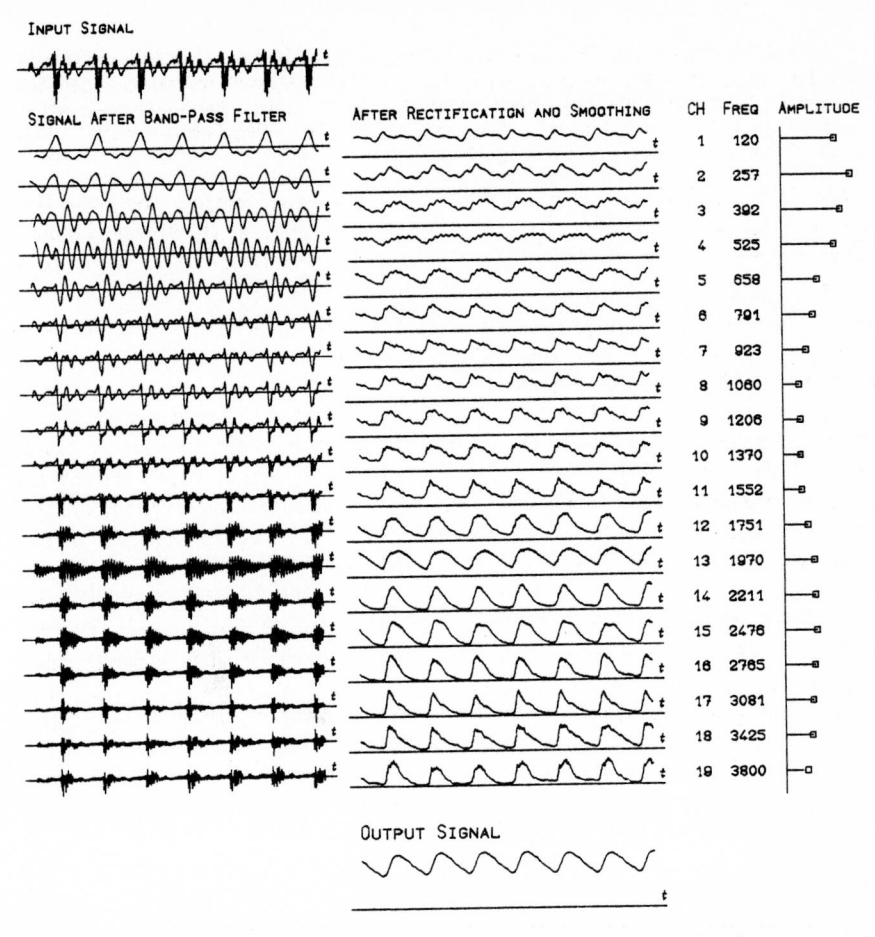

Fig.6.40. PDD by Yaggi (1962): example of performance. (CH) number of channel; (FREQ) center frequency of channel (in Hz). The amplitude is displayed in a linear scale; all the signals were normalized before plotting. Signal: /æ/ from "Algorithms...", speaker MIW (male); undistorted signal. The second-order band-pass filters used for this display had a bandwidth between 114 Hz (channel 1) and 342 Hz (channel 19)

must rely upon the signals gained from the higher formants. A certain reduction in the signal-to-noise ratio, however, might result from this coincidence.

The "vector" PDA by Rader (1964), which emerged from an earlier PDA by Lerner (1959), also uses the bandpass filters of the spectrum analyzer in the channel vocoder. The remainder of the preprocessor is different from Yaggi's approach. The outputs of the bandpass filters are each passed through 90° phase-splitting networks (Hilbert filters), whose realization can be very simple since the bandwidths of their input signals are small. The momentary amplitudes of the 2N output signals (N bandpass filters and N Hilbert filters) are taken as vector components in a 2N-dimensional space. In case of a periodic

INPUT SIGNAL

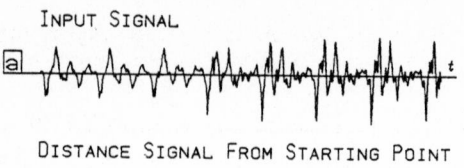

DISTANCE SIGNAL FROM STARTING POINT

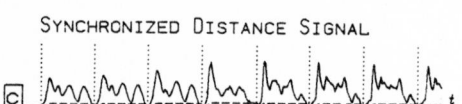

SYNCHRONIZED DISTANCE SIGNAL

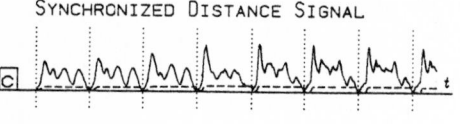

Fig.6.41a–c. Vector PDA by Rader (1964): example of performance. **(a)** Original signal; **(b)** distance function D(n) with arbitrary reference point; **(c)** D(n), threshold and new setup after each period. The filterbank used was the same as in Fig.6.40

signal this time-varying vector describes a closed curve, and the time which elapses during tracing of one complete cycle equals the length of one pitch period. If thus the momentary value of the vector at an arbitrary instant n is taken as a reference, the distance D(n) of the vector from this point will come to zero or close to zero after one period has elapsed; in almost any case there will be a local minimum that is easily detected by a threshold analyzer (see the example in Fig.6.41). A similar algorithm was proposed later by Gikis (1968).

The arbitrary choice of the reference point indicates that the PDA works phase independently; the phase relations between the original signal and the distance function D(n) are lost. Hence this PDA works inherently according to a nonstationary short-term analysis principle (similar to the average magnitude difference function, as discussed in Sect.8.3) unless the phase relations are reestablished by a special synchronization routine (that might be quite simple in the present case, since it has to be invoked only when a new voiced segment has been started, or when the PDA has fallen out of synchronization due to a measurement error).

The discontinuity in the excitation function caused by the abrupt closure of the glottis is the primary cue for the PDAs by Ananthapadmanabha and Yegnanarayana (1975, 1979). The authors called this instant the *epoch* of the signal and define the epoch as follows:

Let f(t) be a function defined over an interval (a,b) and zero outside the interval. Also let f(t) possess continuously differentiable derivatives in the interval (a,b). Then the point of discontinuity of the lowest-ordered derivative will be regarded as an epoch. The epoch therefore can occur either at a or at b or at both a and b. Here by the term "discontinuity of the lowest-ordered (n) derivative" we mean that at $t = t_j$, $f^{(n+1)}(t)$ is discontinuous, but $f^{(n)}(t)$ is continuous. ... The value of (n+1) will be referred to as the order of the epoch.

In the above definition we have restricted the function to possess continuously differentiable derivatives in the interval (a,b). If the function is piecewise continuous, then the interval (a,b) can be further subdivided so that the restriction is satisfied in each of the subintervals. Then the number of epochs could be more than two for a given function. Thus, for example, for a triangular function there will be epochs of order one at the ends as well as at the apex. (Ananthapadmanabha and Yegnanarayana, 1975, p.563)

The authors then showed that, for high frequencies and a bandpass filter whose bandwidth 2B is small compared to its center (radian) frequency ω_c, the filter response to a single epoch at $n = a$ will approximately be

$$f_1(n) = P g(n-a) - Q g_H(n-a) \qquad (6.24)$$

where the functions $g(n)$ and $g_H(n)$ are

$$g(n) = \frac{\sin \frac{Bn}{2}}{\pi n} \cos \omega_c n + \frac{1 - \cos \frac{Bn}{2}}{\pi n} \sin \omega_c n \qquad (6.25)$$

$$g_H(n) = - \frac{1 - \cos \frac{Bn}{2}}{\pi n} \cos \omega_c n + \frac{\sin \frac{Bn}{2}}{\pi n} \sin \omega_c n \ .$$

[Note that $g_H(n)$ is the Hilbert transform of $g(n)$]. In a sampled system this expression becomes simpler when the center frequency f_c of the band-pass filter is chosen equal to the folding frequency (which is half the sampling frequency). In this case we obtain

$$\cos \omega_c n = (-1)^n \qquad \text{and} \qquad \sin \omega_c n = 0$$

for all n, and the function $f_1(n)$ is given by

$$f_1(n) = (-1)^n \left\{ \frac{P \sin \frac{B(n-a)}{2}}{\pi(n-a)} + \frac{Q \left[1 - \cos \frac{B(n-a)}{2} \right]}{\pi(n-a)} \right\} . \qquad (6.26)$$

Reversing the sign of each alternate sample, we get rid of the unwanted oscillation at the folding frequency and obtain

$$f_e(n) = \frac{P \sin \frac{B(n-a)}{2}}{\pi(n-a)} + \frac{Q \left[1 - \cos \frac{B(n-a)}{2} \right]}{\pi(n-a)} , \qquad (6.27)$$

where P and Q are constants that depend on the order of the epoch, the phase conditions at the instant of excitation, and the frequency characteristics of the signal in the passband of the filter. The problem is that both these values are constants for a single epoch, but time variant for the signal in general. Depending on P and Q the waveform of the output signal of the epoch filter is different. For $Q = 0$ we obtain the normal $\sin(n)/n$ waveform with the highest peak at $n = a$; for $P = 0$, however, we obtain a waveform that crosses zero at $n = a$ and possesses an odd symmetry with respect to this point. This waveform is much harder to process by conventional algorithms, like a threshold analyzer. This difficulty can be overcome if we compute the momentary amplitude of the pertinent analytic signal. For this we need the Hilbert transform of $f_e(n)$,

$$f_{eH}(n) = \frac{P\left[1 - \cos\dfrac{B(n-a)}{2}\right]}{\pi(n-a)} - \frac{Q\sin\dfrac{B(n-a)}{2}}{\pi(n-a)}$$ (6.28)

and can then compute the momentary amplitude

$$f_0(n) = \sqrt{f_e^2(n) + f_{eH}^2(n)} = \sqrt{P^2 + Q^2} \cdot \frac{2\sin\dfrac{B(n-a)}{4}}{\pi(n-a)} .$$ (6.29)

If there are several epochs within the analyzed signal, care must be taken that they do not affect each other due to the nonlinear squaring operation when the momentary amplitude of the signal is computed. This is achieved, however, 1) when the transfer function of the band-pass filter is suitably chosen, and 2) when the reciprocal 1/B of the bandwidth of the filter is small compared to the interval between successive epochs, i.e., when the bandwidth B considerably exceeds the fundamental frequency.

Since they started from a finite signal to develop their theory, Ananthapadmanabha and Yegnanarayana (1975) applied a Fourier transform; they carried out the Hilbert transform and band-pass filtering in the frequency domain (Fig.6.42). The sign reversal of alternate samples corresponds to a spectral shift by N/2, if N is the length of the DFT (in the z plane this would correspond to a spectral rotation by 180° which moves the point z = -1 into +1 and vice versa). In order to avoid artificial "epochs" at the beginning of the interval, an appropriate time-domain window is applied before the DFT.

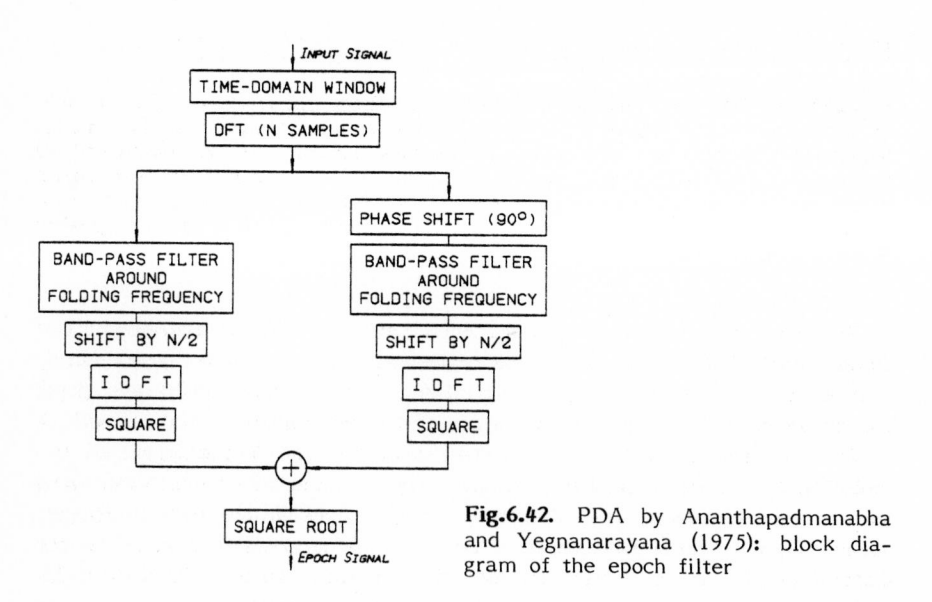

Fig.6.42. PDA by Ananthapadmanabha and Yegnanarayana (1975): block diagram of the epoch filter

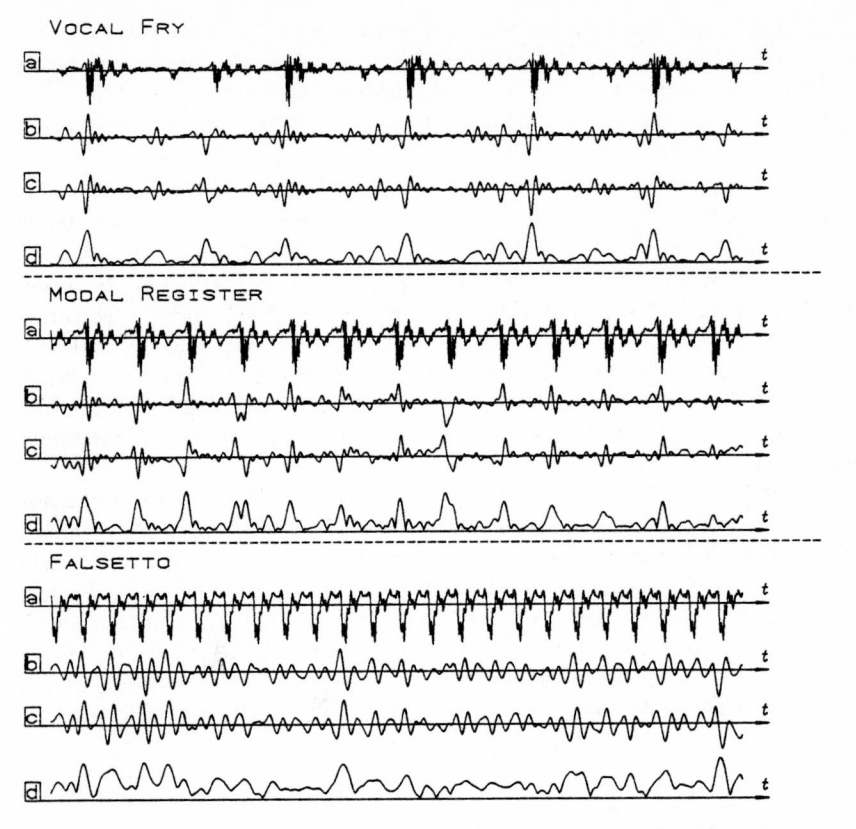

Fig.6.43a–d. PDA by Ananthapadmanabha and Yegnanarayana (1975): example of performance for different modes of phonation (as indicated in the figure). Signals: (a) input speech signal; (b) after band-pass filter; (c) 90° phase shifted signal after band-pass filter; (d) output signal of the epoch filter. The figure clearly indicates that the epoch filter works reliably only when there is a distinct glottal closure in the input signal. Signal: stationary vowel /ɛ/, speaker WGH (male), undistorted

The DFT provides an exact realization of the Hilbert transform for a finite signal. For infinite signals the Hilbert transform can be approximately realized by a linear-phase digital filter. Examples of such filters are given, for instance, by Schüssler (1973), Kunt (1980), or Rabiner and Gold (1975).

In their first paper (1975), Ananthapadmanabha and Yegnanarayana developed the principle of epoch detection and discussed the application using some special examples. They did not evaluate the performance in a practical vocoder. They mentioned, however, that the PDA may run into trouble if the discontinuity of the excitation is weak, i.e., during a voiced fricative, or for a falsetto voice. An example which supports this statement is shown in

Fig.6.43 with the speech material from Figure 3.8. The PDA may also run into trouble if a weakly attenuated formant falls into the passband; it works best when the spectrum in the passband is approximately flat.[27]

This fact may have motivated the authors to apply a spectral-flattening procedure before the epoch filter, a procedure they proposed in their second publication (1979). Since the spectrum of the LPC residual is flat over most of the frequency range, the bandwidth of the filter can be . substantially increased. The shape of the epoch markers in the output signal of the epoch filter is more distinct and more pulselike, which permits the instant of glottal closure to be determined more exactly. Compared to the speech signal, the high frequencies in the LPC residual are enhanced by 6 dB/octave; the residual can thus be regarded as being composed of 1) a sequence of pulses which marks the epochs of the signal, 2) low-frequency components due to glottal zeros at higher frequencies, and 3) attenuated and delayed parts of the original signal due to antiresonances in the vocal tract and inaccuracies in the estimation of the LPC coefficients. Since components 2) and 3) fall off with at least 6 dB/octave versus higher frequencies, the low frequencies are suppressed. Applying a band-pass filter to this task, the high frequencies (above 4 kHz) where the signal-to-noise ratio of the LPC residual is bad, are suppressed as well. Other than in the first approach (1975), the band-pass filter is not centered around the folding frequency with consecutive demodulation, but is realized by a Hann window around 1/4 of the sampling frequency.

Although the performance is reported to be better than in the first approach, there are some problems:

> The method ... decribed here ... relies on the accuracy of LP analysis for satisfactory performance. For sounds like vowel /u/, high-pitched voiced sounds and voiced fricatives, it is noticed that the LP residual cannot be used satisfactorily for the extraction of epochs. Large errors in the estimation of formants and bandwidths occur in the LP analysis of these sounds. In order to produce frication for a voiced fricative, the glottis never closes completely. Moreover, there are two sources simultaneously exciting the vocal tract system. Further, the transfer function of the system contains zeros for these sounds. Hence, it is difficult to analyze fricatives for extracting epochs. (Ananthapadmanabha and Yegnanarayana, 1979, p.314).

Additional difficulties result from the possibility that a) the glottis does not close completely (as cited for voiced fricatives and high fundamental frequencies, but also applying to a soft voice or falsetto register), and b) that there are secondary points of discontinuity in the period, such as the instant of

---

[27] This has been implicitly verified by Larsson (1977) who applied the same principle to low-pass filtered speech and music signals. Larsson theoretically showed that the method works perfectly if a single, sufficiently damped formant is present in the signal. If there are several formants, as in speech, they will heavily interfere due to the nonlinear squaring operation. Larsson reported that this PDA failed systematically for such signals.

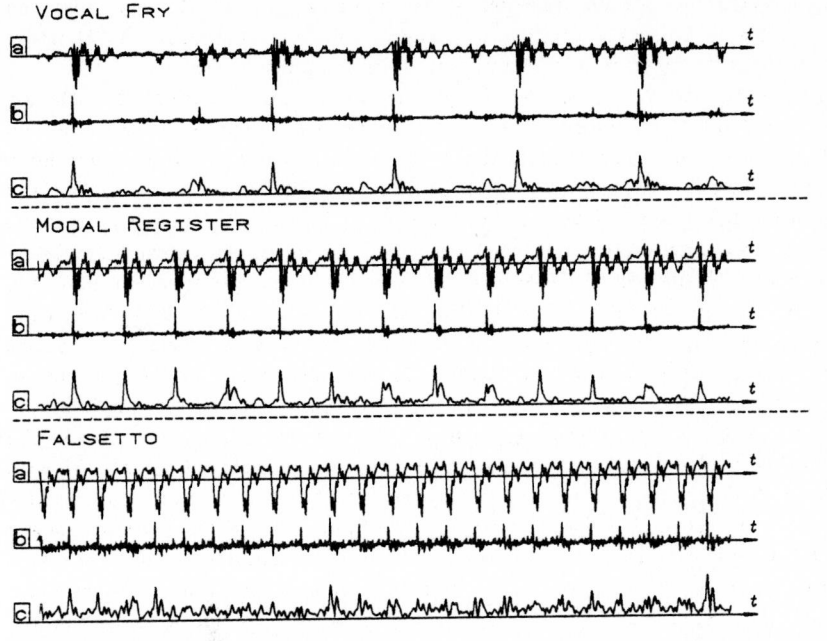

Fig.6.44a-c. PDA by Ananthapadmanabha and Yegnanarayana (1979): example of performance for different modes of phonation (as indicated in the figure). Signals (same utterance as in Fig.6.43): **(a)** input speech signal; **(b)** LPC residual signal; **(c)** output signal of the epoch filter. The figure clearly indicates that the epoch filter works reliably only when there is a distinct glottal closure in the input signal

glottal opening (which is assumed to be smooth in nearly all models, but which can also result in an epoch in real speech). As Fig.6.44 shows, the basic difficulties are the same as for the earlier PDA (Ananthapadmanabha and Yegnanarayana, 1975; cf. Fig.6.43). The problems just mentioned make these PDAs sensitive; they are thus better suited for voice research and exact analysis of high-quality signals than for use in a vocoder with possibly bad input conditions.

Exactly the opposite approach has been developed by Berthomier (1979) with a PDD built in analog technology. For all the PDAs discussed up to now in this section, a strong first harmonic tends to degrade the performance; at least it does not contribute to the quality of the output signal of the preprocessor. In contrast to this, Berthomier's PDD isolates the first harmonic and processes it as follows.

Consider a point M on a plane, following some trajectory around an origin O. The movement of this point can be described by its x and y co-ordinates or by its distance from the origin, E, and the polar angle ϕ. In the case of a circular trajectory with uniform motion ($x = A \cos 2\pi F_0 t$, $y = A \sin 2\pi F_0 t$) we find that

$E = A$ and $\phi = \arctan y/x = 2\pi F_0 t$,

and the derivation of y with respect to time is given by

$$\phi' = \frac{xy' - yx'}{E^2} = 2\pi F_0 . \tag{6.30}$$

The value of $\phi'/2\pi$ gives the frequency of rotation of the point M about the origin. The idea behind the method of calculation is that the result, F_0, could have been obtained by considering either x or y and performing a harmonic analysis of one of these functions. In other words, in the case of uniform circular motion (and strictly speaking only in this case) the frequency of rotation as derived from the motion of the point on the plane is identical to the frequency as derived from the harmonic analysis of one or other of its components. (Berthomier, 1979; original in French, cf. Appendix B)

A block diagram is shown in Figure 6.45. An 8th-order filter performs the Hilbert transform on the low-pass filtered signal. The estimate of F_0 directly results from the computation of the rotation speed of the vector in the x-y plane. In Berthomier's paper the low-pass filter remains unspecified; if higher harmonics are insufficiently damped, however, they come out as unwanted additive oscillations in the output signal.

Like all PDAs and PDDs that rely upon the presence of the first harmonic, Berthomier's PDD falls into trouble when this harmonic is weak or absent. The attractive feature, however, is the elegant way how the time-domain signal is directly converted into an indication of frequency with a linear scale since the output voltage of the basic extractor linearly depends on the fundamental frequency F_0 of the input oscillation.

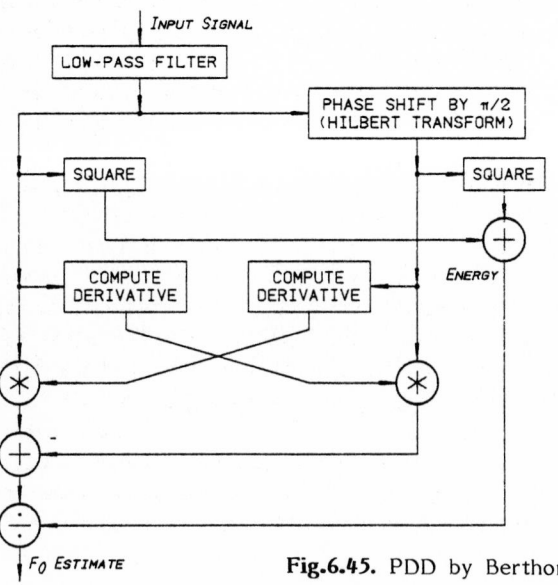

Fig. 6.45. PDD by Berthomier (1979): block diagram

Berthomier's PDD is treated in this section because its preprocessor corresponds to the epoch detection PDAs reported here. The way the basic extractor is realized, however, makes this PDD difficult to categorize into one of the time-domain groups. The PDD is best classified into the group PiTLO (time-domain PDA with basic extractor omitted) and will be further discussed in Section 6.6.2.

The PDA by De Mori, Laface, Makhonin, and Mezzalama (1977), the last PDA to be discussed in this section, uses five band-pass filters with center frequencies between 1.3 to 3 kHz. No Hilbert transform is computed in this PDA; the epoch information is rather extracted using a momentary squared amplitude estimation, the autocovariance function, and subsequent nonlinear processing. The rest of this PDA will be discussed in Sect.6.4 since the five independent channels require a multichannel basic extractor.

Each PDA principle is as valuable as the feature it stems from. Most of the PDAs discussed in this section detect the discontinuity of the second derivative of the glottal waveform which corresponds to the instant of glottal closure. derivative of the glottal waveform. This instant has to be regarded as the real point of excitation of all the formants and is thus the real beginning of the pitch period (Fant, 1979a-c). There are different methods to detect this instant. One can get rid of phase by a phase-splitting network or a Hilbert transform, by a filter bank with rectification in the individual channels, or by a (very) short-time energy estimation, and one can extract the epoch information from the signal after it has passed the filter bank or a band-pass filter around the folding frequency (Ananthapadmanabha and Yegnanarayana, 1975) or a moderate band-pass filter after LPC inverse filtering (same authors, 1979) or - just the opposite - a narrow-band filter centered preferably around formants (De Mori et al., 1977). With respect to robustness, the extraction of the information from a formant region is preferable since the formant regions carry most of the signal energy. To perform this, however, the formants must either be known, or a number of fixed filters has to be set up in the hope that at least one of them traces a formant (and with a multichannel basic extractor behind it that extracts the right information). As far as the performance of these PDAs is evaluated by their authors, it is reported good. Problems, however, arise if there are secondary excitations in the periods, or if the discontinuity in the glottal waveform is weak. These problems are avoided when the analysis centers around a higher formant whose natural damping always guarantees the success of the method. In this respect these PDAs are closely related to the envelope detection PDAs discussed in Sect.6.2.1, and the high-pass filter applied in one of those PDDs (Schunk, 1971) is finally justified. In fact, the output signals of the epoch filter and the nonlinear envelope detector look very similar; the epoch detector, however, seems more reliable, especially if the input signal centered around formants. This point will again be taken up in Sect.6.5.2 when the problem of detecting the instant of glottal closure is discussed in more detail.

The two PDAs by Rader (1964) and by Berthomier (1979) remain. Essentially they apply a time-domain preprocessor in conjunction with a short-term analysis technique. They have been discussed here because the implementation of the preprocessor is practically equal to those PDAs which apply epoch detection in the time domain.

6.4 Parallel Processing in Fundamental Period Determination. Multichannel PDAs

It is an old engineering principle that a circuit should be realized several times if the single circuit is not sufficiently reliable. All the PDAs reviewed up to now in this report have their shortcomings; the combination of several channels, whether in preprocessor, basic extractor, postprocessor or any of them, should result in an improved performance.

Multichannel PDAs can be classified according to one of the following principles.

1) The principle of *main and auxiliary channel*. A main channel which is accurate but requires previous tuning is provided with this tuning by an auxiliary channel which is inaccurate but works without tuning. One can also interpret this principle as the combination of a crude (but reliable) PDA and a more accurate (but less reliable) PDA which then serves as a vernier.

2) The *subrange* principle. The PDA consists of several similar or identical PDAs, each of which covers a subrange of the total measurement range of F_0. This is particularly helpful when the partial PDA would not be able to cover the whole measurement range alone.

3) The *multiprinciple PDA*. In this device, the channels are completely independent. Each channel processes different parameters, or it processes the same parameters according to different criteria. The multichannel operation mode need not extend over the whole PDA: any part of the preprocessor, the basic extractor, or the postprocessor may be commonly used.

A number of these PDAs have already been discussed elsewhere in this book. For instance, the PDAs by C.P.Smith (1954) and Yaggi (1962; both discussed in Sect.6.3.2) as well as the one by Gold and Rabiner (1969; Sect.6.2.4) may be counted among the multiprinciple PDAs. PDAs and PDDs with tunable filters operating with an open-loop control circuit follow the principle of main and auxiliary channel. They have been discussed in Section 6.1.4. A small number of PDAs combine a short-term analysis PDA and a time-domain PDA. In this case the basic performance is that of a short-term analysis PDA; the time-domain channel is only used as a vernier of for synchronizing selected periods. Since these are about the only PDAs applying that principle, a further discussion of the concept of main and auxiliary channel in the present section is not necessary. We will thus deal in Sect.6.4.1 with the PDAs which follow the subrange principle, i.e., which apply multichannel preprocessors. In Sect.6.4.2 the remainder of the multiprinciple time-domain PDAS (which are not too many) will be discussed.

Two problems arise in connection with multichannel pitch determination: channel selection on the one hand, and marker synchronization on the other hand. The channel selection problem exists in any multichannel PDA (except for those which follow the main-and-auxiliary-channel concept). One rarely finds the ideal case that only that channel is active which carries the correct information, and that the remainder of the channels are inactive. Usually the "correct" channel has to compete against other channels which are also active but possibly incorrect. How this selection is performed depends on the individual configuration of the PDA, and heuristic elements tend to play a major role.

If the individual channel processes a subrange of F_0, it is frequently found that all channels are inactive whose upper cutoff frequencies are situated below the actual value of F_0. In this case the selection can be made in favor of the lowest active channel. In multiprinciple PDAs we can frequently observe a certain error behavior of the individual basic extractors and conceive the selection routine according to this behavior. For instance, basic extractors of type PiTL/N tend towards higher-harmonic tracking. In a PDA that applies several basic extractors, one channel can be selected which carries the lowest frequency (minimum-frequency selection principle). If such a criterion is not applicable, selection must be performed according to some matching process (Gold and Rabiner, 1969) or even according to syntactic rules (De Mori et al. 1977). In general, however, the selection subroutine does not significantly contribute to the overall number of errors.

The problem of marker synchronization does not arise in every PDA. When it does arise, however, it is quite complicated and difficult to solve. First of all, one has to consider whether the individual markers are needed at all. If not, one can disregard the problem and be satisfied with the T_0 contour as it is at the output of the channel selector. In this case, however, one may end up with a PDA which is no longer able to track individual periods although its principle is still time domain. In any other case the synchronization task is still feasible when all channels yield their markers at the same instant. (This may apply, for instance, if one channel is correct and another tracks the second harmonic.) Things become difficult, however, when there is no clear phase relation between the markers in the individual channels. If the output of such a PDA (the one by Gold and Rabiner is this kind) were required in the form of individual markers, the synchronization could be nearly as complicated as the pitch determination algorithm itself.

6.4.1 PDAs with Multichannel Preprocessor Filters

A direct consequence of the problems with single-channel PDDs (PiTL) with tunable low-pass or band-pass filters was the design of PDDs in which one filter is selected from several different fixed preprocessor filters. Figure 6.46 shows the block diagram of a proposal by Peterson and Peterson (1968). Its basic extractor input signal is supplied by a preprocessor consisting of a bank of low-pass filters. The speech signal is first nonlinearly distorted (optional, to be chosen by a manual switch) and then high-pass filtered (also optional) to remove low-frequency spurious signals. After that the signal reaches the filter bank. The cutoff frequencies of the preprocessor filters are spaced about two thirds octave apart. The output signal of each preprocessing filter LP (except the one for the highest cutoff frequency) represents the input signal of a control filter LC whose cutoff frequency is situated somewhat below that of the associated preprocessing filter. The output signal of the control filter controls a diode gate G which blocks the preprocessing filter(s) with higher cutoff frequencies when a signal is present at the control input C of the gate. A preprocessor filter is thus connected with the summing amplifier at the output only as long as there is no signal on the associated control line. In total, this filter behaves like a tunable low-pass filter with the fundamental frequency in the vicinity of the upper cutoff frequency. Consequently, at the input of the basic extractor there is nothing but the waveform of the fundamental. The advantage, compared to a tunable filter, is that the approach is straightforward and does not require feedback. Hence, a preset is not necessary, and there is no danger that the device might lose track and become unstable. A minor criticism from the point of view of Chap.7 (to anticipate this investigation again) is that it would be preferable to place the switchable high-pass filter for low-frequency noise suppression before the nonlinear element and to add a fixed offset suppression filter thereafter.

A PDD with four channels covering a range of F_0 of about three octaves was proposed by Dibbern (1972). This PDD applies four low-pass filters (preceded by a dynamic compressor and some nonlinear circuitry) with cutoff frequencies two-thirds octaves apart. A level discrimination logic ensures correct channel selection. Phase distortions due to channel switching are corrected by a special compensation logic.

Such a four-channel PDD covering a three-octave range, although operating with band-pass filters, had already been described in 1963 by Righini. Hamm (1977) reported a five-channel PDD (for investigation of guitar sounds) with a cascade of switches for channel selection. This PDD has a measuring range of four octaves. A three-channel PDD for a measuring range of two octaves with nonlinear preprocessing (band-pass filtering, amplitude modulation, and full-wave rectification, in that sequence) was reported by Kubzdela (1976).

An eight-channel PDD for phoniatric research and musical investigations covering an extremely wide measuring range (50-4800 Hz) has been published by Fedders and Schultz-Coulon (1975). For the channel selection unit this PDD applies a frequency-matching rather than a minimum-frequency selection principle. Following the low-pass filter, each channel is equipped with a Schmitt trigger and a monostable multivibrator (monoflop) whose pulse duration is somewhat above the highest value of T_0 for which that particular channel is to serve. The monoflop is triggered whenever the output of the Schmitt

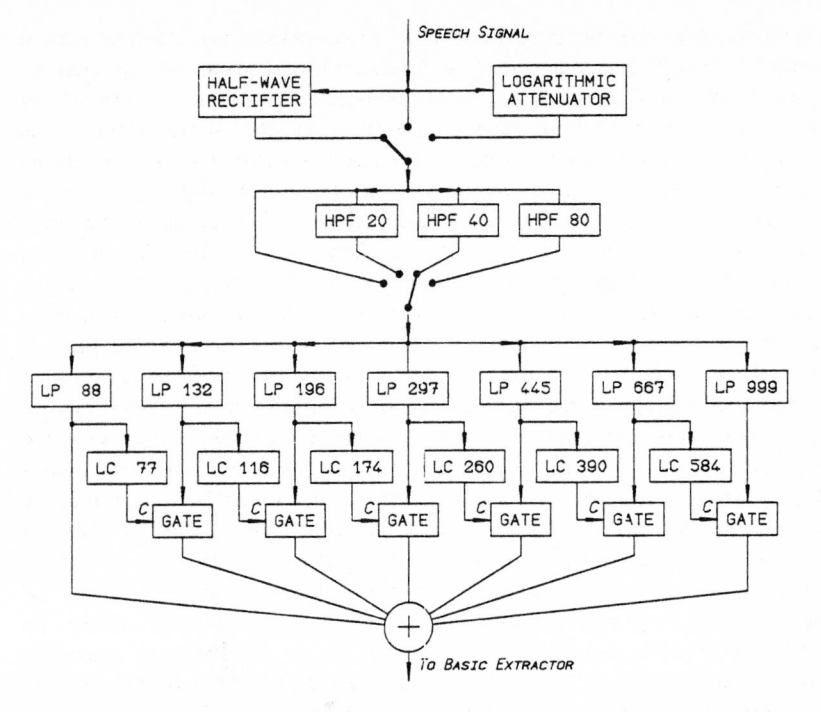

Fig.6.46. PDD by Peterson and Peterson (1968): block diagram. (LP) Low-pass preprocessor filter;   (LC) low-pass control filter;   (HPF) high-pass filter; (GATE) control gate; (C) control input of gate. The numbers denote the cutoff frequencies of the filters. Each gate is closed when there is a nonzero signal at the control input. The high-pass filter and the nonlinear preprocessor components are optional and can be changed by manual switching

trigger switches from 0 to 1 (i.e., from low to high voltage); it can also be retriggered when it is in its unstable state. This combination of Schmitt trigger and monoflop thus serves as a threshold discriminator, a TABE, and a channel selector at the same time. Due to its hysteresis, the Schmitt trigger does not switch until the signal level exceeds a certain threshold; hence, in the channels where there is no signal, the monoflop remains in its stable state (for convenience, a logical value of zero is assigned to this). In the channel which corresponds to the actual value of T_0 the Schmitt trigger acts as a TABE (with hysteresis). The monoflop is triggered regularly; since its pulse length slightly exceeds T_0, it will remain consistently in its unstable state (corresponding to a logical 1) as long as the signal is present. In those channels where the cutoff frequency of the low-pass filter exceeds $2F_0$, higher harmonics of the signal will also be present. There the Schmitt trigger will usually switch at irregular intervals causing the subsequent monoflop to switch to and fro between 0 and 1. The logic thus selects that channel whose output shows an uninterrupted logical 1. In the case of a dominant first formant at some higher harmonic there may be more than one channel with a consistent 1 at the output; in this case a minimum-frequency selection principle must be applied again.

The selection logic of this PDD causes a delay of (at least) one pitch period, since a decision can only be made after that amount of time as to whether a 1 at the output of a monoflop is consistent or not.

The use of the Schmitt trigger as a TABE is very simple but may be the cause of temporary measurement inaccuracy. If the switching threshold is fixed and a sinusoid of variable amplitude is processed, amplitude changes of the sinusoid cause phase shifts of the switching instants and thus a phase modulation of the pulses at the output (see the discussion in Sect.6.1.1, Fig.6.7).

A three-channel PDD with nonlinear processing was proposed by Di Toro et al. (1955). Each channel consists of a band-pass filter (50 to 90 Hz, 90 to 160 Hz, and 160 to 290 Hz, respectively), a limiter, and a differentiating circuit, in that sequence. The output signals of the three channels are attenuated with a ratio of 1:2:4 and added. The selection always turns out to be in favor of the low-frequency channel since this channel is least attenuated. The performance of this type of channel selection is reported to be strongly sensitive to amplitude variations of the input signal (which directly influence the slope of the sinusoid at the output of the bandpass filter and therefore strongly affect the amplitude of the peaks at the differentiating circuits). A similar PDD was proposed later by Nagata and Kato (1960).

Other PDDs with several preprocessor channels have been reported by Halsey and Swaffield (1948), Petrov (1957), Hollien (1963), Yaggi (1962), Barney (1958), and R.L.Miller (1951); the last three PDDs have already been discussed in Section 6.1.4.

One problem accompanying the application of an adaptive bank of low-pass or band-pass filters is that of the transients when one of the filters is switched off or on. In the PDAs which follow this principle, the fundamental waveform always passes the filter in the vicinity of a cutoff frequency. Unless the filter is phase compensated (or linear phase), the output signal of the preprocessor might be grossly phase distorted even if there is little change in the fundamental signal itself. For a time-domain PDD, however, phase distortions, especially time-variant phase distortions, always signify a source of possible errors. Kubzdela (1976) who quantitatively investigated this issue, found that the error may become effective during the one or two periods when the channels are actually switched. For this reason, Dibbern's PDD (1972) momentarily stops measurement during such a period. Peterson and Peterson (1968) avoided the problem by 1) providing more filters than necessary, and 2) not exploiting the whole passband of the low-pass filters in the individual channels. In any low-pass filter phase distortions tend to accumulate in the vicinity of the cutoff frequency due to the accumulation of poles and zeros of the transfer function around the transition band. In the PDD by Peterson and Peterson each channel is switched off (and on) when F_0

is situated about 10% below the cutoff frequency (cf. Fig.6.46). At that point, however, the phase distortions are still negligible. The PDD by Fedders and Schultz-Coulon avoids such errors by providing each channel with its own basic extractor.

It must be noted, however, that frequency-dependent phase distortions - whatever the design of the PDD may be - are not tolerable when the locations of individual periods are to be determined. On the other hand, the problem of phase distortions is of minor interest if only the pitch contour is desired, i.e., if the individual period limitations are not needed in the respective applications of the PDA. Hence, this principle is very suitable for phonetic and linguistic research. A good example is the *Toronto Melodic Analyzer* by Martin (Léon and Martin, 1969; Martin, 1972b). This PDA has been realized in hybrid technology. The preprocessor as well as parts of the basic extractor are analog, whereas the channel selection logic and the postprocessor are computer routines (PDP-8). This realization allows real-time operation. Considerable detail is devoted to the display of the T_0 contour which is very important for phonetic research (see Sect.9.4.2 for a discussion of this point).

Like Dibbern's PDD, the PDA by Martin uses four preprocessor input channels with four low-pass filters. The remainder of the PDA differs from earlier proposals in the way that the channel selection is not performed by level criteria, as in the PDAs by Dibbern or by Peterson and Peterson, but rather by the output of the basic extractor itself. [This PDA thus operates in a similar way as the one by Fedders and Schultz-Coulon (1975); it will have the same problems with measurement inaccuracies due to amplitude changes of the input signal.] In each channel the output signal of the filter is processed by a trigger with a fixed nonzero threshold. This trigger generates a pulse as soon as the threshold is exceeded; it thus serves as a signal-silence discriminator and a basic extractor (TABE) at the same time. If the signal in the individual channel is so small that the threshold of the trigger is not exceeded, the trigger does not generate a pulse for a time longer than the maximum period duration; consequently, it indicates silence in that particular channel. Hence each channel has its own basic extractor which is still realized in analog technology. The pulse sequences are then transferred into the computer. For each channel the intervals between successive pulses are determined and stored. The lowest active channel is selected for further processing. The subsequent processing steps pertain to the postprocessor tasks. A three-point median smoother (see Sect.6.6.4 for a detailed discussion) removes small hops in the T_0 contour. The smoother is disabled, however, when the actual estimate of T_0 is obviously erroneous; it is then replaced by a list correction routine which uses the last 32 T_0 estimates. This routine corrects gross errors such as holes or large hops due to a change of the preprocessor channel.

This PDA is to serve for phonetic and linguistic research. It admittedly requires the presence of the first harmonic (principle PiTL). On the other hand, it can process a range of F_0 from 75 to 500 Hz without needing a manual preset of an expected estimate. A further development in microprocessor technology is now available on a commercial basis (Frøkjaer-Jensen, 1980; "pitch computer").

A quantitative performance evaluation of the channel selection routine was carried out by Baronin (1965) for a three-channel PDD whose basic arrangement is similar to the PDD by Peterson and Peterson (1968). If we characterize the short-term amplitudes of the signals in the three channels by $A_1(q)$, $A_2(q)$, and $A_3(q)$, the minimum-frequency selection principle, as it is applied in these PDAs, can be formulated as

$$\underline{if}\ A_1(q) > TH(1)\ \underline{then}\ SEL(q) = 1;$$
$$\underline{else}\ \underline{if}\ A_2(q) > TH(2)\ \underline{then}\ SEL(q) = 2;$$
$$\underline{else}\ SEL(q) = 3;\ . \tag{6.31}$$

In this statement $SEL(q)$ denotes the selected channel at the instant q; $TH(i)$ is the channel-specific selection threshold. In the majority of PDAs implemented this threshold is a constant for each channel; the value of the threshold has mostly been determined empirically. A graphic display of the selection scheme according to (6.31) in an A_1-A_2 plane is shown in Figure 6.47a. Regarding the selection problem as a pattern recognition task with three classes and a two-dimensional pattern (the three classes are represented by the three selection possibilities, and the patterns are represented by the respective values of A_1 and A_2), Baronin found that 1) the adjustment of the thresholds $TH(i)$ [i = 1, 2] is quite critical, and 2) that a constant threshold for each channel does not represent the optimum, i.e., the selection criterion yielding the minimum selection error rate. In Baronin's experiment the optimal adjustment with constant thresholds gave a selection error rate of 3.4%. Optimizing the two thresholds so that they depend on both A_1 and A_2 decreased the error rate to 2.4%. (A further reduction of the error rate might have been possible if the amplitude A_3 in the third channel were also taken into account.) The location of the optimized thresholds is shown in Figure 6.47b. The pertinent PDD (Baronin, 1965, 1974a), realized in analog hardware, is rather complicated. Using a digital or a hybrid implementation, like the one applied in Martin's PDD (Léon and Martin, 1969; Martin, 1972b), such a

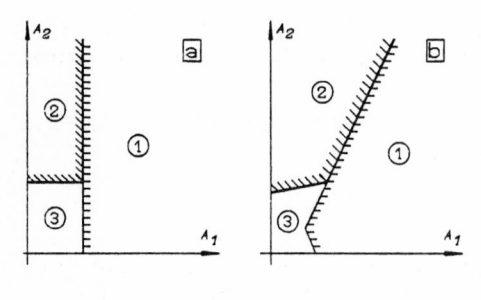

Fig.6.47a,b. Channel selection in a three-channel PDA with (**a**) constant thresholds and (**b**) optimized thresholds according to the investigation by Baronin (1965). (A_1) Amplitude of the signal in channel 1; (A_2) amplitude of the signal in channel 2. (1, 2, 3) Regions in the A_1-A_2 plane associated with the selection of the pertinent channel. Channel 1 corresponds to the low-frequency end of the measuring range

selection scheme can be implemented without great effort. Replacing the selection logic by limiters in the two lower channels, followed by an adding circuit, Baronin (1974a) himself proposed a greatly simplified but suboptimal analog version of this PDD.

In conclusion, a bank of low-pass filters with different cutoff frequencies permit applying the principle PiTL with a simple basic extractor (ZXABE) over a wide measuring range without serious implementation problems. The question of frequency-dependent phase distortions requires some attention. The PDA can be realized with a common basic extractor or with individual basic extractors for each channel. In both cases a channel selection routine is necessary. This routine has been realized in two ways: 1) a minimum-frequency selection principle which selects the lowest active channel and blocks all the higher ones, and 2) an optimum frequency-matching principle which selects the correct channel according to a logical condition (Fedders and Schultz-Coulon, 1975; Martin, 1972b). From a theoretical point of view the minimum-frequency selection principle may do a better job since it is not as sensitive to higher-harmonic tracking and does not cause a significant processing delay. It may be more sensitive, on the other hand, to low-frequency spurious signals and rapid transitions with strong low-frequency components.

Besides the experimental investigation of a particular channel selection routine, Baronin (1974a) investigated the evaluation of such channel selection schemes in general. He found that

> ... it is difficult to properly judge the performance of multichannel PDAs for various reasons.
> 1) In order to work properly, an estimate not far from the correct value is supposed to be available at the output of at least a certain number of the elementary PDAs.[28] If this does not apply, the momentary estimate of F_0 will strongly deviate from the real value. ... However, ... one could think of a scheme that interpolates local estimates from the neighbors, and in principle it should be possible to construct a multichannel PDA that works correctly even if the estimates are momentarily wrong in all channels.
> 2) It is evident that the elementary PDAs do not perform uniformly throughout the whole measuring range. Existing systems ... do not take this into account.
> 3) The selection logic discards part of the information that might be useful. Information about F_0 is also contained in those channels which are in error, for instance when they are tracking the second harmonic.
> 4) The auditory perception of erroneous results is totally disregarded in theexisting systems. (Baronin, 1974a, p.171, original in Russian, cf. Appendix B)

Based on earlier work (Pirogov, 1963; Baronin, 1965), Baronin (1974a) developed a probabilistic channel selection routine that tries to minimize the risk of

---

[28] Baronin assumes that each channel of the investigated multichannel arrangement contains a PDA of its own (with a preprocessor and at least a rudimentary basic extractor).

errors. Since this routine is of more general significance for error correction
and postprocessing, it will be further discussed in Section 6.6.3.

The PDA by De Mori et al. (1977) meets one important point of Baronin's
criticism, i.e., it is able to draw relevant information from channels that are
temporarily in error. The PDA is based on the epoch-detection principle, as
discussed in Section 6.3.2. Since the epoch information is extracted from
higher formants, narrow-band filters have to be applied in the preprocessor,
and here the use of several filters does not serve to establish a subrange of
F_0, but to increase the overall reliability. Instead of performing formant
determination, the PDA applies five fixed filters in the region between 1.3
and 3 kHz expecting that at least one formant always falls into the passband
of one of the filters.

The epoch filter supplies a peak at the instant of excitation, i.e., at the
instant of glottal closure. Due to a mismatch of the formants and individual
filters and to secondary excitations at other points in the periods there are a
number of false alarms in each channel. If a true epoch occurs, it is
registered at the same instant in all those channels where it is registered.
Hence, no synchronization is necessary when the selected channel changes;
there is even no genuine channel selection in this PDA, but only a check of
the major peaks (which may occur in any channel) as to whether they represent genuine glottal pulses or false alarms.

This starting point enables the basic extractor to be built up in a unique
way (Fig.6.48). First of all, the decision rules in the basic extractor are
formulated as a finite-state grammar. In principle, all those PDAs which
involve a heavy decision process (these are especially the PDAs of the type
PiTS) can be formulated this way. The rules of the grammar establish a

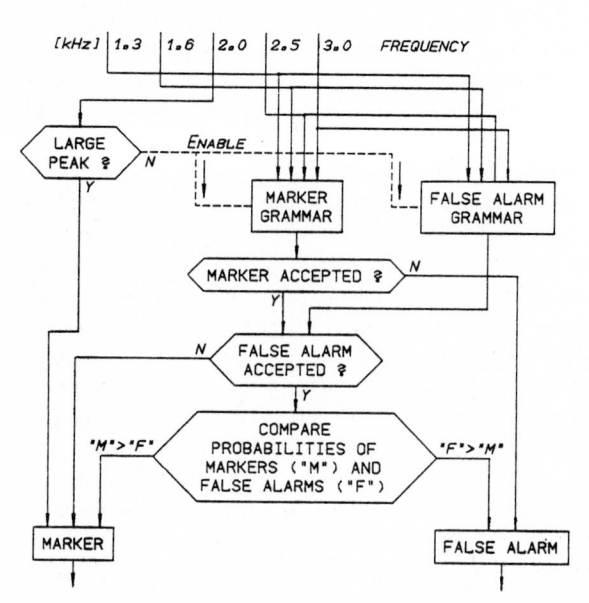

Fig.6.48. PDA by De Mori
et al. (1977): flow chart
of basic extractor. In this
flow chart it is implicitly
presupposed that in at
least one channel a peak
has occurred which ex-
ceeds the fixed thresh-
old(s) of the system

finite-state automaton or a transition network from which the pitch markers (or period durations) are extracted in such a way that an individual candidate is labeled a marker if and only if it is able to pass the automaton or the transition network by a legal path. In this case we would have the binary decision that the candidates are labeled markers if they pass the automaton, and rejected if not. De Mori's approach is different. De Mori et al. provided their automaton with transition probabilities (or, as one could say since the rules were experimentally established, they established *degrees of plausibility*) and assigned a plausibility coefficient to their markers. Even this would not have changed the procedure if done alone. But De Mori et al. established such a grammar not only for the true markers ("marker grammar"), but also for the false alarms ("false-alarm grammar"). Thus a decision can be made even in those ambiguous cases when a certain combination of peaks at the input is accepted by both the marker grammar and the false-alarm grammar. In most cases, however, one can suppose that the situation is accepted by the two grammars with clearly different degrees of plausibility, according to which the basic extractor can then make its decision.

This principle looks very attractive especially for the algorithms of the type PiTS where large efforts are made towards isolating true markers and rejecting false alarms, but practically nothing is done towards isolating false alarms which could otherwise be detected as "true markers" unless they are overridden by a better match with the false-alarm condition.

6.4.2 PDAs with Several Channels Applying Different Extraction Principles

Multichannel time-domain PDAs that apply different principles in the individual channels are rather rare. Two of those that combine a PDA of type PiTL and a PDA of type PiTN will be discussed in this section. The PDA developed by the present author (Hess, 1979a,c,d) applies three different nonlinear functions (and thus combines three elementary PDAs of group PiTN); it will be dealt with in Chapter 7. Baronin (1974a) proposed combining the principles PiTL and PiTE, but has not realized his proposal. The combination of a time-domain and a short-term analysis technique has been realized several times (Moser and Kittel, 1977; Bruno et al., 1982); it has also been used in the interactive PDA by McGonegal et al. (1975; Sect.9.2.2). This multichannel arrangement tries to combine the reliability of the short-term analysis PDA and the capability of the time-domain PDA to synchronize with individual periods. In this respect one can group such PDAs among the principle of main and auxiliary channel (see the initial discussion in Sect.6.4). Most of the time, however, these PDAs operate as short-term analysis PDAs and not really as time-domain PDAs. The short-term analysis PDA yields a reliable estimate; the time-domain PDA is then used only as a vernier or as a means for synchronizing the estimate with a particular period within the frame; in contrast to genuine time-domain PDAs the time-domain channel is not used to provide a train of markers which yield a contiguous image of the pitch contour within a voiced segment. Hence, the time-domain channel in such

PDAs is usually rather primitive, and there is no need to further discuss it in this section. These PDAs will be briefly mentioned in connection with the respective short-term analysis channel (Chap.8).

Two PDDs have been proposed by Fant (1962b) which are worthy of discussion since they are typical for the problems associated with multichannel analysis. The first PDD, based on earlier work by Risberg (1960a,b, 1961a) and further developed by Floyd (1964), is similar to that by Peterson and Peterson (1968), as discussed in the previous section. The preprocessor consists of three band-pass filters in three parallel channels which cover the measurement range of F_0 (in this PDD, 55 to 320 Hz). Each filter is followed by a basic extractor (ZXABE) whose output signal has to pass a gate before it reaches the selection logic. The gate performs a threshold discrimination and blocks the channel if the level of the signal is too low. At the output of each ZXABE, as long as it is active and not blocked, we can thus expect F_0 itself or a higher harmonic thereof, but no subharmonic tracking (unless the signal is grossly disturbed by low-frequency noise and spurious signals). Consequently, the channel selector applies the *minimum-frequency selection* principle. Among the active channels it selects the one which displays the lowest frequency. This principle is quite common in multichannel PDAs when the single channel tends towards higher-harmonic tracking, and when it can be blocked against subharmonic detection. Since the individual band-pass filters are fixed, no

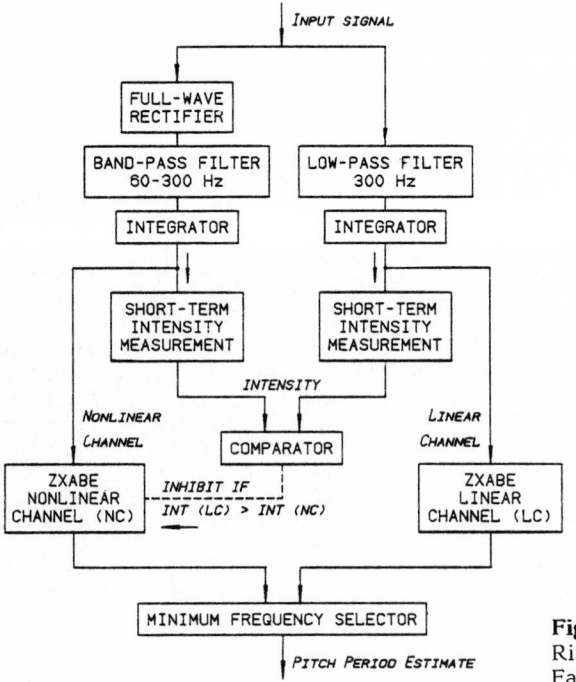

Fig.6.49. Nonlinear PDD by Risberg (1960a,b, 1961a) and Fant (1962b): block diagram

time-variant phase distortions occur in this PDD. The fundamental harmonic, however, must be present in the signal since the PDD is linear.

The second PDD proposed by Fant (1962b), based on earlier work by Danielsson and Gustrin (1961) and by Risberg (1960a,b, 1961a), is a further development of the first one towards application of nonlinear functions (Fig.6.49) and thus applicable to band-limited signals. This PDD takes account of the fact that no single NLF is ideal for any signal (see Sect.6.1.3 and Chap.7 for more details). For this reason one linear and one nonlinear channel are combined. The linear channel consists of a low-pass filter followed by a single integrator (6 dB/octave attenuation starting at 50 Hz). The nonlinear channel consists of a full-wave rectifier, a band-pass filter (60 to 300 Hz), and an integrator, in that sequence.

> The low-pass filter channel works best for high fundamental frequencies or signals with weak overtones, and the other channel works best on signals in which $F1/F_0$ is high, i.e., at low fundamental frequencies. (Fant, 1962b; p.20).

The two channels are again combined in a minimum-frequency selector. In addition, a gate is switched between the basic extractor and the selector in the nonlinear channel. This gate blocks the nonlinear channel when the amplitude of the filtered signal in the linear channel, relative to the corresponding amplitude in the nonlinear channel, exceeds a certain threshold.

The results of this PDD are very promising. More than 97% correct F_0 indication for a wide variety of voices has been reported. As to the minimum-frequency selection principle, Fant remarked that

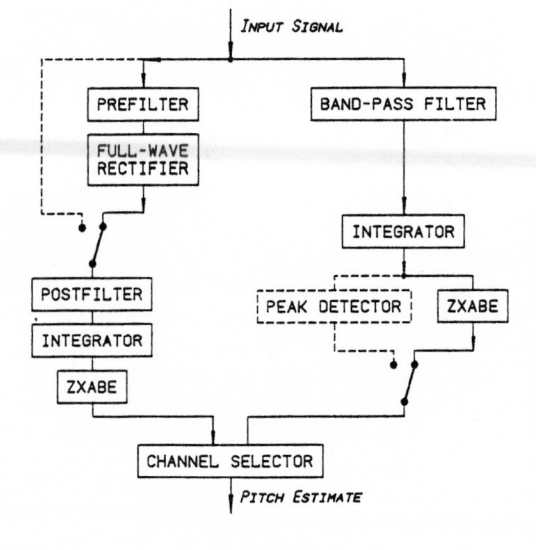

Fig.6.50. PDD by McKinney (1965), realized by McKinney and Carlson (1976): block diagram. The figure shows the original proposal by McKinney (1965). The paths depicted in dashed lines were not realized in the later implementation. Furthermore, the channel selector realized by McKinney and Carlson deviates from the minimum-frequency selector proposed originally. In the final version the selector selects that channel whose output has been most stable during a short time interval

... frequency is measured in several parallel circuits each designed for optimum result over a limited range of fundamental frequencies. The frequency that is most probable to be the fundamental frequency is then obtained by selecting the smallest of the signals from the parallel circuits. These circuits must then incorporate a device that inhibits frequency values lower than the fundamental frequency. (Fant, 1962b, p.20).

If one wants to implement a reliable wide-range PDA of type PiTL/N, this realization of several parallel channels seems to be the only promising way out of the limited performance of the individual NLFs, including linear processing. Apparently, some of the shortcomings of the PDDs reported by Fant were due to the nonideal circuitry available at that time, especially for the function generators necessary to nonlinearly distort the signal. Fant as well as Risberg, Möller, and Fujisaki (1960) were about the first to systematically discuss the effect of different NLFs on PDDs of type PiTN. This issue was further pursued by McKinney whose proposal (realized by McKinney and Carlson, 1976) contains about the same elements (see Fig.6.50 for a block diagram). After that time, there was more and more silence about this type of PDA. The idea, however, is still attractive and deserves a further investigation, in particular since much more ideal circuitry now exists for the realization of NLFs. This includes the facilities of digital signal processing where the implementation of a nonlinear function shrinks to a single arithmetic statement or a table-lookup procedure. This argument has led the present author to carry out a digital investigation of different NLFs in order to find out which one is best suited for which type of input signals. Details of this experiment and the subsequent development of a three-channel PDA with three different NLFs will be reported in Chapter 7.

6.5 Special-Purpose (High-Accuracy) Time-Domain PDAs

Pitch determination is one aspect of voice source investigation. As defined in Sect.2.5, pitch describes one component of the source of voiced signals, namely, the generator $P(z)$ of the (quasi-)periodic excitation pulse sequence. From the discussion in Sects.6.1-4 it follows that most time-domain PDAs are able to track the signal period by period with an (implicitly or explicitly) defined relationship between the position of an individual marker and the position of the pertinent laryngeal pulse. In the current section algorithms will be discussed that investigate the voice source in more detail. In this respect there are two tasks: 1) exact location of the beginning and end of each laryngeal pulse, and 2) reconstruction of the laryngeal excitation signal from the speech signal, also referred to as *glottal inverse filtering*. It is obvious that task 2) requires task 1) to be solved. For didactic reasons, however, it appears better to first discuss the algorithms that yield an estimate of the glottal excitation signal (Sect.6.5.1). These algorithms will be referred to as *glottal-waveform determination algorithms* (GWDAs).

Since these algorithms perform a high-accuracy task, they have usually been conceived without regard to the computational effort they require. Some of them, especially those which perform glottal inverse filtering, need human interaction or work only for selected speech signals. All algorithms, however, require that the signal be completely undistorted, i.e., neither phase distorted nor band limited. (Of course, nonlinear distortions are not tolerable either.) In this respect, phase distortions are more difficult to avoid; this requirement even excludes the use of high-quality analog tape recorders. In most cases, phase distortions do not cause these PDAs to fail; however, they are preserved in the output signal and may thus yield useless results. The question of input signal distortions is taken up again in Sect.9.2.1 together with a more general discussion on data acquisition. Here we assume that the signal is undistorted, disregarding this issue.

Section 6.5.2 deals with algorithms that give an estimate of the location of individual laryngeal pulses. The end of the laryngeal pulse, i.e., the *instant of glottal closure* (IGC) coincides with the instant where all formants are excited simultaneously; this point can thus be defined as the *beginning of the pitch period*. Determination of the position of laryngeal pulses is thus mostly confined to detection of the instant of glottal closure; such algorithms will be referred to as *instant-of-glottal-closure determination algorithms* (IGCDAs).

6.5.1 Glottal Inverse Filtering

Glottal inverse filtering is the approximative reconstruction of the excitation signal (the *glottal waveform*) from the speech signal. From the linear model of speech production we know that the voiced speech signal $x(n)$ can be thought of as being generated by the pulse generator characterized by its z transform $P(z)$. The pertinent pulse sequence $p(n)$ passes the glottal shaping filter $G(z)$, at the output of which we have the glottal excitation signal $s(n)$. This signal excites the supraglottal system consisting of the vocal tract $H(z)$ and the radiation component $R(z)$. In terms of transfer functions we obtain

$$X(z) = P(z) \ G(z) \ H(z) \ R(z) \ A \ , \tag{3.10}$$

where A represents the overall amplitude. A PDA, in this model, can be defined as a device which determines $P(z)$ from $X(z)$. For glottal inverse filtering the task would then read

$$S(z) = P(z) \ G(z) = X(z) \ / \ [\ H(z) \ R(z) \ A \ ] \ . \tag{6.32}$$

Thus a filter has to be applied whose transfer function reverts the influence of the vocal tract and the radiation component.

In speech production the radiation component is the low-impedance load which terminates the vocal tract; the volume velocity of the air flow at the lips (and the nose) is converted into sound pressure in the distant field. In a first approximation, which is valid for lower frequencies where the wavelength is large compared to the diameter of the mouth opening, this conversion involves a differentiation, causing a zero at zero frequency. In the inverse filter this zero is reverted by an integrator component, i.e., by a first-order recursive filter with a pole at or near $z = 1$. For reasons of stability, the pole must take on a value slightly below 1; nevertheless, the signal at zero frequency is amplified by a factor $(1-z_{pr})$ if z_{pr} is the location of the pole.[29] Together with the condition that the signal may not be phase distorted even at low frequencies, this poses rather high requirements on the quality of the recording equipment. In the earlier analog realizations, FM tape recorders and special phase-compensated high-pass filters for suppression of spurious signals at frequencies below 10 Hz had to be applied[30] (R.L.Miller, 1959; Holmes, 1962b,c; Lindqvist, 1965). Digital implementations do this job more easily. Like FM recordings, digital recordings are free from intrinsic phase distortions; linear-phase high-pass filters remove the spurious low-frequency components; and temperature and time drift of components which might cause a slight displacement of the pole of the integrator (which then could lead to instability) does not occur with digital filters.

The main component of the inverse filter models the reciprocal of the vocal-tract transfer function H(z). To design this filter, we must first know the transfer function H(z) itself; i.e., a formant analysis is necessary. The inverse filter not only requires an estimate of the formant frequencies, but one of the formant bandwidths as well. Out of the many methods developed for formant analysis (Flanagan, 1972a), we will discuss only those which have been applied in an inverse-filter implementation:

1) individual determination of the different formants, mostly in an interactive way (R.L.Miller, 1959; Lindqvist, 1965; Holmes, 1962b,c; Nakatsui and Suzuki, 1970; Hunt et al., 1978; Rothenberg, 1972);

2) automatic formant measurement by nonstationary linear-prediction (LPC) analysis during the closed-glottis interval (Markel and Wong, 1976; Wong et al., 1977, 1979, Hunt et al., 1978); and

3) cepstrum techniques (Derby, 1978; Yegnanarayana, 1979a,b, 1981).

In the classical method (R.L.Miller, 1959; Holmes, 1962b,c; Lindqvist, 1964, 1965, 1970; Hunt et al., 1978), which is carried out in an interactive way, an antiresonance circuit (i.e., a second-order filter with a complex zero) is provided to cancel each formant individually. The input signal is confined to stationary vowels with significant high-frequency components and formants

---

[29] The question of stability and low-frequency enhancement is important for the design of the integrator, but is not the only reason for keeping the pole somewhat below $z = 1$. The recordings have mostly been made with the microphone about 30 cm away from the speaker's lips. At this distance, however, the acoustic near field around the speaker has not been completely left, so that a low-frequency enhancement by 6 dB/octave, as done by the integrator, is not fully justified.

[30] Because of these problems, Rothenberg (1972) developed his pneumotachographic mask (see Sect. 5.2.2) which renders the integrator unnecessary.

that are well separable (/a/, /ɛ/). These vowels are recorded on a tape loop so that they can be repetitively played. A crude formant analysis provides reasonable initial estimates. Then the antiresonance filters are manually adjusted to the frequencies and bandwidths of the individual formants.

In the first investigation of this kind (Miller, 1959) the inverse filter only contained one adjustable formant; in the case of front vowels like /ɛ/ the analysis was thus confined to frequencies below 1.5 kHz. The system developed by Lindquist (1964, 1965 and 1970) provided four tunable filters [developed by Fant (1961)] to cancel the first four formants; fixed low-pass filters attenuate the higher formants (Fig.6.51). The same applies to the digital system developed by Holmes (1962b,c).

> The inverse filter used in the experiments described here consists of five cascaded similar networks whose transfer functions each have one complex conjugate zero pair adjustable in frequency and bandwidth, together with a higher pole pair which is well above the frequency range of interest. The correction for higher zeros is included in the inter-network couplings. The low-frequency pole which corrects for the radiation at the mouth is included in the formant-one zero circuit, and may be switched out if the differentiated larynx wave is required. Each adjustable zero network is arranged to give a low-frequency transmission which is independent of the frequency of the zero, and to have facilities for disconnecting the zero circuit without affecting the low-frequency gain. This latter facility is useful for showing the effect of inverse filtering all except one of the formants, so that the excitation pattern of each formant can be seen separately. (Holmes, 1962c, p.1)

Data by Fant and Sonesson (1962) comparing the inverse filter to the output signal of a photoelectric PDI (Sonesson, 1960; Sect.5.2.5) show that there is good temporal agreement between the two estimates of the instant of glottal closure, as provided by the inverse filter on the one hand and the PDI on the other hand. The inverse filter used in this investigation was limited to frequencies below 1400 Hz; it contained adjustable filters for two formants. Since the microphone was placed at a distance of 5 cm from the speaker's lips, no integrator was necessary for compensation of R(z).

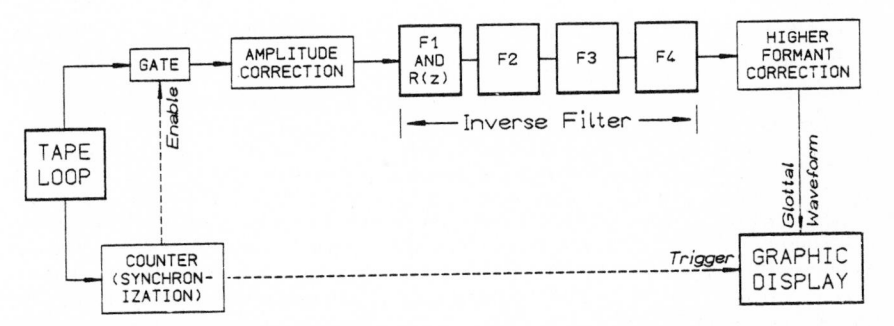

Fig.6.51. Instrumentation for interactive glottal inverse filtering: block diagram (Holmes, 1962b,c; Lindqvist, 1964)

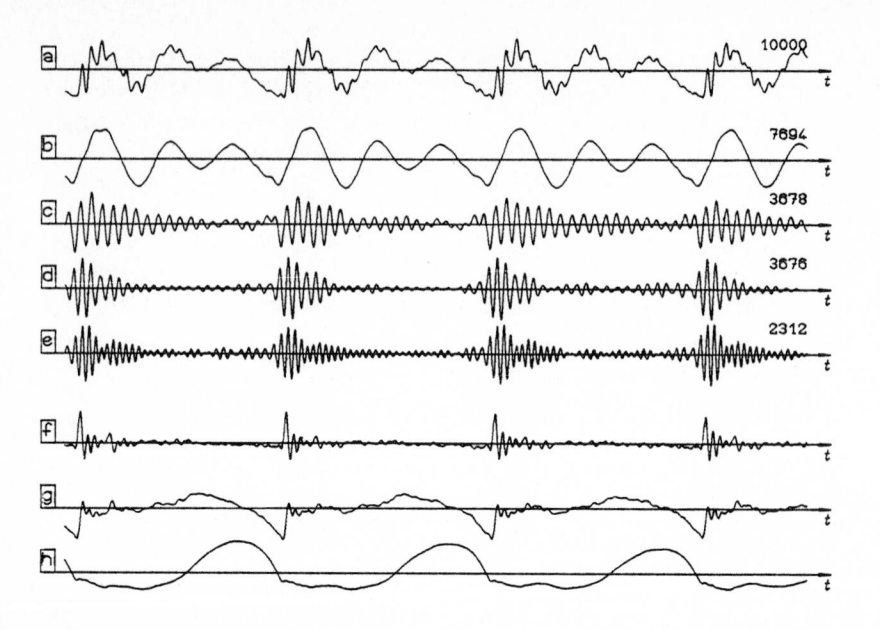

Fig.6.52a–h. Inverse-filter analysis. (**a**) Signal: sustained vowel /e/, speaker LJB (male), 32 ms per line; (**b**) waveform of the formant Fl; (**c–e**) same as (b), this time for the formants F2-F4; (**f**) differentiated output signal of the inverse filter; (**g**) output signal of the inverse filter; (**h**) reconstructed glottal excitation signal, filtered by the inverse filter and the integrator. The inverse filter was tuned to the following formant frequencies and bandwidths: F1 = 357 Hz, F2 = 2056 Hz, F3 = 2493 Hz, F4 = 3500 Hz; B1 = 26 Hz, B2 = 40 Hz, B3 = 150 Hz, B4 = 250 Hz. The transfer function of the integrator filter used is $1 / H_i(z) = 1 - 0.995z^{-1}$. All signals were normalized before plotting. The numbers on the right-hand side of (a-e) indicate the amplitude of the signal and the individual formants; the amplitude of the signal was normalized to a value of 10 000

The criterion for the adjustment of the filter is based on the postulate that during the closed-glottis interval the excitation is almost zero. Therefore the output signal of the inverse filter should be maximally flat during this time:

> For normal nonnasalized voice samples, we do not encounter any difficulty in finding an optimum setting of the antiresonance filters. The frequencies and bandwidths of the zeros are simply set to produce minimum formant frequency ripples in the output waveform. (Lindqvist, 1964, p.2)

> The frequencies and bandwidths of the zeros are then adjusted to produce minimum formant-frequency ripple in the output waveform. During vowels this condition is usually well defined, and provides, incidentally, an accurate measure of the formant frequencies. The waveform with all the formant ripple removed represents the volume velocity at the larynx. (Holmes, 1962c, p.1)

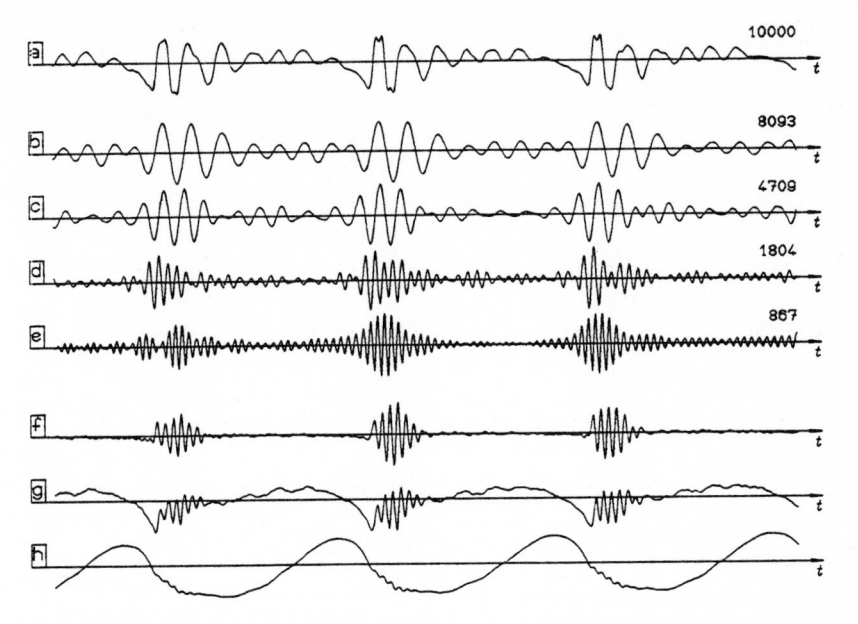

Fig.6.53a–h. Inverse-filter analysis. (a) Signal: sustained vowel /a/, speaker LJB (male), 32 ms per line; (b–h) as in Fig.6.52. The inverse filter was tuned to the following formant frequencies and bandwidths: $F1 = 745$ Hz, $F2 = 1225$ Hz, $F3 = 2268$ Hz, $F4 = 3650$ Hz; $B1 = 60$ Hz, $B2 = 70$ Hz, $B3 = 150$ Hz, $B4 = 400$ Hz. The proper adjustment for this vowel was not possible with 4 formants since an additional weak resonance was encountered in the vicinity of 3000 Hz

Data by Holmes (1962b,c, 1976) and Hunt et al. (1978) show that this is not always true and that there may be a secondary excitation at the instant of glottal opening or even during the closed-glottis interval. This issue will again be taken up later in this section.

Besides secondary excitation, other signal-dependent features may spoil the results. Holmes (1962c) mentioned that careful selection of the vowels to be analyzed is necessary in order to avoid nasalization (which leads to additional formants and antiformants in the signal). An additional pole-zero pair in the inverse filter may compensate for this effect if the frequency range of the analysis is restricted (Lindqvist, 1964). Another problem may arise from the coupling of the subglottal and the supraglottal system during the open-glottis interval (Lindqvist, 1964). For these reasons, interactive inverse-filter analysis without use of a PDI has almost exclusively been confined to male voices and chest register phonation where the closed-glottis interval is well defined and not too short. Figures 6.52-54 show some examples.

A GWDA using linear-prediction (LPC) analysis was proposed by Markel and Wong (1976) and further developed by Wong, Markel, and Gray (1977, 1979). Similar algorithms were published by Wood (1977) and by El Mallawany (1976a,b).

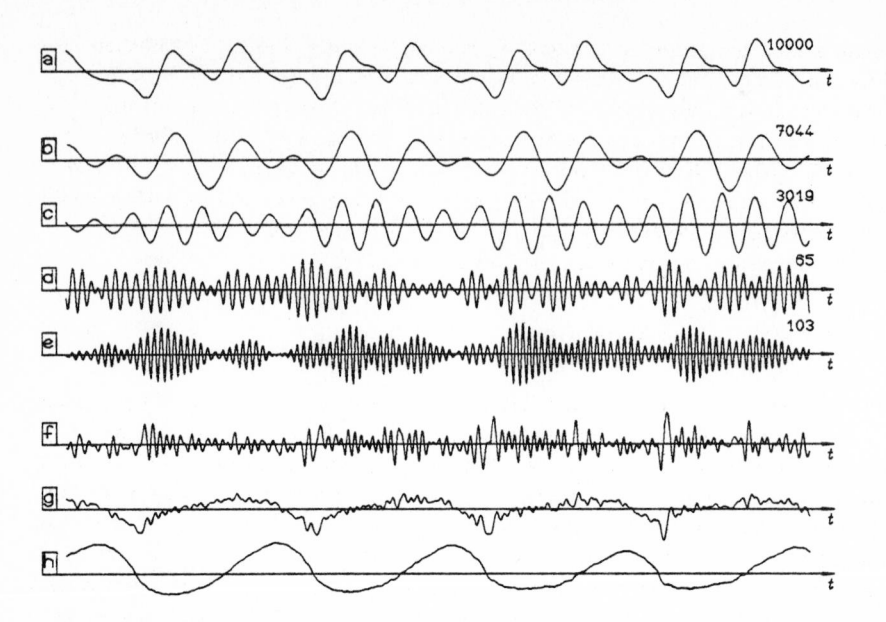

Fig.6.54a-h. Inverse-filter analysis. **(a)** Signal: sustained vowel /o/, speaker LJB (male), 32 ms per line; **(b-h)** same as Fig.6.52. The inverse filter was tuned to the following formant frequencies and bandwidths: F1 = 357 Hz, F2 = 709 Hz, F3 = 2480 Hz, F4 = 3520 Hz; B1 = 55 Hz, B2 = 20 Hz, B3 = 500 Hz, B4 = 300 Hz. This vowel represents a rather peculiar waveform. The formant F2 has an extraordinarily narrow bandwidth; F3 is hardly above the noise level (-45 dB with respect to the signal amplitude); and F4, although somewhat higher in amplitude, is also rather noisy. It can be expected that the instant of glottal closure for this signal is not easy to determine other than by inverse filtering

As already pointed out in Sect.6.3.1, LPC models the speech tract as a digital all-pole filter,

$$x(n) = \sum_{i=1}^{k} a_i x(n-i) + e(n) , \qquad (6.18)$$

and determines the filter coefficients in such a way that the filter optimally matches the structure of the signal. "Optimally," in this respect, means that the filter has been optimized according to a given criterion. The criterion mostly used involves minimizing the short-term energy of the prediction error, i.e., the energy of the residual signal e(n) within the frame analyzed.

Although LPC is well known and well documented in the literature (see the reference list in Sect.6.3.1), some mathematical details must be added in order to understand the GWDA by Wong et al. and the difference between this algorithm and an ordinary LPC analysis. Equation (6.18) says that the sample x(n) can be approximately predicted as the weighted average of the k previous sample of the signal x; e(n) will be the prediction error at the instant n. From the speech production point of view, if x(n) is the speech signal, and if the filter is to serve as a model for the speech tract, then $e(n)$ represents some kind of excitation signal; however, $e(n)$ is usually not identical with the glottal waveform. From digital-filter theory we know that the coefficients a_i, $i = 1\,(1)\,k$ represent a purely recursive k^{th}-order digital filter (Fig.6.38a) whose transfer function $A(z)$ has no zeros (except in $z = 0$) and k real and/or complex poles. We also know that the inverse filter (Fig.6.38b) with the transfer function $1 / A(z)$ can be realized using the same coefficients a_i and computing the residual signal $e(n)$ from $x(n)$,

$$e(n) = x(n) - \sum_{i=1}^{k} a_i x(n-i) \ . \tag{6.19}$$

LPC analysis can be used in a GWDA when the algorithm is modified in such a way that $e(n)$ represents the glottal waveform itself or at least a waveform which has a defined relation to the glottal waveform. The most straightforward way to achieve this is to verify that the LPC filter transfer function $A(z)$ represents the transfer function $H(z)$ of the vocal tract; in this case the residual signal $e(n)$ represents the glottal waveform except for the radiation component, whose reciprocal must be supplied in the form of the first-order integrator filter already known from the interactive GWDA discussed earlier in this section.

If the optimization criterion postulates minimizing the energy E of $e(n)$ with respect to the predictor coefficients a_i, $i = 1\,(1)\,k$,

$$E = \sum_{n} e^2(n) = \sum_{n} \left[\sum_{i=1}^{k} -a_i x(n-i) \right]^2 , \tag{6.33}$$

where $a_0 = -1$. The coefficients a_i, $i = 1\,(1)\,k$ represent a least-squares estimate of the transfer function of the speech tract. Minimizing E means equalling the derivative of E with respect to the predictor coefficients a_i to zero. In practice the partial derivatives $\partial E / \partial a_m$, $m = 1\,(1)\,k$, must be set equal to zero,

$$\partial E / \partial a_m = 2 \sum_{n} x(n-m) \sum_{i=0}^{k} a_i x(n-i) = 0, \quad m = 1\,(1)\,k \ . \tag{6.34}$$

With $a_0 = -1$ and the substitution

$$c_{im} := \sum_{n} x(n-i)\,x(n-m) \tag{6.35}$$

we obtain the linear equation system for the coefficients a_i,

$$\sum_{m} a_m c_{im} = c_{0i} \ , \quad i = 1\,(1)\,k \ , \tag{6.36a}$$

which gives in matrix form

$$
\begin{pmatrix}
c_{11} & c_{12} & c_{13} & \cdots & c_{1k} \\
c_{21} & c_{22} & c_{23} & \cdots & c_{2k} \\
\vdots & \vdots & \vdots & & \vdots \\
\vdots & \vdots & \vdots & & \vdots \\
c_{k1} & c_{k2} & c_{k3} & \cdots & c_{kk}
\end{pmatrix}
\begin{pmatrix}
a_1 \\
a_2 \\
\vdots \\
\vdots \\
a_k
\end{pmatrix}
=
\begin{pmatrix}
c_{01} \\
c_{02} \\
\vdots \\
\vdots \\
c_{0k}
\end{pmatrix}
, \qquad\qquad (6.36b)
$$

from which the predictor coefficients can be estimated. The question of the frame length for the computation of the coefficients c_{im} according to (6.35) plays an important role when the method is to be used in a GWDA; it will be discussed a little later.

As already pointed out in Sect.6.3.1, the LPC algorithm models the signal according to a global optimization criterion without taking into account the place in the speech tract where a particular pole originates. When applying the method in a GWDA it is therefore necessary to be certain that the poles of the filter transfer function A(z) represent formants only and nothing else. This leads to a modification of the LPC algorithm which involves the following two steps.

1) In order not to miss a formant, the filter order k must be more than twice the number of the formants within the frequency band analyzed. This means that the poles of A(z) have to be explicitly determined after the analysis; poles that do not pertain to a formant must be excluded from the inverse filter. In practice this means that the direct structure (see Sect.2.3) of the LPC filter, corresponding to (2.12), must be converted into the cascade structure as defined by (2.17). Routines which perform this task are standard in most scientific program libraries. [Note that for LPC analysis the numerator in (2.12,17) will reduce to a constant times z^k since the filter is purely recursive.] Once the poles are explicitly known, one can easily assign them to the formants as far as possible and exclude the remainder. One can also exclude a whole frame from further processing if the LPC algorithm has obviously missed a formant (this happens, for instance, when two real poles are supplied instead of a low-frequency or high-frequency formant).

2) In order to represent the vocal-tract transfer function H(z) as accurately as possible, the LPC analysis should be carried out during the closed-glottis interval only. During the open-glottis interval the subglottal system and the vocal tract are coupled via the glottis. In contrast to the model assumption (Fig.2.13) where the influence of coupling between the voice source and the vocal tract has been neglected, this coupling in reality affects the transfer function of the supraglottal system: subglottal formants and antiformants are added to the overall transfer function, and the frequencies and bandwidths of the vocal-tract formants are slightly changed (Wakita and Fant, 1978; Guérin and Boe, 1979; Flanagan, 1972b; Ananthapadmanabha and Fant, 1982). For normal LPC analysis the global estimate is sufficient; for the GWDA, however, greater accuracy is required.

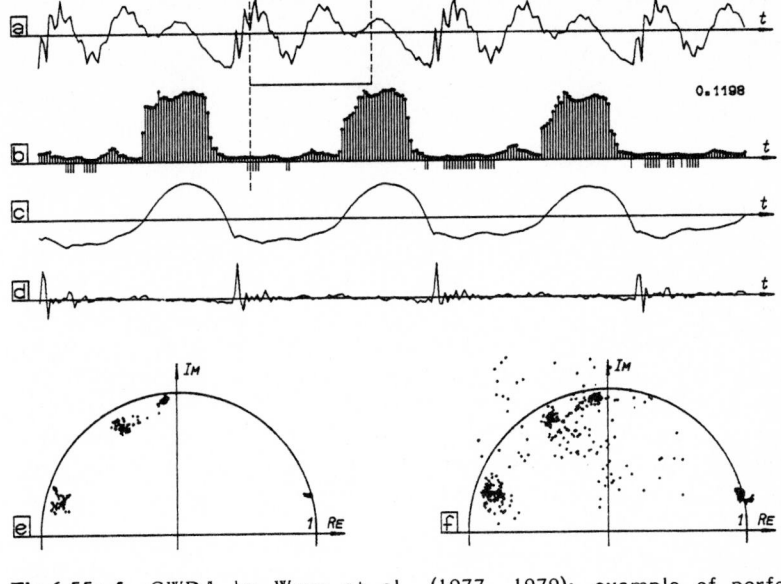

Fig.6.55a–f. GWDA by Wong et al. (1977, 1979): example of performance. **(a)** Signal: sustained vowel /e/, male speaker, 32 ms per line; same signal as in Fig.6.52; **(b)** prediction error depending on the starting point q of the frame with the maximum of the normalized error indicated on the right-hand side; **(c)** reconstructed glottal waveform (the integrator being the same as in Fig.6.52); **(d)** differentiated output signal of the inverse filter; **(e)** locations of the poles of A(z) in the z plane for those cases where A(z) was found appropriate to serve for use in the inverse filter; **(f)** locations of the poles in A(z) for all other cases; (-----) frame selected for computation of the inverse filter. All the frames which pertain to (e) have been marked in (b) by a short continuation line below the baseline. Formant frequencies and bandwidths for the inverse filter applied: F1 = 352 Hz, F2 = 2081 Hz, F3 = 2652 Hz, F4 = 3733 Hz; B1 = 10 Hz, B2 = 109 Hz, B3 = 193 Hz, B4 = 246 Hz. The constraints of the LPC analysis to separate those frames which are suited for selection for the inverse filter (e) and the remainder (f) are rather simple. A frame was excluded from selection when 1) the pertinent LPC filter was not stable, 2) less than 4 formants were detected in the frame, 3) the frame contained a formant frequency below 250 Hz, or 4) one or several formants had excessively large bandwidths. Although there is some variance in the estimates in (e), the formant frequencies and bandwidths are determined rather consistently for the pertinent frames

The last issue leads back to the question of the frame length in (6.35). In order to be able to minimize the energy E, the signal-dependent coefficients c_{im} must be evaluated on a short-term analysis basis. For ordinary LPC analysis it is appropriate to apply the stationary principle of short-term analysis (see Sect.2.4) where the signal is forced to zero outside the interval $n = q(1) q+K-1$. In doing so, we can extend the summation interval in (6.35) to infinity and obtain the well-known autocorrelation method of LPC analysis (Makhoul and Wolf, 1972; Markel and Gray, 1976). For analysis over the

closed-glottis interval, however, the use of the autocorrelation method is
unsuitable; the truncation of the signal would cause severe distortions since
the interval to be processed is rather short. Hence the nonstationary principle
of short-term analysis has to be used; it leads to the covariance method of
LPC analysis (Atal and Hanauer, 1971; Markel and Gray, 1976). In this case
the summation in (6.35) involves a constant number of elements,

$$c_{im} := \sum_{n=q}^{q+K-1} x(n-i)\,x(n-m) \ . \tag{6.37}$$

The total number of samples of $x(n)$ involved in the computation of (6.37)
equals $k+K+1$ since both i and m vary within $0(1)k$. To guarantee that the
linear equation system (6.36) is well defined, K must be greater than k so
that at least $2k+1$ samples are involved in the analysis. For the speech signal
this means that the frame must be longer than 2 ms since there is usually
one formant per 1000 Hz which requires two complex poles.

To determine the closed-glottis interval, any IGCDA may be used (see
Sect.6.5.2). This interval, however, can also be determined via the LPC
analysis itself, since it need not start exactly at the IGC. If the assumption

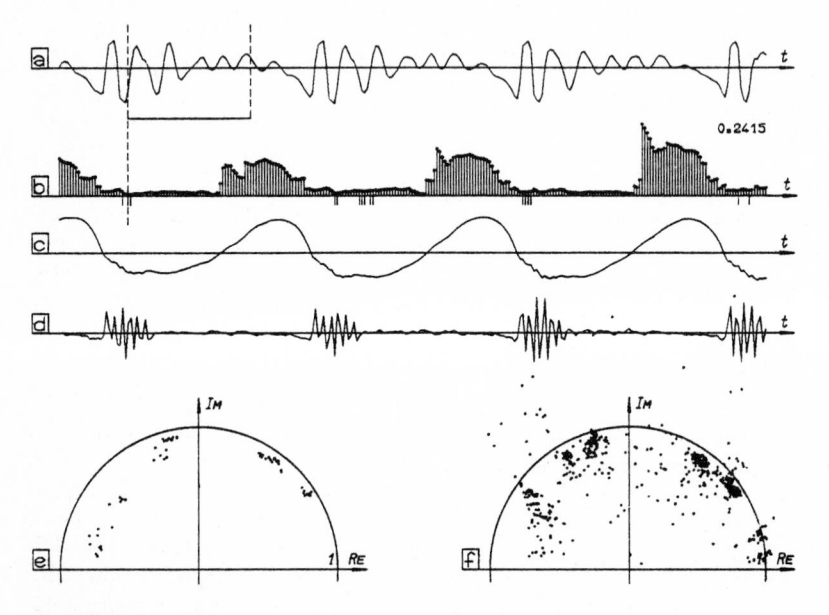

Fig.6.56a-f. GWDA by Wong et al. (1977, 1979): example of performance.
(a) Signal: sustained vowel /a/, male speaker, 32 ms per line; same signal as
in Fig.6.53; **(b-f)** same as in Fig.6.55. Formant frequencies and bandwidths for
the inverse filter applied: F1 = 745 Hz, F2 = 1221 Hz, F3 = 2340 Hz,
F4 = 3727 Hz; B1 = 120 Hz, B2 = 118 Hz, B3 = 172 Hz, B4 = 502 Hz. Corresponding to Fig.6.53, the signals as well as the formant estimates show that there
is an additional weak resonance near 3000 Hz

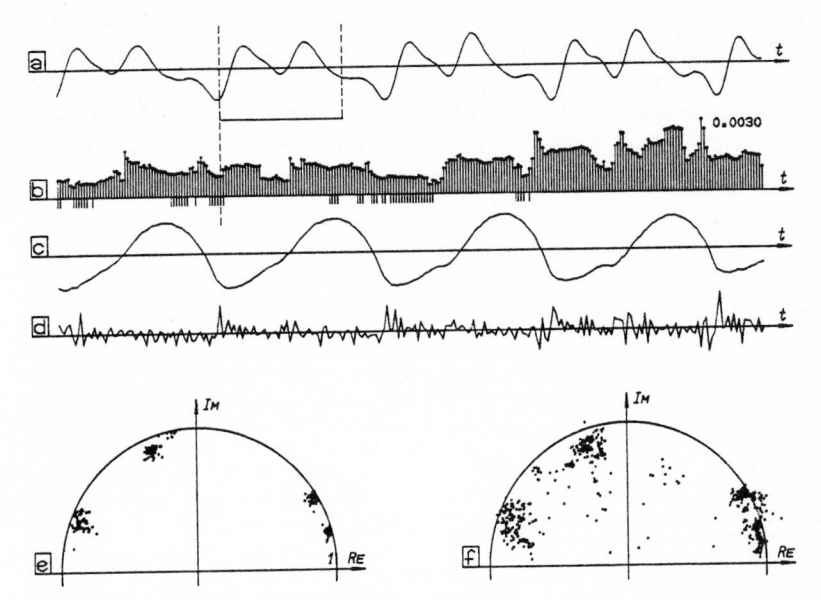

Fig.6.57a-f. GWDA by Wong et al. (1977, 1979): example of performance. (a) Signal: sustained vowel /o/, male speaker, 32 ms per line; same signal as in Fig.6.54; (b-f) same as in Fig.6.55. Formant frequencies and bandwidths for the inverse filter applied: F1 = 359 Hz, F2 = 709 Hz, F3 = 2484 Hz, F4 = 3529 Hz; B1 = 133 Hz, B2 = 123 Hz, B3 = 344 Hz, B4 = 350 Hz. According to Fig.6.54, the estimates for the bandwidths of the formants F1 and F2 are not correct

holds that the vocal tract is not excited during the closed-glottis interval, the prediction error will be very low in this case since the vocal tract then represents a linear passive all-pole system. To determine the closed-glottis interval therefore the LPC analysis (using the covariance method and a frame length K which guarantees that k+K+1 does not exceed the length of the closed-glottis interval) must be carried out at each sample individually (i.e., using a frame interval equal to the sampling interval of the signal). Low prediction error then indicates that the frame is totally embedded in the closed-glottis interval (Figs.6.55-57). Performing this analysis requires the frame length K to be substantially greater than k in order to obtain a prediction error sufficiently different from zero (Wong et al. applied a frame of 4.75 ms duration). This GWDA is thus confined to male voices in modal register (or vocal fry) where the closed-glottis interval is long enough.

The algorithm is much easier to handle than the interactive algorithms discussed earlier in this section. It automatically yields estimates of the formants which are approximately correct in most of the cases. Yet the algorithm cannot be left completely without manual control. For fundamental reasons the analysis has to be confined to nonnazalized vowels and nonnasalized stationary consonants, i.e., voiced signals that do not have antiformants. In addition, the assumption that there is no excitation during the closed-glottis interval does not always hold (Holmes, 1976), and sometimes the prediction error is lower during the open-glottis interval than during the closed-glottis interval.

In a comparative evaluation of an interactive GWDA (as discussed earlier) and the linear-prediction GWDA (as implemented by Hunt, 1977), Hunt, Bridle, and Holmes (1978) discussed this question in detail.

> Airflow through the glottis is interrupted by closure of the vocal cords, and in normal phonation the cords might typically remain closed for about half of the glottal cycle. Acoustic excitation of the formants predominantly takes place at the instant of glottal closure, and it is frequently assumed that there is no excitation during the closed phase.
> If this last assumption is valid it should be possible in principle to design an all-zero inverse filter which would reduce the energy during the closed period of a nonnasalized vowel to exactly zero, and the standard methods of designing inverse filters for glottal airflow investigations are based on determining the filter settings which minimize energy during the closed period. Early attempts at inverse filtering using analog hardware (Miller, 1959) or computer-implemented digital filtering (Holmes, 1962c) required human interaction to decide what filter settings to try and to judge the results. More recently (Hunt, 1977), the covariance method of linear-prediction analysis has been applied to closed-glottis sections of speech waveforms in order to design inverse filters completely automatically. Under the assumptions listed above linear-prediction analysis properly applied should specify the perfect inverse filter and so produce an inverse-filtered waveform with no energy in the closed period. If linear-prediction analysis gave this result there would be little case for considering any alternative method. But it does not; and because it does not we feel there are strong arguments for retaining the possibility of human interaction in inverse filtering. (Hunt et al., 1978, p.15)
> Although the glottal waveforms obtained by applying the filter specified by the linear-prediction analysis look quite credible, they rarely indicate a close approximation to zero effective glottal airflow in the closed period, and taking the second differential of the computed flow makes the activity in the closed period quite evident [see Figs.6.55-57]. This activity is not random noise because it shows considerable correlation and gradual evolution from cycle to cycle. It could conceivably result merely from incorrect settings of the zeros of the inverse filter, but we reject this possibility since linear-prediction analysis is known to work extremely well on idealized signals, and in any case our interactive method failed to find any alternative filter settings which could completely remove the closed-glottis features. We are left with two possibilities: either the all-pole model is invalid in nonnasalized vowels or there is substantial glottal excitation (i.e., effective glottal flow) during the closed period. For reasons set out elsewhere (Holmes, 1976) we favor the second explanation. (Hunt et al., 1978, p.17)

In their final conclusion, Hunt et al. (1978) found that, from the speech production point of view, both these approaches are justified and each one has its application area.

> We feel that linear-prediction analysis used with discretion is an extremely powerful tool in inverse-filtering investigations. It is quick and almost effortless, and provides, in terms of an objective - if debatable - criterion, an optimum result. It is clearly the method to adopt for work requiring the inverse filtering of large amounts of speech.

On the other hand, in any investigation of the fine details of glottal airflow and vocal-cord behavior, we are convinced of the importance of an interactive capability. We contend that the ability to appreciate the full range of possible waveforms, and the ability to incorporate external constraints on the filter settings and to determine the sensitivity of the waveform to those settings are all assets that should not be discarded lightly. (Hunt et al., 1978, p.18)

Faris and Timothy (1974) proposed a modification of the linear-prediction method for use within short intervals in the form that the interval over which the analysis is carried out need not be contiguous. This means that one might take samples from the closed-glottis intervals of consecutive pitch periods and compute the predictor coefficients from this sequence of noncontiguous intervals. This works as long as the closed-glottis interval exceeds 2 ms, and it increases the safety of the estimate. Since the LPC analysis cannot be used for the location of the closed-glottis interval if it is that short, other algorithms must do the job in this case.

Another modification proposed by Deller (1982), based on earlier work by Berouti (1977), takes account of the fact that the minimum of the LPC error does not reliably indicate the closed-glottis interval. [This agrees with findings by Hunt et al. (1978) as well as with the example displayed in Figs.6.55-57.] If there is excitation during the closed-glottis interval, or if the excitation signal is not really pulselike so that it is well predictable, the minimum of the prediction error may be situated in the open-glottis interval. (Note that linear prediction, in particular when the covariance method is used, models an exponentially rising oscillation equally well as an exponentially decaying oscillation.) Deller's criterion is the consistency of the estimates of the formant frequencies, and it is established interactively.[31] The range within which the starting point q may vary is delimited by the operator.

The algorithm ... repeated computation of the coefficients a_i as q proceeded through the selected region, at each step displaying the glottal waveform estimate by using the computed values a_i in an inverse-filtering structure. When a reasonable and similar waveform resulted for several sequential values of q, the corresponding set of coefficients a_i in this small region was assumed to be the correct identification of the vocal tract and was used as the inverse filter throughout the case.

The rationale for the procedure ... is that, in the absence of a true closed period ... the covariance analysis is allowed to seek a region of best identification of the formant structure ... which is not necessarily the point of minimum prediction error. (Deller, 1982, p.760)

One could easily think of performing this selection of the correct inverse filter automatically.

---

[31] Deller's criterion is implicitly verified in the examples displayed in Figs.6.55-57. The frames that lead to selectable estimates (as shown in Figs.6.55-57e) form small contiguous islands within the signal where the estimates are much more consistent than for the remainder of the frames.

An alternative criterion for the selection of the closed-glottis interval is the stability of the modeled filter A(z). During the closed-glottis interval the waveforms pertaining to the formants always decay; in this case the LPC filter A(z) will be stable. On the other hand, an instable filter A(z) indicates that there is strong excitation within the analysis interval.

Recent algorithms for vocal-tract modeling permit determining zeros and poles of the speech tract transfer function simultaneously. If LPC analysis is involved, the filter design is carried out iteratively (Steiglitz, 1977; Konvalinka and Mataušek, 1979) or in several approximations (Atal and Schroeder, 1978). Mataušek and Batalov (1980) proposed a GWDA using pole-zero modeling of the vocal tract on a basis of an iterative LPC algorithm. Another possibility for obtaining a pole-zero model of the speech signal is homomorphic decon-volution (Kopec et al., 1977; Oppenheim and Schafer, 1975), i.e., deconvolution via the complex cepstrum (see Sect.8.4.1). Glottal inverse filtering using such an algorithm was carried out by Yegnanarayana (1981).

Processing zeros in a GWDA is always dangerous and must be done with great care. First of all, there is the question of stability. A zero in the transfer function of the vocal tract will result in a pole of the inverse filter; this pole must be inside the unit circle. This means that a stable inverse filter can only be constructed for a *minimum-phase filter*, i.e., a filter which has both zeros and poles inside the unit circle. Algorithms that apply homo-morphic deconvolution always assume a minimum-phase system (Yegnanarayana, 1981). Second, all the algorithms for pole-zero modeling use comparatively long frames since they require the short-term analysis to be performed accor-ding to the stationary or the synchronous principle (cf. Sect.2.4), whereas the nonstationary principle would be the appropriate one for taking into account the intraperiod variations of the formants. Last but not least, one must always verify whether a complex zero in the overall transfer function really pertains to the vocal tract. This is a nontrivial problem. For nonnasalized vowels and nonnasalized voiced consonants the vocal-tract transfer function H(z) is supposed only to have poles (Fant, 1960, 1970); zeros always signal an extraordinary situation, such as nasalization or failure of the speaker to completely close the glottis. Such conditions occur quite frequently in the investigation of pathologic voices where GWDAs using all-pole models of H(z) consistently fail for individual speakers. For this kind of application GWDAs using pole-zero models of the vocal tract may thus prove extremely useful.

In summary, glottal waveform determination centers around two algorithms: 1) interactive analysis with individual formant adjustment, and 2) LPC analysis over the closed-glottis interval using the (nonstationary) covariance method of linear prediction. Due to the necessity of human interaction, the first method is more laborious, but more versatile at the same time (Hunt et al., 1978). The two methods are not contradictory so that a combination may increase the effectivity. In some special cases, such as permanent nasalization, the addition of a complex pole pair to the inverse filter may prove necessary.

6.5.2 Determining the Instant of Glottal Closure

Among all events that characterize the pitch period the *instant of glottal closure* (IGC) occupies a key position. Due to the Bernoulli force exerted on the vocal cords by the air flow in the glottis during the open-glottis interval, the vocal cords are so strongly forced together that they close abruptly and remain closed for about half the glottal cycle (for details see the discussion in Sect.3.1). The air flow is abruptly terminated; this causes a discontinuity in the time derivative of the glottal volume velocity. All formants, particularly the higher ones, are thus simultaneously excited at the IGC (Miller, 1959, Holmes, 1962b,c, 1976; Flanagan, 1972a; Fant, 1979a-c; and many others). It is thus justified from the speech production point of view to define the *beginning of the pitch period* in the speech signal to coincide with the IGC.

The IGC is rather prominent in normal phonation, i.e., modal register and medium voice effort. It is rather prominent during vocal fry as well. For soft voices as well as for the falsetto register glottal closure still occurs, but somewhat more gradually. In some special cases (breathy voice, certain voice pathologies) the glottis never closes completely. This kind of speech is characterized by weak higher formants. On the other hand, the instant of glottal opening, which passes rather smoothly most of the time, tends to exhibit a second discontinuity (and thus tends to become a second point of excitation) when the voice effort is high.

We can thus expect that the IGC usually represents the most significant and - at the same time - the most easily detectable event within the pitch period when a reference point with respect to the excitation function is required. In spite of this the task of IGC determination is not at all trivial.

There is still another reason which makes determination of the IGC desirable. When the glottis is closed, there is in principle no interaction between the subglottal and the supraglottal system, and the speech signal represents the response of the vocal tract to the excitation pulse. The transfer function of the vocal tract is thus most accurately derived from the closed-glottis interval. To locate this interval, however, one must know the IGC.

Scanning the PDAs discussed up to now in this chapter, we see that the algorithms of type PiTA (Sect.6.3) are best suited for IGC determination. In principle most time-domain PDAs place their markers at positions which have some defined relation to the excitation signal. But in many cases this relation is time variant since it depends on the momentary state of the vocal tract. In addition, IGC determination implies the detection of a discontinuity, which is wide-band information, and which is thus masked both by narrow-band formants and high-frequency attenuation in the signal. From this it follows that PDAs which derive pitch from the fundamental harmonic can hardly be used in the present task. PDAs of the algorithmic-structural-analysis-type PiTS have their problems as well. Mostly they derive their information from isolated points such as maxima, minima, or zero crossings. All these points, however, depend on the momentary phase relations between the excitation signal and the transfer function of the vocal tract and are thus time variant; they cannot be unambiguously related to the IGC.

Hence two time-domain PDA principles remain for IGC determination: 1) the *nonlinear* approach, represented by the principle PiTE, and 2) the *linear* approach, represented by the principle PiTA, especially by the PDAs of the event-detector type. The requirements on an IGCDA are a little different from the requirements on an ordinary PDA. As for the GWDA, we can presuppose for the IGCDA that 1) the signal is undistorted and has a high signal-to-noise ratio; 2) the measuring range can be restricted if necessary since the speaker is usually known; and 3) the algorithm can be designed without regard for on-line performance or computing effort. Moreover, the algorithm is allowed to fail in certain situations. Usually an IGCDA is used in pitch-synchronous speech or voice analysis systems that process a limited amount of selected data (creation of a reference data base for a speech synthesis system, for instance, or analysis of irregularly excited speech signals) and permit manual interaction. On the other hand, the IGCDA must place its markers as accurately as possible. This means that an IGCDA may produce chirps and extra markers (which can be removed by a correction routine or by manual interaction), but it may not produce hops, i.e., misplaced markers, and it should not produce holes unless a systematic failure under specific conditions can be tolerated in the specific application. In special cases, such as the analysis of irregularly excited voiced signals (vocal fry as well as certain pathologic voices), it is important that the IGCDA does not apply a regularity criterion, but rather detects each IGC for itself, independently of its neighbors.

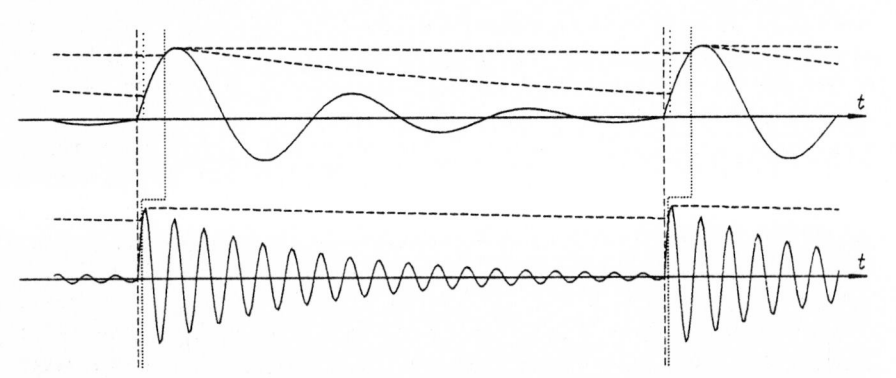

Fig.6.58. Performance of an IGCDA of type PiTE. The signal displayed is the response of a second-order recursive digital filter to a pulse train; the pulses are to represent the IGCs. (Dashed vertical line) Correct location of a pulse; (·····, ·····) estimates of the pulse location for different time constants of the peak detector circuit. The figure indicates that the error 1) is proportional to the period of the decaying oscillation (which represents the formant used for measurement), and 2) decreases with decreasing time constant of the peak detector circuit, i.e., increasing bandwidth of the formant

A PDA of type PiTE can easily be converted into an IGCDA. PDAs of type PiTE model the instantaneous envelope of the signal by nonlinear means (see Sect.6.2.1). One can easily realize from Fig.6.19 that the minimum before the principal peak of the envelope represents the IGC. If the PDA applies a peak detector followed by a differentiating circuit (which is the usual sequence in this type of PDA), the circuit enhances the ascending part of the waveform between the minimum and the principal peak and thus gives an estimate of the IGC. The only problem is that usually the oscillation of the previous period has not yet completely decayed when the new excitation pulse occurs. This causes a slight shift in the location of the pulse estimate which is proportional to the period duration of the dominant formant in the signal and also depends on the bandwidth of the formant. High-frequency formants with large bandwidths are best suited for this analysis (Fig.6.58). In an IGCDA using the principle PiTE, the signal is thus first high-pass filtered so that the first formant is always removed. The remainder of the signal is then processed as described in Section 6.2.1. It is clear, however, that a low-pass filter cannot be used in the preprocessor. For reasons of accuracy, the number of peak detector circuits should be kept low. The greatest accuracy is given after the peak detector of the first stage; the IGC there corresponds to the minimum before the principal peak. If several stages are used, as is frequently done in an ordinary PDA of this type, the output signal of the last stage can serve as a means to distinguish between correct markers and false alarms, and to give a crude estimate of the exact location of the IGC which is then derived from the signal after the first peak detector.

This IGCDA works well for signals with significant high-frequency components (such as front vowels). For back vowels and nasals where most of the energy is confined to the low-frequency band, the high-frequency noise (in the environment as well as in the recording and processing equipment) must be kept low so that the higher formants still exceed the noise level.

Linear-prediction analysis is used by the majority of IGCDAs which follow the principle PiTA. The simplest and most straightforward idea is to look at the residual signal (Atal and Hanauer, 1971). The spectrum of the residual is approximately flat so that a pulselike waveform should be expected, and the instants of the pulses should coincide with the IGCs. If there is sufficient high-frequency energy in the signal, this will be the case (Fig.6.59), regardless of whether the predictor coefficients have been determined by one global approach or by individual optimization using nonstationary analysis over short frames [as done, for instance, in the GWDA by Wong et al., 1979].

The underlying idea is that the excitation peak which occurs at the IGC causes a significant peak in the residual signal with pulselike shape. There are a large number of signals for which this principle works perfectly (see, for instance, the vowels /e/ and /a/ in Fig.6.59), but there are other signals and speakers where it fails completely (see, for instance, the vowel /o/ in Fig.6.59). The main reason for such a failure, as Gold (1977) pointed out, is that, in addition to the information on the vocal tract, the information on the excitation signal is modeled into the coefficients and thus disappears from the residual. In this case, however, the prediction error is unusually low.[32]

Since the normalized prediction error is defined as the ratio of the energy of the residual signal to that of the input signal, it is easy to see that the signal-to-noise ratio of the residual, compared to that of the input signal, is reduced by the same amount. For instance, if the normalized prediction error is 0.05, this means that the signal-to-noise ratio of the residual is reduced by 26 dB. It is clear that, under these circumstances, even a high-quality signal with a S/N ratio of 50 dB or more may be converted into a noisy residual when the prediction error is too low.

There are several reasons for the prediction error to take on a low value:

1) The signal does not contain significant high-frequency components. This applies to all back vowels, like /o/ or /u/, which are frequently reported to cause trouble (e.g., Strube, 1974a,b), and to soft and breathy voices.

2) A major formant coincides with a harmonic. In this case the onset of the new period is rather smooth.

3) A narrow-band formant dominates the signal. Due to the small bandwidth, the waveform of the formant hardly decays until the beginning of the next period.

The IGCDA by Ananthapadmanabha and Yegnanarayana (1979; discussed in Sect.6.3.2) raises the question of the phase of the excitation signal. The ideal case is given when the excitation pulse has a unipolar peak. If the excitation signal is phase shifted by 90°, the IGC coincides with a zero crossing of the excitation pulse, and the amplitude of the pulse is much reduced. This difficulty is overcome by investigating the instantaneous magnitude of the signal which is pulselike when the spectrum of the signal investigated is approximately flat. Ananthapadmanabha and Yegnanarayana (1979) applied this principle to the LPC residual. Compared to the investigation of the residual as it is, the performance of this IGCDA is significantly better; however, the intrinsic problem of the residual (bad S/N ratio when the prediction error is low) is not overcome by this solution.

In their earlier paper, Ananthapadmanabha and Yegnanarayana (1975) applied the epoch-detection principle, i.e., event detection by determining the instantaneous magnitude of the signal, to the unprocessed speech signal. Since the method only works for a signal whose spectrum is approximately flat, Ananthapadmanabha and Yegnanarayana (1975) filtered the signal by a high-pass filter (or band-pass-filter) prior to instantaneous-magnitude determination. The method was found to work when the high-frequency components of the signal

---

[32] In the example displayed in Fig.6.59, the normalized prediction error was 0.273 for /e/, 0.331 for /a/, but only 0.024 for /o/. The covariance method of LPC analysis was applied to a 40-ms frame; the LPC filter had 9 poles, and the sampling frequency was 8 kHz. Prior to the analysis the signal was preemphasized using a first-order filter with the transfer function $H(z) = 1 - 0.9z^{-1}$, which gave a high-frequency emphasis of 20 dB at 4 kHz.

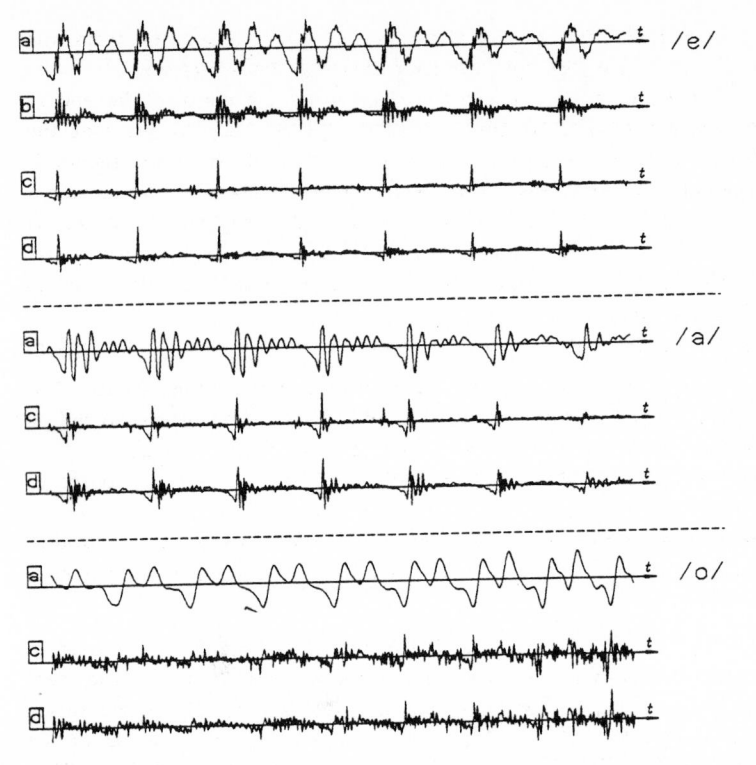

Fig.6.59a–d. Investigation of the LPC residual for detection of the instant of glottal closure. **(a)** Signal: sustained vowels as indicated on the right-hand side of the figure, speaker LJB (male), 64 ms per line, sampling frequency 8 kHz; **(b)** signal after preemphasis with a first-order filter with the transfer function $H(z) = 1 - 0.9z^{-1}$ (displayed for the vowel /e/ only); **(c)** LPC residual after an individual LPC analysis for each frame, using the covariance method and a short interval [as applied in the GWDA by Wong et al. (1979); see Fig.6.52)]; **(d)** LPC residual after global LPC analysis using the covariance method and a single 40-ms frame. The figure shows that the method works well for a front vowel, like /e/, moderately for an open back vowel, like /a/, and clearly fails for back vowels like /o/ or /u/

were sufficiently large, but suffered from a bad S/N ratio for back vowels and other speech signals where the energy is concentrated in the low-frequency band.

It was already mentioned by Ananthapadmanabha and Yegnanarayana (1975) and exploited later by De Mori et al. (1977) that the method also works when the signal is limited to a frequency band which centers around a major formant. In this case the measurement centers around an energy maximum and should thus be more noise resistant than the ordinary measurement in a frequency band where the spectrum is flat. For both Ananthapadmanabha and Yegnanarayana (1975) as well as De Mori et al. (1977), however, the effort for

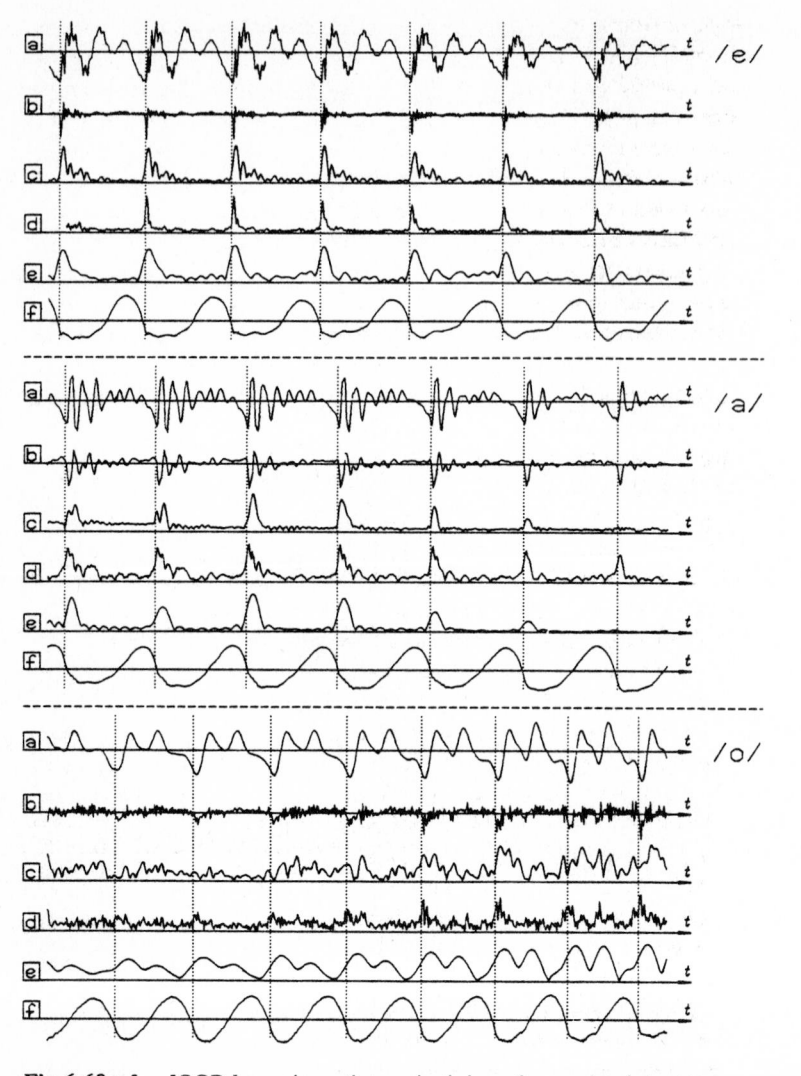

Fig.6.60a–f. IGCDAs using the principle of epoch detection: examples of performance. (a) Signal: three sustained vowels as indicated in the right-hand corner of the figure, speaker LJB (male), 64 ms per line, sampling frequency 8 kHz, signals identical to those displayed in the previous figures; (b) LPC residual (autocorrelation method, 10 poles); (c) epoch signal, directly calculated from a flat part of the spectrum (Ananthapadmanabha and Yegnanarayana, 1975); (d) epoch signal, calculated from the LPC residual (Ananthapadmanabha and Yegnanarayana, 1979); (e) epoch signal, calculated from a major formant; (f) output of glottal inverse filter using the GWDA by Wong et al. (1979). The figure shows that signal (d) determines the IGC most accurately, whereas (e) appears more reliable; possibly a combination of the two methods may improve the results. In principle, method (d) only fails when there is no high-frequency information available in the signal

a complete formant analysis was too great to be justified for pitch detection alone; for use as a PDA, DeMori et al. (1977) therefore suggested the multichannel solution, as discussed in Sects.6.3.2 and 6.4.1. In the application as an IGCDA, however, where an increased effort may be tolerated, the principle, as outlined above, again becomes attractive; only one major formant with not too low a bandwidth (preferably F3 or F4) has to be determined; this is suitably done by some (nonstationary) LPC analysis. The preprocessor band-pass filter can then be adjusted to the momentary value of the selected formant. Figure 6.60 shows some examples.

An IGCDA related to, but not solely based on linear prediction, was presented by Strube (1974a,b) with reference to earlier work by Sobakin (1972). Sobakin investigated the (nonnegative) determinant of the autocovariance matrix of the signal [as given by the matrix at the left-hand side of (6.36b)],

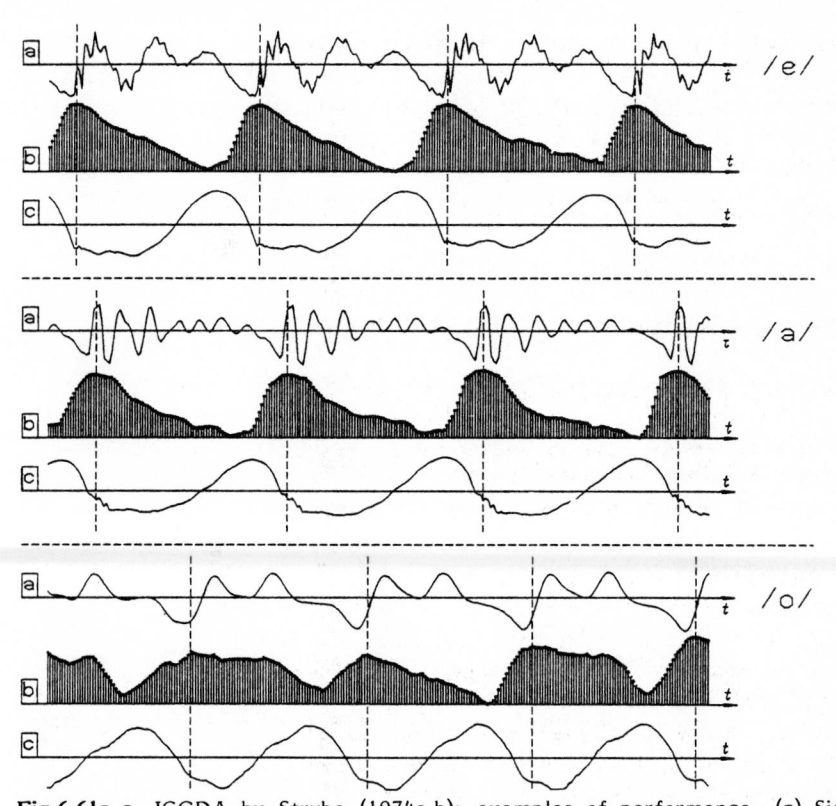

Fig.6.61a–c. IGCDA by Strube (1974a,b): examples of performance. (a) Signal (sustained vowel as indicated on the right-hand side); (b) logarithm of the determinant of the autocovariance function; (c) result of LPC glottal inverse filtering [using the GWDA by Wong et al. (1979)]. The input signals displayed (32 ms per line) are identical to those shown in Fig.6.59

$$\begin{pmatrix} c_{11} & c_{12} & c_{13} & \cdots & c_{1k} \\ c_{21} & c_{22} & c_{23} & \cdots & c_{2k} \\ \vdots & \vdots & \vdots & & \vdots \\ c_{k1} & c_{k2} & c_{k3} & \cdots & c_{kk} \end{pmatrix} \qquad\qquad (6.36c)$$

taken from a segment of speech which is considerably shorter than a period. Sobakin showed that the determinant vanishes for zero prediction error since there is a linear dependence between the samples involved. If the minimum number of samples is exceeded so that the prediction error can no longer become zero, the minimum value of the determinant indicates that there is a minimum of excitation in the signal. This point, however, is admittedly not the IGC. Strube reinterpreted the method so that it was finally converted into an IGCDA (Fig.6.61). The value of the determinant is influenced by 1) the energy of the signal within the frame investigated, and 2) the quality of the prediction, i.e., the value of the normalized prediction error; for both these parameters it holds that the value of the determinant increases with rising parameter values. In the speech signal segments with high energy and low prediction error (which applies to the closed-glottis interval) and segments with low energy and high prediction error (during the open-glottis interval) alternate. The IGC is the only point where both the prediction error and the signal energy are high; thus the value of the determinant will be a maximum at this point. Strube's IGCDA (1974a,b) computes the determinant of the autocovariance matrix of short frames (2 ms). The frame interval equals the sampling interval so that for each sample a value is provided; the maximum of the determinant indicates the IGC. Since the determinant can vary over several orders of magnitude, Strube investigated the logarithm rather than the determinant itself. Figure 6.61 shows an example. The algorithm is somewhat sensitive to global amplitude changes in the signal; such changes result in a temporary low-frequency spurious component in the signal which is gained from the determinants. Nevertheless, Strube's IGCDA is more reliable than the algorithms which only investigate the residual signal since it is based on two criteria, that is, the energy and the prediction error.

Besides glottal inverse filtering, determination of the IGC is the second task for which special-purpose high-accuracy PDAs are used. The methods applied concentrate on the two principles PiTE and PiTA. All methods require the presence of high-frequency components in the signal. Most of these algorithms need considerable computational effort, and there is a certain mismatch between this effort and the effectiveness of the algorithms when compared to their most serious competitor, the electric PDI (Sect.5.2.3), which determines the instant of glottal closure accurately and reliably.

6.6 The Postprocessor

The postprocessor in a PDA has various tasks. More than within the other parts of the PDA, these tasks depend on the individual application. Time-domain PDAs are able to mark the boundaries of every individual fundamental period. Often this form of output is not necessary, and one needs only the contour of F_0 or T_0. In this case a *time-to-frequency conversion* may be necessary. The conversion itself does not pose any problems; problems, how-ever, could arise when errors are incorporated in the measurement.

Another task of postprocessors is *global error correction.* Determination of individual period markers is the main task of the basic extractor. In some algorithms, however, especially in those of the group PiTS, the output of the basic extractor is not reliable enough to be used as the final output of the PDA. The algorithms by Gold (1962a,b) and by Reddy (1966, 1967) are classical examples. Such algorithms need a global correction routine which scans a whole utterance, at least over a whole voiced segment between two pauses and/or voiceless segments. This routine may or may not require a second scan of the signal; if not, it merely processes a list of extracted period markers and is thus referred to as *list correction routine.* List correction is defined to pertain to the tasks of the postprocessor although, for individual algorithms, it may influence the basic extractor by feedback. Parts of individual correction routines have already been described in connection with the pertinent PDAs (see, for example, the description of Reddy's algorithm in Section 6.2.2).

Smoothing is an alternative to global error correction, although it can only be applied to pitch contours, not to a list of individual period markers. We distinguish between *linear* and *nonlinear* (median) smoothing. Linear smoothing is simply the low-pass filtering of an obtained pitch contour; it diminishes the effect of "measurement noise," i.e., small inaccuracies of the unsmoothed result. Median smoothing, on the other hand, is a straightforward procedure for removing isolated gross errors in the pitch contour.

This section will start, however, with a discussion of the problems of display and time-to-frequency conversion (Sect.6.6.1). This will include a dis-cussion of simplifications that can be made in case only the pitch contour is needed and individual period markers are not required (Sect.6.6.2). Error correction will come next (Sect.6.6.3), and Sect.6.6.4, finally, will deal with smoothing routines.

6.6.1 Time-to-Frequency Conversion; Display

Time-domain PDAs and PDDs, with few exceptions, can determine each funda-mental period individually. For many applications, however, the investigators are not interested in that feature. What they rather need is the display of the F_0 (or, respectively, the T_0) contour. In this and the following section, we assume an arbitrary basic extractor to provide a train of constant-length pulses which represent the sequence of markers. An individual pulse is defined to occur whenever a new pitch period begins. From this sequence, a sequence of T_0 estimates is directly derived; the actual estimate is simply the elapsed time between the two most recent pulses provided by the basic extractor. If the sequence of T_0 estimates, for its part, is regarded as a signal, it is always (even if the basic extractor is analog) a sampled signal since a new estimate is always provided when a new period starts. The nonuniform sampling interval of this signal equals T_0, and, correspondingly, the nonuniform sampling rate equals F_0. Thus, in terms of digital signal processing, the conversion of the output of the basic extractor into a display of the fundamental frequency would be identical to a digital-to-analog conversion if the overall system is analog, and a mere arithmetic division if the whole PDA is digital. Since the latter is trivial, we can confine the following considerations to analog PDDs.

The problem of this display conversion is twofold. Firstly, there may be individual errors, such as higher-harmonic tracking, which definitely disturb the course of a F_0 contour. Secondly, there is the problem of removing the higher-frequency parts of the marker signal which come from the pulses generated by the individual markers. In the following, the two circuits of period-to-voltage conversion are described that are most commonly applied in the older PDDs. The output signal of the basic extractor in this form is directly taken to drive the converter.

The *interval-to-voltage converter* (McKinney, 1965) can be treated as a simple digital-to-analog converter whose principle is displayed in Figure 6.62. The conversion network is driven by a buffer register which serves to hold the actual value of pitch period duration. This value is actually determined in a counter which counts at a relatively high frequency (according to the required accuracy). Each pulse (of defined polarity) coming from the basic extractor causes the contents of the counter to be transferred to the buffer

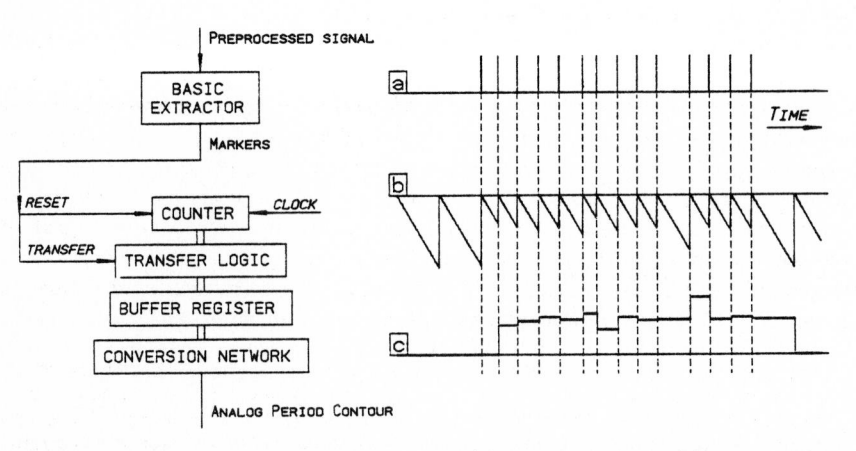

Fig.6.62a–c. Interval-to-voltage converter. **(a)** Pitch markers; **(b)** state of counter; **(c)** indication of period duration or fundamental frequency. Single lines: analog data; double lines: digital data

register and, simultaneously, the counter to be reset to zero. If a display in terms of fundamental frequency is needed rather than in terms of period duration, the conversion time to frequency is either performed by an arithmetic division in the transfer logic or in an analog way in the DAC network (the simplest way to do this is just to invert the duration scale if the investigator who interprets the display does not mind having a nonlinear frequency scale). Otherwise the analog signal at the output will be an estimate of period duration. If the whole is to be carried out in analog technology, the counter may be replaced by a sawtooth waveform generator and the buffer register by a sample-and-hold circuit. The analog signal, of course, is an unsmoothed step function and must be further processed in order to represent a smooth estimate of pitch. It can also be combined with a logic for regularity check which acts as a local correction routine (Admiraal, 1970).

If the buffer register as well as the transfer logic are omitted, the conversion network is directly driven by the counter. In this case the sawtoothlike step function represented by the actual status of the counter is converted to the analog domain. The positive peak of this function then represents the period duration (in the figure, it is the negative peak since the scale has been inverted). This method of display gives a very convenient indication of voicing. If the basic extractor is prevented from yielding impulses if no fundamental component is present, the counter is never reset. Hence, during voiceless or silent intervals the step function will have maximum amplitude and minimum frequency and can thus easily be recognized. The main advantage of this method is that it allows an easy visual interpolation and performs an instantaneous indication instead of causing a delay of at least one period due to the normal interval-to-voltage converter.

If the conversion network has a logarithmic characteristic, and if the polarity of the analog output signal is inverted, the display directly indicates the fundamental frequency with a logarithmic scale. This principle [which had first been used in the mechanical postprocessor by Meyer (1911), cf. Sect.5.1.1] was applied - of course in analog technology - by Grützmacher and Lottermoser (1937, 1938, 1940). Figure 6.63 shows the block diagram of this PDD. The preprocessor (type PiTN) uses a squaring circuit as the nonlinear element and a moderate low-pass filter in tandem. The following limiting amplifier performs the function of the basic extractor: it converts the signal at the output of the low-pass filter which is more or less a sinusoid at the fundamental frequency F_0 with extremely varying amplitude, into a pulse signal of constant amplitude. This pulse signal then drives a thyratron that is used as a triggerable sawtooth generator. The frequency of this waveform generator is chosen low (20-50 Hz) compared to the fundamental frequency.

The decisive element in the procedure ... consists of a thyratron which produces relaxation oscillations in the well-known way by means of a capacitor and a resistor - their size determining the frequency of oscillation. The frequency is chosen to be low compared to the speech frequencies - about 20-50 Hz.

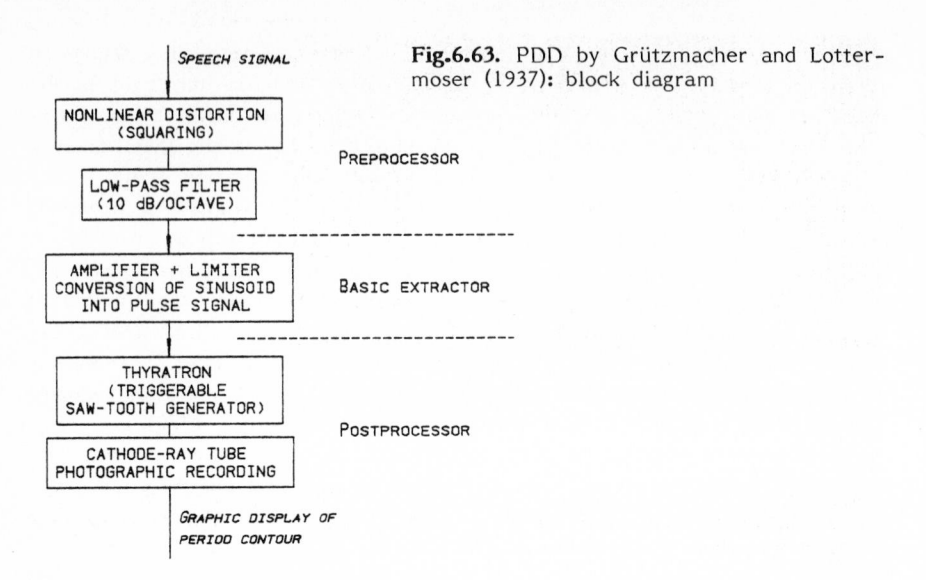

Fig.6.63. PDD by Grützmacher and Lottermoser (1937): block diagram

The relaxation oscillations are made visible by a cathode-ray oscilloscope, appearing as a line of given length. By stretching it along the time axis it can be fixed as a triangular (or similar) waveform. If an alternating voltage whose frequency is higher than that of the relaxation oscillation is connected to the control grid of the thyratron, the thyratron will discharge when the grid has received a certain charge which can be almost independent of the peaks in the control voltage. The higher the frequency of the voltage connected to the control grid, the faster the discharges are repeated; this means, however, that the charge built up previously by the relaxation oscillation is interrupted sooner, so that the line on the screen of the cathode ray oscilloscope becomes shorter with increasing frequency. With this setup the fundamental frequency of the speech wave can be used as well to control a constant relaxation oscillation. The procedure is completely inertia-free so that the problem of providing a reliable display of rapidly changing processes can be regarded as solved. (Grützmacher and Lottermoser, 1937, p.243; original in German, cf. Appendix B)

Figure 6.64 shows the performance of this PDD for the German word "Leben." This principle of fundamental frequency display (which is referred to as *Tonhöhenschreiber* in the original publication) has been widely used and improved (Kallenbach, 1961, 1962; Fant, 1957, 1958; Carré et al., 1963; Grützmacher, 1965; Eras, 1967; Admiraal, 1970; Kubzdela, 1976), especially for phonetic and linguistic research. It provides a simple, easily understandable, and at the same time accurate representation of the course of F_0. The accuracy of the original PDD (Grützmacher and Lottermoser, 1937) has been found to vary between 3% and 6%. Later it was improved by a device (1940) which was used as a vernier to accurately measure small frequency variations. It must be mentioned that this device was not designed solely for speech applications, but for any short-term (fundamental) frequency measurements of time-variant signals, such as the sounds from musical instruments, or machine noise.

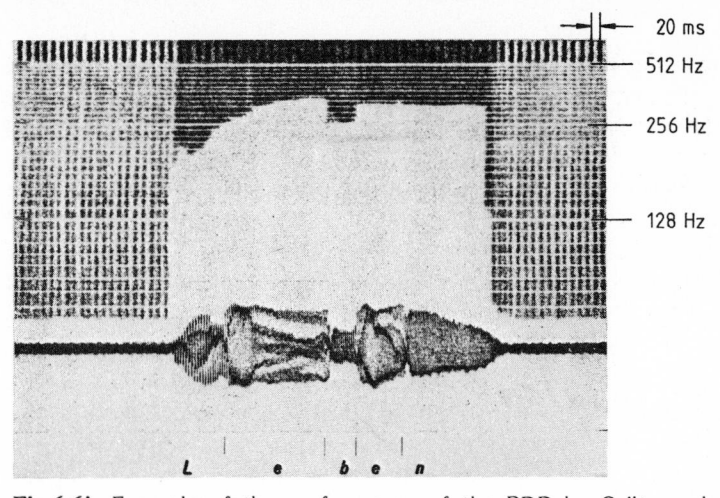

Fig.6.64. Example of the performance of the PDD by Grützmacher and Lotter-moser. Signal: German word "Leben", female speaker. From (Grützmacher and Lottermoser, 1938, p.185)

6.6.2 F_0 Determination With Basic Extractor Omitted

The short-term energy of the output signal of the basic extractor is directly proportional to the fundamental frequency provided that the pulses at the output of the basic extractor are of equal length and shape. Thus a "pulse rate converter" (McKinney, 1965) simply consisting of a low-pass filter is sufficient to convert this signal into an indication of frequency. Under these conditions, the basic extractor can be totally omitted. It may be replaced by the configuration given in Figure 6.65. In this case the output signal of the preprocessor (assumed to be a sinusoid or some other regular waveform at fundamental frequency) passes a constant-output amplifier where its amplitude is made constant. For the moment, it will remain unspecified whether this is done by limiting or by quasi-linear dynamic-range compression. To replace the basic extractor, two ways are possible.

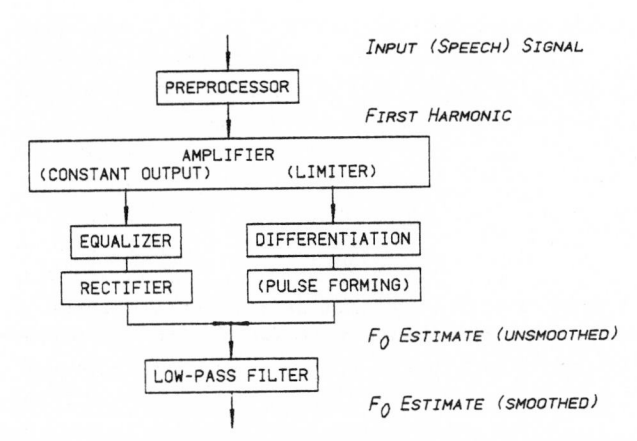

Fig.6.65. Block diagram of a PDD with basic extractor omitted (type PiTLO and PiTNO, respectively)

1) An equalizer filter (preferably following a dynamic-range compressor) which has an attenuation definitely increasing or decreasing with frequency converts the constant amplitude into a frequency-dependent one. This signal is then rectified.

2) A differentiation circuit (preferably in connection with a limiting amplifier) is followed by a pulse former to yield pulses of constant length and amplitude. The signal is finally passed through a low-pass filter in order to obtain the average and to remove the higher frequency components resulting from previous pulse forming or rectification. At the output of the low-pass filter, a signal is obtained which is a direct estimate of F_0 ("F_0 contour signal").

Although this PDD may be completely identical to a normal PDD (PiTL/N), from the functional point of view the basic extractor has been omitted. The position of individual period boundaries is not determined; only the overall contour of F_0 is of interest. Probably for this reason, McKinney (1965) and others (Gruenz and Schott, 1949) have referred to all PDAs in the categories PiTL,N as "frequency-domain" PDAs. The particular PDD described above as well as the others of this group are clearly time-domain devices in the sense of the definition given in Chap.2; the output signal of the preprocessor is a time function which allows for extraction of individual periods. Since this facility is not used, this device is referred to as time-domain PDD *with basic extractor omitted*[33] (type PiTLO and PiTNO, respectively).

There are serious reasons for the criticism directed at this realization. They were clearly formulated by Grützmacher and Lottermoser (1938) with respect to the PDD (PiTLO) by Obata and Kobayashi (1937, 1938). The main objection is that the F_0 contour signal, as available at the output of the low-pass filter, does not correctly represent the F_0 contour at the transitions from and to unvoiced segments. The preprocessor does not yield any signal when the incoming speech signal is unvoiced. Consequently, the F_0 contour signal is forced to go to zero immediately when voicing ends. Such discontinuities in the slope of F_0, however, are incompatible to the use of a low-pass filter in that path (see Fig.6.66). For that reason, this postprocessor is today of merely historical interest if used in a single-channel PDD. A practical use of this arrangement may be given in a multichannel PDD when the output of this PDD is used to provide the preset for a second, more accurate PDD which needs a preset to operate correctly [see, for example, the system by Barney (1958), Sect.6.1.4].

------

[33] With the same justification these PDAs could be treated as *short-term analysis* PDAs since they have abandoned the ability to track the signal period by period in correct phase. Nevertheless, they still operate in a period-synchronous mode. Coming from the short-term analysis side, one would have to make up a new category of short-term analysis pitch determination with these PDAs which would then be labeled *nonstationary short-term analysis PDAs without spectral transform*. PDAs of this category are further discussed in Section 8.3.3.

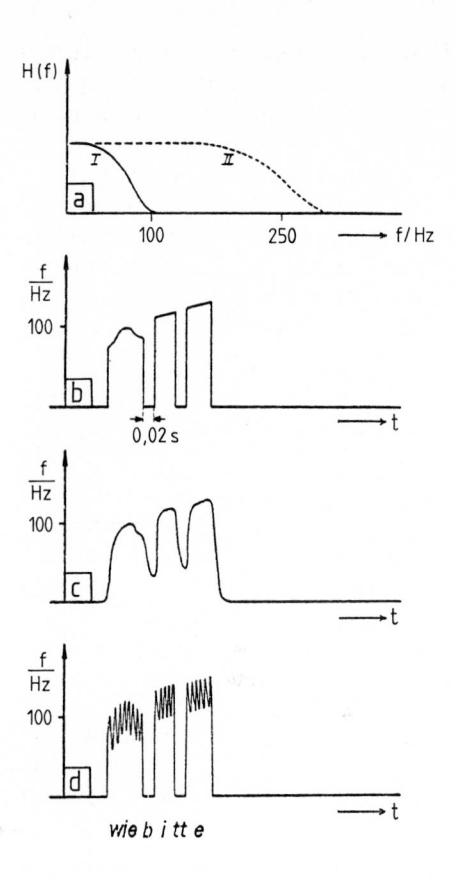

Fig.6.66a–d. Performance of a pulse-rate converter. Signal: utterance "wie bitte" (German, male speaker). (**a**) Frequency response of low-pass filters; (**b**) pitch contour as desired; (**c**) pitch contour after use of filter 1, smeared; (**d**) pitch contour after use of filter 2, with remaining ripple. From (Grützmacher and Lottermoser, 1938, p.186)

An alternate approach which falls into this category was proposed recently by Berthomier (1979). As already discussed in Sect.6.3.2, this PDD (in analog technology) measures the speed of the rotation of the vector with the complex oscillation

$$f(n) = x(n) + jy(n) = \exp(2\pi jF_0) \qquad (6.39)$$

as the derivative of the phase function of f(n). If the higher harmonics are sufficiently suppressed, no low-pass filtering is necessary at the end of the basic extractor since F_0 is directly indicated; if not, the necessary low-pass filtering is done more easily than in those devices where all the high-frequency components of the marker pulses have to be removed. In voiceless segments the behavior is undefined; the output signal will not go to zero unless forced to do so by the absence of low-frequency energy. This PDD avoids the difficulty associated with low-pass filtering of the pitch contour since it directly converts the time-domain waveform into an indication of frequency.

Together with a preprocessor of the type PiTL or PiTN, this basic extractor might thus represent the ideal combination for applications where only the F_0 contour is needed, provided that the first harmonic is present in the signal.

6.6.3 Global Error Correction Routines

In many algorithms, especially in those of the group PiTS, individual period markers are frequently misplaced (hop), left out (hole), or added (chirp) where they should not be (Gold, 1962a,b; Reddy, 1966, 1967; Hess, 1972b, Secrest and Doddington, 1982). These errors in measurement usually cannot be corrected by the basic extractor which only performs a local period marker determination. One possibility to detect and correct local errors is *list correction*. The list correction routines accomplish the corrections in a more global way since they perform a scan of all the markers of a whole utterance or at least of a whole voiced segment. A good example of such a routine is given by Reddy (1967); it has been discussed in Sect.6.2.2 (Fig.6.25). All these list correction routines, however, have a common basic structure which may be realized (in total or as a subset) in the flow chart of Figure 6.67.

The figure needs some further comments. One advantage of a list correction routine is that it can process the markers with and against time. In particular, it can start and exit at any point in the list[34]. One crucial problem of almost any speech analysis system is given by the bad starting conditions when the signal is processed only in a time-sequential manner. Usually the onset of voicing is the most critical point for starting an analysis that presumes some regularity in the signal, such as pitch determination, or even for starting short-term measurements, such as intensity determination. The majority of irregularities and the most rapid parameter changes occur at the onset of voicing. Most speech analysis systems, however, including the adaptive ones, work best when the signal is in a steady-state condition or when it slowly varies with time. For voice onset, this condition is definitely not fulfilled. Adaptive PDAs that use a tracking procedure easily get into trouble when they commit an error at the onset of a voiced segment. Such analysis systems would work better if they could operate from an entry point situated in a stationary part of the signal. For formant analysis as well as for pitch determination by short-term analysis (see Chap.8) this principle can easily be realized with moderate delay. (Usually a suitable entry point can be found within the first 50 ms after voice onset.) For time-domain analysis, especially for PDAs of the group PiTS, however, this realization is difficult to carry out since all the conditions, in this case, have to be reversed in time. Therefore it is convenient to maintain the time-sequential processing mode in

---

[34] Processing with and against time from well-defined entry points in order to avoid bad starting conditions was also performed in various formant analysis systems, such as the one by McCandless (1974) or the one by Hess (1972b, 1976b).

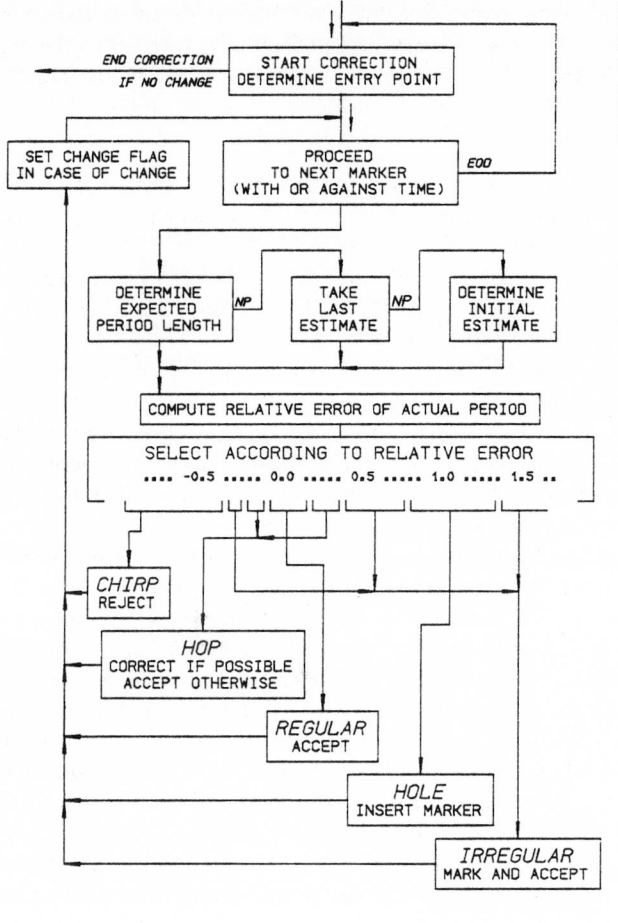

Fig.6.67. Flow diagram of a sample list correction routine in a time-domain PDA. (EOD) End of data; (NP) not possible

the basic extractor and to leave the correction of these initial errors to the list correction routine.

Another problem associated with the starting point is that of expected period length. It should not be necessary to scan a whole utterance in order to get a useful initial estimate of the average period duration. Provided that it is not required to tune the basic extractor to some expected value of T_0 before it can work (which unfortunately does not hold for every PDA), the initial estimate can be generated from the marker list by scanning it for a sequence of several (e.g., five to ten) regular markers. "Regular" means in this case that the relative deviation of the length of an individual period compared to the average period length within that particular sequence does not exceed a given threshold.

Once an estimate of period duration has been found, the algorithm can perform corrections. What is actually done depends on the relative error of the respective period length compared to the respective estimate (which, of course, will be updated as the algorithm proceeds). It proves very useful to perform these corrections in several steps.

1) Check all the markers, accept those found regular, and mark the others for correction.

2) Scan all the markers again. Leave the regular ones unchanged. Correct chirps, hops, and holes where they can be corrected, and label the corrected markers regular where they fulfil the condition of regularity with respect to the expected period length. Repeat this step until there is no more change in the list. Label those markers irregular which do not fit into the correction scheme.

Both steps may be started from regular "islands" and proceed with and/or against time towards the boundaries where measurement errors have occurred.

In any case, care must be taken that the algorithm does not become unstable. This is not always easy since a number of operations are contradictory. The main reasons for instability are 1) corrections of markers which have already been labeled "regular" in a previous run; 2) reinsertions of markers that have already been removed as chirps; and 3) removal of markers that have been inserted in the course of a hole correction. Of course, it depends on the individual algorithm whether a sequence of contradictory operations can be permitted. Usually the correction operations of an individual PDA are heuristically balanced against the particular error scope of the actual basic extractor. Throughout the PDAs which apply list correction, however, the operations are fairly consistent since the errors are about the same everywhere.

The decision whether a correction will be applied or not is made according to the relative deviation of the duration of the actual period with respect to the expected period length at that place. This decision always requires a threshold analysis, and the result of the correction routine is definitely influenced by the thresholds which delimit the individual correction operations as well as by the window which determines the length of the marker sequence that is used to extract the expected value of period length. A simple example may further clarify this consideration. The frequent error of second-harmonic tracking causes a sequence of markers to be generated (at twice the fundamental frequency) with a stronger jitter than the correct marker sequence shows unless the first formant directly coincides with the second harmonic of the signal (see, for example, Fig.6.31, where a weakly attenuated artificial "formant" is moved from 1.4 to 2.6 times the fundamental frequency). If the threshold of regularity is small enough, this jitter causes the erroneous markers to be labeled "irregular" and thus permits correcting them. Otherwise, there is no means to detect this error when it occurs for a sequence longer than the given length of the window used to determine the expected period.

Title: Pitch determination of speech
signals : algorithms and devices /
Wolfgang Hess. -

ID: 31187013702335
Due: Thursday, September 15, 2011

Title: Digital speech : coding for low bit
rate communication systems / A. M.
Kondoz.

ID: 31187027324712
Due: Thursday, September 15, 2011

Total items: 2
6/24/2011 4:32 PM

University of Waterloo

If higher-harmonic tracking remains undetected, i.e., if the erroneous markers spaced at half the real period length are labeled regular, there will be at least one uncorrectable large hop at the boundary between these markers and the real regular ones. If the algorithm is prevented from correcting markers once they have been labeled regular, both sequences, for example, remain unchanged, and the error cannot be corrected. If the correction of markers is permitted even when they have previously been labeled regular, it is unpredictable whether all the erroneous markers will be removed as chirps or all the right markers will be "corrected" as holes. This holds unless the correction routine is biased towards the one or the other individual step. For example, if the investigator knows that the basic extractor of his PDA tends towards second-harmonic tracking, he may inhibit hole correction at places already labeled regular, but not correction of chirps at such locations.

The main difficulty of list correction routines in connection with markers is that the correction must exactly fit into the whole sequence. The algorithm can remove and insert markers, and it can shift individual markers when they are misplaced. However, it cannot change the length of an individual period by shifting the rest of the (preceding or following) markers in order to fit the corrected period into the sequence. Such a correction would destroy the temporal relations between the positions of the markers and the individual periods in time. In some cases this restriction may lead to uncorrectable errors which must then be marked as irregular.

In short-term analysis PDAs as well as in those tasks where a time-domain PDA has been implemented but only a pitch contour is needed, a modified (and simplified) global error correction can be applied. In this scheme correction always means the substitution of an individual estimate by a different value according to the environment. To yield results which agree with reality, a feedback to the basic extractor is necessary in order to verify whether or not an individual correction is justified. This means that the basic extractor has to look once more at that particular signal segment (or, in the case of a short-term analysis PDA, at the short-term spectral function of that particular frame) where the erroneous situation has been found. For the following considerations, let $\hat{p}$ be the global estimate, i.e., the expected period duration as provided by the correction routine, and p_L the local estimate originally provided by the basic extractor. In principle, three possible actions can be taken as a result of the renewed check by the basic extractor.

1) The signal (or the short-term spectral function) provides a strong pitch indication at p_L but no pitch indication at $\hat{p}$. In this case the correction is not justified, and p_L is accepted in spite of being an outlier.

2) Pitch indications of comparable salience are provided at both p_L and $\hat{p}$. In this case the correction is performed, and p_L is replaced by the best estimate in the vicinity of $\hat{p}$.

3) The salience of the pitch indication is low for both p_L and $\hat{p}$. This means that at the particular instant no safe estimate of pitch can be provided. What to do now depends on the particular error and the particular routine. Two actions, however, appear suitable:

a) The ratio of p_L to $\hat{p}$ indicates an octave error or other subharmonic or higher-harmonic tracking. In this case a local voice perturbation may have occurred, or the PDA may have failed locally due to a dominant first formant or a rapid transition. The correction is likely to improve the overall pitch contour and should thus be executed.

b) No harmonic error is indicated. In this case the routine should have the possibility of rejecting the current point; later steps can interpolate a value or substitute for the point by smoothing, or check whether the frame is voiced at all.

List correction routines have been implemented in many short-term analysis PDAs. Frequently the "list" is confined to a short window of three or five consecutive frames (Noll, 1967; Bristow and Fallside, 1982). The postprocessor by Specker (1982), developed for use in connection with the time-domain PDA by Tucker and Bates (1978, cf. Sect.6.2.4), obtains a local reference estimate from the histogram of the period durations of a short segment (50-200 ms). In a first pass all estimates are removed which significantly deviate from the reference value. Then the postprocessor gives the control back to the basic extractor which is now invoked with a narrowed measuring range and sub-stitutes correct estimates at the places where the incorrect ones had been removed previously. As Specker (1982) showed, such correction routines, in conjunction with a feedback to the basic extractor, can decrease gross errors by as much as an order of magnitude (from 21% to 2% in Specker's experi-ments).

Similar results were derived by Secrest and Doddington (1982). Their algo-rithm applies dynamic programming for postprocessing of pitch contours. Dynamic programming and dynamic pattern matching are well-known and powerful techniques for searching for an optimal path through a transition network or for optimally matching two sets of templates. In speech research dynamic programming was first applied in automatic recognition of isolated words (Velichko and Zagoruyko, 1970; Sakoe and Chiba, 1971); nowadays it is widely used in both speech and speaker recognition. A good introduction is found in (Hadley, 1964), or (Sakoe and Chiba, 1978). In this book dynamic programming will be discussed in more detail in connection with the PDA by Ney (1982; Sect.8.3.2). In pitch determination dynamic programming provides an ideal means for combining temporal features (regularity) and local, intra-frame features (salience of individual pitch estimates). Using dynamic program-ming, not only the most likely estimate can be taken into account, but also the second-best and possibly even the third-best choice. The basic extractor, however, must provide some measure for the salience, i.e., the "strength" of each of these estimates. Short-term analysis PDAs usually are able to do so; a number of time-domain PDAs, such as the one by Gold and Rabiner (1969), can easily be modified this way. The dynamic programming routine now checks the temporal sequence of all these estimates. A "score of penalty" is assigned to each suboptimal solution, i.e., to each second- or third-choice pitch estimate as well as to each irregular temporal sequence of estimates.

Tracking a voiced segment the algorithm now pursues a number of possible paths through the complex pattern which is built up by the sequence of all these estimates. For each path the penalty scores of all the points touched are added. Highly unlikely paths may be abandoned before the end of the computation. The pitch contour finally selected corresponds to the path with minimum penalty. Dynamic programming thus enables the user to select a path that is finally optimal in some global sense although it need not be optimal at any place on the way.

Global error correction routines like these are only applicable when on-line applications of the PDA are not intended. Most linguistic research, speech recognition, or speaker verification tasks do not directly require on-line performance (although real-time performance in a more global sense, i.e., with reference to the processing time for a whole utterance, is desirable). On the other hand, on-line operation is needed for vocoder applications. In this case, the delay and the necessary storage capacity within the vocoder are directly proportional to the length of the marker list required by the correction routine. For these applications, PDAs are thus preferred which do not need such a routine at all or which need only a small local subset thereof.

A probabilistic postprocessor for pitch contours was presented by Baronin (1966, 1974a). Originally designed as a channel selection routine for multichannel PDAs[35] to be applied in vocoder systems, the principle has a more general significance in the postprocessing of pitch contours and is therefore discussed in this section. A similar proposal was made recently by Bruno et al. (1982).

Let T_0 be the correct period duration[36] at the current frame q, and let a multichannel PDA with K channels, each channel having a basic extractor of its own, yield the momentary estimates m_i, $i = 1\,(1)\,K$. In this case one can regard the deviations $m_i(q)-T_0(q)$ as additive noise. The task of the postprocessor consists in removing this noise; thus the postprocessor acts as a noice-cancelling filter. If the noise were normally distributed, the filter would be linear. But this is definitely not the case (see Fig.6.68), and so the filter will be nonlinear. The investigation is simplified when the errors in the individual channels and at individual frames are regarded as mutually independent.

Baronin's postprocessor starts from a modified minimum-risk classifier formulation[37] where the risk R is defined by

---

[35] This kind of channel "selection" routine has already been briefly discussed in Section 6.4.1.

[36] The same approach can be expressed in frequency-domain terms with $F_0(q)$ as the correct value. It will lead to corresponding results.

[37] A modification of the minimum-risk classifier known from pattern recognition approaches is necessary for the present application. In pattern recogni-

Fig.6.68. Schematic diagram of the probability density function $g_c(p \,|\, T_0)$ of the estimates of an arbitrary basic extractor. (T_0) Correct value of the period duration; (p) estimate of the basic extractor. The most common errors are 1) higher-harmonic tracking and 2) sub-harmonic tracking. Hence there are secondary peaks in the distribution at $p = T_0/2$, $2T_0$, etc., so that the distribution is definitely not Gaussian

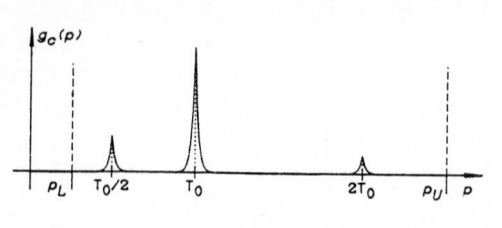

$$R = \int_{T_0} L\,[T_0, \hat{p}]\; g_p\,[T_0 \,|\, p]\; dT_0 \;, \tag{6.40}$$

In this equation, g_p is the a posteriori probability density for T_0; the vector **p** represents the estimates p_i, $i = 1\,(1)\,K$ of the individual basic extractors; and $\hat{p}$ is the global estimate as provided by the channel selector and the postprocessor. The integration is performed over the measuring range of T_0. [Of course this formulation is also valid for a single-channel PDA; in this case we will have $K = 1$.] The task of the classifier is to minimize the global risk of the problem. The "cost function" L represents the degree of freedom and must be specified in such a way that the solution of (6.40) using statistical methods at the same time represents an optimal solution from the viewpoint of the original task. In pattern recognition the usual way of defining L is to assign a value of zero to L whenever the decision is correct, and some nonzero value otherwise. For the present problem, where a defined boundary between "correct" and "wrong" does not exist, a squared-error function may be appropriate. Baronin's definition of L comprises several consecutive frames and, in addition, takes account of the changes in T_0 between subsequent frames,

$$L(q) = \sum_{i=2}^{q} c_1\,[T_0(i) - p(i)]^2 + c_2\,\{\,[T_0(i) - T_0(i-1)] - [p(i) - p(i-1)]\,\}^2 \;. \tag{6.41}$$

The weighting constants c_1 and c_2 permit switching between an *instantaneous* and an *incremental* mode of operation. If $c_1 \gg c_2$, emphasis is put on the correctness of the instantaneous estimate of T_0. On the other hand, if $c_2 \gg c_1$, the correctness of the transition between the new estimate and the previous one is the main criterion.

Using (6.41) to minimize (6.40), the optimized estimate $\hat{p}(q)$ is computed as follows:

---

tion we usually have a small number of discrete classes and a decision that is either correct or wrong. Here the "decision" extends over the measuring range of T_0 which is a continuum, and a small deviation of the estimate from the correct value may be tolerable. The classifier can cope with both these situations provided that the "cost function" L is suitably defined.

$$\hat{p}(q) = C_1 k_1 \int_{T_0} T_0(q) g_a[T_0(q)] \cdot \prod_{i=1}^{K} g_{ci}[p_i(q)|T_0(q)] \cdot dT_0(q)$$

$$+ C_2 \hat{p}(q-1)$$

$$+ C_2 k_2 \int_{T_0} \int_{T_0} [T_0(q)-T_0(q-1)] g_a[T_0(q)]$$

$$\cdot \left\{ \prod_{i=1}^{K} g_{ci}[p_i(q-1)|T_0(q-1)] g_{ci}[p_i(q)|T_0(q)] g_{tr}[T_0(q)|T_0(q-1)] \right\}$$

$$\cdot dT_0(q-1) \, dT_0(q) \,. \tag{6.42}$$

In this equation, $g_a(T_0)$ is the a priori probability density for T_0; $g_{ci}(p_i|T_0)$ is the conditional probability density that the ith channel gives the estimate p_i when T_0 is the correct value (see Fig.6.68); and g_{tr} is the transitional probability density, i.e., the probability density that the period duration at frame q is $T_0(q)$ when it is $T_0(q-1)$ at the previous frame. C_1 and C_2 represent the two weighting constants c_1 and c_2 from (6.41); they are obtained as follows:

$$C_1 = c_1/(c_1+c_2) \quad \text{and} \quad C_2 = c_2/(c_1+c_2) \,. \tag{6.43}$$

The values k_1 and k_2 serve for normalization; k_1, for instance, must be determined such that the weighted integral over the probability densities,

$$k_1 \int_{T_0} g_a(T_0) \prod_{i=1}^{K} g_{ci}(p_i|T_0) \, dT_0 \,, \tag{6.44}$$

equals unity for an arbitrary frame q when the current measured estimates $p_i(q)$, $i = 1\,(1)\,K$ are substituted (Bruno et al., 1982). The integrals in (6.40,42,44) have to be evaluated over the measuring range of T_0.

The postprocessor is interpreted as follows. The first term in (6.42) represents the contribution of the instantaneous estimates $p_i(q)$, weighted by the factor C_1. To evaluate (6.42) for the current frame q, the actual estimates p_i, $i = 1\,(1)\,K$ must be substituted. If most of the estimates p_i are correct, the product of the conditional probability densities will have a large peak at T_0, thus yielding the integral a value close to T_0. If a number of estimates are wrong, the first term will give a correct estimate only if the errors are not biased, i.e., if about the same number of channels perform higher-harmonic tracking and subharmonic tracking; otherwise the estimate will deviate from the correct value.

The second and third terms of (6.42) represent the incremental mode of the postprocessor. The second term simply is the optimized estimate from the previous frame, and the third term is the increment between the previous and the actual frame. This increment is a function of two different quantities. First, it depends on the correctness of the measurements at the frames q and q-1, represented by the same conditional probabilities as in the first term of

(6.42) and depending on the elementary PDAs in the individual channels. Second, it depends on the transitional probability density g_{tr} that $T_0(q)$ will follow $T_0(q-1)$. Both the a priori probability density g_a and the transitional probability density g_{tr} are characteristic for an individual voice or for a number of speakers, whereas the probability densities g_{ci}, $i = 1\,(1)\,K$ represent the performance of the individual PDAs.

If the individual estimates p_i are correct for both the frames q and q-1, the increment will take on the correct value so that the actual estimate given as the sum of the second and third terms in (6.42) will be about equal to T_0. If there are gross errors in the individual estimates, the product of the conditional probability densities will become rather diffuse so that the value of the third term in (6.42) will be close to zero. This means that in the incremental mode the pitch contour tends to become monotonic when the elementary PDAs fail systematically. According to Baronin (1974a) this feature is desirable since jumps and discontinuities in the pitch contour during voiced segments are most annoying to the human listener (Pirogov, 1963). This has implicitly been confirmed by McGonegal et al. (1977) who showed experimentally that a smoother in the postprocessor in most cases improves the perceptual quality of a pitch contour. Baronin (1966, 1974a) thus postulated that the postprocessor should emphasize the incremental mode, i.e., that the weighting constant C_2 be significantly greater than C_1. If the incremental mode is overemphasized, however, errors which occur at the beginning of a voiced segments may remain undetected and propagate throughout the whole segment.

For practical implementation, detailed statistical knowledge about the performance of the individual basic extractor(s) of the (elementary) PDA(s) as well as about the characteristics of the speakers (transitional probability, a priori probability) is required.[38] In a first approximation, however, rather crude assumptions are sufficient. The elementary PDAs can be modeled by a probability density like that displayed in Fig.6.68. The a priori probability density $g_a(T_0)$ represents a characteristic of a single voice or a number of speakers whose utterances are being investigated by the PDA. Here one can start from investigations known from the literature (see Sect.4.3) or even from a uniform distribution within the measuring range if nothing is known about the speaker. In addition, the a priori probability density can be updated and adapted during the measurement. This may temporarily increase the reliability of the PDA as long as that particular speaker is speaking; if the speaker changes, however, the PDA may be led into failure when wrongly adapted. The transitional probability density function g_{tr}, finally, may be approximated by a normally distributed function with zero mean or by experimentally obtained data, as they have been published in the literature (see Sect.4.3). The transitional probability density function g_{tr} is characteristic for an individual

---

[38] For more details on statistical methods in general, the reader is referred to a textbook, such as the one by Papoulis (1965).

voice or for a number of speakers. In contrast to the a priori probability density function, however, g_{tr} can be regarded as time invariant and can thus be approximated by some well-known distribution, as described before.

Baronin (1966, 1974a) realized his postprocessor in analog technology (!) for a multichannel PDD. Unfortunately, quantitative results on the performance are not available. From a theoretical point of view, the main advantage of this postprocessor is that it is able to draw relevant information from erroneous individual estimates. The drawback is that an individual erroneous estimate p_i may influence the global estimate $\hat{p}$ in such a way that $\hat{p}$ finally takes on a value which equals neither p_i nor the correct value T_0, but is situated somewhere in between. This case is similar to that of a linear smoother if there is a single outlier in a pitch contour which is correct otherwise. The same effect may occur permanently when the elementary PDAs have a clear bias toward higher-harmonic tracking or subharmonic tracking so that the pertinent probability density function becomes nonsymmetric with respect to T_0. Of course this is undesirable.

The PDA by Bruno et al. (1982) combines two elementary PDAs by a probabilistic postprocessor which corresponds to (6.40,41) but does not use an incremental mode (i.e., $c_2 = 0$). Instead, this PDA continuously updates the a priori probability density function $g_a[T_0(q)]$, and the initial values for g_a are determined in a training phase. Bruno et al. evaluated the performance of the PDA using a reference which was generated by a mechanic PDI (see Sect. 5.2.2); working with a small preliminary data base, they found that the combination of the two elementary PDAs in connection with the probabilistic postprocessor is likely to improve the overall performance.

It is possible to think of modifiying Baronin's approach in such a way that, instead of minimizing the "cost" function $L(q)$, the a posteriori probability $g_p[T_0(q)|\mathbf{p}(q)]$ is maximized. If we formulate g_p for the present problem in a way corresponding to (6.42), we obtain

$$
\begin{aligned}
g_p[T_0(q)|\mathbf{p}(q)] = & \; C_1 \, g_a[T_0(q)] \prod_{i=1}^{K} g_{ci}[p_i(q)|T_0(q)] \\
& + C_2 \, g_a[T_0(q)] \cdot \Bigg\{ \prod_{i=1}^{K} g_{ci}[p_i(q)|T_0(q)] \\
& \qquad \cdot \int_{T_0} g_{tr}[T_0(q)|T_0(q-1)] \, g_{ci}[p_i(q-1)|T_0(q-1)] \, dT_0(q-1) \Bigg\} \\
= & \; g_a[T_0(q)] \cdot \Bigg\{ \prod_{i=1}^{K} g_{ci}[p_i(q)|T_0(q)] \\
& \cdot \Big\{ C_1 + C_2 \int_{T_0} g_{tr}[T_0(q)|T_0(q-1)] \cdot g_{ci}[p_i(q-1)|T_0(q-1)] \, dT_0(q-1) \Big\} \Bigg\} \; .
\end{aligned}
\tag{6.45}
$$

The final estimate $\hat{p}(q)$ is given as the value for which the a posteriori probability density $g_p[T_0(q)|\mathbf{p}(q)]$ reaches its maximum. This expression is simpler than (6.42) and avoids deviations in the final estimate due to a bias of the elementary PDAs toward a certain type of errors.

A simple example may illustrate this point. Let a single-channel PDA yield the estimates $p(q-1) = 10$ ms and $p(q) = 4.8$ ms at the output of the basic extractor, and let the basic extractor behave in such a way that it gives correct estimates 70% of the time, 20% estimates equal to $T_0/2$, and 10% equal to $2T_0$. Then the probability[39] $P(p|T_0)$ is 0.7 for $p = T_0$, 0.2 for $p = T_0/2$, 0.1 for $p = 2T_0$, and zero otherwise. For simplicity the a priori probability density is assumed to be uniformly distributed so that it does not contribute to the results. Within frame $q-1$, with $p = 10$ ms, the period duration will thus be $T_0 = 10$ ms with 70% probability, $T_0 = 20$ ms with 20%, and $T_0 = 5$ ms with 10%. According to (6.45) we will thus have $\hat{p}(q-1) = 10$ ms. If the frame q were regarded in isolation, it would yield $\hat{p}(q) = 4.8$ ms in the same way. Let us now regard the incremental mode alone, i.e., let us assume $c_1 = 0$ and $c_2 = 1$. Let us further assume that the transition probability P_{tr} is ten times greater for a transition with no change of T_0 than for a transition with a change of about one octave, and that changes of more than one octave between subsequent frames do not occur. Applying (6.45) we then obtain

$$k_c P_p(4.8\text{ ms}) = 0.7\,(1 \cdot 0.1 + 0.1 \cdot 0.7) = 0.119 \;,$$

$$k_c P_p(9.6\text{ ms}) = 0.2\,(0.1 \cdot 0.1 + 1 \cdot 0.7) = 0.142 \;.$$

The a posteriori probability formulation thus leads to the decision that the original estimate of the frame q is wrong; the estimate is corrected to $\hat{p}(q) = 9.6$ ms. (The constant k_c is necessary since for P_{tr} only ratios, not real probabilities have been assumed.) To continue the example, let us assume $p(q+1) = 4.6$ ms. Then we have two consecutive estimates that are about equal, and the postprocessor will vote for $\hat{p}(q+1) = 4.6$ ms. If $p(q+1) = 9.2$ ms, however, then the postprocessor, in spite of the original estimate $p(q) = 4.8$ ms, will yield $\hat{p}(q+1) = 9.2$ ms. Hence, in this approach a bias of the basic extractor toward higher-harmonic tracking automatically invokes some minimum-frequency selection principle in the postprocessor.

This example should not suggest that the approach given by (6.45) is optimal in some way. It should only serve to show how statistical methods could be introduced to solve the task of error correction in the postprocessor. Most global error correction schemes implemented are based on experimental data and follow heuristic strategies. They are thus not easily transferable from one PDA to another whose basic performance might be different. Statistical methods, on the other hand, provide a universal means to evaluate and to

---

[39] For reasons of simplicity we assume that there are no fine measurement inaccuracies so that the probability density g_c is nonzero only for three discrete values of T_0. Strictly speaking, we therefore have to change the expression (6.45) for the continuous case into the corresponding expression for the discrete case where the functions are no longer probability densities but the corresponding probabilities.

correct pitch contours with occasional errors, as most basic extractors supply them. Of course the use of statistical methods requires a substantial amount of work to be done to evaluate the performance of the individual components, otherwise the probability densities and functions may not be representative for the data and yield results far from reality. The difficulties in obtaining representative evaluative data on PDAs are discussed elsewhere in this book (Sects.9.2,3). They may have been a reason why statistical methods have rarely been applied in PDAs. Baronin (1966, 1974a) had a relatively large data base at hand which was obtained by hand labeling and comparative evaluation of several PDAs. An alternative to creating such a data base is given by the use of a reliable PDI; the recent investigation by Bruno et al. (1982) has gone this way. Nevertheless, in this field there still are a number of open questions, and statistical methods may prove to be a viable alternative to simple nonadaptive smoothers on the one hand and heuristically oriented correction routines on the other hand.

6.6.4 Smoothing Pitch Contours

In spite of all the corrections, the signal carrying the F_0 information (i.e., the fundamental frequency or fundamental period contour) is noisy and sometimes afflicted with isolated errors, such as second-harmonic tracking. On the other hand, as discussed in the previous section, F_0 contours exhibit sharp discontinuities which must be preserved since they indicate the beginning and end of voicing.

Smoothing is a way of removing the effects of noise and isolated errors. However, it is applicable only to contours (of F_0 or T_0, respectively), not to a sequence of markers. The reason is easy to see; by smoothing one changes period durations. Applying a smoothing routine to a marker sequence would thus force the markers out of synchronization with the true beginnings of the individual periods.

The most usual way of smoothing is the convolution of the given signal by the smoothing function. Since the smoothing function (window) usually is of very short length, this convolution is reduced to the weighted addition of a few samples. Since this convolution is linear, we speak of *linear smoothing*. If the smoothing function is a three-point Hann window, for example, this reads as

$$w(q) = 0.25\ x(q-1) + 0.5\ x(q) + 0.25\ x(q+1)\ , \qquad (6.46)$$

where y(q) represents the smoothed and x(q) the unsmoothed contour. Linear smoothing is nothing but the application of a nonrecursive digital filter which in this case always has low-pass characteristics.

From the discussion in the previous section, however, we know that the application of low-pass filters to unsmoothed pitch contours is questionable. The low-pass filter removes much of the local jitter and noise, but it does not remove local gross measurement errors, and, in addition, it smears the intended discontinuities at the voiced-unvoiced transitions. Hence, some kind of nonlinear smoothing might be more appropriate.

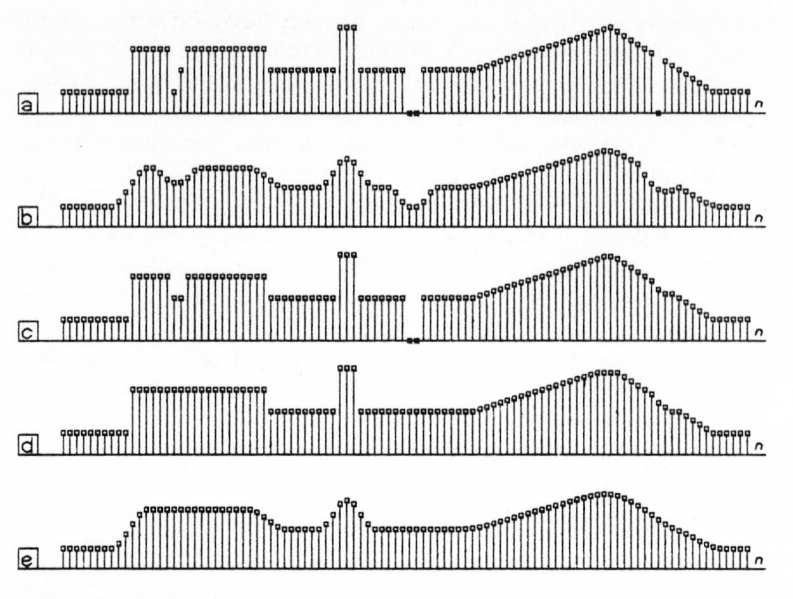

Fig.6.69a–e. Test example for the performance of median and linear smoothing. (a) input contour with isolated errors; (b) after linear smoothing with a 5-point Hann window; (c) after nonlinear smoothing with a 3-point median smoother; (d) after nonlinear smoothing with a 5-point median smoother; (e) after combined smoothing with a 5-point median smoother and 5-point Hann window (in that sequence). After Rabiner et al., (1975)

In a paper by Rabiner, Sambur, and Schmidt (1975), based on an earlier paper by Tukey (1974), *median smoothing* is proposed as an adequate nonlinear method. The K-point median of K consecutive samples x(0) to x(K-1), where K is always an odd number, is defined to be the middle sample if the K inputs are ordered in value,

$$\text{med}\,(x, K) = x\,[(K-1)/2]\ \text{if}\ x(i) \leqslant x(i+1)\,;\ i = 0\,(1)\,K-2\ . \tag{6.47}$$

Figure 6.69 shows a test example of the performance of the different smoothers; Fig.6.70 displays a disturbed sample T_0 contour with one case of higher harmonic tracking and several errors in voiced-unvoiced discrimination.

K/2 is computed from K by integer division. With an appropriate choice of K, this kind of smoothing removes short errors, i.e., such errors whose length is less than K/2; longer discontinuities are preserved. Such a smoother thus performs well when single gross errors are to be removed. Jitter and noise are barely affected. For this reason, Rabiner et al. recommended a combination of median and linear smoothing in that order.

Although median smoothing preserves sharp discontinuities in the data, it fails to provide sufficient smoothing of the undesirable noise-like components for which the smoothing was originally designed. A fairly good solution is a smoothing algorithm based on a combination of running medi-

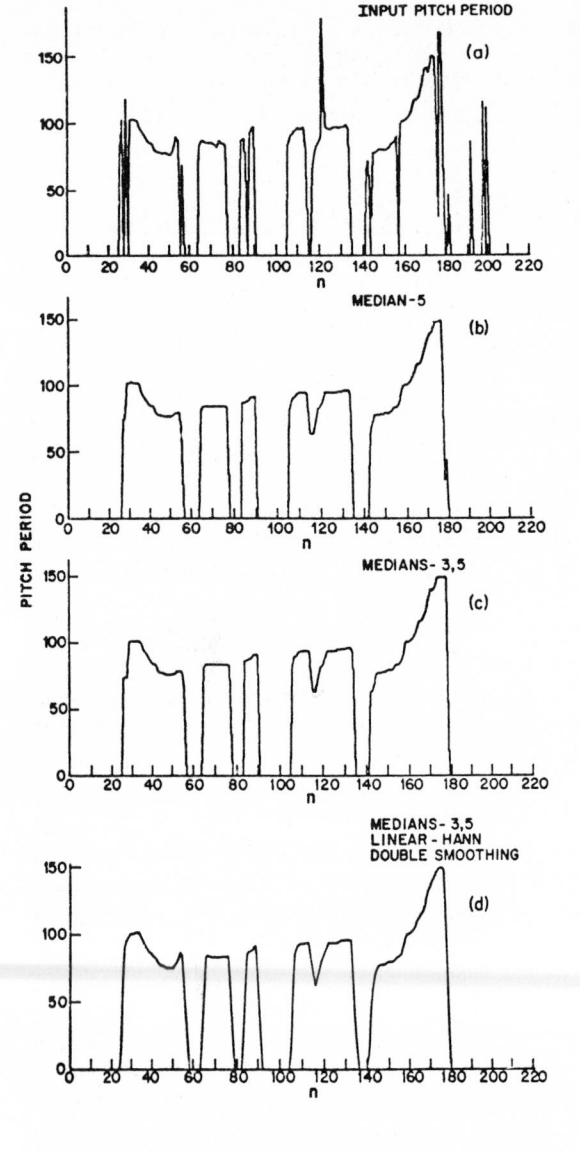

Fig.6.70a–d. Comparison of two median smoothers and the combination smoother for a pitch contour with gross measurement errors. (**a**) Input contour as is; (**b**) after median smoothing (five-point median); (**c**) after double median smoothing (three- and five-point median, respectively); (**d**) after double median smoothing as in (c) and linear smoothing by a three-point Hann window, in that sequence. From (Rabiner et al., 1975, p.555)

ans and linear smoothing. Since the running medians provide a fair amount of smoothing already, the linear smoothing can consist of a fairly low-order system and still give adequate results. Tukey (1974) proposed the use of a three-point Hann window [see (6.46)] as one candidate for the linear smoother. (Rabiner et al., 1975, p.553)

Besides the gross pitch determination errors, the discontinuities at the beginning and the end of voicing are the crucial points when a pitch contour is smoothed. It is convenient to assign a value of zero to pitch outside the voiced intervals. If the pitch contour, however, is stored and processed as a

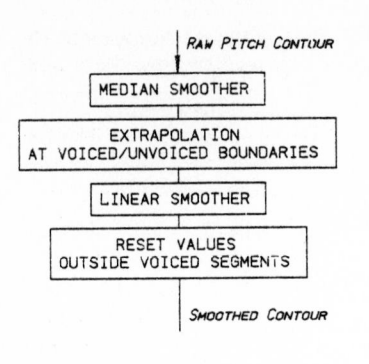

Fig.6.71. Sample of a smoother (after Rabiner et al., 1975) which preserves the values of the pitch contour at the edges of voicing

signal, this signal has gross discontinuities due to these zero samples, and the contours are smeared in an undesirable way at the edges of voicing when linearly smoothed. Yet, do the beginning and the end of a voiced segment necessarily have to be characterized by zeros in the pitch contour when this is not implied by the technological realization of the system (in a PDA or PDD where the voicing function is determined separately this is not the case!)? Instead, any arbitrary value can be assigned to the pitch outside the voiced segments, and voicing can be controlled by the separately determined voicing function. Thus nonzero dummy values can be assigned to the pitch outside the voiced segments which do not affect the contour at the edge of voicing while passing the smoother (Fig.6.71). The median smoother preserves the discontinuities. The nonzero dummy values should be assigned before application of the linear smoother, and they can be reset thereafter.

A smoother can be applied in addition to list correction routines; the smoother may even completely replace the error correction. With respect to isolated outliers, median smoother and list correction routine act similarly although the smoother functions in a much less "intelligent" way. Compared to sophisticated list correction routines the median smoother is thus less complicated but also less powerful. Linear smoothing, on the other hand, reduces the influence of fine measurement inaccuracies which are not at all touched by list correction and hardly influenced by the median smoother. Since its influence on gross errors is unfavorable, linear smoothing is usually applied as the last step, i.e., when the gross errors have already been removed as far as possible.

6.7 Final Comments

In this chapter a survey of time-domain PDAs and PDDs has been given. As defined in Chap.2, a time-domain PDA is a device which processes a signal in its basic extractor whose time base is identical to that of the original signal. This signal ranges from the more or less pure fundamental harmonic to the speech signal itself with its many temporal structure patterns. In summary, we arrive at the various categories of time-domain PDAs as listed in Table 6.4.

The preprocessor in a time-domain PDA is a filter which may or may not contain nonlinear distortion of the signal. The basic extractor usually is built up according to one of the following principles: zero crossings analysis, non-zero threshold analysis (nonzero triggering), or (algorithmic) determination of all significant points whatsoever.

Periodicity presents itself in more than one way: in the presence of a harmonic structure, in the presence of a lowest frequency which is expected to equal the fundamental harmonic, and in the presence of a more or less regular pattern of certain significant samples in the signal, such as peaks or zero crossings. Accordingly, the methods for pitch period extraction are manifold. A common feature of all time-domain PDAs is their ability to determine the boundaries of any individual period (although generally not in correct phase). Not all of the PDAs, however, really use this feature. Those which do can be applied to tasks which require pitch-synchronous speech analysis.

Most time-domain PDAs take on one of two almost contradictory positions. The first group applies a simple basic extractor and an elaborate preprocessor. This basic extractor necessarily tends towards detection of the fundamental harmonic and not of a structural pattern. If the preprocessor filter is purely linear (type PiTL), such a PDA will necessary fail when the first harmonic is weak or totally absent in the signal. If those detectors do work for a given application, they are attractive because of their simplicity. As one main field for PDAs of this type the present author recommends investigating female

Table 6.4. Principles and labels of time-domain PDAs

Label	Description
PiTL	Linear device: detection of the fundamental frequency or waveform in the signal. The fundamental is necessarily enhanced by a linear filter, such as a low-pass filter or a chain of integrators
PiTN	Nonlinear device: detection of the fundamental frequency or waveform in the signal after its enhancement by a nonlinear procedure or function
PiTE	Modeling of signal envelope and search for discontinuities which mark the beginning of individual periods
PiTS	Direct investigation of the temporal structure by algorithm; search and extraction of anchor points from which periodicity is derived
PiTA	Investigation of the temporal structure in a simplified way after adaptive filtering
---M	Multichannel PDA
---O	... with basic extractor omitted

voices (see, for example, Simon, 1977) which have a relatively prominent first harmonic due to the closer vicinity of the fundamental harmonic and the first formant.

Nonlinear distortion (as in the group PiTN) is one means to enhance the first harmonic when it is weak or absent. Best results for band-limited signals are achieved when the applied nonlinear function is even, such as full-wave rectification. The same NLFs which do so well with band-limited signals affect signals with a predominant first harmonic in a most undesirable way. They generate a strong second harmonic and attenuate the first harmonic in favor of the second one. Apparently no single NLF is really optimal. More details about this problem will follow in the next chapter, and a general discussion on preprocessing methods is to be read in Section 9.1.

The PDAs belonging to the other extreme group perform structural analysis (group PiTS). These algorithms, if implemented digitally, extract the principal peaks (or other significant points) of the individual periods step by step from the local peaks. If implemented in an analog way, they perform envelope periodicity detection (group PiTE). These PDDs take account of the fact that the individual period is a truncated version of the impulse response of the vocal tract, i.e., the sum of a finite number of simultaneously excited exponentially damped sinusoids. All these algorithms have the fact in common that equal or at least similar steps are carried out several times (parallel or in series). With few exceptions (Gold and Rabiner, 1969; Tucker and Bates, 1978) the digital PDAs of the type PiTS need a global correction routine which scans at least a total voiced segment. The analog PDDs of type PiTE cannot apply such a step since they do not have a memory available. Thus the frequency range of these PDDs is restricted or must be manually preset. Accordingly, we find these PDDs used preferably for educational applications such as training for the deaf, where they may perform an excellent job.

The PDAs which take their place between the two extremes perform some kind of adaptive temporal structure simplification (group PiTA). Usually this principle leads to some crude approximation of the inverse filter or to an enhancement of the discontinuity in the laryngeal excitation function at the point of glottal closure. This principle is attractive especially if such an analysis, e.g., linear prediction analysis, has to be carried out in the same device for other purposes as well. The dangerous point in the group PiTA is more related to the signal itself than to the method: 1) coincidence between F_0 and the formant F1 - no problem to the group PiTL/N nor to PiTE/S - arises problems here, and 2) the absence of a distinct point of excitation or, on the other hand, the presence of secondary points of excitation in the glottal signal (without the requirement that the excitation signal itself becomes irregular) may cause trouble.

Parallel and multichannel processing appear likely to avoid the shortcomings associated with any single-channel PDA. Parallel channels are used in three ways.

1) A single-channel PDA uses a second auxiliary channel as a vernier to increase the overall accuracy.

2) An auxiliary channel yields a crude estimate of F_0 which is used as a preset for the main channel. Thus a manual preset for the PDA can be avoided.

3) Different channels realize different principles, apply preprocessors with different input conditions, or subdivide the range of F_0. The channels are combined in the postprocessor (or maybe even in the basic extractor) to select the most likely estimate.

The last case appears to be the most important one. A frequently used selection principle in this case is the minimum-frequency selector. This principle is applicable when the basic extractor tends towards higher-harmonic tracking, as is the case for the categories PiTL and PiTN. The principle is usually not applicable to PDAs of the category PiTS (digital as well as analog) since these PDAs produce both chirps and holes at a comparable rate. For the category PiTS, the selection is achieved by choosing the most likely candidate (this principle is most clearly realized in the matrix selection of the PDA by Gold and Rabiner). The channel selection is carried out according to a majority vote (Gold and Rabiner, 1969), a regularity criterion (see Chap.7), or by syntactic plausibility rules (De Mori et al., 1977). The incremental relation between T_0 and the location of individual period markers generates new problems when a multichannel PDA is used to determine individual periods.

As already stated, time-domain PDAs have the advantage that they can locate any pitch period individually. On the other hand, if they fail to find a periodicity, they are generally not yet able to decide whether the marker(s) detected to be irregular result from a measurement error or from a truly irregular signal. This weakness is a feature which is more or less common to all time-domain PDAs and represents one important question still open in this field.

7. Design and Implementation of a Time-Domain PDA for Undistorted and Band-Limited Signals

This chapter describes the design and implementation of a time-domain PDA for undistorted and band-limited signals. The PDA applies a simplified inverse filter in the preprocessor and a threshold analyzer with hysteresis in the basic extractor. This configuration (Hess, 1972b, 1976a,b), which is discussed in Sect.7.1, corresponds to principle PiTL (see Sects.6.1,4). The performance evaluation, however, has shown that this PDA fails systematically with band-limited signals and, in addition, that the PDA tends to fail occasionally when the formant F1 coincides with F_0. Reconstruction of the first partial by nonlinear distortion is one possible way out of this problem. Since the opinion in the literature is not uniform as to what kind of nonlinear distortion is suitable to optimally enhance the first partial, this question has been experimentally investigated (Sects.7.2,3). Experiments show that no single nonlinear function is optimal for any signal. A combination of three functions (odd, even, and rectified single-sideband modulation), however, gives good results in most of the cases (Sect.7.4). This result leads to the implementation of the three-channel PDA discussed in Section 7.5. The three channels result from the application of the three nonlinear functions mentioned above. The nonlinear distortion represents the first step of the PDA, and the subsequent steps of the preprocessor and the basic extractor, which are performed for each channel separately, correspond to the linear implementation. The last step performs the selection of the channel which is most likely to contain the correct information on pitch. The selection routine applies a short-term regularity check and makes its decision according to the minimum frequency selection principle. In the three-channel version the PDA can process both undistorted and band-limited speech signals uttered by male and female speakers.

7.1 The Linear Algorithm

One method for simplifying the signal structure is inverse filtering. PDAs of this group have been discussed in Sect. 6.3.1. The problem of this approach is the great effort of realization if done exactly. For pitch determination, however, a very crude approximation of this filter, whose implementation effort is much lower, is sufficient. The linear algorithm under consideration here applies such an approximation of the inverse filter. Implementation and evaluation of this PDA are discussed in detail in this section.

The inverse filter is implemented in two steps. A fixed low-pass filter suppresses the higher formants. The first formant is then adaptively cancelled by a time variant band-stop filter. From the filtered signal pitch period markers are derived by a TABE with hysteresis. This linear implementation tolerates phase distortion and attenuation of the first harmonic, but it fails when the first harmonic is completely missing.

7.1.1 Prefiltering

First the higher formants are removed by a low-pass filter (Fig.7.1). For reasons of computing speed, a double gap-function low-pass filter is used which does not need multiplications and operates with linear phase. The linear phase response is important when the capability of time-domain algorithms to determine each pitch period individually is taken advantage of. The low-pass filter does not need a sharp cutoff frequency; the filter shown in Fig.7.1 damps the higher formants sufficiently. In the "prefiltered" signal (PFS) at its output only the formant F1 is going to play an important role, except for those vowels where F2 also falls into the passband of the filter. This special case, however, cannot be easily solved and will be discussed later.

7.1.2 Measurement and Suppression of F1

The frequency range of the first formant is too great for a fixed filter to cope with any situation; hence a time variant adaptive filter is needed to cancel this formant. To implement such a filter, it is necessary to perform

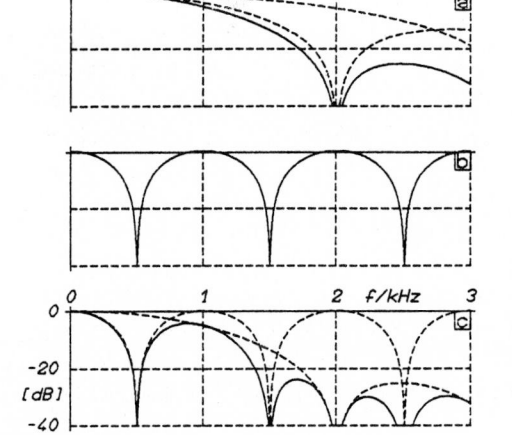

Fig.7.1a–c. Frequency response of the realized approximation of the inverse filter. (a) Low-pass filter (F2, F3, F4); (b) adjustable comb filter (F1, plotted: 500 Hz); (c) total frequency response. (-----) Components of two-step filter

at least a crude measurement of F1. The waveform of F1 dominates the temporal structure of the PFS, and the formant frequency measurement can thus be carried out in time domain using the short time distribution of distances between adjacent zero crossings (Fig.7.2).

For this special application (cancelling F1) a comb filter proved more suitable than a band-stop filter. With respect to the sampling frequency, the stopband is situated fairly low, and the filter is hard to realize, at least with linear phase. If realized crudely (e.g., as a second-order filter with a single zero at the formant frequency F1), such a filter tends to attenuate the lower passband so that it behaves like a high-pass filter after all. Of course this is undesirable. The comb filter, on the other hand, has a neutral frequency response, and it is easier to implement. One possible realization of the comb filter is the two-pulse filter (TPF) whose impulse response consists of two impulses of equal size and equal or opposite sign with a pulse distance of d samples. According to (2.24,25), the zeros of the transfer function of this filter are uniformly spaced on the unit circle of the z plane:

$$z_{0m} = \pm \exp (j\pi m/d) \qquad m = 1 \; (2) \; d\text{-}1 \; \text{(if d greater 1)}$$
$$z_{0d} = -1 \qquad\qquad\qquad d \; \text{odd} \qquad\qquad\qquad\qquad\qquad (7.1)$$

if the two pulses have equal signs, and

$$z_{00} = 1$$
$$z_{0m} = \pm \exp (2\pi jm/d) \qquad m = 1 \; (1) \; \frac{d\text{-}1}{2} \; \text{(if d greater 2)} \qquad (7.2)$$
$$z_{0d} = -1 \qquad\qquad\qquad k \; \text{even}$$

if they do not.

7.1.3 The Basic Extractor

If the first zero of the TPF according to (7.1) is set to the formant frequency F1, the higher zeros will be situated at 3F1, 5F1 etc., i.e., mostly outside the interesting frequency range. At the output of the comb filter all formants are expected to be attenuated; the filter has performed a spectrum flattening, at least in the interesting region below the cutoff frequency of the low-pass filter. The corresponding "filtered" signal FS, if correctly processed, shows one significant maximum per pitch period. The basic extractor determines the pitch markers by threshold analysis. Every peak exceeding a certain threshold is a candidate for a forthcoming pitch marker. This raw marker, however, is not positioned to the peak where it is detected but rather to the subsequent zero crossing since the latter point gives the most accurate phase relation between the FS and the original signal (Fig.7.3). The threshold function necessary for this analysis shall be discussed in more detail in Section 7.4.

To give an idea how accurately F1 must be determined, Fig.7.4 shows some speech signals processed by two-pulse filters with different pulse distan-

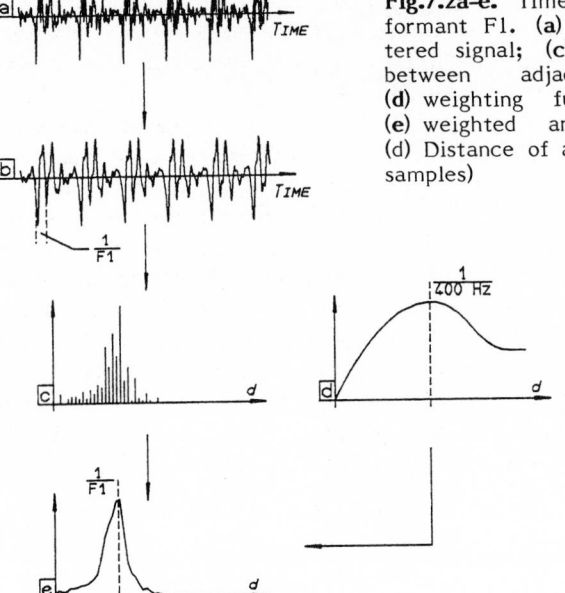

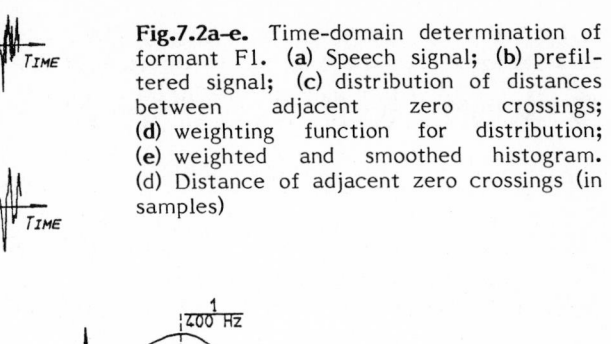

Fig.7.2a–e. Time-domain determination of formant F1. (a) Speech signal; (b) prefiltered signal; (c) distribution of distances between adjacent zero crossings; (d) weighting function for distribution; (e) weighted and smoothed histogram. (d) Distance of adjacent zero crossings (in samples)

ces k. If k is too low, that means, if F1 is determined too high, then the attenuation of F1 will be insufficient; this results in a greater tendency towards detecting chirps. If k is too high, on the other hand, then the phase response is reversed before F1 is reached. Again F1 is insufficiently attenuated; in addition it is phase shifted by 180°. Thus, the phase relation between original signal and FS is temporarily disturbed; in consequence, individual markers will be misplaced (this event cannot be observed in Fig.7.4 since the signals there are taken from the experimental study of Sect.7.3 and thus are strictly periodic). A catastrophic error finally occurs when the zero of the

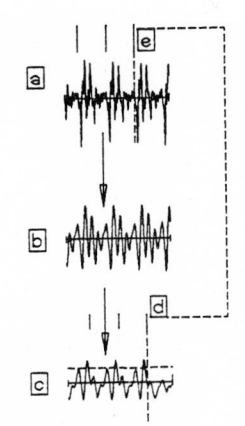

Fig.7.3a–e. Determination of pitch markers. (a) Speech signal; (b) prefiltered signal (PFS); (c) filtered signal (FS) and threshold function; (d) markers in preliminary position; (e) markers in final position

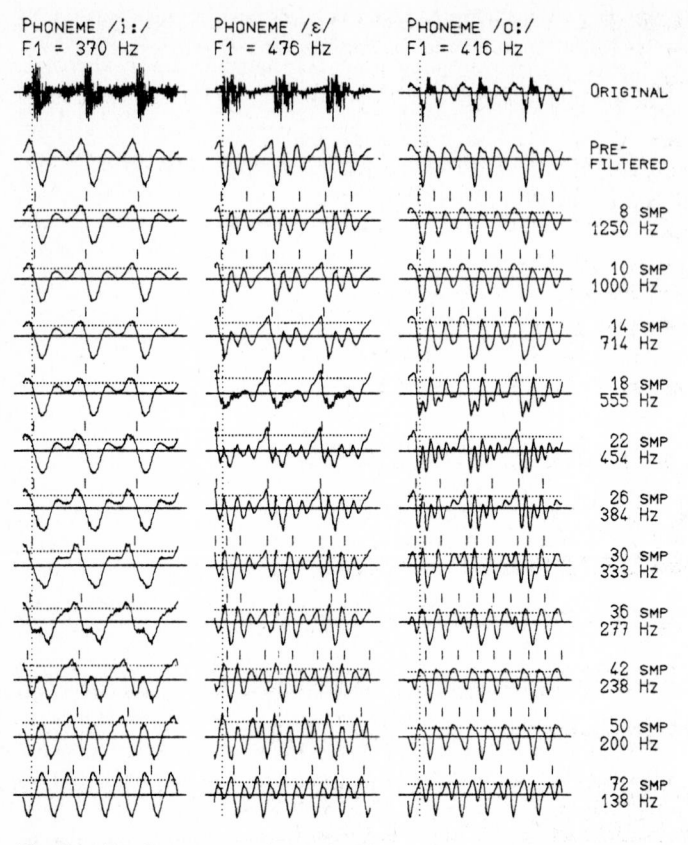

Fig.7.4. Performance of comb filters for the attenuation of F1. A single two-pulse filter with equal polarity of the pulses was used. The rest of this figure corresponds to Fig.7.5

filter coincides with the fundamental itself. In this case all the odd harmonics are removed; F_0 cannot be determined any more. This problem (which is common to all inverse filter approaches) will play an important role in the further discussion. In a preliminary statement it can be concluded that an error of 25% in the determination of F1 can easily be tolerated if F1 is estimated too high, whereas the algorithm is more sensitive to errors in the other direction. For this the weighting function included in the F1 measurement emphasizes the frequencies around 400 Hz. Higher frequencies are favored anyhow by the greater number of zero crossings per time interval. As to the temporal behavior of the algorithm, it is laid out to yield a new value of F1 almost immediately, that is, within one or two pitch periods so that the filter adjustment is able to follow rapid formant changes.

7.1.4 Problems with the Formant F2. Implementation of a Multiple TPF

There remains one problem: the second formant. This formant is either situated well above 1500 Hz or quite close to F1. In the first case it can be forgotten since the low-pass filter removes it. The latter case applies to all back vowels and some consonants, like /r/. The comb filter attenuates F1 by a single zero and has therefore a very narrow stopband at this point; it will not attenuate F2 properly even when this formant is situated closely above F1. As long as F2 is relatively weak, such as in a vowel like /o:/ (Fig.7.4, right part), its influence can be disregarded. For such vowels as /o/, however, the amplitude of F1 is comparable to that of F1, and a single TPF will not give reasonable results any more (Fig.7.5). In this case, F2 must be removed separately. As

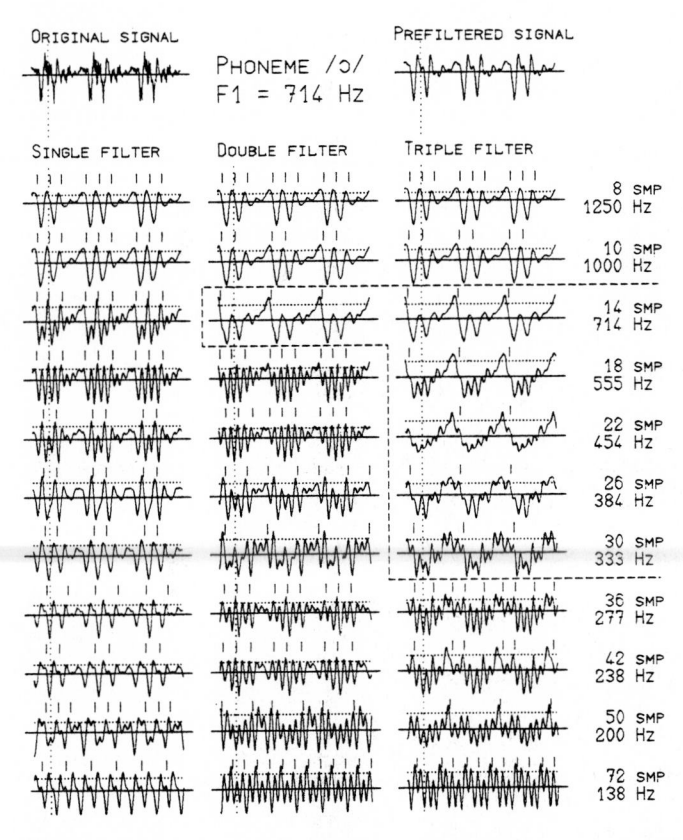

Fig.7.5. Performance of different comb filters for the attenuation of F1. Numbers at right-hand side: pulse width (samples) and frequency of first zero (Hz). Sampling frequency: 10 kHz. Lowest figure: zero of filter coincides with the fundamental frequency. The signal was generated by repetition of one (manually extracted) pitch period, taken from the experiment to be reported in Section 7.3. The dotted vertical line indicates the beginning of the period. The threshold for marker detection was set at 65% of the signal maximum

the experiment shows, however, a separate measurement of F2 is not necessary. An additional zero of the filter at the frequency where F2 is expected
when it is in the low position solves the problem. The frequency of this zero
(depending on the estimate of F1) was experimentally determined as

$$f_{z2} = \max (1.3 \; f_{z2}, \; 500 \; \text{Hz}) \; . \tag{7.3}$$

The lower limit was fixed to 500 Hz since F2 never takes on values below
that limit (Fant, 1968; Hess, 1976b). The zero is realized in form of a second
TPF in series to the first one; it is applied to all signals. In general this
second zero enlarges the work area of the formant determination; it permits
a cruder measurement and decreases the sensitivity towards too low estimates
of F1 at the same time. A third zero slightly above the second one, as indicated in the right part of the figure, offers an additional degree of safety.
At least for rapid speech, this advantage cannot be neglected so that finally
a triple time variant TPF is applied. The filter can be realized in one step;
in this case it needs 8 additions per sample. No multiplications are required.
Modification of the pulse width k requires 8 address computations whenever
this parameter is changed.

   All these TPFs operate with linear phase. Hence the phase relation between the original and the filtered signal is maintained also for the double
and the triple TPF. In the computer simulation the noncausal version can be
implemented which has zero phase response. In a hardware device one has to
prevent the linear phase response from becoming time variant. This can be
done by adding a constant delay d which exceeds the maximum pulse width
used in the TPF.

7.1.5 Phase Relations and Starting Point of the Period

The zero crossing where the raw pitch marker is positioned is the safest point
to determine a pitch period limitation but not the laryngeal pulse time where
the period actually starts. From the speech signal itself the latter point is
hard to extract (cf. the discussion on epoch extraction in Sect.6.3.2). It is
normally not accompanied by a zero crossing which can only be expected
there when F1 coincides with a harmonic of the signal. As can be seen from
the examples in Appendix A (Figs. A.1,3,5), the formant frequency F1 is rather
frequently situated between two harmonics of the signal. In the present algorithm the laryngeal pulse time can thus only be extrapolated from the preliminary marker position. This is done in the following way: the glottal waveform
starts the new pitch period with a positive signal since the glottal pressure
is increased. After glottal closure the new period of the speech signal is
started in the same direction (Pinson, 1963). Since F1 forms the dominant
waveform, one can suppose that half a period of F1 has passed when the preliminary marker position is reached. Hence, if the marker is shifted back
(against time) by half a period of F1, it will arrive in its proper position.
The results of this procedure, however, will obviously be directly influenced
by the accuracy of the determination of the first formant.

7.1.6 Performance of the Algorithm with Respect to Linear Distortions, Especially to Band Limitations

The algorithm was originally conceived to investigate undistorted signals in a speech recognition system (Hess, 1972b, 1976b). Phase-distorted signals can be processed if the marker detection threshold is able to change its sign. In this case the algorithm will put the preliminary marker into the nearest zero crossing from positive to negative values of the FS so that the actual sign of the threshold does not influence the final marker position. Obviously a proper starting point of the period cannot be defined any more when the signal is phase distorted (Holmes, 1975). Nevertheless, the shiftback of the markers by half a period of F1 is maintained since it optimizes the marker position at least with respect to this formant.

With band-limited signals, however, the algorithm gets into serious trouble. From the signal processing point of view the only difference between this PDA and a usual algorithm of type PiTL is that the preprocessor filter is time variant. Like in any PDA of this group, the structure of the FS is thus so strongly influenced, if not dominated, by the first harmonic that the algorithm can hardly work without its presence. This drawback represents the first of two significant disadvantages of this algorithm.

The second disadvantage has already been discussed briefly in the previous section. If the zero of the comb filter coincides with F_0, then the fundamental harmonic is canceled and cannot be determined any more. If this were only an error of measurement, it could be easily avoided. But the problem is inherent in the signal. In the course of female utterances it occurs quite often that F1 equals F_0 so that the algorithm would then cancel F_0 although F1 had been correctly determined. Since this cannot be tolerated, the algorithm must set a lower bound to the zero of the comb filter. This bound obviously cannot be a fixed frequency. A suitable value would be the second harmonic of the signal. To perform this correctly, however, a valid estimate of F_0 must be available to the algorithm. Once a correct estimate has been determined, this is no problem. At the beginning of a new utterance, however, when the result of the measurement is not yet available, the algorithm needs a crude preset of pitch (within the range of an octave). The application of the triple TPF for filtering helps considerably to increase the maximum difference between this arbitrary preset and the actual pitch which the algorithm is still able to process. Nevertheless, it must be admitted that the necessity of such a preset severely restricts the possible applications of this algorithm.

In summary, the linear algorithm has two disadvantages which prevent it from being used as a universal PDA: 1) it cannot process band-limited signals, and 2) it needs a preset for the expected value of pitch. In order to solve this problem we will have to take a closer look at the problem of nonlinear distortion in time-domain PDAs. This will be done in the following three sections.

7.2 Band-Limited Signals in Time-Domain PDAs

As shown in the previous chapter, the majority of the digital time-domain PDAs can be split up into the main categories as summarized in Table 6.4. In the following considerations, however, we will only distinguish between the principles PiTL, PiTN, and PiTS; all PDAs that do not pertain to principle PiTL or PiTN will be regarded like "structural-analysis PDAs."

The algorithms of category PiTL are purely linear. In the preprocessing step there is no nonlinearity, and the fundamental frequency F_0 is measured directly from the input signal of the basic extractor. This requires that the fundamental be either the most prominent or the lowest frequency inherent in the signal. Algorithms of this type will necessarily fail when the fundamental is absent (for instance, because of band limitation in a telephone channel).

In the algorithms of category PiTN a step of the form

$$y(n) := f[x(n)] \tag{7.4}$$

is included. In this equation $y(n)$ represents the output signal, $x(n)$ the input signal, and $f(x)$ a nonlinear correspondence between the momentary amplitude of the input and output signals. A great variety of such nonlinear functions $f(x)$ are implemented in existing PDAs. The positive effect of the nonlinear step with respect to the processing of band-limited signals is clearly demonstrated. On the other hand, there have been relatively few attempts to investigate the influence of $f(x)$ upon undistorted signals of high pitch, especially upon those of female voices.

The algorithms of category PiTS are nonlinear in a different way. Investigation of the signal structure in the time domain requires statements of the form

if C then A else B;

in this case C stands for a condition which may or may not be fulfilled, i.e., *true* or *false*. A simple example for C would be to determine whether a certain sample $x(n)$ of the input signal x is a maximum or not. Depending on C the algorithm will either process A or B, which stand for certain statements, procedures, or subroutines. Since C forms a (binary) decision element, it performs a threshold discrimination between A and B. Such a system cannot be described as piecewise linear since C may change from any sample $x(n)$ to the next one. Hence any algorithm of this group must be regarded as essentially nonlinear, and these algorithms are expected to be able to cope with band-limited signals. Since there exists an even greater degree of freedom in the design of algorithms for this group than for the group PiTN, a general a priori statement as to the influence of band limitation upon a particular algorithm of this type cannot be expected.[1]

---

[1] For the more elaborate PDAs of this category (i.e., the ones which are actually applied in speech communication systems), the applicability to band-

Table 7.1. Signal categories with respect to cutoff frequency f_c of band limitation and fundamental frequency F_0

Number	Label	Description
1	SiLU	F_0 is situated below f_c, and there is no band limitation. The first harmonic is thus included in the signal.
2	SiLL	F_0 is situated below f_c, and the signal is band limited. Hence the first harmonic is not included in the signal.
3	SiH	F_0 is situated above f_c. This means that the first harmonic is included in the signal regardless whether there is a band limitation or not.

7.2.1 Concept of the Universal PDA

Let x(n) be a speech signal which is possibly phase distorted and band limited, but undisturbed otherwise. If the cutoff frequency f_c of the band limitation falls into the measuring range of the PDA, as is the case for the cutoff frequency of 300 Hz in telephone channels, then the signal x(n) is subdivided into three categories with respect to F_0 and f_c; these categories are listed in Table 7.1.

With respect to the signal, there is no difference between the categories SiLU and SiH, whereas the category SiLL represents the truly band-limited signal. However, there are considerable differences between the categories SiLU and SiH with respect to the implementation of a PDA.

If a PDA is able to process signals belonging to any of these three categories over the whole range of possible pitch and without the necessity of a manual preset of range or category, it is labeled a *universal* PDA (UPDA). This UPDA need not be ideal; it may commit errors, but it must not fail systematically for certain signal categories, phoneme classes, or ranges of pitch. From this it follows, for instance, that a PDA of group PiTL is never "universal" since it cannot process the band-limited signal of category SiLL. The decision whether a PDA of group PiTS is a UPDA must be made individually. For the PDAs of group PiTN, however, some general statements seem possible which will be further discussed in the following sections.

---

limited signals has been experimentally verified. This holds, e.g., for the PDA by Gold and Rabiner (1969) as well as for the one by N.Miller (1974, 1975) where it has been proved by Rabiner et al. (1976). It also holds for the PDA by Tucker and Bates (1978) and - in the opinion of the present author - to that by Reddy (1966, 1967). Simpler PDAs, such as the one by Baudry and Dupeyrat (1976), may get into trouble, and the PDA by Tillmann (1978), the simplest one, definitely needs the presence of the first partial.

7.2.2 Once More: Use of Nonlinear Distortion in Time-Domain PDAs

Most of the nonlinear functions proposed in the literature are of the kind given in (7.4). These functions can be purely odd or purely even; if neither, they will be called *mixed*. According to McKinney (1965), a number of researchers have given preference to even nonlinear functions since these functions perform a better amplification of the first harmonic than the odd or mixed ones which were generally judged unsatisfactory for low relative amplitudes of the first harmonic. On the other hand, it is a well-known fact that the application of an even nonlinear function to a sinusoid of the frequency f_S will completely cancel this frequency; the resulting spectrum will only consist of even harmonics of f_S including zero frequency. In female voices, however, the first formant often coincides with the fundamental frequency. In this case up to 95 percent of the total signal energy is concentrated in the first harmonic. Hence, application of an even nonlinear function to a female voice is expected to lead to catastrophic errors in pitch detection. As can be seen from Table 7.2, neither an odd nor an even nonlinear function will provide a universal PDA. Applying a mixed nonlinear function, such as one-way rectification, might be thought of. According to McKinney, however, a majority of researchers regard the mixed function only as a compromise which may cause the PDA to fail as soon as the signal gets difficult to process.

The proposal by Risberg, Möller, and Fujisaki (1960) to apply a SSB modulation, full-wave rectification, and low-pass filter to the signal can also be put into group PiTN. According to the authors this function avoids the boosting of the second harmonic which is often associated with even functions, but yields a good amplification of the first harmonic which makes it applicable to female voices. Just as for even functions, however, the application to a pure sinusoid with frequency F_0 will make this frequency completely disappear in the resulting spectrum, so that a PDA applying rectified SSB modulation may also fail systematically with female voices. A simple example will underline this effect. (Ideal) single-sideband modulation (SSBM) of a signal x(n) results in a spectral shift of all components X(m) with m greater than zero by a constant C which equals the carrier frequency,

Table 7.2. Expected performance of PDAs applying a nonlinear function with respect to the signal category

Signal Category	Nonlinear Function		
	Odd	Even	Mixed
SiLU	as linear	good	possible but
SiLL	poor	good	usually not
SiH	good	poor	satisfactory

$$Y(m+C) = X(m) \qquad m \text{ greater } 0 \qquad\qquad (7.5a)$$
$$Y(m-C) = X(m) \qquad m \text{ less } 0 \qquad\qquad (7.5b)$$
$$Y(k) = 0 \qquad\qquad k = -C \ (1) \ C. \qquad\qquad (7.5c)$$

Equation (7.5b) is required since both the input signal $x(n)$ and the modulated signal are real (and not complex); to fulfil (7.5c) it is assumed that the carrier C is completely suppressed, and that the spectrum $X(m)$ disappears for $m = 0$. If an ideal SSBM is assumed for a waveform $x(n)$ which is purely sinusoidal with the frequency f_S, then a spectral line will appear at (f_S+C) and $-(f_S+C)$, respectively. Hence, the resulting signal is again a pure sinusoid, this time with the frequency (f_S+C). If this waveform is full-wave rectified, the resulting signal will consist of the frequencies zero, $2(f_S+C)$, $4(f_S+C)$, etc., leading again to complete failure of the PDA.[2] Applying a nonideal SSBM, for instance, with only partial carrier suppression will avoid this problem; in this case, however, the performance would correspond only to that of a mixed nonlinear function and might represent another unsatisfactory compromise.

7.3 An Experimental Study Towards a Universal Time-Domain PDA Applying a Nonlinear Function and a Threshold Analysis Basic Extractor

This section describes an experiment in order to verify the hypothesis that a single NLF is not sufficient for the implementation of a universal PDA which uses principle PiTN. From three speakers (two male, one female) 48 pitch periods were selected in such a way that they could be regarded as representative for German spoken phonemes. These periods were nonlinearly distorted by 16 nonlinear functions, and the relative amplitude of the first harmonic with respect to the strongest higher harmonic was measured. The results show that at least two different NLFs are necessary when a universal PDA is to be realized with principle PiTN.

7.3.1 Setup of the Experiment

According to the considerations of the previous section, the following hypothesis is set up:

A universal time-domain pitch detection algorithm applying a nonlinear function and threshold analysis basic extraction is not possible when only one single nonlinear function is used. (7.6)

---

[2] Other than for an ordinary even NLF in this case, there will be no signal within the measuring range of the PDA when full-wave rectified SSB modulation is applied. The algorithm could thus be able to automatically detect its own failure - a significant advantage!

To verify this hypothesis an experimental study was carried out which will be discussed in the following. The greater the predominance of the first harmonic in the filtered signal, the better will be the performance of a PDA pertaining to group PiTL or PiTN. This "predominance" may be defined, for instance, as the relative amplitude of the first harmonic with respect to the largest higher harmonic below a certain cutoff frequency well above the measuring range of the PDA.[3]

$$RA1 := 20 \lg \frac{Y(1)}{\max [Y(i)]} ; \qquad \begin{array}{l} i = 2, 3 \ldots \\ (\text{below } 1500 \text{ Hz}) \end{array} \qquad (7.7)$$

Hence, if the values of RA1 before and after application of a nonlinear function are compared, we obtain a measure for the performance of this nonlinear function. The measurement is greatly facilitated if the speech signal at the input is reduced to single pitch periods. In this case the amplitude of the individual harmonics can be exactly measured by applying the (discrete) Fourier transform. This idea was realized in the experiment whose block diagram is shown in Figure 7.6. The results are representative for the performance of a certain nonlinear function in the case that the processed pitch periods are representative for the input speech signal of the corresponding PDA. To achieve this, 48 individual pitch periods were manually selected from prerecorded utterances by three speakers (2 male, 1 female). The material had been recorded in an anechoic chamber and had been directly digitized so that it could be regarded als free from distortions. The 48 periods (16 from every speaker) were selected in such a way that they represent the most frequent German vowels (including the schwa) and the consonants /n/, /m/, /l/, and /w/. The fundamental frequency of the male speakers centered around 150 Hz, whereas the pitch of the female speaker varied from 220 to 300 Hz. After selection the signals were systematically distorted in several ways. A list of the applied distortions is compiled in Table 7.3. Phase distortion was performed by a second-order digital all-pass filter with the pole-zero pair situated around 300 Hz. This kind of phase distortion roughly corresponds to that of an ordinary tape recorder (Holmes, 1975). The low-pass filter was the one applied by Hess (1972b) in the linear PDA (Sect.7.1); here it was realized in a noncausal way in order to operate with zero phase response. The cutoff frequency of the band limitation was set to 220 Hz in order to obtain signals of all three categories from the given samples. The band limitation itself was done by discrete Fourier transform (DFT) in order to avoid further phase distortions. After nonlinear processing all periods were reduced to an amplitude of 1000 (in digital representation). In order to investigate the influence of amplitude changes on the results of the subsequent nonlinear processing, the band-limited but otherwise undistorted signal was also processed with a higher

_____

[3] This definition takes into account that it is usually a single higher harmonic which dominates the temporal structure of the waveform. The upper cutoff frequency contained in the definition was arbitrarily set to 1500 Hz.

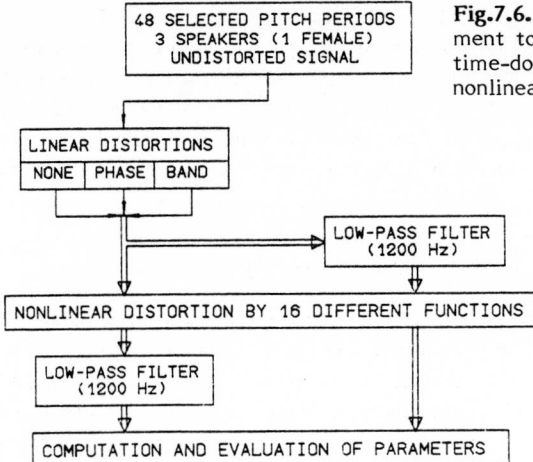

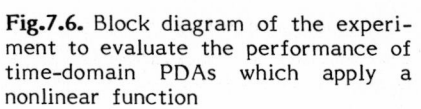

Fig.7.6. Block diagram of the experiment to evaluate the performance of time-domain PDAs which apply a nonlinear function

(+3 dB) and lower (-3 dB) amplitude. The same procedure was repeated for the 48 periods after a slight preemphasis (6 dB per octave starting at 1 kHz) so that finally each of the 48 periods was represented by 16 different versions. This material was then used as the data base for application of the nonlinear functions. The amplitude and phase characteristics of the linear distortions are depicted in Figure 7.7.

According to the proposals made in the literature, 16 functions were selected and applied. These functions are listed in Table 7.4. Thirteen of them, including linear processing, fulfill (7.4), the rest consist of SSB applic-

Table 7.3. Type of distortions applied to the sample pitch periods. The amplitudes (dimensionless integer values) are given in form of a 12 bit representation. The values can thus range from -2048 to +2047

Number	Label	Type of Distortion
1	SgU	Reduced to amplitude 1000 (quantization steps); undistorted otherwise.
2	SgP	Phase distorted by a second-order digital all-pass filter (pole frequency around 300 Hz, see Fig.7.7).
3	SgF	Low-pass filtered by linear-phase gap-function digital filter (see Fig.7.7).
4	SgL	Band limited by Fourier transform; cutoff frequency: 220 Hz (see Fig.7.7)
5	SgLP	Band limited and phase distorted
6	SgLF	Band limited and low-pass filtered
7	SgLH	as SgL; reduced to an amplitude of 1400
8	SgLL	as SgL; reduced to an amplitude of 700

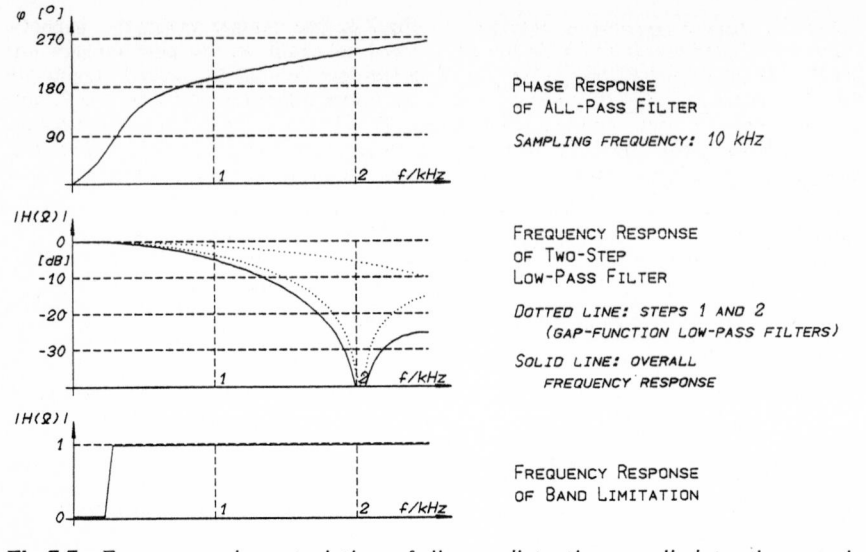

Fig.7.7. Frequency characteristics of linear distortions applied to the sample pitch periods

ations. As the good performance of even functions has been emphasized in the literature, four even functions of the form

$$y = |x|^a \tag{7.8}$$

with different exponents a were analyzed. According to the proposal by Rabiner (1977) for autocorrelation PDAs, center clipping was also examined in three different versions. Finally, in the three SSB functions investigated, ssb(x) is defined as SSBM with completely suppressed carrier, whereas in sst(x) a carrier with half the amplitude of the original signal was added. The SSB was simulated according to Figure 7.16. The characteristics of the nonlinear functions used for this investigation are shown in Fig.7.8 except for the SSB functions where an instantaneous relation between x(n) and y(n) cannot be given. In Appendix A the reader will find figures of all 48 sample periods before and after band limitation as well as their harmonic spectra and auto-correlation functions (Figs.A.1-5).

7.3.2 Relative Amplitude and Enhancement of First Harmonic

In the previous section it was found that the relative amplitude RA1 of the first harmonic in a given frequency range yields good evidence for the performance of an individual nonlinear function. Therefore RA1 was computed for all combinations of pitch periods, types of distortion, and nonlinear functions applied in the experiment. In order to evaluate the effect of a given nonlinear function Fu, the relative enhancement RE1 of the first harmonic was derived from the relative amplitude as follows,

Table 7.4. List of nonlinear functions applied to the sample periods. All functions are listed in FORTRAN notation (except that no capital letters are used). (x) Input signal. In correspondence to the labels introduced in Chap.6 for the individual PDAs, here we introduce labels for the individual methods of preprocessing using nonlinear functions. This helps to formalize the expressions and to make the text more easily readable. The labels for NLFs will not be used outside Chap.7 and Appendix A. See text for more details

Number	Label	Function (y=...)	Mode	Comments
1	FuL	x	odd	linear
2	FuR03	abs(x) ** 0.33	even	
3	FuR05	abs(x) ** 0.5	even	
4	FuR	abs(x)	even	full-wave rectification
5	FuR20	abs(x) ** 2	even	square
6	FuE20	1 - exp(-x*x)	even	
7	FuH	max (x, 0)	mixed	one-way rectification
8	FuL05	sgn(x) * abs(x)**0.5	odd	
9	FuL20	x * abs(x)	odd	squaring with sign of x maintained
10	FuM	abs(x) + x/2	mixed	nonideal rectification
11	FuCCR	abs(clc(x))	even	center clip, compress, and rectify
12	FuCPR	abs(clp(x))	even	center clip and rectify
13	FuCP	clp(x)	odd	center clipping
14	FuSH5	max(ssb(x),0)**0.5	SSB	SSB modulation
15	FuSR5	abs(ssb(x))**0.5	SSB	
16	FuST5	abs(sst(x))**0.5	SSB	SSB modulation with DC component added

$$RE1 \ (Fu, \ Sg) = RA1 \ (Fu, \ Sg) - RA1 \ (FuL, \ SgU) \ , \qquad (7.9)$$

where FuL represents linear processing as defined in Table 7.4, and SgU represents the undistorted pitch period. Fu and Sg denote an arbitrarily distorted and preprocessed signal within the framework of Tables 7.3,4. Figure 7.9 shows the relative amplitude RA1 for an undistorted signal and linear processing, separately for speakers 1 and 2 (male speakers, low pitch) and for speaker 3 (female speaker, high pitch). The two signal types are significantly different. The average value of RA1 for the male speakers was found around -8 dB, and the values of RA1 for the individual pitch periods ranged from -23 dB (phoneme /a/, speaker 2) to +3 dB (phoneme /n/, speaker 1). The corresponding values for the female speaker are +0.3 dB for the average, and -14 to +18 dB for the range of the individual periods. The procedure of averaging was done in a somewhat arbitrary way. In order to reduce the influence of individual periods far outside the common range, the range of RA1 was

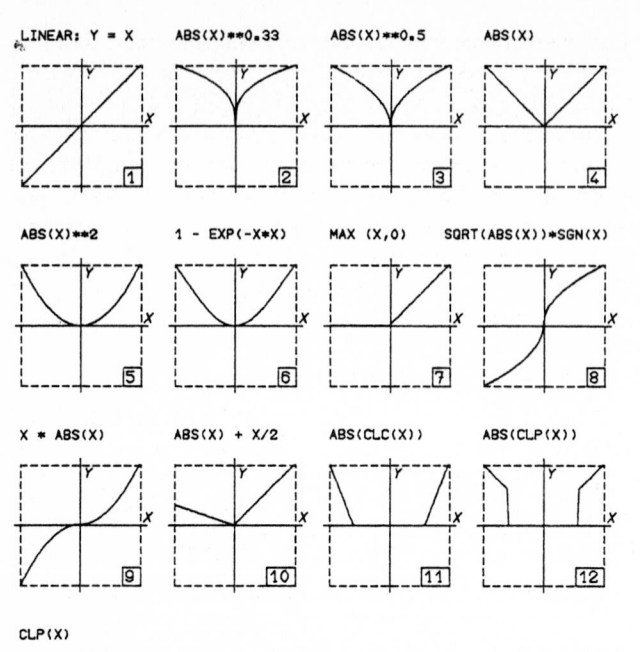

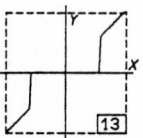

Fig.7.8. Characteristics of nonlinear functions used for PDA evaluation. (x) Input signal; (y) output signal. Note that the nonlinear functions (in all plots) are listed in FORTRAN notation: (*) denotes multiplication and (**) exponentiation. The numbers correspond to Table 7.4

limited to -20 to +10 dB. Averaging was then performed in the decibel scale using the limited value of RA1. This "average" is thus not to be used as a true average in the statistical sense, and the evidence it yields with respect to the performance of a PDA is rather a qualitative one. It is good enough, however, to show the overall tendency induced by a particular (linear) signal distortion as well as by a particular nonlinear function.

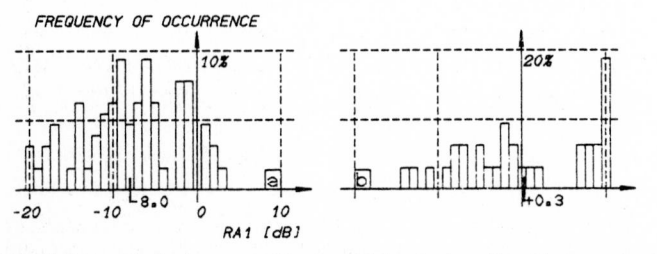

Fig.7.9a,b. Relative amplitude RA1 of the first harmonic in the sample pitch periods. (a) Two male speakers; (b) one female speaker. (Number below each figure) Average of RA1 (averaged over the dB scale)

Figure 7.10 gives a comprehensive view of the relative amplitude RA1 for all the nonlinear functions. Obviously the results verify hypothesis (7.6). For the signal type SiLU (upper left figure) as well as for the band-limited signal SiLL the odd functions show least enhancement of the first harmonic. It must be noted, however, that all the nonlinear functions, on the average, enhance the first harmonic even for the undistorted signal. Among the nonlinear functions with good performance are all the even and SSB functions, whereas the mixed functions show only a moderate enhancement. For male speakers, therefore, the experiment confirms the statement by McKinney and others that even functions do a better job than odd or mixed ones.

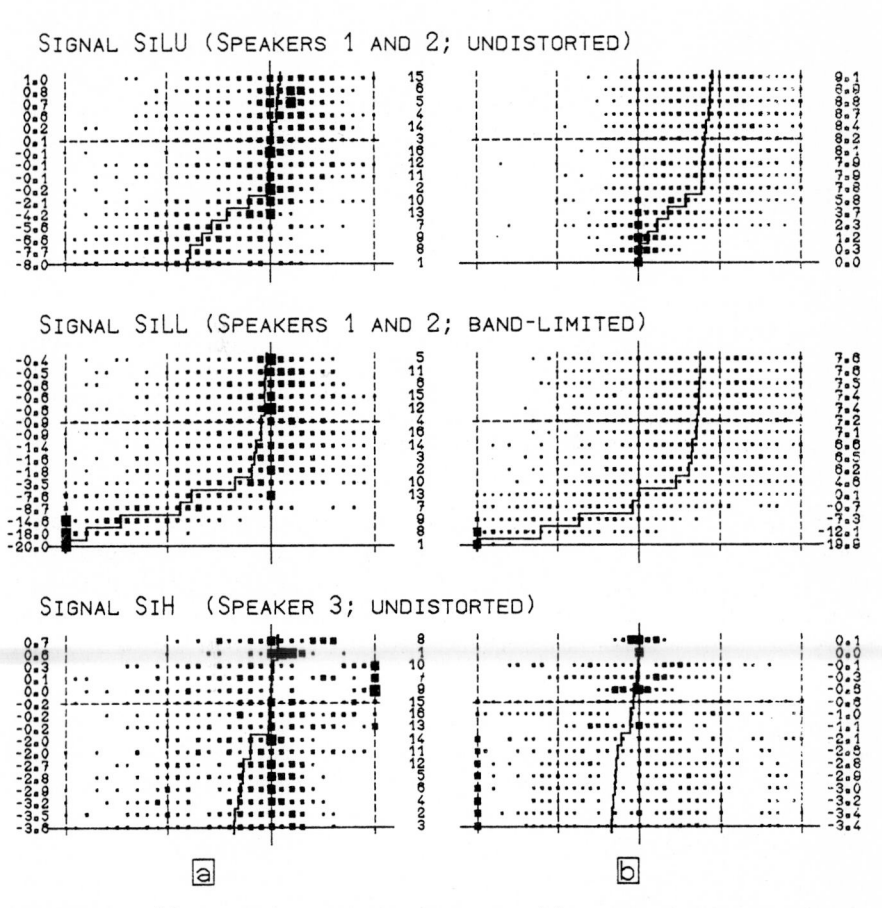

Fig.7.10a,b. (a) Relative amplitude RA1 and (b) enhancement RE1 of first harmonic. Relative amplitude RA1: -20 to +10 dB along the abscissa, enhancement RE1: -20 to +20 dB along the abscissa. Numbers at left and right margins: averages (averaged over the dB scale); numbers between figures: type of function used (see Table 7.4 for specification)

For the female voice, however, a different behavior was obtained. Here linear processing shows best performance. Mixed and odd functions behave about equally well, whereas SSB modulation and especially the even functions clearly deteriorate the performance of a PDA. A universal PDA which applies only a single nonlinear function therefore does not seem reasonable. This is demonstrated more clearly when the enhancement RE1 of the first harmonic is displayed (Fig.7.10b, more detailed view in Appendix A, Fig.A.15). Even and SSB functions, which do a good job for low-pitched voices, may attenuate the first harmonic of a high-pitched voice by 20 dB (or even more). In practice this would mean that a PDA, when using an even nonlinear function, will fail systematically when it processes certain classes of pitch periods uttered by a female speaker. It should not be forgotten that every pitch period of this experiment stands for a certain class of phonemes which frequently occur in real speech. One predictable failure at this point therefore signifies that a PDA applying this particular nonlinear function will fail for a whole class of signals.

The results of this experiment with respect to the relative amplitude and relative enhancement of the first harmonic thus lead to the following conclusions:

Even functions and SSBM show a good performance in conjunction with low-pitched signals whether these are band limited or not, whereas odd functions are less, and mixed functions only moderately effective. (7.10)

Nonlinear processing by even functions as well as by SSBM can cause catastrophic attenuations of the first harmonic when applied to high-pitched signals, whereas odd and mixed functions have little effect. (7.11)

The enhancement of odd functions in the case of band-limited signals is not sufficient for a PDA (PiTN) to operate with. Application of even functions or SSBM leads to failure for high-pitched voices. A universal PDA of this group is therefore not possible by only using a single nonlinear function. (7.12)

A unified view of the behavior of the nonlinear functions is gained when the value of RA1 or RE1 after nonlinear processing is displayed depending on the value RA1 of the undistorted signal (Fig.7.11). Only signals which are not band limited have been taken into account for this figure. For the undistorted signal, the amplitude RA1 of the first harmonic depends mainly on the ratio of the fundamental frequency F_0 to the formant frequency F1; it will take on a low value when these two are far from each other, and an extremely high value when F1 and F_0 coincide. (From this point of view, it seemed justified to limit the range of RA1 as it was done in this experiment.) The significant difference in this value between male and female speakers is mainly due to the fact that, for the latter, F1 and F_0 are usually much nearer to each other. Hence, if the value RA1 of the undistorted signal is taken as the independent variable of the diagram, the separation of male and female

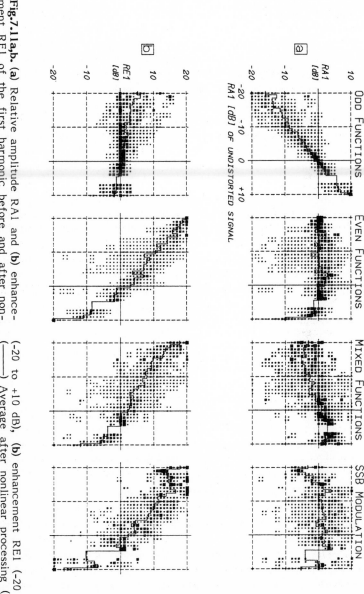

Fig.7.11a,b. (a) Relative amplitude RA1 and (b) enhancement RE1 of the first harmonic before and after nonlinear processing. (Abscissa) RA1 of undistorted signal (-20 to +10 dB); (ordinate) (a) relative amplitude RA1 (-20 to +10 dB), (b) enhancement RE1 (-20 to +20 dB). (———) Average after nonlinear processing (averaged on the dB scale). Only signals without *linear* distortions were taken into account for this figure

voices is not necessary. Both figures - relative amplitude as well as relative enhancement - clearly show that there exists a "class behavior" of the non-linear functions. Odd functions (linear processing was excluded from this figure) show relatively little deviation from linear processing. Even functions as well as SSBM amplify the first harmonic as long as the original value of RA1 is below 0 dB. In case RA1 takes on a value above 0 dB (this means that the first harmonic has become the strongest), they attenuate the first harmonic more and more. Mixed functions, finally, are better than even ones or SSBM for originally high values of RA1; on the other hand, however, they seem to add the disadvantages of both odd and even functions so that their application can hardly be justified. From this we postulate that

> *a universal PDA of group PiTN should use an even or a SSB non-linear function when the relative amplitude of the first harmonic is below 0 dB, and an odd nonlinear function otherwise.*   (7.13)

Further investigations have to show whether statement (7.13) can be realized in this categoric way.

7.4 Toward a Choice of Optimal Nonlinear Functions

According to statements (7.11-13) at least an even and an odd function are required for the UPDA. The SSB function may be optional. To gain insight into the performance of these functions, three (odd, even, and SSB) functions with identical amplitude characteristics (square root) were selected and further investigated (see Figs.A.12-14 in Appendix A for an example). Figure 7.12 shows the relative amplitude RA1 for the function which was actually the best of these three functions, depending on RA1 of the undistorted signal. For all three signal types the average of RA1 after nonlinear treatment is found above 0 dB. (Due to the display procedure any value in the diagram outside the normal range is usually marked by a quadruple of points.) The figure shows that the number of periods whose RA1 remains below -5 dB is extremely low. The male voice was best processed by the even and the SSB function. The odd function was never selected with band-limited signals, whereas it was preferred to process the female voice. To give an idea how much better the actually selected function performs, Fig.7.13 displays the difference (in dB) of RA1 between the selected function[4] and the other two functions. The independent variable is again the relative amplitude RA1 for the undis-

---

[4] A function is regarded as *selected* (or, equivalently, *marked applicable*) if the relative amplitude RA1 takes on 1) the highest value of all functions, or 2) a value less than 1 dB below the highest value, or 3) a value above 0 dB so that the first harmonic is the strongest regardless whether there is a function which performs better or not. Due to this procedure several NLFs could be selected for the same signal.

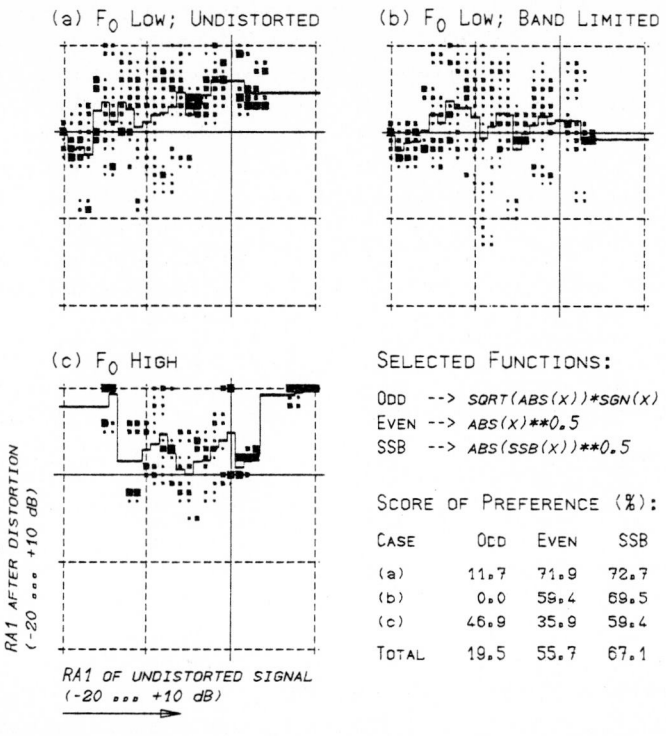

(a) F_0 Low; Undistorted   (b) F_0 Low; Band Limited

(c) F_0 High

RA1 AFTER DISTORTION
(-20 ... +10 dB)

RA1 OF UNDISTORTED SIGNAL
(-20 ... +10 dB)

Selected Functions:

Odd --> SQRT(ABS(X))*SGN(X)
Even --> ABS(X)**0.5
SSB --> ABS(SSB(X))**0.5

Score of Preference (%):

Case	Odd	Even	SSB
(a)	11.7	71.9	72.7
(b)	0.0	59.4	69.5
(c)	46.9	35.9	59.4
Total	19.5	55.7	67.1

Fig.7.12a–c. Relative amplitude RA1 of the first harmonic: optimal performance out of three nonlinear functions. (———) Average after nonlinear processing (averaged over the dB scale). A suboptimal performance was also taken into account for the score or preference when it was less than 1 dB below the optimal value, or when it rendered RA1 above 0 dB

torted signal. As expected, the odd function is mostly selected when the amplitude RA1 of the undistorted signal is high. In this case the odd function yields a significant improvement of the PDA performance when compared to the even or the SSB function. The same holds for the even or the SSB functions with respect to the odd function. The differences between even and SSB functions are generally not very large; the average is situated around 5 dB for both of them. Some more investigations are therefore necessary in order to answer the question whether the effort to implement the SSBM is justified. A UPDA, however, can already be built with two nonlinear functions (odd and even) applied simultaneously.

7.4.1 Selection with Respect to Phase Distortions

Phase distortions influence the temporal structure of the signal thoroughly. As long as processing is linear, they do not influence the amplitude spectrum even after filtering. Any linear filter can be divided into a linear phase filter

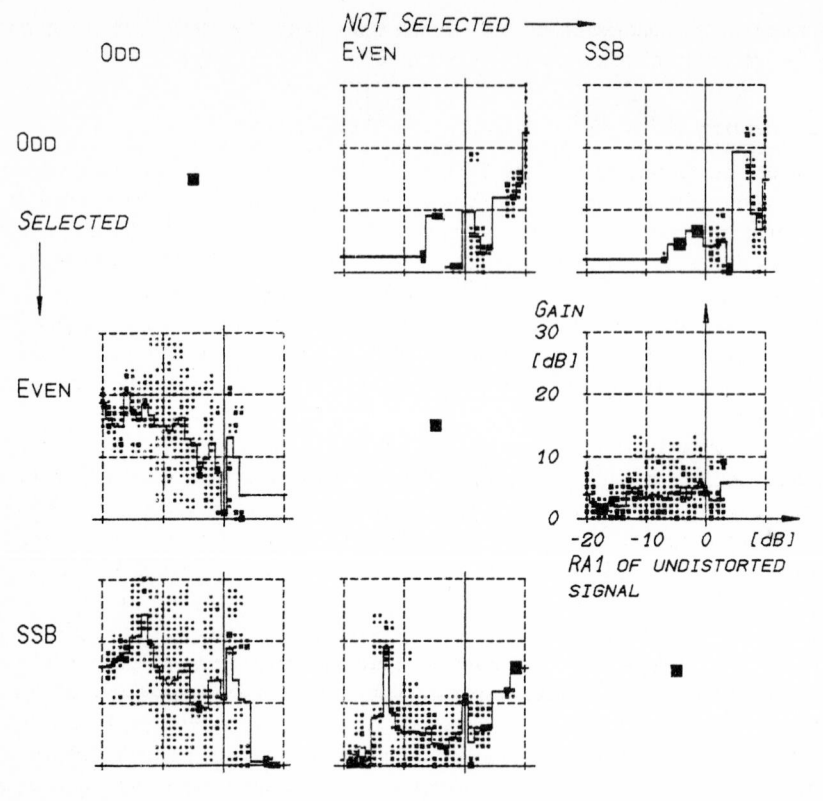

Fig.7.13. Relative amplitude RA1 of first harmonic: Gain by choice of optimal function compared to the functions which were not selected. Top to bottom: selected function (same as in Fig.7.11). For details of the selection procedure, see footnote 4 on the previous page). Left to right: base of comparison (function which was not selected)

with a certain amplitude response and, in series to that, an all-pass filter supplying the phase distortions. Things are different when processing is non-linear. To investigate the influence of phase distortions, especially low-frequency phase distortions, on the relative amplitude RA1, the value

$$RP1 \ (Fu) = RA1 \ (Fu, SgP) - RA1 \ (Fu, SgU) \hspace{2cm} (7.14)$$

was computed for all nonlinear functions (see details in Appendix A, Fig.A.23). The average differences are negligible (1.2 dB for an even function, 1.3 dB for SSB, -0.9 dB for odd functions), but individual values vary much more widely. According to these figures SSB and even functions are more phase sensitive than the odd ones; the general tendency of even and SSB functions with respect to phase distortions is contradictory. This might explain the average difference of 4 to 5 dB between even and SSB functions shown in Figure 7.13. Moreover, this effect yields an argument in favor of SSB since this function is supposed to supply good results when the even function might fail due to phase distortions, and vice versa.

7.4.2 Selection with Respect to Amplitude Characteristics

An important parameter for the PDA is the amplitude range within which it can operate. Disregarding the influences of prosody and recording conditions, a dynamic range of about 30 dB between the weakest and strongest voiced phonemes is to be expected. In addition, there is the 20-dB interphonemic difference of the relative amplitude RA1. Certainly the weaker phonemes tend to take on a higher value of RA1, so that the overall amplitude change of the first harmonic will be less than the sum of the individual changes. Nevertheless, short-term amplitude changes of up to 40 dB must be taken into account for the first harmonic.

As to the nonlinear function, it can be chosen in such a way that it attenuates part of these amplitude changes. It was discovered very early that squaring, as a nonlinear function, gives better results than cubing, and, again, that full-wave rectification gives better results than squaring (Kriger, 1959; McKinney, 1965). In some of the older statements made in connection with analog devices, the limitations of the physical realization should not be overlooked. For instance, it was a technological problem to build a full-wave rectifier with ideal characteristics over an aplitude range of 40 dB or more before the era of operational amplifiers. In the computer, however, internal amplitude range is a matter of word length (or not even that!), and the choice of nonlinear functions can be made by simple insertion or exchange of a table. Thus the nonlinear function should be chosen in such a way that it does not increase the intrinsic amplitude changes of the signal.

In Fig.7.9 the value of RA1 was found to be fairly independent from the amplitude characteristic of the selected function provided that these functions belonged to the same class (odd, even, or SSB). On the one hand, this confirms the statement made in the literature concerning full-wave rectification as opposed to squaring or cubing; on the other hand, it suggests that characteristics of the form

$$y(n) = |x(n)|^a \tag{7.8}$$

with an exponent a less than one could give even more favorable results. Figure 7.14a shows the difference between the relative amplitude RA1 for two even functions (square root of absolute value and square) and the value RA1 for full-wave rectification as reference. The average difference is low: 0.3 dB for the square and -0.4 dB for the square root. Even the cubic root still gives good results (-0.6 dB). From this the square root and cubic root even appear preferable to full-wave rectification. The relative amplitude RA1 is hardly touched by the exponent a, whereas the amplitude changes due to the dynamics of the input signal are drastically influenced; squaring would increase the 30 dB of original dynamic range to 60 dB, whereas square root

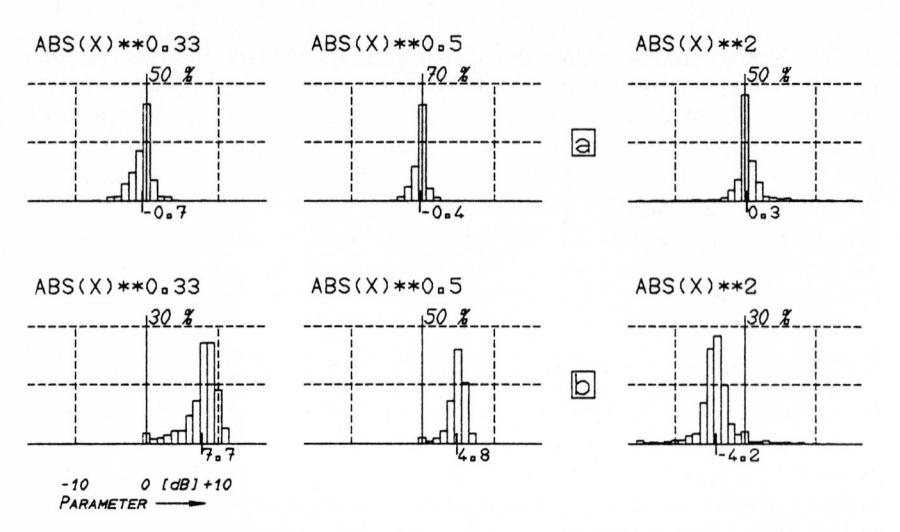

Fig.7.14a,b. Comparison of performance for even nonlinear functions with different amplitude characteristics: difference (in dB) between value of parameter for indicated function and value of the same parameter for full-wave rectification. **(a)** Relative amplitude RA1 of first harmonic; **(b)** relative amplitude RA0 of offset. (Ordinate) Percentage of occurrence

would pull it down to 15 dB, and cubic root to 10 dB. The downward limitation of the exponent a is mainly given by the fact that the amplitude of the first harmonic, with falling exponent a, gets smaller and smaller compared to the amplitude at a frequency of zero. McKinney (1965) has measured that just the step from squaring to full-wave rectification causes a decrease in the relative amplitude of the first harmonic with respect to the amplitude at zero frequency (average offset) by 3 to 6 dB. Therefore the (high-pass) filter which suppresses zero frequency, and which is indispensable for any nonlinear function other than odd, gets more and more ineffective when the exponent a is decreased. Just as the original signal, the filtered signal and in particular the zero-order harmonic is subject to the amplitude changes of the input signal. This means that for low values of the exponent a short-term variation of the overall amplitude of the signal tends to produce time-varying offsets at the filter output which get more and more difficult to control. To quantitatively investigate this influence, the parameter RA0 is defined and determined as the ratio of the amplitude A(0) of the offset and the amplitude A(1) of the first harmonic,

$$RA0 := 20 \lg \frac{A(0)}{A(1)} .$$
(7.15)

The detailed view of RA0 is displayed in Appendix A (Fig.A.16). Figure 7.14b displays the performance of the indicated NLFs for the parameter RA0 in the same way as (a) for the parameter RA1.

The course of RA0 suggests that there is a relatively flat optimum for the exponent a between the two limitations a) dynamic range of the distorted signal, and b) amplitude of the offset. This optimum is found slightly below a = 0.5 where the decrease in the overall dynamic range equals the increase of RA0 when the exponent a goes down.[5] In addition, it seems unnecessary to decrease the amplitude range of the input signal to a lower value than to that of the interphonemic differences of RA1 which are not influenced by the exponent a of the nonlinear function. An exponent a somewhere around 0.5 (square root) thus appears reasonable.

7.4.3 Selection with Respect to the Sequence of Processing

In most PDAs low-pass filtering is implemented somewhere in the course of processing. Even in autocorrelation PDAs (see Rabiner, 1977) it is usual to perform a low-pass filtering with a cutoff frequency well above the possible range of pitch, i.e., 1000 Hz or more. For the PDAs of group PiTN there is no uniform opinion in the literature as to whether to implement the nonlinear function before or after the low-pass filtering step (see Sect.6.1 for more details on this issue). Figure 7.15 shows the relative enhancement of the first harmonic when the nonlinear function is applied after low-pass filtering, compared to the value of RE1 for the reverse order of processing. There is also support for the application of the nonlinear function *before* any low-pass filtering. The effect is greater for band-limited signals than for undistorted ones, and greater for even functions than for odd or mixed ones (SSB functions are disregarded in this consideration since they require low-pass filtering after nonlinear processing in any case). That means that the higher formants with their greater bandwidths significantly contribute to the reconstruction of the first harmonic, in particular, when one considers that the low-pass filter applied in this experiment (Fig.7.7) was fairly moderate. As a control experiment the same values were computed after slight enhancement of the higher frequencies (+3 dB at 1300 Hz and +9 dB at 5000 Hz). The results (see Fig.A.22 in Appendix A) point in the same direction, at least for the even functions. That means:

> Application of the nonlinear function in a pitch detector of group
> PiTN should be done in any case before low-pass filtering. A slight
> enhancement of the higher formants before the nonlinear step also
> shows a positive effect when cancelled by a low-pass filter there-
> after.   (7.16)

---

[5] According to Fig.7.14, RA0 goes up by 2.9 dB when the exponent a is decreased from 0.5 to 0.33. At the same time the overall dynamic range goes down from 15 to 10 dB. This condition alone would thus suggest an optimum below 0.33 for the exponent a. On the other hand, the interphonemic differences of RA1 extend over more than 20 dB (Fig.7.11). This condition would suggest an optimum of the exponent a around 0.7. A value of a = 0.5 thus appears reasonable.

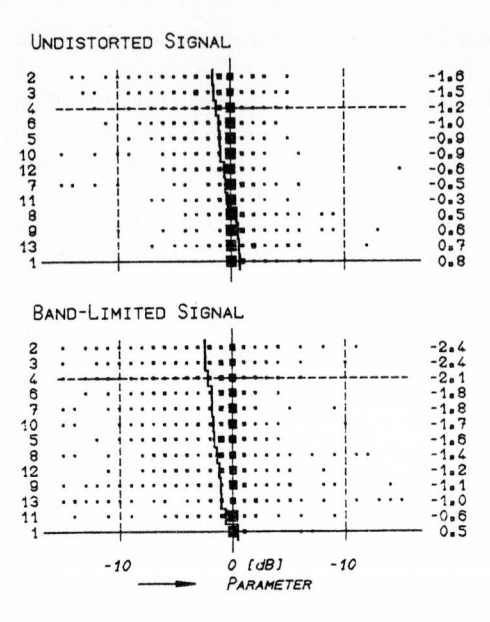

UNDISTORTED SIGNAL

BAND-LIMITED SIGNAL

-10   0 [dB]   -10
━━━▶   PARAMETER

Fig.7.15. Influence of processing sequence on the enhancement RE1 of the first harmonic. (Abscissa) Difference of enhancement RE1 computed from the signal when nonlinear processing was performed before low-pass filtering, and RE1 from the signal when low-pass filtering was done first. (Range) -15 · to +15 dB. (Ordinate) Type of function used (marked left; see specifications in Table 7.3). (———) Average difference (averaged over dB scale, numerical value noted on the right-hand side of the figure). The SSB functions were not taken into account for this figure

7.5. Implementation of a Three-Channel PDA with Nonlinear Processing

The results of the previous sections have provided us with information on the behavior of nonlinear functions in the preprocessor of a PDA. To implement a universal PDA that uses the principle PiTN, at least two NLFs are necessary. In the present PDA a three-channel version with three NLFs (odd, even, and rectified SSB, each with square-root amplitude characteristics) is implemented. As suggested by the experiments, the nonlinear distortion is performed first, and the signal is split up into three independent channels. The following steps, which correspond to the linear algorithm discussed in Sect.7.1, are performed separately (but with nearly identical subroutines) for the three channels. If the adaptive comb filter which is used to suppress the dominant waveform (F1 in the linear case) is set to a common zero for the three channels, the critical case of coincidence between F1 and F_0 can be detected and appropriately handled. For this purpose a "minimum frequency rejection" principle is applied in the determination procedure for this zero. After the basic extractor the channel that is most likely to carry the correct information on pitch is selected according to a minimum frequency selection principle. The algorithm works well with both undistorted and band-limited signals without requiring an a priori estimate of pitch.

7.5.1 Selection of Nonlinear Functions

Nonlinear processing enhances the first harmonic optimally when the following conditions (concluded from Sects.7.2-4) are observed:

1) three nonlinear functions with equal amplitude characteristics (odd, even, and SSB) shall be applied;

2) the nonlinear step is to be performed first, i.e., before low-pass filtering;

3) an amplitude response such as square root gives good results with respect to the intrinsic dynamic range of the signal; and

4) there is no possibility of preselection of a certain NLF so that a short-term selection algorithm must do this job in the basic extractor.

According to these conditions the original speech signal is now first processed by three nonlinear functions (in the notation of Table 7.4),

$$y_{odd}(x) = \text{sqrt (abs}(x)) * \text{sgn}(x) , \qquad\qquad [7.4\text{-}8]$$

$$y_{even}(x) = \text{sqrt (abs}(x)) , \qquad\qquad [7.4\text{-}3] \qquad\qquad (7.17)$$

$$y_{ssb}(x) = \text{sqrt (abs(ssb}(x))) . \qquad\qquad [7.4\text{-}15]$$

The numbers in square brackets refer to the definitions in Table 7.4. In the following discussion these functions and the respective signals will be labeled *odd, even,* and *SSB.* In the equations, y and x stand for an arbitrary sample y(n) or x(n) of the signals involved; *sgn* is the sign function which takes on a value of +1 or -1, according to the sign of its argument. The SSB function is easily obtained in a digital system (Fig.7.16, illustrated with a real speech example). First the sampling frequency is increased by a factor of three; the missing samples are set to zero. The spectrum of the new signal now contains three periods of the (periodic) spectrum of the original signal. From this the SSB signal is obtained by high-pass filtering. The next step is the nonlinear one: full-wave rectification and square rooting. The signal is then low-pass filtered; at the same time the sampling frequency is reset to its original value. The high-pass and low-pass filter need not be linear phase for this purpose.

As discussed in Sect.7.4, an a priori criterion for short-term selection of one of the nonlinear functions cannot be given. For any signal, whether undistorted or band limited, all the nonlinear functions can be selected and actually were selected (Fig.7.12). Hence there is no choice for the present implementation except separate processing of the three partial signals which result from the application of the three NLFs. (The remainder of the preprocessor as well as the basic extractor can be identical for the three channels.) The first subsequent step therefore is the low-pass filter which removes the higher formants. A first-order digital filter is added to suppress the offset occurring in the SSB and the even signals.

7.5.2 Determination of the Parameter for the Comb Filter

As in the linear algorithm, the determination of F1 and the application of the comb filter come next. Only with the odd NLF will the algorithm find a value which corresponds to the real formant F1. In the other two partial

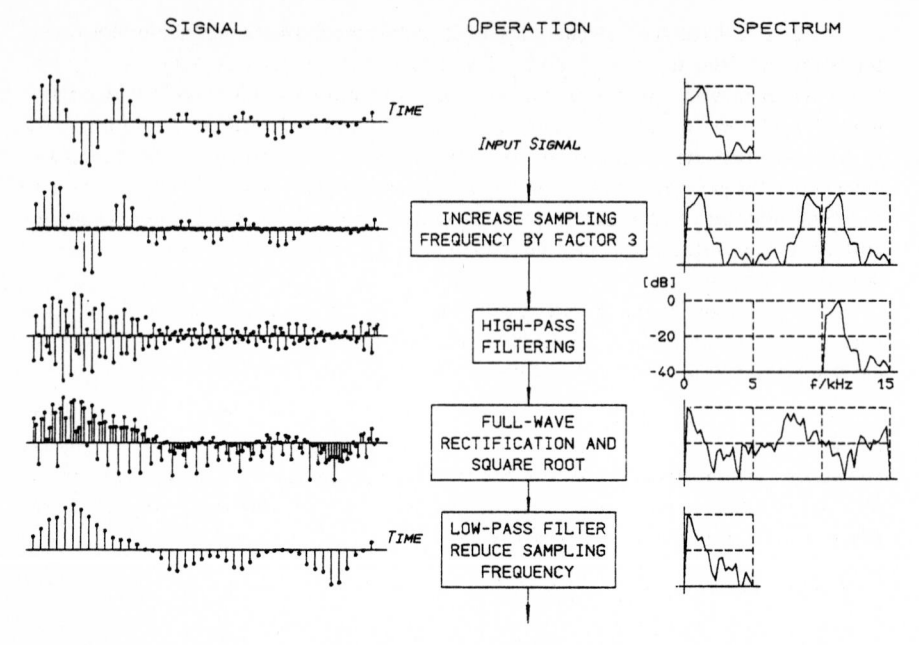

Fig.7.16. Flow chart and example of SSB implementation. Example: phoneme /a/, female speaker

signals the structure of F1 has been destroyed, and a strong signal is found either at F_0 or at a frequency of 2F1. Nevertheless, the algorithm used to determine F1 is applied to each of the three channels because the three values of "F1" resulting from this measurement yield a criterion that enables the algorithm to detect the critical case of coincidence between F_0 and F1. This will be the case if the comb filter is set to a common value for all the partial signals. The following examples clarify that determination of maximum frequency (minimum pulse width) because the zero of the comb filter gives correct results:

1) F1 is well above F_0 but relatively weak. In this case the partial PFSs derived from the SSB and the even function will be dominated by the first harmonic. The F1 histogram which is built up independently for each partial PFS will yield F1 for the odd NLF and F_0 for the others. Since the comb filter must not destroy F_0, its zero must be set to the higher of the offered frequencies, in this case to the real formant F1.

2) F1 coincides with F_0. The SSB and the even functions will then generate strong second harmonics which dominate the temporal structures of these two partial PFSs. The odd NLF will maintain the real formant F1. After determination of the dominant frequency the value of the odd function will be F1 (which equals F_0); for the two others the value will be 2F1. In order to avoid errors, the filter zero is set this time to the pseudo-formant 2F1 - again to the higher of the measured dominant frequencies.

3) F1 is well above F_0 (say, at $2F_0$) and strong. By the SSB and the even function this frequency is doubled. The first harmonic is also emphasized, but the waveform with the frequency 2F1 will dominate the signal. The algorithm may thus determine 2F1 for SSB and even; for the odd signal, it detects F1. If the zero of the comb filter is now set to the highest of the measured "formant" frequencies, the filter is mistuned for the odd signal but optimal for the others. Yet this does not cause damage; in this case the odd signal will never be selected since the first harmonic of the other partial PFSs is definitely stronger.

4) F1 is slightly above F_0, but lower than the second harmonic. In this case F1 may pass all the nonlinear functions unchanged and may be correctly determined for all the partial signals.

These four rules practically cover all the possible positions of F1 with respect to F_0. Two of them require that the highest of the three estimates be taken - or, at least, that the lowest one not be taken. The two others do not oppose this requirement. The frequency f_{ZC} of the zero of the comb filter is thus determined as follows:

$$f_{ZC} = \max (F1_O, F1_E, F1_S) \tag{7.18a}$$

$$f_{ZC} = \max [f_{ZC}, 2 \cdot \min (F1_O, F1_E, F1_S)] \tag{7.18b}$$

where $F1_O$, $F1_E$, and $F1_S$ represent the individual "formant" frequencies derived from the partial PFSs. The second part of condition (7.18) was imposed by rule 4. The general underlying hypothesis is that at least in one of the partial signals the first harmonic (whether equal to F1 or not) is so strong that it dominates the determination of "F1" in this channel. (Again this does not apply to the signals which are covered by rule 3, but this rule does not oppose the hypothesis.) This means that the problem has been solved once the zero of the comb filter is limited to a value greater or equal to two times the lowest "F1" estimate. This hypothesis was verified in the experiment so that the common adjustment of the comb filter according to (7.18) makes the preset of an expected estimate of pitch unnecessary.

7.5.3 Threshold Function in the Basic Extractor

The results of the marker detection procedure in the basic extractor strongly depend on the applied threshold function. A ZXABE appeared undesirable since it requires too extensive low-pass filtering (see Sect.6.1). If the threshold is nonzero, however, a constant threshold definitely is not sufficient; the threshold must be adapted to the course of the envelope of the filtered signal. For this the algorithm determines the positive maximum of the FS within a given frame that contains at least one complete pitch period. In order to assure that the procedure is able to follow rapid intensity changes, the frame interval must be considerably shorter than the frame length. In the implementation the frame length is set to 25.6 ms, and the frame interval to 6.4 ms. (This is done for implementation, since the sampling frequency is 10 kHz and the operating system of the computer (PDP 11/45) supports a block size of 256 words for disk input-output operations.) To calculate the threshold function itself, the envelope is then multiplied by a constant k_T less than one. The value of this constant is limited by two conditions. Firstly, not all the peaks in the FS are of equal size (although the short-term fluctuations of amplitude are much less for the FS than for the original speech signal). If k_T is chosen too high, the algorithm may thus miss peaks it should not miss. This applies mainly to passages of major intensity variations. This first condition thus gives an upper limit for k_T. Secondly, the advantage of this algorithm with respect to the effort necessary for filtering consists in the fact that several zero crossings per pitch period are permitted in the FS. The threshold analysis therefore must use a nonzero threshold. In addition the comb filter is not always optimally adjusted. The second condition thus defines a lower limit for k_T.

The algorithm can only operate in a reasonable way if the work area, i.e., the range of k_T between the lower and the higher limit is large enough (at least for the great majority of signals). In the linear algorithm of Sect.7.1, a value between 0.6 and 0.7 was found experimentally. This value already provides a certain amount of safety against missing a peak due to amplitude fluctuations; on the other hand it proved high enough to suppress most of the spurious peaks. For the nonlinear algorithm the experiment in Sect.7.3 can give an answer. Besides the relative amplitude RA1 of the first harmonic, the relative amplitude RASM of the largest spurious peak was defined and measured for this purpose:

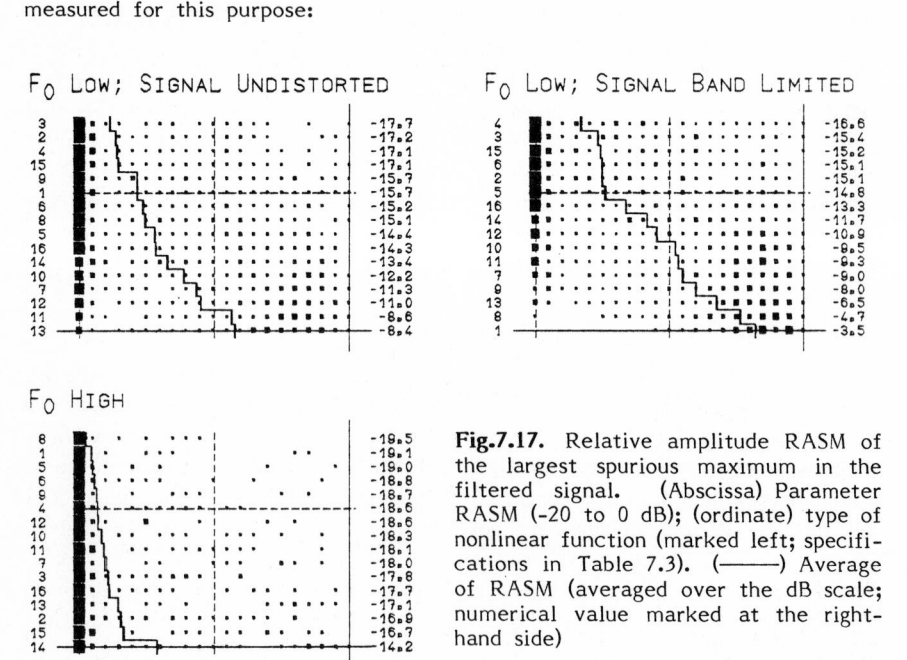

Fig.7.17. Relative amplitude RASM of the largest spurious maximum in the filtered signal. (Abscissa) Parameter RASM (-20 to 0 dB); (ordinate) type of nonlinear function (marked left; specifications in Table 7.3). (———) Average of RASM (averaged over the dB scale; numerical value marked at the right-hand side)

$$\text{RASM} := \max\left(-20,\ 20\ \lg\ \frac{|\ x(n_s)\ |}{|\ x(n_p)\ |}\right). \tag{7.19}$$

In this equation, n_s is the address of the spurious peak and n_p that of the principal peak. In order to enable a meaningful averaging over the decibel scale, the parameter was downward limited to -20 dB. "Largest spurious peak" means the largest peak (maximum or minimum) whose sign equals that of the principal peak of the period, and which is separated from the principal peak by at least one sample of opposite sign. The sample periods (after linear and/or nonlinear distortion) were filtered by the filter applied in the linear algorithm (Fig.7.1). This means that the first zero of the triple comb filter was set to F1 (always the value measured for the linear signal so that for some NLFs the filter could be mistuned) and limited to a value greater than twice the fundamental frequency (examples are displayed in Appendix A, Figs.A.7-14). The results are shown in Fig.7.17 (overview) and in Appendix A (Figs.A.21-24 where this parameter is labeled "amplitude threshold"). The method may be judged safe as long as the value of RASM does not exceed -6 dB (corresponding to $k_T = 0.5$). This condition is fulfilled for linear proces-

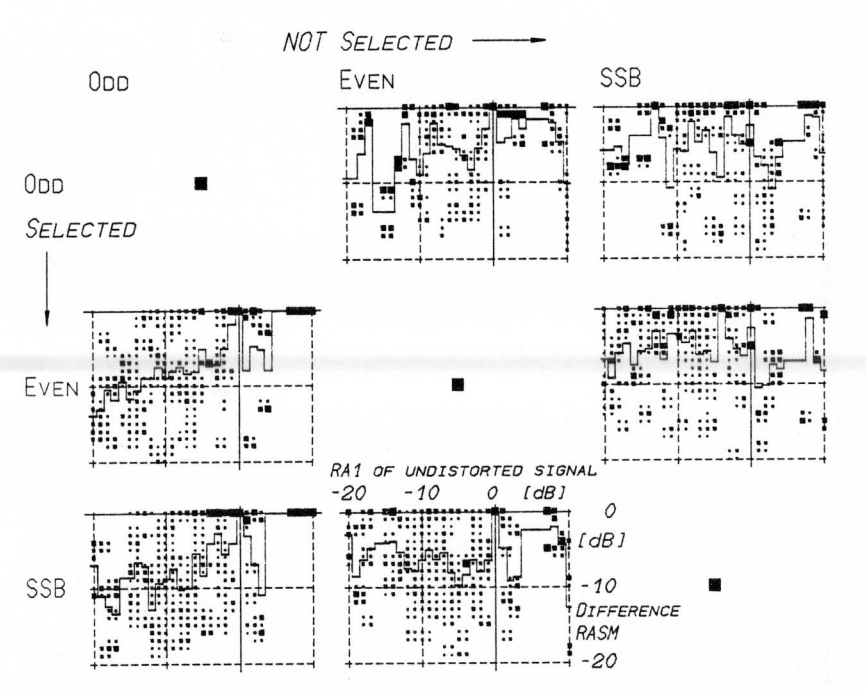

Fig.7.18. Relative amplitude RASM of largest spurious maximum: Gain by the chosen optimal function compared to the functions which were not selected. (Top to bottom) Selected function; (left to right) base of comparison (function which was not selected). The gain takes on negative values since the algorithm shows best performance when RASM is low

sing. For band-limited signals, the even functions give best results but are not free from occasional failure. Linear processing and the signals treated by odd NLFs are severely influenced by band limitation, whereas for the other NLFs the average influence is relatively small (Fig.A.24 in Appendix A).

Especially in connection with even NLFs, this parameter is sensitive to the processing sequence; there is an average loss of 2 dB or more (particularly for the band-limited signal) when the nonlinear function is applied after low-pass filtering. For the even NLFs this difference even takes on a value of 6 dB! The parameter RASM thus supports the processing sequence "perform the nonlinear step first" even more strongly than the relative amplitude RA1 of the first harmonic (Fig.A.21). An additional enhancement of the high frequencies, on the other hand, shows little effect (Fig.A.22).

When the parameter RASM is measured for the optimal function out of the three nonlinear ones implemented in the algorithm (the same as in

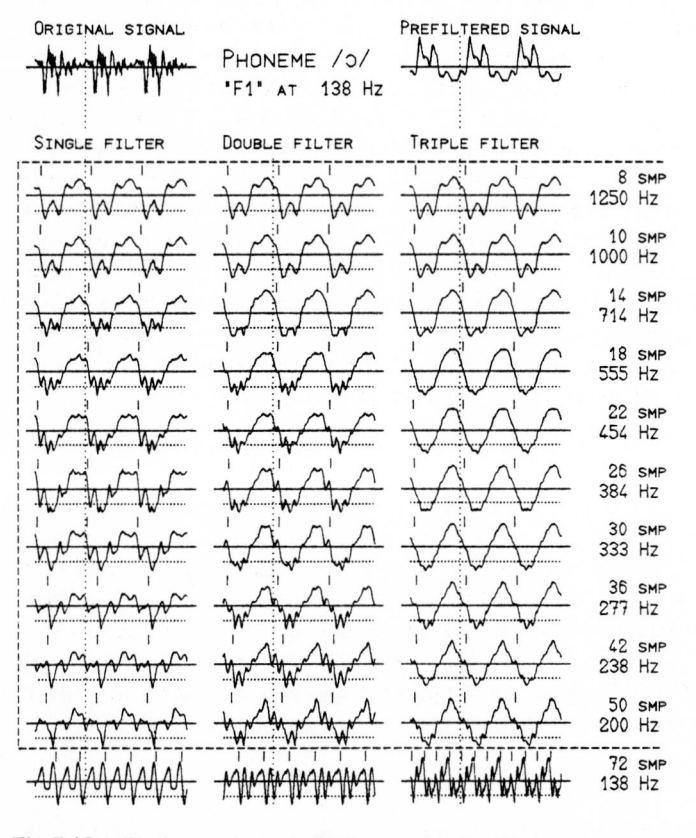

Fig.7.19. Performance of different comb filters for the attenuation of F1. The signal was nonlinearly distorted by an even function (with square root amplitude characteristics) before preprocessing. The rest of this figure corresponds to Fig.7.5

Fig.7.12), the result shows that the spurious peak, if existing at all, can always be kept below -8 dB, and for more than 90% of the investigated signals even below -15 dB. (The corresponding figure is not displayed since it is rather uninformative.) Figure 7.18, like Fig.7.13 for the relative amplitude RA1, displays the gain (in dB) when the function which is actually best is selected with respect to the other two NLFs. A gain of 0 dB which is quite frequent signifies that for more than one NLF the spurious secondary peak has disappeared.

In conclusion, nonlinear processing together with selection of the NLF which is actually best makes the value of k_T uncritical. In the implemented algorithm, it was left at 0.6, the value which had already been assigned to this constant in the linear algorithm. As another consequence, the triple TPF is not necessary; a double one is completely sufficient. This reduces the necessary additions per sample to four.

The procedure of marker determination, as displayed in Fig.7.3, could be transferred from the linear algorithm without change, at least for the odd NLF. For the FSs resulting from the SSB and even NLFs, the phase relations to the original signal are somewhat different. Here the reconstructed first partial was found exactly in phase with the original first partial of the signal. That means that in these partial FSs the proper beginning of the period is marked by that very zero crossing which was so hard to extract from the linearly processed signal. The old marker determination procedure from the linear PDA can therefore be kept for the new algorithm when the sign of the FS is reversed for the SSB and the even functions, and when the markers are left in the position they have been shifted to. They are already in their proper place. Figure 7.19 shows an example for the performance of the algorithm for a partial signal resulting from the even NLF.[6] The dotted vertical line indicates the beginning of the second of three periods (determined manually) and is in good agreement with the measured pitch markers.

7.5.4 Selection of the Most Likely Channel in the Basic Extractor

As in any multichannel PDA, a final channel selection routine is necessary to select that channel which is most likely to momentarily carry the correct information on pitch. Three strings of pitch marker candidates (PMCs) are available to this selection routine. No external criterion has been found which permits reliable selection of one of the partial signals for further processing. The logic of the basic extractor must thus make the decision by means of the internal criterion of regularity: the algorithm selects the one PMC which is found most "regular," i.e., which gives the best estimate of periodicity. As

---

[6] The reader is referred to Fig.7.4 where the same signal is displayed with linear processing. There F1 is determined to be 714 Hz. According to (7.18) the filter zero f_{ZC} would be set to 714 Hz which gives identical results for the odd and for the even signal.

in most selection routines of multichannel PDAs, the approach is mainly heuristic and verified by experiment.

Two principal methods are applied in existing time-domain PDAs to check for regularity. The first one could be called the *frame method.* This method takes a short frame of the signal which contains a small number of pitch periods, and checks it independently from its vicinity. If this method fails for one frame, there is no means to correct this error.[7] The subsequent frame, however, can be correctly measured. On the other hand, the method does not depend on starting conditions as is the case for the second method, the *sequential* or *tracking method.* Here any period is checked after the previous one. The tolerance interval within which an individual period shall be labeled regular is determined by some criterion derived from its predecessors, e.g., from their length average. Simple list correction algorithms tend to apply this method. It gives a good track of regular pitch periods and can well correct local errors. On the other hand, it may fall into trouble if the error occurs just at the beginning of a voiced section, or if the signal is irregular per se. In the present algorithm there is the additional problem that the logic has to choose between three marker estimates. This last point strongly favors the application of the frame method here. Due to reasons of computer implementation the input signal is subdivided into frames of 25.6-ms duration. Since this frame length is not long enough to be used for regularity check, the algorithm also regards the vicinity (10 ms on either side) of the frame under process.

A pitch period may be marked "regular" when its length differs from that of its neighbors by less than a certain percentage (usually somewhere between 10% and 20%). Hence the range of possible pitch (50 to 830 Hz in this algorithm which corresponds to a period length between 12 and 200 samples at a sampling frequency of 10 kHz) is subdivided into tolerance intervals whose center frequencies differ by 20%. For every frame and each string of PMCs, the algorithm takes the distribution of period lengths with respect to these tolerance intervals (Fig.7.20). The expected value of pitch for a PMC string then is the center frequency of the tolerance interval for which the histogram has its maximum. If all the markers pertaining to one frame can be grouped into the same tolerance interval, the marker string is labeled "regular". This check is carried out individually for each of the three PMC strings. After that the decision as to which of the three NLFs gives the most likely marker string is ready to be made. It is based on a set of experimental rules which are discussed in the following. In Fig.7.21 typical examples for the first three rules are displayed.

1) If at least two marker strings are regular and grouped into the same or into adjacent tolerance intervals, one of them is selected in such a way that

---

[7] The selection routine in the PDA by Gold and Rabiner (1969, see Sect.6.2.4 for a discussion) is a good example for this method.

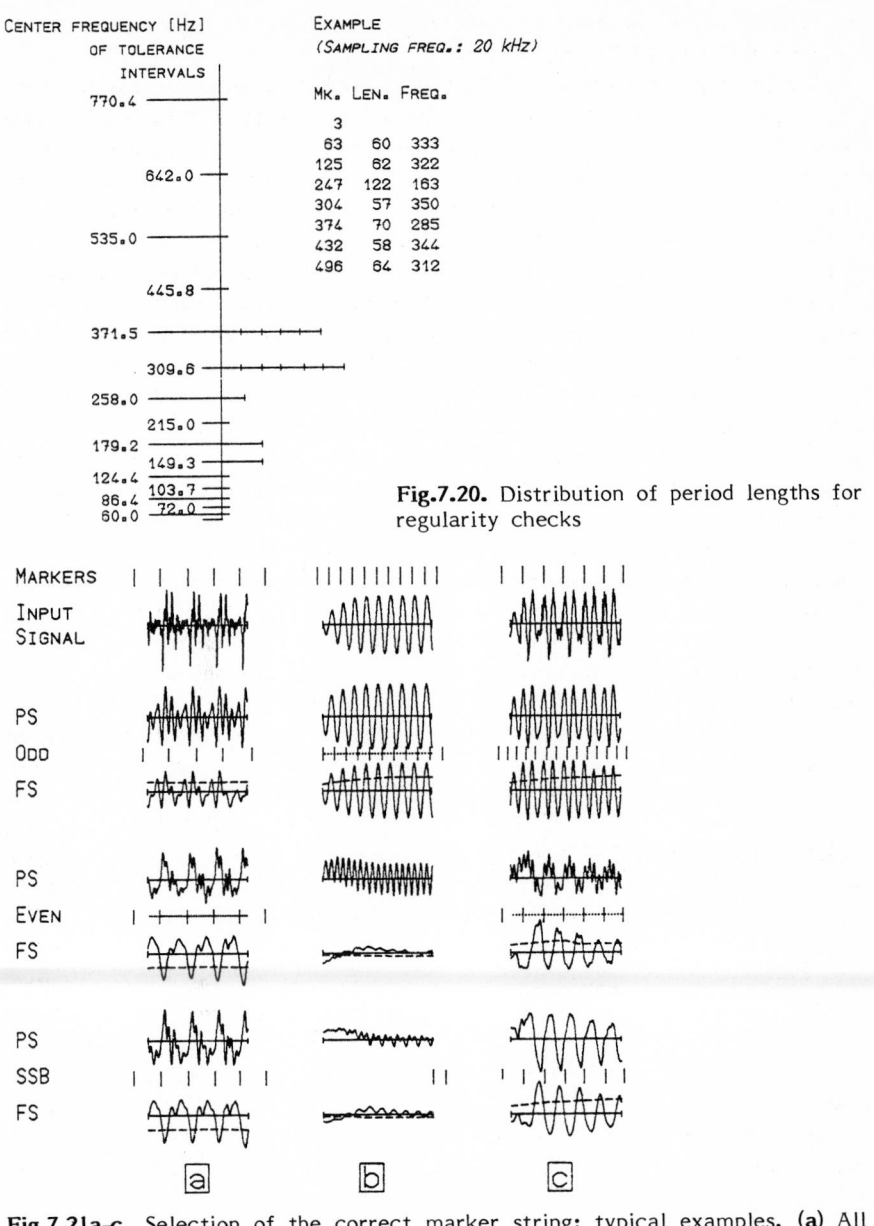

CENTER FREQUENCY [Hz]
  OF TOLERANCE
    INTERVALS

EXAMPLE
(SAMPLING FREQ.: 20 kHz)

770.4 ————

Mk. Len. Freq.

    3
   63   60   333
  125   62   322
  247 122   163
  304   57   350
  374   70   285
  432   58   344
  496   64   312

642.0 ——

535.0 ————

445.8 ——

371.5 ————

309.6 ——

258.0 ————

215.0 ——

179.2 ——
149.3 ——
124.4 ——
103.7 ——
86.4
72.0
60.0

Fig.7.20. Distribution of period lengths for regularity checks

MARKERS

INPUT
SIGNAL

PS
ODD
FS

PS
EVEN
FS

PS
SSB
FS

a   b   c

Fig.7.21a–c. Selection of the correct marker string: typical examples. **(a)** All strings regular: prefer the even NLF; **(b)** only one string regular (2nd harmonic cancelled by comb filter); **(c)** F_0 in one string, $F1$ in the other one: prefer the one which has the longer periods (minimum frequency selection principle). (PS) Prefiltered signal; (FS) filtered signal

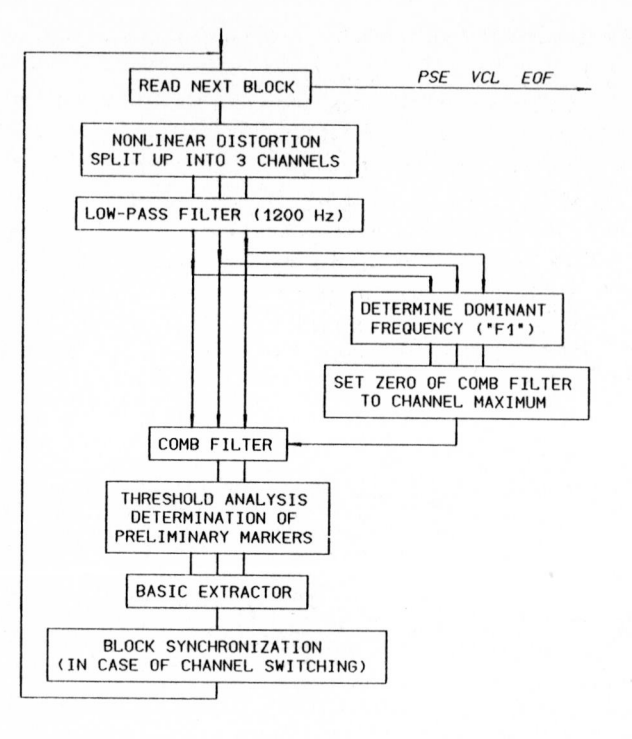

Fig.7.22. Flow chart of the PDA

the even NLF is preferred to SSB, and SSB is preferred to odd since the even NLF provides the best phase relation between pitch markers and the proper beginning of the period (Fig.7.21a).

2) If only one PMC string is regular, the algorithm checks the other two strings as to whether at least 80% of their PMCs are grouped into one tolerance interval. If this applies to a string, it is also marked regular, and the algorithm returs to rule 1. If not, the one regular marker string is selected (Fig.7.21b).

3) If two or three marker strings are regular but not within the same tolerance interval, the algorithm checks whether the period lengths of the strings take on a ratio of 2 to 1 (or, more rarely, 3 to 1). If this ratio is considerably less than 2, rule 1 is applied; if it is considerably greater than 3, the frame is treated as irregular (rule 4). In all other cases the algorithm selects that PMC string which yields the *longer* periods (*minimum frequency selection principle*, see Fig.7.21c).

4) None of the marker strings is regular. In this case the algorithm checks first whether one of the strings is regular within the 80% limit. If yes, and if rule 4 has not been enforced by rule 3, the algorithm retries rules 1 through 3. Otherwise it checks whether the previous frame was regular. If yes, it selects that PMC string which can be joined to that of the previous frame in a regular way. If not, a marker string is selected which is expected to fit regularly to that of the forthcoming frame.

Rules 1, 2, and 4 are rather self-explanatory, whereas rule 3 needs some further discussion. The most common error in PDAs of type PiTL and PiTN is the determination of the second harmonic instead of the first. The fact that three PMC strings are available, however, helps to detect and to avoid this error. Second harmonic tracking preferably occurs in two particular cases:

a) An utterance is given where F1 coincides with F_0. As shown in previous discussions, the odd NLF will then yield the right value whereas the others might track the second harmonic and, in spite of this error, indicate regularity. The error is detected when two regular marker strings differ in their period lengths by a factor of 2. In this case the string with the longer periods is correct. (Usually this case is already prevented by the filtering steps. If the comb filter is adjusted to the second harmonic, it erases the PFS completely for the SSB and the even function so that no regular PMC string can result from these functions. See Figure 7.21b.)

b) Especially for band-limited signals the odd NLF may track a regular second or even third harmonic instead of the correct value. This case is not necessarily ruled out by the comb filter since the filter might be adjusted to even higher values than these harmonics due to a dominant higher harmonic in the SSB or even channels. In this case, however, the even and SSB channels substitute the first harmonic so that two marker strings are available whose

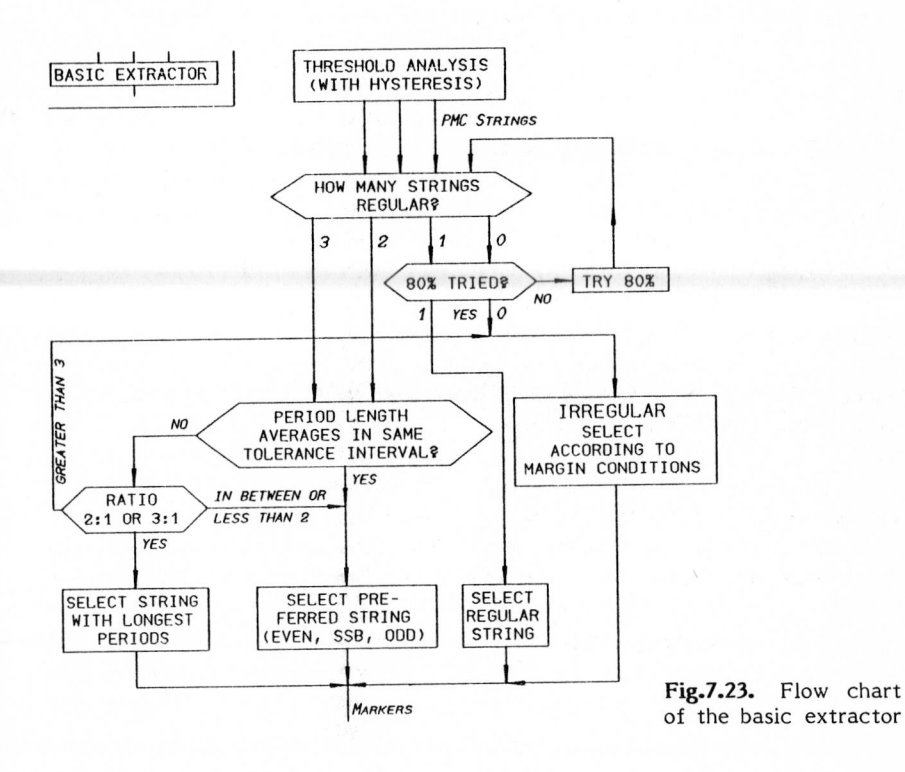

Fig.7.23. Flow chart of the basic extractor

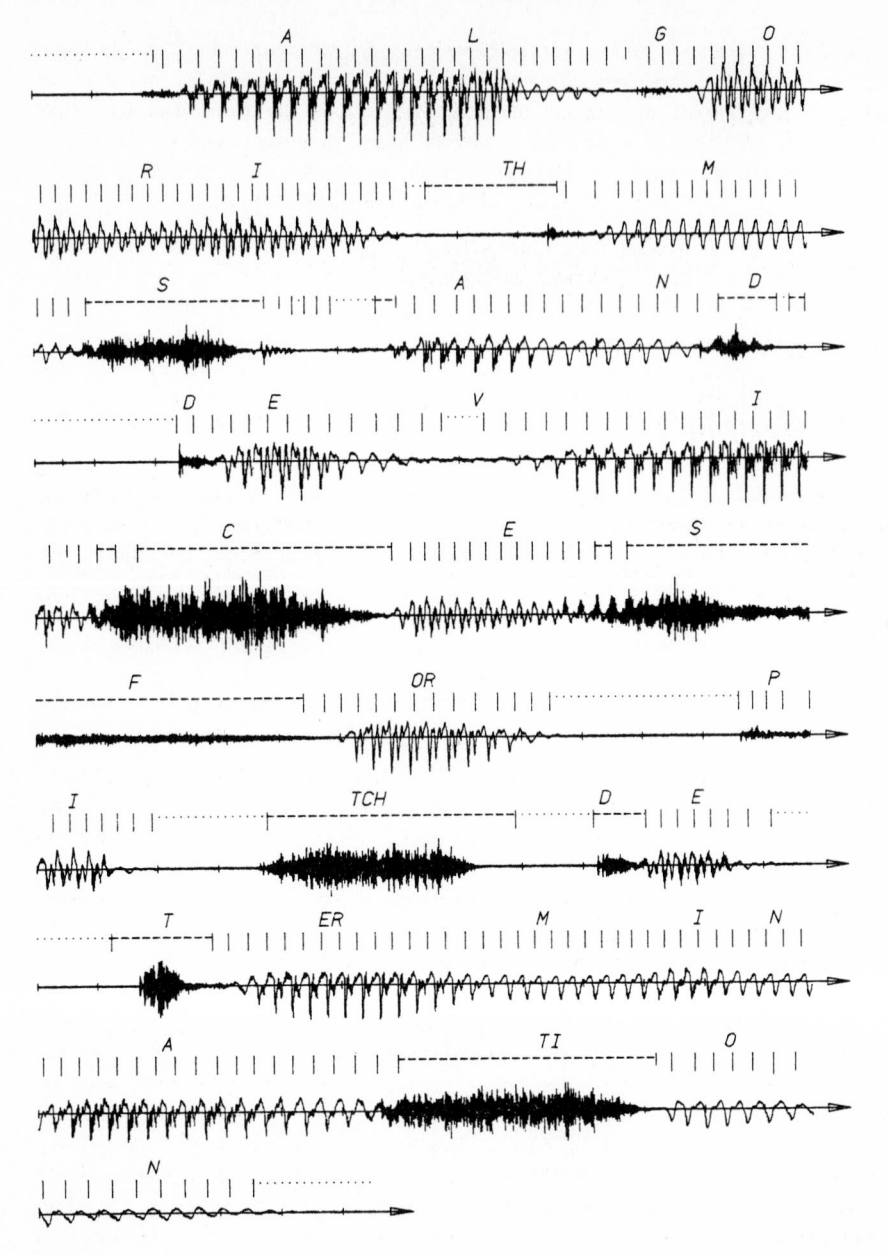

Fig.7.24. Example of a speech signal after pitch determination. Utterance: "Algorithms and devices for pitch determination" (same signal as in Fig.2.8); speaker: MIW (male); undistorted signal. (——) Voiceless; (·····) silence; scale: 320 ms per line. Markers displayed by short lines were classified irregular

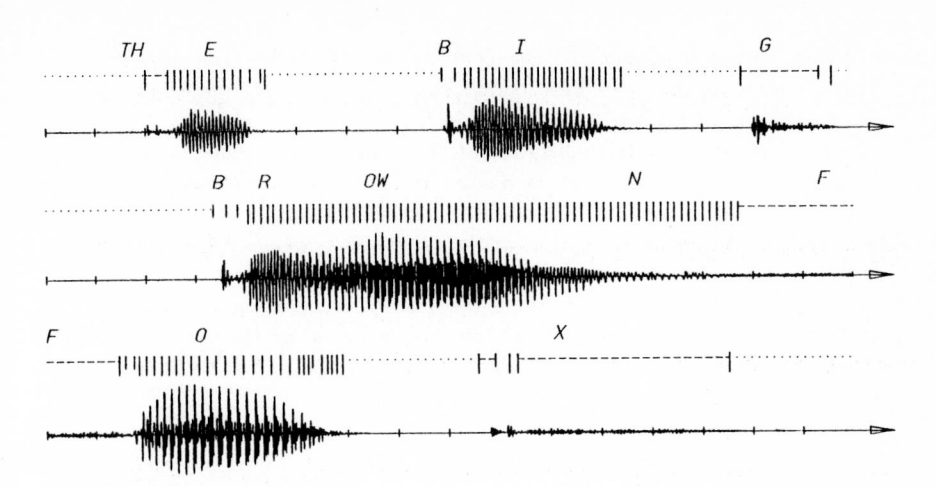

Fig.7.25. Example of a speech signal after pitch determination. Utterance: beginning of the sentence "The big brown fox jumped over the lazy dog;" speaker: JNC (female; fundamental frequency relatively low). The signal was band limited to telephone channel bandwidth. (-----) Voiceless; (.....) silence. scale: 500 ms per line. Markers displayed by short lines were classified irregular

period lengths take on a ratio of two to one. Again the string with the longer periods is the correct one.

Although these two errors have different sources, and although they occur at different NLFs, they affect the PMC strings in an identical way so that they can be ruled out by the same rule, i.e., by a minimum frequency (maximum period) discrimination routine, as has been proposed by various researchers (for details see Sect.6.4). The only real error might occur when for any reason whatsoever the correct PMC string is not regular. In this case, however, the irregularity is mostly confined to one single marker; this error can still be corrected by the 80% clause in rule 2.

With the selection and decision logic of the basic extractor the algorithm has produced a sequence of pitch markers which correspond to the starting points of the individual pitch periods. Few things remain to be done: mainly, the algorithm has to synchronize the markers at the frame boundaries where the choice of the NLF has changed. This task, however, is not exceedingly difficult. Since each of the three channels is expected to synchronize its markers with the proper beginnings of the periods, one could in principle concatenate marker strings from different channels without any further synchronization. In reality, however, there are phase shifts between the individual channels; this is mainly due to the fact that the markers in the odd channel are shifted back by half a period of F1 and are thus sensitive to measurement errors of this formant, whereas in the other channels the markers, once they have been determined, remain where they are. Problems can thus arise at boundaries where the channel selection switches from even or SSB to odd and vice versa; here the markers of one channel, preferably the odd one, must be shifted in such a way that they match the markers of the other channel at these boundaries.

Figure 7.22 shows a flow diagram of the algorithm; a more detailed flow diagram of the basic extractor is displayed in Figure 7.23. Examples of the performance can be found in Figs.7.24,25. In the present form the algorithm works without a preset over the whole range of possible pitch and for all categories of signals. Hence it can be classified as a *universal* PDA according to the definition in Section 7.2. Since the frames are processed independently, occasional errors occur especially at the beginning or at the end of a voiced section. These errors, however, usually do not cross the boundary of an individual frame.

8. Short-Term Analysis Pitch Determination

The short-term analysis PDAs leave the time domain by some short-term transform which characterizes the individual methods. They provide a sequence of average estimates of fundamental frequency F_0 or fundamental period T_0 which emerges from the short-term intervals (frames).

Table 8.1 and Fig.8.1 present an overview of the prevailing short-term analysis PDAs. These PDAs are further grouped according to the individual short-term transformation they apply. The sequence of operations is quite similar for all these PDAs. After an optional preprocessing step (which can be a moderate low-pass filter, an approximation of the inverse filter, or an adaptive center clipper), the signal is subdivided into short segments (frames) whose typical duration ranges between 20 and 50 ms. This value is chosen because most of the short-term transforms (with the only exception of the AMDF, see below) require several pitch periods within a single frame. After the formation of the frames, the short-term transform is performed on every frame individually. This transform is intended to perform in such a way that it concentrates all the available information on periodicity into a single principal peak which is then detected by a peak detector and labeled as the pitch estimate for that frame. For each transform, the figure shows a typical example (computed from the 50-ms frame displayed in the upper part) of the short-term function from which the estimate of pitch is derived. Not all the

Table 8.1. Grouping of short-term analysis PDAs. As in the previous chapters, labels are introduced to characterize the different PDAs

Domain	Label	Explanation
Correlation	PiC	
	PiCA	Autocorrelation
	PiCAL	... with linear adaptive preprocessing
	PiCAN	... with nonlinear preprocessing
	PiCD	Distance functions (AMDF)
Frequency	PiF	
	PiFH	Direct harmonic analysis
	PiFM	Multiple spectral transform (cepstrum)
Time	PiE	Periodicity estimation (maximum likelihood)

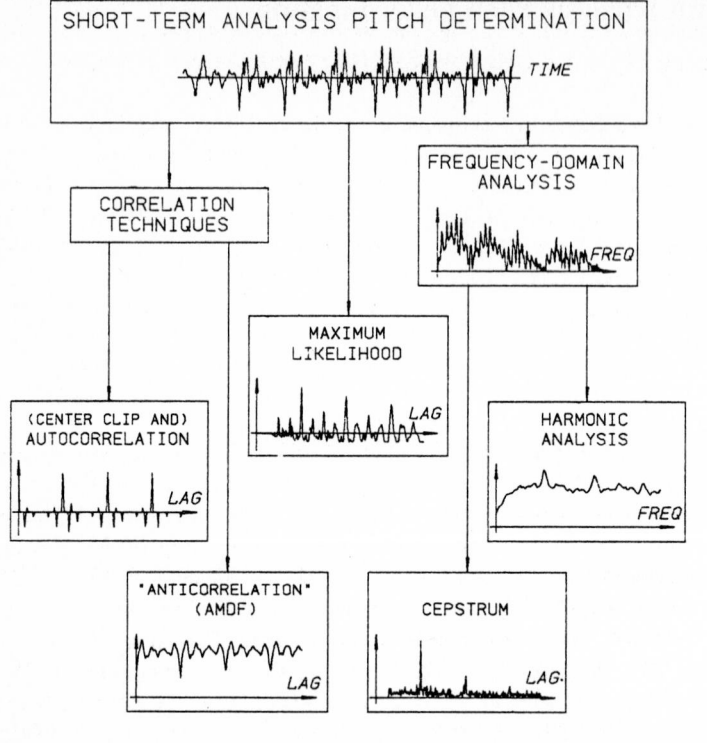

Fig.8.1. Grouping of short-term analysis PDAs. Signal: part of the sustained vowel /ɛ/, male speaker, undistorted recording. For definition of the term *lag*, see Section 8.2.1

known short-term transformations, however, behave in the desired way. Among those which do, we find the well-known short-term autocorrelation technique whose performance is greatly improved when the signal is preprocessed by center clipping. Autocorrelation PDAs will be discussed in Section 8.2. The counterpart of autocorrelation analysis is the average magnitude difference function (AMDF), which represents a special case of a distance function and is evaluated using nonstationary short-term analysis. PDAs that apply one of these principles will be dealt with in Section 8.3. The well-known cepstrum analysis (discussed in Sect.8.4) as well as harmonic analysis (Sect.8.5) are frequency-domain methods. Both are derived from the (logarithmic) power Fourier spectrum. The cepstrum PDA transforms this spectrum back into the time domain, thus generating a large peak at T_0. One of the several methods of harmonic analysis is spectral compression. The log power spectrum is compressed along the frequency axis by integer factors $k = 2, 3, 4$, etc. Adding all these compressed spectra causes the harmonics to contribute coherently to the distinct peak at F_0. The maximum likelihood PDA (to be discussed in Sect.8.6), finally, represents the mathematical procedure of detecting a periodic signal with unknown period T_0 in a noisy environment. A detailed discussion of the consequences of applying the short-term analysis principle will precede the survey of the individual algorithms and will form the first section of this chapter.

Short-term analysis PDAs are principally unable to track individual periods since the phase relations between the signal and the short-term function are lost during the transform. This loss, on the other hand, makes these PDAs insensitive to phase distortions of the signal. Moreover, since they focus all the information on periodicity into one single peak, they tend to be resistive against noise and signal degradation. The disadvantage of large computing effort is gradually losing its importance due to the current progress in fast digital circuitry.

8.1 The Short-Term Transformation and Its Consequences

The essential component of short-term analysis (STA) PDAs is the short-term transformation which is part of the preprocessor. The main properties of short-term signal analysis have been briefly reviewed in Chapter 2. The short-term transformation causes the time domain to be left in favor of other domains (frequency, autocorrelation, etc.).

The signal is divided into consecutive frames which are then further processed separately (see Fig.2.10). Within a single frame, one estimate of pitch is calculated which is representative of that frame. Hence, we do not obtain the boundaries of individual periods, not even the lengths of individual periods, but rather an estimate of the average period length or fundamental frequency within a given frame. To detect periodicity at all, at least two periods must be situated within one frame; otherwise the information of periodicity is lost. Thus a minimum frame length of two maximum-duration pitch periods must be observed. On the other hand the STA principle implies that the signal is quasi stationary, i.e., the extracted parameters can be assumed constant within the frame. Thus the frame length must not be too large, otherwise the natural change of pitch in the signal may become significant and may spoil the intraframe estimate of this parameter. For speech signals, these two conditions are just compatible; they do not yet really conflict. Usual values for the frame length range between 20 and 50 ms, according to the actual value of F_0 in the signal under consideration.

Parameter extraction is one form of data reduction. The parameter *pitch* is represented by the average period duration or fundamental frequency estimate for the individual frames. Hence this parameter, regarded as a signal, has a bandwidth much smaller than that of the original signal, and it can thus be sampled at a much lower rate, i.e., the frame rate. For pitch determination, the frame interval, as the reciprocal of the frame rate, takes on values between 10 and 40 ms. Seemingly this statement conflicts with the fact found in Chap.6 that a pitch contour has a bandwidth much greater than

25 to 100 Hz. This increased bandwidth, however, is due only to the discontinuities at the voiced-unvoiced transitions. If the parameter *pitch* is undersampled at a voiced-unvoiced transition in a short-term analysis PDA, this will mean that there exists an uncertainty of about half a frame interval as to the exact instant where voicing begins or ceases.[1] In contrast to the time-domain PDAs where low-pass filtering of the pitch contour smears the contour at the edges of voicing, the marginal values are not spoiled. If the PDA succeeds in getting an estimate in the last frame at the voiced-unvoiced boundary, it will get it correctly; if not, it tends towards complete failure, thus suggesting that the frame be labeled unvoiced.

The discussion up to now already suggests that the majority of the short-term analysis PDAs go digitally. In fact, the analog PDDs of this category - not very numerous anyhow - have mostly been outperformed by their digital "colleagues." Exceptions, such as the PDD by R.L.Miller (1970, see Sect.8.5.1), confirm the rule. The difference in effort between the analog realization of such a PDD and a time-domain one usually is too great. Owing to the possibility of low-rate parameter sampling, this difference for digital PDAs is much smaller, although the computing effort of most digital short-term analysis PDAs is still an order of magnitude greater than that of comparable time-domain PDAs. Again there are exceptions, such as the PDA by Dubnowski et al. (1976) or the one by Seneff (1978).

Most of the computing effort has to be put into the short-term transformation. From (2.32) we know that in the digital domain (presuming a finite window) most short-term transforms can be seen as a matrix multiplication of the signal vector $\mathbf{x}$ by the transformation matrix $\mathbf{W}$,

$$\mathbf{X} = \mathbf{W} \mathbf{x} , \qquad\qquad\qquad (2.32)$$

where $\mathbf{X}$ denotes the *short-term spectrum* of the transformation given by the matrix $\mathbf{W}$. For the following considerations we will label the vector $\mathbf{X}$ as the *spectrum* of $\mathbf{x}$ regardless of what the matrix $\mathbf{W}$ looks like.

The multiplication of the vector $\mathbf{x}$ by the matrix $\mathbf{W}$ is a mathematically simple procedure which is nevertheless associated with high computation effort, more or less due to the vast number of arithmetic operations, in particular multiplications, involved. The number of multiplications which determines the basic computational complexity to evaluate (2.32) is in the order of N^2 if N is the transformation interval, i.e., the length of the vector $\mathbf{x}$. There are several possible methods, mostly in connection with particular transformations, to reduce the computation effort.

1) A number of transforms permit a more effective implementation in the form of a "fast" algorithm. The matrix $\mathbf{W}$ is decomposed into a chain of partial matrices which are connected in a multiplicative way,

---

[1] In resynthesized speech undesirable clicks use to occur in consequence of these uncertainties (Rabiner et al., 1976; Gold, 1977).

$$\mathbf{W} = \mathbf{W}_{pk}\mathbf{W}_{p,k-1} \cdots \mathbf{W}_{p2}\mathbf{W}_{p1} . \qquad (8.1)$$

An example is the radix-2 FFT where $k = \operatorname{ld} N$ partial matrices are obtained if the transform interval N is an integer power of two. In these partial matrices, the majority of elements take on a value of zero so that, for a single matrix multiplication $\mathbf{W}_{pi}\mathbf{W}_{p,i-1}$, the number of multiplications is reduced to one per line and column (instead of N when all the elements are expected to be different from zero). If N is an integer power of two, the total number of multiplications is thus reduced from N^2 to $(N \operatorname{ld} N)$ where N is the number of samples in the frame, and $(\operatorname{ld} N)$ its dual logarithm. Among the transformations which allow for this fast algorithm, we find the well-known fast Fourier transform (FFT) as the fast variation of the discrete Fourier transform (G-AE Subcommittee, 1967). The usual price that has to be paid when applying such a fast algorithm is that some freedom of choice of the transform interval N is lost (see Sect.2.4 for more details).

2) A considerable reduction in computing effort is achieved if part or all of the multiplications can be omitted by setting the coefficients of $\mathbf{W}$ to 1, 0, or -1. Usual spectral transformations, such as the discrete Fourier transform, do not permit this simplification. After special nonlinear distortion (infinite clipping), however, the autocorrelation method can be varied in such a way that the multiplications are avoided (Dubnowski et al., 1976). The average magnitude difference function (AMDF, see Sect.8.3.1) also falls into this category. The AMDF is a short-term transformation that is not computed via (2.32); it does not need multiplications either.

3) On a digital computer calculations such as matrix multiplications are known as *array processing* or "stupid arithmetics." They do not need any decisions, and the address computations are simple. Thus a general purpose computer is too "intelligent" for this job and does it too slowly. Fast signal processors developed in recent times allow real-time computation of transformations like the FFT and make short-term analysis PDAs attractive even for real-time applications (Krasner et al., 1973; Seneff, 1978; Sluyter et al., 1980; Martin, 1981).

4) The transformation interval N greatly influences the computational effort since the number of multiplications, in the basic version, is proportional to N^2. Hence one should keep N as low as possible. However, N is a function of a number of external conditions. The most important condition, which must never be violated, is that pitch remains measurable in the spectral domain. One can directly apply the sampling theorem and postulate that the sampling rate in the spectral domain be such that the information on pitch is completely represented. For instance, in a frequency-domain PDA the spectral resolution must be better than half the lower-end frequency of the measuring range, i.e., less than 25 Hz. The requirement that pitch be determined with high accuracy - which particularly involves low quantization error in the measurement - often demands a spectral resolution which is much better (down to 1 Hz; see Sect.8.5). If this increased resolution is obtained directly in the short-term transform, it augments the computing effort by increasing

N. Here one must carefully check whether it is really necessary to provide the full measurement accuracy at once, or whether a basic accuracy may be increased by interpolation in the spectral domain or in the basic extractor. This issue is particularly important to frequency-domain PDAs (it will thus be further discussed in Sect.8.5.4); in PDAs that operate in the lag domain (see Sect.8.2.1) other method of effort reduction may be more effective.

5) If points 1 and 3 fail, partial computation of the transform may still help. The "fast" algorithms do not permit partial computation; they require all the spectral samples to be computed simultaneously. In the direct computation of (2.32), however, each spectral sample X(m) can be computed individually. Only those samples that are needed for pitch determination will then be computed. Using a good strategy, one can reduce the computational effort by as much as an order of magnitude.

Figure 8.2 shows the block diagram of a sample short-term analysis PDA. The short-term transformation may or may not be preceded by a linear or nonlinear filter that operates similarly to time-domain PDAs. In the spectral domain, some algorithms need further processing (weighting, absolute of complex spectral values, etc.). The basic extractor is nearly always a peak detector (Fig.8.3). Together with spectral-domain processing, one can say that the short-term transformation acts like a concave mirror; all the indications of periodicity, which are scattered over the whole frame in the original signal, are focused into the one peak (maximum or minimum, depending on the individual transformation). Hence, if periodicity is present, and if the signal is regular, there will be a strong indication of pitch which makes the algorithms of this category extremely reliable, especially under bad recording conditions such as noisy environment or reverberation. In particular, pulse-like

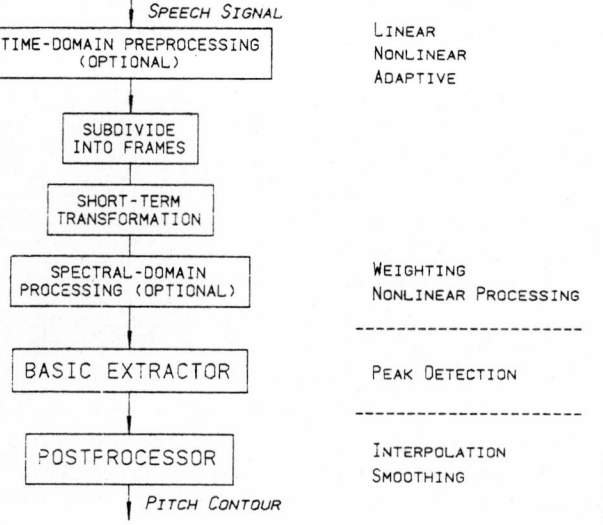

Fig.8.2. Block diagram of a sample short-term analysis PDA

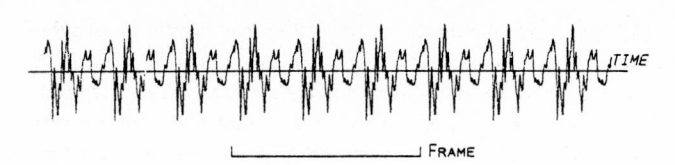

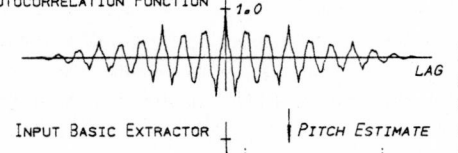

Fig.8.3. Basic extractor of a short-term analysis PDA: principle of operation. Example: autocorrelation. For the definition of the (short-term) autocorrelation function, see Sect.8.2.1

spurious signals are tolerable as long as they do not exceed the length of a whole frame.[2] In addition, these algorithms are insensitive to phase distortions and band limitation.

On the other hand, immediate failure can occur when the speech signal itself is momentarily irregular. Consider a local perturbation present in form of an extra laryngeal pulse (Fig.8.4). Trying to predict what happens in the PDA, one can think of three possibilities which depend on the ratio of fundamental period duration T_0 and frame length K.

1) The frame is short; it contains only the perturbated period(s) and the beginning of the consecutive one (Fig.8.4a). In this case the algorithm acts like a time-domain PDA and follows the perturbation.

2) The frame is somewhat longer so that it contains the perturbated period(s) and an about equal number of correct ones. The PDA then cannot decide within that single frame whether the period length of the perturbated signal or the length of the regular period is correct (Fig.8.4b). Presumably it will indicate both of them, thus significantly reducing the amplitude of the principal peak. Most of the short-time spectral functions used in PDAs, however, yield a reference relative to which the real strength the indicated periodicity can be judged. Short-term analysis PDAs thus usually apply a threshold relative to the reference peak (for autocorrelation analysis, for example, the peak at zero lag may serve as a reference; see Fig.8.3) which must be exceeded in order to be recognized as an indication of pitch. If the (reduced) amplitude remains below that threshold, there must be something wrong with the signal in the frame under consideration.

3) The frame is long enough to contain more than one regular period besides the perturbated one. In this case the regular periods overrule the

---

[2] Every time-domain PDA would fail with such signals unless it corrects the error by some external correction routine, such as list correction, or nonlinear contour smoothing.

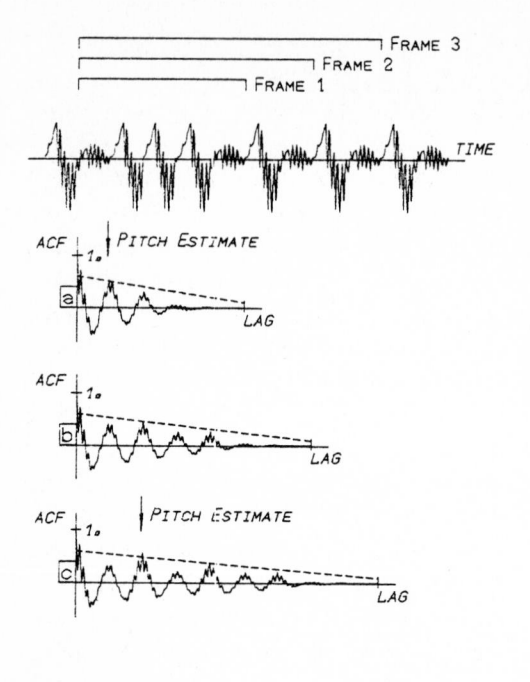

Fig.8.4a-c. Typical performance of a short-term analysis PDA with perturbated signals (depending on the frame length). **(a)** Short frame: the influence of the perturbated period prevails; **(b)** medium length frame: ambiguous situation in the basic extractor; **(c)** long frame: the influence of the correct periods prevails. Signal: One period of the vowel /i:/, male speaker, repeated. Perturbation by doubling the repetition rate. For the definition of the term *lag*, see Sect.8.2.1

perturbated one, and the perturbation will not be registered; it is smoothed out (Fig.8.4c).

As this comparatively long initial discussion shows, short-term analysis PDAs have a number of common features which separate them from time-domain PDAs. The last of these features to be discussed here is the loss of the phase relations between the signal and the pitch estimate during the short-term transformation. Thus, on the one hand, the PDA is insensitive to phase distortions; on the other hand, it yields an indication of pitch which has nothing to do with the individual instants of glottal excitation, even when the estimate is given in time-domain terms, i.e., in the form of fundamental period duration. Hence, the short-term analysis PDA is in principle unable to locate pitch markers. If nevertheless forced to do so, it will place them in an arbitrary way, certainly at a correct distance, but without any real relation to the actual signal.

In the following sections, we will discuss the individual algorithms grouped according to the short-term transformation applied: *correlation* and *"anticorrelation"* (distance function) algorithms, *frequency-domain* algorithms including those which apply a double short-term transformation, like the cepstrum, and, finally, *direct periodicity estimating* algorithms such as the maximum-likelihood approach.

8.2 Autocorrelation Pitch Determination

Correlation is a measure of similarity, i.e. of the degree of agreement between two input functions. In the case of autocorrelation both input sequences are derived from the same original signal: the input signal is correlated with itself. Between the two input channels there is the delay (lag) d which forms the parameter of the autocorrelation function (ACF). If the signal is periodic or quasi periodic, there are great similarities, i.e., high correlation coefficients when the lag d equals one period or a multiple thereof. This principle looks simple, nearly trivial, so that the earliest experiments with short-term analysis PDAs were made in this domain. On the other hand, this approach raises delicate detail problems which had to be overcome by sophisticated preprocessing methods. The ACF is the inverse Fourier transform of the power spectrum of the signal. That means, on the one hand, that all the complications due to unknown phase relations have been overcome since the phase is forced to zero. On the other hand, if there is a distinct formant structure in the signal, it is maintained in the ACF. Thus, one is not safe from such well-known errors as higher harmonic or subharmonic detection.

The section starts with a comparatively broad discussion on the ACF itself, in particular with respect to short-term analysis. A short survey of the historical development of the application of AC methods in pitch detection follows. The remainder of the section deals with the more recent autocorrelation PDAs which compensate for the shortcomings of the ordinary ACF (which retains the formant structure) by elaborate preprocessing methods. Two competitive methods were found to be the most powerful ones: 1) nonlinear preprocessing, i.e., center clipping, and 2) adaptive linear preprocessing, i.e., inverse filtering. Both these methods greatly improve the performance of the method, particularly since there are arithmetically simple implementations which cut down the computing effort drastically.

8.2.1 The Autocorrelation Function and Its Relation to the Power Spectrum

The correlation function[3] of two sampled signals u(n) and v(n) is defined as follows:

$$\text{corr}(d) = \lim_{N \to \infty} \frac{1}{2N+1} \sum_{n=-N}^{+N} u(n)\, v(n+d) \ . \tag{8.2}$$

---

[3] Since all the signals are regarded as digital, the autocorrelation function will also be a digital signal. Hence it appears more appropriate to speak of an autocorrelation *sequence* instead of an autocorrelation *function* (Oppenheim and Schafer, 1975) in order to emphasize its discreteness. Most technical papers on autocorrelation PDAs, however, use the term *function.*

The special case of autocorrelation (AC) is given when the two input signals are identical,

$$r(d) = \lim_{N \to \infty} \frac{1}{2N+1} \sum_{n=-N}^{+N} x(n)\,x(n+d) , \qquad (8.3)$$

no need to be finite energy

provided that u(n), v(n), and x(n) are defined for all n. The parameter d is the *lag* or *delay time* between the immediate and the delayed signal. The properties of the autocorrelation function (ACF) are well known (Lee, 1960; Schroeder and Atal, 1962). For analog systems the relation corresponding to (8.3) holds provided that the time-domain signal x(t) has finite power (Schroeder and Atal, 1962). This means that, for digital systems, (8.3) holds when the samples of x(n) have finite amplitudes for all values of n.

If the signal x(n) is of finite duration, i.e., if the samples of x(n) are zero outside an interval n = 0 (1) N-1, then, for positive values of the delay d, (8.3) is written as follows,

$$r_N(d) = \frac{1}{N} \sum_{n=0}^{N-d-1} x(n)\,x(n+d) . \qquad (8.4)$$

The upper summation index in (8.4) is given as (N-d-1), since for n = N-d the index of the second term in (8.4) is already greater than the nonzero range of x(n). For the finite signal of (8.4), the autocorrelation function can be expressed according to (2.32) so that its matrix representation would be

$$
\begin{pmatrix} r(0) \\ r(1) \\ r(2) \\ \vdots \\ r(N-1) \end{pmatrix}
= \frac{1}{N}
\begin{pmatrix}
x(0) & x(1) & x(2) & \cdots & x(N-1) \\
0 & x(0) & x(1) & \cdots & x(N-2) \\
0 & 0 & x(0) & \cdots & x(N-3) \\
\vdots & \vdots & \vdots & \vdots & \vdots & \vdots \\
0 & 0 & 0 & \cdots & x(0)
\end{pmatrix}
\begin{pmatrix} x(0) \\ x(1) \\ x(2) \\ \vdots \\ x(N-1) \end{pmatrix}
. \qquad (8.5)
$$

Since the ACF is an even function, one can rewrite (8.4,5) extending the ACF towards negative values of the lag d. In this case, we obtain

$$\mathbf{r}_N = [r(-N+1) \ \ldots \ r(0) \ \ldots \ r(N-1)]^T$$

$$= \frac{1}{N}\,\mathbf{W}_{AC}\,[0 \ \ldots \ 0 \ x(0) \ x(1) \ \ldots \ x(N-1)]^T \qquad (8.6a)$$

with the transformation matrix

$$\mathbf{W}_{AC} =$$

$$\begin{pmatrix}
x(0) & x(1) & x(2) & \text{.......} & x(N-2) & x(N-1) & 0 & \text{.......} & 0 & 0 \\
0 & x(0) & x(1) & \text{.......} & x(N-3) & x(N-2) & x(N-1) & \text{.......} & 0 & 0 \\
0 & 0 & x(0) & \text{.......} & x(N-4) & x(N-3) & x(N-2) & \text{.......} & 0 & 0 \\
\vdots & \vdots & \vdots & \vdots\ \vdots & \vdots & \vdots & \vdots\ \vdots & \vdots \\
0 & 0 & 0 & \text{.......} & 0 & x(0) & x(1) & \text{.......} & x(N-1) & 0 \\
0 & 0 & 0 & \text{.......} & 0 & 0 & x(0) & \text{.......} & x(N-2) & x(N-1) \\
x(N-1) & 0 & 0 & \text{.......} & 0 & 0 & 0 & \text{.......} & x(N-3) & x(N-2) \\
\vdots & \vdots & \vdots & \vdots\ \vdots & \vdots & \vdots & \vdots\ \vdots & \vdots \\
x(2) & x(3) & x(4) & \text{.......} & 0 & 0 & 0 & \text{.......} & x(0) & x(1) \\
x(1) & x(2) & x(3) & \text{.......} & x(N-1) & 0 & 0 & \text{.......} & 0 & x(0)
\end{pmatrix}.$$

$$(8.6b)$$

This represents a (cyclic) convolution of the signal $x(n)$, $n = 0\,(1)\,N\text{-}1$, extended by zeros until length $2N$ (in order to enable the computation of the ACF by cyclic convolution without overlap), and the same signal in reverse order of index, i.e., $x(n)$, $n = N\text{-}1\,(\text{-}1)\,0$. According to the convolution theorem for discrete convolution,[4] $r(d)$ can be computed via the discrete Fourier transform,

$$\mathbf{r} = \mathbf{W}^{-1}\ [\ (\mathbf{Wx})\ \cdot\ (\mathbf{W}^{-1}\mathbf{x})\ ] = \mathbf{W}^{-1}\ (\mathbf{X}\cdot\mathbf{X}^{*}) = \mathbf{W}^{-1}\ |\mathbf{X}|^{2} \qquad (8.7)$$

In this equation, $\mathbf{W}$ denotes the transformation matrix of the discrete Fourier transform, as defined in (2.32); $\mathbf{W}^{-1}$ denotes the inverse of the matrix $\mathbf{W}$ and thus the transformation matrix of the inverse DFT; $\mathbf{x}$ is the signal vector as defined in (8.6), $\mathbf{X}$ represents its (complex) DFT spectrum, and $\mathbf{X}^{*}$ is the conjugate complex spectrum. The multiplication point signifies that an element-by-element multiplication of the two vectors has to be performed; the other multiplications indicated in the equation are matrix multiplications. Equation (8.7) is the digital representation of the well-known fact that the autocorrelation function is the (inverse) Fourier transform of the power spectrum of the signal $x(n)$.

Sometimes in the literature another definition of the autocorrelation function is given (Schroeder and Atal, 1962),

$$\phi(d) = \sum_{n=-\infty}^{\infty} x(n)\,x(n+d)\ . \qquad (8.8)$$

Should be of *finite energy.*

---

[4] For details, see a textbook on digital signal processing, such as Rabiner and Gold (1975), Oppenheim and Schafer (1975), Achilles (1978), or Kunt (1980).

This definition requires that the signal be of finite energy, i.e., that the sum

$$\phi(0) = \sum_{n=-\infty}^{\infty} x^2(n) \qquad (8.9)$$

be finite. Equation (8.8), in connection with (8.9), is valid for infinite signals as well as for finite signals.

The ACF r(d), according to definition (8.3), has its highest peak at $d = 0$ which equals the average power of the signal x(n); for a finite signal, according to (8.4), the peak indicates the average power during the interval $n = -N(1)N$,

$$r(d) \leq r(0) . \qquad (8.10)$$

If the signal x(n) is periodic with the period duration T_0, its ACF will also be periodic,

$$r(d+kT_0) = r(d) \quad \text{if} \quad x(n+kT_0) = x(n); \quad \text{k integer} . \qquad (8.11)$$

Together with (8.10), (8.11) shows that the ACF of periodic signals has a significant peak at $d = kT_0$,

$$r(kT_0) = r(0) . \qquad (8.12)$$

For a quasi-periodic signal (8.12) holds approximately so that the peaks at $d = kT_0$, especially the one at $d = T_0$, will have an amplitude comparable to that at $d = 0$.

To process nonstationary signals like the speech signal, we can define a short-term ACF as follows,

$$r(d,q) = \sum_{n=q}^{q+N-1} x(n) x(n+d) w(n-q) . \qquad (8.13)$$

In this equation q is the starting sample (the *address*) of the frame under consideration, whereas w(i) is a window which forces the signal to zero outside the interval $i = 0(1)N-1$ and takes on arbitrary (but finite) values within this interval. N is the number of samples used for the computation of r(d,q); this number is not necessarily equal to the window length K. Note that in the strict sense of (8.3) this function is not an autocorrelation but an *autocovariance* function since the two input signals, although they have the same origin, are not identical. In particular, the relation (8.10) will become invalid. For segments with rising intensity, this could lead to spuriously high peaks at multiples of the period duration T_0, causing subharmonic detection in fundamental period analysis as a consequence. This effect is avoided in the following definition of the short-term ACF (Rabiner, 1977a):

$$r(d,q) = \sum_{n=q}^{q+N-d-1} x(n)w(n-q)\ x(n+d)w(n+d-q) \qquad (8.14a)$$

$$= \sum_{i=0}^{N-1-d} [x(q+i)w(i)]\ [x(q+i+d)w(i+d)]$$

$$= \sum_{i=0}^{N-1} [x(q+i)w(i)]\ [x(q+i+d)w(i+d)] \qquad (8.14b)$$

with $w(i)= 0$ for $i = 0\,(1)\,N-1$.

With this definition, which is implicitly given and always obeyed in Rabiner's fundamental paper (1977a), the short-term ACF again fulfils the basic definition (8.4) for the autocorrelation function of finite signals. Equation (8.14) applies the *stationary* principle of short-term analysis, whereas the approach defined by (8.13) is nonstationary.

To demonstrate the difference between definitions (8.14) and (8.13), Fig.8.5 shows the ACF according to both definitions for a periodic signal and for a signal with rising intensity. Definition (8.14) has the effect of gradually tapering the autocorrelation function to zero when the lag d increases (since fewer and fewer samples of $x(n)$ are involved in the computation). This can have an undesirable effect on fundamental period determination when strong secondary peaks due to the formant structure are present. Since that influence is predictable, however, possible higher harmonic tracking due to this error can be avoided.

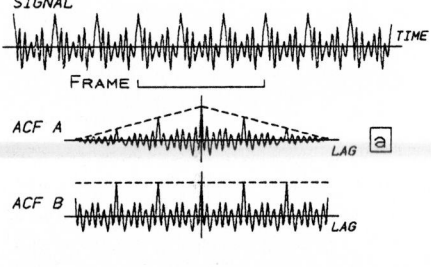

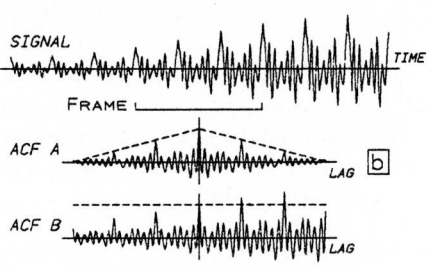

Fig.8.5a,b. Two definitions of the short-term autocorrelation function. (a) Periodic signal, (b) quasi-periodic signal with rising intensity. (ACF A) Stationary short-term analysis; autocorrelation function according to (8.14); (ACF B) non-stationary short-term analysis, autocorrelation (more exactly: autocovariance) function according to (8.13)

The independent variable of the ACF is time. AC analysis, on the other hand, is not a time-domain analysis in the sense of Chap.6 since running time, i.e., the independent variable of the signal x(n), has been left at the moment the signal is subdivided into frames starting at samples k·q, whatever these values are. To avoid confusion, the independent variable of the ACF will henceforth be called *autocorrelation lag*, or simply *lag*.[5]

According to the above discussion, the main task of the basic extractor is to detect the significant peak of the ACF at $d = p$, when p equals the fundamental period duration. This is expected to be the highest peak within the range of possible pitch (Fig.8.3). As stated in the general discussion on STA-PDAs, most short-term functions used in pitch determination yield a reference which may help to discriminate between voiced and unvoiced segments or between regular and irregular segments. Here, this reference is given by the ACF at $d = 0$, and it is usual to calculate the ACF relative to $r(0)$ thus confining its samples to values between +1 and -1. Using this reference, a threshold can be defined in order to label a frame unvoiced if the highest peak of the ACF for a nonzero lag d does not exceed this threshold. On the other hand, a frame can be labeled irregular if an appropriate threshold is exceeded by more than one peak.

In the following discussion, we will first briefly dwell on the analog realizations which have been largely suppressed by their digital successors because of the complicated way how an analog correlator has to be realized. After that the "ordinary" AC analysis, as already outlined above, will be further considered. A discussion of the various preprocessing methods proposed will conclude this section.

8.2.2 Analog Realizations

In analog systems the autocorrelation function is defined as

$$\phi(d) = \lim_{T \to \infty} \frac{1}{2T} \int_{-T}^{T} x(t)\, x(t-|d|)\, dt \qquad\qquad (8.15)$$

and, correspondingly, a short-term ACF as

$$f(d,t_0) = \int_{-\infty}^{t_0} x(t)\, x(t-|d|)\, w(t_0-t)\, dt \ . \qquad\qquad (8.16)$$

The weighting function w(t) is defined to be zero for negative values of t. The main obstacle in the way of an analog implementation is the fact that the calculation depends on two independent variables: the lag d and the integration variable t; t_0 need not be specially regarded since it represents

---

[5] The term *lag* will also be used in conjunction with other short-term transformations where a spectral function has time as its independent variable (cf. Fig.8.1 and the definition in Sect.2.4).

running time. To calculate $\phi(d,t_0)$ within finite time, we have to sample the signal for one of these variables, preferably for the lag d since the signal x(t) itself is present as a continuous signal. The block diagram of an analog correlator realization is given in Figure 8.6. Since (8.16) represents a convolution integral, the short-term ACF is obtained at the output of a low-pass filter whose impulse response equals the window w(t). $\phi(d,t_0)$ is only computed for one value of lag d at a time. To sample the ACF with respect to the lag d, two ways are possible:

1) *Sequential processing*, as proposed by Stevens (1950). This requires an analog memory (in form of a magnetic tape loop) and a variable delay line. On-line processing is prohibited since the loop must be completely scanned for each value of d.

2) *Parallel processing* using a tapped delay line. This requires the configuration of Fig.8.6 to be built up separately for each value of d for which the ACF is to be obtained. If such a PDA is to serve for any voice, it has to have at least 60 channels (Baronin, 1974a), each one connected to a tap of the delay line, and equipped with an analog multiplier and a low-pass filter to evaluate r(d).

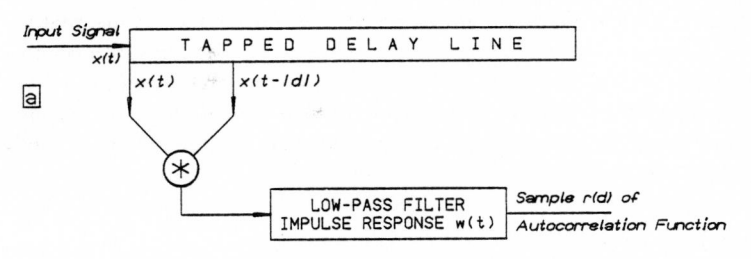

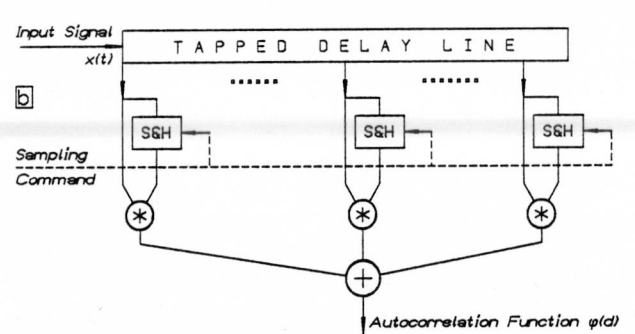

Fig.8.6a,b. Analog evaluation of the short-term autocorrelation function (ACF): block diagrams. **(a)** Evaluation of the ACF for a single value of the delay d when the ACF is sampled (after Stevens, 1950); **(b)** evaluation when the input signal is sampled (after Baronin, 1974a). (S&H) Sample-and-hold circuit; [w(t)] time-domain weighting function

Except for the PDD by Raisbeck (1959) and a number of investigations in the Soviet Union (Pirogov, 1958, 1963; Baronin, 1961; Sapozhkov, 1963) no PDD using this principle has been published. The PDD by Miller and Weibel (1956), realized in a similar way, uses a difference function as the criterion for pitch period indication (see the description in Sect.8.3.1). Since the analog realization has proved too complicated, the PDDs of this type were realized in a digital way even when no computer was used.

Analog evaluation of the ACF requires time *or* lag to be sampled in order to be able to realize the PDD. Although sampling the lag d is straightforward, this is not at all necessary; on the contrary, it is advantageous to sample the input signal and not the ACF. Such a PDD has been realized by Baronin (1964, 1974a) and Sobolev and Baronin (1968). As mentioned in Sect.8.1, the computational effort (as well as, in this case, the hardware effort) is often due to the requirements on measurement accuracy and is not necessary in order to correctly represent the information on pitch in the spectral domain. If the delay d is quantized, i.e., if the ACF is sampled, the sampling rate directly influences the accuracy of the measurement. If the input signal is sampled, however, one can evaluate the ACF without sampling it, thus being able to maintain the basic accuracy. The principle, which has some similarities with the digital realizations by Gill (1959) and by Stone and White (1963), works as follows (Fig.8.6b). The signal is low-pass filtered (in Baronin's PDD the cutoff frequency was 800 Hz) and fed into a tapped delay line. This time the delay line only serves to delimit the current frame. At the moment t = q the momentary signal value at all taps is sampled and stored in the sample-and-hold circuits. The stored value is then multiplied by the momentary value at the tap, and the output signals of the multipliers at all taps are added. As time goes on, the output signal of the adder directly is the desired autocorrelation function $\phi(d)$; the current value of d equals the difference between the current time t and the instant q where the sample-and-hold circuits were loaded. In this case $\phi(d)$ is not sampled although it is being evaluated using a sampled input signal. When the signal is low-pass filtered before entering the delay line, sampling can be rather crude. In Baronin's PDD the signal was sampled at 1.65 kHz so that 20 channels were sufficient for a 13-ms frame. (Since this PDD implicitly applies the nonstationary principle of short-term analysis, the effective frame length is d + 13 ms.)

Baronin proposed a simplification; evaluating $\Sigma\, x(t)\, sgn\,[x(t)]$ instead of the true autocorrelation function, he could replace the sample-and-hold circuit by a flipflop which just stores the sign of x(t) at t = q, and the multiplier by a simple switch controlled by the flipflop. Nevertheless, this PDD failed; however, it did not fail due to the fact that the effort was too great or that the implementation was unreliable; it failed just because of the basic problem arising when the ACF is taken for pitch determination without suitable preprocessing (see the discussion in Sects.8.2.3,4). A corresponding PDD using the AMDF (see Sect.8.3.1) was more successful.

8.2.3 "Ordinary" Autocorrelation PDAs

The term *ordinary* (used by Rabiner, 1977a) says that for this analysis the incoming speech signal is more or less unprocessed except perhaps some low-pass filtering in order to reduce the effects of higher formants. Digital PDAs of this type were reported by Fujisaki (1959, 1960, 1963), Sugimoto and Hashimoto (1962), David and Pierce (1963), and Maezono (1963). The results of this ordinary AC analysis were not very encouraging, and consequently Schroeder (1970d) judged with particular respect to the PDDs by Raisbeck and by Miller and Weibel:

> The results of this well-meant application of detection theory were most discouraging: the performance was often worse than of existing simpler devices. ... The most common error of pitch detection by autocorrelation analysis is that it often gives a value for the fundamental period of $T_0 \pm T_F$ rather than T_0, where T_0 is the true fundamental period and T_F the period of a major formant. This type of error is easily traced to the fact that the position of the major peak in the short-time autocorrelation function is often influenced by the formants, particularly during rapid pitch variations. (Schroeder, 1970d, p.707)

Researchers thus readily arrived at proposals using some more sophisticated preprocessing in order to reduce the influence of the formants, especially of the first formant.

8.2.4 Autocorrelation PDAs with Nonlinear Preprocessing

There have been two "generations" of digital PDAs (PiCAN) using nonlinear preprocessing. The first generation, in the late fifties and early sixties, was realized digitally in order to yield a computationally simple solution which could lead out of the calamity with the analog ACF realization. Among these PDAs we find the one by Gill (1959) and the one by Stone and White (1963, again applying a distance function). After that, this principle was not further investigated. The second generation started with the paper by Sondhi (1968) who, according to a proposal by E.E.David (after Schroeder, 1970d), presented new methods for successful AC pitch determination. This paper anticipated the breakthrough of this analysis which then culminated in the recent digital hardware PDDs by Gillmann (1975), Dubnowski et al. (1976), and the fundamental and possibly final investigation by Rabiner (1977a).

Let us consider the first generation for a moment. The PDD by Stone and White (1963) applies a difference function rather than a correlation and thus is to be dealt with in Section 8.3. The PDD by Gill (1959), in contrast, applies a nonlinear autocorrelation analysis principle. This PDD is remarkable for its technical realization. Gill implemented a four-channel principle using four different preprocessors: 1) 400-Hz low-pass filtering, 2) 400- to 1000-Hz band-pass filtering, and 3) and 4) rectified SSBM of two different subbands of the speech signal. The correlation circuit is displayed in Figure 8.7. The signal is infinitely clipped and sampled (1 bit) every 180 μs. The resulting 1-bit sample is then entered into a magnetostriction delay line which acts as

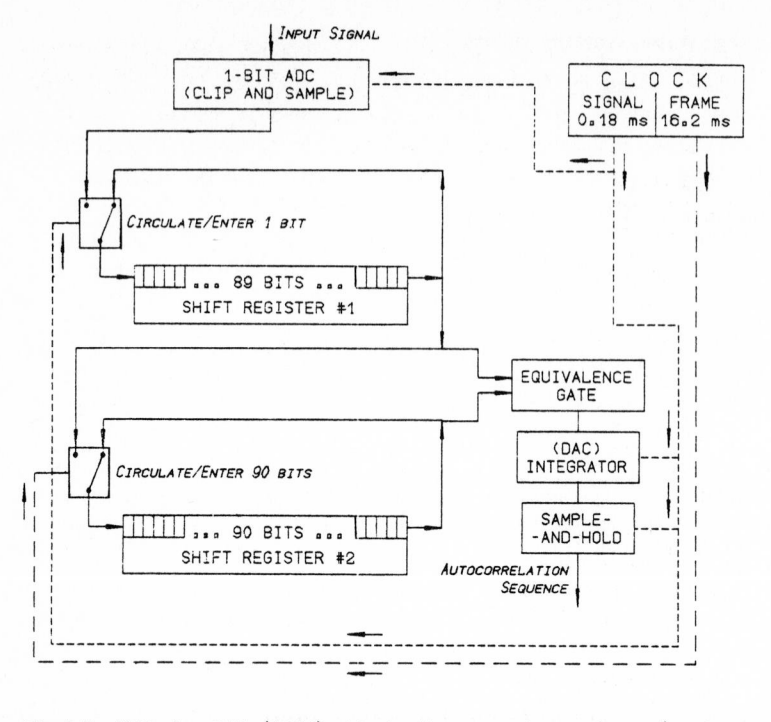

Fig.8.7. PDD by Gill (1959): block diagram of correlator (analog realization!)

a dynamic shift register (register 1 in the figure), and which contains the last 89 samples of the clipped input signal. The contents of this register are correlated (equivalence gate and integrator in tandem) to the contents of the data reference register (register 2) which is one sample longer (90 bits). The clock frequency of the two registers is adjusted in such a way that the data reference register performs one complete cycle of the data every 180 µs (i.e., the sampling rate of the input signal). This means that one sample of the ACF is calculated within that time. Consider now the first 89 bits of the two registers to be identical. For the current cycle the circuit first calculates the ACF for zero lag. In the following cycle the last bit of the input sequence is replaced by the actual new one. The signal of the data reference register then appears delayed by one sample with respect to register 1. The total short-term ACF [according to definition (8.13)] is thus calculated within 90 cycles (16.2 ms). As soon as the calculation of the ACF for the current frame has been completed, the data in the reference register (register 2) is replaced by the contents of the shift register (register 1) as a whole, and the circuit starts to calculate a new frame.

   At this point, let us jump ahead almost 20 years and have a look at the technical realization of the hardware PDA by Dubnowski, Schafer, and Rabiner (1976). The comparison shows that the two realizations are in prin-

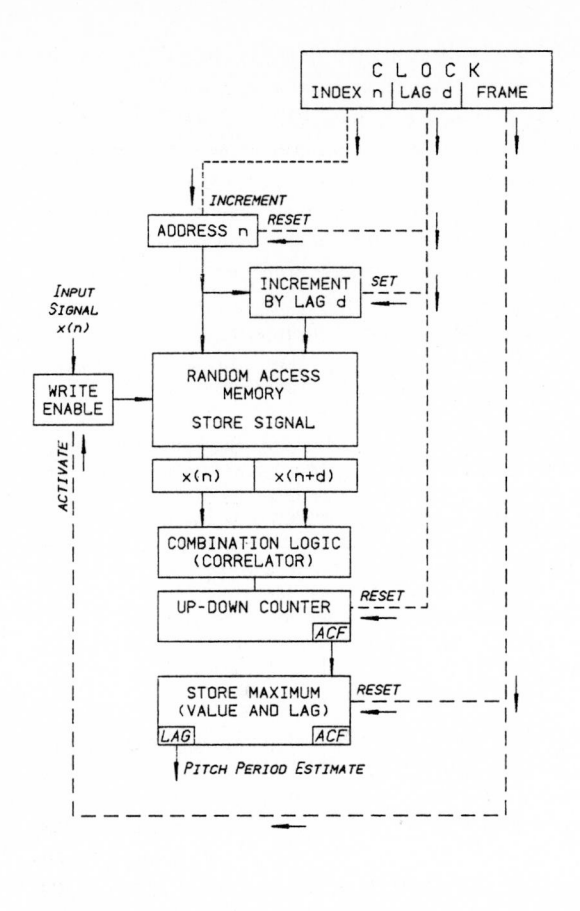

Fig.8.8. PDA by Dubnowski et al. (1976): block diagram of the correlation circuit

ciple not as far from each other as one might imagine at first glance. In the PDA by Dubnowski et al. the actual frame is stored in a register which is realized as a 512×2-bit random access memory (RAM). More flexible than a mere shift register, the RAM permits the computation of the ACF to be confined to those values of the lag d which are really needed for determination of the fundamental period. In this approach, two bits per sample are needed since every sample is permitted to take on the values 1, 0, and -1 (Fig.8.8). The combination logic, which replaces the multiplicative part of the correlation, drives an up-down counter where the actual value of r(d) is calculated; this counter performs the additive part of the arithmetic. Together with the accompanying value of the lag d, the maximum value of the ACF is stored in the maximum ACF register which, finally, contains the actual estimate of T_0. From the functional point of view, and apart from the technological differences, these two correlators are not too dissimilar. And yet there are large differences in performance. Gill's PDD needs four channels to cope with all signals possible in the "real world"; that of Dubnowski et al. needs only one channel, and systematic failures are not reported.

The difference results from the way the signal is preprocessed. To explain this, let us go back to Sondhi (1968). As known from previous attempts (see the citation by Schroeder in the previous section), the shortcomings of ACF pitch determination result from the formant structure, especially from the first formant. Hence Sondhi suggested spectral flattening as a possible way out of this problem. Spectral flattening is a technique which equalizes the speech spectrum in such a way that the spectral peaks (which represent the formants of the speech signal) are removed. Sondhi proposed two methods of spectral flattening, one linear and the other nonlinear. He described the linear method first and found that, once the formant structure has been removed, AC pitch determination works extremely well.[6] He then turned to the non-linear technique of adaptive center clipping:

> Autocorrelation of a speech signal, after removal of the formant struc-ture, is a powerful method of pitch extraction. We have successfully tried another way of removing formant structure, namely, center clipping. ... Center clipping of speech was first used by Licklider and Pollack (1948) in an experiment in which they showed that, whereas speech that has been infinitely peak clipped is highly intelligible, even a few percent of center clipping drastically reduces intelligibility. The explanation is not hard to find. Whereas infinite peak clipping retains the formants of the speech signal, ... center clipping destroys formant structure, while retaining the periodicity. It is the removal of the formant structure that is so important for a good pitch extractor (Sondhi, 1968, p.264)

Eight years later, the PDA by Dubnowski et al. (1976), which was mentioned above with regard to a technical detail of its realization, successfully com-bined the knowledge about center clipping resulting from Sondhi's paper and the computational simplification achieved by infinite clipping as realized by Gill. The signal is low-pass filtered to 900 Hz bandwidth and sampled at a sampling rate of 10 kHz. After that the signal is subdivided into frames of 30 ms duration; the frame interval is set to 10 ms. The characteristic of the subsequent clipping operation is shown in Fig.8.9 (function "sgc"); an example of the performance is given in Figure 8.10 later in this section.

> The first stage of processing for each section is the computation of the clipping level for that section. Because of the wide dynamic range of speech, the clipping level must be carefully chosen so as to prevent loss of waveform information when the waveform is either rising in amplitude or falling in amplitude during the section. ... The way in which the clip-ping level is chosen is as follows. The 30-ms section of speech is divided into three consecutive 10-ms sections. For the first and third 10-ms sections the algorithm finds the maximum absolute peak levels. The clip-ping level is then set as a fixed percentage of the *smaller* of these two maximum absolute peak levels. (Dubnowski et al., 1976, p.3)

---

[6] This method will be discussed in the next section together with other PDAs that apply linear preprocessing.

For this fixed percentage, a value of 80% has experimentally been found appropriate. The other components of the PDA have already been discussed. More details on the performance of this PDA will be presented in Sect.9.3.1 since it was included in the comparative evaluation by Rabiner et al. (1976). As a main cause for occasional gross errors the authors mentioned the fixed frame length which proved to be too small for extremely low-pitched voices.

The PDA by Gillmann (1975) which applies center clipping but no peak clipping, reduces the computational effort by downsampling the signal to a 2-kHz sampling frequency before computing the ACF. Gillmann's PDA thus combines the nonlinear preprocessor proposed by Sondhi (1968) with the computation of the ACF as implemented by Markel (1972b; to be discussed in Sect.8.2.5). The raw values for T_0 as obtained from the ACF are then refined by interpolation.

Infinite peak clipping, but no center clipping is applied in the autocorrelation PDA by Galand et al. (Galand, 1974, 1976; Galand et al., 1976). The ordinary ACF of the peak-clipped signal was found insufficient; hence, additional signals were processed: the differenced signal (after infinite peak clipping as well) and two ternary features which indicate whether the signal has a peak, and whether the slope of the clipped signal is positive, zero, or negative. The ACFs of all these signals are added and yield an accumulated ACF which serves as the input signal for the basic extractor. For a given instant n and delay d, the partial value of the accumulated ACF depends on 8 bits; it is not computed but read out of a stored table. This PDA thus does not need multiplications; it has been implemented as a real-time digital hardware device. The performance, however, shows the known sensitivity to dominant first formants and is thus inferior to that of autocorrelation PDAs with center clipping.

The PDD by Lukatela (1973) uses cubic nonlinear distortion instead of center clipping prior to computation of the ACF. The performance of this way of preprocessing may be comparable to that of center clipping since, by cubic distortion, the spectrum is flattened, and the formant structure is heavily degraded. Compared to center clipping, however, cubic distortion has two disadvantages: 1) the effect of distortion is amplitude dependent unless an adaptive nonlinear function is applied; and 2) the algorithm cannot be speeded up by crude quantizing; infinite peak clipping, for instance, would completely annihilate the effect of the cubic distortion.

The issue of preprocessing was further pursued by Rabiner (1977a) who investigated a modified autocorrelation PDA design. Two different signals, both derived from the input signal, can be correlated there. Linear processing as well as three different types of nonlinear distortion can be selected as optional modes of preprocessing (Fig.8.9). From these Rabiner derived 10 individual correlators whose performance was evaluated using the speech material from the comparative evaluation (Rabiner et al., 1976). Some examples are shown in Figure 8.10. From the statistical error analysis, Rabiner drew the following conclusions:

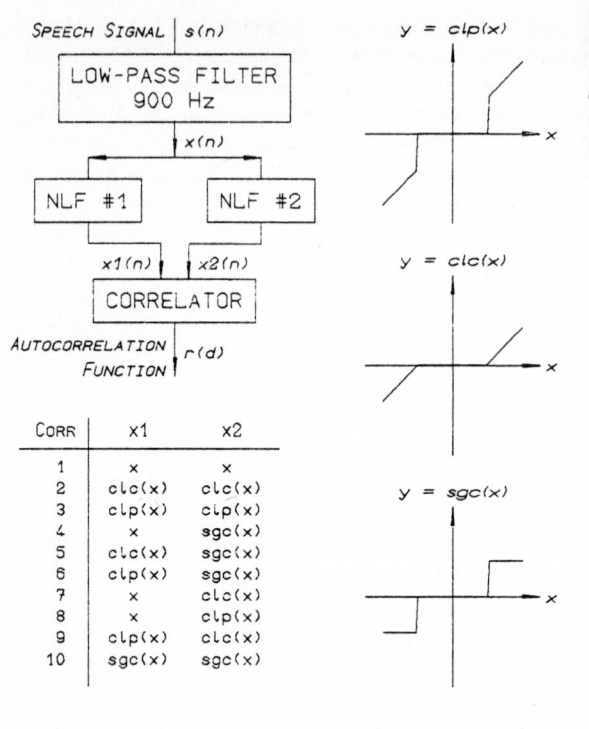

Fig.8.9. Autocorrelation PDA by Rabiner (1977a): Block diagram of preprocessor, characteristics of the applied nonlinear functions, table of correlators investigated

Corr	x1	x2
1	x	x
2	clc(x)	clc(x)
3	clp(x)	clp(x)
4	x	sgc(x)
5	clc(x)	sgc(x)
6	clp(x)	sgc(x)
7	x	clc(x)
8	x	clp(x)
9	clp(x)	clc(x)
10	sgc(x)	sgc(x)

1) For high-pitched speakers the differences in performance scores between the different correlators are small and probably insignificant. It is for this class of speakers that any type of correlation measurement of pitch period tends to work very well.

2) For low-pitched speakers fairly significant differences in the performance scores existed. Correlator number 1 (the normal linear autocorrelation) tended to give the worst performance for all utterances in this class. Correlators number 4, 7, 8 [the ones involving an unprocessed x(n) in the computation] were also somewhat poorer in their overall performance. ...

3) Differences in the performance among the remaining six correlators were not consistent. Thus, any one of these correlators would be appropriate for an autocorrelation pitch detector. (Rabiner, 1977a, p.30)

It is interesting to note that Rabiner's correlator number 10 (actually the one used by Dubnowski et al. where both input signals are center and peak clipped), in spite of the low information content preserved there, yields results equivalent to those correlators processing signals that were only center clipped.

The final point of investigation in Rabiner's paper is the frame length which is adapted to contain about three to four periods once an estimate of fundamental period duration has been established.

The PDA proposed by Rabiner was implemented by the present author. Results were found corresponding to those obtained by Rabiner. Autocorrelation analysis was also investigated as a by-product of the experiment discus-

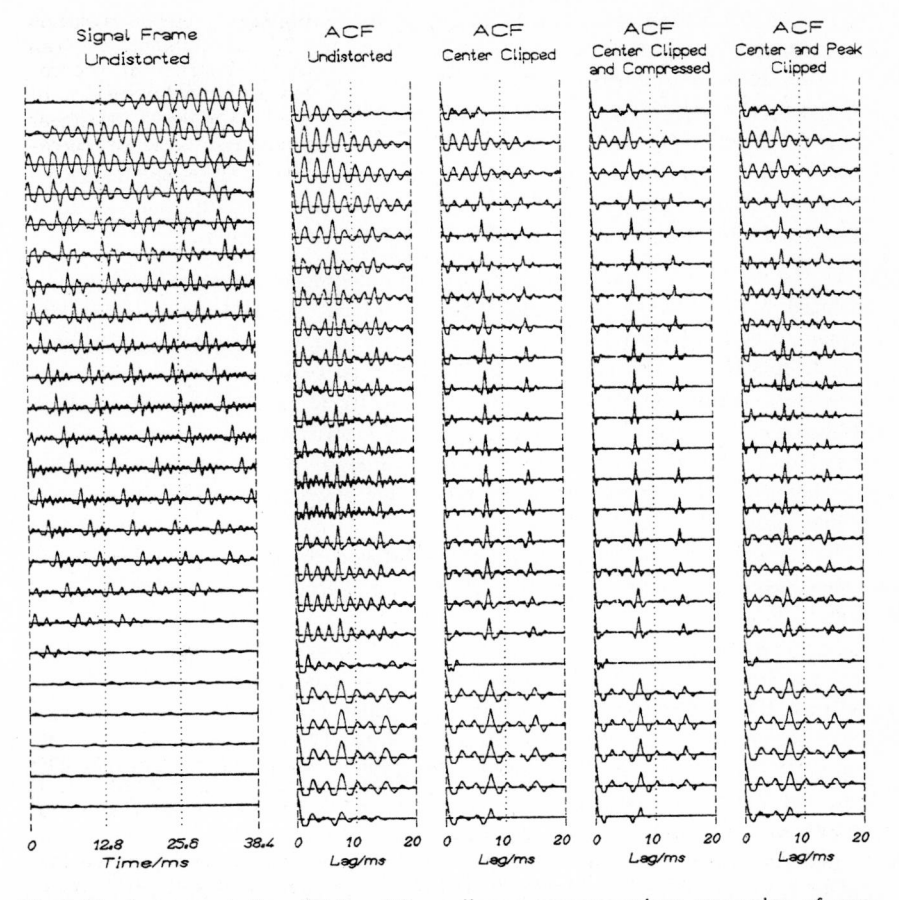

Fig.8.10. Autocorrelation PDAs with nonlinear preprocessing: examples of performance. Signal: utterance "brown", part of the sentence "The big brown fox"; male speaker, band-limited signal, sampling frequency 8 kHz, frame interval 12.8 ms, frame length 38.4 ms. The signals involved in the computation of the four displayed autocorrelation functions were preprocessed as follows (from left to right): 1) undistorted, (leftmost ACF display, corresponding to correlator #1 in Fig.8.9); 2) center clipped without compression (corresponding to correlator #3); 3) center clipped and compressed (correlator #2); 4) center and peak clipped (rightmost ACF, correlator #10 in Fig.8.9). The figure shows clearly that the unprocessed signal gives the worst results, especially in the first few frames where a strong first formant coincides with the third harmonic. In this extreme case, center clipping with compression (function "clc") performs best since it yields an additional attenuation of the weakly damped formant. In the rest of the figure the three correlators with center clipped input signals behave about equivalently. The center clipping threshold was set to 40% of the signal maximum within the frame. In case of the transition /ow-n/ (sixth frame from bottom), this proved to be too much so that all periods except one were clipped for that particular frame

sed in Chapter 7. The main results of this investigation are found in Appendix A: ACFs for the original periods (Figs.A.1-5), ACFs for center clipped speech (Fig.A.6), relative amplitude RASM of largest spurious maximum for the ACF (Fig.A.19), and the effect of band limitation (Fig.A.24) as well as processing sequence (Fig.A.21) on this value. The relative amplitude of the largest spurious peak (defined for the ACF in the same way as discussed in Chap.7 for a time signal) gives a good measure of the quality of the analysis to be expected. In the following the results are briefly summarized.

1) The performance of the NLF, i.e., the degree of spectral flattening, is indicated by the value of RASM. Since the input signal is assumed strictly periodic, the peak at a lag equal to the period duration T_0 takes on a value of one, relative to the peak at zero lag. For all phonemes and for all signal categories, center clipping gives the best results (Fig.A.19a, function 13). On the average, the results were between 4 and 8 dB (i.e., by a factor 1.5 to 3) better for center clipping than for ordinary analysis (same figure, function 1). No other NLF performed nearly as well. This strongly confirms Sondhi's and Rabiner's results.

2) The influence of band limitation was found to be moderate for center-clipped speech (a degradation by 1.6 dB, i.e., around 20% for the average value of RASM).

3) The result of the procedure is improved by another 1.4 dB (about 16%) on the average if the NLF is applied before low-pass filtering (Fig.A.21). This confirms the statement found already in Chap.7 that the contribution of higher formants to periodicity is nontrivial (cf. also the discussion by Friedman, 1978, on this point). This applies especially to band-limited signals where often the weakly attenuated waveform of the first formant is the only information available for periodicity detection after low-pass filtering. The validity of this statement in connection with the combined center and infinite peak clipper, however, was not tested. A disadvantage for the realization would be that it is difficult to obtain a simple two-bit representation of the preprocessed signal when the clipping is performed before the low-pass filter.

8.2.5 Autocorrelation PDAs with Linear Adaptive Preprocessing

In opposition to time-domain analysis by principle PiTL, autocorrelation analysis does not require the presence of the fundamental harmonic. Thus it is possible to apply some of the linear adaptive methods discussed in Chap.6 in connection with autocorrelation analysis, even when the signal is band limited.

Again the existence of PDAs of this type starts more or less with Sondhi's investigations (1968). Besides the nonlinear spectrum flattener, Sondhi proposes a linear spectrum flattener that consists of a set of band-pass filters (of approximately 100 Hz bandwidth) covering the frequency range of the signal (Fig.8.11). In each channel, the short-term amplitude of the signal is

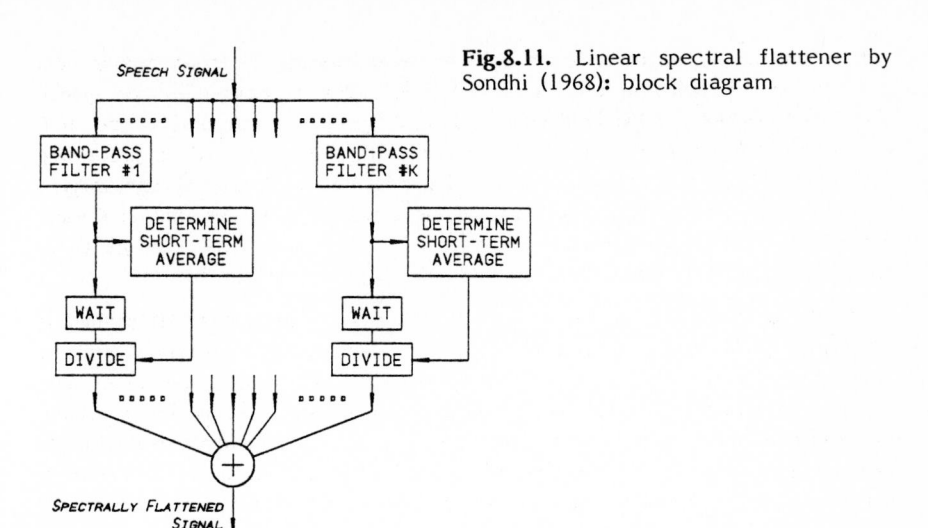

Fig.8.11. Linear spectral flattener by Sondhi (1968): block diagram

measured (full-wave rectification and smoothing), and the (delayed[7]) channel signal is divided by its short-term amplitude. At the output of the spectral flattener the formant information is practically removed, but the periodicity still remains prominent. This flattener is not an inverse filter although the principle is similar. It must be noted that this filter was designed before linear prediction entered the field. Dubnowski et al. (1976) objected that this flattener may introduce unwanted noise in those channels which do not carry a speech signal. Sondhi himself stated that this kind of preprocessing is inferior to center clipping in case the signal consists of very few harmonics. This approach, however, was about the first to apply linear adaptive preprocessing in an autocorrelation PDA.

The PDA by Fujisaki and Tanabe (1972) exploits the fact that the frequency domain is touched when the (ordinary) autocorrelation function is computed via (8.7) using the discrete Fourier transform. Spectral flattening, in this case, can be performed directly in the frequency domain. After a thorough discussion on the relation between the autocorrelation function and the cepstrum, the PDA applies a deconvolution (see the discussion in Sect.8.4.1) of the excitation function and the vocal tract transfer function, as given by (8.44). As usual in the computation of the ACF, a window with the length K is applied which contains several pitch periods. The corresponding Fourier spectrum $X(m)$ which is representative of the (short-term) speech signal $x(n)$ exhibits strong peaks at integer multiples of F_0. The estimate $H(m)$ of the vocal tract transfer function is evaluated using a shorter time-domain window;

---

[7] Delaying the channel signal is necessary to compensate for the delay caused by the short-term amplitude measurement.

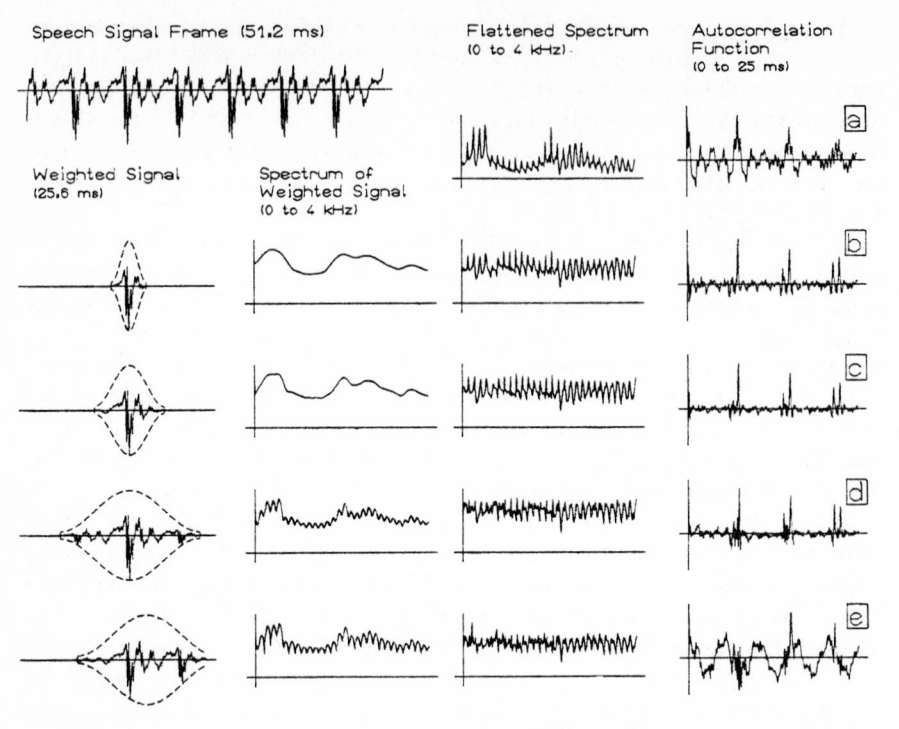

Fig.8.12a-e. PDA by Fujisaki and Tanabe (1972): example of performance. **(a)** Spectrum and autocorrelation function of the unprocessed speech signal; **(b)** weighted signal, spectrum of the weighted signal, flattened spectrum, and autocorrelation function for a window length of 5 ms; **(c)** same as (b) for a window length of 10 ms; **(d)** same as (b) for a window length of 20 ms; **(e)** same as (d), but with a badly positioned window. Signal: sustained vowel /ɛ/, speaker WGH (male). A Hamming window was used for this display. Due to the spectral flattening, the algorithm tends to somewhat decorrelate the spectrally flattened signal so that the peak at T_0 becomes smaller with respect to the peak at zero lag. To allow for this, the ACF in this figure was normalized with respect to the peak at $d = T_0$ and not to the peak at $d = 0$

its length is chosen approximately equal to the pitch period T_0. Note that signal weighting is a multiplication in the time domain; in the frequency domain this leads to a convolution of the two spectral functions involved (i.e., the spectrum of the unweighted signal and that of the window) which acts as a spectral smoother if the time window is short. Figure 8.12 shows an example. Applying (8.44), which is valid for both complex and power spectra, to the two short-term spectra, a flattened spectrum is obtained which is then converted into an autocorrelation function via (8.7) by executing an inverse DFT.

The choice of the length K' of the short window is uncritical as long as K' is small compared to the frame length K. If K' is smaller than T_0 the situation is uncritical as long as the window is placed in such a way that the principal peak of a pitch period falls into the interval $n = q'+K'/2-1$ (1) $q'+K'/2+1$. If K' is greater than T_0 the division of the two spectra will remove part of the harmonic structure; as long as K'≪K holds, however, sufficient information on pitch will be preserved.

In 1972, Markel presented the "SIFT" algorithm (SIFT stands for simplified inverse filter transformation). Figure 8.13 shows the block diagram of this PDA. The signal is first low-pass filtered and then sampled at a rate of 2 kHz. Since there are never more than two vocal tract resonances within the frequency range of 0 to 1 kHz, the following LPC analysis only needs four poles to model the formant structure. The formants are removed by the pertinent nonrecursive four-zero inverse filter, thus converting the input signal into the residual signal (for more details of the LPC analysis, see Sect.6.3). Markel then applied ordinary autocorrelation analysis to the signal.

Due to the low sampling rate, the computation of the ACF is done quite fast, but the measurement of T_0 is still too inaccurate to meet the linguistic requirements, as specified in Chapter 4. Markel thus performed an interpolation in the postprocessor. If one regards the autocorrelation sequence as a (finite) signal, the interpolation of new samples is achieved just by upsampling,

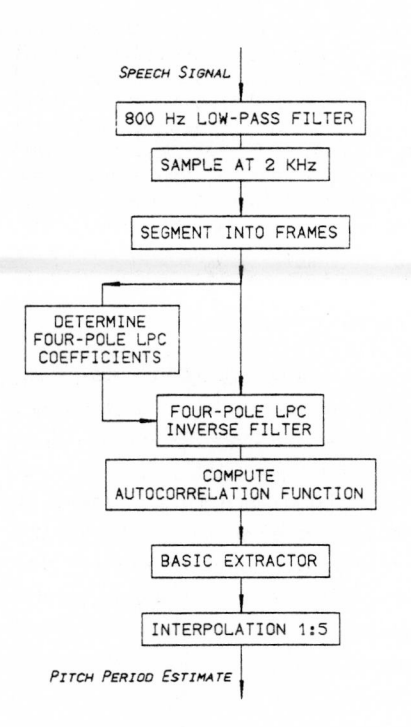

Fig.8.13. SIFT PDA by Markel (1972b): block diagram

i.e., by increasing the sampling frequency. Upsampling is a well-known procedure in digital signal processing. For those readers who are not familiar with
it, the principle is briefly displayed (Fig.8.14). The Fourier spectrum of a
sampled signal is known to be periodic with $2\pi/T$, if T is the sampling rate
(cf. Sect.2.1 for more details). If the sampling frequency is increased by an
integer factor of k, just by inserting (k-1) zeros between the original samples,
k periods of the original spectrum will now be within the range of the new
sampling frequency. The interpolation is completed by removing the unwanted
spectral components by a linear-phase low-pass filter which must have a sharp
cutoff frequency practically equal to the old sampling frequency (Fig.8.14,
lower part). In order not to distort the signal, this low-pass filter must be
close to ideal. The *ideal* low-pass filter, however, which has a rectangular
frequency and zero phase response, is generally not realizable since it is noncausal and has an infinite impulse response. In this special case, however,
since the input sequence r(d) is finite, this filter *can* be realized! One can
apply the discrete Fourier transform, set the unwanted spectral components
to zero, and transform the signal back. Since only very few interpolated
samples are needed, this interpolation can also be performed in the time
domain directly. For real even functions, such as the ACF r(d), we then
arrive at the following interpolation formula given by Markel. If $M = 2N$ is the
total length of the ACF signal (including the samples at negative values of
the lag d), and if

$$M' = kM; \quad d' = iM/M'; \quad i = 0 \ (1) \ M'-1$$

are the length and index[8] of the new signal r(d'), then, together with the
auxiliary expressions

$$a = (d+d')\cdot 2\pi/M \quad \text{and} \quad b = (d-d')\cdot 2\pi/M \ ,$$

the interpolation is given by

$$r(d') = \frac{1}{M} \sum_{d=0}^{M/2-1} r''(d) \left[\frac{\sin (M-1)a/2}{\sin a/2} + \frac{\sin (M-1)b/2}{\sin b/2} \right] . \tag{8.17}$$

This interpolation corresponds to the truncated impulse response of the ideal
digital low-pass filter. The function r''(d) equals r(d) for all values of d except
$d = 0$, where it equals r(0)/2. For special cases, and if the exact interpolation
is not needed, the formula can be further simplified.

Markel's PDA was included in the comparative evaluation by Rabiner et al.
(1976). In the original paper, Markel compared his PDA with cepstrum processing (see Sect.8.4) and found the two procedures about equivalent. There is,
however, one danger to be seen from a theoretical point of view. As already
stated in Sect.6.3, the label *inverse filter* in connection with linear predic-

---

[8] Note that the index d' of the upsampled signal r(d') proceeds in steps
of 1/k.

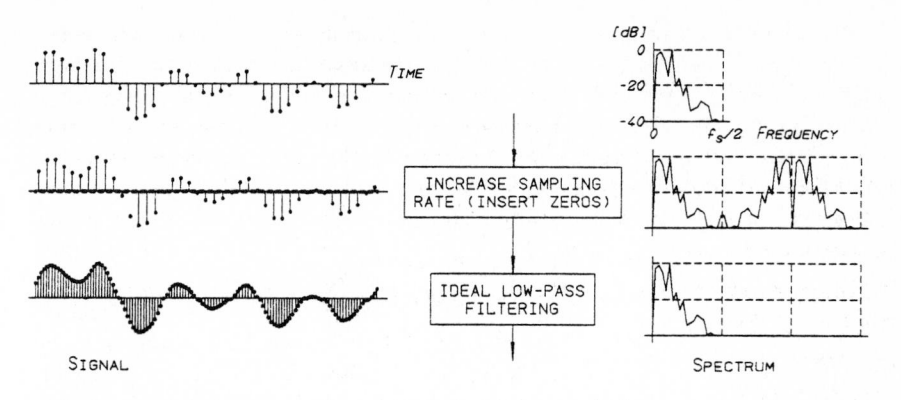

Fig.8.14. Upsampling as a method of signal interpolation

tion may be misleading. Linear prediction does not automatically model the vocal tract transfer function and separate it from the voice source transfer function, but it minimizes the mean square, i.e., the energy of the residual signal. In the worst case of a long frame in connection with a high-pitched voice, the first complex zero of the inverse filter may happen to be adjusted to the first harmonic, and the second zero to the second harmonic, thus cancelling the whole periodicity information available in the downsampled signal. The results obtained while analyzing such a voice in the comparative evaluation (Rabiner et al., 1976) suggest that this failure does really occur.

A modification of the algorithm, using the comb filter applied in Chap.7 instead of LPC, was investigated by the present author as a by-product of the experiment reported in that chapter. The comb filter, however, was prevented from cancelling the first harmonic in the case of coincidence between F_0 and the formant F1, so that the results of this "SIFT" algorithm and that of Markel are not really comparable. In spite of this, the results give some insight into the performance of an adaptive linear filter as preprocessor in an autocorrelation analysis PDA. The parameter of interest is again the relative amplitude RASM of the largest spurious maximum, this time measured for the ACF, relative to the principal peak exhibited at a lag of T_0. This value is displayed in Fig.A.19 (Appendix A) for the three signal types (see Table 7.1 for details) and, in addition, for various kinds of nonlinear processing which are not implemented in Markel's algorithm. Furthermore, in Appendix A, the influences of band limitation (A.24) and applied nonlinear function (A.20) are displayed. The latter two figures present a direct comparison of ordinary AC analysis and SIFT AC analysis, depending on the additionally applied nonlinear function. The results are summarized as follows.

1) For undistorted signals and linear processing, SIFT AC analysis roughly corresponds to nonlinear direct AC analysis. The SIFT procedure used in this investigation, however, exhibits a considerable sensitivity to band limitation.

2) The results can be improved by nonlinear distortion before application of the inverse filter. Full-wave rectification performs well, even for high-pitched voices. Center clipping in connection with linear inverse filtering has a detrimental effect on the performance. This is obvious since center clipping is the only NLF to remove the formant structure so that an inverse filter cannot be designed any more.

3) The comparison between direct AC and SIFT AC analysis is performed in two ways:

a) as displayed in Fig.A.20 (Appendix A) using the *same* NLF for both methods. This shows, for instance, that the inverse filter exhibits a considerable gain for linear processing (except in the case of band-limited signals), and that center clipping and inverse filtering applied together usually cause a loss of performance. (Note that in this figure negative values indicate that the relative amplitude RASM is lower for SIFT analysis, i.e., that SIFT performs better.)

b) The comparison of both procedures applying the *optimal* NLF independently for each of the methods first shows that, for more than 70% of all the cases, both methods work so well that a difference is of no more interest. (The pertinent figure was suppressed because the display proved to be rather uninformative.) In the rest of the cases, the difference is not too great. There is the clear preference towards center clipping in ordinary AC analysis. In SIFT analysis, full-wave rectification as well as linear processing without any nonlinear distortion tend to give good results.

8.3 "Anticorrelation" Pitch Determination: Average Magnitude Difference Function, Distance and Dissimilarity Measures, and Other Nonstationary Short-Term Analysis PDAs

Correlation is a measure of similarity, i.e., a measure of "agreement" between two input functions. Functions which are similar do not exhibit great differences. Thus a possible approach to detect periodicity would be the investigation of the global distance between two functions.

Among the distance measures used for pitch determination, the *average magnitude difference function* (AMDF) has become the best known. Section 8.3.1 deals with the PDAs that are based on this principle. Besides the AMDF there are a few other distance or dissimilarity measures that have been used to design a PDA. These approaches will be discussed in Section 8.3.2.

As we will see from the discussion of the individual distance functions, their computation does not directly involve short-term transformations in the sense of (2.32) since they do not require a matrix multiplication. However, the PDAs based on the application of distance functions undoubtedly pertain to the short-term analysis category since the signal is subdivided into frames and since the output is an estimate of pitch period duration and has nothing to do with the actual position of individual periods. In contrast to the autocorrelation method, however, the distance functions, and in particular the AMDF, consistently use the *nonstationary* principle of short-term analysis. It can easily be seen, for instance, from the definition (8.18) of the AMDF, that more than K samples (K being the frame length) are involved in the computation of a distance function for a single frame, and that the signal is not assumed to be zero or periodic outside the K samples that make up the frame, as it would be if the stationary or the synchronous principle of short-term analysis were applied.

Besides the PDAs that use distance measures there are a number of PDAs which are located at the border between time-domain and nonstationary short-term analysis. These algorithms are difficult to categorize. With time-domain PDAs they share the ability to process the signal period by period. Unlike genuine time-domain PDAs, however, they are not able to synchronize with a defined instant within an individual period. The reasons for this behavior are manifold and depend on the particular device. PDAs of this type will be discussed in Section 8.3.3.

8.3.1 Average Magnitude Difference Function (AMDF)

Correlation is a measurement of similarity, i.e., a measure of "agreement" between two input functions. Functions which are similar do not exhibit great differences. Thus a possible approach to detect periodicity would be the investigation of the global deviation between two functions. For this, a short-term difference function is defined as follows:

$$ \text{AMDF}(d) = \frac{1}{K} \sum_{n=q}^{q+K-1} | \ x(n) - x(n+d) \ | \qquad (8.18) $$

This function is called the *average magnitude difference function* (AMDF). The AMDF is expected to have a strong minimum when the lag d becomes equal T_0. This minimum would be exactly zero in case the input signal $x(n)$ were exactly periodic. Equation (8.18) implies a nonrecursive digital filter which has the form of a two-pulse filter (TPF), as discussed in Sects.2.3 and 7.1 where the TPF was found to be a comb filter. If the two pulses have opposite sign, the frequency response is given by

$$ H_0(\Omega) = 2 \ \sin \ \Omega d/2 \ . \qquad (2.25) $$

If the pulse distance d (which corresponds to the lag in autocorrelation analysis) is set equal to the length of one pitch period, the harmonic structure of the signal is totally cancelled so that the output of the filter becomes zero. Thus the AMDF can be realized by a short-term measurement of the absolute average at the output of a TPF with opposite sign of the two pulses. The estimate of T_0 then equals the pulse distance d where this average becomes a minimum.

It can easily be seen that this method is phase-insensitive since the harmonics are removed without regard to their phase. On the other hand, this function is sensitive to intensity variations, noise, and low-frequency spurious signals, all of which directly influence the magnitude of the principal minimum at T_0. For this reason, for instance, the AMDF, unlike other short-term functions, does not readily provide a reference which could serve as an indication for a discrimination between voiced and unvoiced or between regular and irregular.

About 10 PDAs have been reported which can be grouped in this category: the two early PDDs by Miller and Weibel (1956) and by Stone and White (1963); the algorithms by Sobolev and Baronin (1968), Moorer (1974), Ross et al. (1974), Un and Yang (1977), Sung and Un (1980), Hettwer and Fellbaum (1981); and, in addition, the analog PDD by Christiansen (1981). The PDA by Ney (1982), which uses the AMDF in connection with dynamic programming, pertains to this category at least to some extent; it is discussed, however, in Sect.8.3.2 together with the PDAs that apply a generalized distance measure. In addition, there are a number of publications where this principle is applied without further modification or development.

The PDA by Miller and Weibel (1956), an analog realization, mainly follows the analog correlator principle given by Stevens (1950) using a tapped delay line (Fig.8.6). The multiplication, which was necessary for autocorrelation, has been replaced by subtraction and full-wave rectification. A similar PDD, in analog technology as well, was developed by Christiansen[9] (1981). This PDD measures the peak amplitude of the difference function rather than its magnitude. Differential amplifiers at the taps of a tapped delay line compare the original and the delayed signal; their output signal is fed into peak detection circuits consisting of the diode-capacitor-resistor combination as described in Section 6.2.1. A multiplexing switch examines the output signals of the individual peak detectors. The location of the significant minimum which indicates T_0 if then transferred to an output register. This PDD shows that it is possible to build a short-term analysis PDD completely in analog technology; since the peak detectors at each tap of the delay line cannot be multiplexed, however, the hardware effort is considerable.

The PDA by Stone and White (1963), a digital hardware implementation, is interesting for its realization since it applies the AMDF in a pitch-synchronous time-domain version, similar to some of the nonstationary short-term analysis PDAs to be discussed in Section 8.3.3. This PDA must thus be placed at the border between time-domain and short-term analysis techniques. With the time-domain techniques, it shares the capability to place individual markers; on the other hand, it is unable to find a phase relation between the markers and the signal; the markers will be placed in correct distance, but in

---

[9] This PDD, patented in 1981, had already been developed before 1970. According to Baronin (1974a), a similar PDD was developed in the Soviet Union in the early sixties.

arbitrary phase according to some initial condition. Like Gill (1959), Stone and White designed their PDA this way because of the need for a realization which could escape from the dilemma with analog correlators. The correlator of this PDA strongly corresponds to the one by Gill except that this device counts the number of discrepancies between the contents of the two shift registers. This number then represents the current value of the AMDF. Once this value falls below a given threshold, the logic decides that the lag between the registers now equals a pitch period, and sets a marker. At the same time, the data in the shift register are transferred to the reference data register, and a new frame is started.

By the infinite clipping the sensitivity of the AMDF to intensity changes is largely removed so that a fixed threshold for the minimum detection appears justified; it has always been easier to state that a variable goes to zero than to determine its actual value. Whether a fixed threshold offers enough safety against subharmonic or higher harmonic detection remains an open question. Subharmonic detection in this case is dangerous since it directly influences the frame interval. On the other hand, in order to prevent a signal from remaining in the reference data register for too long a time, a new frame is forced to start as soon as the lag exceeds the measuring range for T_0. A weakness of this PDA (which could easily be avoided at the expense of some overlap between adjacent frames) is the fact that a marker is set as soon as the AMDF takes on a value below the given threshold; it is not asked whether this value already represents the minimum. Thus the periods might become too short by a few samples if the AMDF has a very deep minimum.

The first computer implementation of the AMDF PDA was probably that by Sobolev (Sobolev and Baronin, 1968; Baronin, 1974a). Sobolev and Baronin (1968) implemented both the AMDF and the ACF, and Baronin (1974a) wrote about the experimental comparison of the two methods:

> With optimal parameter values (0.1-ms sampling interval, 70 samples used for evaluating the shift function[10]), the accuracy of the PDA using the shift function is approximately the same as that of the autocorrelation PDA. However, the shift-function method is more suitable for both analog and digital implementation. Besides that, the duration of the signal segment necessary for evaluating the shift function can be a factor 2 or 3 shorter than that for evaluating the ACF. (Smoothing and averaging filters are unnecessary; therefore it is even possible to measure T_0 for a single period.) (Baronin, 1974a, p.200; original in Russian, see Appendix B)

In 1974, computer simulations of AMDF PDAs were developed independently by Ross, Shaffer, Cohen, Freudberg, and Manley as well as by Moorer (and published in the same issue of the same journal 23 pages apart). The two PDAs are equal in their basic approach but different in the realization of

---

[10] "Shift function" is Baronin's term for the AMDF. The frame length of 70 samples corresponds to 7 ms. The autocorrelation PDA used for comparison (Sobolev and Baronin, 1968) was implemented without center clipping (which was never used in these investigations in the Soviet Union).

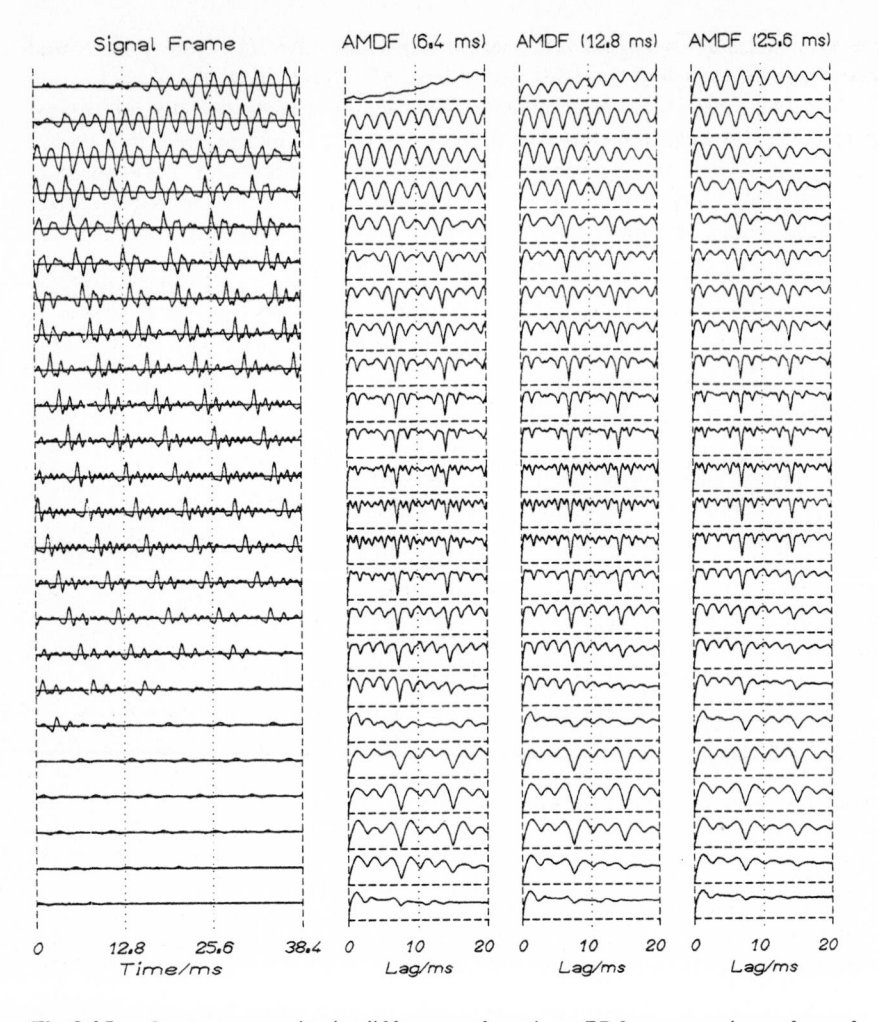

Fig.8.15. Average-magnitude-difference-function PDA: examples of performance. The signal is identical to that in Fig.8.10 (band limited to the telephone channel, male speaker, strong first formant on the third harmonic in
the beginning). The upper limit K of the summation according to (8.18) was
set equivalent to 6.4 ms (leftmost AMDF display), 12.8 ms (middle display),
and 25.6 ms (rightmost display). This yields an effective frame length of
(K+d) for the respective value d of the lag, and a maximum frame length of
(K+20 ms) for the maximum lag of 20 ms. The figure shows that the short
frame is more sensitive to intensity changes (especially at the beginning and
in the transition /ow-n/, 6th frame from the bottom). On the other hand one
might expect that a long frame is sensitive to changes of the fundamental
frequency. This cannot be shown in this figure since T_0 remains approximately
constant for this example

the final decision logic. Both of them compute the AMDF of the unprocessed speech signal according to (8.18). An example is shown in Figure 8.15.

Ross et al. discussed the similarity between the AMDF and autocorrelation approaches. Starting from the inequality

$$\frac{1}{N} \sum_{n=0}^{N-1} |x(n)| \leqslant \sqrt{\frac{1}{N} \sum_{n=0}^{N-1} x(n)^2} \quad , \tag{8.19}$$

they established a relation to the ACF which holds for a stationary process, i.e., a process where the signal has constant energy (and, in consequence, a time-invariant value of the autocorrelation function at zero lag). In this case they found that

$$AMDF(d) = b(d) \sqrt{2 [r(0)-r(d)]} \quad . \tag{8.20}$$

This means that for stationary signals the AMDF behaves like the square root of the difference between the ACF values at zero lag and at the nonzero lag d; $b(d)$ is a signal dependent factor which varies slowly with d. For nonstationary signals, especially during rapid variations of energy, this relation is no longer valid so that the results obtained by the AMDF are different from those obtained using the ACF. Compared to the ACF, the square-root relation has the effect that the minima of the AMDF are relatively sharp, sharper than the comparatively broad peaks of the (ordinary) ACF.

The AMDF does not as readily provide a reference which could serve to indicate failure of the method as do other STA techniques, such as autocorrelation. Thus a logic must be incorporated which excludes ambiguities. A simple possibility to realize this consists in a tracking logic which selects the minimum in best agreement with the previously established fundamental period estimates:

> The normal path for sustained voicing employs a feature for locking onto the true pitch and tracking it. A tracking window of 12 samples on either side about the last measured pitch period is defined within which the logic looks for a minimum. This minimum is then compared with the minimum in the entire AMDF search range. Normally, the tracking minimum is selected as the pitch but the logic will change to the nontracking position if the amplitude outside the tracking range is less than 1/2 the tracking amplitude minimum. For higher frequencies, more local minima are present in the AMDF, so a minimum outside the tracking window is required to be less than 1/8 the minimum in the tracking window to be chosen. (Ross et al., 1974, p.357)

The application of a tracking logic appears much less dangerous for this PDA than in time-domain approaches (cf. Sect.6.1). Time-domain PDAs that apply the tracking filter in the preprocessor may lose the correct value of T_0 completely if the PDA locks onto a wrong value of T_0 due to an occasional mistuning. In all those cases the tracking filter is implicitly or explicitly incorporated in a feedback loop. In the actual PDA this cannot happen. The tracking logic is incorporated in the basic extractor; it gives the basic extractor the characteristic of a hysteresis, and a solution for a mistuning is possible as soon as the obviously wrong situation has been registered.

At the time the algorithms came out, the main advantage of this PDA - reported both by Moorer (1974) and by Ross et al. (1974), was the increased speed, compared to conventional AC analysis (Ross) and to cepstrum (Moorer). This is due to the fact 1) that the AMDF does not require multiplications, and 2) that, as a procedure that applies *nonstationary* short-term analysis, it can be confined to a relatively short frame[11] (Sobolev and Baronin: 7 ms; Ross: 9 ms; Moorer: 1 period). The fast AC algorithm by Dubnowski et al. (1976) was not yet available at that time. The performance was found to be more or less comparable to that of cepstrum and AC analysis. The PDA by Ross et al. was included in the comparative evaluation by Rabiner et al. (1976).

The principal difference of the PDA by Un and Yang (1977) from its predecessors is the fact that the residual of linear prediction, i.e., the local error signal e(n), is used as input for the AMDF computation, and not the original speech signal. This involves the advantage that the formant structure has already been removed, and in addition that a cue for the voiced-unvoiced discrimination appears possible. On the other hand, the use of the LPC residual always reduces the signal-to-noise ratio, especially when the frame used for evaluation of the LPC parameters is long and the number of poles is high (for comparison, see the examples in Sect.6.5.1). Figure 8.16 shows the block diagram of this PDA. Again the AMDF is computed via the nonstationary formulation of short-term analysis, the number of involved samples not being a function of the lag d,

$$\text{AMDF}(d) = \frac{1}{E} \sum_{n=q}^{q+K-1} \left| e(n) - e(n+d) \right| , \quad d = 1 \ (1) \ K , \tag{8.21}$$

where $E = \sum e(n)^2$ is the energy of the residual signal in the frame.

Among the details of their basic extractor, the authors found that, using a sampling rate of 6.8 kHz, a frame length of 20 ms (K = 136 samples), and calculating the AMDF between 2.5 and 15 ms (18 to 102 samples),

---

[11] The fact that the AMDF follows the nonstationary approach of short-term analysis (see Sect.2.4) is easily seen from (8.18). Unlike the autocorrelation function according to definition (8.4), there is no defined frame i = q (1) q+K-1 in the AMDF outside of which the samples are forced to be zero. If we take the definition (8.18), set the upper limit K of computation to one period for any value of the lag d, and limit d between zero and 20 ms, then we can easily see that the real *frame*, i.e., the total number N of samples involved in the computation of one pitch estimate, is always greater than one pitch period of maximum duration, as is the case with the other short-term analysis PDAs. The advantage of this approach is that only K+d, and not all the N samples, needs to be involved in the computation of the AMDF for a single value of the lag d. If one were to compute the AMDF corresponding to definition (8.14) for the ACF, the computation would involve fewer and fewer samples for great lags and thus force the AMDF to zero. Of course this is undesirable.

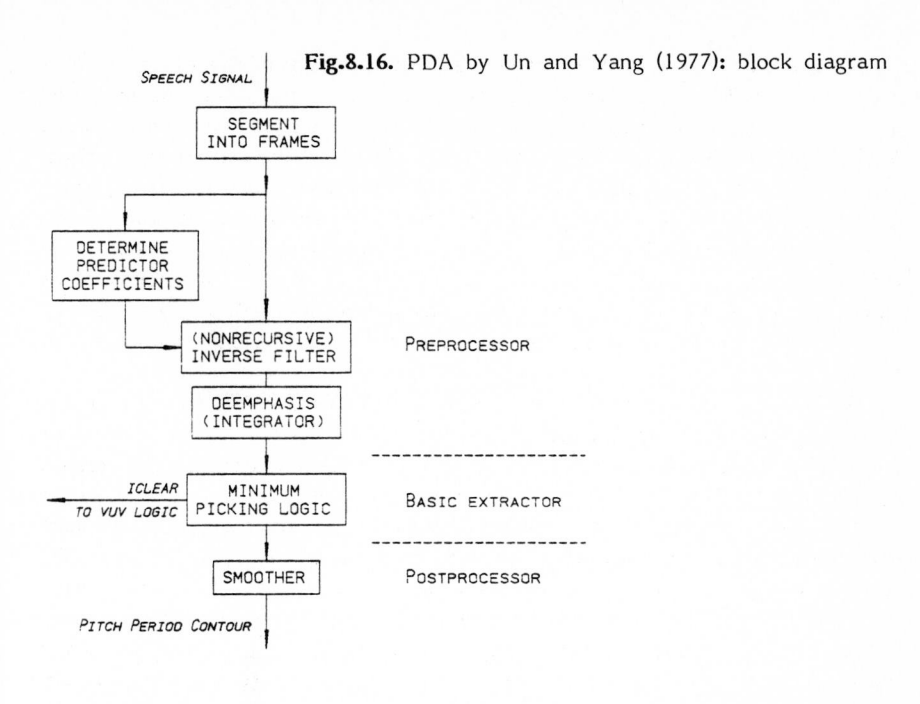

Fig.8.16. PDA by Un and Yang (1977): block diagram

for a pitch period within the range of 2.5 to 15 ms, at least one local minimum[12] but not more than six local minima will show up in the AMDF waveform. These minima should be separate from one another by at least 2.5 ms (or 17 points in AMDF) and tend to remain constant in size.

On the basis of the above observations, the process of selecting candidates for a significant minimum involves two threshold values, TH1 and TH2. As shown in Fig.8.17, the absolute minimum MIN is the first candidate; the local minima that fall in the region of MIN and MIN+TH1 and are at least TH2 points apart from any other candidate are selected as one of the significant minimum candidates. The threshold values TH1 and TH2 not only provide the reliable selection of candidates, but also drastically reduce the effort of searching for local minima.

Since the d value of the leftmost significant minimum candidate usually represents the pitch period for a residual signal with high periodicity, we choose the leftmost candidate and test to see whether it is a deep and clear minimum. For this purpose we use a third threshold value, TH3. A minimum is said to be clear (ICLEAR = 1) [see the block diagram in Fig.8.16] if, within 17 points for each side of the minimum, the difference between the maximal AMDF value and AMDF(d) exceeds TH3. Otherwise, ICLEAR = 0, and the past history of the pitch period which is updated whenever a pitch period corresponds to a clear minimum is used to pick the correct significant minimum among the candidates. (Un and Yang, 1977, p.568)

---

[12] Un and Yang, as well as Ross et al., label these minima *nulls*. Since this term is used nowhere else in the literature, the citations have been changed in such a way that the term *null* was replaced by *minimum*, *local minimum*, or *significant minimum*, according to its respective meaning.

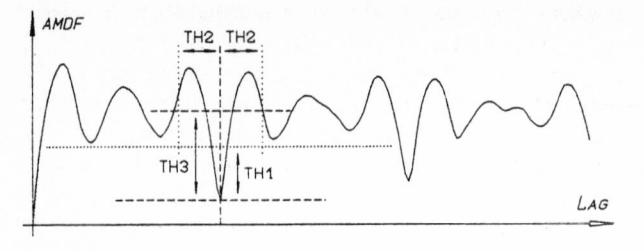

Fig.8.17. PDA by Un and Yang (1977): operational principle of the basic extractor. (TH1) Limit for secondary minimum to be taken under consideration as possible period indication; (TH2) smallest distance between adjacent minima, (TH3) smallest depth of minimum in order to be labeled "clear"

So far, the tracking logic applied in this PDA is about the same as that in the PDA by Ross et al. (1974) which operates with the AMDF of the unprocessed signal. Compared to that PDA, however, the tracking thresholds experimentally found and implemented by Un and Yang appear much more moderate. This may serve as a indication that the local performance of the AMDF is better for the LPC residual than for the unprocessed speech signal; the better the local extraction, the less the past history must be troubled. This is to be expected, since the residual signal has no formant structure left. The deemphasis prior to computation of the AMDF consists of a first-order low-pass filter (integrator) which gives the residual signal an overall attenuation of -6 dB/octave, similar to that of the original speech signal.

The main field of application for this PDA is LPC analysis and synthesis or any speech communication system where LPC coefficients are required for other purposes. The performance is reported to be very good. Gross errors or systematic failures, after extensive testing, are not mentioned. Unlike Markel's SIFT PDA (1972b; see previous section), this PDA apparently does not suffer from the occasional failures due to a coincidence between F_0 and the formant F1.

Although the computation of the AMDF is quite fast due to the short frame and the lack of multiplications, it can be further speeded up when the AMDF need not be evaluated for all possible values of T_0. To reduce the number of points for evaluation of the AMDF, however, a preliminary estimate of T_0 must be available. A convenient way to provide such an estimate is to use the estimate from the previous frame, and to evaluate the AMDF only in the vicinity of this estimate. This approach is dangerous, however. As we have already seen from the discussion of tunable band-pass filters (Sect.6.1.3) which apply a similar strategy, the PDA can easily lose its track without a chance to recover once a gross error has been committed or when the excitation signal is temporarily irregular. A much better way to realize such a principle is to implement a multiprinciple PDA. A primitive but fast PDA supplies a crude estimate (or even several estimates, in order to take into account possible gross errors) of T_0 around which the AMDF is then evaluated.

Such an algorithm has been implemented by Sung and Un (1980). To obtain the preliminary estimates, a simple time-domain PDA of the type PiTE is used which is realized in much the same way as the parallel processors in the PDA by Gold and Rabiner (1969, cf. Sect.6.2.4). Using the preliminary estimates, only few points remain where the AMDF must be evaluated. The computing effort is further reduced by choosing an increment m greater than one in the computation of the AMDF (and thus effectively downsampling the signal during evaluation of the AMDF without losing accuracy of the T_0 estimate[13]):

$$AMDF(d) = \sum_{k=0}^{K/m-1} |\, x(q+mk) - x(q+mk+d)\,| \ \ .\tag{8.22}$$

Taking these actions, Sung and Un arrived at a PDA that operates in real time on a standard 8-bit microprocessor.[14] The performance was found comparable to that of the standard-AMDF PDA and a little, but not significantly inferior to the earlier algorithm (Un and Yang, 1977).

The question of signal windowing prior to the computation of the AMDF was investigated in a paper by Fette et al. (1980). They applied rectangular and exponential windows to the signal. As can be expected, the application of an exponential window increased the error rate when the window was too short, and even a long window did not improve the performance significantly. This is consistent with the fact that the AMDF uses the nonstationary principle of short-term analysis which does not really provide the framework for the application of a window; there is no doubt that a windowed signal usually is less "periodic" than an unwindowed one.

The idea of combining center clipping and the AMDF was realized by Hettwer and Fellbaum (1981). They verified that moderate center clipping reduces the problems associated with a strong first formant for the AMDF, as well. Combined with peak clipping, it also speeds up the algorithm since it replaces the necessary additions by simple counting operations. An example is shown in Figure 8.18.

In conclusion, the AMDF is a variation of autocorrelation - or, rather, its antonym. For the PDAs of both categories, the following can be summarized:

---

[13] This is the digital counterpart to the statement by Baronin (see Sect.8.2.2), made for the analog domain, that if crude sampling is necessary due for some reason of implementation, it is advantageous to crudely sample the signal rather than the spectral function where T_0 is actually measured.

[14] Sung and Un (1980) implemented their PDA on a Zilog Z-80 microprocessor using a sampling frequency of 8 kHz and an increment m = 4, and evaluating the AMDF for seven possible T_0 candidates per frame. The computation time needed was 0.6 times real time.

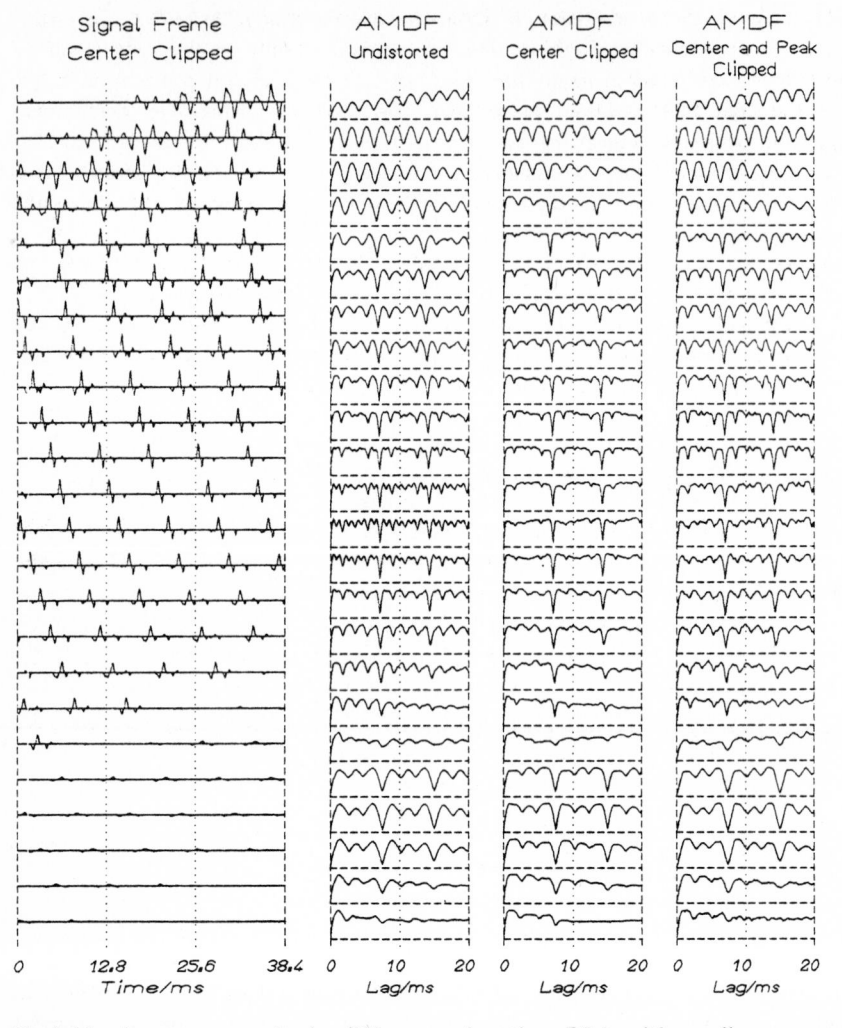

Fig.8.18. Average magnitude difference function PDA with nonlinear preprocessing: examples of performance. The signal is identical to that in Fig.8.10
and Fig.8.14 (band limited to the telephone channel, male speaker, strong first
formant at the third harmonic in the beginning). The upper limit K of the
summation according to (8.18) was set equivalent to 12.8 ms. This yields an
effective frame length of (d + 12.8 ms) for the respective value d of the lag,
and a maximum frame length of 32.8 ms for the maximum lag of 20 ms. The
figure shows that nonlinear processing, in particular simple center clipping, is
advantageous in case of a dominant first formant, although it does not appear
as necessary for the performance as in case of the autocorrelation PDA (see
Fig.8.10)

1) Autocorrelation and AMDF PDAs are antagonistic in behavior, but comparable in performance. Highly correlated signals have a low minimum of the AMDF and vice versa;

2) the significant minima of the AMDF are usually sharper than the corresponding peaks of the ACF;

3) the computation of the AMDF is faster than that of the ordinary AC since AMDF needs no multiplication and allows for a comparatively short frame;

4) due to the nonstationary STA principle, the AMDF is sensitive to intensity changes, whereas here the ACF is fairly insensitive;

5) both ACF and AMDF are sensitive to a dominant formant structure; and

6) linear preprocessing (inverse filtering) as well as nonlinear processing (moderate center clipping) have a favorable effect on the performance of AMDF PDAs.

8.3.2 Generalized Distance Functions

Among the PDAs that apply distance functions other than the AMDF, the PDA by Nguyen and Imai (1977) as well as that by Sánchez (1977a,b, 1979) start from the generalized distance function

$$D(d,k) = \left\{ \sum_{n=q}^{q+K-1} [\ | x(n) - x(n-d) \ | \ ]^k \right\}^{1/k} , \tag{8.23}$$

of which the AMDF is a special case ($k = 1$). Regardless of the value of the exponent k, this distance function exhibits a significant minimum at $d = T_0$. A normalized distance function is obtained as

$$D_N(d,k) = \frac{D(d,k)}{\left[\sum_{n=q}^{q+K-1} | x(n) \ |^k \right]^{1/k}} . \tag{8.24}$$

For a stationary signal $D_N(d,k)$ does not exceed a value of 2. For $k = 1$ we thus obtain an AMDF which is normalized with respect to the signal magnitude; for $k = 2$ we obtain a squared difference function normalized with respect to the signal energy. These normalized distance functions have the advantage that they provide some measure of the salience of the pitch estimate.

Nguyen and Imai (1977) experimentally investigated the influence of the exponent k for stationary and nonstationary voiced signals. They found that

... the distance function for small k is appropriate to the almost perfectly periodic signal, in which the minute parts of the signal must be considered to recognize the repeating.

The distance function for large k is appropriate to the weakly stationary periodic signals, in which large components play the most important role in representing the periodicity of the signal.

Any value of k will also be appropriate to the stationary signals, in which the consideration of the minute part for recognizing the fundamental period is not required.

Consequently, since a signal like the speech signal essentially has both strongly stationary and strongly nonstationary parts, the most appropriate value of k is about 2; however, k = 1 and 3 are also available. (Nguyen and Imai, 1977, p.16)

Hence their PDA uses the normalized *average squared difference function* (ASDF; k = 2). The computational effort, which is much higher than for the ordinary AMDF due to the squaring operations, is reduced by evaluating the distance function for crudely spaced values of the delay d first and computing it in full accuracy only around the minima found from the crude evaluation.[15] Nevertheless, the advantage of this distance function over the AMDF is not great enough to really justify the increased computing effort.

In analogy to autocorrelation and autocovariance functions, the PDA by Sánchez Gonzalez (1977a,b, 1979) uses a distance function called *autodissimilarity*. Sánchez first defined an auxiliary function

$$U[k,a(n)] := \left\{ \sum_{n=q-K/2}^{q+K/2} \frac{1}{W} w(n-q) \left[|a(n)| \right]^{k} \right\}^{1/k}, \tag{8.25}$$

where w(i) is an arbitrary window having the average magnitude W and its nonzero values within i = -K/2 (1) K/2. [With w(i) = 1 within these limits, and a(n) = x(n)-x(n+d) or a(n) = x(n-d/2)-x(n+d/2), we are back at the generalized distance function defined by (8.23).] In Sánchez's definition, the autodissimilarity function (ADS) then reads as

$$ADS(d,k) := \frac{U[k, x(n-d/2)-x(n+d/2)] - \beta \, | \, U[k, x(n-d/2)] - U[k, x(n+d/2)] \, |}{U[k, x(n-d/2)] + U[k, x(n+d/2)]}.$$
$$\tag{8.26}$$

In this equation the first term in the numerator represents the distance function as defined in (8.23). The second term - with the factor β as an additional degree of freedom - is a correction term which decreases the value of the ADS when the signal is nonstationary, and vanishes for stationary signals. The denominator represents a rigorous normalization taking into account the nonstationarity of the signal. Since $U[k, a(n)]$ is a nonnegative function, ADS(d,k) will never exceed unity for positive values of β. Since it can be shown (Sánchez Gonzalez, 1979) that the inequality

$$| \, U[k, a(n)] - U[k, b(n)] \, | \, < U[k, a(n)-b(n)]$$
$$< U[k, a(n)] + U[k, b(n)] \tag{8.27}$$

---

[15] Like in other short-term analysis PDAs that involve raising a signal value to some power as the only operation which requires multiplications, a table-lookup routine can drastically reduce the computational load in an implementation where multiplication is slow, and where sufficient random-access memory is available. For more details, see the discussion in connection with the maximum-likelihood approach in Section 8.6.3.

always holds, ADS(d,k) will be nonnegative for values of ß between 0 and 1. This rigorous normalization allows introducing a fixed threshold for the validity of the pitch estimate found.

The performance of the PDA is comparable to that of the AMDF PDAs except for the evaluation of the salience of the estimate which is easier here. The computational effort is considerable in the general case (arbitrary exponent k, arbitrary window, and the frame length K depending on d); there the number of multiplications can be in the order of K^3. Using a rectangular window and a good programming strategy which allows the terms $U[k, a(n)]$ to be computed recursively, however, the computing effort is not considerably greater than for the ordinary AMDF or the distance function according to (8.24). Note that (8.27) consistently obeys the nonstationary principle of short-term analysis; even if a nonrectangular window is applied, it is applied after the signals x(n-d/2) and x(n+d/2) have been arithmetically combined. In this respect definition (8.25) is analogous to (8.13) for the ACF.

The PDA by J.E.Miller (1982) applies a gap-function low-pass filter instead of the comb filter applied in the AMDF and ASDF PDAs. First the filtered signal

$$y(n,d) = \frac{1}{d} \sum_{i=n}^{n+d} x(i) \qquad (8.28)$$

is computed. In contrast to the AMDF and ASDF PDAs which both use two-pulse filters with zeros in the transfer function at all multiples of the trial fundamental frequency p = 1/dT (T being the sampling interval) including zero frequency, the gap-function low-pass filter (see Sect.7.1 for more details) has no zero at zero frequency and is thus sensitive to low-frequency noise. This sensitivity, however, is cancelled by using the short-term variance of the signal y(n,d)

$$V(d) = \sum_{n=q}^{q+K} [y(n,d) - \bar{y}(q,d)]^2 \qquad (8.29)$$

as the distance function exploited in the basic extractor instead of the short-term magnitude or energy. The short-term average $\bar{y}(q,d)$ is given as

$$\bar{y}(q,d) = \frac{1}{K} \sum_{n=q}^{q+K} y(n,d) . \qquad (8.29a)$$

The performance of this PDA is comparable to that of the AMDF PDA. Miller's PDA has the advantage that there is less interaction between higher formants due to the gap-function low-pass filter.

The PDA by Ney (1982) combines the principle of (generalized) distance functions with dynamic time warping. The outcome is a unique PDA using a slightly modified AMDF and a simple dynamic-programming routine. In principle it could also work with a squared difference function as defined in (8.23) for k = 2, and - with restrictions - with an autocorrelation function. The algorithm, by the way, is a good example to see how dynamic programming works.

Let the problem be defined as follows. Given a signal frame $x(n)$, $n = q(1)q+K-1$, another frame is sought starting at $n = q+d_1$ and ending at $n = q+K-1+d_2$ which is as similar to the first frame as possible. *Similar* means that the distance between the two frames in the sense of a given error criterion is minimum. The main difference of this approach to ordinary short-term analysis PDAs is that the delay between corresponding samples of the two frames *need not be a constant*. Thus the procedure takes account of the fact that T_0 may vary within a frame, and yields a powerful method to minimize the distance between two time-variant signals.

Strictly speaking, the problem involves the global minimization of the error $E[x(n), d(k)]$ for the given frame $x(k)$, $k = 0(1)K-1$ (which may start at $n = q$, corresponding to $k = 0$), and a variable delay $d(k)$ which is a function of the position k of the current sample in the frame. This is a two-dimensional problem which can be reduced to a one-dimensional one by imposing viable constraints on the function $d(k)$. For example, the most rigorous constraint would be to postulate that $d(k)$ be independent of k, i.e., $d(k) = d(0) = d$. This would reduce the problem to the strictly one-dimensional case, as given for any ordinary short-term analysis technique, such as the AMDF PDA. In this case, however, nothing is gained; hence we must look for a possibility to vary $d(k)$ depending on k, but to impose constraints on this dependence such that the algorithm remains realizable. This task is performed using a dynamic-programming routine. Dynamic programming

> ... is a technique for solving nonlinear optimization problems, especially combinatorial optimization problems. The optimization problem is broken down into a sequence of optimization steps, such that the optimization can be performed sequentially step by step. From the solutions for each individual step, the global solution is then pieced together. Speaking in mathematical terms, this approach leads to a recurrence relation which is to be evaluated sequentially step by step. (Ney, 1982, p.386)

The sequence of steps pieced together then forms the *optimal path* through the input data.

For *dynamic time warping* as a special case of dynamic programming, the input data consist of two time-variant signals (or two sequences of parameter sets) which have to be optimally matched. In the case of pitch determination the first signal (the reference signal) is the given frame $x(k)$, $k = 0(1)K-1$, where $k = 0$ corresponds to $n = q$. The second signal, the trial signal, (to be matched to the first one) is the same speech signal $x(m)$, $m = 0(1)...$, starting at $m = 0$ (which corresponds to $n = q$ as well). Any valid path through these data now starts at some point (k_1, m_1) with $k_1 = 0$ and $m_1 = d(0)$ and ends at the point (k_2, m_2) with $k_2 = K-1$ and $m_2 = d(K-1)$, where $d = m-k$ is the current value of the delay $d(k)$. To characterize the sequence of points in the (k,m) plane which form the optimal path, a warping function is introduced

$$W := w(0), w(1), \ldots w(i), \ldots w(L-1) , \tag{8.30}$$

where

$$w(i) = [k(i), m(i)] \tag{8.31}$$

characterizes a point in the (k,m) plane which is touched by the path (Fig.8.19).

Four constraints are now imposed on the path.

1) *Beginning and End.* The path must start at $k = 0$ and end at $k = K-1$, as ready stated. In terms of the index i of the warping function this means

$$k(i=0) = 0 , \quad k(i=L-1) = K-1 . \tag{8.32}$$

2) *Boundary Condition.* The entire path must never leave the measuring range for F_0, i.e., the delay

$$d(i) = m(i) - k(i) \tag{8.33a}$$

can only vary within

$$d(i) \geqslant d_{min} \text{ and } d(i) \leqslant d_{max} \text{ for } i = 0(1)L-1 . \tag{8.33b}$$

3) *Condition of Monotony.* The chronological sequence of the samples must be retained:

$$k(i-1) \leqslant k(i) \quad \text{and} \quad m(i-1) \leqslant m(i) . \tag{8.34a}$$

4) *Condition of Continuity.* No signal sample may be omitted:

$$k(i) - 1 \leqslant k(i-1) \quad \text{and} \quad m(i) - 1 \leqslant m(i-1) . \tag{8.34b}$$

Constraints 1) and 2) delimit the region in the (k,m) plane where a valid path can exist at all. The constraints (8.34a,b) of monotony and continuity can be combined to a continuity constraint between two consecutive points of the path W,

if $W(i-1) = (k,m)$
    then $W(i) = (k+1,m)$ or $W(i) = (k,m+1)$ or $W(i) = (k+1,m+1);$. \tag{8.35}

We can easily see from these constraints that $L \geqslant K$.

The optimization procedure for the path W is now carried out as follows. Given the error criterion E, a local error $e(k,m)$ can be defined and evaluated for each point in the permitted region of the (k,m) plane. In the actual PDA the error criterion is the magnitude difference. In this case we simply obtain

$$e(k,m) = |x(m)-x(k)| \tag{8.36}$$

This error defines a local "score of penalty," i.e., the price to be paid for the permission that the path may pass through (k,m). For an arbitrary point j of the path W, the total error equals the sum of the individual contributions of all the points passed,

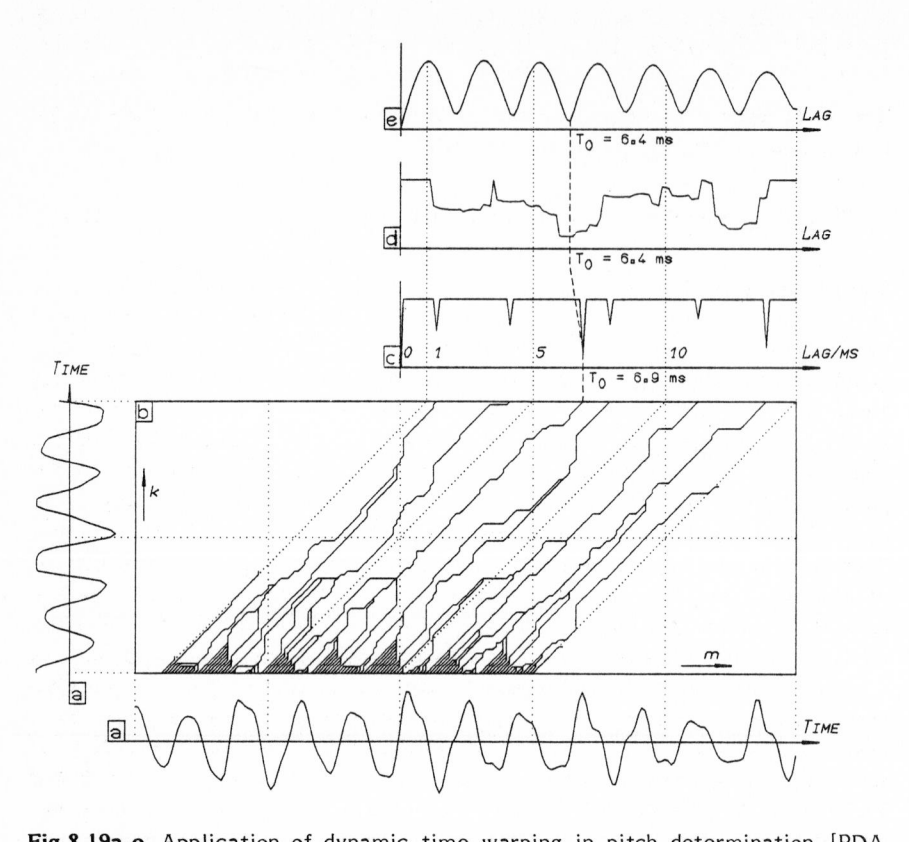

Fig.8.19a–e. Application of dynamic time warping in pitch determination [PDA by Ney (1982)]. (**a**) Speech signal (as in Figs.8.15,18, third frame); (**b**) display of the possible paths after local optimization; (**c**) accumulated distance function depending on the endpoint of each path (final result); (**d**) accumulated distance function depending on the starting point of each path; (**e**) ordinary AMDF (for comparison). Sampling frequency: 8 kHz; frame length (shown along the vertical axis): 10 ms. (k, m) Indices as indicated in (8.31). Paths going beyond the measuring range have been abandoned in this example; and for the accumulated distance function in (c) only complete paths were taken into account; the others have been set to maximum distance. Note that, although at each point within the measuring range (1–15 ms) a path is started, all the locally optimized paths end at one of only six points. The distance function displayed in (d) thus has only six significant points (and was set to maximum at the remainder of the points for reason of display). The difference in the estimates between (c,e) on the one hand and (d) on the other hand results from the fact that for (d) the value of the lag d at the endpoint of the frame is taken, whereas (c) takes the lag a the beginning of the optimal path, and (e), as the ordinary AMDF, takes an average over the whole frame. These results indicate that T_0 increases within the frame displayed

$$E(W,j) = \sum_{i=0}^{j} e\,[k(i), m(i)] = \sum_{i=0}^{j} |\,x\,[m(i)] - x\,[k(i)]\,| \quad . \tag{8.37a}$$

The total error committed on the path W is computed from (8.37a) for $i = L-1$, i.e., for the last point of the path,

$$E(W) = \sum_{i=0}^{L-1} e\,[k(i), m(i)] = \sum_{i=0}^{L-1} |\,x\,[m(i)] - x\,[k(i)]\,| \quad . \tag{8.37b}$$

The optimal path W is the one for which $E(W)$ becomes a minimum.

How can we obtain this path? One possibility is to check all possible paths in the permitted region and to select the optimal one. This strategy would guarantee that the solution obtained is really the optimal one. In the worst case, however, the computational effort grows exponentially with the frame length K and makes the strategy unfeasible. The only situation where such a strategy is realistic is given when such heavy constraints are imposed on the path that branching during the computation is practically impossible. The constraint that the delay d be independent of k, for instance, is such a case. There the number of possible paths reduces to $N = d_{max} - d_{min}$, i.e., the number of possible (integer) values of T_0 in the measuring range, and the strategy reduces to evaluating $E(W)$ for all these paths, i.e., evaluating AMDF(d) for all integer values of d between d_{min} and d_{max}.

It can easily be seen that even the constraint (8.35) with three possible branches at each point of the path, does not permit such a strategy. The next best strategy would be to evaluate all possible paths but to abandon them as soon as they become unlikely, i.e., as soon as their error score $E(W,j)$ becomes so large compared to the (momentarily) best path that they will probably never again be able to compensate for it. The criterion for this may be a threshold, or the number of paths pursued may simply be limited by forcing the algorithm to abandon the worst path as soon as the maximum number of permitted paths is exceeded. Such strategies are sometimes applied when the length K of the reference signal is not too great.

Dynamic programming provides a different solution. According to the strategy that the global minimization is to be split up into a sequence of local minimizations, the path at a given point (k,m) will always be pursued in the direction where the minimum penalty has to be paid. For this, we have to define a partial error criterion for the point (m,k):

$$P(k,m) = \min_{W} \left\{ \sum_{i=0}^{j} e\,[k(i), m(i)] \right\} \tag{8.38a}$$

for all those paths where the jth point w(j) equals (m,k). Decomposing this criterion into the contribution of the most recent step and that of all previous steps, we obtain

$$P(k,m) = e(k,m) + \min_{W} \left\{ \sum_{i=0}^{j-1} e\,[k(i), m(i)] \right\} \qquad (8.38b)$$

again with $w(j) = (k,m)$. According to (8.32-34) we obtain the initial condition

$$P(0,m) = e(0,m) \qquad (8.39)$$

and, to enforce the boundary condition,

$$P(m,k) = \infty \qquad (8.40)$$

if the delay m-k exceeds the permitted measuring range. The continuity condition (8.35) leads to the recurrence relation

$$P(k,m) = e(k,m) + \min \,[\, P(k-1,m), \; P(k,m-1), \; P(k-1,m-1)\,] \; . \qquad (8.41)$$

Using (8.39-41), and starting at $k = 0$ and $m = d_{min}$, the algorithm evaluates $P(k,m)$ for each point (k,m) in the permitted region of the (k,m) plane. The optimal path is not known before the endpoints $(K-1,m)$ for all values of m within the measuring range have been evaluated. The optimal endpoint $(K-1,\hat{m})$ with the lowest error score is the endpoint of the desired optimal path. The path itself must then be backtracked down to $k = 0$ using (8.41).

The information on pitch is implicitly contained in the warping function W. In principle one need not know the optimum path in detail; the information can even be obtained from the position of the endpoint alone:

$$T_0 = \hat{m}-K+1 \; . \qquad (8.42)$$

The error scores $P(K-1,m)$ can serve as a generalized distance function for evaluating the salience of the estimate.

Backtracking the path may increase the accuracy of the estimate,

$$T_0 = \frac{1}{L} \sum_{i=0}^{L-1} m(i) -k(i) \; . \qquad (8.43)$$

Besides some arithmetic effort, backtracking needs a considerable amount of memory since the partial error measures must be saved (or recomputed) for all points (m,k) which might have been touched by the path.

The algorithm can be used as an ordinary short-term analysis PDA on a frame-by frame base. It can also give an estimate of T_0 on a sample-by-sample basis.

There is no need to segment the speech signal into frames. Instead the dynamic properties of continuous speech can be exploited so as to force the estimation procedure to take into account the smoothness of the contour of period estimates. Due to the continuity constraint (8.35) imposed on the optimum warping path, the resulting sequence of period estimates will be relatively smooth. Thus the recursion (8.41) is performed once for the speech signal as a whole, and only one determination of the optimum path is performed in this matrix. Since a period estimate is computed for each signal sample, some type of low-pass filtering and downsampling may be used. The main drawbacks of this method are the storage requirements and that no on-line computation is possible since the optimum path can be determined only after all speech samples have been acquisited. By introducing a certain time delay, say 100 ms, and storing only the equivalent part of the matrix P(k,m), we can expect that the optimum path will lock onto the true fundamental period and track it, thus allowing virtually on-line computation. Anyway a significant improvement upon the second method is achieved due to the following reason. Considering only the current column of the matrix P(k,m) as in the second method does not guarantee global optimality since the minimum of P(m) may be attained at positions different from the a posteriori optimum path. (Ney, 1982, pp.387-388)

8.3.3 Nonstationary Short-Term Analysis and Incremental Time-Domain PDAs

In Sect.2.6 we have categorized the PDAs into the two categories *time-domain* and *short-term analysis* PDAs. The criterion for this categorization was the independent variable of the signal at the input of the basic extractor. If this signal has the same time base as the input signal, the PDA works in the time domain; otherwise it is a short-term analysis PDA. During the discussions in Chaps.6-8 we found that time-domain PDAs are usually able to track the signal period by period, and, in addition, to synchronize and lock onto a reference point in the period (which depends on the particular method applied). They are thus able to act as a phase-locked circuit with respect to the first harmonic, a principal peak, the instant of glottal excitation, or another significant event in the period. Short-term analysis PDAs, when applying stationary short-term analysis, neither process the signal period by period nor are able to lock onto a certain event in the period. There are, however, a number of time-domain PDAs which process the signal period by period but cannot synchronize with a defined event within an individual period. On the other hand, there are short-term analysis PDAs which can apply an (arbitrarily) short frame and are thus able at least in principle to get away from the frame-by-frame processing mode towards processing in a period-by-period mode.

From this short discussion we can conclude that time-domain and short-term analysis PDAs are separated by two features: 1) the ability to process the signal period by period, and 2) the ability to synchronize with a certain event in an individual period. A genuine time-domain PDA is able to do both, and a genuine (i.e., stationary) short-term analysis PDA has none of these abilities. Since 2) requires 1) to be given, there is only one more possibility: PDAs that can process the signal period by period but cannot synchronize with a certain event in the period. These PDAs form a group of their own between the two categories. We will label them *nonstationary short-term analysis* PDAs or *incremental time-domain* PDAs. Seen from the time-domain analysis point of view, these PDAs operate incrementally. They cannot synchronize by themselves, so they must be synchronized externally. Once this has been done, the PDA takes the instant of the external synchronization as the reference event and operates from there in a period-synchronous way until it falls out of synchronization due to a momentary failure or until the voiced segment ends. Once out of synchronization, the PDA must be externally synchronized again. The "vector" PDA by Rader (1964; see Sect. 6.3.2) is typical for this kind of operation.

The majority of PDAs pertaining to this group have been discussed elsewhere in this book. From the time-domain PDAs these are 1) all the PDAs of type PiTLO and PiTNO, i.e., all the PDAs with basic extractor omitted, for reasons of implementation; 2) all the multichannel PDAs where synchronization problems are associated with the channel selection (cf. Sects. 6.4.1,2); 3) other PDAs with parallel processors, such as the one by Gold and Rabiner (1969; cf. Sect.6.2.4); and 4) individual PDAs that operate in an incremental mode, such as the vector PDA by Rader (1964) mentioned above. Of the short-term analysis PDAs, all those which apply nonstationary short-term analysis can operate in this mode (although most of them do not). Examples of synchronous short-term analysis PDAs are the PDAs by Stone and White (1963; Sect. 8.3.1) and by Sung and Un (1980) which follow the AMDF principle; another example is the period histogram PDA by R.L.Miller (1970; to be discussed in Sect. 8.5.1).

Few algorithms thus remain to be discussed in this section. One of them is the PDA by Friedman (1979). Having developed two algorithms which are based on the maximum-likelihood principle (Friedman, 1977b, 1978; cf. Sects.8.6.1,2), Friedman looked for a computationally more efficient but still sufficiently reliable algorithm. He kept the preprocessor of the multichannel maximum likelihood (ML) approach (Sect. 8.6.2; Figs.8.43,44) and replaced the ML estimators by simpler basic extractors. The algorithm thus operates as follows. The frequency band between 0 and 5 kHz is split up into 16 subbands (channels) of equal bandwidths. In each channel the instantaneous envelope of the signal is computed. So far this PDA corresponds to Friedman's former algorithm (1978) as well as to the PDAs by Smith (1954, 1957) and Yaggi (1962). The zero frequency component is then removed in each channel by a (digital) high-pass filter, and a ZXABE is applied. In principle this PDA could act like a time-domain PDA (for an example, see the signal in Fig.6.36). However, it measures T_0 by a majority vote in a short-term analysis mode. The ZXABEs in the individual channels pass their estimates of T_0 to the central basic extractor which forms a histogram of them. As can be seen from the example in Fig.6.36, the instantaneous envelope oscillates with the frequency F_0 in all channels. It thus has two and only two zero crossings per period, and all the partial estimates coherently contribute to the histogram to form a significant peak at T_0 (after some smoothing). After determination of an estimate the histogram is reset to zero. According to the author the performance of this PDA is comparable to that of the cepstrum (cf. Sect.8.4.2) or the ordinary ML PDA (cf. Sect.8.6.1), but inferior to the multichannel ML PDA (cf. Sect.8.6.2). An advantage of this PDA is that the frame length can be dynamically varied according to external conditions: one period is the minimum, and the PDA thus is principally able to analyze signals with irregular excitations; if the signal-to-noise ratio is bad, the frame can be made longer, and the reliability of the final estimate can be increased by averaging the estimates of several periods.

The PDA by Moser and Kittel (1977), designed for research in logopedics and phoniatrics, applies the AMDF together with a time-domain principle. The speech signal is crudely sampled at a 5.5-kHz sampling rate, and the AMDF is applied in the usual way (as described in Sect.8.3.1) to determine the fundamental period T_0. The raw estimate obtained by the AMDF analysis is then refined by a time-domain technique. Starting from a significant zero crossing in the signal within the frame used by the AMDF, the algorithm advances by one period, as indicated from the raw estimate, and looks for another significant zero crossing of the same polarity. The accuracy of the two zero crossings is increased by interpolation, and their distance is taken as the refined estimate of T_0. The PDA was reported to be accurate to within 0.2% at $F_0 = 100$ Hz (which would correspond to a sampling rate of 50 kHz without interpolation). Since the estimate is based on only one period, the PDA can be used to measure signals with pitch perturbation and aperiodic excitation, as given in hoarse and pathologic voices.

Atal (1975, unpublished) proposed a PDA which is based on high-accuracy linear-prediction (LPC) analysis. A brief description of this PDD has been presented in the comparative evaluation by Rabiner et al. (1976, see Sect.9.3.1); a few more details have been added in a paper by Schroeder (1981b). The principle of this PDA is as follows (Fig.8.20). First the signal is

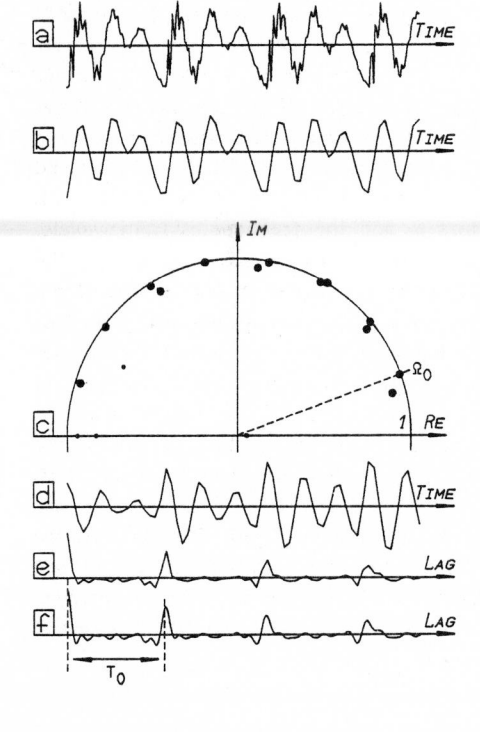

Fig.8.20a-f. PDA by Atal (1975, unpublished): example of performance. (a) Input signal (sustained vowel /e:/, male speaker, sampling frequency 8 kHz); (b) input signal after low-pass filtering and downsampling to a sampling rate of 2 kHz; (c) location of the poles of the LPC filter in the z plane; (d) impulse response of the LPC filter; (e) impulse response of the filter after elimination of all poles that were either real or had too large bandwidths [indicated by the small circles in (c)], and after the Newton transformation; (f) same as (e) with a sampling frequency of 16 kHz. The LPC analysis was performed using the covariance method with 41 poles. The impulse response (d) shows that this filter is unstable. In (c) stability was achieved by forcing all poles inside the unit circle. (Ω_0) Location of the fundamental frequency F_0 on the unit circle in the z plane

low-pass filtered and downsampled to a sampling frequency of 2 kHz. A 41th-order LPC analysis is then carried out on a 40-ms frame of the low-pass filtered signal. In contrast to usual LPC analysis where the number of poles is fairly low compared to the frequency range scanned by the analysis, the LPC algorithm here models the fine structure of the signal. In other words, we can expect that a complex pole pair is assigned to each harmonic of the signal. The order of the LPC analysis guarantees that this holds even at the lower end of the measuring range. (For $F_0 = 50$ Hz, for instance, 20 harmonics can be in the signal below 1000 Hz. If each harmonic is represented by a complex pole pair, 40 poles are necessary for the analysis.) A Newton transformation is then performed on the predictor filter; it converts the LPC filter into another digital filter where the residues at the locations of all poles equal unity. Hence all the damped oscillations corresponding to the individual poles obtain equal amplitude and zero phase. The impulse response h(n) of this filter will thus have sharp peaks at $n = 0$ and $n = kT_0$, $k = 1 (1) \ldots$.

To better understand how this algorithm works, let us briefly regard a purely recursive second-order digital filter with a complex pole pair. The impulse response of this filter is an exponentially damped sinusoid of the form

$$h_1(n) = A_1 \exp(-a_1 n) \cos(f_1 n + \phi_1) ,$$
(8.44)

where A_1 is the amplitude, a_1 the attenuation factor, f_1 the frequency, and ϕ_1 the phase. If we have a higher-order filter which is assumed to only have complex poles, then the impulse response of this filter is a sum of $k/2$ exponentially damped sinusoids, k being the order of the filter,

$$h(n) = \sum_{i=1}^{k/2} h_i(n) = \sum_{i=1}^{k/2} A_i \exp(a_i n) \cos(f_i n + \phi_i) .$$
(8.45)

It can be easily shown that, if we add a nonrecursive part to the recursive filter (i.e., if we add zeros to the transfer function which was hitherto all-pole), the amplitudes A_i and the phases ϕ_i of the individual waveforms $h_i(n)$ will change. The attenuation factors a_i and the frequencies f_i, however, are uniquely determined by the locations of the poles and will thus remain unchanged. By the Newton transformation the transfer function of the filter is now modified in such a way that the amplitudes A_i are forced to unity, and the phases are forced to zero, so that we obtain

$$h_N(n) = \sum_{i=1}^{k/2} \exp(a_i n) \cos(f_i n) .$$
(8.46)

If the poles are equally spaced along the unit circle of the z plane, i.e., if the filter represents a harmonic structure with the fundamental frequency F_0, ths impulse response is approximately

$$h_N(n) = \sum_{i=1}^{k/2} \exp(a_i n) \cos(i n F_0) .$$
(8.47)

This waveform, however, will have its principal peak at $n = 0$ and at multiples of T_0, and a peak detector is able to easily locate T_0 from this impulse response. If the order of the LPC analysis is higher than twice the number of harmonics in the signal, the superfluous poles will either lock onto the harmonics as well, or they will be real (i.e., not complex), or they will represent resonances with exceedingly large bandwidths. In this last case the pertinent oscillations will rapidly decay; they no longer contribute to the peak at $d = T_0$.

Given the recursive LPC filter, one can in principle compute the coefficients of the pertinent nonrecursive filter so that (8.47) is fulfilled. According to the theory of functions, A_i and ϕ_i are determined by the residues of the transfer function $H(z)$ at the pertinent pole pair $z = z_{pi}$ and $z = z_{pi}^*$ which can be evaluated depending on the predictor coefficients (i.e., the coefficients of the recursive filter which are known) and on the unknown coefficients of the nonrecursive filter. Forcing the residues to unity then yields a linear equation system by which the nonrecursive part of the filter can be determined. In practice, however, it is unnecessary to determine the nonrecursive filter which forces the residues of the LPC filter to unity. If the poles of the LPC filter are explicitly known, we also know the respective frequencies f_i and attenuation factors a_i so that $h_N(n)$ can be directly generated via (8.44,46). To determine the poles of the LPC filter, standard routines are available in most mathematical program libraries. In a practical implementation one could thus proceed as follows.

1) Perform the 41-pole LPC analysis on the low-pass filtered and down-sampled signal, as specified in the original algorithm.

2) Determine the roots of the predictor coefficient polynomial, i.e., the poles of the transfer function of the LPC filter.

3) The high-order LPC analysis assigns the poles of the LPC filter to the harmonics of the input signal. Superfluous poles will be irregularly spaced on the real axis of the z plane, or they will have large bandwidths. Such poles can be excluded from further processing once they are explicitly known. In addition, stability of the filter is enforced by keeping all poles within the unit circle.

4) For each complex pole pair that has not been excluded, generate a waveform $h_i(n)$ as given by (8.44), with the frequency f_i and the attenuation factor a_i as specified by the location of the pole pair in the z plane. Add all these waveforms to obtain $h_N(n)$.

5) Determine T_0 from $h_N(n)$ as the location of the first significant peak.

This algorithm has a number of advantages. First, one can expect that all harmonics coherently contribute to the information on pitch, whereas spurious components add incoherently or rapidly decay. The algorithm thus becomes very reliable. Second, this algorithm is one of the rare examples of PDAs working in the lag domain without having trouble with quantization of the lag. Since the z plane is not sampled but represents a continuum, the locations of the poles are not quantized (assuming that the representation in the computer is accurate enough). We can thus synthesize the impulse response $h_N(n)$ with an arbitrarily fine sampling interval and reduce the measurement inaccuracies due to quantizing to almost zero.

The computational effort of this algorithm is fairly high (Rabiner et al., 1976). Once a reliable estimate of pitch is known, however, the order of the LPC analysis can be reduced if the momentary value of F_0 is sufficiently far away from the low-frequency end of the measuring range.

Atal's PDA was included in the comparative evaluation by Rabiner et al. (1976) where it consistently received the best ranking (objectively and subjectively) with respect to gross and fine pitch determination errors. More details about its performance will be added in Section 9.3.1.

8.4 Multiple Spectral Transform ("Cepstrum") Pitch Determination

The principle of cepstrum pitch determination culminates in just one algorithm (Noll, 1964, 1966, 1967, 1968b, 1970; Noll and Schroeder, 1964). The cepstrum PDA was the first short-term analysis algorithm that proved realizable on a digital computer, and it paved the way for the whole area of short-term analysis PDAs since it was reliable to a hitherto unimaginable extent. For a long time this algorithm therefore has served as a calibration standard for other, simpler PDAs.

The cepstrum is the inverse Fourier transform of the logarithm of the (short-term) power spectrum of the signal. By the logarithmic operation, the voice source and the vocal tract transfer functions are separated. Consequently, the pulse sequence originating from the periodic voice source reappears in the cepstrum as a strong peak at the "quefrency" (lag) T_0, which is readily discovered by the peak-picking logic of the basic extractor.

In the beginning, an extreme computational burden accompanied the increased reliability. (Remember that this PDA came out before the fast Fourier transform was rediscovered.) Therefore a number of modifications and proposals were made, partly by Noll himself (1968b), partly by others, in order to speed up the algorithm. Nowadays real-time cepstrum analysis is possible. Nevertheless these modifications are of more than historical interest since they elucidate the role of certain preprocessing methods, such as center clipping.

8.4.1 The More General Aspect: Deconvolution

According to the simplified model of speech production (see Sect.2.5) the speech signal is built up from three processes: 1) periodicity, i.e., a periodic pulse train, 2) the glottal source characteristics except periodicity, and 3) the vocal tract transfer function, including radiation characteristics. Neglecting the mutual dependencies, one can think of the three sources of influence to be switched in cascade so that the respective transfer functions have to be multiplied,

$$X(z) = P(z) \ S(z) \ H(z) \ , \tag{8.48}$$

where $P(z)$, $S(z)$, and $H(z)$ represent the transfer functions of the pulse train, glottal source, and vocal tract, respectively. In the time domain this denotes that the three respective time functions have to be convolved,

$$x(n) = [p(n) * s(n)] * h(n) \ . \tag{8.49}$$

The task of pitch determination, briefly expressed, is to isolate $p(n)$ or $P(z)$ from the other two terms. For (8.49) this means reversal of the convolution, i.e., the deconvolution of the speech signal $x(n)$ in order to obtain $p(n)$.

Deconvolution is a more general task in digital signal processing. Provided that one of the transfer functions is known or well estimated, its influence can be removed by frequency domain division, e.g.,

$$P(z) \ S(z) = X(z) \ / \ H(z) \ . \tag{8.50}$$

In the case of the vocal tract transfer function $H(z)$ this process, defined by (8.50), is known as "inverse filtering." If $H(z)$ represents a linear filter, it is the quotient of two polynomials, according to (2.11). Thus the inverse filter also is linear, and provided that the original filter has no zeros on or outside the unit circle, it is stable. For the vocal tract transfer function, after Fant (1960, 1970), an all-pole model is assumed to be accurate for at least vowels and nonnasal voiced consonants. Thus the inverse filter will become an all-zero filter, i.e., in digital terms, a nonrecursive filter (which is unconditionally stable). Linear prediction (Atal and Hanauer, 1971; Itakura and Saito, 1968; Makhoul and Wolf, 1972; Markel and Gray, 1976) is one of the short-term methods used to estimate the coefficients of this filter. Of course these coefficients are time variant since $H(z)$ itself is time variant. PDAs using inverse filtering in one or the other approximation have already been discussed in this book (Sects.6.3 and 8.2.5). Equation (8.50) has been directly implemented in the PDA by Fujisaki and Tanabe (1972, cf. Sect.8.2.4).

Inverse filtering, however, is only one of several possibilities for carrying out deconvolution. A different approach is the use of the (complex) logarithm in the frequency or the z domain in order to convert the multiplicative relation (8.48) into an additive one,

$$\log X(z) = \log P(z) + \log S(z) + \log H(z) \ . \tag{8.51}$$

If $\log X(z)$ is transformed back to the time domain, the signal obtained is drastically different from the original time-domain signal. A sequence in which the sources of influence are superimposed in an additive way results from this computation, and the individual components are expected to be much more easily separated than in the original signal.

According to the generalization of a term created by Bogert et al. (1963), the inverse Fourier transform of $\log X(z)$ is called the complex *cepstrum* or, according to a definition by Oppenheim (see Oppenheim and Schafer, 1975), a special case of *homomorphic processing*. The pertinent signal processing technique is called the *cepstrum technique*. For pitch determination the approach is not needed in this general form. But for didactic reasons it appears useful to add some more details to this aspect. Most of the following discussion is drawn from the tutorial paper by Childers, Skinner, and Kemerait (1977).

If this approach is valid for the whole z plane as far as the z-transform $X(z)$ exists, it is in particular valid for the unit circle. The condition that $X(z)$ has to exist on the unit circle implies that the three functions $P(z)$, $S(z)$, and $H(z)$ are all regular there. For $S(z)$ and $H(z)$ this is obvious whereas $P(z)$, in the case of a signal of infinite duration, represents the z-transform of an undamped time function which has poles on the unit circle. As soon as the signal becomes finite, however, the pertinent z-transform again becomes regular on the unit circle. For speech signals, and especially for (short-term) signal frames, the validity of (8.51) on the unit circle of the z plane can thus be assumed, and (8.51) can be rewritten in terms of the discrete Fourier transform. This change is possible since the convolution theorem is valid for the Fourier transform as well as for the z-transform,

$$\log X(m) = \log P(m) + \log S(m) + \log H(m) , \tag{8.51a}$$

where $A(m)$ (A stands for X, P, S, and H, respectively) represents the DFT of the signal $a(n)$. With this approach it is possible to implement the cepstrum technique using the FFT. For speech processing further simplification is possible. Equation (8.51a) is likewise valid for the power or the amplitude spectrum,

$$\log |X(m)| = \log |P(m)| + \log |S(m)| + \log |H(m)| . \tag{8.52}$$

Table 8.2. Some terms of cepstrum processing and the corresponding frequency-domain terms. After Bogert et al. (1963) and Childers et al. (1977)

Starting point	Time-domain signal $x(n)$	Frequency-domain signal (spectrum) $X(m)$
Result after Fourier transform	spectrum $X(m)$	cepstrum $x(d)$: Fourier transform from $\log [X(m)]$
Independent variable	frequency	quefrency[a] (= lag!)
Filter terms	filtering	liftering
	harmonic	rahmonic
	low-pass	short-pass
	high-pass	long-pass

[a] Since the *quefrency* is a measure of time, it is equivalent to the lag in the autocorrelation or AMDF domain. For reasons of unification it would be appropriate to label the independent variable of the cepstrum *lag* as well (and there are already some publications that use this term). To remain compatible with the majority of the published literature on cepstrum processing, however, the term *quefrency* is maintained here.

Since amplitude and power spectra are real nonnegative functions, the complex logarithm is replaced by the ordinary logarithm. The power spectrum, in addition, is an even function, so that the transformation back into the time domain (leading to the *power cepstrum*) can be done by a real (cosine) Fourier transform. In the following, the considerations will be confined to this power cepstrum which will henceforth be simply called *cepstrum* (as in the publications by Noll who first applied this technique to speech processing).

One more word about the terminology. In the digital signal representation, i.e., when the signal is stored in computer memory, there is absolutely nothing which forces us to assume that this is a time signal. What we have in the computer is a sequence of numbers which, by convention, are samples of a function sampled in regular intervals of an independent variable which happens to be time. It does not make any difference to the signal representation if this independent variable is frequency or something else. In other words, the methods of digital signal processing (filtering, weighting, DFT, etc.) can be applied to a spectrum in the same way as with respect to a (finite) time signal. To avoid confusion, and to give a clear distinction as to the domain in which these methods are applied, Bogert et al. (1963) have introduced a special terminology, the main terms of which are listed in Table 8.2. This terminology, built up by syllabic mutation of usual English frequency-domain terms (and thus untranslatable!), was completed by Childers et al. (1977).

8.4.2 Cepstrum Pitch Determination

Let us now go back to the problem of pitch determination. The principle once known, the details of the pertinent PDA are presented rather quickly. If a periodic signal of period duration T_0 consists of many adjacent harmonics, the

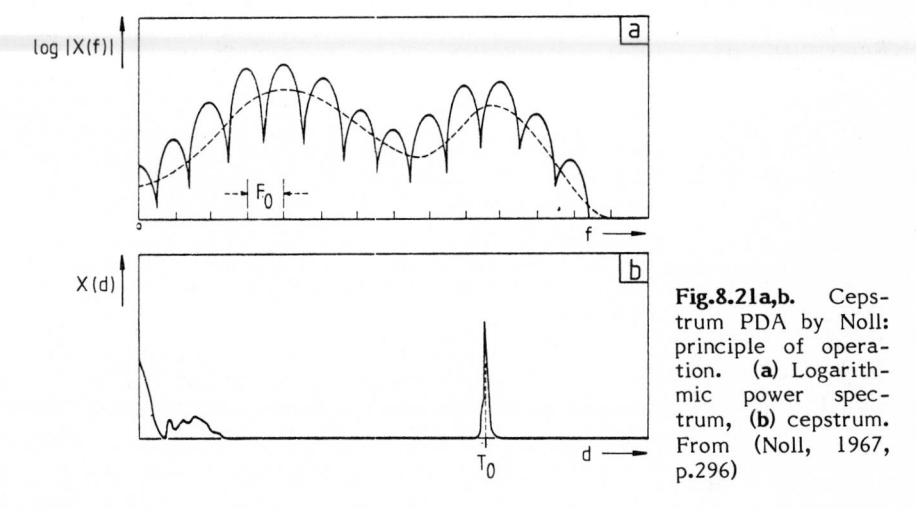

Fig.8.21a,b. Cepstrum PDA by Noll: principle of operation. (a) Logarithmic power spectrum, (b) cepstrum. From (Noll, 1967, p.296)

corresponding short-term spectrum exhibits a ripple due to its harmonic structure, provided that the analyzed frame contains several periods. This is valid
in particular for the logarithmic power spectrum where this ripple gets a
cosinelike shape (Noll, 1967), and where the dynamic range within the spectrum is greatly reduced. The IDFT of this signal, i.e., the cepstrum, exhibits
a strong peak at a quefrency d equal to the period duration T_0 (Fig.8.21).
According to (8.52), we obtain

$$x(d) = p(d) + s(d) + h(d) \tag{8.53}$$

in the cepstrum domain so that the three signals convolved in the time signal
$x(n)$ are now connected in an additive way. Since $p(n)$ is defined to be a pulse
train with no inherent information other than its periodicity, this function is
transformed in the cepstrum into another pulse train starting at quefrency
$d = 0$. This is only approximately valid since the multiplication of the time-
domain signal by a window in order to apply short-term analysis leads to the
already mentioned cosine-shape form of the spectral ripple.[16] Thus the cepstrum contains only few significant "rahmonics" of the ripple at $d = T_0$, $d = 2T_0$,
etc. (Fig.8.21).

Detection of the significant peak at $d = T_0$ in the cepstrum $x(d)$ forms the
principle of the cepstrum PDA, developed by Noll (1964, 1967) according to
an idea by Noll and Schroeder (1964). Figure 8.22 shows the block diagram of
this PDA. The unprocessed speech signal is segmented into 51.2-ms frames
(yielding 512 points for the FFT); the frame interval is set to 10 ms. Each
frame is weighted by a Hann window and then transformed into the cepstrum.
The peak picking basic extractor searches for the principal peak in the range
1-15 ms after appropriate weighting of the cepstrum. Figure 8.23 shows an
example of the performance.

> The examples of cepstra indicate that the cepstral peaks are clearly de-
> fined and are quite sharp. Hence, the peak-picking scheme is to determine
> the maximum value in the cepstrum exceeding some specified threshold.
> Since pitch periods of less than 1 ms are not usually encountered, the
> interval searched for the peak in the cepstrum is 1-15 ms.
>   Since the cepstral peaks decrease in amplitude with increasing que-
> frency, a linear multiplicative weighting was applied over the 1-15 ms
> range. The weighting was 1 at 1 ms and 5 at 15 ms. The Fourier transform
> of the power spectrum of the time window equals the convolution of the
> time window with itself. ... Thus, the higher-quefrency components in the
> power spectrum decrease as the time window convolves with itself. Al-

---

[16] This ripple is due to the correspondence multiplication-convolution and
to the fact that the signal is usually weighted in the time domain before
application of the Fourier transform. In the spectral domain the time-domain
weighting (which means multiplication) results in a convolution between the
spectrum of the original time function and that of the window. Usually the
spectrum of a finite window has some cosine-shape main lobe around zero
frequency (see Chap.2 for more details). The spectrum which is a pulse train
in the case of an infinite periodic time function is thus converted into a
sequence of cosine shaped pulses.

though the mathematics becomes unwieldy for an exact solution, it is reasonable to expect the higher-quefrency components in the logarithm of the power spectrum to decrease similarly, thereby explaining the need of weighting of the higher quefrencies in the cepstrum. The linear weighting with range of 1-5 was chosen empirically by using periodic pulse trains with varying periods as input to the cepstrum program.

The cepstral peaks at the end of a voiced-speech segment usually decrease in amplitude and would fall below the peak threshold. The solution is to decrease the threshold by some factor two over a quefrency range of ±1 ms of the immediately preceding pitch period when tracking the pitch in a series of voiced-speech segments. The threshold reverts to its normal value over the whole cepstrum range after the end of the series of voiced segments (Noll, 1967; p.303).

The following correction logic is typical for high quality short-term analysis PDAs. This logic is mainly designed to correct local errors and to watch the cepstra at the voiced-unvoiced transitions. The most frequent local error to occur is "pitch doubling," i.e., subharmonic tracking at half the fundamental frequency. A simplified block diagram of the correction logic (all the checks dealing with voiced-unvoiced decision are excluded) is shown in Figure 8.24. The threshold THR, whose determination is unspecified except for the fact that it is set to some predetermined value, plays an important role. It is used to decide whether the principal peak of the cepstrum is likely to represent an estimate of period at all.

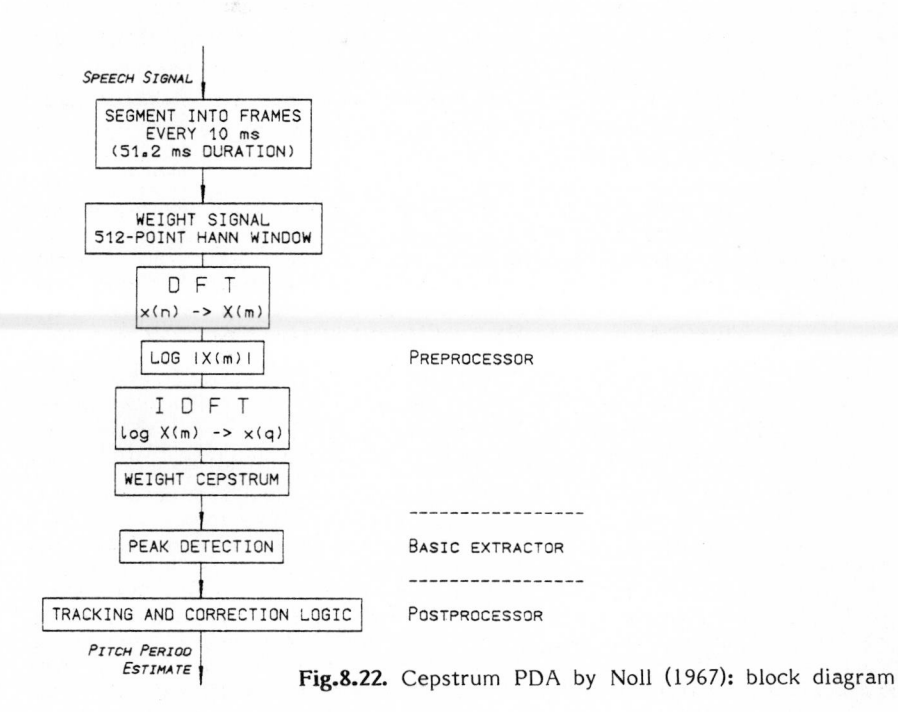

Fig.8.22. Cepstrum PDA by Noll (1967): block diagram

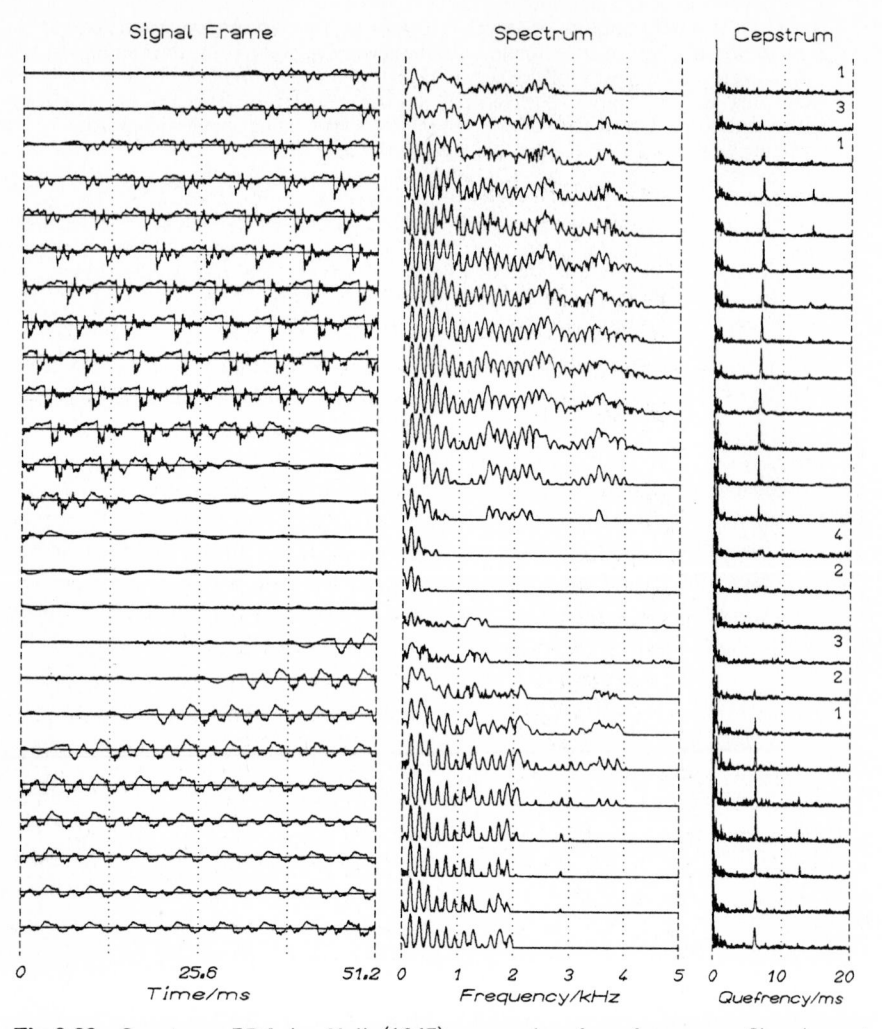

Fig.8.23. Cepstrum PDA by Noll (1967): example of performance. Signal: part of the utterance "Algorithms and devices for pitch determination," male speaker, undistorted recording. Time runs from top to bottom with a 12.8-ms interval between two successive frames. The numbers written at the right-hand side of a part of the cepstra indicate how many spurious peaks (maxima exceeding 70% of the amplitude of the principal maximum) have been encountered within the measuring range of T_0 (1-20 ms). In the spectra a dynamic range of 40 dB has been displayed; all spectral values further below have been set to -40 dB

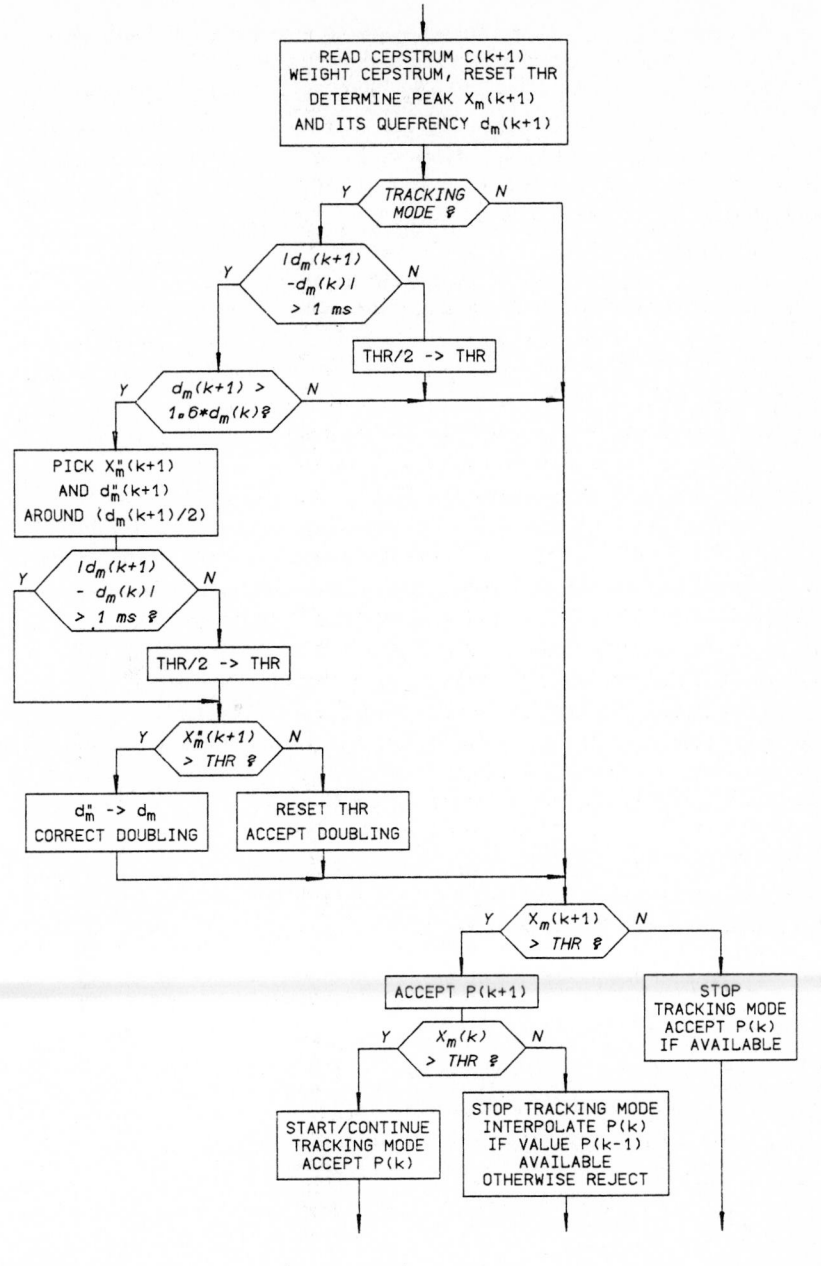

Fig.8.24. Cepstrum PDA (Noll, 1967): Simplified tracking and correction logic (the part dealing with the voiced-unvoiced decision has been left out). (THR) Adaptive threshold which must be exceeded in order to provide a clear estimate for T_0

Before deciding about the "present" cepstrum, knowledge about the preceding and following cepstrum is required for the algorithm used to eliminate another problem, namely, pitch doubling. ... The second rahmonic of a cepstral peak sometimes exceeds the fundamental, and the second rahmonic should not be chosen as representing the pitch period. Thus, the peak-picking algorithm should eliminate false pitch doubling caused by a second rahmonic but should also allow legitimate pitch doubling. For legitimate doubling, there is no cepstral peak at a one-half quefrency, but for erroneous doubling, there is such a peak at one-half quefrency since this is the fundamental. The algorithm capitalizes upon this observation by looking for a cepstral peak exceeding the threshold in an interval of ±0.5 ms of one-half the quefrency of the double-pitch peak. If such a peak is found, then it is assumed that it represents the fundamental, and the double-pitch indication is wrong. The threshold is reduced by a factor of 2 if the maximum peak in the ±0.5-ms interval falls within ±1.0 ms of the immediately preceding pitch period. Pitch doubling has occurred whenever the cepstral peak exceeding the threshold is at a quefrency of more than 1.6 times the immediately preceding pitch period (Noll, 1967, p.305).

For simplification of the cepstrum PDA, Noll (1968b) proposed the *clipstrum* where the time signal as well as the logarithmic spectrum are infinitely peak clipped before the FFT. The performance is inferior to the cepstrum (see Fig.8.25 for an example), and the gain in computing speed is modest. The development of the clipstrum must be understood historically. Like the cepstrum, the clipstrum PDA was developed at a time when the FFT had not yet been rediscovered. In the ordinary discrete Fourier transform (the "slow" DFT), infinite peak clipping allows the multiplications to be completely replaced with additions, so that the gain in computation speed is considerable. In the FFT, however, only the first of (ld N) steps (N is the transformation interval) can be accelerated that way.

The primary motivation for developing the clipstrum was to simplify the hardware requirements of the spectrum analyzers. However, the fast Fourier transform (FFT) has greatly reduced the number of digital multipliers, but a FFT clipstrum would only eliminate the very first stage of digital multipliers. Therefore, it becomes quite debatable whether the clipstrum can be justified from the standpoint of hardware savings in the FFT age. But, the clipstrum might nevertheless offer certain hardware simplifications. ... (Noll, 1968b, p.1591)

From today's point of view, the most important statement of that paper could be the one about center clipping:

It has been known for some time that center clipping of speech nearly completely destroys intelligibility by eliminating much of the formant structure of the speech signal (Licklider and Pollack, 1948). However, the center-clipped speech is still periodic, and in fact Sondhi (1968) has reported that center clipping can greatly improve many simple forms of pitch detectors.
        The amount of center clipping is specified by a threshold ... which ... varies with time and is specified as some constant fraction of the absolute maximum of the speech signal within the analysis interval. ... The center-clipping fraction was 70% of the absolute maximum of the speech signal

within each 40-ms interval[17]. ... The clipstrum of the center-clipped speech shows considerable improvement over the straightforward clipstrum and actually gives the pitch for intervals missed even by the cepstrum.

The cepstrum of center-clipped speech was investigated by computer simulation a few years ago when Sondhi first suggested center clipping for removing the formant structure of the speech in order to improve the operation of an autocorrelation pitch detector. However, the conventional cepstrum was a trifle superior to the cepstrum of center-clipped speech (Noll, 1968, p.1590).

These results suggest that the difficulties in autocorrelation analysis and cepstrum PDAs with the formant structure are similar. Yet this is not quite true. Ordinary AC analysis fails when there is a predominant formant (especially F1) at some higher harmonic. Cepstrum analysis, in contrast, fails as soon as the presumption is violated that the signal contain many adjacent harmonics. The cepstrum PDA would thus completely fail if the input signal were a pure sinusoid at the fundamental frequency, a case AC analysis easily copes with. The problem of the cepstrum is the absence of adjacent harmonics rather than a predominant formant structure. A predominant formant structure, on the other hand, can easily be removed in the frequency domain by a "long-pass lifter," as is necessarily done in the clipstrum PDA before clipping the spectrum. The reason for the good performance of center clipping for both AC and cepstrum pitch determination is that center clipping does both: it removes the formant structure and, at the same time, adds a great number of higher harmonics to the signal.

At the time it was published, the cepstrum PDA marked a milestone in the development of PDAs. It was about the first PDA which could reliably cope with the whole range of F_0 without requiring a manual preset. For a long time this PDA has served as a standard for calibration of other newly developed PDAs, especially time-domain ones (see, for instance, Liedtke, 1971; Markel, 1972a,b; Moorer, 1974; Friedman, 1978; or Martin, 1981, 1982). Cepstrum processing has - or at least had until most recently - a serious disadvantage: the computing speed, or, rather, the computing slowness. Noll himself devoted considerable detail to an analog realization proposal using a

---

[17] Compared to later investigations (Dubnowski et al., 1976; Rabiner, 1977a) this center clipping threshold is so high that a number of periods were probably missed during rapid intensity changes. This might partially explain the disagreement between Noll's findings and what one might expect according to Sondhi's statement (1968), i.e., that (moderate) center clipping increases the number of harmonics and improves the overall performance of the PDA. On the other hand, the examples in Fig.8.25 show that the spectrum tends to get noisy when heavy center clipping is performed. This is probably due to the fact that no low-pass filtering is applied, and that interactions between major formants affect the results of center clipping in an unpredictable way. This again supports Noll's view that center clipping does not really improve the performance of the cepstrum PDA.

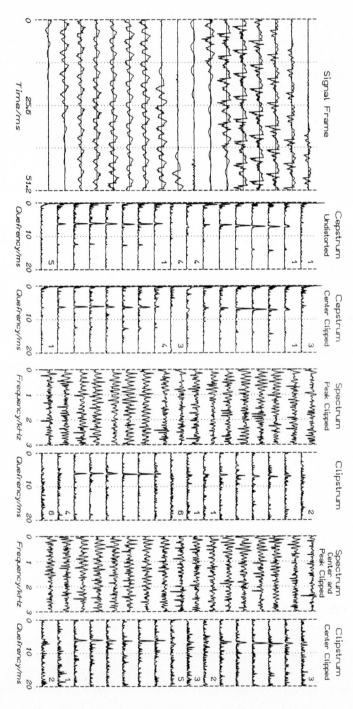

Fig. 8.25. Cepstrum and Clipstrum PDA (Noll, 1967, 1968b): comparative example of performance. The signal is the same as in Figure 8.21. Time runs from top to bottom; frame interval 25.6 ms. Prior to clipping, the spectra have been long-pass liftered. The spectra leading to the ordinary cepstrum and the cepstrum of center-clipped speech are not displayed. For further details, see the caption of Fig. 8.23

Signal Frame — Time/ms 0 25.6 51.2

Cepstrum Undistorted — Quefrency/ms 0 10 20

Cepstrum Center Clipped — Quefrency/ms 0 10 20

Spectrum Peak Clipped — Frequency/kHz 0 1 2 3

Clipstrum — Quefrency/ms 0 10 20

Spectrum Center and Peak Clipped — Frequency/kHz 0 1 2 3

Clipstrum Center Clipped — Quefrency/ms 0 10 20

sampled-data system.[18] Kelly and Kennedy (1966) reported a study using one of the sampled-data analog spectrum analyzers available at that time. A similar PDD was proposed by Weiss et al. (1966) with the additional feature that the logarithmic operation can be suppressed. They claimed that this processing is more appropriate for extremely noisy signals. In this case, however, one should not start the inverse transform from the power spectrum, which would result in the (short-term) autocorrelation function whose short-comings for pitch detection are well known. Spectra which have been nonlinearly processed in some other way will do a better job. Sreenivas (1981, see Sect.8.5.2), for instance, proposed the use of the fourth root of the amplitude spectrum instead of the logarithmic spectrum [i.e., $|X(m)|^{1/4}$ instead of $\log X(m)$], mainly for an arithmetic reason, however. It often happens that individual spectral samples become zero. In this case the logarithm would yield $-\infty$. The pertinent arithmetic alarm is easily avoided by adding a small constant to each sample of the power spectrum before calculating the logarithm. However, the advantage of the fourth-root amplitude spectrum is that, regarded as a whole, its characteristic is very similar to that of the logarithm, whereas for small amplitudes the fourth-root amplitude spectrum will be close to zero instead of going to $-\infty$. As the example in Fig.8.26 shows, there is indeed little difference between the performance of the cepstrum and that of the inverse Fourier transform of the fourth-root amplitude spectrum.

As the example in Fig.8.26 further suggests, it is not the effect of deconvolution which makes the the cepstrum PDA so successful, but the spectral flattening which is performed by taking the logarithm. (The aspect of deconvolution is more important for tasks such as cepstral smoothing, but not so significant for pitch determination.) The figure shows the performance of double-transform PDAs[19] which involve a different degree of amplitude compression in the spectral domain: no compression at all for the power spectrum, moderate compression for the amplitude spectrum, and severe compression for the fourth-root amplitude spectrum and the logarithmic spectrum. The greater the amplitude compression in the spectrum, the more pronounced and the narrower is the principal peak at T_0 in the pertinent IDFT. Cepstrum and

---

[18] Due to the recently developed charge-coupled circuits (CCD technology) which permit analog sampled-data hardware realizations without great cost and effort, these considerations may regain interest. In addition there is the general trend towards more powerful and at the same time cheaper digital hardware.

[19] PDAs of this type require two (discrete) Fourier transforms with nonlinear processing in the spectral domain. They will thus be called *double-transform* PDAs. The individual PDA is characterized by the way the spectrum is nonlinearly processed. Since it is not at all necessary that the nonlinear processing has a logarithmic characteristic, the cepstrum PDA represents one special case of this more general category. Except for the proposal by Weiss et al. (1966), however, the cepstrum PDA has been the only one of this category which has become widely known.

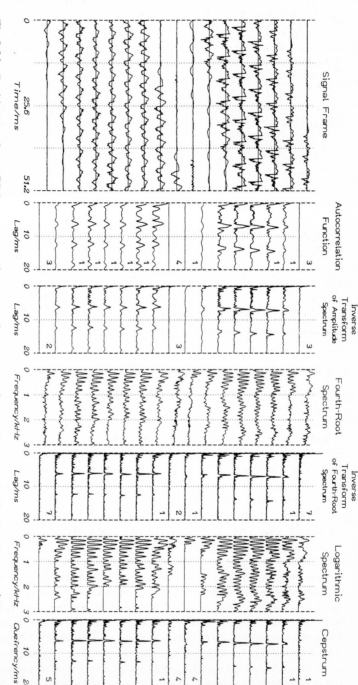

Fig. 8.26. Double-transform PDAs: comparative examples of performance. The signal is the same as in Figure 8.21. Time runs from top to bottom; frame interval 25.6 ms. Four double transforms have been displayed. The inverse transform of the power spectrum leads to the autocorrelation function; the inverse transform of the logarithmic spectrum leads to the cepstrum. For all inverse transforms the magnitude is shown. (The power and amplitude spectra were not displayed because they are rather uninformative.) The numbers at the right-hand side of a part of the inverse transforms indicate how many spurious peaks (maxima exceeding 70% of the amplitude of the principal maximum) have been encountered within the measuring range (1–20 ms)

IDFT of the fourth-root spectrum behave similarly; the autocorrelation function (emerging from the power spectrum) shows the well-known problems, as discussed in Sects.8.2.2,3. The performance of the IDFT of the amplitude spectrum is found somewhere in between. This experiment also supports Noll's findings that center clipping does not greatly improve the reliability of the cepstrum PDA. Although the methods are completely different, center clipping of the signal and logarithmic processing of the spectrum both achieve spectral flattening; therefore, if the spectrum has already been flattened by one of these methods, the other method is no longer able to significantly improve the overall performance.

The example in Fig.8.26 does not say anything about the noise resistance of the double-transform PDAs since the signal investigated has a high signal-to-noise ratio. From experiments by Noll (1970) it is known that the cepstrum PDA is not optimal with respect to noisy signals. Thus, as stated by Weiss et al. (1966), it becomes likely that the PDA will be more noise resistant when the spectrum is not as amplitude compressed as results from taking the logarithm. It is an open question, however, as to which degree of spectral amplitude compression represents the optimal tradeoff between the sensitivity to dominant formants, as given in the autocorrelation function, and the sensitivity with respect to noise, as given in the cepstrum.

8.5 Frequency-Domain PDAs

The obstacle to entering the frequency domain, which is presented by the Fourier transform, seems to give frequency-domain PDAs the reputation of being clumsy and nonversatile approaches to pitch determination. However, this is not true. Once the frequency domain has been entered, the measurement of F_0 is as easily and accurately possible as in the time domain or by other short-term analysis methods. One advantage of frequency-domain PDAs is that the accuracy of the F_0 estimate can easily be improved by interpolation. In addition, and in contrast to any PDA whose basic extractor works in the time domain or in a spectral domain with an independent variable equivalent to time (such as lag or quefrency), the spectral resolution and with it the accuracy of the measurement do not suffer from time-domain downsampling of the signal.

The easiest way to determine F_0 would be the extraction and identification of a spectral peak at the fundamental frequency itself. In order to yield a spectral peak at that frequency the first harmonic must be present in the signal. This cannot be expected. The way out of this calamity, however, is Thus, direct measurement of the spectral peak at the fundamental frequency F_0 proves unsuitable; the peak is often weak or totally absent, and the accuracy is bad since F_0 is located at the lower end of the frequency scale. Possible ways out of this dilemma are 1) measurement of F_0 as a fraction of the frequency of a higher harmonic, and 2) measurement of F_0 as the distance between adjacent spectral peaks caused by higher harmonics. Practically all the frequency-domain PDAs follow one of these concepts.

The first of the two concepts just mentioned leads to the principle of spectral compression. Again this is performed in two ways: a) compression of the spectrum along the frequency axis and subsequent addition of the original and compressed spectra, and b) search for an optimally matching harmonic structure. An important application is given by the relation of this principle to psychoacoustics. It allows developing PDAs that are oriented towards pitch perception (i.e., residue pitch or virtual pitch).

Section 8.5.1 deals with spectral compression. The principle is first described together with the pertinent histogram techniques (Schroeder, 1968; R.L.Miller, 1968, 1970; Kushtuev, 1969a,b) and then generalized towards the harmonic product spectrum (Noll, 1970; Schroeder, 1968). Section 8.5.2 deals with psychoacoustic PDAs (Terhardt, 1979a,b; Terhardt et al., 1982b; Duifhuis et al., 1978, 1979a,b, 1982) and harmonic matching (Martin, 1981; Sreenivas and Rao, 1979; Sreenivas, 1981).

The second concept, i.e., measurement of F_0 as the distance between adjacent spectral peaks, is discussed in Section 8.5.3. A refinement of this technique results in a combination of the two principles: first a crude but reliable estimate of F_0 is derived from adjacent peaks; then this estimate is used to identify a higher harmonic and to determine its harmonic number. The frequency of this harmonic, in turn, is then taken in order to improve the accuracy of the F_0 estimate. A computationally efficient realization (Seneff, 1978) is discussed in some detail. Section 8.5.4 finally deals with the issue of computational complexity, especially with possibilities of decreasing the computational effort of the FFT.

8.5.1 Spectral Compression: Frequency and Period Histogram,   Product Spectrum

Spectral compression, first proposed by Schroeder (1968), opens the way for accurate determination of F_0 from higher harmonics without requiring the respective harmonic numbers to be known. Given a short-term (power or amplitude) spectrum that contains many harmonics, the individual harmonics are detectable by a peak-picking procedure. Once a peak has been determined, the only disadvantage is that the machine does not know which harmonic it represents (contrary to the human spectrogram reader who will always know!). But this can be tried out: if several harmonics are present, the fundamental frequency is the greatest common divisor of the frequencies of the individual harmonics. According to Schroeder's proposal (1968), a histogram is built up, and the frequencies of all the spectral peaks are noted as its entries; the individual peaks are thought to represent the higher harmonics of the periodic input signal (Fig.8.27). The frequencies of the peaks are then divided by two and added to the histogram. If a first and a second harmonic have been present, the entry at the fundamental frequency (the first harmonic) is now

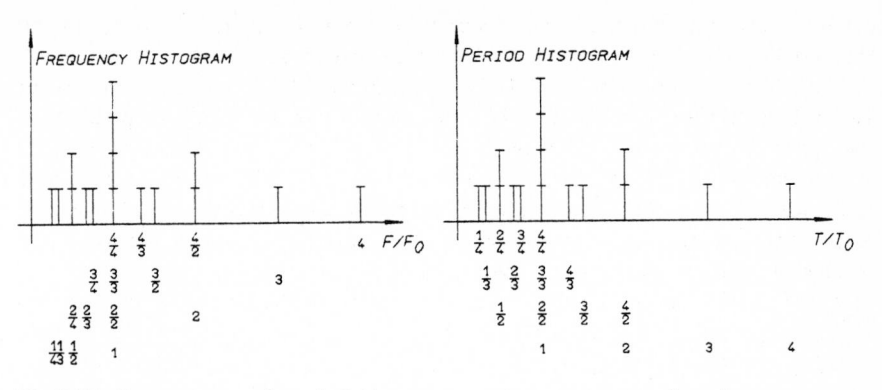

Fig.8.27. Frequency and period histogram. After Schroeder (1968)

occupied twice: by the original F_0 value and by the value of the second harmonic after dividing this frequency by two. The same procedure is repeated with compression factors of 3, 4, etc., so that the histogram finally has a distinct maximum at the fundamental frequency (Fig.8.27). The same procedure is possible in the time domain. The duration of the fundamental period is the smallest common multiple of the period durations of the individual harmonics. If thus the period duration or the waveform of higher harmonics is known without knowing which waveform represents which harmonic, a *period histogram* can be built up where all the possible period durations are entered and added up after expansion by a factor of 2, 3, 4, and so on. The period duration is finally measured as the highest peak in the histogram. To verify this system, the individual harmonics must be separated using a filter bank with equally spaced narrow band-pass filters. These filters, however, do not need to have nearly as good frequency resolution as required for a direct frequency-domain F_0 determination; the band-pass filters must only be spaced in such a way that not more than one harmonic of the signal can be present within a single channel. This means that the individual filters must have bandwidths about equal to the lowest fundamental frequency to be measured.

In this histogram (frequency or period) each entry is marked by just one bit: the fact of its existence. The second proposal by Schroeder (1968) deals with the weighted histogram where the individual entries are weighted by their respective amplitudes:

> The frequency and period histograms do not reflect the *amplitudes* of the harmonics whose frequency or period have been measured. Since, in the presence of noise, a higher signal amplitude usually means a better accuracy and reliability of the frequency measurement, it may be advantageous to "weight" the contribution of each harmonic to the histogram with a monotonically increasing function of its amplitude. Weighting functions that have been found useful are the amplitude and the logarithm of the amplitude of the measured harmonics (Schroeder, 1968; p.831).

Hardware realizations of both the period and the frequency histogram have been reported by Schroeder (1970a) in a patent. Another version of this principle was realized in analog technology by R.L.Miller (1968, 1970) as the *experimental harmonic identification pitch extraction* (HIPEX)[20] system. Note that this PDA is a short-term analysis PDA although it works in the time domain. The HIPEX system (Fig.8.28)

> ... consists primarily of a set of narrow-band filters, each being connected to a period translator and an amplitude weighting circuit (rectifier, low-pass filter, and multiplier). The period translator carries out the function of measuring the period of the filter output signal as given by zero-axis crossings. [in the terminology of this book, the period translator is a time-domain ZXABE which here operates on a higher harmonic.] This measure is then used to control the period of a pulse generator [a voltage controlled oscillator, VCO] linearly related to the period of the harmonic wave from the filter. The amplitude of the pulsed wave is weighted linearly with respect to the measured amplitude of the harmonic wave. An important feature of the various channel period translators is that the output pulse trains can all be initiated at a common point in time by a clock pulse, whenever it is designed to obtain a measure of the pitch period. The output pulse trains are summed together to give what has been termed a period histogram (Miller, 1970, p.1594)

The capability of the basic extractor to synchronize all the pulse generators of the period translators in the individual channels represents the most important feature of this PDD. By virtue of this, the phase relations are thrown off, and the domain of running time is left in favor of the domain of lag, which is the independent variable of the histogram. Figure 8.29 illustrates

> ... how this process acts to give an accurate measure of the pitch period. For purposes of illustration, it is assumed that each channel has a harmonic, which corresponds to the channel number, although in actual practice a harmonic can fall in any channel with similar results. If, as shown in the figure, the different waves have a harmonic relation, then at some point in time, corresponding to the fundamental period, the pulses will all coincide and add together to give a maximum value. We have, in effect, set up a computational process, which covers all possible combinations of the wave periods. It will be observed that the coincidence also repeats again at $2T_0$ with the same amplitude. To obtain a measure of the fundamental period, it is necessary to determine the position of the first major peak. This process is carried out by the peak detector arrangement (Miller, 1970, p.1594)

To correctly determine the peak at the fundamental period, a threshold THR defines the amount a later peak must exceed the previously determined one in

---

[20] Starting from a probabilistic frequency-domain approach (Baronin, 1966; Kushtuev, 1968, 1969a,b), Baronin and Kushtuev (1970b) arrived at a PDD which is almost identical to Miller's PDD, although it was developed independently of Schroeder and Miller (Baronin, 1974a). Explaining the PDD in this section we will widely follow Miller's description. Baronin and Kushtuev's probabilistic approach will be further discussed in Section 8.6. Another PDD of this kind was proposed recently by Arkhangelskiy et al. (1982).

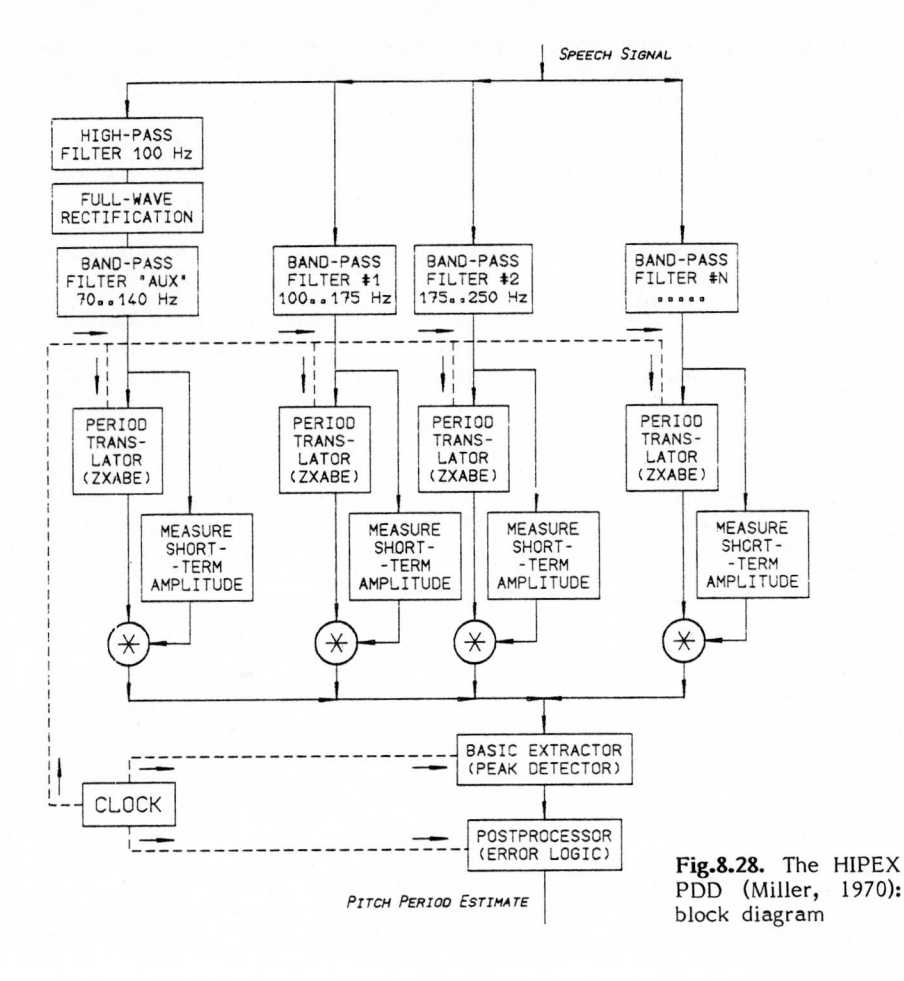

Fig.8.28. The HIPEX PDD (Miller, 1970): block diagram

order to override it as the fundamental period estimate. This threshold was experimentally optimized and found to produce fewest errors when set to a value of 1.1. An additional advantage of the VCO, besides the ability of synchronization, is the fact that the process of building up the histogram can be done faster than in real time:

An obvious arrangement is to adjust the period of the VCO to be the same as the period of the measured harmonic. Since the longest speech periods to be measured are of the order of 15 ms, this would mean that at least this same length of time would be required to obtain a measure of the period from the histogram. It is actually not necessary to keep a one-to-one relation between the period of the measured harmonic and the period of the VCO, but only a constant ratio. It is, in fact, very desirable to make use of a large speed-up ratio, a factor of 15:1 being used in the experimental system. This allows a histogram to be completed in about 1.0 ms. The interval between samples can then be 1.0 ms or longer, depending on the desired rate of pitch samples. The experimental system was arranged to carry out a sample of the histogram every 5 ms. The use of a high speed-up also means that the histogram cannot change appreciably during a sample process, thus simplifying the construction of the period translator. (Miller, 1970; p.1595)

For low-pitched voices there is an auxiliary channel that uses nonlinear distortion (full-wave rectification) before the low-frequency band-pass filter in order to reconstruct and to isolate the first harmonic. This harmonic is then processed in the same way as the other signals, except for the fact that the pulses generated here are somewhat longer. Since this channel never contains a higher harmonic, its output signal is used to identify the principal peak without contributing much to its actual position.

The harmonic product spectrum and the harmonic sum spectrum are generalizations of the principle of spectral compression. Again the basic idea was suggested by Schroeder (1968). This method has no direct time-domain counterpart. In the ordinary frequency histogram only the peak frequencies were compressed by a factor of 2, 3, etc., to yield the histogram. A similar effect, however, is achieved if the whole spectrum is compressed by these constant factors and added up without previous peak location. Noll (1970) has realized this principle in an experimental PDA. The (logarithmic) harmonic product spectrum is computed when the log power spectrum is compressed and added:

$$\log P(m) = \sum_{k=1}^{K} \log |X(km)|^2 = \log \prod_{k=1}^{K} |X(km)|^2 \qquad (8.54)$$

According to Noll, the reasoning for this method is that the pitch peaks in the log spectrum add coherently while the other, uncorrelated components

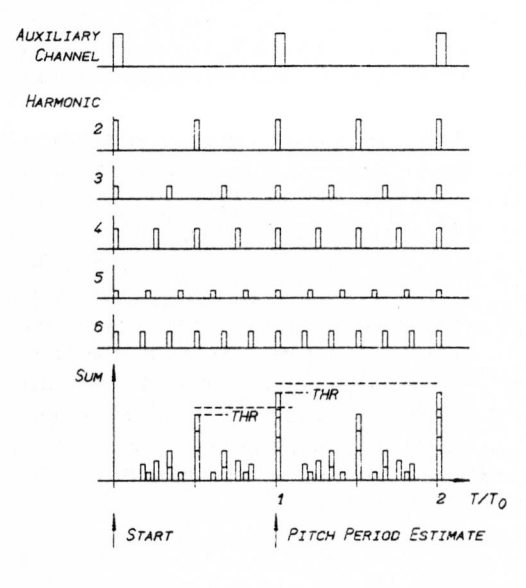

Fig.8.29. The HIPEX PDD (Miller, 1970): example of weighted period histogram

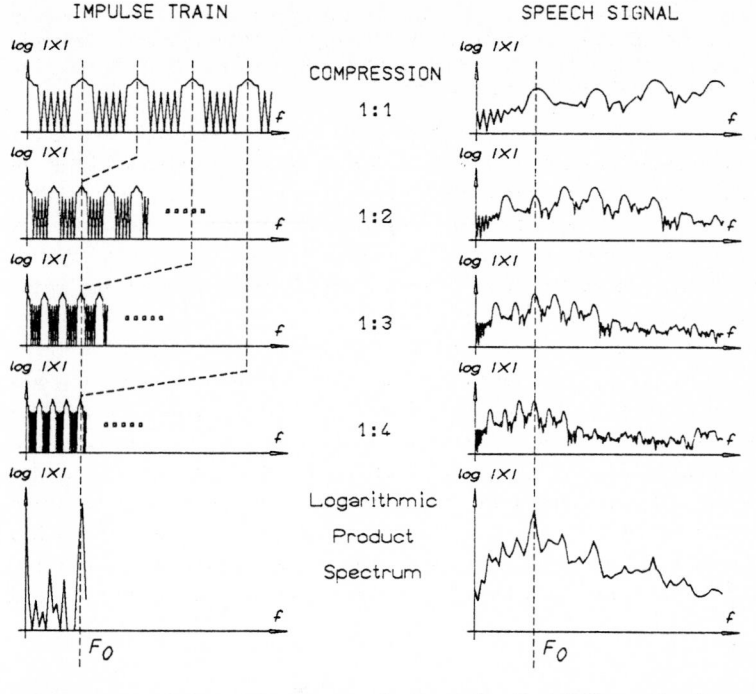

Fig.8.30. Harmonic product spectrum PDA (Noll, 1970): evaluation of the product spectrum

do not. The harmonic product spectrum is obtained taking the antilogarithm of (8.54). Similarly, the harmonic sum spectrum is defined when the compressed amplitude spectra are added instead of the compressed logarithmic spectra. Figure 8.30 shows the principle of compression and adding, whereas Fig.8.30a displays an original example of the performance of the three methods. It can clearly be seen that the harmonic sum spectrum acts much more like the frequency histogram than the harmonic product spectrum does. In the figure, all the spectra have been "long-pass liftered" to remove the formant structure prior to compression, and have been normalized before plotting.

Short-term analysis PDAs, as stated at the beginning of this chapter, tend to focus all the energy of the periodic component into one single peak. For no PDA (except perhaps the maximum likelihood method) is this statement as valid as for this one. Thus the harmonic product spectrum PDA is especially suitable for noisy signals. Figure 8.31b shows the three spectra of the same utterance (female voice), but with noise added to the signal so that the overall signal-to-noise ratio becomes about 0 dB. Cepstrum was found to fail under these circumstances. Yet the harmonic product spectrum method still succeeds in giving a reasonable estimate. (With this much noise added, the speech already becomes unintelligible.)

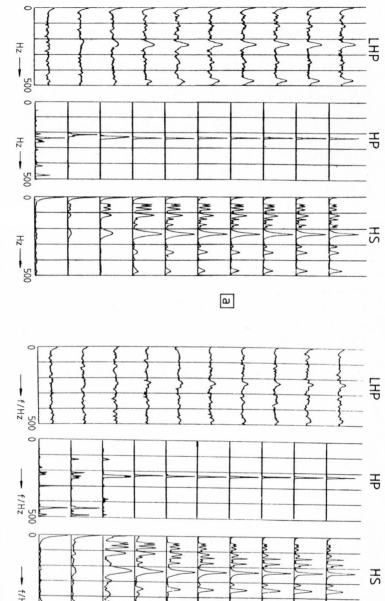

Fig.8.31a,b. Product spectrum PDAs (Noll, 1970): example of performance. **(a)** Normal speech (female voice), **(b)** same speech with added noise (overall S-N ratio 0 dB). (LHP) Log harmonic product spectrum, (HP) harmonic product spectrum, (HS) harmonic sum spectrum.

Frequency scale: 150 Hz per division. Time runs from top to bottom. The amplitudes of the harmonic product and sum spectra have been normalized before plotting. From (Noll, 1970, pp.785,790)

This principle is probably the most reliable way to establish a PDA. It always works whenever a harmonic structure is present. Probably, however, this method is also one of the slowest PDAs in existence. According to Noll (1970), the spectrum must be heavily oversampled in order to yield a reasonable accuracy of measurement. Noll used 2048 points at a time-domain sampling rate of 10 kHz (the frame of 51.2 ms duration is filled up with zeros after weighting), thus achieving a frequency-domain resolution of less than 5 Hz. This resolution requires a voluminous FFT which makes this PDA slower with respect to the cepstrum PDA by at least a factor of 2. It is thus not surprising that the search for a faster frequency-domain PDA continued although this extreme robustness could probably not be achieved.

In more recent times, however, a change in attitude towards this last argument has occurred because of the advances of large scale integration technology in connection with the development of low-cost high-speed digital hardware. Fast signal processors which are able to perform a 1024-point complex FFT in less than 10 ms bring this (and any) PDA close to real-time approximation. According to a footnote by Dubnowski et al. (1976),

> ... some special purpose extremly fast processors such as the fast digital processor (FDP) and the digital voice terminal (DVT) at the MIT Lincoln Laboratories, and the programmable signal processor (PSP) at Sylvania, have been built which are capable of running most pitch detection algorithms in real time. Generally these processors are either expensive, or are not commercially available. (Dubnowski et al., 1976, p.2)

Facing the development of the modern LSI circuits, we have to expect the last statement of the above note to become more and more insignificant, and that other arguments in connection with PDAs, such as a overall delay enforced by some global correction subroutine, will increasingly suppress the aspect of computational effort.

8.5.2 Harmonic Matching. Psychoacoustic PDAs

Five algorithms have to be reviewed in this section. Two of them deal with functional models of pitch perception and apply them to speech and speech-like signals. Among these PDAs, the one by Terhardt (1979a,b) incorporates psychoacoustic facts and models to a greater extent than the PDA by Duifhuis et al. (1978, 1979a,b, 1982) which is oriented more toward speech signal processing. The other three PDAs apply the principle of harmonic matching and arrive at very similar solutions although their starting point is rather different. The PDAs by Martin (1981) and by Paliwal and Rao (1983) exploit the harmonic structure in order to increase the robustness of the measurement. Sreenivas (1981) also presented an algorithm which checks coincidences within tolerance intervals. In this case, however, the tolerance intervals are not established by psychoacoustic facts, but rather by possible corruption of the spectral peaks due to limited frequency resolution, noise, or nonstationarity of the signal.

The PDA by Martin (1981, 1982) applies the principle of harmonic pattern matching in a rather rigorous way. It applies a comb filter to the short-term amplitude spectrum.

> The principle behind the comb method consists in the search for values of the spectrum situated at harmonic frequencies, and whose sum is a maximum for a given frequency interval. The intercorrelation of spectrum and comb amounts to the calculation of the sum of the spectral components corresponding to a given harmonic structure. ... The fundamental corresponding to the harmonic structure giving the largest sum is then taken to be the fundamental frequency of the signal, as long as this sum differs sufficiently from the values obtained for other structures in the same spectrum (which would correspond to the case of voiceless segments). (Martin, 1981, p.224; original in French, cf. Appendix B)

The spectral comb is given as an impulse sequence; the distance of the individual pulses equals the trial fundamental frequency p,

$$C(m,p) = \begin{cases} C(k) & m = kp; \ k = 1 \ (1) \ ... \\ 0 & \text{otherwise.} \end{cases} \tag{8.55}$$

For each value of p the spectrum A(m) is weighted by the spectral comb $C(m,p)$, and the spectral components that pass the comb are added up to form the harmonic estimator function $A_C(p)$,

$$A_C(p) = \sum_{k=1}^{N/2p} A(kp) \, C(kp,p) \tag{8.56}$$

The value of p where A_C reaches its maximum is then taken as the estimate $\hat{p}$ of the fundamental frequency F_0.

To realize this PDA in practice, a number of additional features are necessary. First of all, the individual "teeth" of the comb must be provided with appropriate weights in order to avoid octave errors (i.e., subharmonic or second harmonic tracking). In this respect, it proves most suitable to provide the comb with amplitude weights that decrease with increasing frequency, and not to further normalize $A_C(p)$ with respect to the number of components that pass the comb, or with respect to the number of teeth that are found within the frequency range of the spectrum.

> It is possible to ensure that only F_0 corresponds to the maximum in the intercorrelation function by giving different weights to the different lines of the spectrum such that the higher harmonics are penalized in the calculation of $A_C(p)$. Since the same harmonic peaks in the spectrum correspond to teeth of the comb of a different order for $p = F_0/2$, $F_0/3$, etc., the desired result is achieved by giving the teeth of the comb an amplitude that decreases according, for example, to:

$$C(k) = k^{-1/s} \quad \text{with s an integer} > 1. \tag{8.57}$$

> This decreasing amplitude of the teeth of the comb is preferable to normalizing the intercorrelation function by the number of teeth present in the frequency band used. The latter method leads to equal maxima being

obtained for $p = F_0$ and $p = 2F_0$ if the spectrum only has two harmonic peaks with the first one corresponding to the fundamental. (Martin, 1981, pp.225-226; original in French, cf. Appendix B)

In addition, care must be taken that the spectral values between the individual harmonic peaks that are nonzero due to various effects (background noise, nonstationarity of the signal, truncation and time-domain weighting) do not contribute to the sum A_C. For this reason, only the peaks of the spectrum A(m) and their immediate neighbors are retained, and all the other values are set to zero.

> The amplitude spectrum is not in fact a line spectrum and exhibits nonzero values between its different peaks. Even when the teeth of the comb have decreasing amplitudes there is thus a chance that the value of $A_C(p)$ for subharmonic values of p will be greater than the value obtained for $p = F_0$ since some intermediate teeth of the comb then correspond to the nonzero valleys between the harmonic peaks.
> In order to be sure of obtaining a maximum in $A_C(p)$ for $p = F_0$ it is therefore advisable to eliminate the effect of the nonzero spectral values lying between the spectral maxima by carrying out a frequency-domain weighting around each of the peaks according to a function that is analogous to the windows used for weighting signals in the time domain (Hamming etc.). This weighting is to be preferred to the reduction of the signal to the discrete values corresponding to the maxima alone. The inaccuracy involved in numerical methods of calculation and possible phase distortions can result in the frequencies of the spectral peaks not lying at exact harmonics of the fundamental frequency. The teeth of the comb would then no longer correspond exactly to all of the harmonics. Emphasis in the frequency domain, on the other hand, allows the maximum in the intercorrelation function to be interpreted as the result of performing the best possible adjustment of a given spectrum as constrained by a fundamental frequency with exact harmonics. (Martin, 1981, p.226; original in French; cf. Appendix B)

This nonlinear weighting of the spectrum goes hand in hand with an improvement in the spectral resolution. The whole algorithm thus presents itself as follows (Fig.8.32).

Since only frequency components below 2 kHz are taken into account, the signal is first downsampled in the time domain to a sampling rate of 4 kHz (in the hardware PDD that was built according to this design, the A-D conversion of the signal is already performed with a sampling rate of 4 kHz). A frame of 32 ms length (128 samples) is then windowed and transformed into the frequency domain. In the amplitude spectrum all the values are set to zero except the peaks that exceed a threshold of -35 dB relative to the global maximum of the spectrum, and their immediate neighbors. The original spectral resolution of 33 Hz guarantees that the information on fundamental frequency is present in the spectrum for values of F_0 down to 70 Hz. This resolution is then increased to 1 Hz (!) interpolating the missing points by convolution of the remaining nonzero samples and an appropriate window (parabolic interpolation in the actual implementation). The comb function is then applied to this fine spectrum yielding the spectral histogram that serves as the basis to estimate F_0.

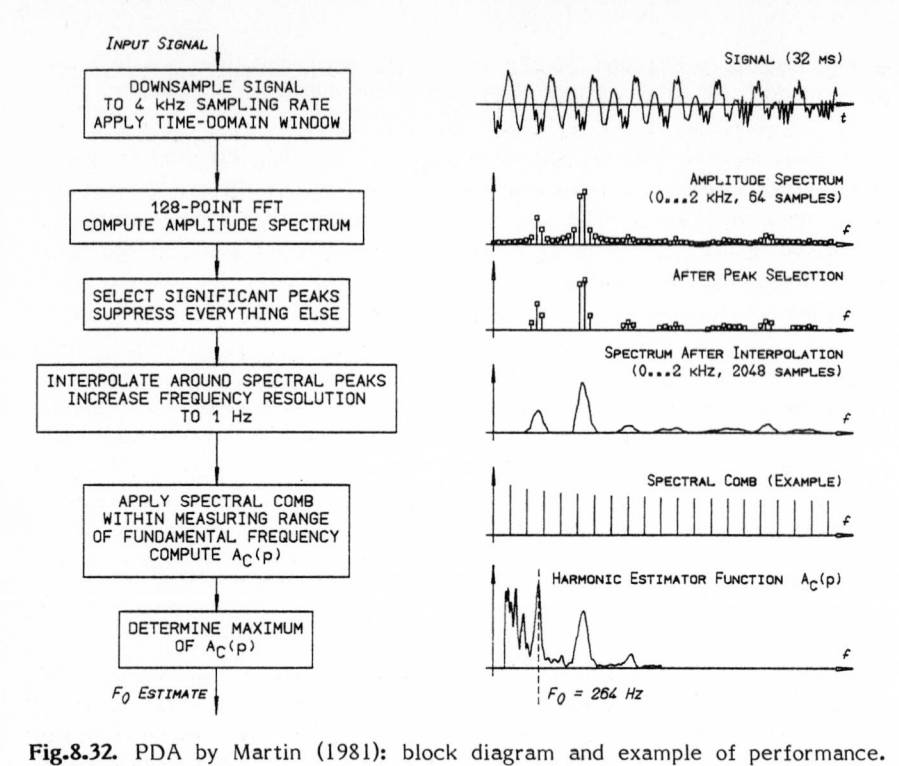

Fig.8.32. PDA by Martin (1981): block diagram and example of performance. Signal: /i/ from "big", female speaker, band limited to telephone channel. The interpolation was carried out with a window of cosine shape

Martin compared his PDA to cepstrum pitch determination and found that it performs well with signals where the cepstrum PDA may fail systematically.

> The cepstrum method depends on the determination of a periodic component in the spectrum in order to be able to measure the fundamental. The intercorrelation function exhibits a maximum for F_0 even if this frequency is the only component in a spectrum in which every harmonic is absent, or if only two consecutive harmonics are present. With the cepstrum method this result cannot be guaranteed. (Martin, 1981, p.226; original in French, cf. Appendix B)

In addition, Martin presented some preliminary quantitative results in the form of global error rates without further specifying what kind of error occurs. In spite of the frequency-domain approach which should be insensitive with respect to low-frequency band limitation, the overall error rate rises from 1% to 10% for the same speaker when telephone quality speech is compared to good quality speech. With only the preliminary data that have been published up to now, and with no specification of the applied error criteria, one cannot say whether this performance is typical for frequency-domain PDAs in general. On the other hand, even such a reliable concept apparently still has its problems with degraded or low-frequency band limited speech. Here, further evaluative work is necessary.

The relation between Martin's PDA and the one by Paliwal and Rao (1983) is in some way comparable to the relation between ACF and AMDF PDAs. Martin's PDA maximizes the energy of the components that pass the spectral comb. The PDA by Paliwal and Rao, on the other hand, minimizes the difference between the input spectrum and reference spectra that depend on the trial fundamental frequency p.

Figure 8.33 shows the block diagram of this PDA. The input spectrum is computed from a 40-ms segment of speech via a 512-point FFT. In the frequency domain the spectral resolution is increased to 1 Hz by interpolation. The reference spectra are synthetically generated in two steps. First, a LPC analysis is performed on the input frame; from the predictor coefficients a smooth spectrum $H(m)$ is computed (and refined to a frequency resolution of 1 Hz as well). For each trial fundamental frequency p the reference spectrum is then defined as

$$R(m,p) = \begin{cases} H(m) & m = kp, \ k = 1 \ (1) \ \dots \\ 0 & \text{otherwise}, \end{cases} \qquad (8.58)$$

i.e., $R(m,p)$ represents a spectral comb which is computed from the LPC spectrum $H(m)$ by 1) assuming a strictly periodic excitation whose spectrum equals 1 at the multiples of the fundamental frequency p and is zero otherwise, 2) taking $H(m)$ as the combined voice source and vocal tract transfer function, and 3) applying the model defined in Section 2.6. The spectrum $R(m,p)$ is then compared to the input spectrum $X(m)$ by taking the spectral distance function

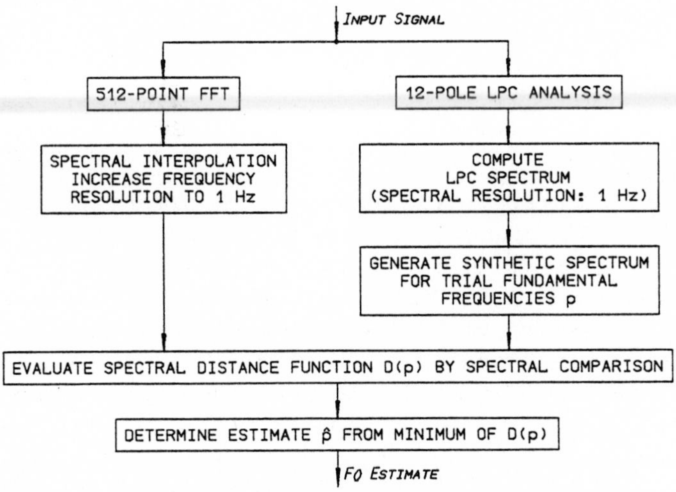

Fig.8.33. PDA by Paliwal and Rao (1983): block diagram

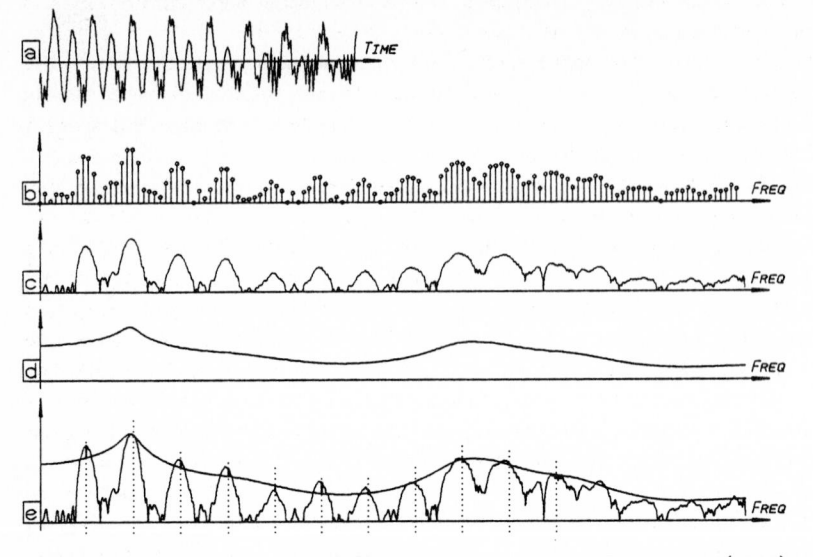

Fig.8.34a–f. PDA by Paliwal and Rao (1983): example of performance. **(a)** Input signal (same as in Fig.8.32); **(b)** logarithmic spectrum X(m) after 256-point Fourier transform (0–4 kHz, 50 dB); **(c)** same spectrum after interpolation; **(d)** LPC spectrum H(m), obtained by a 13-pole LPC analysis using the autocorrelation method; **(e)** example of the evaluation of D(p) for the optimal estimate $\hat{p} = F_0$; **(f)** distance function D(p). (·····) Spectral comb for the final estimate; the number of harmonics involved was limited to 11. Due to an implementation restriction, the spectral resolution of the spectra in (c–f) is about 1.8 Hz

$$D(p) = \frac{1}{K} \sum_{k=1}^{K} |\log R(kp,p) - \log X(kp)| \; ; \tag{8.59}$$

K is the number of harmonics involved (unless otherwise restricted, this is simply the number of harmonics falling into the frequency range of the spectrum). If p equals the real fundamental frequency F_0, there is no great difference between the two spectra at multiples of p; and D(p) takes on a distinct minimum.

To work properly, the algorithm needs careful normalization of the LPC spectrum with respect to· the input spectrum[21]; in addition, it may be advisable to limit the number of harmonics involved in the evaluation of D(p) at the lower end of the measuring range since for low values of F_0 the

---

[21] Since D(p) is evaluated on the basis of a logarithmic spectrum, normalization means *adding* a constant to the individual difference log R(m,p) – log X(m).

contribution of the higher harmonics (above 2 kHz) becomes insignificant, and the overall performance of the algorithm may be degraded if the LPC spectrum does not match the input spectrum at the higher frequencies well. The high spectral resolution of 1 Hz is necessary to provide sufficient accuracy; the minimum is rather sharp, and it is hardly possible to apply (8.58,59) when the value of p is not an integer.

Figure 8.34 shows an example of the performance. Since X(m), as the spectrum of the speech signal, is the output spectrum of the whole speech tract (including the vocal tract), and since H(m) is the LPC estimate of the vocal tract transfer function, by taking the difference of the logarithmic spectra, (8.59) acts like a spectral flattener and removes the influence of the vocal tract on the distance function D(p). D(p) is thus no longer subject to interaction with dominant formants. This is an advantage over the PDA by Martin[22] (1981), and a simple comparison of Fig.8.32 and Fig.8.34 shows that there are fewer spurious peaks in the distance function of the latter figure.

A PDA that is based on a functional model of speech perception (Goldstein, 1973; Goldstein et al., 1978), but ends up with a very similar procedure as the PDA by Martin, was developed and realized in software by Duifhuis, Willems, and Sluyter (1978, 1979a,b, 1982). Sluyter et al. implemented (1980) and improved (1982) it as a digital hardware PDD. Bosscha and Sluyter (1982) tested it in a vocoder. The basic idea of the model of pitch perception

> ... was reported by Goldstein et al. (1978). They found that the pitch of a stationary sound, containing a few successive harmonic components, is perceived as the fundamental frequency of a pure harmonic template which best fits the harmonic content of the sound. Although the pitch extractor ... is mainly based on Goldstein's model, some essential changes had to be made: in speech we have to cope with continuously varying pitch and with sounds which generally do not only contain successive harmonic components. We have solved the latter problem by the concept of the "harmonic sieve" which provides the proper processing algorithm. (Sluyter et al., 1980, p.45)

The "harmonic sieve," which represents the harmonic matching routine, is a spectral comb[23] which excludes spurious peaks and retains the optimal harmonic structure in the spectrum. Figure 8.35 shows an example. First, a limited number P (implemented: P = 8) of significant spectral peaks is determined starting from the low-frequency end of the spectrum; each peak that exceeds a given threshold (implemented: 26 dB below the highest spectral peak) is

---

[22] Paliwal and Rao (1983), besides comparing the performance of their PDA to that of the cepstrum algorithm (see Sect.8.4.2), directly compared it to Martin's PDA. They found the performance of their own PDA superior; their implementation of Martin's PDA committed some isolated gross errors due to interaction with dominant formants.

[23] For reasons of uniformity, the term "spectral comb" is applied in the discussion of both the PDAs by Martin (1981) and by Duifhuis et al. (1978, 1980, 1982).

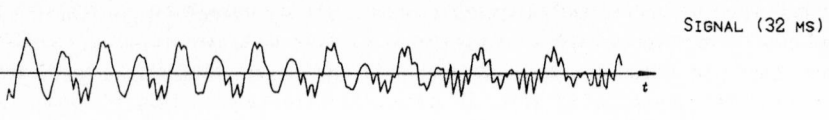

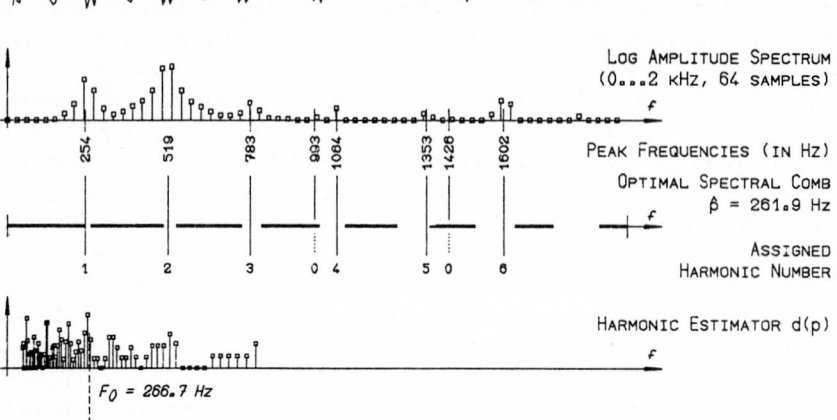

Fig.8.35. PDA by Duifhuis et al. (1978, 1979a,b, 1982): example of performance. Signal: /i/ from "big", female speaker, band limited. The signal and its spectrum are identical to those displayed in Figs.8.32,34. For reasons of display, the parameters were chosen somewhat different from Sluyter's implementation. The tolerance interval of the spectral comb was set to 4%, the number H of harmonics checked by the comb was limited to 7, and 80 trial values of p were computed within the measuring range between 50 and 800 Hz. Besides the principal maximum at $p = F_0$, the estimator function d(p) shows strong maxima at the 2nd and 4th subharmonic of F_0

regarded as significant.[24] After peak selection, the amplitude of the peak is thrown off; this agrees with Goldstein's model which judges the amplitude information insiginificant once a spectral peak has been found to contribute to the perception of pitch. The measurement accuracy of the peak frequency is improved by parabolic interpolation; i.e., for the ith peak P_i at the m_i^{th} sample of the spectrum, the pertinent frequency is interpolated as follows:

$$f_i = \Delta f \left[m_i + 0.5 \frac{A(m_i+1) - A(m_i-1)}{2A(m_i) - A(m_i-1) - A(m_i+1)} \right] \qquad (8.60)$$

where A(m) are the spectral amplitudes, and Δf is the frequency interval (in Hz) between adjacent spectral samples. Once the peaks have been determined, the spectral comb C(p) is invoked. In contrast to Martin's PDA, where

---

[24] In the most recent version (Duifhuis et al., 1982) a simplified masking threshold is checked and applied. The present PDA only takes into account masking of individual spectral components by other individual spectral components. Accumulation of the contributions of several maskers, like in Terhardt's PDA (1979b), has not been implemented. For more details on masking, see the discussion in Section 4.2.1.

we had smoothed peaks and a pulselike shape of the comb, here we have very "pulselike" information about the peaks (i.e., only their frequencies) and are thus obliged to establish small tolerance intervals in the spectral comb around the first H harmonics. (This gives a substantial difference to Martin's approach since there the bandwidth of the interpolation window is independent of frequency, whereas here the width of a tolerance interval is a constant fraction of its center frequency and thus increases with increasing frequency.) For a trial fundamental frequency estimate p, the frequencies f_i of the significant spectral peaks P_i, $i = 1\,(1)\,P$ are checked for whether they fit into the tolerance intervals. If a frequency f_i does, the harmonic number corresponding to the current tolerance interval of the comb is assigned to it; otherwise it does not contribute to the actual measurement (the pertinent harmonic number is set to zero). The trial period p is varied over the whole measuring range in (logarithmically growing) comparatively crude steps. (In the implemented PDD we have $K = 8$, the width of the tolerance interval is 1/16 of its center frequency, and the measuring range of F_0 is scanned with 40 trial values p in the range 50 to 500 Hz.) The optimal match $\hat{p}$ is found when the expression

$$d(p) = k(p) \,/\, [\; q(p) + c(p)\;] \,, \tag{8.61}$$

which is a modified distance measure for a minimum distance classifier, is maximized. In this equation, k is the number of peaks that fall into the tolerance intervals, i.e., that pass the spectral comb; q is the number of the peaks whose frequencies are situated within the measuring range of the current spectral comb, i.e., below the upper frequency limit $K \cdot p \cdot (1+b)$, and c(p) is the number of tolerance intervals situated below the spectral peak P_i for $i = p$.

The fundamental frequency F_0 is then computed according to Goldstein's model (1973) as the least squares estimate from all the spectral peaks that contribute to the formation of pitch, i.e., in the actual case, that pass the optimum spectral·comb,

$$F_0 = \frac{\displaystyle\sum_{i=1}^{P} f_i \cdot h_i(\hat{p})}{\displaystyle\sum_{i=1}^{P} h_i^2(\hat{p})} \,. \tag{8.62}$$

In the improved version, Sluyter et al. (1982) introduced some quality criteria for the peaks similar to those applied by Sreenivas (1981, cf. Fig.8.39). In addition, the selection routine of the optimal spectral comb was made more reliable. Instead of considering all configurations possible, the most recent version of the PDA only keeps those spectral combs which contain the two highest spectral peaks. Among those combs the decision is made according to (8.61). If there are several combs with equal selection scores, the one with the highest value of the F_0 estimate is selected.

Both the software PDA and the hardware PDD are reported to perform very well. Duifhuis et al. (1979b) stated that the performance of their PDA is superior to that of a sophisticated autocorrelation PDA. This may be because the present PDA is extremely reliable in the case where a harmonic structure is present, and because it computes F_0 as the weighted average of a number of subharmonics to the individual partials of the signal whose accuracy had already been improved by the parabolic interpolation. It thus minimizes the fine measurement errors which have been found to contribute significantly to the degradation of the subjective quality of the speech in a vocoder system (Fellbaum, 1982, private communication). This statement is supported by the fact that evidently no smoothing routine is needed in the postprocessor (Duifhuis et al., 1979b). With the claim, however, that *"existing techniques to date show weaknesses which in our opinion are caused by an inadequate approach to the problem"* (Sluyter et al., 1980, p.45), the present author cannot agree. One could conclude from this statement that the frequency domain approach must have a superior performance because it is *perception oriented.* Yet there is no evidence for such a statement, and it is not necessary that a perception-oriented parameter perform better than other parameters when it is, for instance, used to drive the synthesizer in a speech communication system (which is necessarily oriented toward speech production). If such a PDA is judged to perform very well in a speech communication system (and there is no reason to question that), the argument that one error source, i.e., fine errors and noisiness of the pitch contour, is practically abolished may contribute more to the overall improvement than the fact that the PDA is oriented towards perception.

The proximity of the frequency histogram methods to the basic ideas of pitch sensation and perception in the ear has also been demonstrated by Terhardt (Terhardt, 1979a,b; Terhardt et al., 1982a,b). From his model of virtual pitch, Terhardt (1972a,b, 1974) derived a PDA which is closely related to that by Schroeder (1968) and also corresponds to the PDA by Duifhuis et al. (1978, 1979a,b, 1982). Terhardt's theory of pitch perception (the concept and model of *virtual pitch*) has been briefly outlined in Section 4.2.1. In summary, perceived pitch, or, as Terhardt labeled it, virtual pitch,

> ... is obtained by a specific matching process of subharmonics pertinent to certain determinant spectral pitches. This process is dependent on a previous perceptual mode of learning, in which the knowledge of harmonic pitch intervals has been acquired. (Terhardt, 1979b, p.156)

The principle of the calculation routine for virtual pitch is shown in Figure 8.36. The individual harmonics of a periodic sound are, as usual, determined as the significant peaks in the (short-term) power or amplitude spectrum. To each harmonic is then assigned the pertinent *spectral pitch* which is computed via some kind of "psychoacoustic transformation" that takes account of frequency-domain nonlinearities of the human ear, such as pitch shift or musical interval stretching (Terhardt, 1972a, 1979b). At a first approximation, however, these features are negligible so that the output measure of this

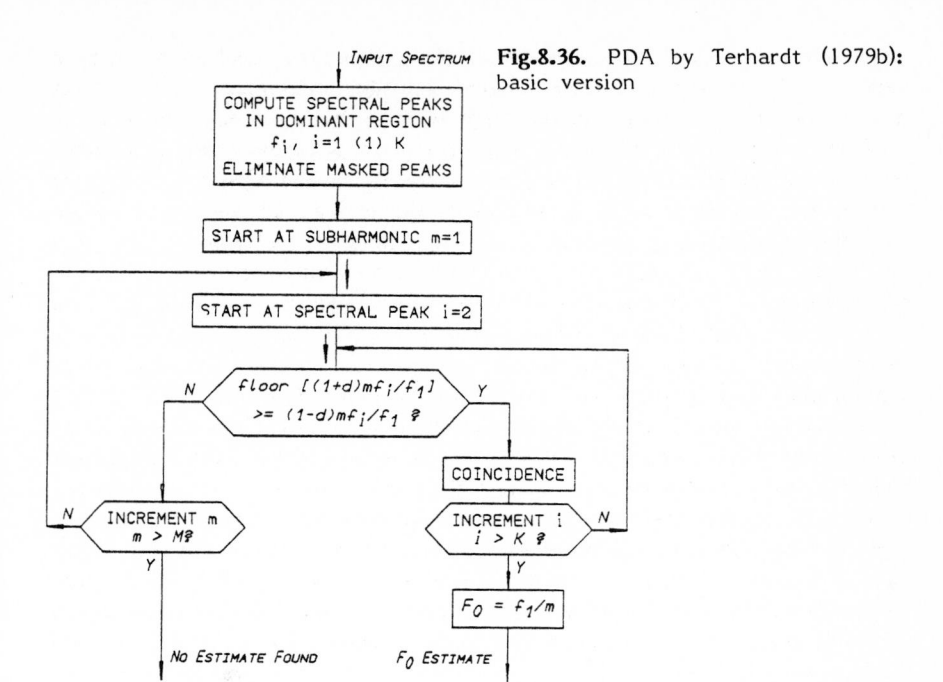

Fig.8.36. PDA by Terhardt (1979b): basic version

transformation, the *pitch unit* ("pu") can be set equivalent to the unit of frequency, the hertz (Terhardt, 1979b), and the spectral peaks can simply be characterized by their respective frequencies. (To be consistent with Terhardt's terminology, we would then compute the *nominal virtual pitch* instead of the true virtual pitch; to understand the principle of the algorithm, however, this difference is insignificant.) The pertinent selection mechanism of the spectral peaks that actually contribute to the formation of virtual pitch is far more important. Terhardt's PDA is the only one that applies a significance threshold for the peak selection rigorously oriented towards perception. A spectral peak can evoke a spectral pitch if and only if it is audible, i.e., if its amplitude exceeds the masking threshold at that particular frequency (see Sect.4.2.1 for more details); in addition, the peak must be located within the *dominant region* (this is, crudely speaking, the frequency region between 0.3 and 3 kHz). Otherwise the peak has to be rejected and must not contribute to the formation of virtual pitch. Virtual pitch is then computed as the optimal match between the subharmonics of the individual spectral pitches that have survived the selection procedure. In this respect, Terhardt's PDA is algorithmically somewhat different from the one by Sluyter et al. (1980), but in principle the basic extractor is the same. A significant difference between the two algorithms is again found in the final computation of the pitch estimate. Sluyter et al., in agreement with Goldstein's model (1973), applied a least squares estimator to compute F_0 as the optimal match between all the harmonics, whereas in Terhardt's model the formation of virtual (or nominal) pitch is dominated by the first significant partial above 300 Hz. In the example displayed in Figs.8.32,33,35, for instance, the final estimate would be computed in Terhardt's PDA as the frequency of the second-order subharmonic of the second partial which is situated at the beginning of the dominant region. The numerical value would thus be $518.8/2 = 259.4$ Hz (pitch shifts and octave stretching not taken into account).

The harmonic matching routine has already been discussed in Sect.4.2.1; it has been described by (4.7,8). If condition (4.8) is fulfilled by a spectral peak f_i, the integer subharmonic f_i/n (n unspecified) falls into the tolerance interval and indicates coincidence. The value f_1/m for which all the peaks f_i, $i = 2\,(1)\,K$ signal coincidence, is then taken as (virtual or nominal) pitch estimate.

This simple version of the PDA expects that the harmonic pattern be undisturbed; it will thus fail if any of the selected spectral peaks are spurious, especially when f_1 is a spurious peak. In this case, a more complex and speech-oriented version (Fig.8.37) must be applied. The latter PDA allows for the existence of spurious peaks in the spectrum. Accordingly, the F_0 estimate[25] is that subharmonic f_i/m which shows the largest number of (near) coincidences. In a practical implementation, this number, compared to the number K of peaks present, could even serve as a voicing indicator.

The PDA has been proposed for arbitrary stationary sounds, but has not yet been tested for time-variant speech signal processing. A reliability comparable to that of the frequency histogram, however, has to be expected, especially if the procedure is compared to other PDAs that have been tested with speech signals (Sreenivas, 1981; Martin, 1981; Sluyter et al., 1980). In the opinion of the present author, the main merit of this PDA is that it executes pitch determination rigorously according to a psychoacoustic model, in particular, that it applies a unique criterion for selection and rejection of spectral peaks. However, further work has to be done in order to refine the peak selection routine. Since masking a complex harmonic sound by itself, i.e., masking individual partials by the remainder of the sound, is among the difficult problems in psychoacoustics, the peak selection routine is still rather rudimentary.

A question which is still completely open is that of *temporal* masking in connection with speech signals. It has been known for some time that masking

---

[25] In Fig.8.37 (and 8.36 as well), the processing is intended to be terminated as soon as a sufficiently reliable pitch estimate has been found. For speech-domain applications this is sufficient since only one F_0 has to be determined. The ear, however, can register several virtual pitches simultaneously. To perform such a measurement with this algorithm, the check of the number of coincidences (in Fig.8.37) has to be moved from its place in the figure to the end of processing. In this case all possible coincidences are processed, and the algorithm finally assigns the attribute "virtual pitch" to all those estimates where the number of coincidences exceeds the given threshold k.

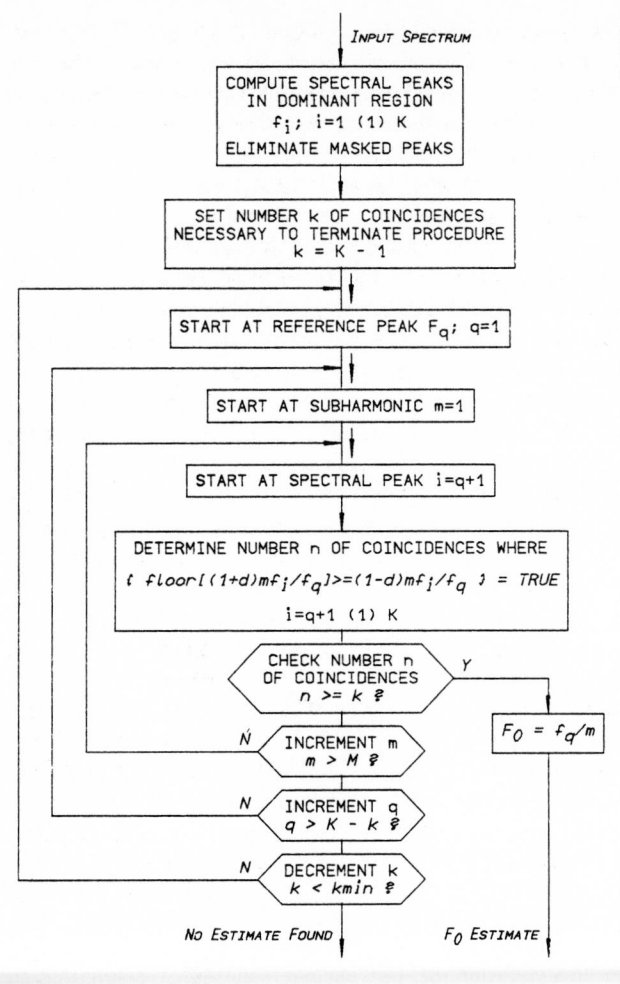

Fig.8.37. PDA by Terhardt (1979b): fault-tolerant version

in the human ear is a two-dimensional process that involves frequency as well as time (Fastl, 1976, 1977, 1979). The intensity changes in speech, especially before a plosive, can be so rapid that parts of the signal may very well be masked and thus inaudible. We know that rapid intensity variations are a source of problems for a great number of PDAs. If one could prove that in these regions the signal is widely masked, one could ignore the errors made by PDAs during such segments, and change the strategy towards interpolation or extrapolation of the pitch contour from more stable points. However, temporal masking of nonstationary complex sounds is far more delicate to investigate than frequency-domain masking of stationary harmonic complex sounds. Hence, a final answer to this question cannot be expected in the near future.

The PDA by Terhardt, Stoll, and Seewann (1982b) is a further development of the earlier PDA (Terhardt, 1979b) discussed above. It has been designed for any complex sound which is able to invoke pitch perception in the human auditory system. The main difference to the earlier PDA (Terhardt, 1979b) is that now weights are introduced where the earlier PDA made a binary decision. Using these weights, a measure for the salience of a given virtual pitch estimate can be derived from the spectral information. This serves to discriminate between competitive virtual pitches that are perceived simultaneously. Since the PDA has been designed to model pitch perception in general, it explicitly allows for several virtual pitches to occur simultaneously, and it decides in favor of one of them only if the salience of that virtual pitch considerably exceeds the salience of the others.

The weights are introduced at three points in the algorithm:

1) A weight is assigned to each spectral pitch in order to mark its "importance." This weight depends on the SPL excess of that spectral pitch and on its frequency with respect to the dominant region (see Sect.4.2.1).

2) A weight is assigned to each (near) coincidence. It replaces the binary decision established in the earlier algorithm by definition (4.7) of the tolerance interval.

3) A weight is assigned to each virtual pitch according to its frequency. This weight takes into account the fact that virtual pitch sensation is practically confined to the range of F_0 in speech.

The PDA by Sreenivas (1981), a preliminary version of which was published earlier (Sreenivas and Rao, 1979), attacks the problem of harmonic analysis from a completely different starting point. Seeking a PDA that is reliable even when the signal is strongly degraded, Sreenivas arrived at the principle of harmonic analysis with tolerance intervals around the significant spectral peaks. These tolerance intervals, however, are not deduced from a perceptual criterion, but from possible corruption of the spectral peaks by noise, signal degradation, intrinsic aperiodicity, and insufficient spectral resolution. Sreenivas dealt with an investigation of the various sources of corruption in great detail. In particular, he examined the effect of white noise, change of fundamental frequency, and change of amplitude within the frame. Here, he found that, using some global measure of spectral peak corruption,

> ... a 4% period transition, a 20-dB amplitude transition and white noise at 0 dB SNR have equivalent effect on the spectrum ... and ... that the effect of simultaneous occurrence of different aperiodicities is in general cumulative with the exception of amplitude transition reducing the effect of other transitions. (Sreenivas, 1981, p.II-27)

Starting from these results, Sreenivas developed a PDA which is by far the most complex one to be discussed in this section. Figure 8.38 shows the block diagram. The algorithm can be subdivided into four major steps: 1) selection of significant peaks, 2) search for harmonic coincidences, 3) determination of a preliminary estimate for F_0 from that frequency where the majority of har-

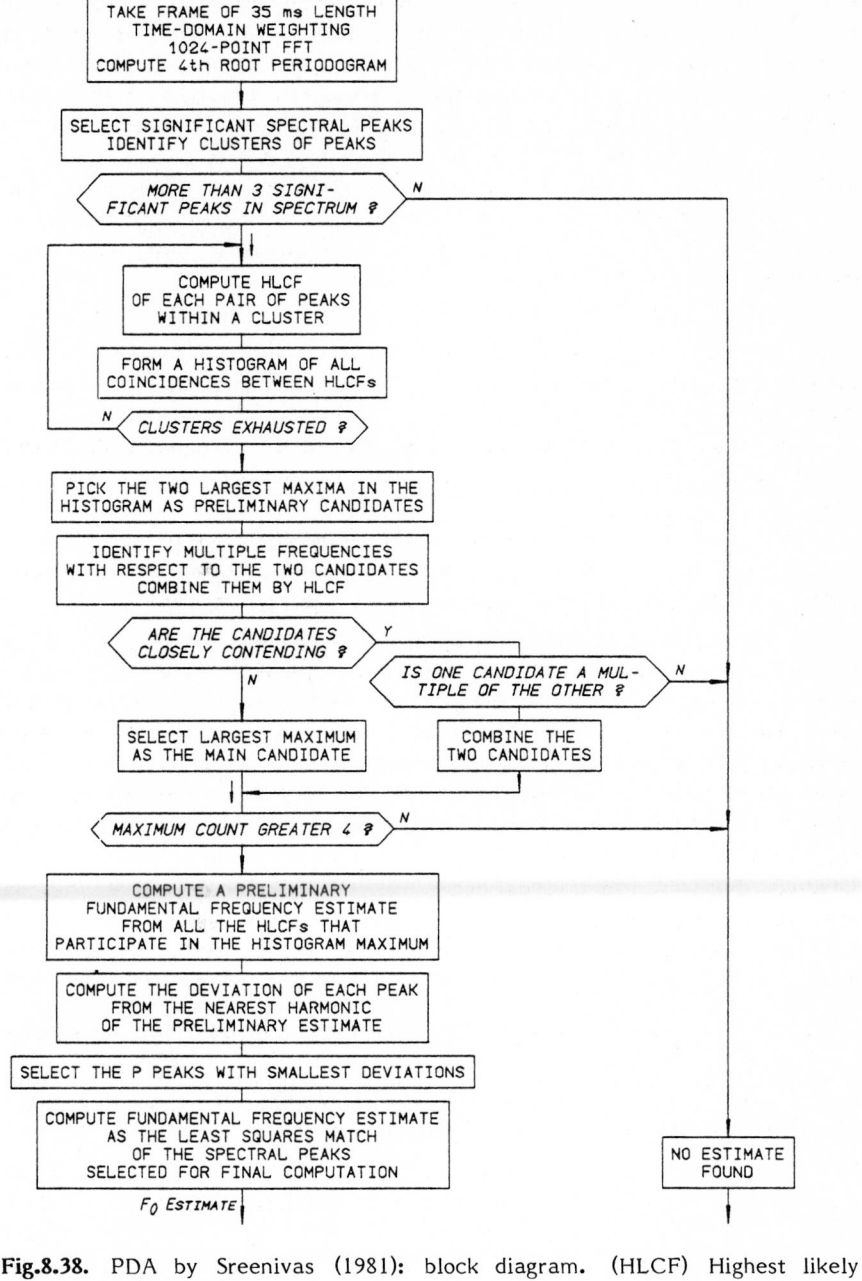

Fig.8.38. PDA by Sreenivas (1981): block diagram. (HLCF) Highest likely common fundamental

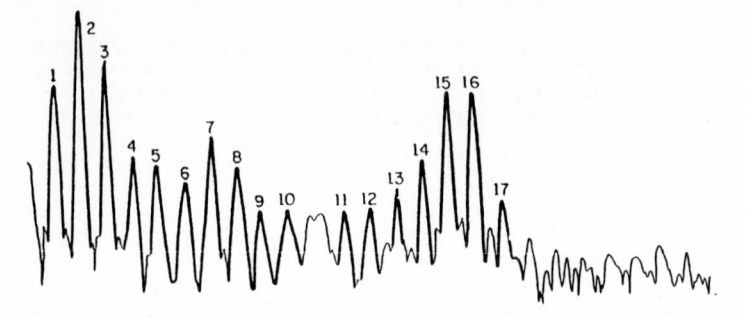

Fig.8.39. The peak selection routine in the PDA by Sreenivas: example of performance. The numbered peaks have been selected as significant. From (Sreenivas, 1981, p.V-5)

monic coincidences occur, and 4) computation of the improved estimate of F_0 from the least corrupted harmonics using the maximum likelihood estimate by Goldstein (1973) which is given in (8.62).

To understand how this algorithm works, we have to go into some further detail. Sreenivas's PDA takes a frame of 35 ms and computes a 1024 point FFT which corresponds to a spectral resolution of better than 8 Hz (the sampling frequency of the signal is 8 kHz). In the first major step, the peak selection routine, only those peaks are kept that are *clear*, i.e., such peaks that are accompanied by deep minima on both sides (a minimum being labeled "deep" when it is situated below a given threshold relative to the peak under consideration). An original example is given in Figure 8.39. Sreenivas investigated neither the amplitude nor the log amplitude spectrum; he rather computed what he called the fourth-root periodogram, the samples of which are equal to the fourth root of the samples of the amplitude spectrum. The advantage is that this periodogram (which will nevertheless be labeled "spectrum" in the following), like the logarithmic amplitude spectrum, has a greatly reduced dynamic range, but, like the amplitude spectrum, is a nonnegative function. Adjacent peaks are grouped together into clusters when they are separated by less than the maximum expected value of F_0, and the following harmonic coincidence checks are preferably carried out within these clusters.

In the next step the main harmonic match is performed. Sreenivas defined his own calculus to handle corrupted harmonics which then results in the so called HLCF routine (highest likely common fundamental). In his investigation of spectra that result from signals with changing fundamental frequency, Sreenivas found that the spectral peaks may be shifted due to the aperiodicity of the signal so that they take on inharmonic positions; in this case, however, they form clusters within which the difference of adjacent peaks still equals the fundamental frequency. So the HLCF algorithm has to process both harmonic structures and the difference of adjacent (or nonadjacent) peaks.

Let a corrupted harmonic be represented by a doublet (A,a), meaning to say that the exact harmonic frequency lies in the range A_1 to A_2, where

$A = (A_1+A_2)/2$ and $a = (A_2-A_1)$. In other words, if p is a fundamental frequency, h_1 is a harmonic number, e_1 is the unknown corruption that has got added to the harmonic frequency h_1p, and d_1 is the bound on the magnitude of corruption, the corrupted harmonic can be represented as

$$(A,a) = (h_1p+e_1, d_1) \text{ where } |e_1| < d_1/2 \tag{8.63}$$

The error or corruption e_1 could be positive and negative; p and d_1 are real numbers and usually $d_1 \ll p$; h_1 is an integer.

Subtraction between two corrupted harmonics (A,a) and (B,b) is defined as follows. Let $(A,a) = (h_1p+e_1, d_1)$ and $(B,b) = (h_2p+e_2, d_2)$, where $h_2 > h_1$. In such a case (B,b) is said to be *greater than* (A,a). The subtraction is then defined as

$$(B,b) - (A,a) := (B-A, b+a) , \tag{8.64a}$$

which means that subtraction of a number in the range $(A-a/2)$ to $(A+a/2)$ from another number in the range $(B-b/2)$ to $(B+b/2)$ would result in a number in the range $[B-A-(b+a)/2]$ to $[B-A+(b+a)/2]$; this is intuitively quite clear. *Addition* is defined in analogy to subtraction,

$$(B,b) + (A,a) := (B+A, b+a) . \tag{8.64b}$$

Two corrupted harmonics are said to be "*equal*," represented as (A,a) :=: (B,b), iff[26] there is an overlap in their respective ranges, i.e.,

$$(A,a) :=: (B,b) \text{ iff } |A-B| < (a+b)/2 . \tag{8.65}$$

When two corrupted harmonics are "equal," an operation of *intersection* between them is defined as follows,

$$(C,c) := (A,a) \cap (B,b) \quad \text{iff } (A,a) :=: (B,b) , \tag{8.66}$$

where

$$\begin{aligned} C &= [\min (A+a/2, B+b/2) + \max (A-a/2, B-b/2)] / 2 , \\ c &= [\min (A+a/2, B+b/2) - \max (A-a/2, B-b/2)] . \end{aligned} \tag{8.67}$$

The present task is now to compute the best possible estimate of p. When the errors e_1 and e_2 are zero and known to be so, p can be exactly computed ... . But when the errors are nonzero and unknown, the problem becomes very different. However, given at least the bounds on the errors, d_1 and d_2, it should be possible to estimate p within some error bound. (Sreenivas, 1981, pp. V-8/9)

With these definitions, Sreenivas constructed an algorithm that can cope with both a harmonic structure and an inharmonic sequence of peaks if the distance of adjacent peaks equals the fundamental frequency. The algorithm always investigates a pair of spectral peaks. Let (A,a) and (B,b) represent such a pair of harmonic peaks whose (unknown) harmonic numbers are assumed to be prime. The subtraction of (A,a) from (B,b) using (8.64) leads to a new "harmonic" (C,c) which has a lower frequency but increased uncertainty. (C,c) then replaces the greater frequency of (A,a) and (B,b); let it be (B,b) in this example. The same step is repeated for (A,a) and (C,c) and results in a new

---

[26] "Iff" is a standard expression in computer science and means "if and only if."

value (D,d) which may now replace (A,a). The highest likely common fundamental (HLCF) of (A,a) and (B,b) is found when the two "harmonics" have finally become "equal", as defined in (8.66). The algorithm always terminates; if the peaks under consideration are really harmonics with mutually prime harmonic numbers, it ends up with the fundamental frequency; if the harmonic numbers have a common divisor, it ends up with the fundamental frequency times this divisor; if the peaks under consideration are spurious, the algorithm may terminate at an arbitrary value, or one may abort it once the lower limit of the measuring range has been exceeded. Of course the algorithm works only when the increased corruption bounds after the successive subtractions do not exceed the value of the fundamental frequency itself. In the realized PDA, if an upper error boundary d_1 is not explicitly known, the maximum error induced by limited spectral resolution is taken as the upper boundary.

All the HLCFs from pairings of significant peaks in the spectrum are gathered in a histogram and checked whether they are "equal" in the sense of (8.65). If two HLCFs from different pairs of peaks are found "equal," a "coincidence count" is incremented in the histogram.

In the third major step the HLCF algorithm is used to compute a preliminary estimate of F_0. The histogram is scanned for the two frequency values that show the largest and the second-largest number of coincidences. The following checks are performed in order to recognize and to clear ambiguous situations. The two candidates are checked against all the other HLCFs in order to find those HLCFs that are multiples of the candidates. If a multiple is found, the HLCF routine is used again to merge it with the pertinent candidate. If the one candidate has a frequency which is a multiple of the other, the two are merged. Otherwise the coincidence counts of the two candidates are compared. If their frequencies are not harmonically related, and if the number of coincidences is "closely contending," i.e., about equal for both candidates, the search for F_0 is terminated since the algorithm cannot decide between two competing pitches. The search is also terminated if the number of coincidences for the best candidate is less than four. Once all ambiguous situations have been cleared, the preliminary estimate of F_0 is computed as the average of all those HLCFs that have contributed to form the maximum of coincidences in the histogram.

In the last step the final estimate of F_0 is computed. All the spectral peaks are checked against the preliminary estimate by computing the distance between the frequency of the peak and the nearest multiple of the preliminary estimate. The algorithm then selects a limited number of peaks whose distance to a harmonic of the preliminary estimate is smallest. From these peaks the final estimate of F_0 is computed using the Goldstein maximum likelihood estimator (8.62).

Sreenivas's publication contains detailed evaluative data on his PDA. A part of his results will be presented in Chapter 9. In general, the algorithm is reported to work with less than 5% error for a signal-to-noise ratio better than 5 dB, and with less than 20% error for a SNR better than -10 dB (there the speech is unintelligible, but pitch can still be perceived).

8.5.3 Determination of F_0 from the Distance of Adjacent Spectral Peaks

The easiest way to determine F_0 would be the extraction and identification of a spectral peak at the fundamental frequency itself. In order to yield a spectral peak at that frequency the first harmonic must be present in the signal. This cannot be expected. The way out of this difficulty, however, is easier to find in the frequency domain than in the time domain. Since the harmonics are well separated in the spectrum, it is sufficient to measure the distance between adjacent or even nonadjacent spectral peaks representing the higher harmonics of the periodic or quasi-periodic time-domain signal. This is, for instance, performed by Harris and Weiss (1963) and later by Aschkenazy, Weiss, and Parsons (1974). Measuring this spectral distance, one gets a crude estimate of F_0 (see Fig.8.21 for an illustration). The accuracy of this value is improved by deriving a better estimate from the frequency of a higher harmonic whose harmonic number, in turn, can be determined from the crude estimate. This is the usual way to improve the accuracy of manual spectrogram F_0 determination (see Sect.5.1.2 for details). With their real-time analog spectral analyzer whose output is A-D converted and further processed by a computer, Harris and Weiss achieved relatively fine resolution in the frequency domain.

After the PDD by Harris and Weiss, however, it was very quiet in this field until the era of digital spectral analysis began with the rediscovery on of the FFT (1965, 1967). Since frequency-domain analysis needs only one FFT per frame and is thus twice as fast as cepstrum analysis, the principle again became attractive. Most frequency-domain PDAs, however, then followed the principle of spectral compression (see Sects.8.5.1-3). It was only in 1976 that once again a PDA investigating the distance of adjacent spectral peaks was published (Seneff, 1976, 1978). It is worthwhile presenting some details of this approach since it is rather unconventional from the point of view of its implementation, and a good example how computational effort can be saved by good application of appropriate signal processing methods.

In the previous parts of Sect.8.5 it was shown that frequency-domain PDAs must provide a relatively high spectral resolution in order to achieve reasonable accuracy. Therefore they require voluminous FFT sequences which slow down the analysis. The question is whether the process can be speeded up by reducing the length of the FFT.

Seneff's PDA presents an unconventional solution to this problem. This PDA rigorously excludes all the redundant or irrelevant information before executing the FFT. Thus the number of samples to be transformed is drastically reduced. This is achieved in two steps. The signal is originally sampled at a sampling interval of 130 µs (i.e., a sampling frequency of 7.5 kHz); for the determination of F_0, however, only frequencies up to 1 kHz are used in this PDA. The principle is to detect the fundamental frequency directly in the frequency domain as the distance between adjacent significant spectral peaks. Hence at least two spectral peaks are needed; they require a minimum frequency band of 700 Hz (when the range of F_0 is upward limited to 350 Hz). Since the

PDA is designed for telephone speech, the frequency band below 300 Hz is not occupied and therefore not needed either.

The PDA manages to get rid of the irrelevant frequency components by a spectral rotation procedure. In the first step, the signal is low-pass filtered by a nonrecursive digital filter and downsampled by a factor of three. In contrast to time-domain PDAs, downsampling does not decrease the accuracy of frequency-domain PDAs.

> The first step in the pitch extraction is to filter the speech down to 1260 Hz and downsample, throwing away two out of every three samples. For this purpose a finite impulse response filter seemed to be a good choice. Since FIR filters have only zeros, one need compute outputs only at the downsampled rate, which in our case represents a three to one savings in time. ...
>
> We now have a waveform $s_1(n)$ which contains information from -1260 to +1260 Hz. We know that, since the waveform is real, the negative frequency information is redundant. Digital signal processing theory tells us that if we multiply each sample of the waveform $s_1(n)$ by $\exp(j\Omega n)$, we will cause the spectrum to be rotated by Ω in the z plane. By choosing $\Omega = 90°$, we cause the spectrum to be rotated such that $1260/2 = 630$ Hz is at the origin. Now a second pass of this complex $s_2(n)$ through the same (nonrecursive low-pass) filter, with 3 to 1 downsampling again, will yield a complex waveform $s_3(n)$ containing frequencies up to $1260/3 = 420$ Hz. However, because of the rotated spectrum, 630 Hz in the original waveform corresponds to zero Hz in $s_2(n)$, and thus our doubly downsampled complex waveform contains the information from (630-420)Hz to (630+420)Hz in the original speech, which is in the desired spectral region.
>
> Choosing $\Omega = 90°$ has certain advantages in terms of speed. Multiplication by $\exp(j\Omega n)$ involves only data transfer, since the sine and cosine of multiples of 90° are either ±1 or zero. Furthermore, as a consequence, each sample of $s_2(n)$ is either purely real or purely imaginary. One can therefore choose an implementation method for the FIR filter so that filtering of this complex waveform takes essentially no more time than would filtering a real waveform.
>
> Pitch detection generally requires a long time window of speech in order to assure at least two periods of a low pitched voice. Fortunately, the double downsampled signal consists of samples which are spaced by 132·3·3 µs, or 1.188 ms. Only 32 samples of this waveform are required to yield 38 ms of data, a time window that is sufficient to encompass two periods for pitches of up to 19 ms, or 53 Hz, a very deep male voice. (Seneff, 1978, p.359)

Using a 128-point FFT (which is 1/16 the number of points required in the harmonic product spectrum), this PDA achieves a frequency resolution of 6.6 Hz.[27]

For readers who are not so familiar with digital signal processing procedures, the principle of this spectral rotation is explained in some more detail. The exponential-weighting theorem of the z-transform says that weighting a signal $f(n)$ by an exponential function a^n leads to the z-transform $F(z/a)$

---

[27] With the assumption of a 14th-order nonrecursive low-pass filter the computational saving of this algorithm over a conventional FFT with equal

when $F(z)$ is the z-transform of $f(n)$. It can easily be seen that this relation is true,

$$G(z) = Z\{f(n)a^n\} = \sum_{n=0}^{\infty} f(n)a^n z^{-n} = \sum_{n=0}^{\infty} f(n)(z/a)^{-n} = F(z/a) . \qquad (8.68)$$

This theorem holds for arbitrary values of a, i.e., it also holds when a is complex. If a is now chosen on the unit circle, i.e., $a = e^{j\Omega}$, then the complex division will cause the unit circle of the z plane to be rotated by the angle $-\Omega$ and depicted on itself since the absolute of $\exp(j\Omega)$ equals 1. The whole operation is linear (the superposition principle holds!); i.e., although frequencies are shifted, they are shifted by a constant, and their difference remains unchanged. The advantage of the spectral rotation is that irrelevant spectral information can be discarded both at the low-frequency and the high-frequency end of the scale.

Spectral rotation is equivalent to complex modulation or demodulation of a signal. The resulting signal is complex. For the digital low-pass filter (having real coefficients) this has no effect other than that it has to be run twice: separately for the real and for the imaginary part of the signal.

The amplitude spectrum of the complex signal obviously is not an even function. But that bothers neither the FFT nor the subsequent peak-picking basic extractor. This similarly unconventional routine, which somewhat resembles the channel selection routine in the time-domain PDA by Gold and Rabiner (1969; Sect.6.2.4) as well as the previously discussed frequency histogram technique, derives the fundamental frequency in an iterative way. In a preparatory step, obviously irrelevant and/or erroneous peaks are excluded from

---

spectral resolution yields a factor of more than ten, if the number of multiplications is taken as the basis of comparison. In the original signal 1152 samples per frame would be necessary to yield the required spectral resolution; if a (well-programmed) FFT were available for that number it would need about 8640 complex multiplications, i.e., 34560 simple (real) multiplications. Of course this long transformation interval (more than 150 ms) would contain only about 40 ms of speech. The rest would be filled up with zeros, but this does not really influence the computational complexity of the FFT. However, the downsampling steps can be done before subdividing the signal into individual frames. Hence, if we take the computational load per frame as the basis of comparison and assume a frame interval of 10 ms, we have to take into account 75 samples of the original signal within this interval. The first 1:3 downsampling step with the 14th order filter thus needs $(75/3) \cdot 7 = 175$ multiplications and yields 25 samples. The subsequent complex exponential weighting, due to the choice of $a = \exp(j\pi/2)$, does not need any multiplications. The second downsampling step needs $(25/3) \cdot 7$, i.e., about 63 multiplications per 10 ms and yields about 9 samples. The final FFT which operates on a frame length of 40 ms and extends the frame by zeros to 128 samples, requires $128 \cdot 4 \cdot 4 \cdot 5 = 2304$ multiplications. Altogether the computing effort will be 2542 multiplications which is less than 1/12 the effort of the full-length transform with equal spectral resolution. The question of the computing effort in connection with the FFT will be further discussed in Section 8.5.4.

further consideration (they are recognized as being either too small or too close to larger peaks).

> The peaks that remain after the elimination step are given a rank order according to size. At the first iteration, a single pitch estimate is entered into a table of potential pitch estimates, defined as the distance between the two largest peaks. At the second iteration, the third largest peak is added to the list of peaks under consideration and two new pitch estimates are added to the table, defined as the distance between adjacent peaks, among the three under consideration. At each subsequent i^{th} iteration, the largest peak among those remaining is added to the list of candidate peaks, and i new pitch estimates are added to the growing list of estimates, defined, again, as distance between adjacent peaks. (Seneff, 1978, p.361)

The iteration steps are terminated as soon as there are 8 coincident F_0 estimates (i.e., estimates situated within a common small tolerance interval) or as soon as no significant peaks are left. In the postprocessor, finally, the F_0 estimates are smoothed by a combined 3-point and 5-point median smoother.

The PDA was incorporated into the previously mentioned Lincoln Labs digital voice terminal. It was found to perform better for band-limited or noisy signals than the time-domain PDA by Gold and Rabiner (1969). As to the computing speed, the algorithm runs in real time on this signal processor. It was found to require about twice the time of the Gold-Rabiner PDA and 25% that of an implemented cepstrum algorithm.[28]

This PDA certainly will not reach the reliability of the harmonic product spectrum PDA, and it is certainly not designed for that. It signals the fact, however, that fast and accurate frequency-domain pitch determination need not hide its face in the presence of the more fully exploited STA techniques, such as autocorrelation, AMDF, or cepstrum.

8.5.4 The Fast Fourier Transform, Spectral Resolution, and the Computing Effort

The length and the computing effort of the DFT which defines spectral resolution definitely deserve some further consideration. As just mentioned, Seneff (1978) achieved a spectral resolution of 6.6 Hz. Noll (1970) reported a DFT of 2048 samples at 10 kHz sampling rate, yielding a spectral resolution of 4.95 Hz. Sluyter et al. (1980) confined themselves to a spectral resolution of 25 Hz and executed some refinements of the measuring accuracy by parabolic interpolation after the crude determination of F_0. The same applies to Martin (1981) who improved the basic resolution of 32 Hz to 1 Hz (!). Sreenivas (1981) performed a 1024-point FFT which yields an accuracy of about 8 Hz. Terhardt (1979b) left the question open.

---

[28] The implementation of the cepstrum PDA must be extremely fast in this processor since Rabiner et al. (1976), in their comparative evaluation, found a ratio of about 1:50 for the computing speed between the Gold-Rabiner (1969) algorithm and the cepstrum PDA.

In principle, there are two issues that determine the requirements with respect to spectral resolution. On the one hand, there is the sampling theorem. It has already been mentioned several times that the spectrum, in its digital representation, is a sampled signal, just as the time-domain signal is. If the time-domain signal $x(n)$ is periodic with the fundamental period duration T_0 and contains many harmonics, then the spectrum shows ripples with the minimum width F_0. In order to meet the sampling theorem, the spectral resolution must thus be less than $F_0/2$ which gives a lower bound of $2T_0$ for the frame length K, i.e., at least two complete periods of the signal must be within the frame. Since the usual minimum value of F_0 is about 50 Hz, a minimum spectral resolution of 25 Hz, i.e., a minimum frame length of 40 ms is required. The appropriate transformation interval for application of the FFT is 51.2 ms when the sampling rate is 5 or 10 kHz; this yields a spectral resolution of 19.8 Hz. On the other hand, there is the spectral resolution required for the measurement. In the frequency domain, accuracy is independent of the time-domain sampling rate; it depends rather on the frequency-domain sampling rate, i.e., the spectral resolution. As was just shown, a spectral resolution of 19.8 Hz is sufficient to completely describe the harmonic pattern caused by the periodicity of the signal, but not to measure the fundamental frequency exactly enough; in this case we would commit an error of 20% at $F_0 = 100$ Hz if the first harmonic itself were measured. If the 10th harmonic, for instance, is taken as a reference, the error would decrease to 2%. Not all the frequency-domain PDAs, however, can rely on such a refinement of the initial spectral estimate. Those which cannot (the harmonic product spectrum, for example) need a better spectral resolution for measurement.

The resolution which is ultimately required in order to render the measurement inaccuracies inaudible is the 0.2%-0.5% given by the difference limen for the frequency of sinusoids (valid above 500 Hz), which demands a spectral resolution of 2.5 Hz at a frequency of 500 Hz and of 5 Hz at a frequency of 100 Hz. No PDA has attemped to reach this resolution directly; a number of them, however, obtain it by interpolation of the spectral samples (Martin, 1981), interpolation of the peak frequencies (Duifhuis et al., 1982), or averaging of several crude estimates (Sreenivas, 1981).

One way to achieve a good frequency resolution is to increase the length of the FFT. Since the actual frame length K is upward limited by the condition that the speech signal be rather stationary within the frame, the signal beyond K - in correspondence with stationary short-term analysis - is padded with zeroes to yield the required transformation interval N. Of course the signal must be weighted by an appropriate time-domain window in order not to introduce spurious ripples in the spectrum due to the truncation of the signal. Increasing the transformation interval, however, results in an increase of the computational effort. The computational burden of the FFT is in the order of (N ld N) so that it grows faster than proportionally with N.

For the following examples, let us assume a sampling rate of 10 kHz and a spectral resolution of about 5 Hz. If the first partial above 500 Hz is taken as the reference peak for determination of F_0, the error is then kept below 1% over the whole measuring range of F_0. Without any countermeasures a FFT of 2048 complex samples would be needed to do this job.

Two countermeasures appear possible to reduce the computational effort: 1) optimize the implementation of the FFT, and 2) reduce the length of the FFT as much as possible.

Let us first deal with the optimization of the FFT. For reasons of space we cannot cover the elementary knowledge of the FFT in all detail, so that some familiarity with this transform is required to understand the following remarks. In this respect the reader is referred to the literature (Nussbaumer, 1980; Achilles, 1978; Rabiner et al., 1972; G-AE Subcommittee, 1967). The discrete Fourier transform in its elementary version (2.32) needs N^2 complex multiplications. For certain values of the transformation interval N, in particular for N equal an integer power of two, the transformation matrix $\mathbf{W}$ can be factorized into (ld N) partial transformation matrices $\mathbf{W}_{pi}$, $i = 1 (1) \text{ld N}$,

$$\mathbf{X} = \mathbf{W} \mathbf{x} = \mathbf{W}_{pk} \mathbf{W}_{p,k-1} \cdots \mathbf{W}_{p2} \mathbf{W}_{p1} \mathbf{x} \; ; \quad k = \text{ld N} , \tag{8.69}$$

such that the computation of one of these partial transforms,

$$\mathbf{X}_{pi} = \mathbf{W}_{pi} \mathbf{x}_{pi} , \tag{8.70}$$

where $\mathbf{X}_{pi}$ is the output vector, $\mathbf{x}_{pi}$ the input vector, and $\mathbf{W}_{pi}$ the partial transformation matrix, in principle requires one multiplication per line, i.e., N (complex) multiplications as a whole. Figure 8.40 shows an example for $n = 16$. Since (ld N) partial transformations are necessary to perform the FFT, the overall computational effort is in the order of (N ld N). This is an immense gain, compared to the N^2 multiplications of the ordinary DFT; but for most applications, in particular for the frequency-domain PDAs, this is still too slow or too expensive to be justified for a real-time implementation.

To understand the following remarks, let us go somewhat further into detail. Each partial transform is implemented as a program loop whose significant element, the so-called *FFT butterfly* (Rabiner et al., 1972), connects two input samples $x_{pi}(n_1)$ and $x_{pi}(n_2)$ to form two output samples $X_{pi}(m_1)$ and $X_{pi}(m_2)$,

$$X_{pi}(m_1) = x_{pi}(n_1) w^0 + x_{pi}(n_2) w^k \; ; \quad w = \exp(-2\pi j/N) ,$$

$$X_{pi}(m_2) = x_{pi}(n_1) w^0 + x_{pi}(n_2) w^{k+N/2} = x_{pi}(n_1) w^0 - x_{pi}(n_2) w^k , \tag{8.71}$$

where k is an integer between 0 and N whose actual value is not of interest here. The elementary operation (8.71) needs one complex multiplication since w^0 equals unity. One partial transform with N output samples executes this operation N/2 times, thus requiring exactly N/2 complex, i.e., 2N ordinary multiplications. In total this gives

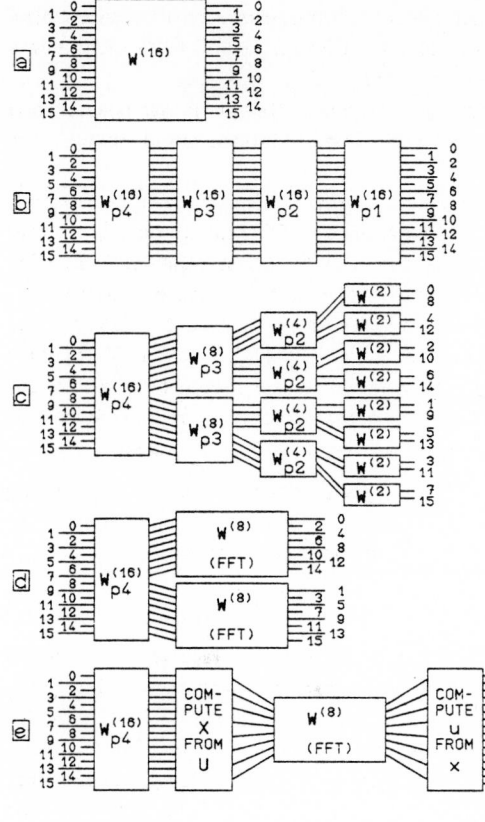

Fig.8.40a–e. Factorization of the discrete Fourier transform. **(a)** Scheme of ordinary DFT; **(b)** scheme of ordinary FFT according to (8.70,71); **(c)** complete factorization according to the decimation-in-time principle (with bit reversal at the input); **(d)** decomposition into two FFTs with length N/2 and the last partial transform; **(e)** computation of **X** from the real input signal **x** using the complex signal **u** and a FFT of length N/2. (N) Transformation interval (in this example, N = 16), (x) input signal, (X) spectrum

$$Z = 2N \ \text{ld} \ N \qquad\qquad (8.72)$$

multiplications for one FFT. In our numerical example, with N = 2048 and (ld N) = 11, this would be Z = 45056 multiplications. Equation (8.72) is valid for the FFT programmed without any optimization; the value of Z for N = 2048 can thus serve as a reference for possible savings of the computing effort.

In the remainder of this section we will discuss the various possibilities to reduce the computing effort without losing the accuracy required for reliable measurement of F_0. Table 8.3 gives an overview of the methods to be used for this, and Fig.8.41 shows the block diagram of a preprocessor for frequency-domain PDAs which has been optimized with respect to the computing effort.

A considerable part of the multiplications can be saved if one looks more closely at the values w^i of the complex exponential function. For the whole transform we need only N values of the complex exponential, i.e.,

$$w^{mn} = \exp(-2\pi jmn/N), \quad m = 0 \ (1) \ N-1, \quad n = 0 \ (1) \ N-1$$

where mn reduces to mod(mn,N) since the complex exponential function is periodic with 2π. These N values are equally spaced at multiples of the exponent $(-2\pi j/N)$. For the partial transforms we need only N/k values, k = 1,2,4,8,

Table 8.3. Comparative evaluation of the effort necessary to compute a FFT spectrum for frequency-domain pitch determination under various aspects of possible algorithmic optimizations.• The number of multiplications per frame is taken as the reference quantity. The computing effort is always given for the displayed optimization alone. The sampling rate is assumed to be 10 kHz, the required spectral resolution to be less than 5 Hz, and the frame rate to be 100 Hz

Optimizing operation	FFT Length	Number of Multiplications			Saving (%)
		FFT	Other Operations	Total	
No optimization	2048	45056	---	45056	0
Optimized FFT Programming	2048	32776	---	32776	29
Exploit the fact that the input signal is real: perform spectral rotation, shift imaginary part of the signal against time, and decompose spectrum	1024	20480	2048	22528	50
All programming optimizations	1024	14344	2048	16392	64
Downsampling to 5 kHz	1024	20480	200	20680	56
Downsampling and programming optimizations	512	5942	1224	7166	84
Limit transformation interval to 51.2 ms and upsample spectrum by factor 4 in the frequency domain	512	9016	6144	15160	66
Time-domain downsampling, limitation of transformation interval, and frequency-domain upsampling	256	4096	3272	7368	83
All optimizations applied together	128	1032	3528	4660	89

... $N/2$ which are then equally spaced at multiples of the exponent ($-2\pi jk/N$). For an implementation according to the principle "decimation in time with bit reversal after transformation" (G-AE Subcommittee, 1967), which allows for the most rigorous exploitation of this feature, this means that the first partial transform with the transformation matrix $\mathbf{W}_{p1}$ needs only the two values w^0 and $w^{N/2}$, which are equal +1 and -1, respectively; it can thus be executed without multiplications. The same applies to the second partial transform where the exponents of the complex exponential function proceed in steps of $N/4$, and where the elements of $\mathbf{W}_{p2}$ can thus only take on values of +1, -1, +j, and -j. In the third partial transform $N/2$ multiplications can still be saved this way; then the saving is divided by two for each subsequent partial transform until it reaches a value of two in the last step. In this special decimation-in-time algorithm, the elements of $\mathbf{W}_{pi}$ which allow the computation of the output samples without multiplication are always found at the beginning of the partial transforms. Hence, here the saving can be fully exploited. In this case the number of multiplications becomes

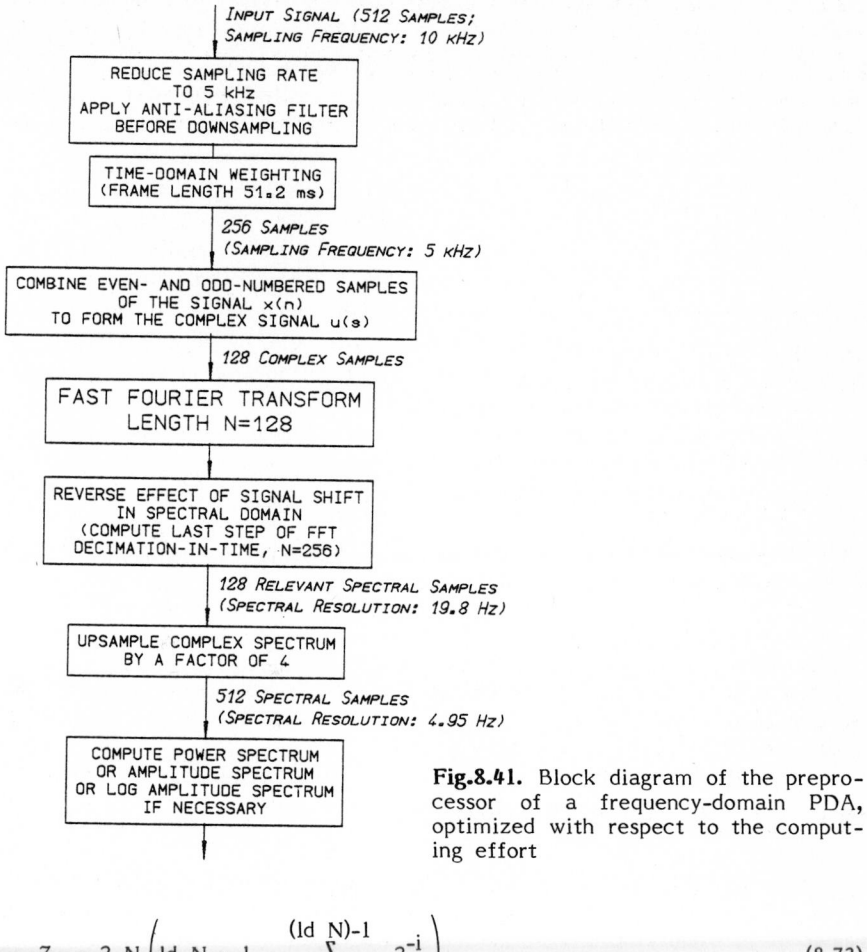

INPUT SIGNAL (512 SAMPLES; SAMPLING FREQUENCY: 10 KHZ)

REDUCE SAMPLING RATE
TO 5 kHz
APPLY ANTI-ALIASING FILTER
BEFORE DOWNSAMPLING

TIME-DOMAIN WEIGHTING
(FRAME LENGTH 51.2 ms)

256 SAMPLES
(SAMPLING FREQUENCY: 5 KHZ)

COMBINE EVEN- AND ODD-NUMBERED SAMPLES
OF THE SIGNAL x(n)
TO FORM THE COMPLEX SIGNAL u(s)

128 COMPLEX SAMPLES

FAST FOURIER TRANSFORM
LENGTH N=128

REVERSE EFFECT OF SIGNAL SHIFT
IN SPECTRAL DOMAIN
(COMPUTE LAST STEP OF FFT
DECIMATION-IN-TIME, N=256)

128 RELEVANT SPECTRAL SAMPLES
(SPECTRAL RESOLUTION: 19.8 Hz)

UPSAMPLE COMPLEX SPECTRUM
BY A FACTOR OF 4

512 SPECTRAL SAMPLES
(SPECTRAL RESOLUTION: 4.95 Hz)

COMPUTE POWER SPECTRUM
OR AMPLITUDE SPECTRUM
OR LOG AMPLITUDE SPECTRUM
IF NECESSARY

Fig. 8.41. Block diagram of the preprocessor of a frequency-domain PDA, optimized with respect to the computing effort

$$Z_p = 2N \left(\text{ld } N - 1 - \sum_{i=0}^{(\text{ld } N)-1} 2^{-i} \right). \tag{8.73}$$

For $N = 2048$ this gives $Z_p = 32816$ which is equivalent to a 30% saving in computational effort due to some sophistication in programming alone.

Another way to save multiplications is sometimes proposed in the literature as *FFT pruning* (Markel, 1971b; Sreenivas, 1981). The FFT is faster than the ordinary DFT because many elements of the partial transformation matrices are zero, and the algorithm knows which ones these are. If the FFT is pruned, multiplications are saved because *signal values* are zero and the algorithm knows which ones, or because the spectral samples are only partially needed. FFT pruning might thus be of interest when a short frame is padded with zeros to yield a long transformation interval. In the ordinary DFT, pruning can save a huge amount of computing effort. In the FFT, since everything is computed simultaneously, only the first or the last steps (in the case of zero signal samples it is the first) are really affected. The just-mentioned decimation-in-time algorithm does not need multiplications in the first partial transform anyhow; so with pruning nothing is gained. One could think of the principle "decimation in frequency with bit reversal before the transformation" which allows for similarly favorable control of the exponential values that need no multiplications, and reverses the sequence of the partial transforms so that the steps without multiplication come last. If just a few percent need to be saved to fulfil real-time conditions, such an algorithm may be justified. We will see, however, that there are better methods of saving computing time than relying on input samples that happen to be zero.

The DFT transforms a finite complex input signal of length N into the complex spectrum $X(m)$. The input speech signal $x(n)$, however, is real; the imaginary part remains zero. Thus 50% of the computing effort is in vain, but the FFT algorithm does not permit computing the transform only from the real part of the complex input signal and reducing the computing effort at the same time. Hence, other countermeasures must be thought of. The usual way to speed up the FFT if the input signal is real is as follows. Let $x(n)$, $n = 0\,(1)\,N\text{-}1$ be the input signal, and let N be an integer power of 2 so that the radix-2 FFT algorithm, as defined by (8.69-71), can be applied. If we use the decimation-in-time principle, the distance

$$n_d = n_2 - n_1 \tag{8.74}$$

between the two samples $x_{pi}(n_1)$ and $x_{pi}(n_2)$ connected by (8.71) depends only on the index i of the partial transform [according to (8.70)] just being applied. For the first partial transform, i.e., $i = 1$, n_d takes on a value of $N/2$; for each subsequent partial transform, n_d must be divided by two until it equals 1 in the last step. This means that all partial transforms except the last one can be performed for the samples $x(n)$ with odd index n and the samples with even index n separately. In addition, the product of the partial transformation matrices $\mathbf{W}_{pi}$, $i = 1\,(1)\,k\text{-}1$, regarded separately for odd and even values of n, yields the transformation matrix $\mathbf{W}$ of the FFT of length $N/2$. The FFT of N samples $x(n)$, $n = 0\,(1)\,N\text{-}1$ can thus be decomposed into two FFTs of $N/2$ samples each, which transform the samples $x(n)$, $n = 0\,(2)\,N\text{-}2$ and the samples $x(n)$, $n = 1\,(2)\,N\text{-}1$ separately, and the last partial transform $\mathbf{W}_{pk}$ which then yields the N output samples $X(m)$, $m = 0\,(1)\,N\text{-}1$. If the input signal is real, these two FFTs with length $N/2$ can be performed simultaneously. From the input signal $x(n)$, a complex signal $u(s)$ is formed such that

$$u(s) = x(2n) + jx(2n+1); \quad n = 0\,(1)\,N\text{-}1,\ s = 0\,(1)\,N/2\text{-}1 \; ; \tag{8.75}$$

this signal is then transformed into the spectrum $\mathbf{U}$ using a single FFT with length $N/2$. The spectrum $\mathbf{U}$ corresponds to the partial transform $\mathbf{X}_{p,k\text{-}1}$ in (8.69,70), and the computation of $\mathbf{X}_{p,k\text{-}1}$ from $\mathbf{U}$ is rather simple if one takes into account that, if the input signal is real, the real part of the spectrum is always even, and the imaginary part is always odd. (The corresponding theorem says that, if the input signal is purely imaginary, the imaginary part of the spectrum is even, and the real part is odd.) The last partial transform $\mathbf{W}_{pk}$ is then carried out as given in (8.69,70). In this manner the number of multiplications is reduced by a factor of two. In total, the computing effort can be cut down by more than 60% due to elaborate programming alone.

Further reductions of the computing effort can only be achieved when the length N of the transform is reduced. Two ways for doing this are feasible: 1) time-domain downsampling, and 2) reduction of the duration of the transformation interval. The latter operation causes a loss in spectral resolution, whereas time-domain downsampling does not. This possibility was extensively exploited by Seneff (1978, cf. Sect.8.5.3). Excessive time-domain downsampling, however, may cause a loss of frequency regions that are important for pitch information (Sreenivas, 1981). If we follow the statement by Terhardt (1979b) and others that frequencies above 3 or 4 kHz do not significantly contribute to pitch perception, then a downsampling of the signal to a sampling rate of (not less than) 5 kHz can be justified. In our example this would cut down the length N of the transformation from 2048 to 1024 points (no other kind of optimization taken into account) and thus reduce the computing effort by a factor of more than two. If only good quality voiced speech were analyzed, one could even omit the antialiasing filter before downsampling. For degraded speech the antialiasing filter is nevertheless necessary. For example, this filter could be realized as a 14th-order linear-phase half-band filter (Bellanger, 1980, Goodman and Carey, 1977). In this case it only needs four multiplications per sample and already operates at the reduced sampling rate. Since the signal can be downsampled before subdividing it into frames, the computing effort for the antialiasing filter would be 200 multiplications per frame (if the frame interval is assumed to take on a value of 10 ms). Compared to the computational effort of the FFT this is negligible.

A final reduction in the computing effort by another factor of four results from the discrepancy between the spectral resolution necessary to satisfy the sampling theorem and the accuracy requirements of the measurement. The spectral resolution of 19.8 Hz is sufficient for maintaining the pitch information during the transform; the spectrum is oversampled only for the sake of an accurate peak determination. Instead of performing a voluminous FFT on an artificially lengthened interval, one could think of transforming the ordinary frame interval of 51.2 ms which yields the spectral resolution of 19.8 Hz, and subsequently upsampling the spectrum by a factor of four in the frequency domain. For upsampling we use the same half-band filter as for downsampling, and again we use it at the lower sampling rate. In our example (without any other optimization) we would now get along with a FFT of only 512 samples and a saving in computational effort of more than 75%. The upsampling is less costly: the 14th-order half-band filter (with four multiplications per sample) needs 2048 multiplications for 256 complex spectral samples in the first step, and twice that number for the 512 spectral samples in the second step. (Note that 50% of the spectral samples are redundant and need not be upsampled.) A number of PDAs, for instance the one by Duifhuis et al. (1979b), provide parabolic interpolation around the crudely determined spectral peaks in order to improve the accuracy. Such operations may replace the spectral upsampling and thus further reduce the computing effort[29]. Nevertheless, they have been disregarded here since this discussion only intended to show how signal processing methods and discarding irrelevant information can lead to algorithmic realizations that yield practically identical spectra, but differ by almost an order of magnitude in their computing effort.

8.6 Maximum-Likelihood (Least-Squares) Pitch Determination

A signal a(n) of finite duration K is assumed to consist of Gaussian noise gn(n) and a periodic component x(n),

$$a(n) = gn(n) + x(n) ; \quad n = 0 \ (1) \ K\text{-}1 \ . \tag{8.76a}$$

Nothing is known about the periodic signal except that it exists over the whole interval $n = 0 \ (1) \ K\text{-}1$ and that its period duration T_0 is shorter than the interval length K. The task of the maximum likelihood $(ML)^{[30]}$ approach is then to reconstruct x(n) and a Gaussian random noise signal gn(n) by a mathematical procedure, i.e., to find a periodic estimate $\hat{x}(n,p)$ depending on p, which is *most likely* to represent the original periodic wavelet x(n). The solution involves maximizing the energy of the estimate $\hat{x}(n)$ as a function of the trial period p.

---

[29] Interpolation of samples in the power or amplitude spectrum is dangerous and may lead to erroneous results. The correct sequence of operations to improve spectral resolution is 1) the FFT, 2) frequency-domain upsampling, and 3) computation of the amplitude or power spectrum. Since operations 1) and 2) are both linear, their sequence can be interchanged. This permits optimizing the FFT and the digital filtering in order to globally minimize the computing effort. Computation of the power or amplitude spectrum, however, involves squaring and is thus a nonlinear operation. Interchanging the sequence of operations 2) and 3) thus leads to principally different results. A correct method of selective interpolation around spectral peaks could thus be the following. 1) Compute a crudely sampled power or amplitude spectrum and determine the significant peaks; 2) perform an interpolation around these peaks using the pertinent complex spectrum, and 3) compute the power or amplitude spectrum of the samples interpolated and determine the location of the peak therefrom with better accuracy.

[30] As to the terminology, different terms for this method are found in the literature. The use of the term *maximum likelihood* is due to Noll (1970) and Slepian (see Noll, 1970). Friedman (1977) called the method *pseudo maximum likelihood* since it does not perform an exact maximum likelihood task (for definition of that, see publications such as Gueguen and Scharf, 1980). Irwin (1979) called this method *least-squares* approach. Throughout this section we maintain the original term *maximum likelihood* as the major concept since it agrees with the majority of publications on this subject.

The mathematical formulation of the principle, as outlined above, is due to Slepian (unpublished; see Noll, 1970, for details) and was first applied by Noll (1970) to develop a PDA. The performance was found to be excellent, comparable to the harmonic-product spectrum PDA; some trouble, however, was reported in connection with multiple peaks and a trivial maximum at $p = K$. The paper by Wise et al. (1976), as well as the first paper by Friedman (1977b), modify and improve the approach in order to avoid this trouble.

Another probabilistic approach to the pitch determination problem had been formulated earlier by Baronin (1966). This approach, however, was expressed in frequency-domain terms. Baronin never realized it as a PDA or PDD; however, later investigations (Kushtuev, 1968, 1969a,b; Baronin and Kushtuev, 1970) dealt with so-called suboptimal solutions, one of which is almost identical to the PDD by R.L.Miller (1970; Sect.8.5.1). Only the PDA by Steiglitz et al. (1975), which applies least-squares trigonometric curve fitting, comes rather close to Baronin's original formulation.

Section 8.6.1 outlines and discusses the basic algorithm and the corrections necessary to cancel the trivial peak at $p = K$. A multichannel approach proposed by Friedman in his second paper (1978) is discussed in Section 8.6.2.

Although the ML procedure starts from a probabilistic approach, the final result consists of the intensity measurement of the output signal of a tunable comb filter in connection with stationary short-term analysis. Optimal tuning of the comb filter gives the most likely estimate of T_0. The computing effort can be reduced if a simpler measure for the signal intensity (or a more efficient implementation) can replace the cumbersome squaring of the samples. A PDA simplified in this way was developed by Ambikairajah et al. (1980); it determines the peak amplitudes of the signal as intensity estimates. Section 8.6.3 deals with such modifications and simplifications of the maximum-likelihood principle.

8.6.1 The Least-Squares Algorithm

As initially stated, it is assumed that the signal $a(n) = gn(n) + x(n)$, or, in vector notation

$$a = gn + x \qquad\qquad (8.76b)$$

with the vectors defined in the same way as in Chap.2, is given. The Gaussian noise is assumed to have zero mean and the variance σ^2, i.e., it obeys the normal distribution $N(0,\sigma^2)$.

The task is now to estimate 1) the period duration T_0 of the periodic component x, 2) the samples $x(n)$ of x, and 3) the variance σ^2 in the most likely way, i.e., in such a way that x and gn are optimally reconstructed from the composite signal a.

The following deduction is drawn from the paper by Wise et al. (1976). Since gn is assumed to be Gaussian noise, one can write the conditional probability density function for the composite signal a as follows:

$$g\ (\mathbf{a}|p,\mathbf{x},\sigma^2) = \frac{1}{(2\pi\sigma^2)^{K/2}}\ \exp\left\{\frac{1}{2\sigma^2}\sum_{n=0}^{K-1}[a(n)-x(n)]^2\right\}\ . \tag{8.77a}$$

The conditional probability density function[31] g becomes a maximum if the estimated values of the signal $\hat{\mathbf{x}}$, period duration $\hat{p}$, and noise variance $\hat{\sigma}^2$ equal the original values $\mathbf{x}$, T_0, and σ^2. This means that (8.77a), expressed for the estimates $\hat{\mathbf{x}}$, $\hat{p}$, and $\hat{\sigma}^2$, must be maximized. The task is easier if, instead of the probability density function g itself, its logarithm (ln g) is maximized,

$$\ln\ g\ (\mathbf{a}|p,\mathbf{x},\sigma^2) = -\frac{K}{2}\ln\ (2\pi\sigma^2) - \frac{1}{2\sigma^2}\sum_{n=0}^{K-1}[a(n)-x(n)]^2\ . \tag{8.77b}$$

The search for the global maximum of (ln g) with respect to the three parameters to be estimated involves setting the partial derivatives to zero. This is suitably done first for the samples x(n) of the signal vector. Since $\mathbf{x}$ is periodic with the period p, we have

$$x(n) = x(n+kp)\ ,\quad k\text{ integer and }n+kp\ \varepsilon\ [\,0(1)K-1\,]\ . \tag{8.78}$$

After rewriting the sum in the second term of the right-hand side of (8.77b), we obtain the partial derivative with respect to the samples x(n) for an arbitrary period p as follows:

$$\frac{\partial(\ln\ g)}{\partial x(n)} = \begin{cases} \dfrac{1}{\sigma^2}\displaystyle\sum_{k=0}^{P}a(n+kp)-x(n)\ , & n = 0\ (1)\ K-Pp-1\ , \\[4mm] \dfrac{1}{\sigma^2}\displaystyle\sum_{k=0}^{P-1}a(n+kp)-x(n)\ , & n = K-Pp\ (1)\ p-1\ . \end{cases} \tag{8.79}$$

This approach takes account of the fact that, for an arbitrary value of p, the last period of x(n) is incomplete. P is the number of complete periods contained in the interval $n = 0(1)K-1$. Equating the derivative with zero gives the ML signal estimate $\hat{x}(n,p)$ for a given period p,

$$\hat{x}(n,p) = \begin{cases} \dfrac{1}{P+1}\displaystyle\sum_{k=0}^{P}a(n+kp)\ , & n = 0\ (1)\ K-Pp-1\ , \\[4mm] \dfrac{1}{P}\displaystyle\sum_{k=0}^{P-1}a(n+kp)\ , & n = K-Pp\ (1)\ p-1\ . \end{cases} \tag{8.80}$$

---

[31] Qualitatively explained, this is the probability (density) that the composite signal $\mathbf{a}$ comes out as it is under the condition that $\mathbf{x}$ is the periodic component with the period p, and that $\mathbf{gn}$ consists of Gaussian random noise samples with zero mean and the variance σ^2. Since $\mathbf{a}$ consists of many samples, we obtain a probability density function and not simply a probability.

The estimate $\hat{\mathbf{x}}$ proves to be independent of the noise variance σ^2. Setting now the partial derivative $\partial(\ln g) / \partial\sigma^2$ [using the estimate $\hat{\mathbf{x}}$ from (8.80)] to zero, we obtain

$$\hat{\sigma}^2(p) = \frac{1}{K} \sum_{n=0}^{K-1} [a(n) - \hat{x}(n,p)]^2 . \qquad (8.81)$$

Substituting (8.80,81) in (8.77b), it can easily be seen that the second term in (8.77b) becomes a constant. Maximizing $g(\mathbf{a}|p,\hat{\mathbf{x}},\hat{\sigma}^2)$ with respect to p thus implies minimizing the first term of (8.77b), i.e., minimizing the variance $\hat{\sigma}^2(p)$ of the Gaussian noise estimate. For the assumption that the noise is Gaussian, the problem is thus reduced to a one-dimensional maximization over the trial period duration p.

Equations (8.80,81) can be combined (Wise et al., 1976) as follows:

$$\hat{\sigma}^2(p) = \frac{1}{K} \left[\sum_{n=0}^{K-1} a^2(n) - \sum_{n=0}^{K-1} \hat{x}^2(n,p) \right] . \qquad (8.82)$$

The first term in (8.82) is the energy of the signal a(n) within the frame; the second term is the energy of the periodic estimate $\hat{x}(n,p)$, also for the frame $n = 0\,(1)\,K-1$. Equation (8.82) is minimized when the second term becomes a maximum, i.e., when the energy of the periodic component $\hat{x}(n,p)$, depending on the trial period p, becomes a maximum,

$$E(p) = \sum_{n=0}^{K-1} \hat{x}^2(n,p) = (P+1) \sum_{n=0}^{K-1-Pp} \hat{x}^2(n,p) + P \sum_{n=K-Pp}^{p-1} \hat{x}^2(n,p) . \qquad (8.83)$$

In case K is an integer multiple of p, (8.83) reduces to

$$E(p=K/P) = P \sum_{n=0}^{p-1} \hat{x}^2(n,p) . \qquad (8.84)$$

This can also be expressed in terms of the autocorrelation function

$$E(p=K/P) = \frac{1}{P} \left[r_a(0) + 2 \sum_{k=1}^{P-1} r_a(kp) \right] , \qquad (8.85)$$

where $r_a(d)$ is the ACF of the signal a(n) within the interval $n = 0\,(1)\,K-1$, according to definition (8.14). If x(n) has the original period T_0, and if noise is missing, it is shown that E(p) has peaks of equal amplitude for $p = T_0$, $2T_0$, $3T_0$, etc., and, in addition, a trivial maximum for $p = K$. We must keep in mind that (8.77a,b), and with it the whole deduction, is invalid whenever σ^2 takes on a value of zero, that means, in the total absence of noise. Equation (8.81), on the other hand, remains formally valid also for $\hat{\sigma}^2 = 0$, and we can interpret it in this case in such a way that the estimate $\hat{\sigma}^2$ is always zero when $\hat{x}(n)$ equals a(n) for all n. This forces the noise component gn(n) to zero, and so far the results agree: when there is no noise, the variance of gn(n) cannot be anything else than zero. If we apply (8.80), we can obtain $\hat{x}(n,p) = a(n)$ for $p = kT_0$ (k integer) and, in addition, for $p = K$, so that, according to (8.81), $\hat{\sigma}^2$ is zero for these values of p. In the presence of noise, the estimate for the variance is nonzero for $p = kT_0$, but (8.80) always yields $\hat{x}(n) = a(n)$ for $p = K$. Therefore, the peak at $p = K$ is the global maximum of $E(p)$ which retains its amplitude whether noise is present or not; of course this trivial solution is undesirable and must be excluded. In addition, if the noise is other than Gaussian, the peaks of $E(p)$ at $p = kT_0$ tend to be greater for k greater than 1 than for k equal to 1, so that the ML estimator yields peaks with increasing amplitude at multiples of T_0. This annoying fact almost always occurs with speech since the signal is not strictly periodic. If the speech signal is thought to be split up into a strictly periodic component and a deviation signal (which is the difference between the strictly periodic component and the real signal), then for the ML estimator this deviation signal will pertain to the noise part of the total signal. Since the deviation "noise" due to the imperfect periodicity of the speech signal is not Gaussian, the total model, for speech signal, only operates under suboptimal conditions. There are two ways out: 1) take account of this shortcoming in the basic extractor and the postprocessor of the PDA, and 2) modify the ML estimate so that a better approach is achieved.

Noll (1970) has pointed out the problem but not further pursued it. Wise et al. found an experimental solution for speech signals applying a linear additive weighting which forces the least-squares estimate to zero for $p = K$,

$$E_W(p) = E(p) - \frac{p}{K} r_a(0) . \tag{8.86}$$

With this improved approach, Wise et al. arrive at a ML estimate where the largest peak is situated at the fundamental period duration T_0.

Figure 8.42 shows an example of the two ML estimates; they are displayed for a good-quality stationary speech signal uttered by a male speaker with and without noise added. Especially for the noisy signal, the difference between the two ML signals is significant.

A similar estimator has been deduced by Friedman[32] (1977b) which takes account of the fact that the signal may be windowed before the estimation process so that the weighted mean squared error

$$\sigma_F^2(p) = \sum_{n=0}^{K-1} [a(n) - \hat{x}(n,p)]^2 w(n) \tag{8.87}$$

---

[32] Friedman called his method *pseudo-maximum-likelihood* (PML) approach since the assumption of a truly periodic signal in (truly) Gaussian noise, as he stated, "*does not fit the true state of affairs*".

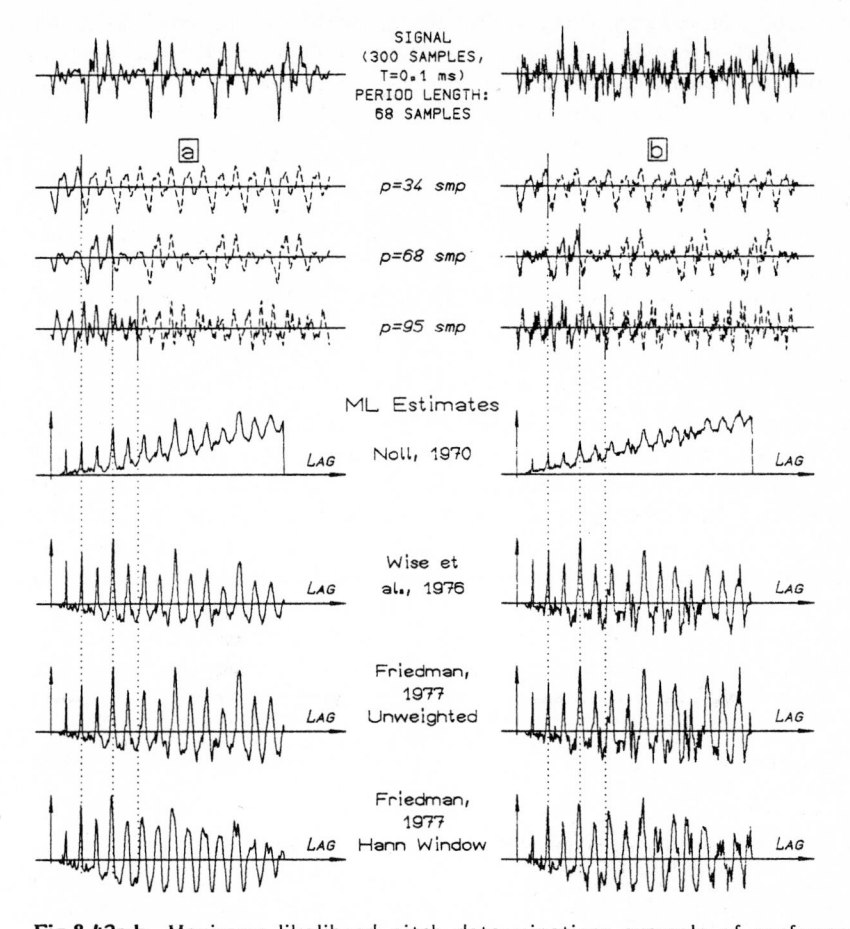

Fig.8.42a,b. Maximum-likelihood pitch determination: example of performance.
(a) Sample speech signal (part of the vowel /ɛ/, male speaker, undistorted);
(b) same signal with noise added (SNR = 0 dB). All signals were amplitude
normalized before plotting. (smp) Samples, (T) sampling interval

is to be minimized. For an arbitrary window w(k) the estimate for $\hat{x}(n,p)$ is
(Friedman, 1977b)

$$\hat{x}(n,p) = \frac{\sum_{k=0}^{P'} a(n+kp)\, w(n+kp)}{\sum_{k=0}^{P'} w(n+kp)} . \tag{8.88}$$

In this equation P' takes on a value of P-1 or P, according to whether the
sample a(n+Pp) is still situated within the frame or not. It is easily seen that,

for a rectangular window, i.e., $w(n) = 1$ for $n = 0\,(1)\,K-1$, we arrive at (8.80). The most likely estimate $\hat{p}$ for the period T_0 of $x(n)$ is that value of p which maximizes the weighted energy of $\hat{x}(n,p)$,

$$E_F(p) = \sum_{n=0}^{K-1} \hat{x}^2(n,p)\, w(n) = \sum_{n=0}^{p-1} \frac{\left[\sum_{k=0}^{P'} a(n+kp)\, w(n+kp)\right]^2}{\sum_{k=0}^{P'} w(n+kp)} . \qquad (8.89)$$

This function is normalized with respect to the weighted energy E_A of the signal,

$$E_{FN}(p) = E_F(p)\, /\, E_A \qquad (8.90)$$

where

$$E_A = \sum_{n=0}^{K-1} a^2(n)\, w(n) . \qquad (8.90a)$$

Friedman pursued the problem of the trivial peak of $E_{FN}(p)$ at $p = K$, which is connected with a gradual rise of the baseline of $E_{FN}(p)$, and solved it from a theoretical point of view. He found that $E_F(p)$ is nonzero for $p \neq kT_0$ even when the input signal $a(n)$ is an impulse train with the pulse interval T_0:

The fact that $E_{FN}(p)$ is nonzero at all for frequencies not harmonically related to the fundamental when the test signal is an impulse train suggests that the source is in the calculation. The expression (8.89) for $E_F(p)$ contains the squared sum

$$\left[\sum_k a(n+kp)\, w(n+kp)\right]^2 ; \quad k = 0\,(1)\,\ldots\,,\quad n = 0\,(1)\,p-1 \qquad (8.91a)$$

which can be written as a double summation

$$\left[\sum_k \sum_m a(n+kp)\, w(n+kp)\, a(n+mp)\, w(n+mp)\right] ; \quad k,\, m = 0\,(1)\,P' . \qquad (8.91b)$$

For k not equal to m, if $a(n)$ is an impulse train with a period harmonically unrelated to p, $a(n+kp)\,a(n+mp)$ will be zero for all n since the phase difference $(k-m)p$ will never be an integral number of periods. On the other hand, for all k equal m the contribution to the sum will be nonzero regardless of p for every signal $a(n)$ not identically zero in $n = 0\,(1)\,K-1$. This implies that the diagonal terms of the double summation are responsible for the rising baseline, and suggests that the energy contribution of these terms be subtracted from $E_F(p)$, and also from E_A, to give a corrected function which will be called E_{FNC}:

$$E_{FNC} = \frac{E_{FN}(p) - E_C(p)}{E_A - E_C(p)} \qquad (8.92)$$

where E_C is the energy contribution of the diagonal terms. (Friedman, 1977b, p.216)

Friedman thus obtained the correction term E_C as

$$E_C(p) = \sum_{n=0}^{p-1} \frac{\sum\limits_{k=0}^{P'} [\, a(n+kp)\ w(n+kp)\,]^2}{\sum\limits_{k=0}^{P'} w(n+kp)} \; . \tag{8.93}$$

For a rectangular window, the correction term $E_C(p)$ proves to be $1/P'$ times the energy E_a of the signal $a(n)$, P' now being the exact (noninteger) number of periods in the frame. Except for the fact that $E_C(p)$ is also subtracted in the denominator of Friedman's estimator formula (8.92), this corresponds exactly to the estimator $E_W(p)$ in (8.86) applied by Wise et al. (1976). The numerical calculation of (8.92) may appear dangerous since the denominator (as well as the numerator) of (8.92) go to zero for $p \to K$. In practice, however, this will never happen since the ML algorithm requires the stationary principle of short-term analysis; thus the frame length will always be at least twice the period duration at the low-frequency end of the measuring range.

Baronin (1966) and Kushtuev (1968, 1969a,b) proposed a statistical approach to the problem of pitch determination in frequency-domain terms. The primary assumption of their approach is the same as that by Noll (1970),

$$a(t) = gn(t) + x(t) \; . \tag{8.76c}$$

(Baronin formulated his approach in the *analog* domain.) In contrast to Noll's formulation, Baronin's approach represents the periodic component $x(t)$ by the amplitudes X_i and the phases ϕ_i of the individual harmonics,

$$x(t) = \sum_{i=1}^{k} X_i \sin(i\omega_0 t + \phi_i) \; , \tag{8.94}$$

where ω_0 is the (radian) fundamental frequency of $x(t)$. Assuming that the noise component $gn(t)$ is normally distributed with zero mean and the variance σ^2, one can formulate a conditional probability density function depending on the trial fundamental frequency p, the variance σ^2, and the values of X_i and ϕ_i in a way similar to (8.77) and optimize it with respect to ω_0. According to Baronin (1974a) one then obtains

$$\ln g_p(p) \approx \ln F(p) + \ln g_a(p) + \sum_{i=1}^{k} K^2(ip)\, X^2(ip) \; . \tag{8.95}$$

In this equation, $g_p(p)$ is the a posteriori probability density for the unknown fundamental frequency p; $F(p)$ is a correction term that takes account of the long-term behavior of the harmonics of $x(t)$ with respect to the noise (for low noise this term will be very close to unity); and g_a is the a priori probability density for p (see the discussion of Baronin's statistical postprocessor in Sect.6.6.3). The term $X^2(ip)$ represents the energy of the ith harmonic for the trial fundamental frequency p, and $K^2(ip)$ represents a kind of long-term signal-to-noise ratio for that harmonic. The optimal estimate $\hat{p}$ for F_0 is that value of p for which the a posteriori probability reaches a maximum.

Without going into too much detail one can see that (8.95) defines a spectral comb where the individual components are weighted by $K^2(ip)$. The a posteriori probability density will be maximized when the spectral comb optimally matches the harmonic structure of the signal, i.e., for $p = \omega_0$. When the noise level is low, the values of $K^2(ip)$ will be rather high so that the two first terms in (8.95) tend to become insignificant.

Baronin showed that the same approach can be formulated in time-domain terms, and that it can be executed not only for the signal itself, but also for the autocorrelation function of the signal. In frequency-domain terms (8.95) directly defines the principle of harmonic compression. Baronin never realized a PDA rigorously following the principle defined by (8.95), probably because the technical effort required was too high. However, at Baronin's laboratory some "suboptimal" PDDs were developed. Among these, the one by Kushtuev (1968, 1969a,b) is almost identical to the PDD by R.L.Miller (1968, 1970; see Sect.8.5.1), and other algorithms implemented differ from this one only in some technical detail (Baronin, 1974a). The PDA by Steiglitz et al. (1975), which approximates the speech signal by a finite Fourier series in a least-squares sense, is perhaps the one which is most closely related to (8.95). Compared to the ML algorithms by Wise et al. (1976) or by Friedman (1977b), Steiglitz's method yields comparable results but is computationally more complex. A similar PDA estimating the frequencies, amplitudes, and phases of the individual harmonics was proposed by van Eck (1981).

As these considerations show, there is a very close relation of the maximum-likelihood PDA (which operates in the lag domain) and the principle of harmonic compression. In both cases the procedure ends up with a comb filter that enhances harmonic structures and is optimally matched to the harmonic structure of the signal. Thus the maximum-likelihood PDA becomes extremely noise resistant. On the other hand, it tends to show a certain sensitivity towards octave errors when a strong first formant coincides with the second or third harmonic. Unfortunately the correction terms in the PDAs both by Wise et al. (1976) and by Friedman (1977b) tend to increase this sensitivity since they emphasize small values of the lag d (in this respect, Friedman's PDA is a little better).

8.6.2 A Multichannel Solution

A later paper by Friedman (1978) deals with an extended version, the *multidimensional pseudo-maximum-likelihood* (MDPML) approach. This PDA was conceived without regard to computational effort and is thus even slower than Friedman's earlier ML algorithm (1977b). The paper is also interesting for the conclusions it draws with respect to preprocessing, especially with respect to inverse and low-pass filtering.

In investigating the PML algorithm (Friedman, 1977b), processing runs were made with the speech signal replaced by the prediction-error signal resulting from linear-prediction inverse filtering applied to the input speech. It was thought that, since the inverse filtering effectively cancels the spectral shaping due to the vocal tract resonances, the prediction error should reflect the true glottal excitation, giving an estimate essentially the same as for an impulse train. The results, however, were rather disappointing. ...

It would seem that the reliability of the pitch estimate - as represented by the height of the peak - is improved by discarding information. This is likely due to the phase uncertainty of the upper harmonics of the recovered excitation function, resulting from the time-varying nature of the speech and the fact that the LPC inverse filter, representing an all-pole analytic model, does not perfectly cancel the actual vocal-tract transfer function. In order to realize the recovered excitation as an impulse train (or other periodic waveform with a flat spectral envelope), a state of phase coherence must exist across the entire spectrum; i.e., the phase uncertainty between any two frequency components at a randomly chosen point in time must be small. If this condition does not hold, the higher frequencies act effectively as noise, masking the periodic component of the waveform. ...

The hypothesis on which the present work is based is that useful pitch information exists in the portion of the speech spectrum above 1 kHz. Linear prediction inverse filtering, which continually corrects the amplitude and phase of the spectral components according to a time-varying analytic model, offers a useful first step in extracting this information. The PML algorithm, however, requires phase coherence over the entire spectrum, which is unattainable in practice, as already shown. A more appropriate pitch-estimation approach for the LPC error signal would be to break up the spectrum into bands such that phase coherence could reasonably be expected over the width of each. This is supported by the observation that in a wide-band spectrogram (typically 300 Hz bandwidth), one can usually resolve the individual pitch pulses as vertical bars extending across most of the spectrum. Inverse filtering would eliminate the formant pattern and leave the pitch-pulse bars. For a given center frequency of the analyzing filter, the spectrograph records the envelope of the filter output when excited by the original (or inverse-filtered) speech. One might therefore consider the following scheme, proposed in the conclusion to the earlier paper (Friedman, 1977b): applying the prediction-error signal to a bank of bandpass filters, and taking the envelopes of the respective outputs as a multidimensional vector signal $x(n)$, to be approximated by a true-periodic vector signal $\hat{x}(n)$, in a generalization of the PML algorithm. ... For this case, the question of phase coherence only applies within the bandwidth of each individual channel filter; each vector-component signal (envelope of one filter output) is treated independently, and gross phase distortion across the spectrum has no effect. (Friedman, 1978, p.187)

Implementing this idea, Friedman arrived at the "multidimensional" (i.e., a multichannel) solution of his PML algorithm. With this PDA, Friedman combined the principles of inverse filtering and epoch detection known from time-domain PDAs with short-term analysis by maximum-likelihood estimation. Figure 8.43 shows the block diagram. The LPC residual signal e(n) is subdivided into 16 equally large subbands. In each channel the short-term envelope is computed. The ML estimator [according to (8.92)] is then applied to each channel envelope to yield a partial estimate, and all the partial estimates are simply added to yield the total ML estimate

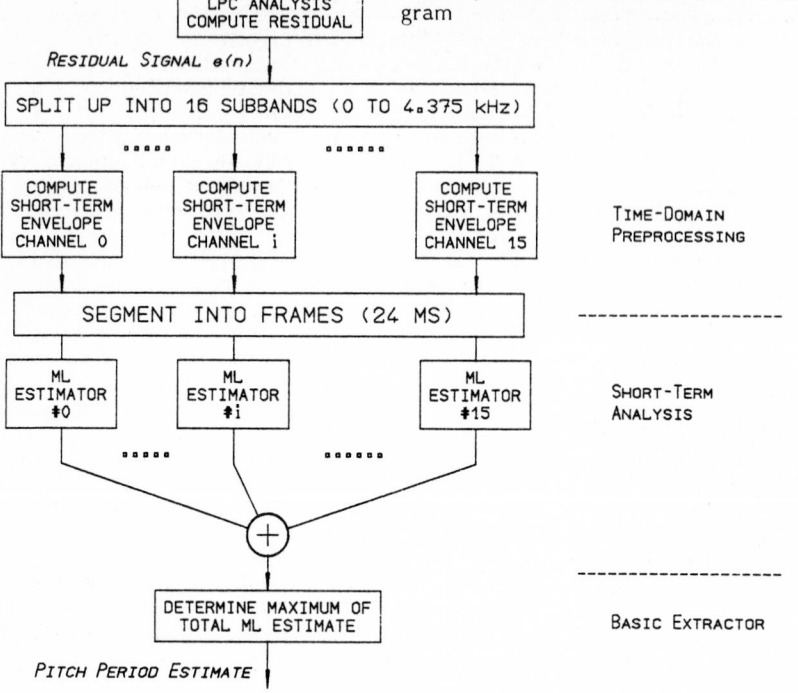

Fig.8.43. Multichannel maximum-likelihood PDA by Friedman (1978): schematic diagram

$$E_{FM}(p) = \frac{\sum\limits_{i=0}^{15} E_F(p,i) - \sum\limits_{i=0}^{15} E_C(p,i)}{\sum\limits_{i=0}^{15} E_A(i) - \sum\limits_{i=0}^{15} E_C(p,i)} . \tag{8.96}$$

To this estimate, a peak picking procedure is applied in the usual way to yield $\hat{p}$, the final estimate for T_0.

The actual implementation (Fig.8.44) of the time-domain preprocessor differs a little from the principal solution shown in Fig.8.43:

> Implementation of such scheme requires ... a bank of band-pass filters with minimum dispersion, so that an impulse train applied to the inputs will appear in the output envelopes with maximum resolution. This calls for a linear-phase design such as a Bessel characteristic, or a symmetrical FIR digital filter derived from one of the standard window functions. One way of digitally realizing a bank of such filters with the advantage of no phase distortion at all, is by a discrete Fourier transform via the FFT algorithm, using a symmetrical time window on the signal. This form of filtering was used to give the realization of the complete pitch-estimation algorithm (Friedman, 1978, p.188)

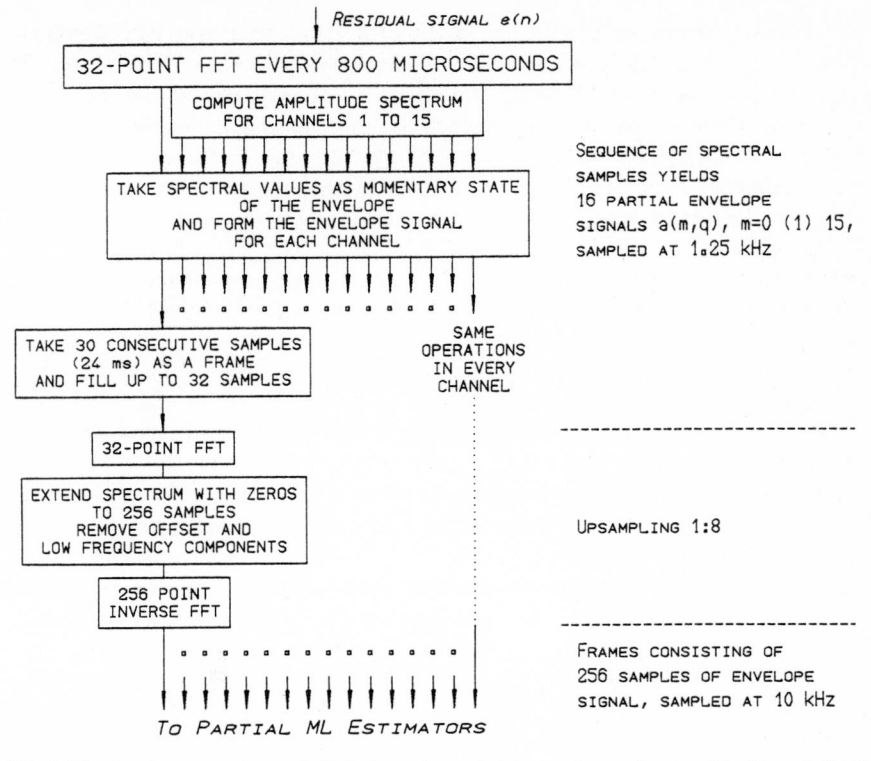

Fig.8.44. Implementation of Friedman's multichannel maximum likelihood PDA (1978): block diagram of the preprocessor

Hence, Friedman applied a 32-point FFT to the signal and computed the amplitude spectrum (in every channel except the one which contains zero frequency) every 800 μs. This yields the sequence a(m,q), m = 0 (1) 15 of partial envelope signals representing the envelope in each of the 16 subbands, sampled at 1.25 kHz. Thirty samples of each envelope signal, corresponding to 24 ms of speech, are then taken as a frame and again upsampled to 10 kHz to yield a sufficient time resolution for the measurement. To emphasize the periodicity of the envelope, the offset and very low frequency components are removed during the upsampling procedure, which is again performed using the FFT.[33]

This design successfully avoids a number of problems arising with envelope periodicity detection and inverse filtering in connection with time-domain PDAs. Firstly, the envelope is computed for each channel except the one

---

[33] Subdivision of a signal into a number of equally large subbands is very suitably realized by a polyphase network (Heute and Vary, 1981). Friedman's PDA, however, was published before this kind of filterbank had been developed.

around zero frequency (which can be directly taken from the FFT spectrum without computing the magnitude since that value is always real) so that a baseband is available within which the signal is original. Hence, the disadvantage that full-wave rectification (which is done in the computation of the amplitude spectrum in the other channels) destroys the fundamental when the signal resembles a pure sinusoid, is avoided in the baseband; it never occurs in higher frequency subbands since the higher formants are more strongly attenuated (Dunn, 1961). The application of a STA procedure makes the implementation of a channel selection routine unnecessary which would arise in a time-domain PDA (see Sect.6.4 and Chap.7); one can simply add up the partial estimates according to (8.96). Since there is a global normalization of the partial estimators with respect to energy, we can expect that those ML estimators which do not have a speech signal envelope in the respective channels do not yield a substantial contribution to the total estimate and thus do not cause errors.

The performance of the multidimensional PDA was found to be superior to that of the simple PML algorithm as well as to that of cepstrum. This is in agreement with the results obtained by Noll (1970) who found that the least-squares approach can tolerate considerably more noise (12 dB) than the cepstrum PDA. This multichannel ML algorithm my be the most robust PDA with respect to signal degradation by noise. It combines the advantages of the optimal periodicity detection with the advantage of fundamental harmonic enhancement by full-wave rectification. The computational burden of the initial 32-point FFT is low (23 complex multiplications per FFT if it is well programmed). Since the claims to the 1:8 interpolation filter are modest, it can be realized by a chain of low-order half-band filters (Schüssler, 1973; Bellanger, 1980) which may need less computational effort than the FFT proposed by Friedman in the original design. Thus even a real-time application of this algorithm appears feasible; possible computational shortcuts to the least-squares estimation algorithm will be discussed in the next section.

One might also think of applying the ML algorithm (like any other short-term analysis algorithm) in connection with a sophisticated preprocessor. In this respect the multichannel processor as realized by C.P.Smith (1954, 1957) and Yaggi (1962) is closely related to the preprocessor applied in the present PDA (cf. Figs.6.39,40 in Sect.6.3.2 for a comparison). This raises the question whether it is really necessary to compute the LPC residual in the present algorithm. As the preprocessor by Smith and Yaggi shows, the principle of extracting the instantaneous envelope can be successfully applied to the original speech signal as well. Processing the original signal instead of the LPC residual could even result in a better performance with respect to ambient noise since the S-N ratio of the original speech signal is always higher than that of the LPC residual.

Friedman's results obtained for degraded speech strongly support the view that there is information above 1000 Hz relevant for pitch extraction.

The large increase in error rate [the MDPML algorithm was tested with degraded speech in two versions: one with full bandwidth of input signal, the other version with 1000-Hz low-pass filtered speech] for restriction to frequencies below 1 kHz indicates that the MDPML pitch estimate is based on the periodic structure of the speech signal as a whole, and not mainly on the fundamental or lower harmonics. This in turn supports a view of pitch in the time domain, based on discrete independent impulse events (glottal closings) which appear as a wideband time-periodic energy distribution in a simultaneous time-frequency representation such as a wideband spectrogram. Each glottal event yields a signature determined by the impulse response of the vocal cavity at the time of occurrence, and even this influence can be largely canceled by inverse filtering with time-varying parameters through linear prediction. The usual view of pitch, in terms of harmonics of a fundamental frequency varying with time, appears as merely a convenience which happens to give a good description of normal speech in which the glottal events have a well-defined repetition rate. (Friedman, 1978, p.195)

This is in agreement with the fact that the ear can distinguish the fundamental frequency of speech much better than pure sinusoids at the same frequency (Flanagan and Saslow, 1958; Terhardt, 1979b; see Sects.4.2.1 and 8.5.2 for more detail), as well as with the fact that the higher formants are usually more strongly attenuated than F1 (Dunn, 1961), and thus are less ambiguous with respect to the location of the excitation pulses in the signal. The results obtained by the present author in the investigation reported in Sect.7.4.3 (with respect to processing sequence) also support this view.

8.6.3 Computing Complexity, Relation to Comb Filters, Simplified Realizations

For the single-channel maximum-likelihood PDA, Friedman (1977b) discussed the computational effort. He found that the minimum computational effort of the least-squares procedure according to (8.92) is

$$n_{add} = N + \sum_{p=P_a}^{P_e} (2p + 4\frac{N}{p}); \quad n_{mul} = 3N + \sum_{p=P_a}^{P_e} p; \quad n_{div} = 2 \sum_{p=P_a}^{P_e} p; \quad (8.97)$$

where n_{add}, n_{mul}, and n_{div} represent the number of additions, multiplications, and divisions per frame, respectively, and p_a and p_e denote the range of fundamental period duration T_0 to be covered. In the case where the whole range of T_0 from 1 to K is scanned (this is the worst case for the computation but the easiest for estimating the computing effort) the computational complexity[34] for all operations, according to (8.97), is in $O(K^2)$. For the PDA

---

[34] The expression $O(X)$ is standard in computer science and denotes the order of the computational complexity of an algorithm. X here means a function of a significant parameter, such as the frame interval or frame length. If the complexity of an algorithm is in $O(K^2)$, for instance, this means that the number of significant arithmetic operations, e.g., multiplications, grows proportionally to the square of the value of the parameter K.

by Wise et al. (1976), the computational complexity for the divisions is in $O(K)$. In this respect the ML algorithm is as fast as ordinary autocorrelation but slower than cepstrum or harmonic analysis when K becomes large.

Although the starting point of the considerations is a probabilistic problem, i.e., maximizing the probability that a finite and (within the given frame) periodic signal x(n), corrupted by Gaussian noise, can be correctly reconstructed, we arrive at a solution from which we can derive two very "deterministic" statements: 1) any periodic component x(n), whatever its period may be, is optimally enhanced by a multiple pulse comb filter with equal pulse distances whose impulses extend over the whole range of the frame; and 2) among all possible comb filters, a given period duration T_0 is optimally enhanced when the pulse distance p of the filter matches (i.e., equals) the intrinsic period duration T_0 of the signal.

These two statements which look rather trivial reduce the complex problem of stripping ambient noise from a periodic wavelet to the rather simple application of a tunable comb filter. The question is whether this more general principle can be exploited to yield computationally more efficient algorithms of comparable reliability and robustness.

It may be dangerous to restrict the range of the trial period p too drastically. If the algorithm is to be implemented in an actual vocoder, its computing facilities must be designed according to the worst-case computational effort, not to a long-term average. So far, the worst case is that the whole range of T_0 (for instance, from 1 to 20 ms) has to be scanned; this happens regularly if the device processes an unknown speaker or an irregular signal. Reducing the range of p may bring the danger of losing the correct track of T_0 once an error remains undiscovered. There are better methods to reduce the computational burden.

The major part of the computational effort is due to the multiplications; the additions are carried out much faster, and the divisions can be kept in $O(n)$. Nearly all the multiplications, however, are due to the squaring operation for the computation of the signal energy. A technological solution that avoids all these multiplications is the implementation of a table-lookup procedure.[35] In this case the PDA could be realized with $O(K)$ divisions, $O(K^2)$ additions, $O(K^2)$ table-lookups, and no multiplications.

In the deterministic interpretation of (8.86) the ML estimate is computed as the energy of the output signal of the tunable comb filter. To reduce the computing effort, one may replace the measurement of the energy of the output signal by the less complex measurement of the absolute average or the amplitude. In the following, we will label these algorithms as *modified maximum-likelihood* (MML) algorithms. In Figs.8.45-49 some examples are given which will be explained later in the text.

---

[35] For readers that are not too familiar with programming, the principle of a table-lookup procedure will be briefly explained. Instead of performing

One could think of implementing the same principle taking the average signal magnitude (the absolute average) as the criterion instead of the squares. This PDA would be the direct counterpart to the AMDF algorithm. There is the strong relation between the AMDF and the correlation pointed out by Ross et al. [1974; see (8.20)]; this MML algorithm, however, is the approximation of the inverse of the AMDF comb filter by another comb filter.[36] The performance of this algorithm would be expected to be comparable to that of the AMDF. Curiously enough, however, this algorithm does not reliably work! As the example shows (Figs.8.45), it tends to commit octave errors and breaks down when noise is added to the signal. (This also applied to the signals in Figs.8.46,47; for that reason the corresponding display was suppressed in these figures. In the example of Fig.8.48 the procedure works slightly better but is still clearly inferior to the other two methods.) A possible explanation is that the signal estimate $\hat{x}(n,p)$ for a mistuned comb filter has a great deal of small but nonzero samples which are well suppressed in the computation of the squares, but inadmissibly taken into account when the average magnitude is measured.

A computationally even simpler, but probably less reliable algorithm emerges when the *amplitude* of the output signal of the filter, i.e., the global maximum or minimum of the output signal within the whole frame, is taken as the intensity estimate. Such an algorithm, though applying nonstationary short-term analysis, was proposed by Ambikairajah et al. (Ambikairajah and

---

a certain arithmetic operation (squaring in the actual case) over and over with a limited range of values, the results of all possible such operations are precomputed and stored in a table. In the actual case of squaring, M table entries would be necessary if the dynamic range of the input signal extends from 0 to $\pm(M-1)$. The table is organized in such a way that consecutive entries are stored in consecutive memory locations. The actual table would therefore contain the square numbers 0, 1, 4, 9, ... $(M-1)^2$ in the consecutive locations 0, 1, 2, 3, ... (M-1). In the normal operational mode the input value x serves as the address for a read-only memory (ROM) and controls the readout of the memory contents which then is the desired output value, x^2 in the actual case. Not only squaring, but any linear or nonlinear function which assigns an output value y to an input value x by the unique relation $y = f(x)$ can be realized this way (presuming that the number of possible values of x is limited). For a 12-bit integer representation of the input signal x (which corresponds to a range from -2048 to +2047 and a signal-to-noise ratio of about 72 dB) and a 24-bit representation of x^2 one would need a memory of 48 Kbit for the table, which is easily realizable. There is no doubt that this realization works faster than a conventional multiplier implementation. A modern high-speed LSI multiplier, however, can outdo the speed of the table-lookup realization. Hence the design of such an implementation is always a tradeoff between the required computing speed and the cost of the circuitry.

[36] The exact reciprocal of the AMDF comb filter, having poles on the unit circle, would be unstable. Any stable recursive approximation of the inverse AMDF filter, on the other hand, would be unrealistic due to its infinite impulse response.

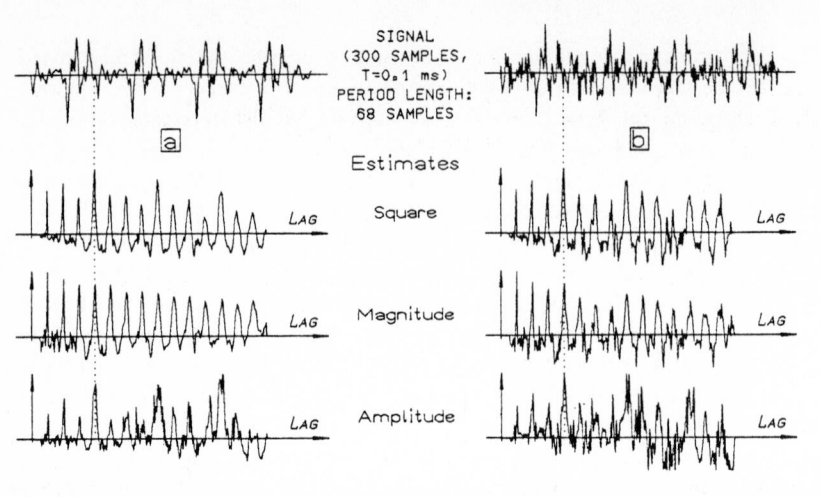

Fig.8.45a,b. Modified maximum-likelihood PDAs: example of performance. (a) Speech signal (part of the vowel /ε/, male speaker, undistorted); (b) same signal with additive noise (SNR = 0 dB). The signals are identical to those in Figure 8.42. (Square) ML estimator after Wise et al. (1976) according to (8.92) and identical to that in Fig.8.42; (Magnitude) same estimator with the squares replaced by the magnitudes of the samples; (Amplitude) same estimator with the squares replaced by the difference between maximum and minimum of the output signal of the comb filter, corresponding to the proposal by Ambikairajah et al. (1980). Stationary short-term analysis; frame length 500 samples (50 ms). The first 300 samples of the frame are shown. All signals were amplitude normalized before plotting

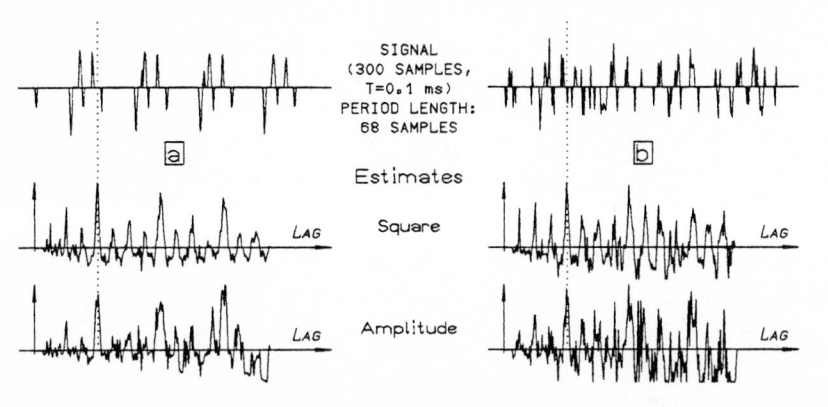

Fig.8.46a,b. Modified maximum-likelihood PDAs: example of performance. (a) Speech signal (part of the vowel /ε/, male speaker, undistorted); (b) same signal with additive noise (SNR = 0 dB). The estimators are identical to those in Figure 8.45. The signals are identical to those in Fig.8.45 except that they were center clipped with the threshold set at 40% of the absolute peak within the frame. For more details, see the caption of Fig.8.45

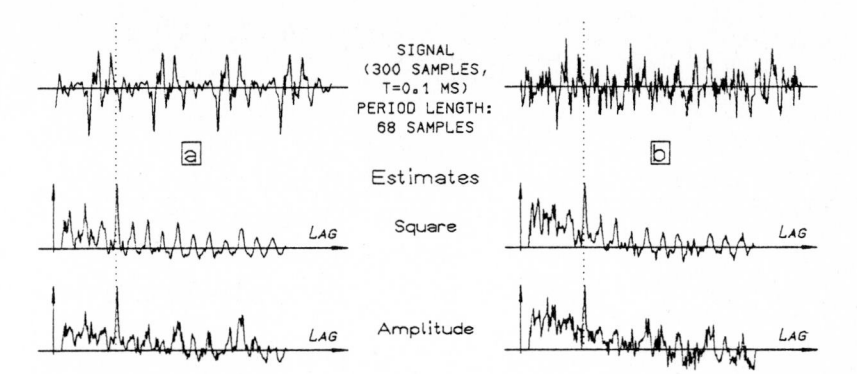

Fig.8.47a,b. Modified maximum-likelihood PDAs: examples of performance.
(a) Speech signal (part of the vowel /ε/, male speaker, undistorted); **(b)** same
signal with additive noise (SNR = 0 dB). The signals are identical to those in
Figure 8.42. In this figure a nonstationary short-term analysis was applied; the
output signal of the comb filter was computed according to (8.84) with P
upward limited to a value of four. The additive correction terms have been
applied according to (8.86) for the nonstationary PDA as well; however, they
proved to be not optimal in this case

Carey, 1980; Ambikairajah, Carey, and Tattersall, 1980). This PDA applies a
comb filter with a fixed number P of pulses (2 to 4) and computes the
difference between the global maximum and minimum of the output signal
within the frame $n = q(1)q+K-1$,

$$y(n,p) = \sum_{k=0}^{P-1} x(n+kp) \; ; \quad A(q,p) = \max_{K,q} \; [y(n,p)] - \min_{K,q} \; [y(n,p)] \; . \tag{8.98}$$

The results are reported to be quite good, even for noisy signals with a S-N
ratio down to 10 dB. The authors complained about some trouble with the
optimal number P of impulses in the comb filter which has been shown to
depend on T_0, especially for high-pitched voices. This problem, however, can
easily be avoided if one enforces some quasi-stationary short-term analysis
where the comb filter has to extend at least over a minimum frame length.
The example in Fig.8.47 was computed this way with a minimum frame length
of 20 ms. A similar algorithm, using P = 3, was recently proposed by Sobolev
(1982).
    One might think of applying center clipping in order to reduce the sensi-
tivity of the algorithm to formant interaction. As the examples in Figs.8.46,49
show, the performance of the algorithm is hereby improved. This can be seen
best from the first four frames in Fig.8.49 (in comparison to Fig.8.48 where
the same band-limited signal is processed without center clipping). In the first
two frames the clipping level is rather low due to the nonstationarity of the
signal, and the PDA fails; in the following frames, however, the signal

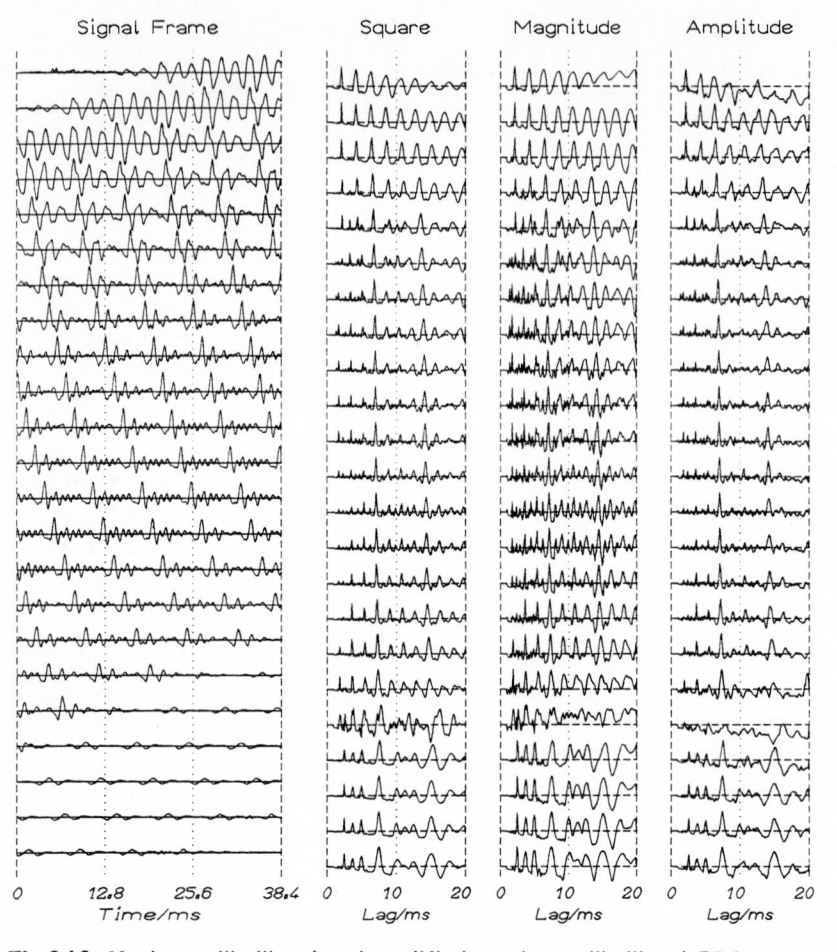

Signal Frame	Square	Magnitude	Amplitude

| 0   12.8   25.6   38.4 | 0   10   20 | 0   10   20 | 0   10   20 |
| Time/ms | Lag/ms | Lag/ms | Lag/ms |

Fig.8.48. Maximum-likelihood and modified maximum-likelihood PDAs: examples of performance. Signal: utterance "brown," part of the sentence "The big brown fox"; male speaker, band-limited signal, sampling frequency 8 kHz, frame interval 12.8 ms, frame length 38.4 ms, stationary short-term analysis. The signal is identical to that used in Figs.8.10,15,18; the algorithms are identical to those used in Fig.8.45. All ML estimator signals were amplitude normalized before plotting. The figure shows that the algorithms which take the energy and the amplitude of the output signal of the comb filter perform similarly (with slight advantages for the energy), whereas the algorithm that takes the magnitude is significantly inferior. All the algorithms, however, show a certain sensitivity with respect to a strong first formant. Together with the correction term given by (8.86), this leads the PDA to failure during the first three frames

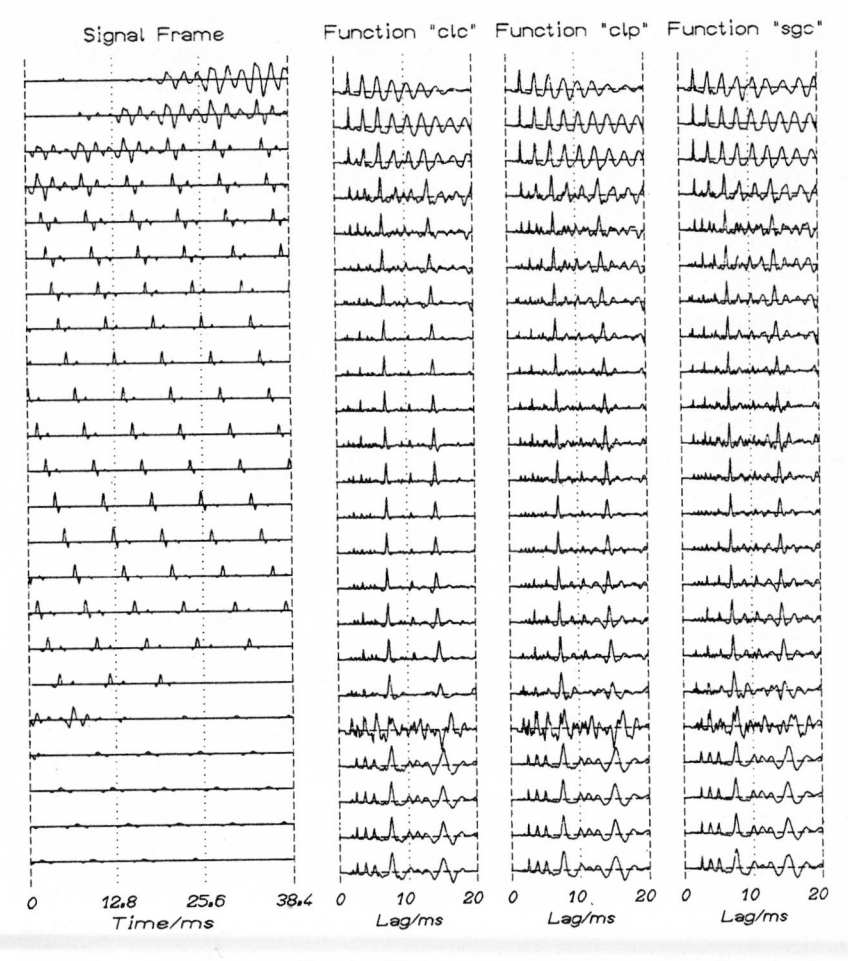

Fig.8.49. Maximum-likelihood PDA operating on center-clipped signals: examples of performance. Signal: utterance "brown," part of the sentence "The big brown fox"; male speaker, band-limited signal, sampling frequency 8 kHz, frame interval 12.8 ms, frame length 38.4 ms, stationary short-term analysis. The input signal is identical to that used in Figs.8.10,15,18,48. The nonlinear functions are the same as illustrated in Fig.8.9; the signal on the left-hand side is plotted for the function "clc" (clip and compress). The center clipping threshold was determined in a way similar to that used in the PDA by Dubnowski et al. (1976; Sect.8.2.4). The results were obtained using the PDA by Wise et al. (1976); all ML estimator signals were amplitude normalized before plotting. The figure shows that the sensitivity of the algorithms to dominant formants has decreased. As in the autocorrelation PDA the function "clc" (clip and compress) gives best results. The function "sgc" (adaptive three-level quantization) is significantly inferior

becomes stationary, and center clipping can be applied to remove the formant structure without removing the principal peaks that indicate periodicity. Although the improvement of the performance is not as drastic as for the autocorrelation PDA, it can be regarded as significant. The function "clip and compress" performs best. In contrast to the autocorrelation PDA, adaptive three-level quantization (function "sgc") cannot be applied. This is obvious, however. If the signal is peak clipped, there is no longer a difference between the energy of the signal and its magnitude so that the ordinary ML PDA will perform like the MML algorithm which uses the magnitude instead of the energy, and which is clearly inferior to the others. From this we can conclude that peak clipping and the application of a maximum-likelihood PDA are incompatible; the ML PDA requires the peak values of the signal to be preserved.

In summary, simplified and computationally more efficient algorithms can be derived from the maximum-likelihood approach when the squaring necessary to compute the energy of the output signal of the comb filter is replaced by a less costly operation. The PDA using the magnitude instead of the signal energy, however, does not work reliably. It is susceptible to higher harmonic tracking and breaks down in a noisy environment. Center clipping (Figs.8.46,49) gives promising results and deserves further investigation. Nonstationary analysis (Fig.8.47) may be useful; nevertheless an additive and/or multiplicative weighting of the estimates (which was applied but not optimized in the examples displayed in this section) seems necessary. In this respect, some work remains to be done; the PDA by Ambikairajah et al. (1980) is a first step toward implementing a PDA which forms a computational shortcut to the maximum-likelihood approach without losing too much of its reliability.

8.7 Summary and Conclusions

The common feature of all short-term analysis PDAs, as stated in the initial discussion in this chapter, is the fact that the time domain is left in favor of some spectral domain whose independent variable may be time, lag (which is essentially the same), or frequency. All these PDAs thus provide a sampled estimate of pitch - sampled by some appropriate frame rate which does not necessarily have anything to do with F_0 itself. Phase information of the original signal is lost. Five categories of PDAs have been reviewed which can be classified into three main groups according to their operating domain (see Table 8.1 at the beginning of this chapter). The fact that the phase information is lost implies that the short-term transformation necessary to leave the time domain is irreversible. This is due to nonlinearity (autocorrelation, computation of power spectrum) or to averaging processes (AMDF, ML).

From a theoretical point of view, almost all the short-term analysis PDAs either maximize a similarity criterion or minimize a distance measure or error criterion, with respect to lag or frequency. This result was succinctly formulated by Ney (1982).

A number of the most common methods of fundamental period estimators can be treated uniformly in terms of a minimum error function. Optimization of this error criterion leads to an estimation procedure for the unknown fundamental period. The approach is as follows.

The problem is to estimate the period of an unknown signal of which the time samples are given and which is corrupted by noise. Let $x(n)$ be the observed signal. The observed signal is modeled as a hypothetical signal $a(n)$, $n = 0\,(1)\,N-1$, with an unknown period p:

$$a(n) = x(n) + v(n) \; , \qquad n = 0 \; (1) \; N-1$$
$$a(n+p) = x(n) + w(n) \; , \qquad n = 0 \; (1) \; N-1 \; . \tag{8.99}$$

The noise samples $v(n)$ and $w(n)$ represent the error between the model and the observation. The basis of the method is to specify a criterion of the overall error and to minimize it with respect to the period p and the signal $x(n)$. One possible choice of the overall error criterion is the sum of absolute values:

$$E_1[p, x(n)] := \sum_{n=0}^{N-1} [\, |v(n)| + |w(n)| \,] \; . \tag{8.100}$$

Using the triangle inequality

$$|a(n) - a(n+p)| \; \leq \; |x(n) - a(n)| \; + \; |x(n) - a(n+p)| \; = \; |v(n)| \; + \; |w(n)| \tag{8.101}$$

and considering that equality is actually attained for $x(n) = a(n)$ or $x(n) = a(n+p)$, we can carry out the minimization with respect to $x(n)$ and obtain

$$E_1(p) = \sum_{n=0}^{N-1} |a(n) - a(n+p)| \; . \tag{8.102}$$

Equation (8.102) is known as the definition of the average magnitude difference function (AMDF) (Ross et al., 1974). The fundamental period is given by the position of the minimum of $E_1(p)$. ...

The sum of absolute values is a well acceptable error criterion. Although squares place too much weight on large errors, they are useful in that they are easier to handle mathematically. Starting with the sum of squares as error criterion, we will develop "different" estimation methods.

The error criterion to be considered is

$$E_2[p, x(n)] = \sum_{n=0}^{N-1} [\, v^2(n) + w^2(n) \,] \; . \tag{8.103}$$

Minimization with respect to $x(n)$ is straightforward and yields

$$E_2(p) = \frac{1}{2} \sum_{n=0}^{N-1} [a(n) - a(n+p)]^2 \; . \tag{8.104}$$

Equation (8.104) may be viewed as the definition of an average squared difference function [see Sect. 8.3.2].

By considering the set of noise samples v(n) and w(n) to be a 2N-dimensional Gaussian random vector it is clear that a maximum-likelihood estimation for the period would lead to the same result as (8.104). Maximum-likelihood estimation techniques in somewhat different formulations have been described by Wise et al. (1976), or Friedman (1977b).

Rewriting (8.104) as

$$E_2(p) = \frac{1}{2} \sum_{n=0}^{N-1} [a^2(n) + a^2(n+p)] - \sum_{n=0}^{N-1} a(n)a(n+p) \tag{8.105}$$

we arrive at an estimation method in terms of a correlation function, since the first sum is virtually independent of p. Strictly speaking, the second term is a cross-correlation function and not an autocorrelation function. The difference arises from the upper limit on the sum in (8.105). Neglecting this difference or, in other words, boundary effects, we can apply Parseval's theorem to (8.104) and obtain

$$E_2(p) = \frac{1}{4\pi} \int_{-\pi}^{+\pi} | A(\omega)(1-e^{n\omega\pi}) |^2 d\omega , \tag{8.106}$$

where A(w) is the Fourier transform of a(n). Some care must be taken to use an appropriate window function for a(n). The term $[1 - \exp(j\omega\pi)]$ defines a comb filter the intertooth spacing of which is 1/p with the sampling period being set to unity.

Thus minimizing (8.104) with respect to p is equivalent to finding the comb filter that optimally nullifies the harmonics of the fundamental frequency. In this view (8.105) provides a clear rule how to determine the fundamental frequency from the harmonic structure of the power density spectrum $|A(\omega)|^2$ of the signal. (Ney, 1982, pp.383-385)

A corresponding error criterion was derived for the cepstrum method.

From the point of view of performance, the average short-term analysis PDA is more reliable than its time-domain counterpart. It is not sensitive to phase distortions and less sensitive to noise and spurious signals. One reason for this is the fact that a frame always contains several periods which are investigated simultaneously instead of individual periods investigated successively as done by time-domain PDAs. Several pitch periods within a single frame can contribute coherently to an accurate pitch indication, whereas uncorrelated noise will add incoherently. For this reason, correcting algorithms are confined to median smoothing or some corresponding "short-term tracking." These operations, however, do not normally extend over more than three consecutive frames. Global correction schemes, which are quite usual in the time-domain PDA category PiTS, are rare, just as multichannel devices. Even the "multidimensional" maximum likelihood PDA by Friedman (1978; Sect.8.6.2) has not more than one basic extractor. These refinements of processing are apparently not needed here.

From the point of view of computational effort, the normal short-term analysis PDA is at least one order of magnitude slower than the average time-domain PDA. Crudely estimated, the computational complexity for a short-term analysis PDA is in $O(N^2)$, if N is the number of samples per frame, unless a mathematical shortcut or a special trick in the implementation can reduce this number. The mathematical shortcut is given for the Fourier transform by the FFT. This necessarily applies to all frequency-domain PDAs and in an optional way to autocorrelation approaches which might make use of it. Tricks of implementation discussed in this chapter are:

-- replacing multiplications and divisions by additions or table lookups (all PDAs using the AMDF and, in principle, also the maximum-likelihood procedures);

-- replacing multiplications by logical operations due to sophisticated preprocessing, in particular due to combined center and peak clipping (see Dubnowski et al., 1976; Rabiner, 1977a, for autocorrelation; Hettwer and Fellbaum, 1981, for the AMDF);

-- downsampling in order to reduce the number N of samples involved in the short-term transformation (Markel, 1972b; Gillmann, 1975; Seneff, 1978; Sung and Un, 1980; Martin, 1981);

-- discarding redundancies and irrelevance before the spectral transformation, again in order to reduce N (Seneff, 1978);

-- confining the operating range of the calculation to samples that are actually needed (nearly all PDAs except those which apply the FFT; there all the spectral samples are computed simultaneously); and

-- adaptive change of the frame length K (Moorer, 1974; Rabiner, 1977a).

It is clear that some information is lost in all these technical shortcuts. As the autocorrelation example shows, however, this need not be a disadvantage for the performance (Rabiner, 1977a). Thus, a number of PDAs already operate in real time or near to real time (Dubnowski et al., 1976; Seneff, 1978; Ross et al., 1974; Sluyter et al., 1980; Martin, 1981; Hettwer and Fellbaum, 1981). Most of these PDAs operate on line since no substantial delays are induced by global correction subroutines or similar delay sources. Thus, the principle of short-term analysis pitch determination will increasingly gain in importance.

Analog short-term analysis PDAs are rare. Among the more recent implementations, we do not find more than one: the HIPEX device by R.L.Miller (1970). This PDD, however, is a device whose performance need not hide its face in the presence of its digital colleagues.

Each of the short-term transformations (except the complex DFT) is irreversible; thus the results obtained by one transform are not readily transferable to the other. Yet these transformations are closely related. Autocorrelation function and power spectrum are tied together by the Wiener relations [Schroeder and Atal (1962), see also (8.6)]; therefore, autocorrelation and cepstrum are not different except for the logarithm applied before the cepstrum transformation. But this "little" difference between the ordinary ACF and the cepstrum is a most striking example of what nonlinear processing can do! There exist, however, strong interrelations between all really successful short-term analysis PDAs. Three ways of processing have proved to be the most reliable: 1) nonlinear preprocessing in connection with autocorrelation and AMDF (center clipping), or cepstrum (taking the logarithm); 2) adaptive linear preprocessing in connection with autocorrelation, AMDF, or maximum likelihood analysis; and 3) the search for the optimum comb filter which forms the underlying principle of AMDF, spectral compression (here applied to frequency-domain processing), and maximum-likelihood. The PDAs based on spectral compression and maximum-likelihood analysis are the most robust and reliable PDAs with respect to degraded speech that are in existence. They are so robust because they manage all the harmonics (i.e., the total energy of the periodic components in the signal) to yield a coherent contribution to the indication of the fundamental.

9. General Discussion:
Summary, Error Analysis, Applications

In the preceding chapters, the prevailing methods of pitch determination have been presented and discussed. The different principles of operation as well as their individual advantages and limitations have been demonstrated as far as possible. In this final chapter, our attention is focused on some important issues dealing with pitch determination from the speech signal in general, such as:

-- Has the problem, in general as well as from the point of view of the respective application, been solved? If not, which basic questions remain?
-- What are the special advantages of the individual principles? Once a special advantage has been found, is it a matter of implementation (and thus transferable), or is it intrinsic to the particular method?
-- What are the weaknesses of the individual principles? Are they a matter of implementation, or are they intrinsic?
-- If an individual PDA or a class of PDAs fail temporarily, do they fail with respect to signal, range of F_0, distortions, noise, spurious signals, or recording conditions? If such a failure is intrinsic, is there another PDA which might be insensitive to this particular situation?
-- Which PDA is best suited for which application?
-- What requirements with respect to output and display facilities arise from the individual applications, and how are they met?
-- What arguments concerning pitch determination principles will gain importance, and what arguments will become obsolete under the impact of modern large-scale integration technology?

This discussion is carried out in several steps. In Sect.9.1 an abridged survey of the prevailing principles of pitch determination is presented, which explains the principles given in Chaps.6 and 8 in a different way. The next two sections deal with the question of performance evaluation and error analysis. In Sect.9.2 the issue of calibration and generation of reference information is discussed. Section 9.3 then deals with the performance evaluation itself. Section 9.4 discusses the various areas of application: phonetics, linguistics, musicology, education, pathology, and - last but not least - speech communication. Section 9.5, finally, tries to sum up the remaining questions (especially for application in speech communication systems) and to present some guidelines for future work in the area of pitch determination.

9.1 A Short Survey of the Principal Methods of Pitch Determination

Due to the preliminary categorization in Chap.2, the time-domain PDAs and the short-term analysis PDAs were treated separately throughout this book. This could give the (wrong) impression that they do not have too much in common. However, there are many features that time-domain and short-term analysis PDAs share. Hence this short discussion will take the opportunity to deal with the PDAs in a different way so that the common features become more evident.

If we regard the categorization in Chap.2 as a "horizontal" categorization, that is, a categorization according to the principle applied, the alternative "vertical" categorization would be according to the functional blocks of the PDA, i.e., the preprocessor, the basic extractor, and the postprocessor. These blocks were introduced at the beginning of Chap.6; each PDA can be subdivided into these categories although one of these blocks may be missing in a given implementation. In the present author's opinion, the "horizontal" categorization is more suitable for detailed presentation of the various principles, whereas the "vertical" categorization is appropriate for the discussion of the interconnections between the individual solutions.

Discussion begins with the presentation of the complete categorization scheme as it has been pursued in this book (Sect.9.1.1); then a brief review according to the "vertical" categorization follows. There we start with the basic extractor (Sect.9.1.2); it is obvious that the strongest differences between the time-domain and the short-term analysis PDAs are in this block. The postprocessor with its application-oriented tasks comes next (Sect.9.1.3); there the respective application is dominant and may mask the differences between the individual categories of PDAs. In principle, the issues discussed in Sect.6.6 are valid for the short-term analysis PDAs too. A corresponding section in Chap.8 thus proved unnecessary, and the review in the present section can be accordingly short. The most important point for discussion is the preprocessor (Sect.9.1.4), where the various processing methods are to be critically elucidated and contrasted. In particular we will check which processing methods can be successfully combined and which ones are mutually exclusive. The final part (Sect.9.1.5) deals with the evolution of algorithms and implementations under the impact of technology, especially in the area of very large scale integration (VLSI).

9.1.1 Categorization of PDAs and Definitions of Pitch

Figure 9.1 and Table 9.1 show the complete categorization scheme as it has been compiled from Figs.2.15, 5.1, 6.2, and 8.1 as well as from the pertinent tables. In this book the gross categorization into the main areas of *time-domain* and *short-term analysis* PDAs was performed according to the time base of the signal at the input of the basic extractor. Within the gross categories further classification was done according to the preprocessing principle applied (in the review of the time-domain PDAs this includes the principle of the basic extractor).

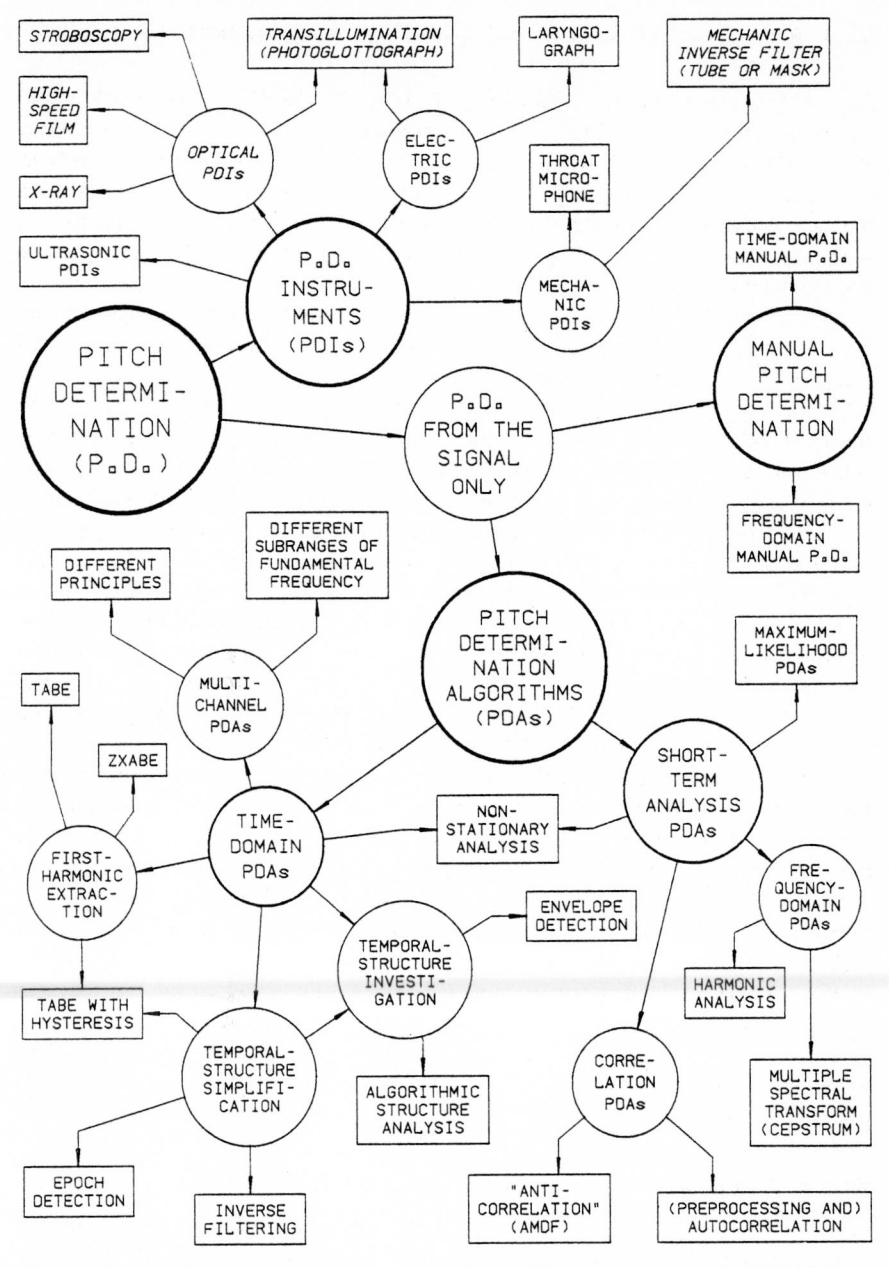

Fig.9.1. Complete categorization scheme of pitch determination algorithms, devices, and instruments. Multiprinciple implementations have been omitted from this figure. For sake of completeness, manual and instrumental methods have been included in this figure although they are not further discussed in the text. The PDIs printed in *italics* prevent the speakers from articulating normally and thus cannot be applied to continuous speech

Table 9.1. Complete categorization scheme of pitch determination algorithms, devices, and instruments. Multichannel implementations have been omitted from this table. For sake of completeness, manual and instrumental methods have been included in the table although they are not further discussed in the text. A "." stands for a preceding or following letter

Label	Description
PiT.	Time-domain PDA
PiTL	Linear device: detection of the fundamental frequency or waveform in the signal
PiTN	Nonlinear device: detection of the fundamental frequency or waveform in the signal after its enhancement by a nonlinear procedure or function
PiTE	Modeling of signal envelope and search for discontinuities which mark the beginning of individual periods
PiTS	Direct investigation of the temporal structure by algorithm; search and extraction of anchor points from which periodicity is derived
PiTA	Investigation of the temporal structure in a simplified way after adaptive filtering
PiT.O	Time-domain PDA with basic extractor omitted
PiC	Short-term analysis PDA based on a correlation principle
Pi..L	with linear preprocessing
Pi..N	with nonlinear preprocessing (center clipping)
PiCA	Autocorrelation PDA
PiCD	Distance function ("anticorrelation") PDA
PiF.	Frequency-domain PDA
PiFH	Harmonic analysis PDA
PiFM	Multiple spectral transform (cepstrum)
PiE	Least-squares estimation PDA (maximum likelihood)
PiM.	Manual pitch determination method
PiMT	... in the time domain
PiMF	... in the frequency domain
PiI.	Pitch determination instrument (PDI)
PiIO	Optical PDI (high-speed photography, stroboscopy)
PiIE	Electric PDI (laryngograph)
PiIM	Mechanic PDI (contact microphone, inverse filter tube, accelerometer)
PiIU	Ultrasonic PDI

Between the gross categories we find a number of PDAs that are either implicitly time domain but do not exploit those features which would make them work according to the time-domain principle (as defined in Sect.2.5), or are implicitly short term but do not perfectly fit into the short-term analysis scheme. Part of these PDAs were labeled "PiT.O," i.e., time-domain PDA *with basic extractor omitted* which was meant in the sense that they have thrown off some of the abilities which would have made them a full member of the time-domain category.

During the presentation of the short-term analysis principle (Sect.2.4) we have seen that there are several possible ways to define fundamental period and fundamental frequency. After the review of all these PDAs we can add some more detail to the definition of T_0 and F_0. Agreeing that the relation

$$F_0 = 1 / T_0 \qquad\qquad (2.34)$$

always holds however T_0 or F_0 are defined, we can understand fundamental frequency and fundamental period in several different ways.

There are three points of view for looking at a speech processing problem (Zwicker et al., 1967; see also Chaps.1 and 4): the *production* point of view, the *signal-processing* point of view, and the *perception* point of view. In the actual case of pitch determination the production point of view is obviously oriented towards the generation of the excitation signal in the larynx; we will thus have to start from a time-domain representation of the waveform as a more or less regular and quasiperiodic train of laryngeal pulses. The signal-processing point of view can be characterized in such a way that (quasi-)periodicity is observed in the signal, wherever it comes from, and that the task is just to extract those features that best represent this periodicity. The perception point of view leads to a frequency-domain representation since pitch sensation corresponds to a frequency and not to an average period or a sequence of periods (Terhardt, 1979a,b; Plomp, 1976).

It appears reasonable to proceed from production to perception. Going in that direction we will start at a local and detailed representation of the parameter *pitch*[1] and arrive at a more global representation in the case of the perception-oriented view. The basic definitions of *pitch* could thus read as follows:

> T_0 is defined as the elapsed time between two successive glottal pulses.[2] Measurement starts at a well-defined point within the glottal cycle, preferably at the point of glottal closure or – if the glottis does not close completely – at the point where the glottal area reaches its minimum. (9.1)

PDAs that obey this definition will be able to locate the point of glottal closure. These PDAs are suited for high-accuracy tasks. To operate properly, they need absolutely undistorted signals (see the discussion in Sect.6.5).

---

[1] As to the use of the term *pitch*, see the comments in Sect.2.1.

[2] It is always assumed that voicing continues between two successive glottal cycles, i.e., that the definitions do not hold for the elapsed time between the last cycle of the present voiced segment and the first cycle of the next one.

> T_0 is defined as the elapsed time between two successive glottal
> pulses. Measurement starts at an arbitrary point within the glottal
> cycle. Which point that is depends on the individual PDA, but for
> a given PDA this point is always located at the same position
> within the glottal cycle. (9.2)

Ordinary time-domain PDAs follow this definition. The reference point can be a significant extreme, a certain zero crossing, an excursion cycle, and so on. This is not necessarily the point of glottal closure itself. Usually, however, it is possible to conclude the point of glottal closure from this reference point when the signal is undistorted. The presence of phase distortions, however, can even destroy this possibility. PDAs that follow this definition usually track the signal period by period in a synchronous way.

> T_0 is defined as the elapsed time between two successive glottal
> cycles. Measurement starts at an arbitrary instant which is fixed
> according to external conditions, and ends when a complete period
> has elapsed. (9.3)

This is the definition of T_0 for those PDAs that are situated between the categories. T_0 is still defined as the length of an individual period, but no longer from the speech production point of view, since the definition has nothing to do with the individual excitation cycle. The synchronous way of processing is maintained, but the phase relations between the excitation function and the pitch period markers at the output of the basic extractor are lost. Once a reference point in time has been established, it will be kept only as long as the measurement is correct and as long as voicing continues. If there is a measurement error, or if voicing ceases, the location of the reference point is lost, and the next reference point may be completely different with respect to its position within the excitation cycle. Among these PDAs we find

-- the time-domain PDAs "with basic extractor omitted" (type PiTLO and PiTNO; Sect.6.6.2);
-- multichannel time-domain PDAs, such as the one by McKinney and Carlson (1976, Sect.6.4.2) or by Martin (1972b, Sect.6.4.1) which do not have a synchronization between the individual channels, or multichannel PDAs that are principally unable to resynchronize with the glottal cycle, such as the PDA by Gold and Rabiner (1969, Sect.6.2.4);
-- short-term analysis PDAs that follow the nonstationary principle of short-term analysis without really performing a short-term transformation. Here we find PDAs that operate according to a modified AMDF principle (Stone and White, 1963; Sect.8.3.3), a nonstationary modified maximum likelihood principle (Ambikairajah et al., 1980; Sect.8.6.3), or a generalized minimum-distance principle, like the vector PDA (Rader, 1964; Sect.6.3.2).

These PDAs miss one feature necessary for "full membership" in the time-domain category: the fixed synchronization with respect to the glottal cycle. The other feature of time-domain PDAs, i.e., the ability to process the signal period by period, has been maintained.

> T_0 is defined as the average length of several periods, i.e., as the average elapsed time between a small number of successive excitation cycles. In which way the averaging is performed, and how many periods are involved, is a matter of the individual method.   (9.4)

This is the standard definition of T_0 for all the short-term analysis PDAs that apply stationary short-term analysis, including the implementations of frequency-domain PDAs. For the frequency domain this definition would read as follows:

> F_0 is defined as the fundamental frequency of an (approximately) harmonic pattern in the (short-term) spectral representation of the signal. It depends on the particular method whether F_0 is calculated as the frequency of a certain harmonic divided by the respective harmonic number m (including $m = 1$), as the frequency difference between adjacent spectral peaks, or as the greatest common divisor of the frequencies of the individual harmonics.   (9.5)

The relation between the time domain and the frequency domain is always established via (2.34).

The perception point of view of the problem leads to a different definition of pitch. Pitch perception happens in the frequency domain. According to the existing theories (Goldstein, 1973; Terhardt, 1974), and with (2.34) still valid,

> F_0 is defined as the frequency of the sinusoid that evokes the same perceived pitch (residue pitch, virtual pitch, etc.) as the complex sound which represents the input speech signal.   (9.6)

This definition is different in principle from the previous ones (Sreenivas and Rao, 1980a, 1981). The main difficulty, however, is that this frequency-domain definition is a *long-term* definition. The pitch perception theories were developed for stationary complex sounds and were only extended toward short pulse trains with varying amplitude patterns and constant frequencies, but not toward signals with varying fundamental frequency. The question of the behavior of the human ear with respect to *short-term* pitch perception is thus still unanswered, and little is known about what kind of short-term "analysis" is executed in the human ear and how it is executed. So we cannot enter the frequency domain other than by an ordinary short-term analysis of the signal as it is known in the signal processing area, i.e., by a discrete Fourier transform (DFT) with previous windowing of the signal. This fact has led the present author to group the frequency-domain PDAs into the short-term analysis category although, from the point of view of their results, they would deserve to be placed in a general category of their own.

Scanning the various surveys of PDAs in the literature, systematic categorizations of the various principles are not too numerous.

Ungeheuer (1963) distinguished between frequency-domain and time-domain PDAs. Among the frequency-domain PDAs he counted 1) isolation and investigation of the first harmonic (PiTL, PiTN, Sect.6.1), 2) short-time spectral analysis using spectrograms (PiMF, Sect.5.1.2), 3) inverse filtering (Sect.6.5), and 4) a procedure called "synchronization with the fundamental," which is similar to the application of a phase-locked loop (Baronin and Kushtuev, 1974). The time-domain PDAs comprised 1) signal distortion with pulselike processing (PiTE, Sect.6.2.1), 2) correlation analysis (Sects.8.2-3), 3) signal processing by computer (PiTS, Sects.6.2.2-5), and 4) manual processing of signal oscillograms (PiMT, Sect.5.1.1). The subdivision of the methods strongly differs from that used in this book, and evidently Ungeheuer did not distinguish between manual and automatic pitch determination.

Léon and Martin (1969) started their categorization from the estimate as given at the output of the basic extractor. If this is a frequency, the PDA works in the frequency domain; otherwise the PDA works in the time domain. The second criterion is whether the PDA uses the whole frequency band of the signal ("complete" analysis) or whether it restricts the analyzed frequency band to the lowest partial(s) ("limited" analysis). Together with the third criterion (manual-versus-automatic processing mode) this leads to the subdivision of the PDAs as given in Figure 9.2. The question whether a PDA is able to track the signal period by period or only on a frame-to-frame basis is not taken into account. From the phonetician's or linguist's point of view (as taken by Léon and Martin) this is totally justified since the phonetician is primarily interested in the global pitch contour unless individual excitation cycles are to be examined. A similar categorization was published by Filip (1972).

A rather detailed categorization scheme for PDAs has been introduced by Baronin and Kushtuev (1974; see Fig.9.3). In the first step the categorization distinguishes between manual methods, "contact" methods (i.e., PDIs), and methods that analyze the signal (PDAs and PDDs). The PDAs are further subdivided into the following categories:

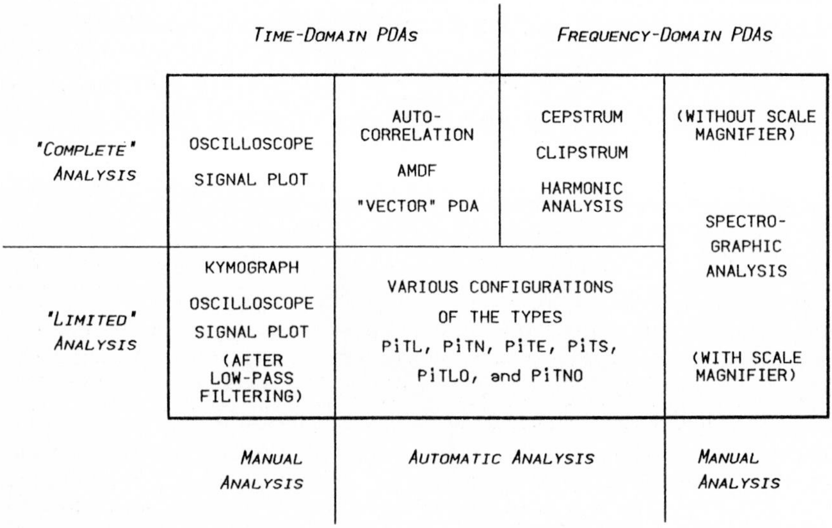

	TIME-DOMAIN PDAs		*FREQUENCY-DOMAIN PDAs*	
"COMPLETE" ANALYSIS	OSCILLOSCOPE SIGNAL PLOT	AUTO-CORRELATION AMDF "VECTOR" PDA	CEPSTRUM CLIPSTRUM HARMONIC ANALYSIS	(WITHOUT SCALE MAGNIFIER)
				SPECTRO-GRAPHIC ANALYSIS
"LIMITED" ANALYSIS	KYMOGRAPH OSCILLOSCOPE SIGNAL PLOT (AFTER LOW-PASS FILTERING)	VARIOUS CONFIGURATIONS OF THE TYPES PiTL, PiTN, PiTE, PiTS, PiTLO, and PiTNO		(WITH SCALE MAGNIFIER)
	MANUAL ANALYSIS	*AUTOMATIC ANALYSIS*		*MANUAL ANALYSIS*

Fig.9.2. Categorization of PDAs by Léon and Martin (1969)

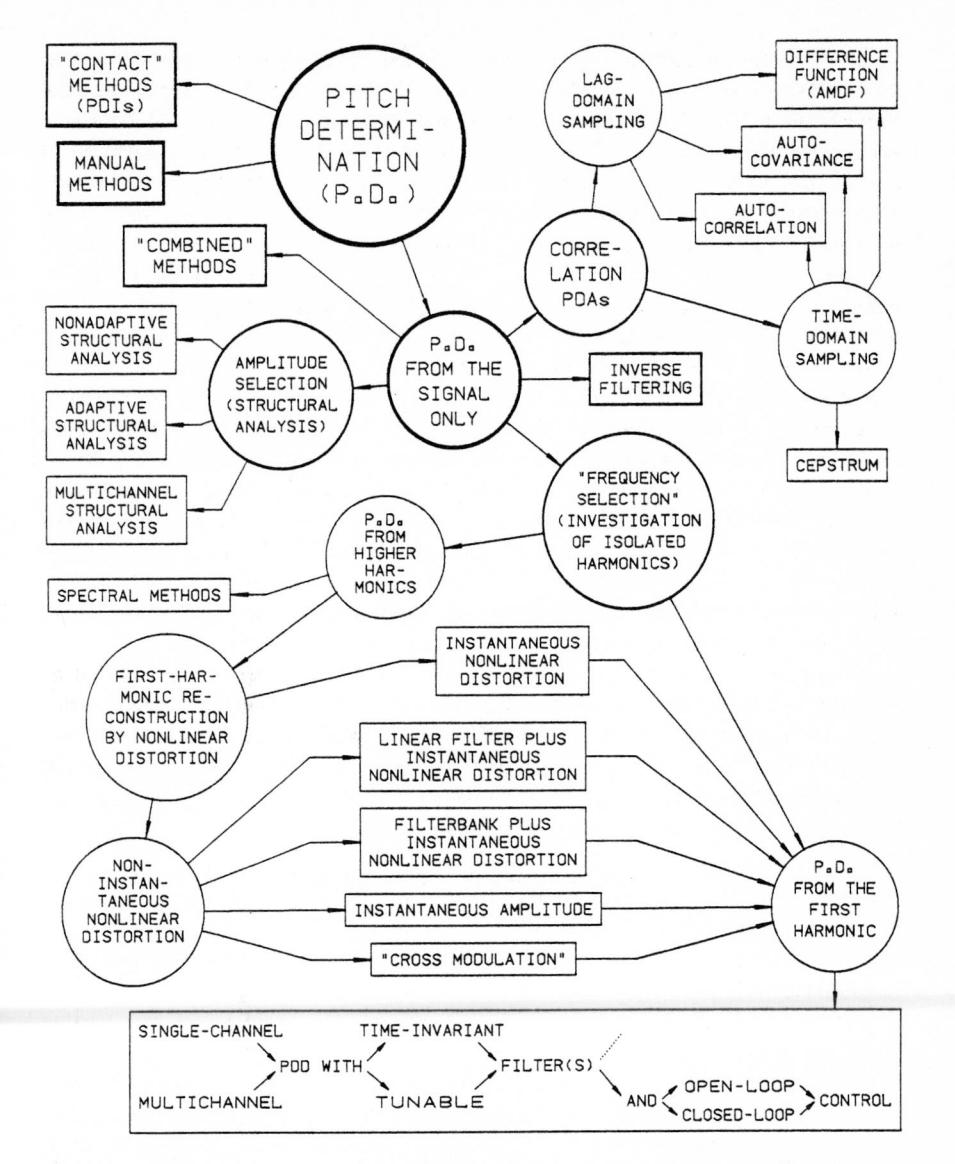

Fig.9.3. Categorization of PDAs by Baronin and Kushtuev (1974). The five main categories are further explained in the text. "Autocovariance" corresponds to the autocorrelation method with nonstationary short-term analysis (see Sect.8.2.1 for more details). Except for the spectral methods, the methods that derive pitch from a higher harmonic reconstruct the first harmonic and determine T_0 therefrom. As to nonlinear distortion, Baronin and Kushtuev distinguish between "instantaneous methods," i.e., distortion where a distorted sample y(n) only depends on the current input sample x(n), and noninstantaneous methods where y(n) also depends on previous samples x(n-k), k > 0. The two categories "filterbank plus nonlinear distortion" and "instantaneous amplitude" correspond to the epoch detectors described in Sects.6.3.1,2. "Cross modulation" has been briefly explained in Sect.6.1.3

1) *Frequency-Selection PDAs.* This category comprises all the PDAs that determine T_0 or F_0 from isolation of individual harmonics; these are the time-domain principles PiTL and PiTN which derive T_0 from the first harmonic or from higher harmonics after nonlinear distortion (including multichannel PDAs) as well as all the frequency-domain PDAs. Note that the simple time-domain basic extractors rank among the frequency-domain principles in this categorization [same as in the surveys by Ungeheuer (1963), by McKinney (1965) and by Lèon and Martin (1969)]. This is due to the fact that in Baronin's survey emphasis is laid upon the determination of the pitch contour, not on the determination of individual period boundaries.

2) *Amplitude-Selection PDAs.* This category contains the principles PiTE and PiTS.

3) *Correlation Methods* (including cepstrum). Except for the cepstrum, all these methods are seen from the point of view of an analog approach. In the analog domain it is necessary to sample at least one independent variable (time or lag); Baronin and Kushtuev distinguished the methods according to which variable is sampled.

4) *Inverse-Filtering Methods.* This category contains simplified inverse-filtering methods for pitch determination as well as sophisticated methods for glottal-waveform detection.

5) *"Combined" Methods.* Any PDA that does not clearly fit into the other categories is grouped into this category. For instance, Rader's vector PDA (1964; Sect.6.3.2), according to Baronin and Kushtuev, combines the principles of frequency selection (due to the filterbank) and correlation analysis (due to the distance function applied in the basic extractor).

In their comparative evaluation of PDAs which comprised seven PDAs (see Sect.9.3.1), Rabiner et al. labeled the autocorrelation PDA (Sect.8.2.4) and the AMDF PDA (Sect.8.3.1) as time-domain techniques since the operating domain of the basic extractor of these PDAs is time (more accurately, the autocorrelation lag). The LPC PDA (Sect.8.3.3) and the SIFT PDA (Sect.8.2.5) were labeled hybrid since they apply (frequency-domain) spectral flattening and a time-(lag-)domain basic extractor. The cepstrum PDA (Sect.8.4.2) was labeled as a frequency-domain method.

Sreenivas and Rao (1980a, 1981) categorized the PDAs according to three functional definitions of pitch. These definitions come rather close to the definitions expressed in (9.1-6). They also correspond to the three points of view for looking at a speech signal, as expressed by Zwicker et al. (1967; see the discussion at the beginning of this section). Sreenivas and Rao distinguished between 1) perceived pitch, 2) pitch as periodicity of excitation, and 3) pitch for laryngeal studies. Perceived pitch corresponds to definition (9.6), and according to the authors all frequency-domain PDAs follow this premise. Periodicity of excitation corresponds to definition (9.4); this category comprises all the short-term analysis PDAs not working in the frequency domain. The third category, pitch for laryngeal studies, corresponds to (9.1,2) and is obeyed by "pattern-recognition" PDAs; this includes all the time-domain PDAs except those which determine F_0 from the first harmonic (those PDAs are not included in Sreenivas and Rao's categorization).

9.1.2 The Basic Extractor

According to the six definitions (9.1-6) that specify how pitch can be measured, we know several types of basic extractors that operate in principally different ways.

The basic extractor of the ordinary time-domain PDA, including the high-accuracy task of determining the instant of glottal closure (IGCDA, Sect.6.5.2), is a typical *event detector*. It sets a marker whenever it detects that a significant event, i.e., a specific feature has occurred. It depends on the particular algorithm how this feature is represented. For the IGCDA the choice of significant features is severely limited: the PDA is obliged to detect the edge in the excitation signal (corresponding to a jump in its first derivative and an impulse in the second derivative). For ordinary time-domain PDAs where pitch is measured according to definition (9.2), there are a wide variety of features that can be used. the The primitive basic extractors, such as the ZXABE, the TABE, and the TABE with hysteresis, pertain to this category as well as the epoch filter (which could also work with definition 9.1), the envelope peak detector, or the algorithmic temporal-structure analyzer which locates and isolates a significant sample, such as a principal peak. Whenever the basic extractor verifies that the significant event has occurred, it sets a marker. This marker is set independently of previous or successive markers. Of course the basic extractor can and will be influenced by the results of previous operations, but this affects only the decision whether a certain event is regarded as significant or not. Once the basic extractor has regarded an event as significant, it positions the marker at the place where the event has occurred, and nowhere else. If the designer of such a PDA specifies events as significant that occur only once per period, he can be sure that his PDA will set its markers synchronously to the individual periods.

The typical basic extractor of the ordinary short-term analysis PDA that measures pitch according to (9.4) is a *peak detector*. The short-term transform forms an image function of the actual signal frame where all the periods are depicted into the interval $(0, T_0)$ in the spectral domain. The basic extractor picks up the peak at T_0 (maximum or minimum, according to the respective PDA) and locates its position with respect to the origin of the spectral function[3]. Of course, the designer must select the spectral transform in such a way that there is really a significant peak in the spectral function from which T_0 can be derived. For the frequency-domain PDAs that follow (9.5), the same applies. In practice, the PDAs that are designed according to (9.6) also apply this type of basic extractor.

---

[3] Remember that, for reasons of uniformity, we speak of a spectral function at the output of the short-term transformation regardless of whether the independent variable at this point is frequency, time, or lag.

The typical basic extractor of the PDAs that measure pitch according to (9.3) works *incrementally* and acts as a *time counter*. This basic extractor operates without a specified reference point. So it must create one. Once this has been done, the basic extractor counts the elapsed time until it again reaches the same status (at least approximately) that it had at the instant when the reference was created. The best example for this way of processing is Rader's vector PDA (1964). There we have a number of band-pass filters and phase-splitting networks in the preprocessor. When the measurement is activated, the current values at the output of all these channels are taken as the actual reference state of the basic extractor, and the PDA starts working. For each instant it then computes the Euclidean distance between the momentary state and the reference state. When the momentary state of the basic extractor becomes sufficiently similar to the reference state (i.e., in the case of the vector PDA, when the Euclidean distance becomes a minimum), the basic extractor assumes that one period has elapsed, sets a marker, and replaces the old reference state by the actual one. The measurement cycle then starts again. Obviously this basic extractor works in an incremental way; after it is positioned according to an external condition (which is at the same time the command to start the first measuring cycle), it counts periods from that point on. If measurement fails, the basic extractor must be repositioned.

Since the positions of the individual markers do not have any relation to the physical locations of the pertinent periods (for instance, to the instants of glottal closure), the user of such a PDA usually throws off the positions of the individual markers and only retains the estimates of T_0 given at these points.

Besides the "synchronous short-term analysis" PDAs that apply nonstationary short-term analysis and follow this principle explicitly, a number of multichannel time-domain PDAs also fall into this category. This applies when the synchronization condition between the individual channels is ambiguous. Definition (9.2) postulates that one can conclude the instant of glottal closure from the reference point of the PDA; hence one should also be able to determine the relation between the reference points of two different basic extractors of this type. However, this holds only for undistorted signals. Even when the signal is only phase distorted, different reference points may shift in a different manner so that we obtain undefined results when switching between the channels. Consequently, again only the period estimate T_0 is a useful result, not the positions of the individual markers. These problems may even arise in simple bipolar structural-analysis PDAs (Sect.6.2.1) with separate channels for the two polarities of the signal.

The basic extractor of the PDA by Gold and Rabiner (1969) pertains to the same type for a somewhat different reason. Six features derived from the significant maxima and minima, each of them determined separately, are exploited. The logic of this PDA permits up to five of them to be grossly wrong; in particular, it is expected that a number of these features are sensed several times per period. Since it is unpredictable which channel is correct when such a feature is detected, the algorithm must work incrementally: each of the six partial extractors just counts the elapsed time between two successive significant events and supplies this value as its actual estimate of T_0.

The PDAs of type PiTLO and PiTNO (time-domain PDA *with basic extractor omitted*) directly combine the basic extractor and the postprocessor; the basic extractor thus no longer appears as a separate component. The extractor itself must obey (9.2) in the case of a pulse rate converter (see Sect.6.6.1 for details). If the extractor obeys (9.3), the valid estimate can be directly converted into a frequency measure, which renders the task simpler. A special case is the PDA by Berthomier (1979) which converts a sinusoid in the time domain directly into an indication of its frequency by just measuring the rotation speed of the complex vector which is established by the sinusoid and its Hilbert transform (i.e., a sinusoid with the same frequency and amplitude but phase shifted by 90°).

The *frequency-domain* basic extractors rely upon the same principles as the extractors that work on a time-domain basis according to (9.3) or (9.4) if previous spectral domain processing (such as harmonic compression) is counted among the preprocessor tasks. If harmonic compression is applied, the pertinent basic extractor has to detect the principal peak at the frequency F_0; it is thus of the same type as those extractors that follow (9.4). If F_0 is detected as the distance between adjacent spectral peaks, the basic extractor operates in an incremental mode and thus in the same way as those extractors that determine T_0 according to (9.3).

In conclusion, we end up with three types of basic extractors, to each of which is assigned a certain definition of F_0 or T_0.

1) The *event detector* marks a well-defined event and sets a pitch period marker whenever the occurence of this event has been encountered. This extractor satisfies (9.1,2).

2) The *counter* works in an incremental mode; it is reset whenever a significant event (which may be the state at the input or the internal state of the basic extractor at a given instant, for instance when an external command is released) has occurred, and stops counting when (approximately) the same event occurs again. This basic extractor satisfies (9.3). The same performance is obtained in multichannel PDAs when the single basic extractors are of the event-detector type, for the case that it is impossible to synchronize the respective significant events.

3) The *peak detector* detects the principal peak (maximum or minimum) in a finite signal and measures its distance from the starting point of that signal. This basic extractor satisfies (9.4).

Frequency-domain PDAs, as shown, work with basic extractors of the peak detector or the counter type.

9.1.3 The Postprocessor

A detailed discussion on the tasks of the postprocessor has been presented in Section 6.6. The following problems were found to represent the essential tasks of the postprocessor: 1) editing and formatting data as well as visualization and preparation of other display; and 2) error analysis and correction as well as smoothing of pitch contours.

Editing and formatting is a matter of the respective application; it will thus be dealt with in Sect.9.4 where the respective applications are discussed. The two remaining points, namely, error correction and smoothing, are closely related. In Sect.6.6 we discussed three methods: 1) linear smoothing, 2) median smoothing, and 3) algorithmic error correction. Let us disregard the case of linear smoothing; this is simply a (zero-phase) low-pass filtering of pitch contours in order to suppress the noisiness of the measurement. Median smoothing is the application of a nonlinear nonrecursive digital filter. It removes isolated maxima and minima from the contour but leaves longer discontinuities unchanged. Global error correction, and in particular list correction, on the other hand, is an adaptive decision procedure designed for certain defined error conditions, usually in a heuristic way. Just as the median smoother, one could regard it as a nonlinear filter, but it executes a different kind of nonlinear filtering. Both procedures do not need the signal to make their decision; thus they do not necessarily have a feedback to the basic extractor (except that they might alter an adaptively changeable preset for future estimates), and they are in some way straightforward. How these routines work was explained in Section 6.6. A more detailed discussion on the various situations arising with the individual errors follows in Sect.9.3.2 in connection with error analysis.

Scanning the postprocessors of short-term analysis PDAs, it can easily be seen that their tasks are more or less the same as in the time-domain PDAs. Usually the error correction logic is less elaborate in short-term analysis PDAs since their preprocessors and basic extractors work more reliably. Thus one can refer to the discussion in Sect.6.6 and close this point here.

There remain two things that might be important. Neither list correction nor smoothing is universally applicable. If a list correction routine is applied, it must reliably detect an error before correcting it. This means that, first of all, the algorithm must reliably recognize the erroneous location as being erroneous. In order to correct the error, the algorithm must then know which value is to replace the incorrect one. Hence a number of correct estimates have to be known before a correction can be carried out. List correction is therefore not applicable to periods at the beginning of voicing unless back-tracking is allowed, i.e., unless the correction of erroneous estimates from their immediate future is made possible. This may cause trouble in speech communication systems where on-line performance is required. Since these systems have to be causal, the list correction routine will cause a substantial delay in processing which can easily reach several hundred milliseconds. Such a delay would be clearly audible and cannot be tolerated in a vocoder. Since this delay is intrinsic to the method (it is the amount of future data that is relevant, not the processing speed of the computer or the hardware), it cannot be compensated and may limit the use of such routines to applications where on-line performance is not needed.

In this respect, median smoothing is much less harmful. It causes a fixed delay which equals half the length of the window of the median. (The same applies to linear smoothing.) This equals about 30 ms and is easily tolerable, even in vocoders, since there the usual frame length is in the same order of magnitude. As opposed to list correction, which can in principle be applied anywhere, smoothing cannot be applied to time-domain PDAs that delimit individual periods. If we smooth a pitch contour, we smooth *period durations*; we are not able to smooth individual markers. The relation between the position of individual markers on the one hand and period durations on the other hand is incremental; if we change the length of an individual period without a compensation somewhere else, we force all future markers within the current voiced segment to be shifted by the amount of that change. The time-domain PDA, however, requires that correctly determined markers delimit individual periods and stay where they are. From this contradiction it follows that smoothing is not applicable to time-domain PDAs when individual periods are delimited.

9.1.4 Methods of Preprocessing

The task of the preprocessor is *data reduction* and, to some extent, *feature extraction*. A useful data reduction enhances the information on the excitation function (and/or on periodicity) and attenuates spurious information.

The voiced speech signal is a quasiperiodic signal. For pitch determination, a number of features can serve as key features for the basic extractor (see also Sect.4.1):

1) a repetitive structural pattern in the signal;

2) the presence of a first partial;

3) a significant peak at the period T_0 in the autocorrelation or a similar spectral function; and

4) the presence of a harmonic structure in the Fourier spectrum, or of a strong peak at F_0.

These features are characteristic for any periodic or quasiperiodic signal. For speech, a view at the mechanism of speech production tells us what the repetitive pattern looks like. The main excitation occurs around the instant of glottal closure; during the rest of the glottal cycle the speech signal equals the impulse response of a linear passive system, at least in first approximation. This means that

5) the instantaneous envelope exhibits its principal peaks at the beginning of the glottal cycle, and becomes smaller toward the end; and

6) the excitation function itself has an approximately triangular shape with a discontinuity in its first derivative at the point of glottal closure.

The last point must be taken with some caution. The statement is valid for (normal) phonation in the modal (chest) register. For the falsetto register the discontinuity is much weaker (see Figs.6.43,44). In vocal fry the excitation function is more pulselike; between the principal excitation pulses there are secondary pulses with lower amplitude.

The main sources of errors are the following:

1) the nonstationarity of the signal;

2) a weakly damped first formant;

3) irregularities in the excitation signal during otherwise stationary segments; and

4) noise and distortions of the recording and/or transmission of the signal.

The error associated with case 3) comes from the fact that most PDAs are unable to distinguish between systematic failures of the method, errors due to distortions, and situations due to irregular excitation signals which look like errors but have been correctly analyzed.

As we have seen, the main task of the preprocessor is data reduction in such a way that the useful features are enhanced, and that the spurious ones are suppressed. A catalog of possibilities to realize this task could thus be made up of the following:

1) The principal peak at the beginning of each period is a good anchor point for a time-domain basic extractor. Thus enhance the principal peaks at the beginning of each period, and suppress the peaks that follow later in the period.

2) T_0 can be measured as the duration of one cycle of the first partial. Thus enhance the first partial in the signal and suppress the others.

3) The formant, F1, if too weakly damped, has a spurious effect. Thus try to artificially attenuate the waveform of F1 within the period or remove it from the signal.

4) In short-term analysis PDAs the short-term transformation is intended to focus the information on pitch from several periods into one significant peak in the spectral function. Thus apply a short-term transformation that has this feature; if possible, give further support by suitable preprocessing methods.

In the following, the prevailing methods of preprocessing are presented in a short survey. The references in brackets indicate where the respective method has been discussed in detail.

Isolation of the First Partial (Sect.6.1). This method looks simple, but the implementation is not trivial at all. First of all, a band-pass filter is needed which eliminates all frequencies outside the measuring range. Within the measuring range, the second partial must be attenuated by 6 to 18 dB with respect to the first partial, depending on the basic extractor. This requires that a dynamic range of 50 dB or more has to be processed unless there is some adaptive shift of the upper cutoff frequency of the low-pass filter within the measuring range. The most promising implementation is the subdivision of the measuring range into subranges of about one octave; this completely avoids the problem with the dynamic range. A further problem arises when the first formant coincides with the second or third harmonic and is weakly attenuated. The method fails systematically when the first partial is not present in the signal. This principle is mostly realized in analog technology. If the measuring range is not subdivided into subranges, the high-frequency attenuation is best obtained by a cascade of first-order integrator filters each of which attenuates the high frequencies by 6 dB/octave.

Moderate Low-Pass Filtering with a cutoff frequency around 1000 Hz (Sects.6.2.4, 7.1, 8.2). This technique is insufficient when used as the only preprocessing step. In conjunction with structural analysis, autocorrelation, or AMDF, however, moderate low-pass filtering removes the influence of higher formants and has a slightly favorable effect. In connection with digital PDAs, moderate low-pass filtering permits reducing the sampling rate and thus helps to reduce the computational effort.

Inverse Filtering (Sects.6.3.1, 7.1, 8.2.5). This means filtering the speech signal with a filter whose transfer function is the reciprocal of the transfer function of the vocal tract. The aim of the method consists in reconstructing or approximating the excitation signal whose temporal structure is much simpler than that of the original speech signal, and in removing the predominance of strong formants. To implement the method, a linear-prediction algorithm can be used, or the method can be combined with moderate low-pass filtering and confined to the formant F1. Since it is linear, this method does not solve the problem of the presence or absence of the first harmonic. This problem is not too severe here, however, because most of the higher partials remain present in the signal. The method may fail systematically when F_0 coincides with the first partial. In this case the first partial is removed by the inverse filter, and the temporal structure is destroyed. Problems may also arise when the speech is distorted by background noise. The inverse filter removes the spectral peaks and thus reduces the overall S/N ratio of the signal.

Comb Filtering (Sects.2.3, 7.1, 8.3, 8.6). Filters pertaining to this special class enhance (or completely suppress) harmonic structures because the maxima and minima of their frequency responses are equally spaced along the frequency axis. To optimally apply a comb filter, however, F_0 should be known beforehand. Since this is not the case, one must try out all possible comb filters in the measuring range of F_0 and obtain the optimum filter parameter as the estimate for T_0. To determine F_0 directly, therefore, this principle is not suited for time-domain PDAs; in short-term analysis PDAs (AMDF, harmonic analysis, maximum-likelihood), where it can be applied in both the lag and the frequency domain, it has to be counted among the most robust and reliable techniques. When used in conjunction with a time-domain PDA, a comb filter may be applied to suppress a low first formant without producing an undesirable high-pass effect (Sect.7.1).

Application of a Filterbank (Sects.6.3.2, 6.4, 8.5, 8.6). This last method of linear filtering is also not useful as the only step of a preprocessor system. It can bring excellent performance, however, in conjunction with envelope detection methods and methods of harmonic analysis (in the time domain). Filterbanks can be applied in two ways: 1) subdividing the measuring range into subranges, and 2) subdividing the significant frequency band of the speech signal into subbands which are then further preprocessed on their own. At the input of the basic extractor, the partial signals may or may not be combined again.

Nonlinear Processing in the Spectral Domain (Sects.8.4, 8.3.3). The purpose of spectral-domain preprocessing is always spectral flattening in order to reduce the influence of strong formants. One way to do this is *amplitude compression* of the spectrum with subsequent transformation back into the time domain. The power spectrum $|X(m)|^2$ leads to the autocorrelation function; in contrast to this, nonlinear spectral-domain treatment of the form

$$Y(m) = |X(m)|^k \tag{9.7}$$

with k significantly less than 2 performs spectral amplitude compression. A special case is the cepstrum where $Y(m) = \log |X(m)|$ is transformed back into the time domain. The other way can be called *harmonic equalization*. Once the individual harmonics are known, their amplitudes can be set to unity, and their phases to zero. Transformed back into the time domain, this yields a waveform with a strong peak at T_0.

Center Clipping (Sects.6.2.4, 8.2.4, 8.3, 8.4.2, 8.6.3). This is one of the instantaneous nonlinear time-domain distortion techniques that destroy the formant structure most effectively without destroying the information on periodicity. It gives an excellent performance together with time-domain structural-analysis, autocorrelation, or AMDF PDAs. Its use may be critical in the case of rapidly varying signal amplitude. Among the various center clipping functions proposed in the literature, the function "clip and compress" (Rabiner, 1977a) gives the best performance since it additionally damps the first formant. Combined with infinite peak clipping (which finally realizes an adaptive three-level quantizer) center clipping substantially reduces the computational effort of the autocorrelation function and the AMDF. The advantage of center clipping is that it works for everything - for any voice and any environmental condition. In some cases, on the other hand, it may not be powerful enough to cope with a dominant first formant coinciding with the second or third harmonic.

Signal Distortion by an Even Nonlinear Function (Sect.6.1, Chap.7, Appendix A). Among the other possibilities of *instantaneous nonlinear distortion* [this technique is called *instantaneous* since the current output sample y(n) is only a function of the current input sample x(n)] the even nonlinear functions have been widely used. The aim of nonlinear distortion consists in enhancing the first partial by formation of difference tones of adjacent harmonics. The method enables PDAs that operate according to the principle of first-harmonic extraction to be used with band-limited signals. The method only runs into trouble when the first harmonic is prominent, especially when F_0 coincides with F1. In this case the signal is almost a pure sinusoid which is converted into a waveform with the fundamental frequency $2F_0$ by the even nonlinear function.

Envelope Detection (Sects.6.2.1, 6.4.2, 8.6.2) is a method of noninstantaneous, or, as one could say, *short-term nonlinear distortion.* Here the current output sample y(n) will depend not only on the current sample x(n) of the input signal, but also on preceding samples x(n-k) in a nonlinear way. The most common envelope detection circuit is a unipolar or bipolar peak detector (preferably in analog technology) with a very small rise time constant and a decay time constant which corresponds to the natural damping of the signal. The circuit is not influenced by the absence of the first partial; on the other hand it gets into trouble with rapid amplitude changes and weakly damped first formants. The performance is also degraded by the unpredictable interference between different formants due to the nonlinear distortion in the peak circuit; previous application of a filterbank with separate peak detection in the individual channels and subsequent summation improves the results considerably. In this configuration the preprocessor also serves as an input for short-term analysis basic extractors.

Instantaneous-Envelope Detection (Sect.6.3.2), besides short-term envelope detection, plays an important role in preprocessing. The instantaneous envelope (or instantaneous magnitude) is the magnitude of the complex analytic signal where the original input signal is the real part, and its Hilbert transform the imaginary part. This technique is able to detect the instant of glottal closure (the "epoch") if the glottis closes abruptly. In this case, the second derivative of the glottal excitation function shows an impulse which is detected by the epoch detector regardless of its phase. Apart from the Hilbert transform which causes some filtering effort, the implementation is simple. The methods get into trouble with soft voices where the glottis does not close abruptly. On the other hand, this preprocessor is ideal for vocal fry. As in the short-term envelope detector, subdivision into as well as separate processing within subbands is advantageous since this method works best when the spectrum in the processed frequency band is flat.

Algorithmic Temporal Structure Investigation (Sects.6.2.2-4) is the final preprocessing technique to be reviewed. This method performs the detection of the relevant information on pitch by rule, i.e., by a finite-state automaton which can be explained by a syntax, or by a set of rewriting rules. The data reduction is carried out in several steps. First a certain feature (e.g., the condition that there must be a maximum) is selected and verified. Any sample without this feature is dropped. Among the remaining samples which exhibit that feature, insignificant samples (insignificant for the subsequent period detection) are suppressed step by step until only one significant sample per period remains which is then the desired marker. Among the time-domain techniques this is probably the most powerful, but also the most complicated one. The method runs very fast but needs a sophisticated algorithm with many decisions and branches. In these PDAs the preprocessor and the basic extractor merge into one block; therefore we will not find algorithmic temporal structure investigation in short-term analysis PDAs. Other preprocessing is not necessary; techniques like moderate low-pass filtering, crude inverse filtering or center clipping, however, simplify the structure (and with it the algorithm) and may improve the performance. Usually this PDA is designed to cope with the absence of the first partial and with a certain amount of background noise; it can suffer from occasional failure when there is a peculiar structure for which no special routine has been provided.[4]

9.1.5 The Impact of Technology on the Design of PDAs and the Question of Computing Effort

When Obata and Kobayashi (1938) published their PDA (PiTLO; Sect.6.6.2), they included a photo of their speech analysis system showing that it had the dimensions of a medium-size cupboard (2 • 1 m). The PDD by Dolansky (1955; PiTE, Sect.6.2.1), in its first version, was still designed in vacuum-tube technology, and the number of active elements and the size of this PDD were comparable to those of a comfortable radio receiver in the mid-fifties. Today, in the era of LSI and VLSI (large-scale and very large scale integration) circuit technology, devices with even much greater complexity are concentrated in a single chip a few centimeters in size. Within these chips most of the space available is used for external connections (pins) so that the "active" area of the integrated circuit is even smaller. The first single-chip digital PDDs have just appeared (Cox and Crochiere, 1982; Bosscha and Sluyter, 1982; Zuidweg et al., 1982), and it is likely that any PDA, provided that a suitable hardware design is performed, can be converted into a digital or - with restrictions - sampled-data PDD that operates in real time; if this does not apply today, then at least within the next few years.

Of course this revolutionary progress in technology has deeply influenced the design of any speech analysis system including the design of a PDA. Looking at the history of speech analysis research, however, it is evident that technology had a deeper influence on the applications of speech processing systems than on basic research. In basic research on speech processing the most significant progress was made after the limitations imposed by the older analog devices had been overcome by the development and the use of digital methods, which, due to the lack of suitable digital circuits, were first implemented on general-purpose computers at the expense of greatly increased

---

[4] Most of these algorithms have been designed after visual inspection criteria. In contrast to the human designer, who is always able to cope with a new and strange situation, the computer can only successfully resolve those situations for which it has been programmed. Apart from that, it is a nontrivial task for the designer to correctly implement the heuristic routines, i.e., to convert empiric knowledge into a reasonable algorithm.

processing time. Nowadays technology permits implementing these methods on special-purpose microcomputers and processors in real time. It thus opens the way for a wide area of applications where real-time operation is necessary which had not even been thought of some years ago. A classical example for this is speech recognition whose future was judged very pessimistically in the late sixties (Pierce, 1969), which was then sponsored and supported in the early seventies, for instance, by the ARPA program (Newell et al., 1971; Klatt, 1977), and which is nowadays increasingly gaining importance in various fields of applications (Martin, 1977; Welch, 1980, Doddington and Schalk, 1981).

In the following, some aspects of the implementation of speech and signal processing systems in general and PDAs in particular are discussed with special respect to the progress in the technology of integrated circuits.

What is Available in Circuit Technology Today? The complexity of VLSI circuits has reached about 100,000 active elements per chip, and 1 million active elements per chip is predicted for 1985 (Hoff, 1980; TRW, 1981). Typical gate delay times (i.e., the time needed for one logical operation) are between 2 and 10 nsec (Hoff, 1980). This yields addition times (for 16-bit fixed-point numbers) of less than 100 ns and - in conjunction with the neces- sary circuit complexity[5] - multiplication times of less than 200 ns. This shortcut in multiplication time is most important for digital signal processing algorithms which need a great number of multiplications, and it indicates that multiplication no longer is the main bottleneck in the design of real-time digital signal processing implementations.

Random-access memory (RAM) and read only memory (ROM) is available at up to 64 Kbit per chip and with access times between 50 and 400 ns. Whereas the ROM is loaded once with information and keeps that information permanently (it is thus exclusively used as program and fixed-table memory),

---

[5] A k $\cdot$ m-bit multiplier requires k times m one-bit full adders and k times m gates if all partial products are realized simultaneously and straightforward- ly (Fig.9.4a). Carry propagation, which represents the critical path of the operation with respect to time, never exceeds a value of k+m, i.e., a carry does not run over more than k+m adders. Since binary addition can be performed as a two-step logical operation, carry propagation involves a max- imum of 2(k+m) gate delay times in series. For a 16 $\cdot$ 16-bit multiplier and gate delay times between 2 and 10 ns this means that a multiplication is safely performed within 128 to 640 ns. As to the number of elements, a full adder can be realized with 9 to 12 gates (according to whether or not it is necessary to invert the input variables) which gives about 3000 gates for the whole multiplier if all adders are explicitly realized. In addition, there are the k $\cdot$ m gates to control the terms to be added for the partial products. To complete the circuit, registers have to be provided for the product as well as for the two factors, and control steps to load the registers and to start the arithmetic operations are necessary. However, this does not greatly increase the general effort, and one can easily see that the complexity of such a multiplier realization is within the scope of realizability on a single chip using today's technology.

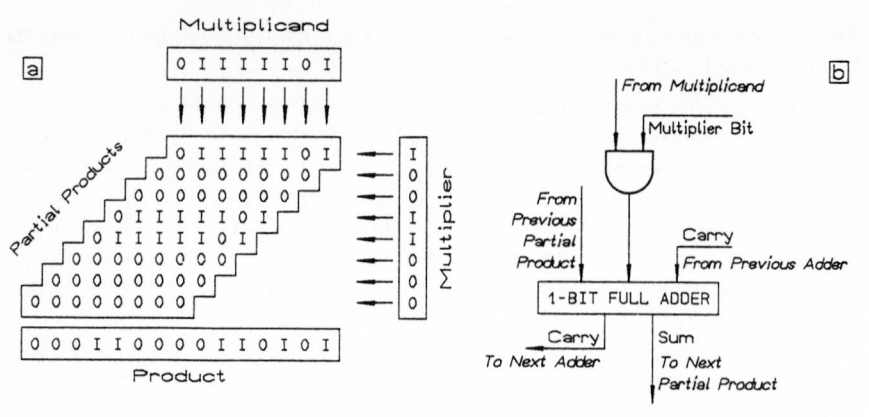

Fig.9.4a,b. Scheme for parallel multiplication of two (positive) binary numbers (a) and principal configuration for computing one bit of a partial product (b). Only the two factors and the (final) product are stored in registers, whereas all places in partial products that are marked by a 0 or 1 are occupied by the arithmetic element as given in (b). Example: 8 · 8 bits; 125 · 25 = 3125

the RAM can be read and written and is therefore suitable as data memory. With a microprocessor, a small number of memory chips, and some special control chips handling input/output operations, a minicomputer can thus be built on a single circuit board.

To terminate this brief inventory with the microprocessors, standard 8-bit microprocessors are not very suitable for speech processing since 8 bits may be sufficient for speech quantizing and transmission, but not for internal signal and parameter representation in speech processing. The more recent 16-bit microprocessors do a much better job. A 16-bit fixed-point representation is sufficient for most applications in speech processing unless extreme accuracy is required. The processing cycle, i.e., the time necessary to execute a standard instruction (add, branch, data transfer, etc.) is comparable to the processing time of arithmetic chips; a high-speed microprocessor can execute between 1 and 5 million instructions per second; this gives an execution time of 0.2 to 1 μs for a microprocessor instruction.

The author is aware of the fact that the numbers given will change even while these lines are being written. However, integrated circuits are usually well documented, and there exist application notes issued by the manufacturers which present the application area a particular circuit has been designed and optimized for, and which give hints how a special algorithm can be optimally implemented.

Where Does the Trend in Integrated-Circuit Development Go? It is evident that the mechanic dimension (besides the type of semiconductor technology) is the most important parameter. Note that, chip size remaining unchanged, the data throughput of a chip - which is the significant parameter of performance - increases with S^3 if the scaling, i.e., the distance between adjacent elements, goes down by a factor of S. The reasons contributing to this are the following (Hoff, 1980):

1) A reduction of the linear dimensions by a factor of S permits increasing the element density - and with it the complexity - by S^2. To give a simple example, if a processor chip, like the digital signal processor developed in the Bell Laboratories (Boddie, 1981), today contains one multiplier which performs a multiplication in one processing cycle, and if the linear dimensions within the chip are reduced by a factor of 2, a successor chip could contain four multipliers of the same type and thus perform a *complex* multiplication within one processing cycle.

2) The processing speed is more or less determined by the parasitic capacitances within the circuit. A reduction of the linear dimensions by a factor of S reduces these capacitances by the same factor, thus enabling the processing speed to be increased by S.

Of course there are a number of limitations which do not permit the linear dimensions to be reduced indefinitely (Hoff, 1980). For instance, to keep the electric fields within the semiconductors and insulating layers of the chip within acceptable bounds, supply voltages must be lowered proportionally to the reduction of linear dimensions. Since thermal noise only changes with temperature, not with the dimensions, the signal-to-noise ratio of the circuit becomes worse as dimensions are reduced. This fact influences analog circuits much more strongly than digital circuits. The error probability of a digital circuit due to thermal noise drops towards $-\infty$ when the signal-to-noise ratio exceeds 20 dB. This means that the linear dimensions of today's digital circuits can be reduced by another order of magnitude before the SNR becomes critical.

Another issue important for integrated circuits is the cost of development. The amount of human work necessary to develop a new integrated circuit comprising some 10,000 active elements is in the order of several man-years even if parts of the circuit can be combined from available standard designs and layouts. This means that integrated circuits only reach a feasible price when they can be sold in great numbers. In speech processing we are in the lucky situation that most of the methods either comprise "intelligent" and flexible but arithmetically rather simple implementations or "stupid" but arithmetically cumbersome "number crunching" for signal processing tasks. In both cases the system designer can widely refer to standard hardware solutions; decision-intensive programs can be run on a general-purpose microprocessor extended by some RAM and/or ROM memory chips and special arithmetic units if necessary. For the "number crunching" there are special one-chip signal processors such as the already mentioned digital signal processor developed at the Bell Laboratories, the NEC µPD7720, or the Texas Instruments TMS320. These chips are specialized in the typical arithmetic tasks as given in digital filtering, in computation of correlation and covariance functions, and execution of the DFT. Furthermore, if a particular processing algorithm is widely used [for instance dynamic pattern matching in isolated-word recognition (Sakoe and Chiba, 1978)], special integrated circuits for this task will be developed within a short time.

Special Processors for Digital Signal Processing. In digital signal processing the operation most frequently required is accumulation of products, as defined for instance in the state equation (2.11) of a digital filter. The increase in multiplication speed has been a very important point. Since multiplication no longer dominates processing time, one must carefully watch the remainder of the operations required. Two issues dealing with hardware architecture are likely to increase the throughput of the system (McDonough et al., 1982): 1) parallelism and 2) minimization of data transfer. Let us again explain this with the aid of a simple example. The state equation of a second-order digital filter

$$y(n) = d_0 a(n) + d_1 a(n-1) + d_2 a(n-2) - g_1 y(n-1) - g_2 y(n-2) \qquad (9.8)$$

requires five multiplications and four additions. As long as each multiplication (in the case of 16-bit operand wordlength) took 16 processing cycles compared to one processing cycle required by any other operation, one could concentrate upon multiplication and neglect everything else. Nowadays multiplication can be completed in one processing cycle as well. Using an ordinary architecture where everything is done sequentially, the execution of (9.8) for one sample of $y(n)$ would need 5 cycles for multiplication, 4 for addition, and 30 for data transfer from and to memory, and the high-speed multiplier would be idle 88% of the time. [Each multiplication would require 4 data transfers: 2 for the operands and 2 for the products which may have 2k-1 bits when k is the word length of the operands. Rounding each product before accumulating the new sample $y(n)$ would render the filter arithmetically unstable and is therefore undesirable.] Changing the architecture in such a way that an accumulator register and an adder are directly connected to the multiplier, the new sample $y(n)$ is built up in the accumulator; the number of data transfers is reduced to 11 (10 for fetching the operands of the multiplications and one for storing the final result), and accumulation (i.e., adding) and data transfer to the input registers of the multiplier can be done in parallel, thus reducing the total number of processing cycles to compute one sample of $y(n)$ from 39 to 16. A further reduction of this number seems possible.

Special signal processors, such as the ones mentioned further above, are optimized in this respect. As an example, Fig.9.5 shows the arithmetic unit of the TMS320 processor (McDonough et al., 1982).

Let us now return from this more general discussion to the specifics of speech analysis, and to pitch determination in particular.

Digital Versus Analog Implementation. The great problem with analog implementations is that speech is a time-variant signal so that a necessary premise for most speech processing systems is the availability of short-term analysis methods. In principle, this renders the problem two dimensional; there is thus the problem of short-term memory with delay time as the parameter

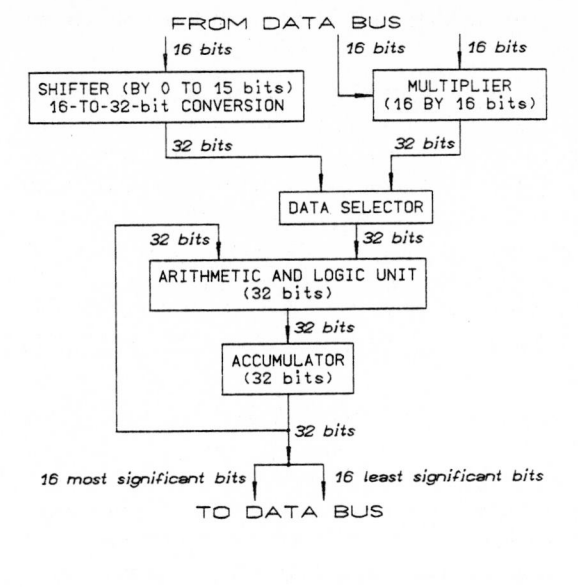

FROM DATA BUS

Fig. 9.5. Arithmetic unit of the TMS320 (simplified). Besides sign extension (conversion from 16-bit 2's-complement to 32-bit number representation), the shifter serves as by-pass of the multiplier and as a tool to perform multiplications and divisions by an integer power of 2. Since division by a value not an integer power of 2 cannot be optimized with respect to processing time, it is not supported on the chip and must be avoided when designing an algorithm to be run on this processor

which is hard if not impossible to be realized in analog technology. The special task of pitch determination is simpler since a PDD may work completely without memory when it is of the event-detector type (see Sect.9.1.2), or with a fixed memory when it operates in an incremental mode [see, for instance, the PDAs by Gill (1959; Sect.8.2.4) or Stone and White (1963; Sect.8.3.1)]. Even here it is very desirable, however, to compare and combine different sources of information on pitch from different instants within one period or from different periods, a feature which is easily realized in a digital system. In digital systems, on the other hand, where principal restrictions with respect to memory do not exist, powerful signal-processing and short-term-analysis methods can be applied, and the most significant progress in speech analysis as well as in pitch determination has been made since the computer was first applied in this area (Schroeder, 1969; Denes, 1970; Flanagan, 1976). The bottlenecks of digital systems, on the other hand, have been real-time processing and hardware complexity. Both these problems, however, have proved less severe than the memory problem in analog systems. Complex hardware has been available - although not very readily accessible - from the beginning on in the form of general-purpose computers, and the lack of real-time processing inhibited applications where man is a part of a closed loop (e.g., man-machine dialogs), but it did not inhibit the development of methodology.

Progress in circuit technology influences both the digital and the analog systems. First of all, as we have seen in the previous discussion, a larger scale of integration and miniaturization reduces the physical size of the apparatus when the hardware complexity is kept constant, or it increases the permitted hardware complexity when the physical size is kept constant.

Second, as components and circuits become smaller, they also become faster and cheaper. The question of circuit speed is especially important for digital systems where it enables real-time operation (which, on the other hand, has never been a problem for anlog systems). In analog systems the memory problem has (at least partially) been overcome by the development of charge-coupled-device (CCD) technology which operates on a sampled-data base. (This means that the signals are sampled but not quantized, and that they are represented as analog voltages and not as digital numbers). Although such systems are not as flexible and not as precise as digital ones, they represent a good compromise between digital and analog technology. A number of short-term analysis PDAs, above all the cepstrum PDA (Noll, 1964, 1967; Sect.8.4.2) were originally intended to run on sampled-data analog hardware (however, with extremely high hardware effort); to develop the methods, the PDAs were simulated on a computer, and sometimes - as in the case of the cepstrum PDA - the digital version became so popular that the (expensive) analog sampled-data version was no longer necessary. Today a sampled-data version of such a PDA seems possible with moderate hardware effort.

Since digital PDAs outperform analog implementations in any respect except cost and operating speed, the current trend toward low-cost and high-speed technology clearly favors digital systems.

Basic Computational Complexity. The basic computational complexity of PDAs, measured as the number of arithmetic operations significant for processing time, is in $O(K)$ for the simplest PDAs and in $O(K^2)$ for usual short-term analysis PDAs. In some multichannel PDAs the computational complexity can even be higher, but it never exceeds $O(K^3)$. K represents important system parameters, such as the frame length, the measuring range of pitch, or, simply, the duration of the signal.[6] Since progress in technology in the last years has made the speed of digital circuits grow nearly exponentially (Hoff, 1980), it seems to be a question of a few years, if this trend continues, before a PDA will be able to run in real time on digital special-purpose hardware and at moderate expense.

It is still usual to judge the efficiency of a digital signal processing algorithm in terms of multiplications required. This has also been done at several places in this book (Sects.8.1, 8.5.4, 8.6.3), and it is fully justified for any software implementation on a general-purpose computer as well as for any conventional hardware realization without special multipliers. On the other hand one cannot ignore that progress in circuit technology has influenced the processing time required for a multiplication to a greater extent than the processing time needed for any other operation. In future implementations, where fast multipliers will be used more widely, a more global optimization

---

[6] This applies to pitch determination. In areas such as speech recognition, the complexity of an algorithm can be of exponential order if the problem is badly formulated (Simon, 1982).

of the computational complexity will thus be necessary, i.e., a minimization of the total number of operations. In such a realization it will no longer be advantageous, for instance, to replace a multiplication by a computation of magnitude or by a table lookup since this will not make the algorithm any faster. On the other hand, it will always be desirable, e.g., to reduce the sampling rate of the signal or the frame length since this indeed reduces the total number of operations required.

How Fast Can Digital PDAs Run Today? Some examples of short-term analysis PDAs will be given here which represent relatively bad cases. Time-domain PDAs usually run faster.

The question of computing effort in connection with frequency-domain PDAs was already discussed in Section 8.5.4. To achieve a spectral resolution of 5 Hz, it was found that one had to apply 1) a 2048-real-point FFT, or 2) a 2:1 decimation in the time domain, a 128-complex-point FFT, and a 1:4 interpolation in the frequency domain, or 3) an implementation somewhere between these two extremes. It is evident that we need at least two chips for a hardware solution, i.e., a processor and a chip for data memory. Many possible hardware configurations and architectures to execute the FFT have been proposed in the literature (e.g., Nussbaumer, 1980; Covert, 1982; Schirm, 1979; Karwoski, 1980a,d). For instance, Schirm (1979) proposed an FFT processor board using a fast multiplier-accumulator which can execute a 1024-point FFT (including data transfer) within 6.8 ms. This means that a frequency-domain PDA with high spectral resolution using the FFT is already possible with relatively low hardware effort. If a more universal digital signal processor, like the TMS320, is used (which is not optimal for this application), a FFT butterfly, conservatively estimated, would need less than 25 processing cycles (including address modification), and thus a 256-complex-point FFT could be managed within 4 ms. The time-domain decimation and frequency-domain interpolation necessary in this case runs faster since the processor architecture has been optimized for this task; it can thus easily be performed within the remaining 6 ms.

This example only takes into account the preprocessor of a frequency-domain PDA. However, the basic extractor requires a nonnegligible computational effort as well, especially when the principle of harmonic matching (Duifhuis et al., 1982; Terhardt, 1979b) is used. This has been verified by Zuidweg et al. (1982) who realized such a frequency-domain PDD with three chips.

As soon as the speeds of multiplication and addition become nearly equal, the difference between autocorrelation and AMDF PDAs (Sects.8.2,3) with respect to computational effort vanishes (except for the different frame length which still favors the AMDF). The basic complexity of both PDAs is in $O(K \cdot N)$ if K is the frame length and N the measuring range of pitch. For instance, with a sampling interval $T = 100$ μs, a frame length of $K = 400$ samples (40 ms) and a measuring range of T_0 between 2 and 20 ms (i.e., 20 to 200 samples), about 54,000 multiplications and additions are required to

evaluate the ACF for one frame. Performing a multiply-and-accumulate operation in 400 ns the TMS320 could do it in 21.6 ms. Two or three such processors would thus be needed to run this PDA in real time. Compared to the
computational load to calculate the ACF, the effort for determining the
principal ACF peak in the basic extractor is negligible. For the AMDF, where
an effective window of 12 ms is sufficient (Ross et al., 1974), the number of
elementary operations (subtract, compute magnitude, and accumulate) would
reduce to 22,000 per frame making this algorithm somewhat faster. For any
other short-term PDA not using a Fourier transform things are similar. We
can thus conclude that any short-term PDA is realizable in hardware with
very few chips (less than 10) operating in real time or at least in 2 to 5
times real time.

Some Realized Examples. Zuidweg et al. (1982) implemented the harmonic-
sieve PDA (Duifhuis et al., 1978, 1982; Sluyter et al., 1980, 1982) on three
custom-designed integrated circuits. The signal enters the first chip, the
segmentation buffer, where it is subdivided into (overlapping) frames of 25.6 ms
duration (the frame interval being 10 ms); the chip then stores the signal for
further processing. The subsequent DFT processor computes the DFT of each
frame (64 spectral samples between 0 and 2 kHz; no FFT is used, only the
ordinary DFT algorithm; signals and auxiliary values for the DFT are rather
crudely quantized). The third component, a special microcomputer, performs
the evaluation of the harmonic sieve. According to the authors, this part of
the algorithm is most critical with respect to real-time operation.

Cox and Crochiere (1982) realized an autocorrelation PDA on the above-
mentioned Bell digital signal processor. (The architecture of this processor is
similar to that of the TMS320; the TMS320, however, which came out later,
is four times faster.) As in Rabiner's algorithm (1977a), the signal is first
low-pass filtered to 1 kHz (without downsampling) by a 7th-order nonrecursive
filter. Since complete evaluation of the ACF is impossible to be realized in
real time on this processor (see the above discussion), the ACF is recursively
updated,

$$r(d,n) = w^k r(d,n-k) + x(n)x(n-d) \,, \quad d = d_{min} (1) d_{max} \,, \quad\quad (9.9)$$

where d_{min} and d_{max} delimit the measuring range of pitch, and w is a
constant less than unity. This way of ACF evaluation is equivalent to applying
an exponential window to the signal. No subdivision into frames is performed,
and compared to direct ACF computation the recursive procedure is considerably faster so that the signal processor alone is sufficient to realize the
PDA; program and data just fit into the processor's internal memory. To
update r(d,n), the update interval k usually equals 1 sample; with the low-pass
filter, however, k can be chosen greater than 1 (k = 4 in the case realized).
For the special application in adaptive speech coding, for which this PDA was
designed, adaptive preprocessing of the signal was not necessary.

In summary, this discussion on hardware realizations can be concluded with the following statements:

1) The progress in integrated-circuits technology permits running most PDAs in (or close to) real time on a digital hardware configuration containing few (usually less than five) VLSI circuits.

2) Digital speech processing devices in general, and digital PDDs in particular, can be realized using standard microprocessors in combination with special-purpose hardware for signal processing, such as fast multiplier-accumulators and digital signal processor chips.

3) Multiplication no longer takes the majority of processing time when fast multipliers are used. Only division cannot be accelerated and should thus be avoided.

4) Besides the application of high-speed circuits, parallel processing and minimization of data transfer are key features for increasing the efficiency of the implementation.

One final word about optimization and testing. Once the requirements with respect to real-time operation have been met, it is principally no longer necessary to optimize the algorithm. However, taking care of a well-structured algorithm is always worth while. Processing time is not the only optimization criterion, although it is probably the most important one in such an implementation. Other criteria are system architecture, use of memory, and the simplicity and conciseness of the program code. Simple and short programs are coded and tested within a short time. Since efficient signal processing programs have to be written in machine code or low-level assembler languages, the programming effort as well as the effort for testing and debugging (which is made even more difficult by the need for testing the hardware and software simultaneously) is considerable. Good design of the hardware architecture and the software structure thus helps saving a lot of human work otherwise necessary for extensive testing.

9.2 Calibration, Search for Standards

Pitch determination is a subproblem of speech analysis and thus a special case of measurement techniques. For any measuring device, data on the performance and accuracy are of primary interest. Performance of a PDA, however, certainly is a complex problem. It may be subdivided into the problems *reliability* and *accuracy*. This subdivision is a generalization of the subdivision found in most published papers, i.e., the discrimination between *gross* and *fine* pitch determination errors. To discuss accuracy first, each PDA has a basic accuracy with respect to the sampling rate, amplifier tolerances, and so on. This basic accuracy may or may not be improved by additional means such as interpolation or a vernier. In addition, and still within the aspect of accuracy, there are the *fine errors*, also known as *jitter* or little *hops* which result from the noisiness of the raw pitch estimate. The other pitch determination errors fall into the category of reliability. These *gross errors* denote a momentary or even long-term failure of the PDA: chirps, holes, large hops, subharmonic tracking, higher harmonic tracking, and - last but not least - voiced-unvoiced (VUV) discrimination errors which have not been discussed in this book. Thorough evaluation of the performance of a PDA must take account of all these points.

How to carry out such an evaluation? Four methods appear possible:

1) objective evaluation with respect to a standard;

2) objective comparative evaluation with respect to other PDAs whose performance may or may not be known;

3) subjective evaluation with respect to a standard; and

4) subjective comparative evaluation and ranking.

Before entering the discussion on comparative evaluation, however, we must think about the possibilities for creating a standard according to which the performance of a PDA can be tested. In the actual case of pitch determination this would be a reference pitch contour. Creating such a contour is a nontrivial problem, even with high-quality signals; it is so serious that many researchers and designers of PDAs have chosen a simpler way to evaluate their PDAs, using the output of a well-known PDA (cepstrum, for instance) as a reference. This may be useful if the researcher knows beforehand that the new PDA will have a performance *inferior* to the reference PDA. However, this is often not the case since there is no PDA which operates free of error. If the performance of the new PDA is comparable or superior to that of the reference, a quantitative statement of how much better the new device works appears hardly possible. Thus we cannot avoid the problem of how to create a reference contour that is free of error and more accurate than the result of any existing PDA. The possibilities and requirements for achieving this will be discussed in the following, whereas we will deal with the performance evaluation itself in Section 9.3.

9.2.1 Data Acquisition

The evaluation of a standard, first of all, requires an adequate acquisition of data. This problem has largely been solved. Analog FM tape drives as well as digital high-quality PCM tape recorders permit recording sample utterances practically free of noise as well as of distortions in amplitude or phase. This recording equipment is today a standard tool of every well-equipped acoustic laboratory. The only disadvantage is that analog FM data are usually not interchangeable since the recording principles of the individual analog FM recorders are incompatible (different carrier frequencies). Transferring digital data, on the other hand, is a minor problem. Normal (not FM) analog tape recordings, however, even when made on high-quality tape recorders, are phase distorted, especially in the low-frequency region (Holmes, 1975; Preis, 1981; see Fig.9.6). This means that an analog recording cannot serve as a

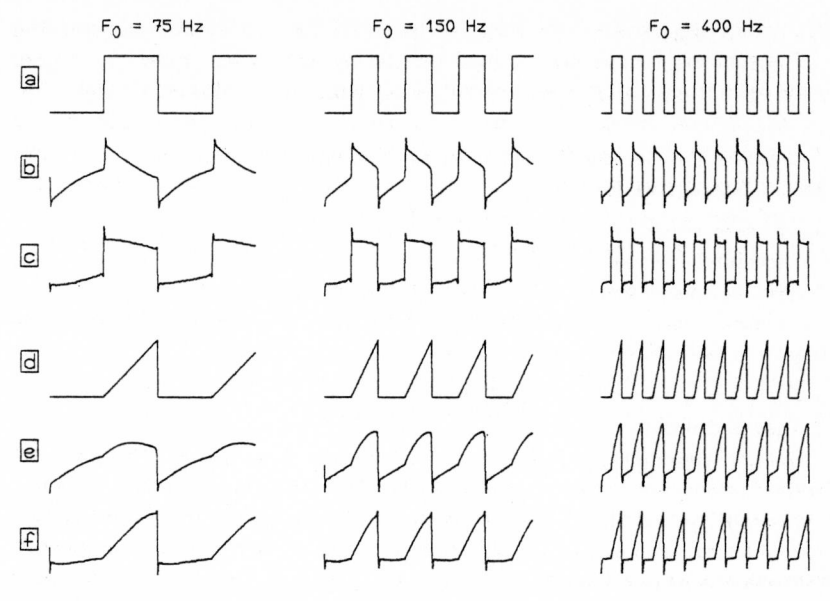

Fig.9.6a–f. Examples of phase distortions in good-quality analog tape recorders due to the frequency response equalizer. Signals: (**a**) rectangular waveforms (frequencies as indicated in the figure); (**b**) same waveforms after recording on the analog tape recorder; (**c**) same waveforms after recording and rerecording on the same tape recorder in backward direction; (**d–f**) same as (a–c), this time for a ramplike time function

standard material for the evaluation of PDAs. If forced to do so anyhow, it may be convenient - if the characteristics of the tape recorder originally used are known - to rerecord the utterance on the same type of tape recorder in the backward direction. Since phase distortions are added in a cascade network (as one could regard this "cascade" of two tape recorders), recording in the backward direction compensates for all the phase distortions so that the second recording is undistorted with respect to phase (Meyer-Eppler, 1950; Beauchamp, 1968; Preis, 1981). Amplitude and nonlinear distortions in this case are tolerable. (Even if the characterictics of the original tape recorder are unknown, rerecording the signal in backward direction will always reduce the overall phase distortions.)

Another way of correcting phase distortions is by using a calibration tone on unknown recordings, as postulated by, among others, Holmes (1975). Such a low-frequency tone consisting of many harmonics in the range of the speech signal (e.g., a 50-Hz rectangular or sawtooth waveform or a pulse train) permits measuring the overall phase distortions of recordings whose provenience is not known, and, accordingly, designing a correcting network such as a (digital) all-pass filter. The issue of phase distortions is important in the evaluation of standards as well as for tasks of voice investigation, such as accurate determination of the closed glottis interval or (exact) inverse filtering to reconstruct the excitation signal. For normal PDAs, phase distortions should play a minor role. To clarify this issue for a new PDA, however, speech material free from distortions is required to generate a suitable standard.

9.2.2 Creating the Standard Pitch Contour Manually, Automatically, and by an Interactive PDA

Manually determined fundamental frequency contours as well as manually determined pitch period boundaries have been the oldest standard with respect to which the performance of a PDA could be evaluated. As far as the manual methods have been discussed (Chap.5), they have three principal short-comings.

1) They are extremely time-consuming.

2) They are usually not accurate enough to serve as a reliable standard for fine errors. If nevertheless forced to do so, visual inspection of the waveform is the only manual method which can be performed with sufficient accuracy. With conventional oscillographic instruments, such as the mingograph, this method requires a considerable additional amount of human work, apart from the material necessary (paper etc.).

3) The results are difficult to synchronize with the results obtained by the test PDA.

Up to the mid-sixties, there was practically no *objective* alternative to visual inspection, and even rather recent PDAs have been evaluated using a manually determined reference contour (Kurematsu et al., 1979). (Starting with the invention of the vocoder, however, it has always been possible to *subjectively* evaluate the performance of a PDA using synthesized speech generated with the test PDA in the pitch determination channel of the vocoder.) The evident difficulties with manual analysis have seduced many designers of PDAs to confine the evaluation of their devices to some (frequently rather vague) declarative statements, so that, for instance, McKinney (1965), with respect to an individual PDA, complained that,

> ... as is typical of descriptions of speech bandwidth compression systems in general and laryngeal frequency analyzers in particular, the evaluation of performance given is limited and only qualitative. (McKinney, 1965, p.52)

The scene changed when the cepstrum PDA (Noll, 1964, 1967) was developed. It provided an easy implementable solution with a degree of reliability hitherto not achieved. Its only disadvantage was that it was slow and did not (yet) permit real-time operation. Thus many researchers have compared their PDAs to the cepstrum PDA. Often the result of this comparison led to the statement *less reliable but much faster* (Reddy, 1967, for instance). Even the cepstrum PDA itself, however, has not been quite perfect (Rabiner et al., 1976), so that the question of a standard for calibration was still open.

The main impulse toward a solution of this question by an interactive system came from the progress in computer philosophy and technology which has been moving further and further away from pure batch-processing methods towards man-machine communication and interaction. New developments, such as moderate-size laboratory computers with multiuser operating systems and interactive graphic terminals paved the way for a solution of this problem. The main obstacle to manual analysis as a standard procedure for the evaluation of PDAs has been its clumsiness and the amount of human working time it requires. Significant progress is achieved - in accuracy as well as in time saving - when an interactive PDA is constructed in the following way. The computer processes a given speech signal frame by frame (or even sample by sample) using a known PDA principle, displays it together with the results, and allows the human investigator to check and correct it if necessary. Such a procedure not only makes the analysis much easier since the human investigator is confined to correcting the results instead of carrying out the whole analysis; it also greatly improves the accuracy of the manual analysis.

> The usual method of manual pitch detection is for the user to mark pitch periods on a period-by-period basis, directly on a display of the speech waveform. Although such a technique is often quite good, there are segments of some speech sounds during which the waveform periodicity is not clearly visible in the waveform due to rapid spectral changes in the sound. During such intervals a rough indication of the pitch period can be obtained from the waveform, but due to the changing spectrum, the pitch period estimate can be off by several samples (McGonegal et al., 1975, p.570).

A first solution,[7] and certainly not the last one, has been presented by McGonegal, Rabiner, and Rosenberg (1975), who developed an interactive PDA for the generation of reference pitch contours. Figure 9.7 shows its block diagram. Three methods of pitch determination are applied: cepstrum, autocorrelation (without center clipping), and a primitive time-domain technique (PiTS). The time-domain basic extractor directly places markers into the first positive zero crossing after the principal maximum of the period.[8] The low-pass filtered waveform, autocorrelation function, cepstrum, markers, estimates, and auxiliary values, such as signal amplitude, are then displayed on a graphic terminal (Fig.9.7).

> The choice of these three displays, i.e., the low-pass filtered waveform, its autocorrelation function, and the cepstrum of the wideband signal, was dictated by the desire to obtain several fairly independent estimates of the pitch period. It was expected that the autocorrelation display would provide a good estimate of the pitch period in cases where the waveform was unreliable due to the changing waveform phase since the autocorrela-

---

[7] An interactive PDA based entirely on a time-domain procedure was proposed earlier by Öhman (1966).

[8] In the opinion of the present author, the negative zero crossing between the principal maximum and minimum provides a better anchor point to place the markers. See the review of the PDA by N.Miller (1975, Sect.6.2.3) and Chap.7 for a more detailed discussion.

tion function is essentially phase insensitive. The low-pass filtered wave-form was used rather than the wideband signal because the high-frequency information present in the waveform tends to mask visual identification of the period rather than aid in the decision. Finally, the cepstrum was used to provide a frequency-domain measurement which again was fairly insensi-tive to changing phase in the waveform. (McGonegal et al., 1975, p.571)

The program computes the period estimate for the current frame as the median of the three individual estimates, and provides the user with the option to change any of these values according to what he (or she) desires. This PDA is designed to yield a fundamental-period contour (T_0 contour), not markers of individual pitch periods. Thus it is more or less based on the short-term analysis principle; little effort is put into the time-domain part. Rather than tracking all the periods the time-domain routine determines the length of that period which happens to contain the center sample of the given frame (in Fig.9.8, the 400th sample out of 600 at 10 kHz sampling frequency), and yields this length as the estimate of the frame.

In case the user corrects one of the individual estimates, he has a cursor at his disposal. He marks the appropriate region in the display where he wants to have his estimate placed. The computer then looks for the nearest peak or positive zero crossing (in the case of the time-domain signal) around the manually set raw estimate. Hence the PDA works with maximum accuracy without forcing the user to determine the exact position of the estimate. If everything fails, the user is even allowed to manually enter a value as the

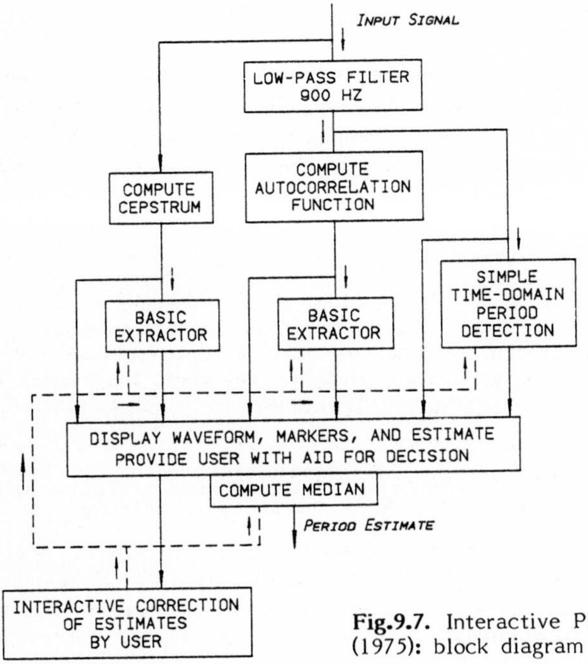

Fig.9.7. Interactive PDA by McGonegal et al. (1975): block diagram

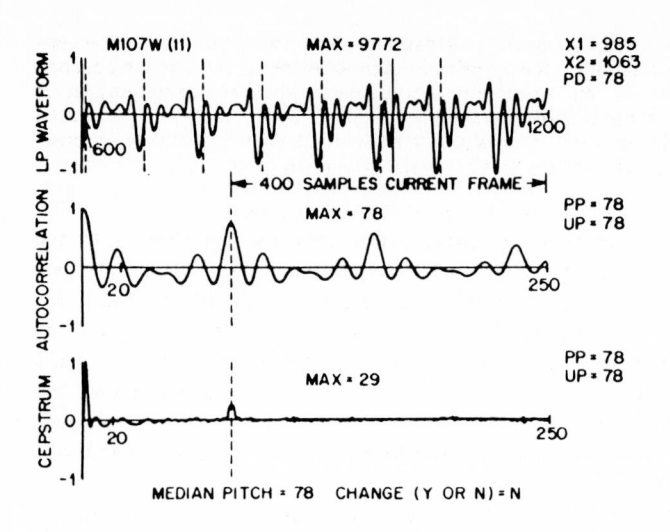

Fig.9.8. Example of the display of an interactive PDA. From (McGonegal et al., 1975, p.571)

final estimate or to label the frame unvoiced. The frame length used is 40 ms (with an additional 20-ms overlap in the display), and the frame interval is 10 ms.

To test the "performance," four users independently processed three utterances by three different speakers. The results differed minimally. Some voiced-unvoiced transitions showed an uncertainty of one frame, and the period estimates of some frames were one sample apart. Thus the contour obtained by this PDA can really serve as a standard contour for calibration of other PDAs.

The interactive PDA by McGonegal et al. (1975) displays every frame and demands approval and/or correction. Accordingly, the processing time is still considerable - 30 min for 1 s of speech. Nevertheless, this concept is extremely promising, especially for linguistic research where large amounts of data must be processed in an extremely reliable way (see Sect.9.3.2). One might think of other interactive systems working on a more automatic basis, e.g., providing a raw F_0 contour which is to be controlled and corrected by the user only at those places where the results appear questionable to the machine.

Dal Degan (1982) proposed an automatic procedure for generating a reference data base. First, a high-quality speech signal is analyzed by a high-quality PDA. The results of this analysis, after postprocessing and (optional) visual inspection, serve as a primary reference pitch contour. The data base for testing the PDAs is generated therefrom by resynthesizing the speech signal in a high-quality vocoder. This procedure has a number of advantages.

1) Since the data base consists of a synthetic signal which realizes a given pitch contour, this contour is 100% correctly represented in the data base, regardless of whether or not errors have occurred in the primary analysis. If a suitable excitation function for the signal is chosen, the testing conditions, according to Dal Degan, are almost the same as for natural speech.

The hypothesis on which the new procedure is based, is that, from the pitch extractor viewpoint, it doesn't matter whether the signal is natural or synthetic, the only importance being, in fact, that the pitch extractor works correctly on the actual signal.

However, some questions immediately arise: 1) are the results of performance evaluations, made using the artificial data base, extensible to a natural-signal data base, and 2) is it important to have a good-quality synthetic signal especially referring to its pitch contour?

The first question could be easily overcome by considering that a synthetic speech is strictly speech-like, at least as far as the analysis-synthesis system uses an adequate model of speech production.

The answer to the second question is that, at least as a first approximation, the quality of the synthetic speech is not too important; in fact the pitch analyzer under test has to extract a pitch contour which matches the reference pitch contour, even if the latter is somewhere incorrect with reference to the original speech signal. (Dal Degan, 1982)

2) By manipulating the pitch contour (shifting it upward or downward by a constant value, multiplying it by a constant factor greater or less than 1, adding noise, or inserting certain irregularities) a wide variety of conditions, "voices," and signals can be generated which would be much more difficult to obtain for natural speech.

One may object that systematic errors in the primary analysis may influence the results of the subsequent PDA performance evaluation. This is true inasmuch as those signals which lead to such errors in natural speech are no longer present in the synthetic data base. This problem, however, can be solved by visual inspection and manual or interactive correction of the primary reference contour. A second objection is perhaps somewhat less trivial. Due to the regularity of the synthetic excitation function, variations of the glottal waveform in amplitude and shape ("shimmer") are excluded from the synthetic signal (yet they may be artificially reintroduced). This problem is insignificant, however, if only the reliability of a PDA is to be tested.

9.2.3 Creating a Standard Contour by Means of a PDI

In contrast to the PDAs, which determine pitch from the speech signal only, the PDI derives this information directly from the vibrations of the vocal cords. Thus the influence of the vocal tract as well as that of environmental conditions is eliminated which is the main source of error for the PDA; using a suitable PDI completely avoids gross pitch determination errors.

From the discussion in Sect.5.2 we know that in such a (hybrid) pitch determination system (PDS) the PDI completely replaces the preprocessor and greatly simplifies the basic extractor. Nevertheless, the basic problem of pitch determination, i.e., the assignment of a period boundary to a discrete time instant, remains present. This is a nonlinear and irreversible process of data reduction. The problem of gross errors can be regarded as solved for such a system. There is, however, the problem of measurement inaccuracies which must be taken care of.

To anticipate the discussion in Sect.9.3.2, measurement inaccuracies, (i.e., "fine errors") result from three main sources: 1) natural fluctuation of the voice, 2) quantizing of pitch in digital signal processing systems, and 3) inaccurate determination of the key feature from which pitch is derived. Source 1 is closely related to the problem of gross errors. Since these errors are avoided, the system does not need an error detection or correction routine. This means that, if irregularities occur, they come from the voice excitation and do not represent failures of the PDS. As to the question of quantization, this problem is relevant only when the output signal of the PDI is analyzed by a time-domain basic extractor, i.e., period by period. Since the output signal of any PDI permits determining the instant of glottal closure, a time-domain basic extractor is applicable and desirable, and the problem of quantization must be solved. Creating a reference pitch contour, however, is never a job which requires on-line or real-time processing. Thus an increased computational effort can be tolerated, and one can increase the sampling frequency (by analog or digital means) until quantization errors due to sampling fall below the auditive difference limen. For the usual measuring range which reaches up to 500 Hz this means that the sampling frequency must be increased to 100 kHz in order to bring the quantization error below 0.5% without interpolation.

Thus the third source of inaccuracy remains: the choice of the key feature from which pitch is derived. This renders the task of accurate pitch determination a nontrivial one even if a PDI is used. Glottal closure is the most abrupt and drastic event during an excitation cycle, but even this event (or, better, its trace in the output signal of the PDI) extends over a finite amount of time. There are several reasons for that.

1) The vocal cords have finite stiffness, and there is always turbulence in the air flow. So, if the vocal cords get contact during glottal closure, they get it gradually and nonuniformly.

2) Even if the event itself happened within an infinitely short time interval, the measurement would always be band limited by (linear or nonlinear) time-variant low-pass filtering effects in the human neck tissue and in the measuring equipment.

3) Even the best PDI has a finite signal-to-noise ratio.

If the excitation is now forced to occur at a single, isolated instant in time, i.e., if a marker is set at the instant determined by the measuring system, the position of this marker will become noisy and uncertain due to interaction of all these sources of inaccuracy. Typically the variance of this "noise" is not more than 100 µs; however, this is sufficient to produce an audible noisiness in the final result.

Hence the determination of a single, isolated time instant as *the point of excitation*, as it is done in most time-domain PDAs certainly is an oversimplification. Part of the measurement inaccuracies, at least with time-domain PDAs, thus result from the failure of this definition to adequately match the facts of speech production (Flanagan and Ishizaka, 1976; Sreenivas and Rao,

1980a). Yet there is no choice but to carefully select the key feature(s) in the output signal of the PDI so that they are resistant with respect to noise and as invariant as possible with respect to slight shifts due to small intra-cycle irregularities (which are due to the *form*, but not to the *instant* of the individual glottal cycle).

Work is in progress by the present author and one of his colleagues (Inde-frey and Hess, 1983) to evaluate the output signal of a laryngograph in order to obtain a reference contour for a comparative evaluation of PDAs. The instant of maximum change of the glottal impedance during glottal closure (i.e., the instant of maximum *speed* of closure) has been (preliminarily) chosen as the key feature of the measurement. This instant, which corresponds to the maximum of the differentiated laryngograph signal, is crudely determined from the laryngograph signal after sampling (at 10 kHz) and digitizing (Fig.9.9c). In the immediate vicinity of this point the differentiated signal is then upsampled to a sampling frequency of 160 kHz using a 368th-order zero-phase digital interpolator filter. Since only few samples are needed, the computational effort for this interpolation is negligible. Differentiation is approximated using a seventh-order digital filter (Parks and McClellan, 1972).

Subjective listening tests are being carried out to verify whether this point is the most suitable anchor point in the laryngograph signal for indication of pitch, i.e., whether the pertinent marker sequence performs best when used to excite a vocoder. Other candidates in the signal are, for instance, the abrupt

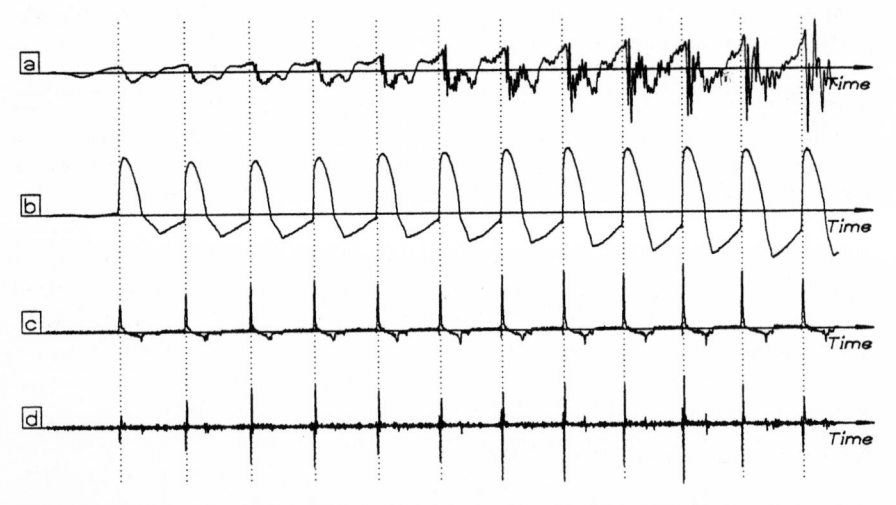

Fig.9.9a–d. Speech signal (**a**), laryngograph signal (**b**), differentiated laryngo-graph signal (**c**), and doubly differentiated laryngograph signal (**d**). The markers delimiting the individual periods were derived from the maxima of (c). Signal: vowel /a/, speaker WGH (male). Sampling frequencies in this example: 80 kHz. Time scale: 51.2 ms per line. The figure suggests that the differentiated laryngograph signal is sufficient for accurately determining the instant of glottal closure

change in the slope of the signal at the beginning of glottal closure, or the instant where the signal crosses the midline between maximum and minimum. The positive-going zero crossing as well as the principal maximum of the laryngograph signal are unreliable; they suffer from spurious low-frequency components (due to larynx movement) which have survived the built-in high-pass filter of the instrument.

The pitch estimates of short-term analysis PDAs have the advantage that they are derived from several pitch periods on the one hand, and, on the other hand, from several features that are distributed over the pitch period. This effect of "focusing the information available into one single peak," which has been mentioned several times in this book, not only increases the reliability of the PDA but also reduces the inaccuracies resulting from an inadequate choice of the key feature. When testing a short-term analysis PDA by means of a PDI, however, nothing prevents the investigator from applying a short-term analysis procedure to the PDI output signal as well. Since the PDI output signal (and especially the laryngograph signal) shows little variation in amplitude and shape, distance function methods (like the AMDF, see Sect.8.3.1) are especially well suited for this task. Using a good strategy the computational effort for providing the spectral-domain resolution necessary to force quantizing errors below the auditory difference limen can be kept low.

The experiments by the present author have shown so far that manual interaction during the generation of the reference contour can be exclusively confined to the voiced-unvoiced transitions (where the PDI is nevertheless very reliable) and to signals with irregular excitation (e.g., vocal fry). Irregular signals, however, are not well suited for testing a PDA, at least not until its performance with regular signals is well known.

In summary, the PDI in general, and the electric PDI in particular, provides a very valuable aid for creating reference pitch contours which can then be used to calibrate a PDA or to evaluate its performance. Although the task of pitch determination using a PDI is nontrivial when high accuracy is required, it is much easier compared to the task of pitch determination from the speech signal alone. Since most methods (especially short-term analysis methods) that work for PDAs can also be applied to the PDI output signal, reliable and good results can be expected. Manual interaction is only occasionally required; a great amount of high-quality data can thus be obtained within a short time.

9.3 Performance Evaluation of PDAs

A number of surveys and critical reviews on PDAs have been published (Halsey and Swaffield, 1948; Meyer-Eppler, 1948; Kriger, 1959; Ungeheuer, 1963; McKinney, 1965; Schroeder, 1966 and 1970d; Léon and Martin, 1969; Erb, 1972b; Schultz-Coulon and Arndt, 1972; Baronin and Kushtuev, 1974; Rabiner et al., 1976a,b; Schröder, 1976; McGonegal et al., 1977; El-Mallawany,

1977c; and others). The majority of these surveys review some or many of the known PDAs with respect to special applications, such as vocoders (Halsey and Swaffield, 1948; Kriger, 1959; Schroeder, 1966; Baronin and Kushtuev, 1974), parameter extraction (Schroeder, 1970d), phonetic and linguistic research (Meyer-Eppler, 1948; Ungeheuer, 1963; McKinney; 1965, Léon and Martin, 1969), and phoniatric problems (Schultz-Coulon and Arndt, 1972). Frequently, these investigations are carried out in order to demonstrate that the implementation of a novel PDA is justified (Halsey and Swaffield, 1948; Kriger, 1959; Léon and Martin, 1970; Erb, 1972; and, with respect to Chap.7, this book also). Except for the reports by Kriger (1959), McKinney (1965), and the NLF evaluation by Risberg et al. (1960), not much practical work has been done to comparatively evaluate the performance of different PDAs. Rabiner et al. (1976) thus complain:

> Because of the importance of pitch detection, a wide variety of algorithms for pitch detection have been proposed in the speech processing literature. ... In spite of the proliferation of pitch detectors, very little formal evaluation and comparison among the different types of pitch detectors has been attempted. There are a wide variety of reasons why such an evaluation has not been made. Among these are the difficulty in selection of a reasonable standard of comparison; collection of a comprehensive data base; choice of pitch detectors for evaluation; and the difficulty of interpreting the results in a meaningful and unbiased way. (Rabiner et al., 1976, p.395)

This section consists of two parts. First, a literature survey of the evaluative work on PDAs will be presented (Sect.9.3.1). After that, the attempt is made to systematically categorize the various errors which occur in pitch determination, and to outline the possibilities for future work in this area.

9.3.1 Comparative Performance Evaluation of PDAs: Some Examples from the Literature

The best known and most comprehensive performance evaluation of PDAs was carried out by Cheng, McGonegal, Rabiner, and Rosenberg at Bell Laboratories. This work was subdivided into an objective study (Rabiner et al., 1976) and a subjective study (McGonegal et al., 1977). Without proposing a new concept, the authors investigated seven PDAs: modified autocorrelation (category PiCAN; Dubnowski et al., 1976; discussed in Sect.8.2.3), cepstrum (PiFM; Noll, 1964, 1967; Sect.8.4), SIFT (PiCAL; Markel, 1972b, Sect.8.2.5), parallel processing (PiTS; Gold and Rabiner, 1969; Sect.6.2.4), data reduction (PiTS; N.Miller, 1974, 1975; Sect.6.2.3), a spectral-flattening LPC method (PiCD; Atal, 1975, unpublished; Sect.8.3.3), and the AMDF method (PiCD; Ross et al., 1974; Sect.8.3.1). Evidently no representative of the categories PiTL and PiTN was among the tested PDAs. As to the scope of their evaluation,

> ... several arbitrary (but hopefully reasonable) decisions were made to limit the scope of the performance evaluation to a reasonable size. ... There are many characteristics of pitch detection algorithms which influence the choice of a set of performance criteria. Among these factors are the following:

1) accuracy in estimating pitch period;
2) accuracy in making a voiced-unvoiced decision;
3) robustness of the measurement - i.e., do they have to be modified for different transmission conditions, speakers, etc.;
4) speed of operation;
5) complexity of the algorithm;
6) suitability for hardware implementation;
7) cost of hardware implementation.

Depending on the specific application, various weights must be given to each of the above factors in choosing a single objective performance criterion. (Rabiner et al., 1976a, pp.400-401)

The authors confined their quantitative investigations to points 1 and 2 of these factors. As a data base they chose a combination of 4 monosyllabic nonsense words and 4 sentences, each spoken by seven speakers (three male, among them one low-pitch whose fundamental frequency managed to drop even below 50 Hz, two female, one four-year old child, and one speaker suffering from diplophonia, a voice disease where the voice strongly tends towards paired pulsing) and recorded under different conditions (telephone line, close-talking microphone, and wide-band recording). A standard T_0 contour (10-ms frame interval) was obtained from this data base by means of the interactive PDA (McGonegal et al., 1975). Then the individual PDAs were evaluated as to the occurrence of four characteristic errors: *voiced-to-unvoiced* error (the PDA classifies a frame as unvoiced which was found voiced in the standard contour), *unvoiced-to-voiced* error (correspondingly), *gross pitch estimation* error (deviation of more than 1 ms between individual estimate and standard contour when both indicate voiced speech), and *fine pitch estimation* error (deviation less than 1 ms).

For the fine errors, the mean value and the standard deviation were calculated. The mean value was found to be almost zero so that the fine errors were not biased. All the other errors were counted and averaged over the utterances of one speaker. Each error was determined from the raw contour and from a median-smoothed version (Rabiner et al., 1975; see Sect.6.6). Usually smoothing substantially improved the performance provided that the error rate was not too high. For more details of the statistical evaluation, the reader is referred to the original publication. About the gross errors, Rabiner et al. wrote:[9]

No single pitch detector was uniformly top ranked across all speakers, recording conditions, and error measurements. ... The results on the gross pitch period errors ... showed that the time-domain and the hybrid pitch detectors had greatest difficulty with the low-pitched speakers whereas the spectral pitch detector (cepstrum) had the greatest difficulty with the high-pitched speakers. The difficulty of time-domain methods for low-pitch speakers are due to the fixed 30-40 ms analysis frame which is generally inadequate for low-pitch speakers. The difficulty of spectral methods for

---

[9] In the study by Rabiner et al. (1976), the grouping of PDAs into categories differs from that used in this book. See Sect.9.1.1 for a discussion.

> high-pitched speakers is due to the small number of harmonics which are
> present in their spectra, leading to analysis difficulties in choosing the
> correct pitch. The poor performance of the SIFT pitch detector on speaker
> C1 [the child] is related to the problem of reliably spectrally flattening
> (by inverse filtering) a signal in which generally only one harmonic occurs.
> (Rabiner et al., 1976a, pp.412-413)

The fine errors were found to be somewhat higher for the time-domain
methods (time domain according to the definition used in this book) than for
the short-term analysis methods. As to the computer time, the fastest short-
term analysis PDA (AMDF, due to a reduced sampling rate in the spectral
domain) ran about 8 times more slowly than the two time-domain PDAs,
whereas the cepstrum PDA, as the slowest, was slower than AMDF by another
factor of 8. It remains unspecified to what extent input-output operations
such as disk read-write influenced the computing speed. The autocorrelation
PDA (Dubnowski et al., 1976; Rabiner, 1977a) which uses the combined
center- and peak-clipped signal and therefore did not need multiplications, was
faster than the SIFT PDA by a factor of 2; the latter needs multiplications
but operates at 1/5 of the sampling rate of the original signal.

A sequel to the objective study was the subjective evaluation by McGonegal
et al. (1977).

> One very basic question arose from this earlier investigation. This is the
> question as to how, and in what manner, the results of the error analysis
> used in the objective evaluation ... are related to perceptual criteria of
> quality in a subjective evaluation. ... Such a subjective evaluation of pitch
> detectors can be obtained by assessing the quality of speech synthesized
> using pitch contours obtained from each of the pitch detectors. Since only
> the pitch contour is being varied, higher subjective quality of a synthetic
> utterance reflects a *better*, or more accurate (in some perceptual sense)
> contour. ...
>   It should be emphasized that the results of a subjective evaluation of
> pitch detectors are applicable primarily to speech analysis-synthesis
> (vocoder) systems. That is, a poor performance in this evaluation does *not*
> preclude using the pitch detector for other applications. (McGonegal et
> al., 1977, p.221)

The investigation was carried out with a subset of the data base used in
the objective evaluation (the four monosyllabic words as well as the diplophonic speaker were excluded). The evaluation was subdivided into two tests:
1) ranking of the seven PDAs plus standard contour (interactive PDA) plus
natural speech using the unsmoothed contours, and 2) examining the influence
of smoothing.

To obtain the data base, the speech was analyzed by a 12-pole LPC
analysis (autocorrelation method, frame interval 10 ms, frame length 30 ms)
and resynthesized using the T_0 contours under investigation. The eight listeners participating in the tests were asked to rank the nine different contours
of a given utterance (recording condition, sentence, and speaker remained unchanged for an individual test run) from low quality (1) to high quality (9)
according to preference. The results, as given in the paper, are displayed in
Figure 9.10. Briefly summarized, the main results are as follows.

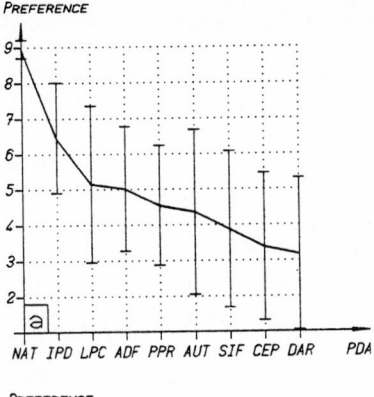

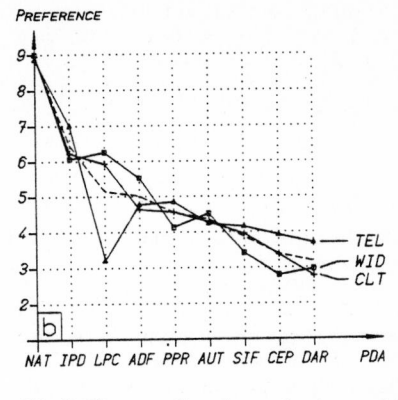

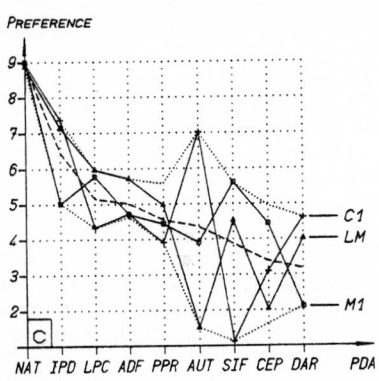

Fig.9.10a–c. Results of the subjective PDA evaluation (McGonegal et al., 1977). (**a**) Mean score of preference and standard deviation; (**b**) performance score depending on the recording conditions; (**c**) effect of speaker (range of pitch) on the mean performance. Three of six speakers are displayed. The dotted lines display the best and worst scores for all the seven speakers. Algorithms: (NAT) natural speech; (IPD) interactive PDA (McGonegal et al., 1975); (LPC) matching of LPC spectra (Atal, 1975, unpublished); (ADF) average magnitude difference function PDA (Ross et al., 1974); (PPR) parallel processing time-domain PDA (Gold and Rabiner, 1969); (AUT) autocorrelation PDA (Dubnowski et al., 1976); (SIF) simplified inverse filtering (Markel, 1972b); (CEP) cepstrum PDA (Noll, 1967); (DAR) time-domain PDA (N.Miller, 1974, 1975). Recording conditions: (TEL) telephone speech; (WID) wide-band recording (0 to 5 kHz); (CLT) close-talking microphone. Speakers: (LM) low-pitched male voice; (M1) male voice; (C1) child's voice

1) The only unanimous decision was that natural speech was ranked best.

2) On the average, the distance in ranking between the standard contour (which was ranked second highest, but not unanimously) and natural speech exceeded the distance between the standard contour and that obtained by the best PDA.

3) The standard deviations of the preference scores exceeded the difference between the rankings except for natural speech. In particular, the mean ranking of the individual PDAs differed much less than the value of their standard deviation. That is, the mean indication of preference still strongly depends on conditions other than the choice of the PDA.

4) The judgment among the listeners was relatively uniform. In the same way, there was little dependence on a particular sentence. The dependence on the recording condition was moderate except for the LPC PDA operat-

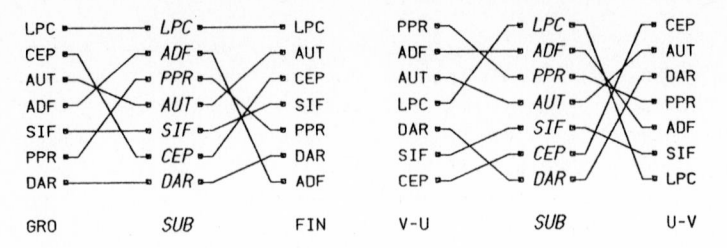

Fig.9.11. Perceptive evaluation of PDA performance (McGonegal et al., 1977): series of comparisons between overall subjective and partial objective ranking of the seven PDAs based on each of the proposed error measures. The PDAs are labeled as in Figure 9.10. (SUB) Subjective ranking; (GRO) objective ranking according to gross pitch determination errors; (FIN) objective ranking according to the standard deviation of fine errors; (V-U) objective ranking according to the number of voiced-to-unvoiced errors; (U-V) objective ranking according to the number of unvoiced-to-voiced errors

ing on telephone speech (because of an excessive number of VUV errors for this constellation). There was a strong dependence on the speaker, i.e., on the individual range of F_0.

> Three of the pitch detectors (autocorrelation, cepstrum, and SIFT) were strongly affected by the pitch of the speaker. The AC method performed worst on low-pitched speakers and best on high-pitched speakers. The SIFT method performed worst on high-pitched speakers and best on low-pitched speakers. The cepstrum method performed worst on some speakers whose pitch was either high or low ... (McGonegal et al., 1977, pp.224-225)

5) For none of the types of errors was the (trial) objective ranking (Fig.9.11) in agreement with the overall subjective ranking of the PDAs. The highest degree of correlation between subjective and objective evaluation is given for the voiced-to-unvoiced discrimination errors, whereas unvoiced-to-voiced errors did not greatly affect the subjective judgment.

Obviously the subjective ranking criteria are different from the objective ones, and many errors which are noticeable and significant in an objective sense tend to be perceptually insignificant, and vice versa.

The second part of this investigation dealt with the influence of smoothers. Here, the general trend showed a strong preference for the smoothed contour when the overall ranking of an individual PDA was low, thus proving the effectiveness of the combined median and linear smoother.

Quantitative data on the performance have been provided by several designers of PDAs (Tucker and Bates, 1978; Friedman, 1979; Somin, 1972). In most of these cases the new PDAs have also been tested with band-limited and/or noisy signals. The results show that band limitation to telephone channel bandwidth degrades the performance of the algorithms more severely than additive white noise. The error rates given in these evaluations are usually higher than those given by Rabiner et al. (1976).

Dal Degan (1982), using synthetic speech (see the discussion in Sect.9.2.2), investigated the influence of various preprocessing conditions (use of different low-pass filters and/or a crude inverse filter; different frame lengths) on the performance of an AMDF PDA (Sect.8.3.1). He found that low-pass filtering before computing the AMDF degrades the performance,[10] and that linear inverse filtering, even if done very crudely (for instance with a second-order filter which cancels just one formant) has a favorable effect.

Based on experimental results by Zurcher (1977), El Mallawany (1977c) evaluated the performance of five PDAs. Four of them pertained to the principle PiTE and were variations and modifications of Zurcher's PDD (Zurcher, 1972, 1977; Zurcher, Cartier, and Boe, 1975; see Sect.6.2.1). For comparison Markel's SIFT algorithm (1972b; PiCA, Sect.8.2.5) was also included in this test. The speech material consisted of five sentences spoken by six speakers (three of them female); the signal was presented 1) without distortions, 2) linearly distorted in order to approximate the frequency response of a carbon microphone, and 3) high-pass filtered at 300 Hz. El Mallawany only considered gross pitch determination errors. The results show that the error rates of all PDAs (including the SIFT PDA) mount by as much as an order of magnitude when the low frequencies are removed from the signal. *"It seems difficult,"* so El Mallawany (1977c, p.30), *"to significantly improve the performance of these detectors without considerably increasing the computing effort and the cost of the device."* The results also show the systematic failure of the SIFT algorithm for high-pitched female voices.

The most interesting result, however, is that El Mallawany - again in cooperation with Zurcher (1977) - quantitatively verified that the bad starting conditions are a main source of errors. It is well known that at the beginning of a voiced segment the PDA works under the worst conditions: no estimate from previous frames is avaliable, and the signal is extremely nonstationary due to rapid amplitude and formant changes. To investigate this effect, El Mallawany distinguished between errors at the beginning of a voiced segment, errors at the end of a voiced segment, and the remaining errors in between. Except for the failure of the SIFT algorithm with female voices (which occurred everywhere), the errors were consistently more numerous in the initial part of voiced segments. For the band-limited signal this discrepancy reached a maximum (about 10% errors at the beginning of voicing, but only 2.4% within the voiced part and less than 1% toward the end of voicing). This investigation only comprised pitch determination errors; no voicing determination errors have been taken into account.

From this investigation it follows that the performance of a PDA can be significantly improved if the starting conditions can be made better. How long the PDA really needs in order to enter a stable operational mode has not yet been determined, and further work is warranted in this respect.

---

[10] This result confirms the findings by Friedman (1978), Terhardt (1979b), and the present author (Sect.7.4.3, Appendix A) that high-frequency information above 2 kHz is significant for pitch detection.

9.3.2 Methods of Error Analysis

In the study by Rabiner et al., the errors of pitch determination were classified into a) gross pitch determination errors; b) fine pitch determination
errors; c) voiced-to-unvoiced errors, and d) unvoiced-to-voiced errors. If we
group errors c) and d) together, we obtain the following three classes of
errors.

1) *Voicing determination errors.* In principle, these errors indicate failures
of the voicing detection algorithm and not of the pitch detection algorithm.

2) *Gross pitch determination errors.* With these errors the question of
reliability of a PDA is raised.

3) *Fine pitch determination errors* which we could also label *measurement
inaccuracies* or simply *jitter.*

In a subjective evaluation where the listener cannot bias his judgment towards
the one or other class of errors, he is forced to judge the global quality of
the pitch contour. As McGonegal et al. (1977) found, there is no simple
relation between an individual error or a class of errors and the subjective
quality when all the errors occur simultaneously. In order to get some idea
how the individual error affects speech quality, we must treat these errors
separately. In the following, the error categories will be discussed in more
detail.

Voicing Determination Errors. These errors were found to correlate highly
with the subjective performance, i.e., a certain class of voicing determination
errors seems to be relatively annoying. In this book, however, the question of
voicing determination has always been kept apart from the question of pitch
determination. Voicing determination is a premise for pitch determination;
pitch can exist only at those instants where a voiced excitation is present
(this statement should not be reversed!). Voicing determination errors, however, are able to turn the attention of the listener away from the very
question of performance with respect to pitch. The best example for this is
supplied by Rabiner et al. (1976, 1977) themselves. One of the PDAs investigated, the one by Atal (1975; unpublished), reached the best ranking (subjectively and objectively) with respect to (gross and fine) pitch determination
errors. Since the voiced-unvoiced discrimination of this algorithm failed systematically for certain recording conditions, the PDA was subjectively ranked
lower (on the average) because of these voicing determination errors (although
it was still ranked best). Just in this algorithm, however, voicing detection
and pitch detection are done by completely separate routines. In the opinion
of the present author, the voicing determination errors should thus be excluded
from a comparative performance evaluation of PDAs (and vice versa), i.e.,
one should only evaluate PDAs using signals with optimal voicing contours,
and VDAs using optimal pitch contours. (Of course, things may be different if
the overall performance of vocoder systems is to be comparatively evaluated.)

The question of evaluating an arbitrary PDA using optimal voicing contours requires that a reliable and accurate voicing detection procedure be applied while the reference contour is generated. In particular, the reference contour must exactly specify the instant where voicing starts and ends. As we have seen, even highly accurate time-domain PDAs are not able to carry out this job without manual interaction. The most suitable device for setting up the reference contour will thus be a PDI which directly measures the opening and closure of the glottis, such as the laryngograph (Sects.5.2, 9.2.3).

Gross Pitch Determination Errors are "drastic failures of a particular method or algorithm to determine pitch" (Rabiner et al., 1976). There is a tendency in the literature to count every parameter sample or marker as grossly in error that deviates from the correct value by more than 1 ms (Rabiner et al., 1976; Tucker and Bates, 1978). The error analysis used by Reddy (1967) for his correction scheme goes into somewhat further detail (cf. Sects.6.2.2, 6.6.3). If we validate his specification, which was made for a time-domain PDA (PITS), also for short-term analysis PDAs, then we arrive at the following error scheme.

1) *Higher-Harmonic Errors.* During one or several periods (or frames), a higher harmonic (preferably the second or third) is taken as the estimate for pitch. This error occurs preferably when a strong first formant is situated at a low-order higher harmonic. In time-domain PDAs this leads to extra markers (or chirps), whereas in short-term analysis PDAs a locally bounded wrong estimate may occur (or no estimate at all if there are conflicting conditions within a single frame).

2) *Subharmonic Errors.* In this case the PDA misses one or two periods and gives a multiple of the period or a fraction of the fundamental frequency as the estimate for pitch. The possible causes for this error are quite complex: rapid amplitude changes, especially in connection with the sound /r/; a slight tendency towards diplophonic voicing (paired pulsing) which even occurs frequently in normal voices; temporary vocal fry; too long frames in connection with short periods, crude quantization of the estimate due to a low sampling rate, and so on. In time-domain PDAs these errors cause missing markers or *holes,* in short-term analysis PDAs again we have a *"runaway,"* i.e., a parameter sample far away from its neighbors.

3) *Nonharmonic Pitch Determination Errors.* These errors result from systematic or nonsystematic breakdowns of the algorithm. Systematic failures of this kind are mostly specific for a particular algorithm. For instance, a PDA which uses inverse filtering tends to fail whenever F1 coincides with F_0. In time-domain PDAs, especially in those of category PiTS where the criterion for marker determination is a maximum or minimum, during rapid formant transitions this criterion quite frequently changes its place abruptly within the period. A *hop* in the marker sequence, as the consequence, may or may not be easily correctable. In lag-based short-term analysis PDAs, a hop may result from taking $T_0 \pm T1$ (where T1 is the period of the formant F1) as the estimate instead of T_0 (Schroeder, 1966).

4) *Correct (or Incorrect) Indication of Irregular Excitation Signals.* From the point of view of the researcher, it may be strange to group this situation among the pitch determination errors. From the "point of view" of the algorithm, however, this is justified since it is not trivial at all for the machine to recognize that the *signal* has caused an irregularity, and not that the routine is ill conditioned. In addition, even some of the definitions of T_0, such as (9.4), do not hold any more when the excitation signal is grossly irregular. We will come back to this point later.

Although gross errors are in general easily correctable (Hettwer et al., 1980), one should, for a number of reasons, not simply leave the issue to the postprocessor, but rather try to suppress at least the predictable systematic failures in the preprocessor and in the basic extractor.

a) A correction routine always needs a certain look ahead in order to correct erroneous parameter estimates at the beginning of a voiced segment. Even if an error is correctable, the correction will always cause a substantial processing delay. If we assume two consecutive pitch estimates to be needed as a minimum in order to substitute an erroneous estimate with the right value (this also holds for median smoothing), then a delay of up to 40 ms will result from the correction alone. This delay is already perceptible when it occurs in a vocoder.

b) An error rate of 5% for gross errors seems usual for moderate signal quality (El Mallawany, 1977c; Tucker and Bates, 1978; Hettwer et al., 1980). If we assume for the moment that the individual errors occur independently of each other, then about every 200th parameter sample (i.e., every 2 s) two errors will immediately follow each other; such error clusters, however, are much more difficult to correct than a single error. Reality is less favorable; the gross errors are correlated with the signal, and they occur preferably at the beginning of voiced segments (El Mallawany, 1977c) and during rapid formant transitions during the voiced segments.

c) "Correction" of a gross error can give wrong results when the signal is the source of the error (and not the algorithm).

From this discussion we see that we have to distinguish between three causes of gross pitch determination errors and errorlike conditions.

1) *Errors That Arise From Adverse Signal Conditions:* strong first formants at low harmonics, rapid amplitude and formant transitions, distortions and absence of the first harmonic. Such conditions arise from a) the vocal tract; b) the fact that the signal is time-variant; and c) the recording and transmission environment. To overcome such errors, good algorithms are needed; however, such algorithms are already available to a great extent.

2) *Errors That Arise From Insufficient Algorithm Performance.* Typical such conditions are: the mismatch of fundamental frequency and frame length in short-term analysis PDAs, the coincidence of F1 and F_0 in inverse-filtering PDAs, or the inability of the epoch detection PDA to cope with soft voices. These errors are very different for different algorithms, and they are usually predictable. Hence a combination of several principles will be right for them.

3) *"Errors" That Arise From the Excitation Signal*: irregularity of the excitation pulses in frequency and shape. This condition can arise at any time in normal speech. It occurs regularly when the voice falls into the vocal-fry register, and vocal fry is often associated with reduced syllables unless the speaker articulates in a very conscious and well-controlled manner. In this situation a great number of PDAs can only fail. Any PDA based on definition (9.4) of T_0 as a short-term average can no longer give reasonable results since T_0 simply becomes undefined with these signals. Definitions (9.1,2) remain valid; we still have discrete excitation pulses although sometimes there are secondary excitation pulses within a period, and it becomes difficult to discriminate between the genuine laryngeal pulses and the parasitory pulses which should be suppressed (see Fig.3.8 for an example of vocal fry). It may be questioned whether (9.6) still holds. Pitch perception definitely exists during irregular signals as well, but there may be a long way from the individual excitation pulses to the output signal of a frequency-domain PDA (and back to the pertinent excitation signal of the generator in the synthesizer part of a vocoder), and the results may be extremely different. A number of PDAs tend to enforce "regularity" even in irregular voiced segments. This is still better than labeling a frame *unvoiced* when a pitch estimate cannot be measured (and thus illegally reversing the statement that pitch can exist only where voicing exists!).

Whereas error sources 1) and 2) seem surmountable, the performance of the PDA when the signal is irregular is still an unsolved problem.

Measurement inaccuracies cause a noisiness of the pitch estimate obtained. These *fine errors* are small deviations; nevertheless, they are well audible and may be extremely annoying to the listener (Fellbaum, 1981, personal communication). In the literature an error threshold separates gross and fine pitch determination errors. Typical values for this threshold are a few percent [Reddy (1967): 12.5%; Tucker and Bates (1978): 10%; Friedman (1979): 25%] or a few samples [Rabiner et al. (1976): 10 samples (equalling 1 ms); Dal Degan (1982): 1 ms]. Every deviation of the measured estimate with respect to the reference contour is regarded as a gross error as soon as it exceeds this threshold.

Just as in the evaluation of gross errors, the separation of the noisiness induced by the voice and that induced by the algorithm renders the task of error analysis and correction difficult. The following three facts (seen from the point of view of the algorithm) act as the main sources of measurement inaccuracies.

1) *Natural Fluctuations of the Voice* from one excitation pulse to the next. Speech is a biological signal; it does not have the regularity of signals generated by a machine. If the inaccuracies caused by little voice perturbations survive the measurement procedure, they contribute to the "naturalness" of the synthetic voice and should thus be preserved.

2) *Quantizing of Pitch* due to sampling of signals and spectra. This error is intrinsic to digital systems. As shown in Chap.4, the error is clearly audible in a vocoder when excitation pulses are permitted only at integer sample addresses.[11] Allowing for higher accuracy, however, either requires higher sampling frequencies or the possibility of interpolation. Higher sampling frequencies are usually not justifiable due to the augmented computational load (remember that, in order to render the error inaudible for a female voice at $F_0=500$ Hz, the sampling frequency ought to be increased to 100 kHz!). Hence some interpolation routine must be provided. At this point the methods differ widely. For a time-domain PDA interpolation is principally possible, but has hardly been attempted.[12] Interpolation is also possible (with slightly increased computing effort) for short-term analysis PDAs with a lag-based spectral function (autocorrelation, AMDF, cepstrum, maximum likelihood). The frequency-domain PDAs, however, are the most suitable ones for accuracy improvement by interpolation. This perhaps explains the enormous success of frequency-domain PDAs in the last few years. Measurement accuracy is increased in a frequency-domain PDA by a factor equal to the harmonic number of a higher harmonic when just this harmonic is taken as the reference from which the estimate of F_0 is derived. Further improvement is possible by exploiting several harmonics. Hence frequency-domain PDAs practically eliminate this source of errors. The same holds for Atal's PDA (see Sect.8.3.3) which constructs the input signal of the basic extractor from individual oscillations whose frequencies are exactly known, and which can thus be generated with arbitrary fine sampling.

3) *Inaccurate Determination of the Key Feature* in the signal or the short-term spectral function. These are the specific measurement errors which should be minimized or eliminated. Errors of this type frequently occur in time-domain PDAs (especially in category PiTS where the high frequencies remain in the signal); in short-term analysis PDAs they arise from a mismatch between T_0 and the frame length. Frequency-domain PDAs again practically do not suffer from this error.

In conclusion, measurement inaccuracies are most annoying in time-domain PDAs, tolerable in short-term analysis PDAs with a lag-based spectral function, and practically not present in frequency-domain PDAs. The fact that the frequency-domain PDAs eliminate the problem of measurement inaccuracies

---

[11] This applies to high-quality vocoders. If the transmission rate of a vocoder is low, pitch information has to be quantized more crudely in order to save bits. For more details, see Section 9.4.4.

[12] The most convenient way to perform an interpolation in a time-domain PDA is the following: 1) determination of a pitch marker at the usual sampling rate, 2) interpolation (i.e., upsampling) of the signal within a narrow interval around this marker, and 3) accurate determination of the marker from the upsampled signal (Indefrey and Hess, 1983; Moser and Kittel, 1977). See Sects.9.2.3 and 9.4.3 for further discussion.

is likely to explain the superior subjective quality of these PDAs which is generally reported by their designers (Martin, 1981; Sreenivas, 1981; Sluyter et al., 1980; Terhardt, 1979b; Paliwal and Rao, 1983). Time-domain PDAs lose at this point. Other short-term analysis PDAs must provide for a suitable inter-polation scheme.

As this discussion shows, there is still a lot of work to be done in the area of performance evaluation of PDAs. The study by Rabiner et al. (see Sect.9.3.1) marked a milestone since it showed for the first time 1) that unbiased error analysis on the basis of a comparative performance evaluation is possible, and 2) how necessary this work really is. Rabiner et al. performed the error analysis from a global point of view. To answer some of the questions their study raised, the signal must be viewed more locally. In order to fight against a particular error, one must first know where it occurs - with respect to the algorithm, the signal, the state of the voice, and environmental conditions - and why it occurs at this place. A detailed investigation of the local performance of PDAs must still be done. It could give us practical hints as to where the problems of PDAs are that can still be solved by better algorithms and, above all, by a better understanding of the speech excitation process at a very detailed level, i.e., within the individual pitch period.

9.4 A Closer Look at the Applications

From the various discussions throughout this book we know the four main application areas of the PDAs, i.e., speech communication, phonetics and linguistics, education, and pathology. In this section a list is compiled of the claims and requirements from the point of view of the respective application. First of all, the provocative question *"Has the problem been solved?"* is used to define *"what the problem is"* and to compile the list of requirements (Sect.9.4.1), which is then discussed in detail in the remainder of this section.

9.4.1 Has the Problem Been Solved?

To anticipate the answer to this most elementary and most general question: definitely not. We do not have a single PDA which operates reliably (and accurately) for all applications, all signals, all speakers, all recording conditions, and all possible quality degradations. We even do not have a single PDA which operates perfectly well for any speaker and a number of well-defined recording conditions (McGonegal et al., 1977). A look at the first or last paragraphs of recent publications on PDAs shows that we are still far from a general solution.

The search for a solution to the problem of obtaining a practical as well as an adequate pitch measurement is as old as the vocoder itself. It has often been termed the "hardy perennial" problem of the vocoder. Useful solutions have, of course, been obtained, particularly when the original speech is of high quality and relatively free of noise. Even here the wide range of possible pitch frequencies (about three octaves) has presented serious problems.

When one attempts to utilize speech waves with noise present (particularly low-frequency noise) or with the fundamental component missing, as often occurs in telephone systems, the difficulty of the problem is greatly increased. (R.L.Miller, 1970, p.1593)

Arguments can be advanced to show that estimation of the voice fundamental frequency is an equivalently difficult problem. For example, this parameter can vary from about 50 to 600 Hz; more than an order of magnitude. Phase shifts in a telephone channel, loss of the fundamental frequency and background noise all create difficulties. Sudden changes in the sound pattern such as the transition from vowel to nasal can result in sudden volume and pitch changes in the speech signals, thus adding to the difficulty of successful decision making. (Gold, 1977, p.1644)

Man easily determines pitch from listening to speech as well as from visual inspection of oscillograms or spectrograms. In contrast to this it is rather laborious at the moment to develop a machine that automatically determines fundamental frequency with low error rate and low processing delay even if the level of noise and spurious components in the signal is low. (Baronin and Kushtuev, 1974, p.113; original in Russian, see Appendix B).

Although pitch extraction has been the subject of a great deal of research since Dudley invented the first vocoder, it is still regarded by many researchers as one of the unsolved problems in vocoder research. A pitch extraction algorithm that is completely reliable and is simple enough for hardware implementation remains to be found. (Un and Yang, 1977, p.565)

A major result is the difference in quality between the synthetic speech using the SAPD pitch detector [the interactive PDA developed by McGonegal et al. (1975)] and that using any other real pitch detector. Again this quality difference was highly nonuniform. However, in general, the real pitch detectors did not approach the quality of the semiautomatic analysis method. Thus, further work on pitch detection methods is warranted by this result. (McGonegal et al., 1977, p.226)

On the other hand, is such a general solution which copes with *everything* really the final goal of research? Is it necessary? It certainly would be necessary if there were only one major application. But this is not the case. As already discussed in Chaps.1 and 4, there are the four main fields of applications which differ widely: 1) *speech communication* (speech analysis, synthesis, and transmission; vocoders; recognition of speech and speakers); 2) *phonetics* and *linguistics* (intonation, tone, microprosody, acquisition of reference data for synthesis-by-rule systems, tone languages, etc.); 3) *education* (instruction aids to the handicapped, language acquisition by foreigners); and 4) *medicine* and *pathology* (phoniatric and psychologic investigations, such as voice disease, voice diagnostics, investigation of emotional state or psychologic stress situations). These four applications deal with speech signals. If other signals, such as musical instrument sounds, are added, there is one more

Table 9.2. Profile of requirements for PDAs depending on the respective application. Applications: (SPC) Speech communication; (PHO) phonetics, linguistics (including musicology); (EDU) education; (PAT) pathology, medicine. Ranking: (1) mandatory; (2) very important; (3) important; (4) marginal; (5) insignificant. The question of a manual range or signal type preset is considered important if the device must *not* have such a preset! Where several rankings are given, the importance depends on the special task to be performed with the PDA. For more details, see the text in Sects.9.4.2-4

	SPC	PHO	EDU	PAT
Reliability	1	1-4	5	3
Accuracy	2-5	1-4	5	5-1
Quality and facility of display	5	1	1	1
Range of fundamental frequency	2	4-1	4	2
Robustness against signal degradation	1	3	5	2
Question of manual preset (F_0 range, signal type)	1	3	5	5
Cost of implementation	3	4	1	3
Real-time (on-line) performance	1-2	5-1	1	5
Ease of operation	5	2	2	3

application to be encountered: 5) *musicology*[13] (analysis and automatic transcription of musical melodies, such as instrument tunes or folk songs).

It appears appropriate to raise the question for each of the particular areas of applications separately. Then one shall see whether the respective applications are so far from each other that they justify an approach of their own.

To obtain an overview, the primary requirements to a PDA according to the various applications are listed in Table 9.2; at the same time an attempt is made to assign a degree of importance to them.

Apparently each application has its own ranking of requirements which make the "ideal" PDAs for the respective applications look very different. In most application areas, surveys on PDAs have been published. To get on with the discussion, some of the conclusions of these surveys will be cited in the following.

9.4.2 Application in Phonetics, Linguistics, and Musicology

The most comprehensive surveys have come from the field of phonetic and linguistic research. The report by McKinney (1965) which mostly deals with time-domain PDAs has already been cited extensively. For this particular application we will thus present the view by Léon and Martin (1969).

---

[13] It seems justified to group the applications in phonetics/linguistics and in musicology together since their profiles of requirements are very similar, perhaps except for the need of an extended F_0 range for musical signals and the way the results are displayed.

While more than a hundred different pitch analyzers have been devised world-wide, phoneticians have really only used two of them, that is, oscillographic analysis and spectrographic analysis - no doubt because they are by and large the most reliable (but also the slowest).

The analysis procedure, particularly when it uses filters, often intoduces delays, phase distortions and various other kinds of distortion whose behaviour is difficult to reduce to rule, particularly as the speech wave is mathematically intractable.

An even more delicate point than the analysis itself, whether spectral (Fourier transform) or periodic (autocorrelation), would seem to be the extraction of the fundamental frequency or the fundamental pitch period. The reduction in the information rate brought about by the analysis often gives rise to serious problems, particularly when the signal is rather noisy or when it is rather aperiodic (rapid formant transitions, sudden change of pitch).

Even if the decision strategy that is applied to all the information in the spectrum - and that consists in the calculation of the highest common denominator of the harmonic frequencies or of the lowest common multiple of the periods of the harmonics - seems in principle resistant to the presence of even large amounts of noise over part of the spectrum, errors can nonetheless occur even when very elaborate procedures such as the cepstrum are employed. This is probably due to the fact that the formant structure (spectral envelope) and the fundamental (spectral fine structure) are inseparable and interact with one another. It is very probably necessary to develop a recognition strategy along pattern recognition lines, which would be applied to the output of the spectral or periodic analysis and would take the future of the analyzed signal into account.

In addition to the extraction problem there remains the problem of measurement and the accuracy of the measurement of the fundamental, the ultimate aim being the best approximation to the frequency of the glottal vibrations. The problem is particularly acute in the most popular methods: "manual" analysis of the oscillogram, which requires enormous (and expensive) paper speeds, and inspection of spectrograms, their accuracy being restricted by the use of filters.

The solutions chosen must obviously be appropriate to the desired goals and the chosen function. While spectrographic analysis may sometimes be acceptable for some tasks, more accurate systems which work in real time are an absolute necessity for applications such as the vocoder and are desirable when large quantities of data have to be processed. (Léon and Martin, 1969, pp.132-133; original in French, cf. Appendix B)

Later they defined the aim of the application more precisely:

This solution is particularly appropriate for our ultimate aims: ... the display of the pitch contour in real time on the oscilloscope screen, and automatic recognition of intonation patterns, both aims being part of what could be described as the "visual" perception of the intonation curve. (Léon and Martin, 1969, p.170; original in French, cf. Appendix B)

This view makes it clear that, for linguistic and phonetic research, overall reliability and accuracy of measurement form the first essential issue. One has to remember that even a frequency change of F_0 by 5 Hz (for a male voice) may override both intensity and duration (McKinney, 1965). For the musical application, just 5 Hz (in the lower range of frequencies, say, around 100 Hz) means a semitone interval. Thus the accuracy of a PDA for linguistic research must be higher. This is valid especially for the fine errors so that such a PDA should contain at least some smoothing and interpolation to make the intonation contour free from noise.

From the short discussion in Sect.1.2 we know that F_0 is a multifunction parameter; it is influenced by microprosody, tone, intonation, by the temporary emotional state of the speaker, and by long-term speaker characteristics (age, sex, range of the voice, etc.). The major application of PDAs in phonetics and linguistics deals with tone and intonation. In addition, the evaluation of statistical data on speakers and voices partly pertains to this area. A further description of these tasks would exceed the scope of this book; the reader is referred to the vast literature published in this field (part of which has been referenced in the bibliography). Compared to the large number of publications in these areas, the investigations referring to pitch as an indicator of the emotional state of the speaker have been rather rare (see, for instance, Fairbanks and Pronovost, 1939; Lieberman and Michaels, 1962; Levin and Lord, 1975; Scherer, 1977, 1981b; Tolkmitt et al., 1981). Except for microprosody whose investigation requires rather accurate measurement, all the significant phenomena in the areas mentioned cause F_0 to vary by more than 5% so that a measurement accuracy better than 2% or 3% is sufficient.

The other essential issue in phonetics and linguistics is the display. The investigation of intonation and prosody requires visual inspection of the signal or its spectrogram, the intensity contour, and the course of F_0. Figure 9.12 shows a narrow-band spectrogram and the corresponding result obtained by the "Toronto melodic analyzer" (Léon and Martin, in Bolinger, 1972). Perhaps the display given by this PDA is in some way ideal for phoneto-linguistic research. The role of the display in linguistic research was recognized very early by Grützmacher and Lottermoser (1937, 1938) as well as by Gruenz and Schott

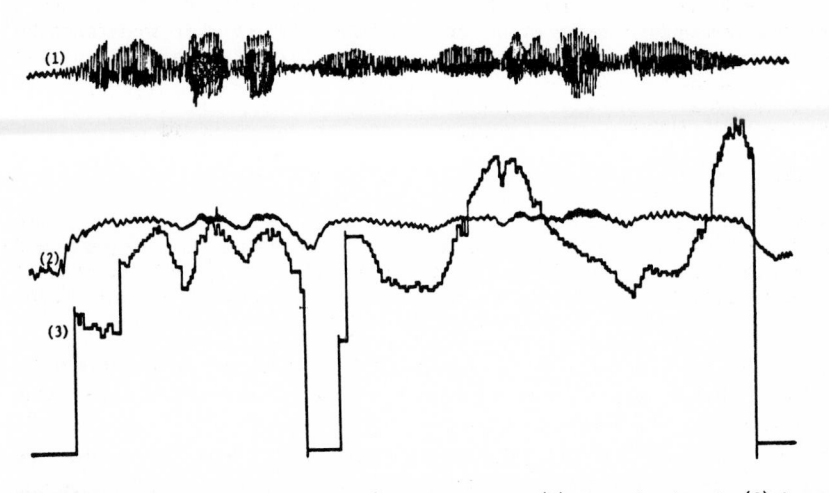

Fig.9.12. PDA by Martin: example of display. (1) Speech signal; (2) intensity; (3) pitch contour. Spoken sentence: Vous nous avez tous vus dans la rue? From (Léon and Martin, 1969, p.173)

(1949), and oscillographic display was one of the strengths of the older analog devices. Most of the older devices designed by phoneticians or used in phonetic laboratories (Kallenbach, 1951; Fant, 1962b) make use of a modified design of the Grützmacher-Lottermoser-Kallenbach type (see Figs.6.47,49 for an example). In the digital domain much less attention was paid to questions of the display. At least until the beginning of the seventies, digital PDAs thus could not contribute very much to the linguistic application area. In addition, digital PDAs were not yet commercially available. These considerations to-gether with the fact that the analog PDAs available at that time were not sufficiently reliable, led researchers like Léon and Martin (1969) to their rather pessimistic view (see above citation). In the meantime, fairly reliable PDAs have been built which also take care of the display (Zurcher, 1977; Léon and Martin, 1972; McGonegal et al., 1975); and graphic terminals are going to contribute to a still wider use of digital PDAs in this field.

To briefly scan the other items listed in Table 9.2, on-line performance is desirable but not mandatory[14] (the application of speech training falls into the group of educational applications). The essential argument in favor of on-line performance is the amount of data to be processed. The least impor-tant point in this field is the wide range of F_0 and the manual preset. For the musical application, a wide range of F_0 is necessary; on the other hand, a temporal restriction of this range by a manual preset might be tolerable. Usually investigators in this field know what they are going to analyze, i.e., they will know the speaker and the range of F_0 to be expected. Thus a manual preset of the range to be analyzed would not be adverse to the aims of the investigation.

As long as the recordings can be made in the laboratory, the robustness against degradation is not very important.[15] The last item (ease of operation) is at least not negligible. Among all researchers who have to do with speech processing, the linguist (as well as the language teacher) is probably farthest away from being skilled to operate complicated technical equipment. Hence this point (it could be regarded as a subset of the issue of display) has to be given some attention.

---

[14] According to Lea (1982), a number of tasks in phonetics and linguistics, especially when they play a role in the preparation of reference data for speech communication systems, require a tradeoff between the parameters *reliability* and *accuracy* on the one hand, and *real-time performance* on the other hand. This means that researchers may have huge amounts of data to be processed with low accuracy in real time (or even faster) on the one hand, and a small number of reference utterances, on the other hand, which need to be processed very accurately without regard to processing time.

[15] Things may change drastically if the recordings cannot be made in the phonetic laboratory. This happens, for instance, when exotic languages or folklore are investigated; in this case the researcher is forced to visit the subjects where they live, and the recording conditions may be very adverse.

Has the problem been solved for this domain of application? If *the pro-blem* were reliability of the analysis, one could say it never existed. Manual F_0 determination, known and practiced as the classical means in intonation research, meets the requirement of reliability, although it does not quite meet that of accuracy. (This holds if the DL of the human ear is regarded as the criterion for accuracy; of course all linguistically significant intonational events can be detected in manual analysis.) Besides the improved accuracy, the main motivation for the linguist and the phonetician towards applying a PDA is the increased ease of data acquisition or further evaluation and the pertinent gain in time and flexibility.

The problem of postprocessing is thus primarily important: the display of a F_0 contour and the facility to bring it into a shape which permits further processing, such as pattern recognition of contour elements, correlation to other parameter contours such as intensity, or statistical evaluation. The accuracy requirements can be met by today's PDAs, at least for the linguistic domain. The reliability requirements certainly are still beyond the current scope, at least when the signal is noisy or degraded. On the other hand, it may not at all be necessary to meet these requirements totally by machine. In the opinion of the present author, the future PDA for linguistic use is an interactive one - interactive, however, perhaps not as much in the way as the one by McGonegal et al. (1975), but rather in such a way that the PDA does the whole analysis automatically as far as the signal is regular, and that it attracts the attention of the researcher only when it seems to have failed for a particular frame, or when it faces a peculiar situation it cannot cope with. Statistical experiments with the PDA implemented in Chap.7 showed that only about one frame out of 20 invokes the correction routine for a high-quality signal (this mostly at the voiced-unvoiced transitions), and less than 2% of the pitch markers are labeled irregular. If any, only these segments would need a manual correction. Compared to the results obtained by McGonegal et al., an interactive PDA could thus be significantly speeded up. In this case, however, the algorithm must learn to recognize its failure. This problem, in the opinion of the present author, is nontrivial, but not exceedingly serious. Coding theory tells us that detecting an error requires much less redundancy and less effort than reliably correcting it. At least the sophisticated digital PDAs, such as short-term analysis PDAs or multichannel time-domain PDAs have correction routines. In Sects.6.6 and 8.4, various correction schemes have been presented, and it would be easy to modify these routines in such a way that they request manual interaction or at least manual approval before carrying out the respective correction.

The main application of PDAs in musicology is automatic notation of played music or singing (Sundberg and Tjernlund, 1970; Askenfelt, 1976, 1979; Larsson, 1977; Filip, 1969, 1970, 1972; Remmel et al., 1975; Moorer, 1975; Philipp, 1979; and others). This is a typical application where masses of data have to be processed with moderate accuracy (better than 1 semitone, i.e., 6%, at least for European music). As an example for this application, let us quote Askenfelt (1979) on the so-called VISA project.

In the Scandinavian countries folk music has been extensively recorded on tape. This has resulted in large collections of recorded melodies, amounting at the present time to approximately 150,000 one-voiced pieces. Card catalogues with associated information about the melodies ... have been assembled. It would also be highly desirable to have access to the melodies in ordinary notation. However, the overwhelming number of melodies gives rise to problems. First, the transcription of 150,000 melodies would occupy a large staff of musicians working many years, hardly a realistic solution. Secondly, how does one find a melody or group of similar melodies with specific melodic characteristics, among 150,000 other melodies?

The VISA project is an attempt to solve the problem of transcribing recorded melodies and of searching for melodies ... with the aid of a computer. The transcription process consists of the following steps. The signal from the tape recorder is fed to a hardware pitch detector outside the computer, which extracts the fundamental frequency of the tones in the melody. The computer records the fundamental frequency and the sound intensity in the melody. Errors in the fundamental frequency can partly be removed by applying a digital filter.

The next step is to define the scale used by the player. If the intervals in the melody resemble the intervals in an equally-tempered major or minor scale, a proper predefined scale may be applied. If, on the other hand, the intervals deviate considerably from the intervals of the equally-tempered scale, the scale must be adapted according to the intervals used in the melody. This is a common situation in folk music. The adaptation is made with the help of a histogram of the frequencies of the scale tones used. The scale tones appear as peaks in the histogram. The operator marks the positions of the peaks with the help of a cursor, and in that way defines the intervals of the scale. ...

After the scale has been defined, the computer can automatically detect transitions between scale tones and thus make a segmentation of the melody in parts, each corresponding to one note. The last step in the transcription process is to let the computer assign note values to the segmented tones. ...

The weakest point in the transcription process at present are in pitch detection and assignment of note values. Pitch detection in real time on music signals is a purely technical problem which will be solved satisfactorily in the near future. The assignment of note values is a more complex problem, because a tone with a given duration does not always correspond to the same note value. Depending on the melodic context, several alternatives might be possible. At this point more research is required. These difficulties merely demonstrate the well-known fact that ordinary musical notation is a very simplified description of the performed piece. It is not a straightforward task to describe to a computer which simplifications are allowed and which are not in the transformation of the acoustic representation of a melody to an acceptable transcription. (Askenfelt, 1979, pp.109-111)

This typical example shows that the requirement with respect to the PDA (as well as the problems with it) are very similar in phonetics, linguistics, and musicology. Only the problems of postprocessing and display are different; in musicology the postprocessor includes a quantization and segmentation routine in order to fit the melody into ordinary musical notation. As the previous quotation shows, this is a rather serious problem. It has not yet been solved in a fully automatic way so that manual interaction is still required.

A special problem in musicology is the automatic transcription of music with two or more voices. Usual PDAs, as applied in speech and one-voiced music, mostly fail since they have not been conceived to determine several simultaneous pitches. An exception are frequency-domain PDAs that operate on a psychoacoustic basis (Terhardt, 1979b; Terhardt et al., 1982a,b; Sect.8.5.2) and multiple-spectral-transform PDAs (Weiss et al., 1966; Sect.8.4.2). Algorithms proposed for this problem have been rare (Moorer, 1975; Philipp, 1979) and rather complex and time-consuming (Askenfelt, 1979; with respect to Moorer, 1975).

9.4.3 Application in Education and in Pathology

There are three mandatory requirements for *educational* PDAs: on-line performance, display facilities, and cost of device. The educational PDA (or PDD) always is an element incorporated into a feedback loop. The subject who uses it as an aid to learn correct intonation represents another element of this loop. The PDD must thus operate on-line and in real time. To realize the ideal state that each person can completely occupy the PDD at least for a certain time, the PDD must not be too expensive. Since data have to be interpreted by a multitude of normally untrained users, the question of (visual) display may be even more important for this application than for the linguistic one. On the other hand, the PDD need be neither too reliable nor too accurate nor too robust with respect to degradation of signals. For example, envelope detection PDDs (PiTE) are ideal for this application (Dolanský, 1955, 1960; Zurcher, 1972, 1977).

Educational PDDs somehow occupy an intermediate position between the areas of linguistic and medical applications. The area can be subdivided into two different applications: 1) aids to the deaf and hard-of-hearing, and 2) teaching intonation patterns to foreigners. Regarding the latter application, the importance of prosody for the intelligibility and the naturalness of speech is known and recognized so that the use of a PDD (or PDI) in language instruction will be of great use for a language training laboratory. McKinney (1965), for instance, made the following comment:

> It has been found that one of the most important contributors to the poor intelligibility and foreign accent of a nonnative speaker is his failure to match the prosodic pattern of his speech to the norms of the language. (McKinney, 1965, p.3)

Nevertheless, at least until a few years ago publications on this matter were rather rare. For instance, Vardanian (1964) used an oscilloscope to display F_0 contours for teaching English intonation. Semmelink (1974) used a speech analyzer and an F_0 indicator in speech training of South African tone languages. Fourcin and Abberton (1976) applied an electric PDI (laryngograph) in foreign language education. The objections to the application of PDAs in this area might have been 1) technical difficulties in selecting, implementing, and running the equipment, and 2) high cost; in addition, 3) it was unclear whether visual feedback of intonation is a help at all in foreign-language teaching (de Bot, 1981b).

Progress in technology is likely to render the first two objections insignificant. Visual displays, wherever they are necessary and desirable, nowadays greatly benefit from computer graphics (Miller and Fujimura, 1979; Myers, 1981) and commercial video electronics. This means that in the future simple, low-priced systems will be developed displaying F_0 contours and/or other speech parameters from recorded or directly from spoken utterances on a standard TV screen in real time. Such systems are then helpful for autodidactic as well as supervised learning of foreign language, or, like the well-known "Speak and Spell," they may even be used as toys.

The question whether visual feedback is useful in teaching intonation has been positively answered by experiments such as the one by de Bot (1981a,b). In language training laboratories it is usual that a student learns correct pronunciation and intonation by repeating phrases spoken by the teacher and recorded on tape. The phrases uttered by the student are also recorded; playing them back the student can acoustically compare his (her) pronunciation and/or intonation with that of the teacher. If there is an additional visual feedback path, both the intonations by the teacher and the student are displayed, for instance, on a TV screen. If the system operates in real time, the student can check his intonation while he is speaking, and he gets a better feeling than by auditory feedback alone as to where exactly he is wrong. Human ability with respect to auditory discrimination between different pitch contours (which are perceived one after another) is rather poor ('t Hart, 1981); this task has proved difficult even for skilled linguists (Lieberman, 1965). As we know from Sect.4.3, this is due to temporal and not to frequency-domain limitations in the auditory system. In visual displays time is converted to an axis on the screen (usually the horizontal one), i.e., the picture stays where it is and displays the whole contour, whereas the acoustic presentation is always in the form of a time function. Hence these limitations are avoided in the visual display. As de Bot (1981a,b) showed, subjects provided with auditory and visual feedback learned intonation of a foreign language better and faster then subjects provided with auditory feedback alone.

The former application - aids to the deaf and hard-of-hearing - is more similar to the application in medicine and pathology. Research has concentrated on the display. Substantial work in this field was done by Dolanský (1955), Anderson (1960), Kallenbach (1962), Grützmacher (1965), Dolanský et al. (1969), Risberg (1968, 1969), Martony (1968), Boothroyd (1972a,b, 1973); Levitt (1972a,b, 1973), Parker (1974, using the laryngograph), Crighton and Fallside (1974), Haton and Haton (1975, 1977a,b, 1979), Heike (1977), Joffe (1977, for articulatory training), and Salles (1980), not to forget the early work on visible speech display (e.g., Gruenz and Schott, 1949). Not all of the work mentioned here was done to display pitch contours. As an example of the requirements for such a display, the following is specified.

1) A portable pitch/intensity display, entirely enclosed in the main frame of a storage oscilloscope;

2) major emphasis on the clarity of the intonation display;

3) simplified operation, so that a child could run the equipment alone;

4) automatic erasing of the pattern and triggering without confusions;

5) sufficient flexibility of equipment operating modes, to allow accomodation of various teaching programs;

6) indication of vocal intensity;

7) a mirror to allow for visual monitoring of the subject's own, as well as the teacher's, facial movements without excessively disturbing his monitoring of the visual pattern;

8) high-level earphone amplifiers for auditory feedback. (Dolanský et al., 1969, p.57)

Salles (1980) showed that the experimental setup for teaching intonation to deaf children is rather similar to that used in foreign-language teaching, as decribed by de Bot (1981). Since the acoustic feedback is blocked, the deaf subject must totally rely on the visual aid. The problem is thus much more basic for this application; it is necessary first to get the voice under control, i.e., to stabilize the average fundamental frequency within an adequate range, and to teach the subjects how to vary the pitch of their voices. Salles's work (1980) contains a detailed discussion of all the phases of such an experiment, including the choice of a suitable PDD (Zurcher, 1977). The results of this study are encouraging; on the other hand it has proved necessary not to confine oneself on teaching intonation alone, but to teach prosody as a whole, i.e., intensity, duration, rhythm, and intonation simultaneously. In this field there is still much work to be done.

Another display possibility is a *tactile display*. This way of transmitting information to a deaf subject is rather old (Gault and Crane, 1925). Much of the work on this display has been referenced in the bibliography by Risberg (1964b). A simple display of this kind has been described, for instance, by Willemain and Lee (1972). The aim of this study is to give deaf persons an everyday tool to control the pitch range of their voices:

> The prototype pitch detector is straightforward. A throat microphone detects voiced speech; its output is amplified, low-pass filtered, and converted to a square wave pulse train by a Schmitt trigger. The pulse train is gated for a fixed time, and pitch frequency is determined from a zero-crossing count on the gated pulse train. The pitch measurement is quantized into one of eight channels. The first seven channels are adjusted to correspond to the range 100-240 Hz in bandwidths of approximately 20 Hz each. The eighth channel corresponds to all pitch frequencies above 240 Hz. Counts in each channel are recorded for analysis of a speaker's pitch distribution. ... The display is also quite simple. Solenoids poke the fingers of the speaker to provide the tactile feedback. The eventual goal of research in tactile pitch feedback is the design of a wearable speech aid, and an important criterion for such an aid is that it be simple and consequently inconspicuous to the user. Therefore, the displays used employed only two and three solenoid pokers. Switching circuits allow the experimenter to assign counts in any of the channels to any tactor. Thus, the eight channels can be grouped for display into "high," "low," and "ok" bands when three pokers are used, or into "high" and "low" bands when two are used. (Willemain and Lee, 1972, p.9)

A comparative evaluation of tactile displays and their use as aids for the deaf was carried out by Spens (1980).

Measurements by Rothenberg et al. (1977) have shown that much information on pitch can be transferred via a single-channel vibrotactile aid by simply changing the frequency of the vibrator. Determining the tactile difference limen for change of the vibrator frequency, Rothenberg et al. (1977) found that there are at least 10 distinguishable steps in frequency in the range 10 to 300 Hz, relatively independent of the vibration site and amplitude. Above 300 Hz the sensitivity of the skin decreases rapidly, and below 10 Hz the oscillation is to slow to convey meaningful information on the time-variant speech signal. Rothenberg and Molitor (1979) carried out an experiment with a single-channel vibrotactile instrument. The pitch contours of six sentences "Ron will win" with different prosodic features (declarative and interrogative, sentence stress on any of the three words) were directly converted into vibrator frequency, and the subjects were asked to judge the prosody from the vibrotactile information alone. This distinction could be well made; the subjects only had some trouble with the temporal judgment in the tactile mode. A vibrotactile display might thus be insufficient if used alone; however, it can be very useful when used together with another aid, such as a visual display or lipreading (Risberg, 1964; Risberg and Lubker, 1978). In addition, vibrotactile displays are useful for a deaf person to control voicing, since the sensation of the vibrator is rather similar to that of the own voice (Rothenberg and Molitor, 1979).

These few and incomplete examples demonstrate some of the problems associated with applications of PDDs as aids for the handicapped.

On the other hand, there is the pure medical application in the investigation and therapy of voice diseases. In this field, the display facilities again play an important role. In addition, the PDD must cope with a variety of irregular signals and conditions. For the medical application instruments are preferred like the laryngograph or other direct-measuring PDIs so that the PDD is not the most important device. If used, principles such as envelope detection (Schultz-Coulon and Arndt, 1972; after the principle by Dolansky, 1955) are preferred, especially since they also indicate T_0 when the glottal waveform is irregular. More recent approaches, for example, use the AMDF and combine it with a time-domain technique (Moser and Kittel, 1977). Many tests in phoniatrics, however, do not need any device since they can be carried out with the ear alone. Investigating a subject's range of fundamental frequency, for instance, is one of these tasks (Schultz-Coulon and Arndt, 1972).

A thorough investigation of the various phonatory skills plays an important part in the examination of patients with phonatory disorders. The usual procedure is that the fundamental frequency of the voice is determined by comparison with notes from the piano or tuning fork. This method of examination has the disadvantage that it requires trained investigators with a good ear; moreover there are quite a number of sources of error so that only approximate values for the voice pitch can be obtained. While this may be enough for the diagnosis and therapy of a phonatory disorder it is insufficient for the purposes of exact documentation and means that comparative examinations are only possible with considerable reservations. (Schultz-Coulon and Arndt, 1972, p.241; original in German, cf. Appendix B)

The reports by Davis (1976a, 1978) present the application of linear prediction analysis to form what Davis called a *voice profile* (see Fig.9.13 for an example). For a sustained vowel, the profile contains the LPC residual signal, the "flatness" of the inverse-filter spectrum (which serves as an indication of the formant bandwidths), the flatness of the residual spectrum (which should reveal strong peaks in order to indicate distinct periodicity), and some statistical measurements on period perturbation. Davis intended the voice profile to indicate mild voice pathologies which might remain undetected in an auditory test.

The normal and mildly pathological subjects ... are the ones for whom acoustic analysis is potentially a valuable addition to existing medical procedures. ... Voice profiles may be useful indicators of voice quality, especially during voice therapy, and it will be necessary to have speech pathologists and physicians evaluate their usefulness. For efficient clinical implementation, tape-recorded voice samples may be sent from a clinic to a central computer facility, either indirectly via the mail system or directly via telephone lines. With the advent of hospital computers and remote terminals in outpatient clinics, an immediate acoustic evaluation of laryngeal pathology is a viable objective. Alternatively, voice samples could be analyzed with a microprocessor-based "black box" built especially for clinical use. (Davis, 1978, p.159)

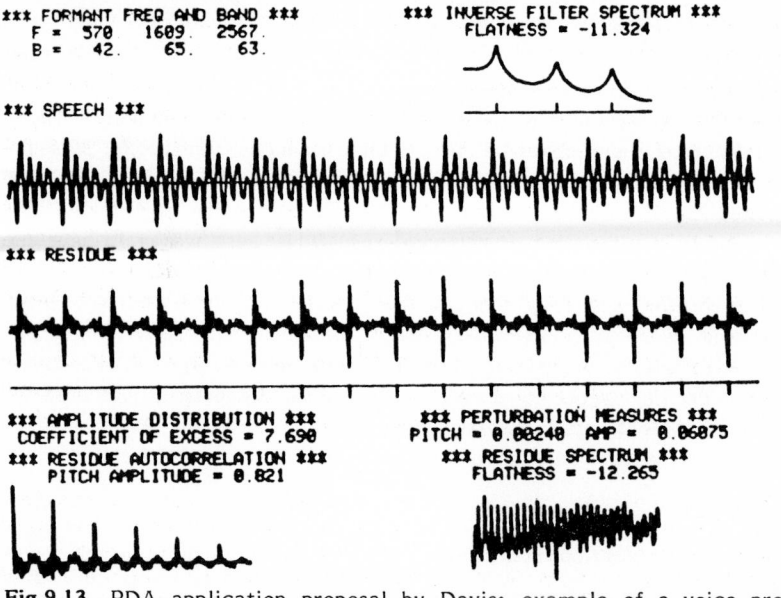

Fig.9.13. PDA application proposal by Davis: example of a voice profile for voice therapy and prophylaxis. The example represents a normal voice. From (Davis, 1978, p.147)

Although some of the specifics remain to be investigated,[16] this application is promising, especially since it could offer a cheap approach to the prophylactic examination of voices.

A task whose requirements with respect to reliability and accuracy are principally different is *voice perturbation measurement*. The question of voice regularity and perturbation has already been discussed in Section 4.3. This application requires time-domain PDAs (Crystal and Jackson, 1970; Horii, 1975, 1979a; Gubrynowicz et al., 1980) or a combination between a nonstationary short-term analysis PDA and a time-domain PDA (Moser and Kittel, 1977). For normal voices where the perturbations are low, the temporal resolution of the signal is critical (Lieberman, 1963; Horii, 1979a). Horii (1979a), for instance, found an overall mean of 36 μs for the jitter of six male voices. With a sampling frequency of 40 kHz (i.e., a sampling interval of 25 μs) even this kind of measurement becomes difficult.

> With a temporal resolution of 25 μs, jitter characteristics may not reliably be examined for phonation above about 210 Hz. ... Perturbation characteristics above 200 Hz need to be investigated with finer temporal resolutions.
> Jitter values reported for various pathological larynges, however, may be large enough to be distinguished from normal phonation by 50-μs or 25-μs resolution. For example, van Michel (1966) reported a mean jitter of 70 μs to 570 μs for sustained /a/ perceptually judged to be pathologic. Data of a severely hoarse voice examined ... had jitter of approximately 300 μs. For sustained /a/, produced by esophagal speakers, mean jitter ranging from 0.62 ms to 5.13 ms was found (Smith, 1977). These relatively large mean jitter values certainly indicate a high probability that their jitter histograms would be quite different from those for normals, even using 50-μs or 25-μs temporal resolution. (Horii, 1979a, p.15)

Since such measurements are usually carried out during sustained phonation, failures of the algorithm due to signal nonstationarity are unlikely. Time-domain PDAs along the PiTS and PiTE principles, which preserve the pulselike character of the excitation signal, are thus best for this application. The same applies to electric or ultrasonic PDIs in connection with a suitable algorithm. As to the question of temporal resolution, here we again have the discrepancy between the required measurement accuracy and the required signal representation given by the sampling theorem. This issue has already been discussed in detail in Sects.4.3, 8.5.4 (in connection with frequency-domain PDAs), and 9.2.3. In this case it is thus unnecessary to excessively increase the sampling frequency during analog-digital conversion; here a sampling rate is sufficient which satisfies the sampling theorem. Increasing the sampling rate to satisfy the accuracy requirements can then be performed

---

[16] A technical objection against this proposal has to be made which seems important. To avoid artefacts by signal distortions, care has to be taken that the signals to be processed are free from any distortions. Thus a telephone line with its unpredictable performance and even a normal tape recorder can have a detrimental effect on the results of such an analysis.

digitally. In this respect, it is again unnecessary to upsample the whole signal. To determine where the period begins, the PDA can operate at the original sampling rate. Once a marker has been found, the sampling rate can be locally increased around this point, and the measurement can be repeated with increased temporal resolution. Any time-domain PDA along the PiTE or PiTS principles which derives pitch from local features can be modified this way. Experiments with such a routine using the output signal of a PDI have been discussed in Sect.9.2.3; this algorithm is very suitable for pitch perturbation measurement as well.

To pose the question again, has the problem been solved for the application in the educational as well as in the medical domain? On the other hand, is this really the problem of PDA development? Except for the relatively hard demand with respect to noise resistance (Laver et al., 1982) - the noise is usually caused by heavy perturbations of the pathologic voice - the requirements on accuracy are rather simple. On the contrary, a PDA originally designed for use in a speech communication system must be prevented from correcting and forcing regularity. It may be a commercial problem to design and manufacture PDDs at low cost so that they can be widely used. On the other hand, from language acquisition to aids for the handicapped and the medical and phoniatric application, the PDD analyzing only the signal faces increasing competition from direct-measuring instruments (PDIs), such as the laryngograph. These PDIs definitely work more accurately than the PDD, and a vast literature has been published describing their use in phoniatric practice (see Sect.5.2). They have only two disadvantages: firstly, they cannot process recorded data so that the subject must be physically present, and, secondly, a part of them cause some inconvenience to the subject. This latter point is of course much more willingly tolerated by a patient in the phoniatric laboratory than by a student who wants to learn a foreign language. Even this point nowadays is of decreasing importance; the pickups of any electric and ultrasonic PDI (Sects.5.2.3,4) as well as of some mechanic PDIs (Stevens at al., 1975; Stilwell et al., 1981; Sect.5.2.2), which have to be worn around the neck, have become so light that a subject can tolerate them for hours without feeling uncomfortable.

Thus the problem in these domains of application seems to be a problem of interdisciplinary collaboration rather than of the PDD itself. Substantial activities exist, and at places where there is a close collaboration between medical and engineering faculties, the results are extremely encouraging [Stockholm (Risberg, 1968; Martony, 1968; and more recent publications in STL-QPSR), Boston (Dolansky, 1955; Dolansky et al., 1960; Dolansky et al. 1969), London and Cambridge (Fourcin and Abberton, 1971; Parker, 1974; Fisher et al., 1978; Crighton and Fallside, 1974; Bristow, 1980), Besançon (André and Lhote, 1977), Nancy (Haton and Haton, 1977, 1979), to name only a few major centers of work]. On the other hand, one cannot ignore the fact that especially in the field of speech acquisition aids to the handicapped the PDA is underrepresented. So this discourse will be terminated by the citation of an abstract from the work by Fisher et al. (1978) which could and should be a challenge to researchers to put more activity into this domain of application:

> Work is in progress ... on various aspects of speech production in hearing impaired children and adults. The main areas are 1) listener evaluation of the speech problem; 2) objective, quantitative analysis; 3) the relationship between perception and production of speech patterns; 4) programs of remediation using visual displays of relevant features. A systematic procedure has been developed for the phonological assessment of abnormal speech, relying on a narrow phonetic transcription in an attempt to determine the distinctions actually produced. Physical measurements are made of various prosodic and speech pattern features, including durations, fundamental frequency contours, formant frequencies, and voice onset time. (Fisher et al., 1978, p.35)

9.4.4 The "Technical" Application: Speech Communication

It is speech communication and in particular speech transmission by analysis-synthesis systems (vocoders) where the hard problems of pitch determination are to be found.

The large discrepancy between the information content of the acoustic speech signal on the one hand and the underlying phonetic message on the other hand has challenged researchers for a long time to look for effective methods of data reduction for speech transmission purposes. PCM-coded telephone-quality speech has a typical information flow of 64 kbits/s (8-kHz sampling rate, 8 bits per sample). High-fidelity speech transmission requires 240 kbits/s or even more (20 kHz; 12 bits per sample). The information flow for the phonetic message, on the other hand, rarely exceeds 100 bits/s even if prosodic information and information on the speaker is taken into account. Effective coding and quantization schemes working on a sample-by-sample basis cannot force the information flow below 8 kbits/s without a substantial loss in speech quality (Flanagan et al., 1979).

Parametric speech transmission permits coding intelligible speech at data rates of 2.4 kbits/s or even less [down to 0.6 kbits/s]. Parametric speech transmission was first applied by Dudley (1937, 1939a-c); in his classic paper *"Remaking Speech"* (1939b) he called the pertinent device a *vocoder* (from "voice coder").

Figure 9.14 shows the principal block diagram of a vocoder. In the analyzer a short-term analysis of the speech signal is performed, and the relevant parameters are quantized, coded, and transmitted. The short-term analysis permits transmitting the data at the usual frame rate which is by two orders of magnitude lower than the sampling rate of the signal (typical values for the frame interval range between 5 and 50 ms). On the receiver side the signal is resynthesized using a speech synthesizer which recovers an intelligible speech signal from the parameters.

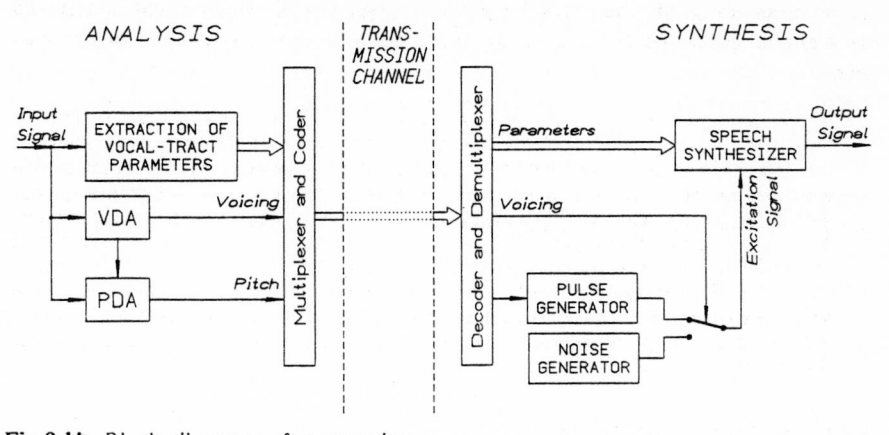

Fig.9.14. Block diagram of a vocoder

Most vocoders use speech production models similar to the one given in Sect.2.5 (Fig.2.15). This means that the analyzer extracts separate estimates for the vocal-tract transfer function $H(z)$ and for the time-variant voice source parameters. There are a number of possibilities to transmit the information on $H(z)$, and the vocoder type is determined by the set of parameters representing $H(z)$. The best-known types of vocoders are the *channel vocoder* (Dudley, 1939b; Gold and Rader, 1967) where $H(z)$ is represented by the short-term amplitudes of the output signals of a filterbank with typically 6 to 15 channels between 0.5 and 4 kHz, the *formant vocoder* (Flanagan, 1955a) where the formant frequencies F1, F2, and (sometimes) F3 represent $H(z)$, and the *LPC vocoder* (Itakura and Saito, 1968) where $H(z)$ is coded in form of linear-prediction coefficients. These and other principles have been abundantly used and varied, and a vast literature has appeared in this field [for surveys, see publications such as the ones by Schroeder (1959, 1966), Flanagan (1972a,b), Pirogov (1974), Gold (1977), Rabiner and Schafer (1978), Viswanathan, Makhoul, and Schwartz (1982), or Crochiere and Flanagan (1983)]. A detailed report on the design of a particular vocoder system - in this case a channel vocoder - has been published, for instance, by Yaggi (1962, 1963).

Whereas a wide variety of possibilities for estimating and transmitting $H(z)$ have been developed, the possibilities for representing the voice source information have been rather restricted. Since there is no way around these parameters (Gold, 1977),[17] each vocoder has its VDA and its PDA in the voice source path and transmits information on voicing and pitch besides the information on $H(z)$ and the overall amplitude.

---

[17] An alternative solution which maintains part of the voice is presented in the next section. This type of vocoder does not need a PDA or a VDA.

Typical data rates per frame in a vocoder are: 1 bit for voicing, 6-8 bits for pitch, 6-8 bits for amplitude, and 30-60 bits for the vocal-tract parameters. With the typical frame rates given further above, such quantization schemes result in an information flow between 1.2 and 9.6 kbits/s. For pitch determination this means that, if the range of F_0 extends from 50 to 500 Hz and is logarithmically quantized, the quantization interval will be 3.7% of the current value of F_0 when 6 bits are available, and 0.98% when 8 bits are available. The problem with the PDA is thus a problem of *reliability* which has absolute priority in the vocoder. Usually the quantization due to the limited transmission rate is cruder than the quantization due to signal sampling so that measurement inaccuracies only become interesting when they exceed a few percent.

In order to resynthesize the speech with minimum redundancy and irrelevance, a number of simplifying assumptions have to be made which mostly concern the time-invariant parameters of the voice source. The original voice source - being that of an individual - is then replaced by a standard voice source model - being that of a machine - with a white-noise excitation for voiceless sounds and a pulse train for voiced sounds. A number of speaker characteristics are thus lost in a vocoder, although one can still recognize the speaker from the prosodic information which has survived the data reduction.

As to the PDA, we again have the assumption of an isolated point of excitation in the synthesizer, regardless of whether the excitation signal itself is represented as a simple pulse train or as a different, more complicated time function which may result in an improved subjective quality (Schroeder, 1963; Makhoul et al., 1978; Rabiner et al., 1976; Gold, 1977; Kurematsu et al., 1979; Un and Sung, 1980; Maitra and Davis, 1980; García Gómez and Tribolet, 1982; Atal and Remde, 1982). As we have seen in Sect.9.2.3, this assumption is an oversimplification; it certainly contributes to the "artificial" quality of vocoder speech. Another possible source of error is the crude quantization of the voicing function. A 1-bit quantization of the voicing function means that the excitation is either voiced or unvoiced. This may be sufficient for stationary phonemes but causes trouble during voiced-unvoiced transitions. Silence may be obtained in such systems by driving the synthesizer with a voiceless excitation and forcing the amplitude of the signal to zero.

If a vocoder is implemented in an actual speech transmission system, it must work for any voice and for (almost) any environmental condition. From the profile of requirements to the PDA, speech communication thus poses the hardest demands: no preset whatsoever as to speakers, expected range of F_0, or recording and transmission channel conditions (except that they may be bad!); no possibility of interactive correction since the job is done in a fully automatic way; and the ear as the final judge of the performance. The last condition undoubtedly is the hardest one. Baronin (1974a) summarized how these conditions and premises affect the choice of a suitable PDA and its performance.

1) The great variety of methods which have been proposed and implemented for pitch detection demonstrate the absence of one such method that, if it were available, would possess absolute advantage over all other methods. For different premises or suppositions it is thus suitable to apply different methods.

2) For pitch detection from high-quality speech signals (dynamic microphone, absence of low-frequency band limitation) from a single speaker, the simplest nonadaptive methods give good results. These methods can operate according to the principle of first-harmonic detection by means of filters with invariant paramters or by amplitude selection [i.e., peak detection using the principle PiTE]. The acoustic quality of synthesized speech using pitch detectors with amplitude selection [see Sect.9.1.1] is somewhat higher because of the lower inertia of such systems, although in the presence of noise and spurious signals the filter methods give better results.

3) If one works with different speakers, the adaptive pitch detectors show noticeable advantages. If the low-frequency components of the signal are present (and undamped), satisfactory quality can be obtained in most cases applying relatively simple systems of amplitude selection with parameters that are automatically adjusted according to the variations of the fundamental frequency, and applying logical correction routines for gross errors. A further improvement of the performance is achieved when one changes over to multichannel analyzers which work in the time (Gold, 1962a,b) or the frequency domain (Kushtuev, 1968, 1969b; Miller, 1970); for low noise levels cepstrum analyzers also give good results. For high noise levels correlation analyzers (Weiss and Harris, 1962, 1963) or multichannel frequency-domain analyzers (Miller, 1970) are preferred.

4) When working with a carbon microphone, one can apply the simple filtering methods. If the first harmonic of the signal is absent, it can be reconstructed by nonlinear processing of the signal. However, the quality of the analysis obtained is quite low when such simple methods are applied. To achieve better results, it is necessary to apply optimal[18] or at least nearly optimal methods (harmonic analysis, correlation schemes with previous adaptive filtering of the signal, etc.). For the calculation of the current estimate of F_0 it is necessary taking into account previous estimates, and it is very desirable that future estimate be taken into account as well. The time delay of the output which emerges from this process, and which usually does not exceed some 10 ms can be tolerated in most cases. (Baronin, 1974a, pp.207-208; original in Russian, cf. Appendix B)

Let us thus drop subproblems such as speech recognition or speaker verification (where one can completely ignore measurement inaccuracies and leave the problem of a grossly erroneous pitch contour to a higher level processing step) or synthesis by rule (where the determination of pitch contours is a matter of phonetics and linguistics), and let us confine the following considera-

---

[18] "Optimal," in Baronin's terminology, means optimized with respect to some statistical error criterion (see the discussions on this issue in Sects.6.6.3 and 8.6.3).

tions to those applications where the ear is actually involved.[19] From Chap.4 we know that the auditory difference limen for fundamental frequency changes is the most sensitive difference limen in the ear to deal with speech signals. This difference limen is smaller than the quantization step for pitch in a vocoder, smaller than the intrinsic accuracy of most PDAs, and smaller than the natural pitch perturbations pertinent to the process of speech production. Thus one might expect that any error made by a PDA will be audible and will degrade the quality of the synthetic speech. As to this point, Gold (1977) has critically commented,

> Superficially, the problem of correctly estimating the excitation parameters in speech analysis-synthesis systems appears to be simpler than that of estimating vocal tract parameters. After all, only a one-bit decision needs to be made to determine whether the excitation should be voiced or voiceless, ... and then a single parameter, the time-varying voice fundamental frequency must be estimated. However, the unfortunate fact is that the speech quality is *critically* dependent on the successful estimation of these two parameters. If the analyzer incorrectly identifies a voiceless sound to be voiced, the listener hears an unpleasant "buzziness" in the synthesized speech; is the analyzer incorrectly identifies a voiced sound (or part of a voiced sound) to be voiceless, the sound suddenly becomes harsh and "reptilian." Mistakes in estimating fundamental frequency of the voice cause comparably highly intrusive unnatural sounds to appear to be incorporated into the perceived speech. These effects are very noticeable even when the analyzer is correct 95% of the time. In difficult environments which cause the analyzer to make a high percentage of mistakes (20%), the effect is to severely lower the overall intelligibility and quality of the speech communications.
>
> Thus the excitation problem is a severe one, and there are several factors that conspire to create major difficulties in attaining the required high performance. ... To further complicate matters, acoustically coupled background noise and preprocessing by a telephone channel introduce further changes into the speech signal which intensify the difficulty. (Gold, 1977, pp.1643-1644)

There is not much to be added to this statement. From the comparative evaluation by Rabiner et al. (1976) and McGonegal et al. (1977) we have evidence that the problem of pitch determination in this area is still far from a general solution. On the other hand, work on PDAs has been focused on this application almost since PDAs began to be developed.

---

[19] Besides vocoding, this includes (at least to some extent) a recent application of PDAs as an aid for the hard-of-hearing: external electrical stimulation of the cochlea (Kiang and Moxon, 1972; Fourcin, 1978; Fourcin et al., 1979). This is possible for patients who suffer from deafness for instance because of a defect in the middle ear or in the tympanic membrane; and it permits transmitting information directly to the auditory nerve. Of course pitch discrimination, if possible at all, is much worse for these subjects than for normal listeners. According to Fourcin et al. (1979), the pertinent difference limens strongly depend on the individual patients; however, no patient was found with a DL for fundamental frequency change below 5% (compared to 0.3%-0.5% in normal listeners). For more details the reader is referred to the original publication (Fourcin et al., 1979).

The problem is unsolved, yes. Yet will we solve it by designing and implementing more and more new PDAs in the hope that one day there will be *the* idea? Should we not rather continue Rabiner's work and put more effort into performance evaluation of existing PDAs in order to learn more about the circumstances of their failure? In addition, isn't it possible to find a way to avoid the problem of pitch determination altogether, i.e., to build a device which serves the desired purpose without the necessity of pitch determination? The remainder of Chap.9 will present some of these topics in more detail.

9.4.5 A Way Around the Problem in Speech Communication: Voice-Excited and Residual-Excited Vocoding (Baseband Coding)

The fact that a bad performance of the PDA and/or the VDA drastically degrades the performance of a vocoder has encouraged research toward the development of parametric speech transmission schemes that do not require PDAs or VDAs in the analyzer part.

In the so-called *pitch-excited*[20] vocoder, as it was described in the previous section, there is no alternative to the VDA and the PDA in the voice source path. These components can only be avoided when the original speech waveform - or some filtered version thereof - is transmitted over the voice source path. Applying this principle one obtains a *semivocoder* (Schroeder, 1959) where the vocal-tract information is transmitted in a parametric way, whereas the voice source information is waveform coded, i.e. quantized and directly transmitted over the transmission channel. Two kinds of semivocoders have been developed: the *voice-excited* vocoder (Schroeder, 1959; Schroeder and David, 1960; David et al., 1962a,b) and the *residual-excited* vocoder (Un and Magill, 1975). Both vocoders are also known under the name of *baseband coders* (Viswanathan et al., 1982). In the following, the two principles will be briefly explained.

Figure 9.15a shows the principle of a voice-excited vocoder. The basic idea is due to Schroeder and David (Schroeder, 1959; Schroeder and David, 1960). The vocal-tract information is parameterized and transmitted as in a usual, pitch-excited vocoder. In the voice source path, however, a part of the original voice of the speaker is preserved and transmitted. To achieve a reasonable amount of data reduction, transmission only comprises a low-frequency band - the so-called *baseband* - which is sufficient for carrying information on pitch and voicing, but not sufficient to produce intelligible speech if used alone.

In the synthesis part, the synthesizer is realized as in a usual pitch-excited vocoder. The excitation signal is generated from the baseband containing the

---

[20] The term *pitch* is used here in its original meaning: *pitch-excited* means excited by a series of pitches, i.e., by a pulse train.

original voice.[21] Since the baseband no longer contains all the frequencies needed to recover an intelligible speech signal, the higher frequencies must be regenerated from the baseband by some nonlinear distortion which preserves the original pitch period and provides an excitation signal with a sufficiently flat spectrum above the baseband.

The vocal-tract information needs only be transmitted above the upper cutoff frequency of the baseband. In the output signal the low-frequency part is best realized by the baseband itself since it contains the original voice. The principle of the voice-excited vocoder is thus best suited to operate with a channel vocoder since there the low-frequency information can easily be suppressed by omitting the low-order channels.

In the original invention by Schroeder and David (1960), the device worked as a "high-fidelity vocoder" which transmitted speech ranging from 0 to 10 kHz over a band-limited telephone channel. The baseband ranging from 80 to 2700 Hz was shifted upward by 600 Hz to fit into the upper band of the telephone channel. For the frequency band from 2.7 to 10 kHz the channel vocoder principle was used; the information was coded and transmitted in three channels which covered the low-frequency end (300-700 Hz) of the telephone line. On the synthesis side the high-frequency excitation signal was recovered, distorting the baseband by a nonlinear function with a zig-zag characteristic and logarithmic compression (Fig.9.16).

The idea to construct a semivocoder for high-fidelity speech transmission over a band-limited channel was in some way untypical for usual parametric-speech-transmission applications. Besides this vocoder, David et al. (1962a,b) thus proposed a voice-excited vocoder (basing on the channel-vocoder principle) for transmission of telephone-quality speech. In this case the bandwidth of the baseband was reduced to 700 Hz. David et al. (1962a,b) found that the total bandwidth required by the voice-excited vocoder could be reduced to 1000-1500 Hz, in contrast to a pitch-excited channel vocoder for which 500-600 Hz would suffice. (These data apply to an analog realization of the vocoder where bandwidth determines the channel capacity.) Golden (1963) postulated a frequency band of 250-920 Hz for the baseband. Gold and Tierney (1963) found that at least the first harmonic of the speech signal or, if the first harmonic is not available, two adjacent higher harmonics must be present in the baseband. So they postulated a frequency range of 150-450 Hz for the baseband of high-quality speech, and 300-900 Hz for telephone-quality speech.

---

[21] In an earlier patent, Feldman (1957) constructed a semivocoder with baseband transmission. In this vocoder, however, the higher frequency part of the excitation signal was not recovered from the baseband but generated in the conventional way as in a pitch-excited vocoder. In another earlier patent (Schmidt, 1942) cited by Schroeder (1961) a narrow baseband containing only the first harmonic was obtained by nonlinear distortion in the analysis part of the vocoder; on the synthesis side the excitation signal was recovered by a frequency multiplication circuit. A similar proposal was made later by Fujisaki and Nagashima (1970).

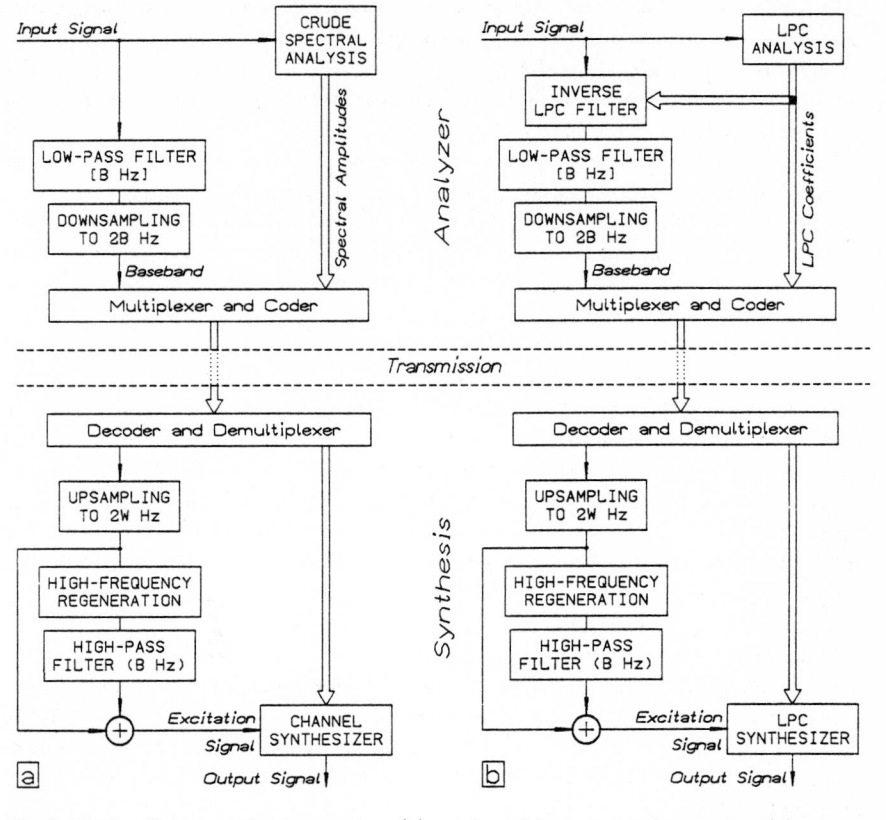

Fig.9.15a,b. Voice-excited vocoder (**a**) and residual-excited vocoder (**b**): principle of operation. (B) Bandwidth of the baseband; (W) bandwidth of the speech signal. In some of these vocoders a low-amplitude noise source is added to the excitation signal on the synthesis side. This source drives the synthesizer during voiceless intervals when there is no signal in the baseband

Fujisaki and Nagashima (1970) transmitted the original baseband from 300 to 450 Hz and created an artificial baseband from 150 to 300 Hz by nonlinear distortion.

Data on a digital voice-excited vocoder were first provided by Yaggi (1962). Yaggi presented a channel voice-excited vocoder operating at 9.6 kbits/s. The baseband ranging up to 900 Hz was sampled at 1800 Hz and quantized to 4 bits per sample yielding a transmission rate of 7.2 kbits/s for the baseband alone; the remaining 2.4 kbits/s were used to transmit the channel information. Gold and Tierney (1963) found that with a quantization of 6 bits/sample there was no audible quality difference between the quantized baseband and an analog baseband transmission. Using nonuniform samp-

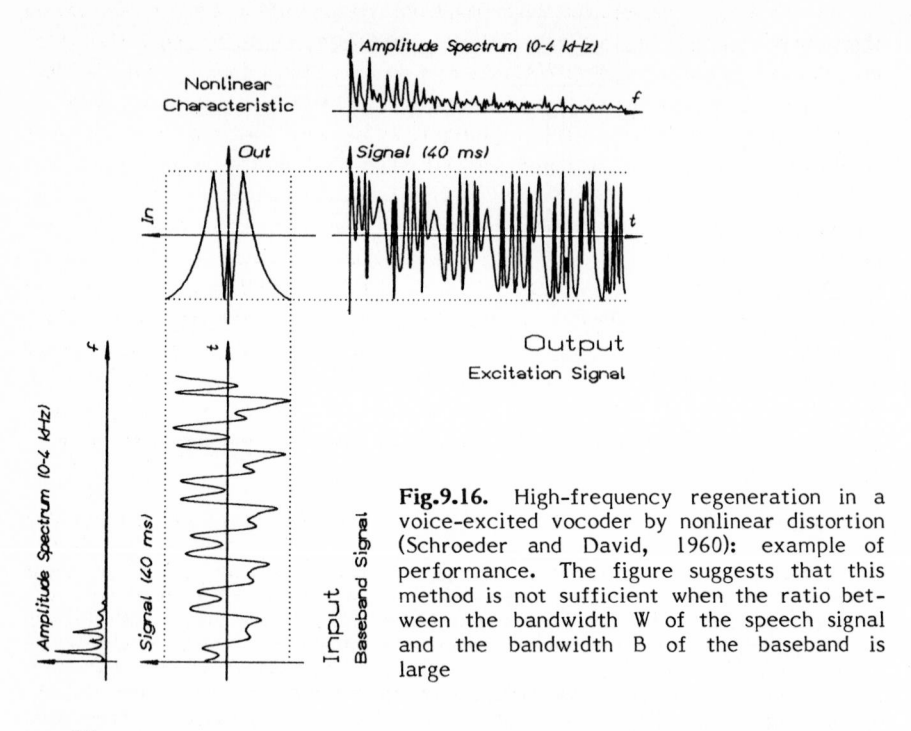

Fig.9.16. High-frequency regeneration in a voice-excited vocoder by nonlinear distortion (Schroeder and David, 1960): example of performance. The figure suggests that this method is not sufficient when the ratio between the bandwidth W of the speech signal and the bandwidth B of the baseband is large

ling,[22] Gold and Tierney reduced the sampling rate for the 300-900-Hz baseband to 1200 Hz and thus achieved a baseband transmission rate of 7.2 kbits/s with a quantization of 6 bits per sample.

From these examples we see that the greater part of the information flow in a voice-excited vocoder is required to transmit the baseband; on the other hand, the quality was found superior to pitch-excited vocoders.

> Our subjective evaluation enables us to say confidently that though the voice-excited vocoder introduces a measurable degradation, it is clearly superior to conventional vocoders. By overcoming the pitch problem and relieving the electrical accent, voice-excitation makes vocoders useful in a variety of contexts, particularly those where input conditions cannot be easily controlled and high-quality output speech is required. (David et al., 1962a, p.6)

---

[22] If a signal is limited to a frequency band of bandwidth B not containing zero frequency, the sampling theorem can be satisfied if the number of samples per second is greater than 2B even if the upper cutoff frequency of the band is considerably higher than B. To realize such a scheme, however, it is necessary either to modulate the signal in such a way that it covers the frequency band from zero to B, or to perform nonuniform sampling, as done in the present case. For further details the reader is referred to the original publication (Gold and Tierney, 1963).

The nonlinear circuit for high-frequency regeneration of the excitation signal was found to be the most critical component in the vocoder (David et al., 1962a,b). Some more details about this will be added later in this section.

Except for an investigation by Flanagan (1959) who constructed a voice-excited formant vocoder, the early investigations on voice-excited vocoders were confined to the channel vocoder principle. After 1964 it became more and more silent about the voice-excited vocoder. At that time research concentrated on improving the quality of conventional vocoders, and significant progress in this direction was made when reliable short-term analysis PDAs were developed (Noll, 1964, 1967; Sondhi, 1968). In the mid-seventies, the principle of baseband coding, as applied in the voice-excited vocoder, regained interest for two reasons: 1) the principle of linear prediction was found very suitable in connection with baseband coding, and 2) the transmission rate of 9.6 kbits/s, which is sufficient (and typical) for voice-excited vocoders, became interesting as a standard data rate for digital data transmission over ordinary telephone lines.

Currently, there is a growing interest in developing speech transmission systems that operate at "intermediate" data rates around 9.6 kbits/s. Several practical reasons exist for this interest, and important among these are: 1) demand on speech quality that is substantially better than currently available from narrowband speech coders (data rates below 4.8 kbits/s); 2) continued unavailability of wideband channels (data rates above 16 kbits/s) due to economical and other factors; and 3) recent availability of modems that operate reliably in the range of data rates around 9.6 kbits/s over regular telephone lines. The types of speech coders that are being investigated for the purpose of 9.6 kbits/s speech transmission include: voice-excited coders or baseband coders, subband coders, adaptive-transform coders (ATC), and adaptive predictive coders (APC). (Viswanathan et al., 1979, p.558)

So the voice-excited vocoder was ready to fill a gap between the pitch-excited vocoders ranging at and below 4.8 kbits/s and the nonparametric waveform coding schemes which operate best at data rates above 9.6 kbits/s (these devices shall not be further discussed here since they are not so much related to pitch determination), and a substantial literature has been published on voice-excited vocoders in recent years (Weinstein, 1975; Un and Magill, 1975; Esteban et al., 1978, 1979; Viswanathan et al., 1979, 1980, 1982; Dankberg and Wong, 1979; Makhoul and Berouti, 1979a,b; Katterfeldt, 1981; Un and Lee, 1982; and others). A number of proposals deal with the application of the voice-excitation principle to the LPC vocoder (e.g., Weinstein, 1975; Viswanathan et al., 1979).

In contrast to the channel and formant vocoders, the LPC vocoder permits a second way of baseband coding which leads to the *residual-excited* vocoder (Un and Magill, 1975; Makhoul and Berouti, 1979a,b). In the residual-excited vocoder the baseband is formed of the LPC residual instead of the original voice. From the discussion in Sects.6.3.1 and 6.5.1 we know that the LPC residual $e(n)$, if transmitted without distortion, serves in connection with the LPC filter to exactly reconstruct the original speech signal. For this reason the LPC residual is extremely well suited to excite a LPC synthesizer even if it is crudely downsampled and/or quantized. Figure 9.15b shows the block diagram of this vocoder. After determination of the predictor coefficients the speech signal passes the inverse LPC filter in order to obtain the residual. As in the voice-excited vocoder the residual is then low-pass filtered, quantized and transmitted and forms the coded baseband of the vocoder; the vocal-tract information is represented by the predictor coefficients. On the synthesis side the high-frequency part of the residual is regenerated by nonlinear distortion; and the original baseband and the reconstructed high-frequency part then serve as the excitation signal of te LPC synthesizer filter.

The residual-excited vocoder was improved and further developed by a number of researchers (e.g., Watkins, 1979; Katterfeldt, 1981; Viswanathan et al., 1982; Kang et al., 1982). Efficient coding schemes have been developed in recent years so that both the voice-excited vocoder and the residual-excited vocoder can satisfactorily operate with transmission rates down to 4.8 kbits/s.

The most critical component of both the voice-excited vocoder and the residual-excited vocoder is the high-frequency regeneration circuit. The baseband has a cutoff frequency of 800 Hz or even less; on the other hand, the excitation signal should have a flat spectrum up to 3.4 kHz. In addition, it is mandatory that the information on pitch transmitted via the baseband is preserved in the high-frequency regeneration circuit. The nonlinear function with the zig-zag characteristic and logarithmic compression, as originally proposed by Schroeder (1959), is one possibility to accomplish this. Another way is (half-wave) rectification and subsequent spectral flattening, which is done, e.g., by double differentiation (Golden, 1963; Un and Magill, 1975) or by LPC inverse filtering (Weinstein, 1975) of the rectified signal. Esteban et al. (1978) showed that part of the remaining roughness in digital voice-excited vocoders and residual-excited vocoders is due to aliasing distortions; the high-frequency regeneration circuit, being a nonlinear device, generates a large number of high-frequency components, and it is likely that some of them violate the sampling theorem. Esteban et al. thus recommended oversampling the excitation signal by a factor of at least two before regenerating the high frequencies, and returning to the original sampling rate of the speech signal at the input of the synthesizer filter. High-frequency regeneration by this way of nonlinear distortion, however, is always critical when the ratio of the bandwidths of the speech signal and the baseband becomes large (see the example in Fig.9.16).

Another possibility of high-frequency regeneration is spectral folding (Makhoul and Berouti, 1979a,b; Dankberg and Wong, 1979). Here the baseband is simply upsampled by interpolation of zero samples without applying an anti-aliasing filter so that the high frequencies are regenerated by aliasing distortions. This procedure is very simple and guarantees significant spectral component throughout the whole frequency range of the signal; however, it introduces audible tonal noise since the dominant spectral components in the regenerated frequency band are usually not located at integer multiples of F_0;

thus the spectrum of the excitation signal becomes inharmonic. To suppress this effect, countermeasures such as random perturbation of selected high-frequency components (Viswanathan et al., 1982) or pitch-dependent frequency-domain regeneration of the excitation signal (Katterfeldt, 1981) have been proposed.

Most investigations agree that both the voice-excited vocoder and the residual-excited vocoder operate much better in a noisy environment than a conventional pitch-excited vocoder (Viswanathan et al., 1982).

In summary, baseband coders, i.e., voice- and residual-excited vocoders, provide a viable alternative to the conventional vocoder, however, at the expense of a certain increase in the transmission rate. For applications where a medium data rate (4.8 kbits/s and more) is tolerable, the baseband coder performs better than a conventional vocoder with respect to robustness against noise and signal degradations; the speech sounds more natural, and the problems with pitch and voicing determination have disappeared.

It is not the aim of this book to discuss aspects of vocoding in more detail; the few examples discussed in this section only serve to show that there are possibilities to circumvent the problems with the PDA and the VDA in speech communication. On the other hand, the examples show that finally there is only one way to avoid these problems: abandoning parametric transmission of the voice cource information and transmitting this information directly by waveform using a suitable method of waveform coding.

9.5 Possible Paths Towards a General Solution

Table 9.2 was the attempt to establish a profile of requirement for the different applications. One might try to set up a similar profile for the individual concepts of existing PDAs and match it with the profile of requirements. There is some difficulty in trying that, however. A number of items, such as the display facility, which is a matter of the postprocessor, can be excluded since they are applicable in any PDA. The question of real-time performance can also be disregarded since even the most complicated or arithmetically cumbersome PDAs may find a hardware implementation which works in real time - if not today, then within a few years. As to the costs of a PDA (e.g., for the educational application where it must be cheap), a considerable cost factor of a portable PDD still is the display facility in the form of an oscilloscope. But, whither an oscilloscope? Today's technology permits preparing the data by one of the commercially available converters for display on a normal TV screen so that such a PDA could be used in connection with any standard TV set.

If we check the remainder of the profile against the existing PDA concepts, it appears that only the combination of robustness, wide measuring range, and high reliability required by the application in speech communication systems forms the problem not met by today's PDAs. All the other areas of applications can be satisfied. Since most points in connection with these applications have already been discussed, we can leave these applications and confine our final considerations to the speech communication problem.

One of the aims of this book, and a main reason for the relatively broad discussion of the older principles from the fifties and even earlier years is the fact that these old PDAs and PDDs exhibit a number of good ideas which failed only because they were not yet practicable at that time and which were never again picked up. For instance, there is the idea of the multichannel PDA. Instead of developing a new PDA, one might combine several existing principles and put the effort in evaluating the cooperation between them in order to improve the overall performance. For example, there is the PDD by Barney (1958) which successfully uses a primitive and inaccurate (but rather reliable) PDD to create the preset for the more accurate and elaborate PDD which has only the disadvantage that it needs a crude estimate in order to start. There are a number of PDAs like cepstrum which might again be realized in analog sampled-data hardware some time in the near future, as originally intended. There is the investigation in Chap.7 leading to the statement that any single nonlinear function in the category PiTN will fail with a certain type of signal, but that an appropriate combination of several nonlinear functions will not. Finally, there is the interactive PDA by McGonegal et al. (1975) which unites three totally different principles (autocorrelation, cepstrum, and time-domain PiTS) in order to provide a basis for the user's decision. These examples show that substantial improvements can emerge from combining different complementary principles. Thus the first statement towards future design of PDAs reads as follows:

Rather than developing new principles of pitch determination, one should take the existing ones and combine them in an appropriate way to yield an overall improvement of performance.   (9.10)

Can even this current review give some proposals as to how to combine PDAs? The first point will deal with correction routines. Coding theory tells us that it is much easier to recognize an error than to reliably correct it. Since any sophisticated PDA has powerful correction routines, it might easily use them to recognize its failure and to pass control to the colleague which might not fail for this particular signal. Thus the second statement reads:

Rather than developing sophisticated correction routines, one should take the existing ones and trim them to reliably recognize an erroneous situation (whether it be due to a failure of the PDA or to a perturbation in the signal) and to pass control to another PDA which might not fail with this particular signal.   (9.11)

An important point in successfully combining several principles is selecting them in such a way that their performance is complementary, i.e., that the one works successfully where the other fails, and vice versa. The PDA discussed in Chap.7 is such an example. Can we transfer this idea to the short-term analysis principle? Definitely. Cepstrum fails when few harmonics are present. Autocorrelation easily copes with this situation but fails when there is a dominant formant structure - a situation easily managed by the cepstrum. Thus the two are complementary at this point. SIFT fails with high-pitched voices whereas it easily copes with low pitches. Ordinary autocorrelation performs just the other way round (Rabiner et al., 1976). Thus the third statement reads:

> *When combining several principles one must take care that they perform in a complementary way so that the one works well where the other fails and vice versa.* (9.12)

For this, further evaluative work is necessary. The evaluation by Rabiner et al. (1976) has marked a milestone since it was about the first to really evaluate PDAs (without designing a new one) on a comparative basis. The main statement if this evaluation - that *all* PDAs have their shortcomings - leads to the open question to find out where they fail and why. Investigating this, one might get an answer as to which PDAs behave in a mutually complementary way. Thus the fourth statement reads:

> *Further work is necessary to evaluate the dynamic performance of PDAs, i.e., to find out which PDA fails at what signal, whether there are situations where most PDAs fail* (labeling an irregular signal as irregular, however, must not be regarded as a failure!), *and whether different PDAs can be combined in a complementary way.* (9.13)

The last statement arises from the discrepancy between subjective and objective evaluation. Apparently it is not yet known how the errors committed by a PDA affect the perception of speech quality. Since the ear is the final judge, the PDA must be designed as to optimally meet the subjective standard. Therefore the fifth statement reads

> *Further work is necessary to correlate the objectively evident errors, such as holes, chirps, hops, or jitter, to the general subjective judgment of performance. In particular, the places in the speech signal must be determined at which the ear is most sensitive to such errors. The aim of this investigation will be to supply criteria which permit* **objective** *evaluation of a PDA in order to optimize it towards a good* **subjective** *performance.* (9.14)

At the end of this chapter, we are back at the questions raised initially. Some of them have been answered. The main problem of pitch determination was found to be the failure of today's PDAs to meet the requirements imposed by high-quality analysis-synthesis speech transmission systems where the ear is the final judge. It is not very likely that a new, revolutionary principle will solve these problems all at once. The combination of several known principles with complementary behavior, in connection with a sophisticated control logic, however, may form a new concept which hopefully will permit research in pitch determination to advance one step further towards a general solution.

Appendix A. Experimental Data on the Behavior of Nonlinear Functions in Time-Domain PDAs

This appendix presents detailed data on the behavior of nonlinear functions (NLFs) in time-domain PDAs. The data result from the experiment discussed in Chapter 7. Appendix A is subdivided into seven parts: 1) a display of the data base; 2) some extreme examples; 3) detailed data on the relative amplitude RA1 and the enhancement RE1 of the first harmonic; 4) detailed data on the relative amplitude RASM of the spurious peak in signal threshold analysis as well as in autocorrelation analysis; 5) data on the influence of band limitation, processing sequence, and high frequency preemphasis; 6) data on the choice of the optimal NLF; and 7) some additional examples on the performance of comb filters. The main parameters are

1) the *relative amplitude* RA1 and the *enhancement* RE1 of the *first harmonic,* as defined in (7.11,13). An individual NLF or a class of NLFs performs best when these parameters take on high values. The same applies to the parameter RS1, defined in (6.2,3) which gives a measure for the applicability of nonlinear distortion in connection with zero crossings analysis;

2) the *relative amplitude* RASM of the *largest spurious peak,* as defined in (7.23). This parameter is important whenever nonzero threshold analysis or peak detection is to be applied. An individual NLF or a class of NLFs perform best when this parameter is low.

All the *signals* displayed in Appendix A are regarded as strictly periodic. Two periods are depicted (three in Figs.A.28-30). The time scale is stretched in such a way that the two periods just fit the space of the figure. The actual length can be taken from the text in Figs.A.1-5. All the signals were normalized before plotting.

The procedure used to obtain the *filtered signal* was described in Section 7.1. To the signals displayed in the figures a double two-pulse filter was applied. The dashed lines indicate the work area of threshold analysis between the principal peak and the second largest maximum (or minimum), provided that at least one zero crossing occurs between these peaks. The second line is omitted when the second largest maximum does not exist or has a relative amplitude below 0.1 (-20 dB). The written-in number gives the amplitude ratio of these two peaks. Due to lack of space the number is omitted when the ratio exceeds 0.75. The polarity of the threshold is chosen where the spurious peak takes on the lower relative amplitude.

All the *spectra* are harmonic spectra computed via the discrete Fourier transform. They display a frequency range from 0 to 5 kHz and an amplitude range of 40 dB.

All the *autocorrelation functions* displayed are periodic since the signal is processed as strictly periodic. The principal peak at $T_0 = 1/F_0$ therefore always takes on a value of 1. The second largest peak is marked by a dashed line and a written number. ACF values below -0.5 were clipped. One period of the autocorrelation function is fit into the space of the figure.

As to the *speakers*, speakers 1 (HRS, male) and 2 (WGH, male) represent the signal categories "F_0 low, undistorted" (SiLU) and "F_0 low, band limited" (SiLL), whereas speaker 3 (IRH, female) represents the category "F_0 high" (SiH).

Differences of performance are defined to be logarithmic differences since all the parameters displayed here are logarithmic. Thus, for instance, the difference of the relative amplitude RA1 for two conditions C_1 and C_2 is defined as follows:

$$DRA1 := RA1(C_1) - RA1(C_2) \quad [dB] .$$

In the captions of all figures where such differences are involved, condition C_1 is named first. According to the definition of the parameter from which the differences are derived, the algorithm fulfilling condition C_1 performs better a) if RA1 or RE1 is the underlying parameter, and the difference is positive, or b) if RASM is the underlying parameter, and the difference is negative.

In some of the comprehensive views (such as Fig.A.18) the *nonlinear functions* are indicated by a number. This number corresponds to the one in the specifications given in Table 7.4. For the convenience of the reader this table is repeated here.

A.1 The Data Base of the Investigation (Figs.A.1-5)

The data base of the investigation contains 48 sample pitch periods uttered by the three speakers mentioned above. The periods were manually determined from isolated words. The recordings were made in an anechoic chamber on an FM tape drive, so that the signal is free from amplitude and phase distortions.

The periods were selected from 12 German vowels and 4 sustained consonants. Since German is a language with a comparatively high number of vowels, and since the extreme positions of the formant chart (corresponding to the vowels /u/, /i/, and /a/) as well as the center position (corresponding to the *schwa*) are well documented, one can regard this selection - although comparatively small - as representative for most German spoken phonemes as well as for the spoken phonemes of other languages with a similar phonetic inventory.

Table 7.4 (repeated from Chap.7). List of nonlinear functions applied to the sample periods. All functions are listed in FORTRAN notation (except that no capital letters are used). (x) Input signal

Number	Label	Function (y=...)	Mode	Comments
1	FuL	x	odd	linear
2	FuR03	abs(x) ** 0.33	even	
3	FuR05	abs(x) ** 0.5	even	
4	FuR	abs(x)	even	full-wave rectification
5	FuR20	abs(x) ** 2	even	square
6	FuE20	1 - exp(-x*x)	even	
7	FuH	max (x, 0)	mixed	one-way rectification
8	FuL05	sgn(x) * abs(x)**0.5	odd	
9	FuL20	x * abs(x)	odd	squaring with sign of x maintained
10	FuM	abs(x) + x/2	mixed	nonideal rectification
11	FuCCR	abs(clc(x))	even	center clip, compress, and rectify
12	FuCPR	abs(clp(x))	even	center clip and rectify
13	FuCP	clp(x)	odd	center clipping
14	FuSH5	max(ssb(x),0)**0.5	SSB	SSB modulation
15	FuSR5	abs(ssb(x))**0.5	SSB	
16	FuST5	abs(sst(x))**0.5	SSB	SSB modulation with DC component added

The limited amount of data, however, necessitates an interpretative approach which may differ from the usual statistical point of view. The aim of the investigation was to gain insight into the behavior of nonlinear functions in connection with different principles of PDAs. If thus the failure of a certain combination (of nonlinear function, PDA principle, and signal) is reported, it is not just a failure which increases some hypothetic error rate, but it is a systematic failure with the consequence that under certain recording conditions a certain principle (or NLF) might not be applicable to a whole class of phonemes.

Figures A.1-5 display the sample periods of the data base. For each spoken phoneme the signal (upper part, two periods), the log amplitude spectrum (0 to 5 kHz, center part; the individual harmonics are indicated by vertical lines), and the autocorrelation function (1 period) are displayed.

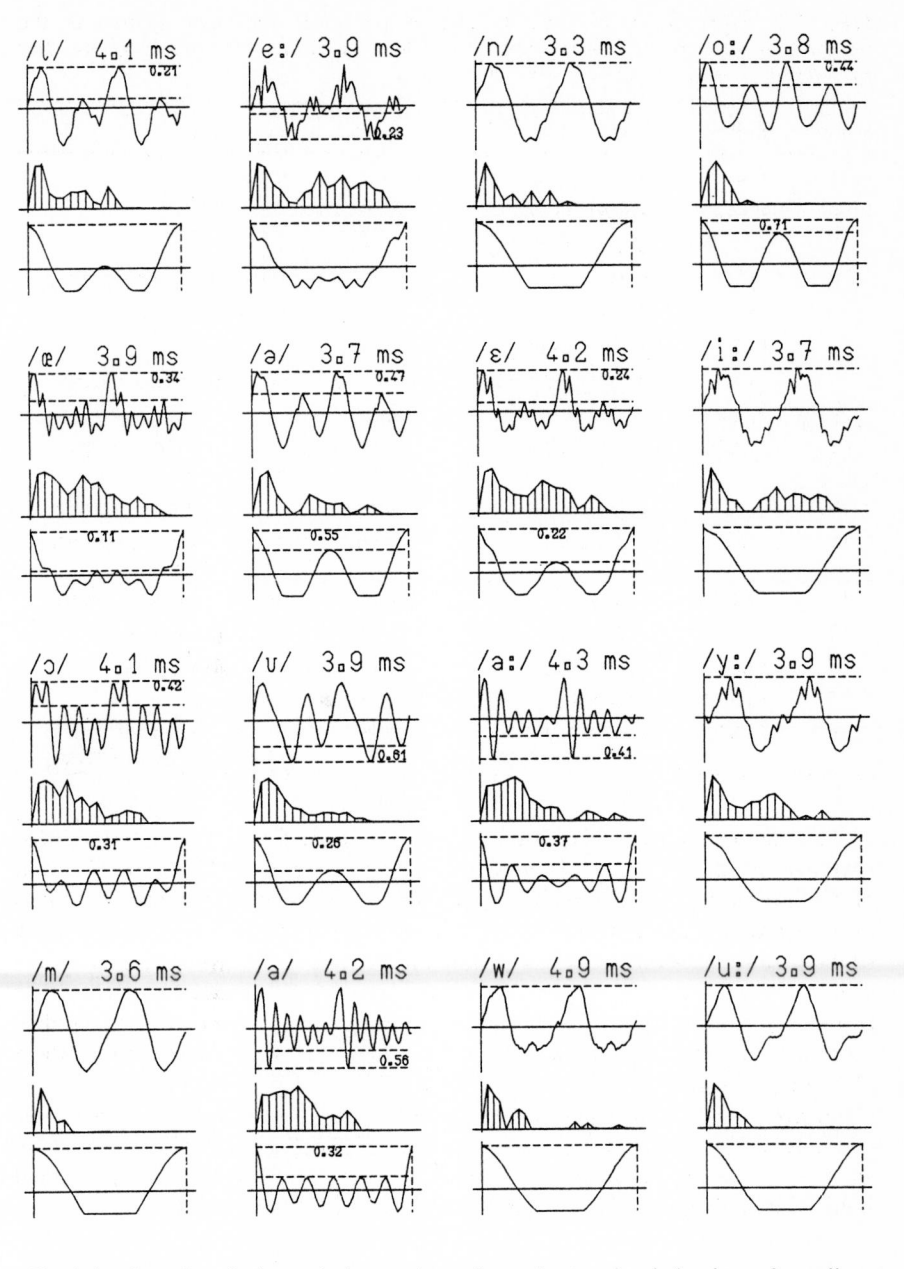

Fig.A.1. Sample pitch periods used to investigate the behavior of nonlinear functions in PDAs. Speaker 3 (IRH, female); undistorted signal. (Upper part of each display) Signal (2 periods); (center part) log amplitude spectrum (individual harmonics marked by vertical lines); (lower part) autocorrelation function (1 period)

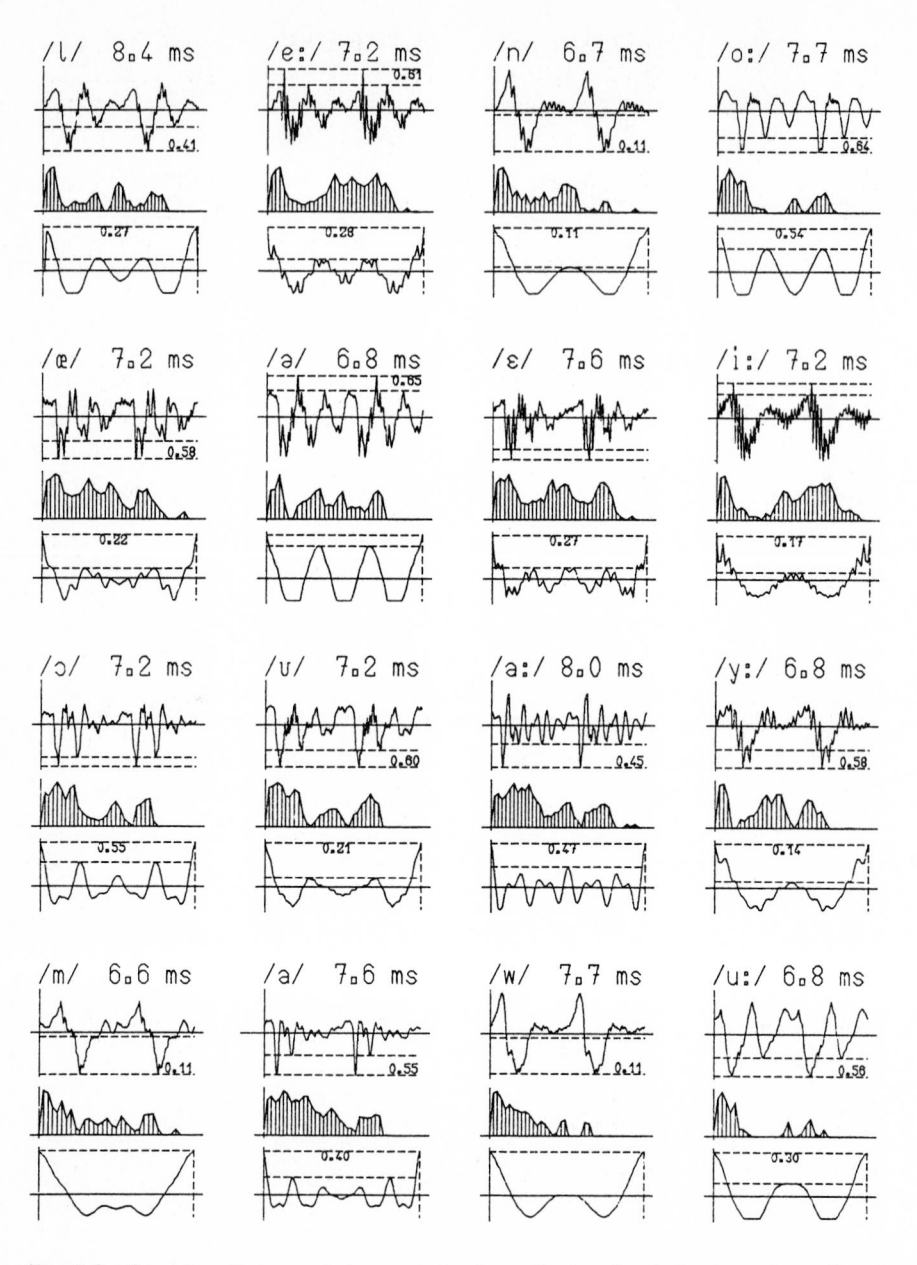

Fig.A.2. Sample pitch periods used to investigate the behavior of nonlinear functions in PDAs. Speaker 1 (HRS, male); undistorted signal. (Upper part of each display) Signal (2 periods); (center part) log amplitude spectrum (individual harmonics marked by vertical lines); (lower part) autocorrelation function (1 period)

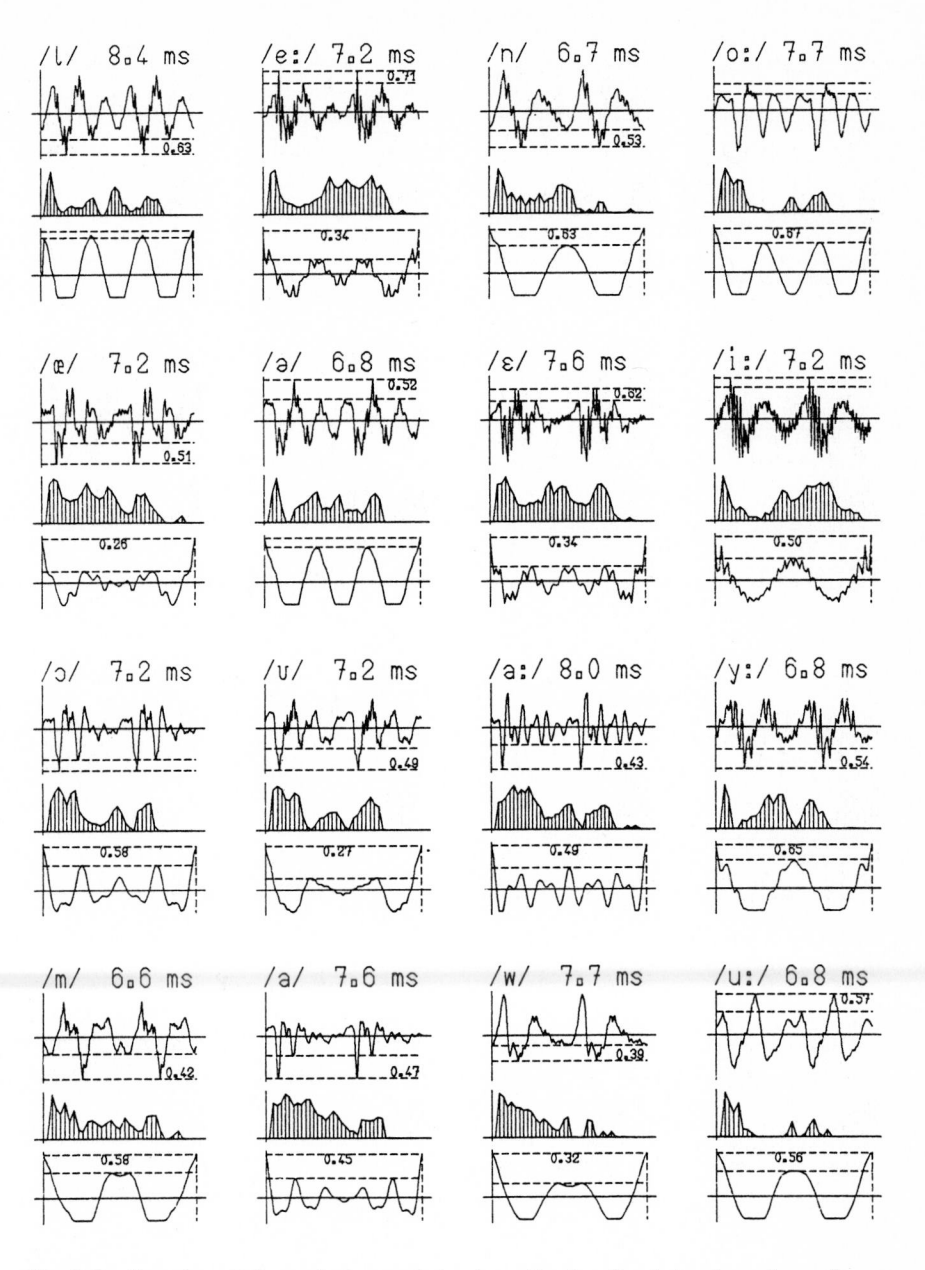

Fig.A.3. Sample pitch periods used to investigate the behavior of nonlinear functions in PDAs. Speaker 1 (HRS, male); band-limited signal. (Upper part of each display) Signal (2 periods); (center part) log amplitude spectrum (individual harmonics marked by vertical lines); (lower part) autocorrelation function (1 period)

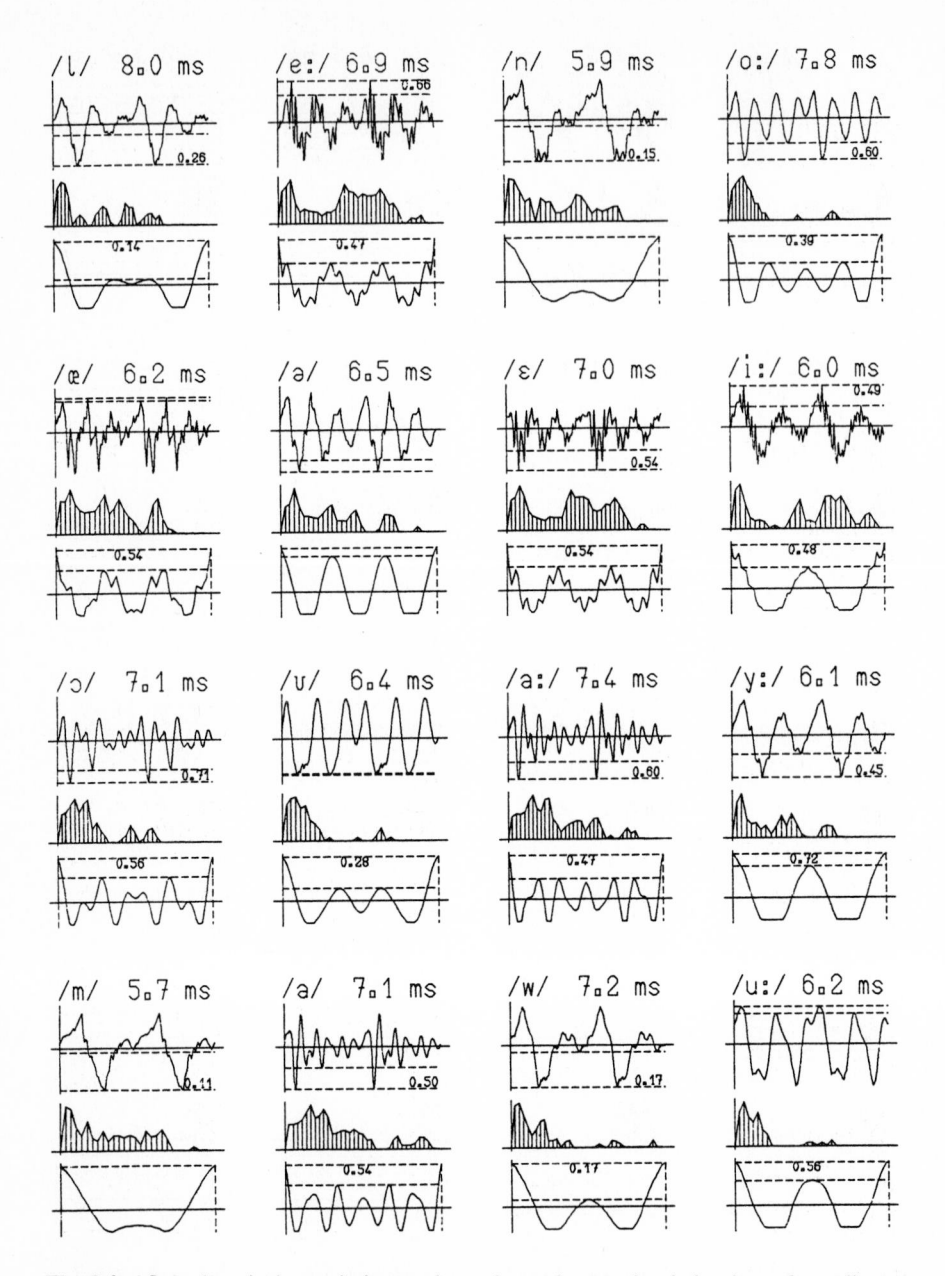

Fig.A.4. Sample pitch periods used to investigate the behavior of nonlinear functions in PDAs. Speaker 2 (WGH, male); undistorted signal. (Upper part of each display) Signal (2 periods); (center part) log amplitude spectrum (individual harmonics marked by vertical lines); (lower part) autocorrelation function (1 period)

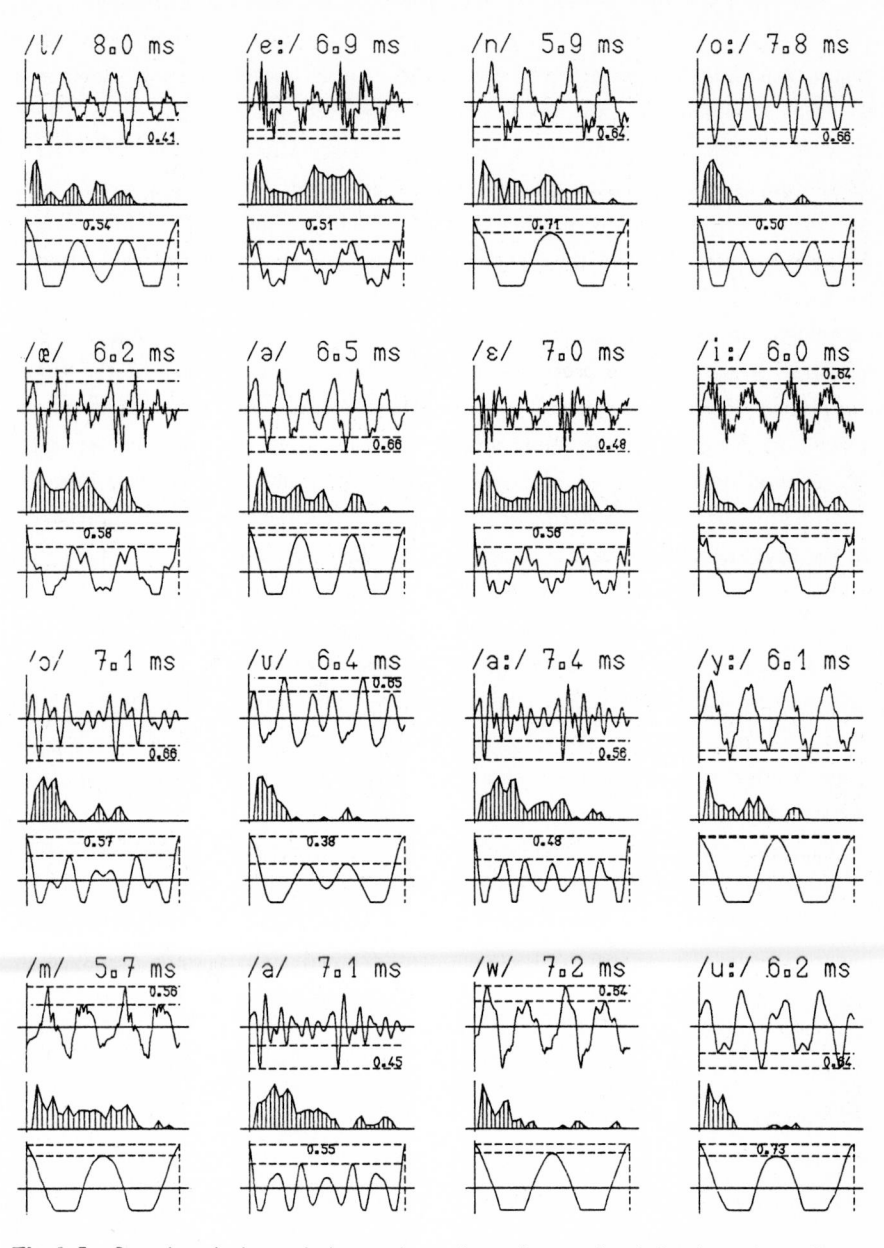

Fig.A.5. Sample pitch periods used to investigate the behavior of nonlinear functions in PDAs. Speaker 2 (WGH, male); band-limited signal. (Upper part of each display) Signal (2 periods); (center part) Log amplitude spectrum (individual harmonics marked by vertical lines); (lower part) autocorrelation function (1 period)

A.2 Examples for the Behavior of the Nonlinear Functions (Figs.A.6–14)

Figure A.6 gives an example for the application of center clipping. Compared to Fig.A.3, which the data were derived from, the spectral flattening effect as well as the favorable effect on the autocorrelation function are clearly seen. Figures A.7-11 display the behavior of individual NLFs in extreme cases: the vowel /a:/ uttered by speaker 1 (HRS, male) as the example of greatest distance between F_0 and the formant F1 (Figs.A.7,8); the vowel /y:/ uttered by speaker 2 (WGH, male) showing the greatest prominence of the second harmonic, and finally the consonant /m/ uttered by the female speaker demonstrating the extreme prominence of the first partial. Instead of the auto-correlation function, however, these figures display the signal after crude inverse filtering (using the filter designed for the PDA described in Sect.7.1). The applied adaptive filter, a time-variant single comb filter, was prevented from destroying the first harmonic by a downward limitation of the frequency of its zero to two times the fundamental frequency. The zero of the comb filter was always adjusted to the *true* formant F1 which had been derived from the undistorted signal, regardless of the applied nonlinear distortion.

The figures show the advantage of the nonlinear distortion by even functions in the case of low F_0; they also show the detrimental effect of the same kind of distortion when F_0 is high. In addition, the incompatibility of linear inverse filtering (linear spectral flattening) and center clipping (nonlinear spectral flattening) is clearly indicated.

The performance of the nonlinear functions selected for the PDA in Chap.7 is shown in Figs.A.12-14 for a band-limited signal. For each sample period the figure displays the signal after distortion (upper part) and after subsequent linear filtering (lower part). The pertinent original signals were displayed in Fig.A.2 (undistorted signal) and Fig.A.3 (band-limited signal).

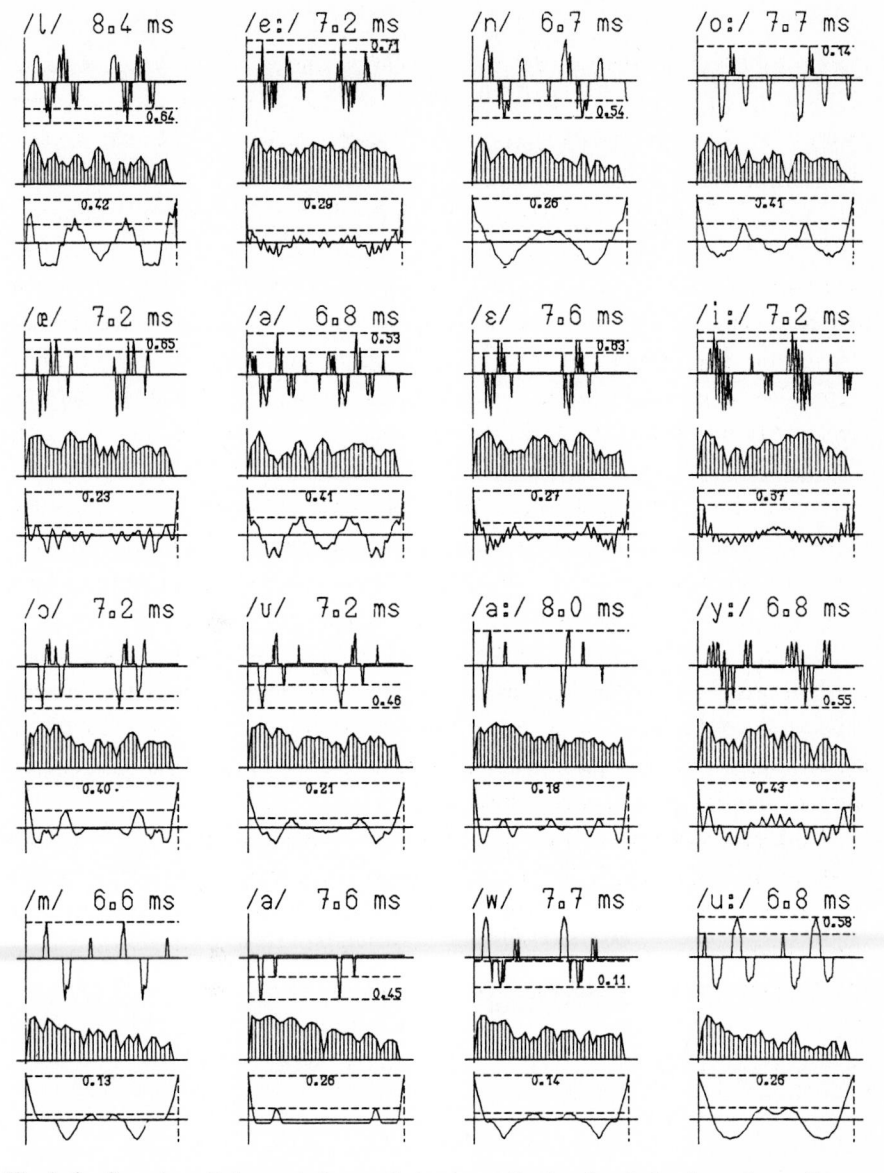

Fig.A.6. Sample pitch periods used to investigate the behavior of nonlinear functions in PDAs. Speaker 1 (HRS, male); signal distorted by center clipping function clp(x). (Upper part of each display) Signal (2 periods); (center part) log amplitude spectrum (individual harmonics marked by vertical lines); (lower part) autocorrelation function (1 period). The clipping threshold was set to 50% of the maximum amplitude

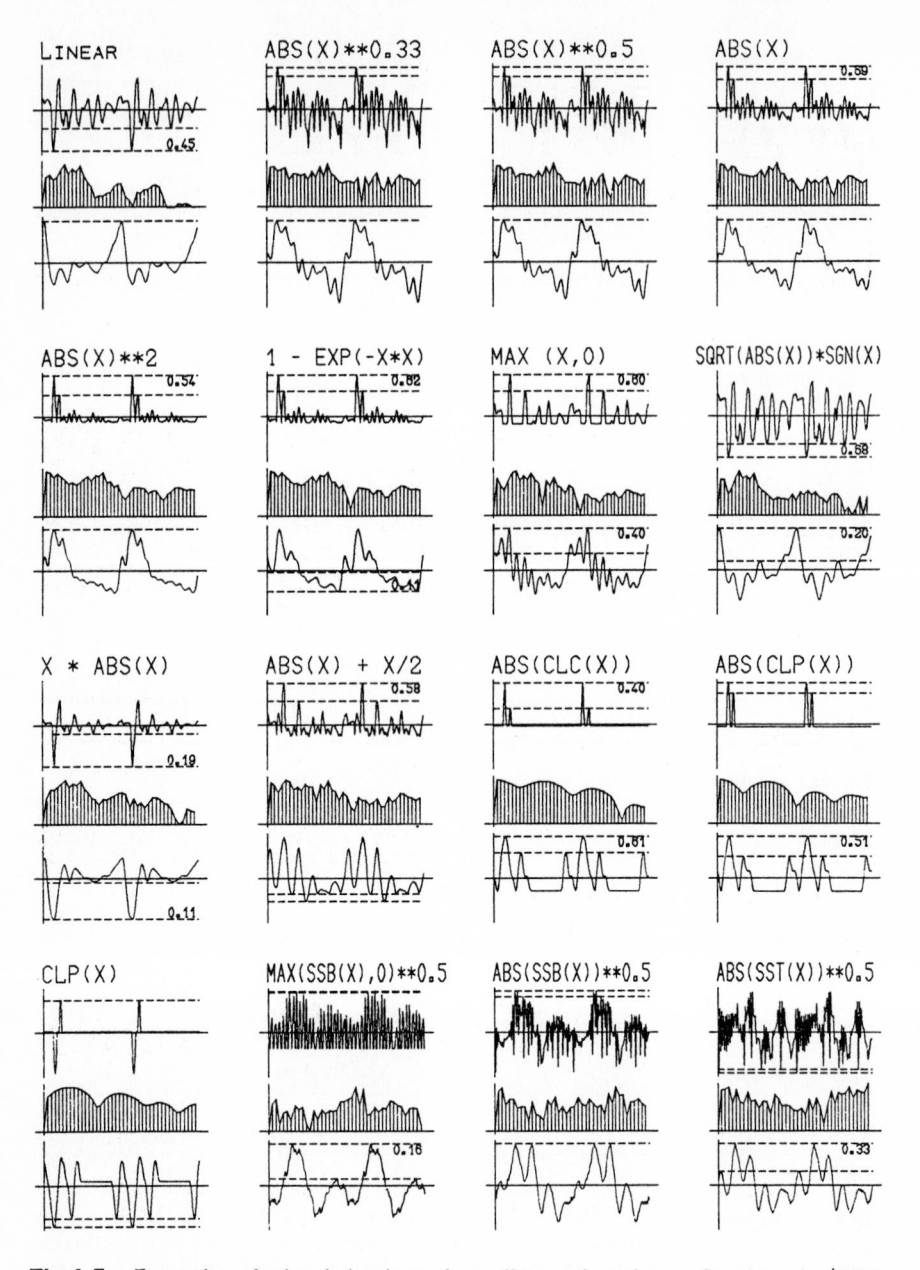

Fig. A.7. Example of the behavior of nonlinear functions. Speaker 1 (HRS, male); undistorted signal; phoneme /aː/. (Upper part of each display) Signal after processing by the NLF indicated (2 periods); (center part) log amplitude spectrum (individual harmonics marked by vertical lines); (lower part) filtered signal (filter from the linear PDA in Chap. 7)

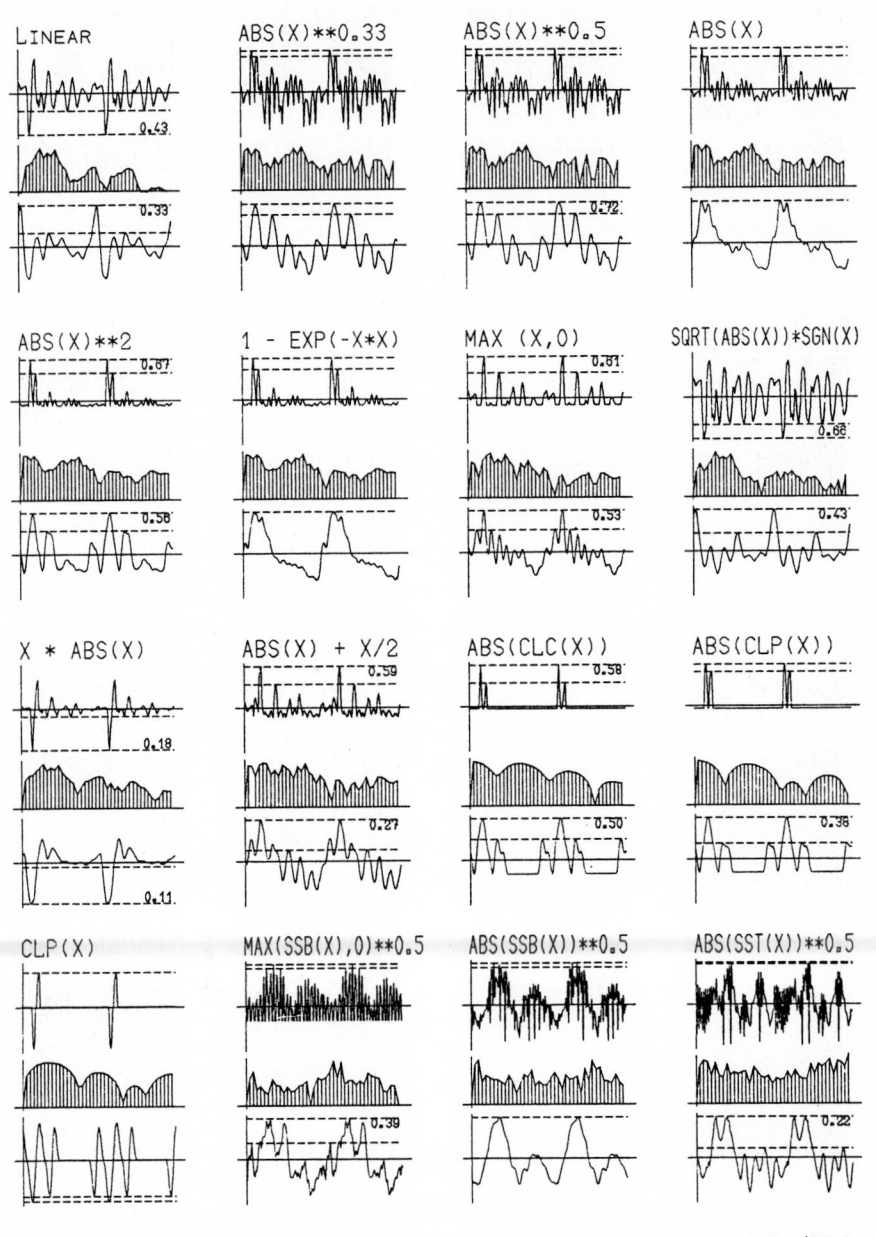

Fig. A.8. Example of the behavior of nonlinear functions. Speaker 1 (HRS, male); band-limited signal; phoneme /a:/. (Upper part of each display) Signal after processing by the NLF indicated (2 periods); (center part) log amplitude spectrum (individual harmonics marked by vertical lines); (lower part) filtered signal (filter from the linear PDA in Chap. 7)

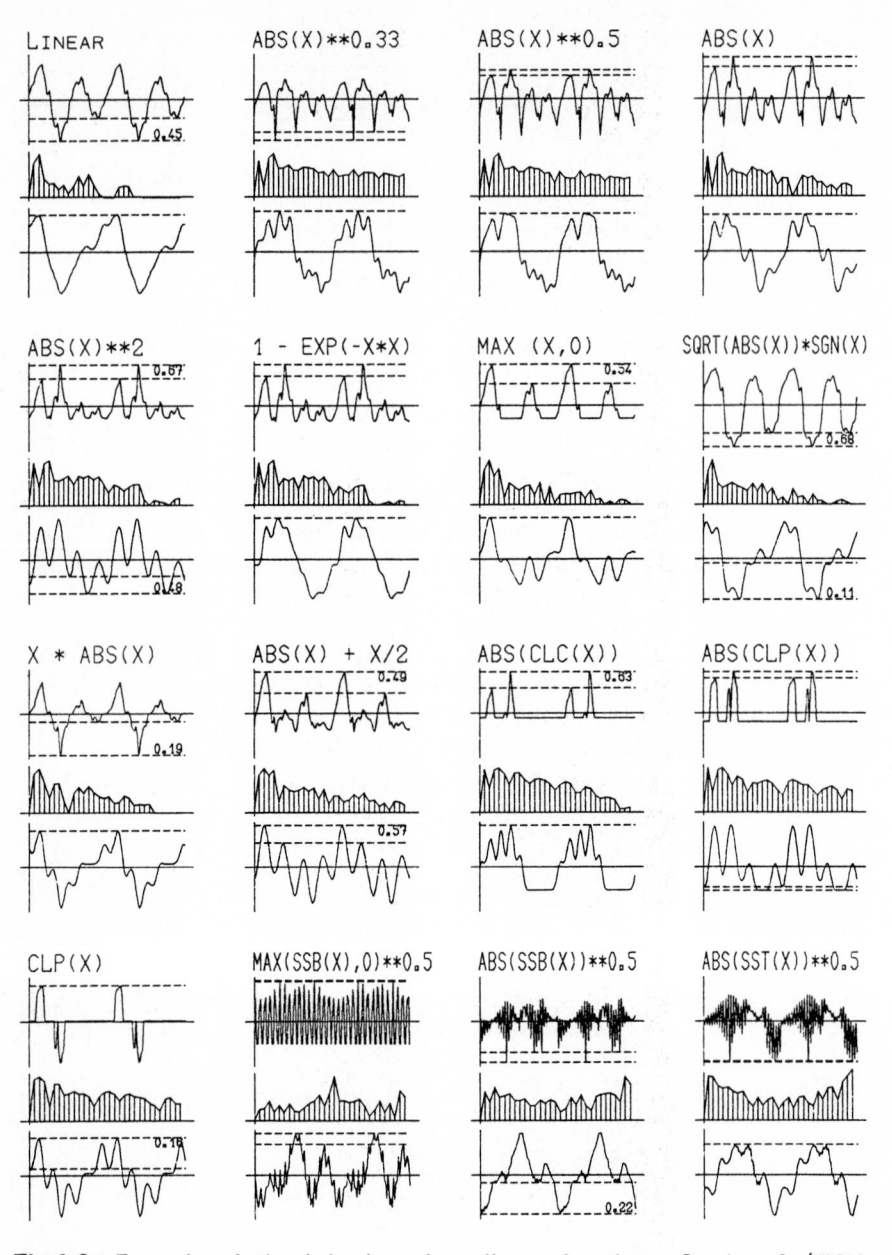

Fig.A.9. Example of the behavior of nonlinear functions. Speaker 2 (WGH, male); undistorted signal; phoneme /y:/. (Upper part of each display) Signal after processing by the NLF indicated (2 periods); (center part) log amplitude spectrum (individual harmonics marked by vertical lines); (lower part) filtered signal (filter from the linear PDA in Chap.7)

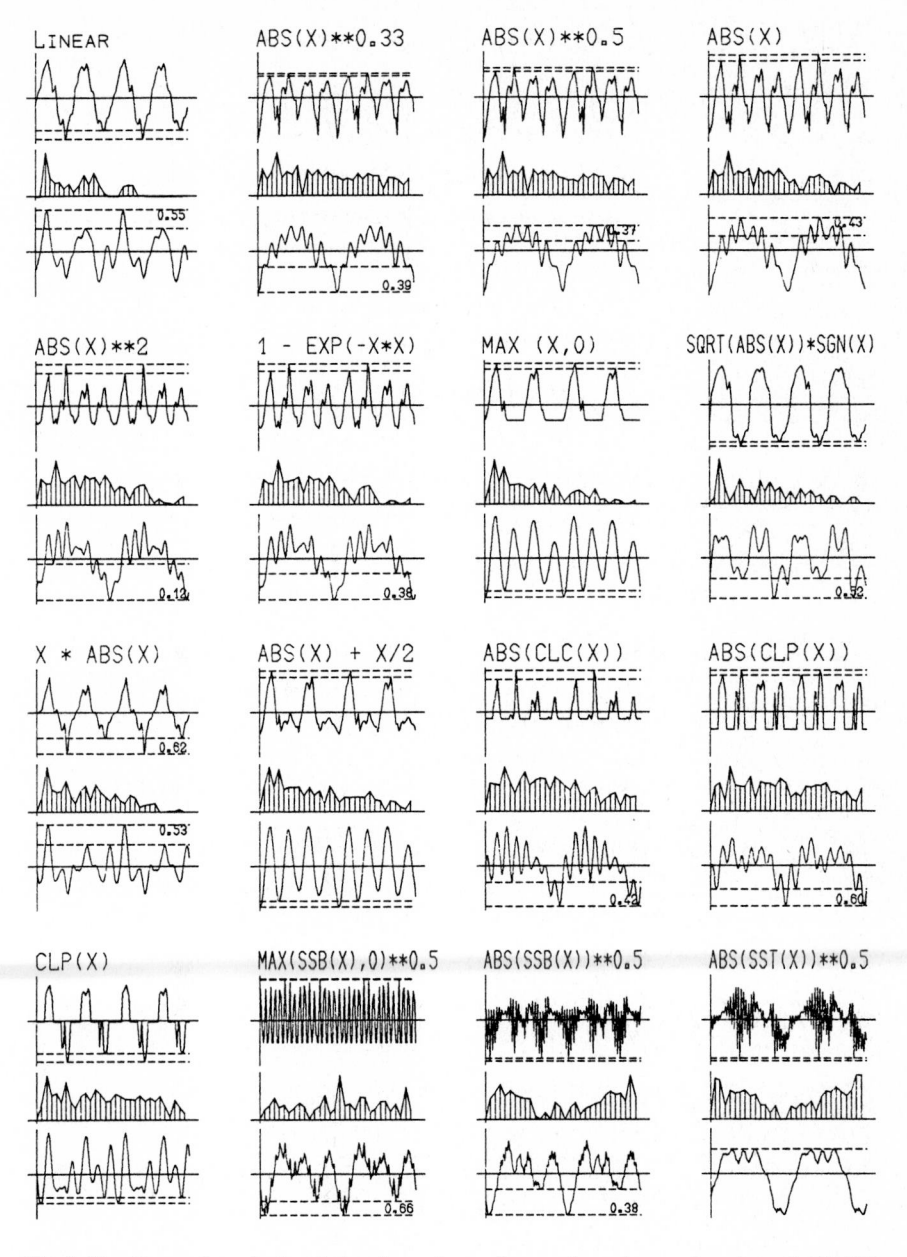

Fig.A.10. Example of the behavior of nonlinear functions. Speaker 2 (WGH, male); band-limited signal; phoneme /y:/. (Upper part of each display) Signal after processing by the NLF indicated (2 periods); (center part) log amplitude spectrum (individual harmonics marked by vertical lines); (lower part) filtered signal (filter from the linear PDA in Chap.7)

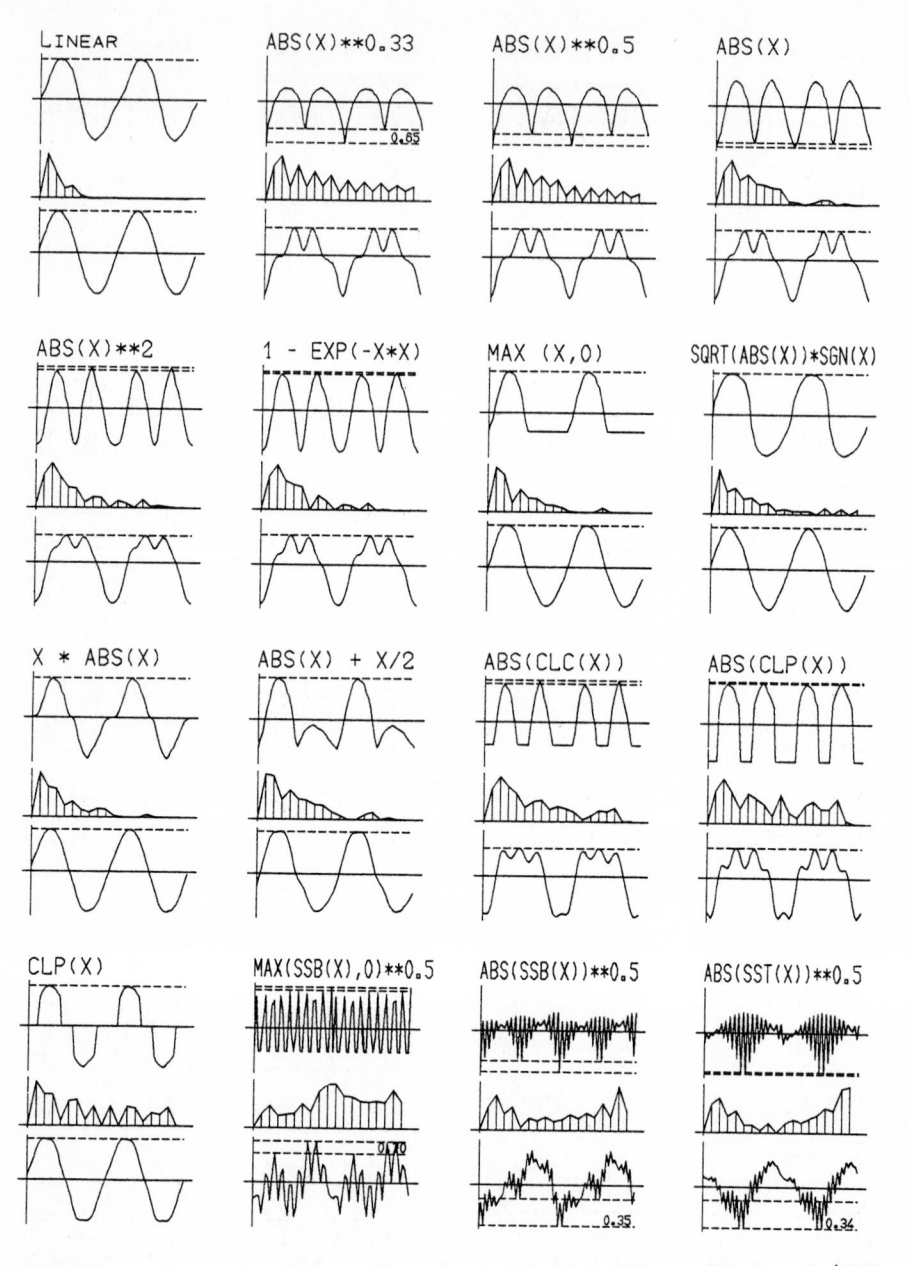

Fig.A.11. Example of the behavior of nonlinear functions. Speaker 3 (IRH, female); undistorted signal; phoneme /m/. (Upper part of each display) Signal after processing by the NLF indicated (2 periods); (center part) log amplitude spectrum (individual harmonics marked by vertical lines); (lower part) filtered signal (filter from the linear PDA in Chap.7)

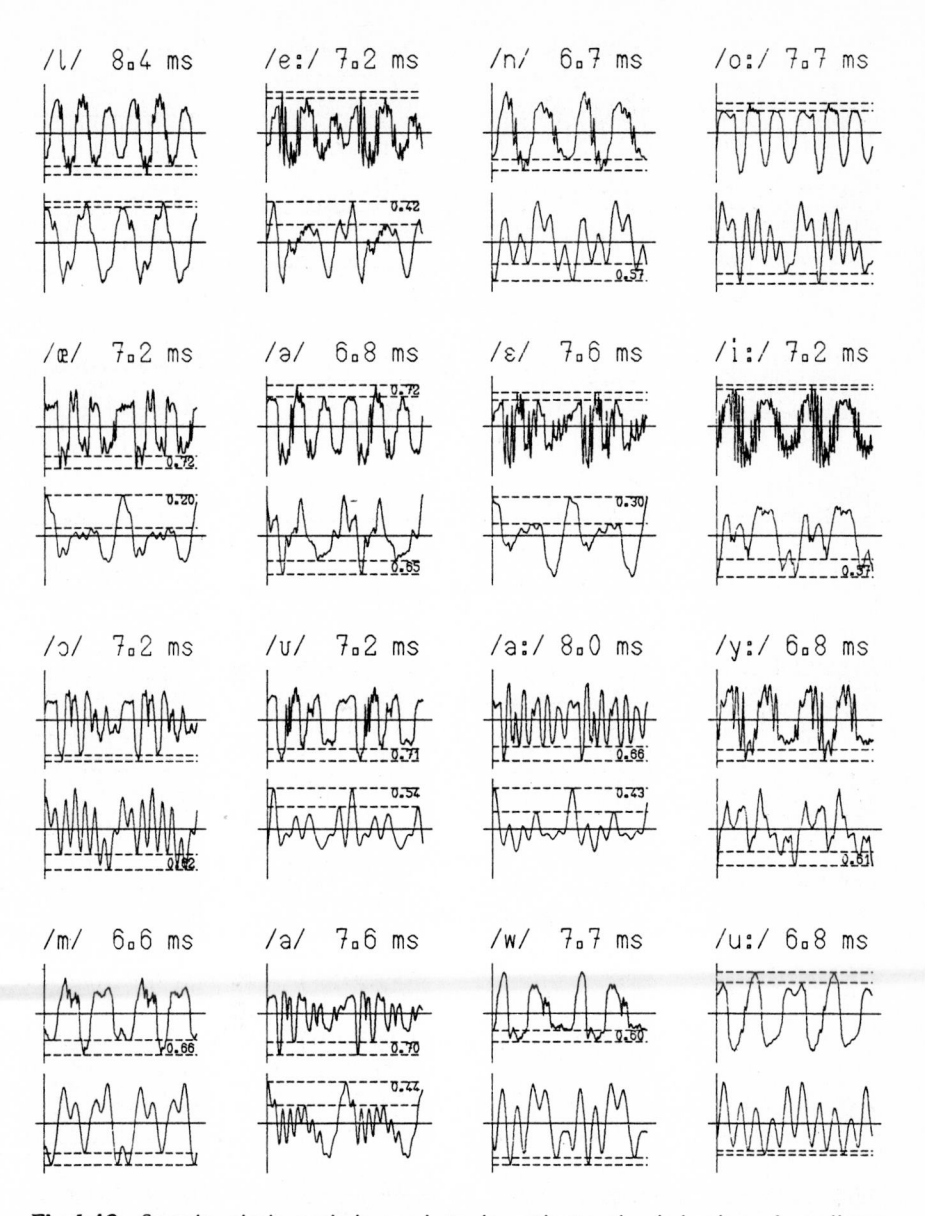

Fig.A.12. Sample pitch periods used to investigate the behavior of nonlinear functions in PDAs. Speaker 1 (HRS, male); band-limited signal. (Upper part of each display) Signal (2 periods), processed by the function y = sqrt(abs(x))*sgn(x), which is the odd function selected for the PDA in Chap.7; (lower part) signal after adaptive linear filtering

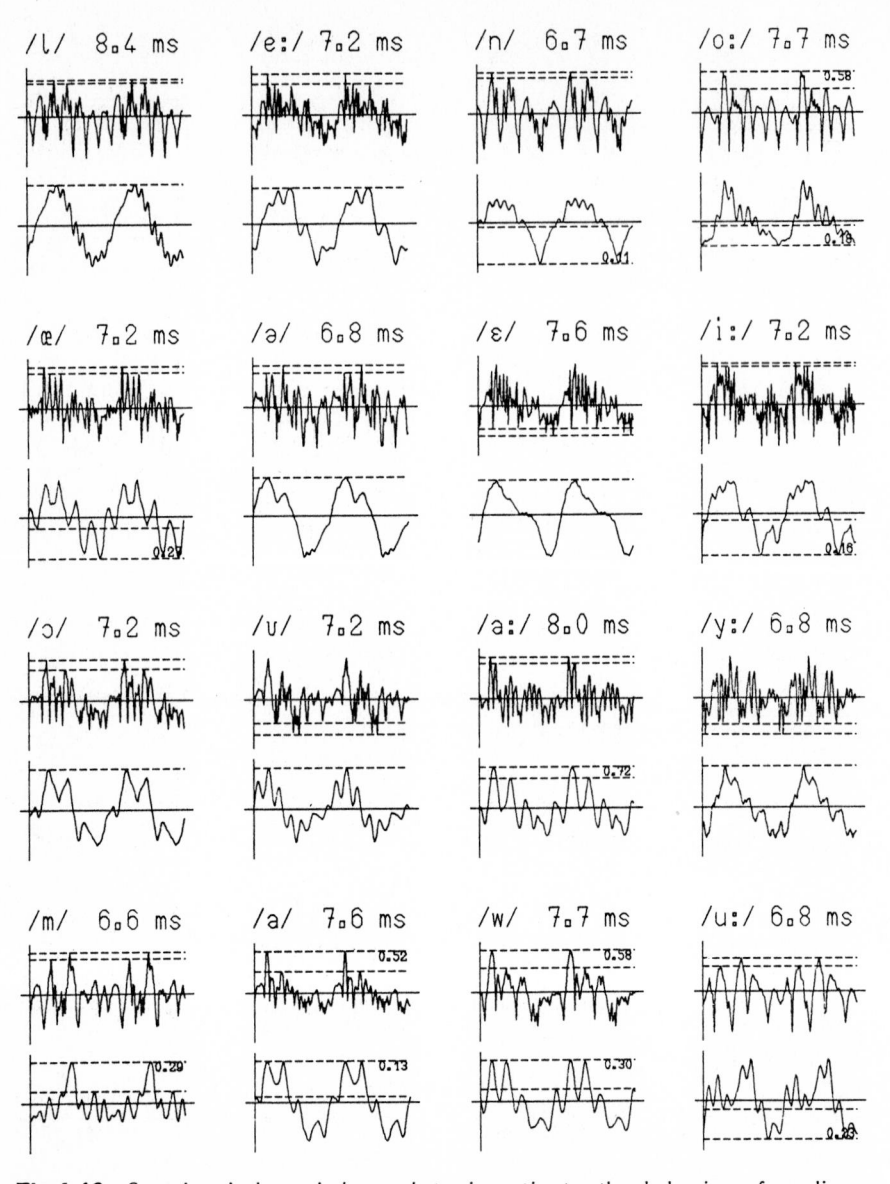

Fig.A.13. Sample pitch periods used to investigate the behavior of nonlinear functions in PDAs. Speaker 1 (HRS, male); band-limited signal. (Upper part of each display) Signal (2 periods), processed by the function y = abs(x)**0.5, which is the even function selected for the PDA in Chap.7; (lower part) signal after adaptive linear filtering

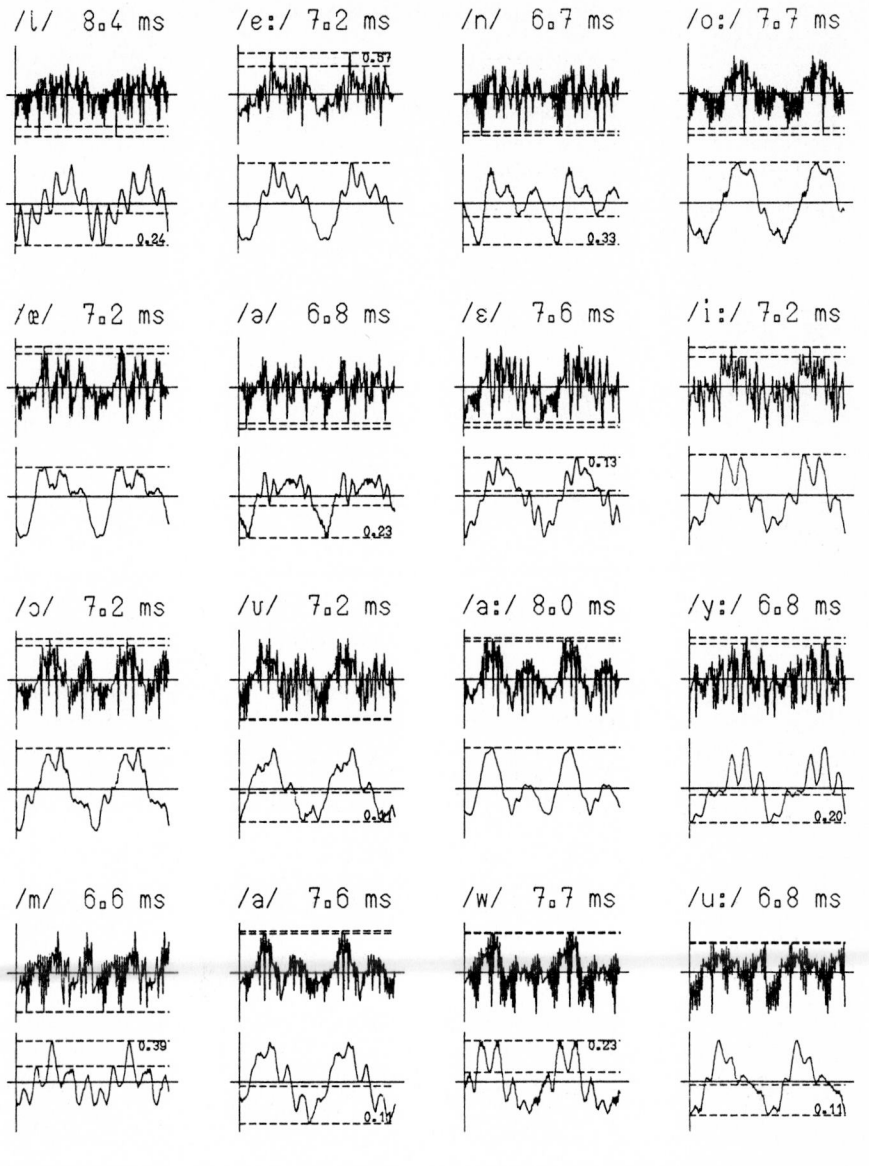

Fig.A.14. Sample pitch periods used to investigate the behavior of nonlinear functions in PDAs. Speaker 1 (HRS, male); band-limited signal. (Upper part of each display) Signal (2 periods), processed by the function y = sqrt(abs(ssb(x))), which is the SSB function selected for the PDA in Chap.7; (lower part) signal after adaptive linear filtering

A.3 Relative Amplitude RA1 and Enhancement RE1 of the First Harmonic (Figs.A.15-17)

The two parameters RA1 and RE1 were defined in (7.11) and (7.13). The distribution of the relative amplitude RA1 for the undistorted signals of the data base was displayed in Figure 7.9. The distribution of RA1 after distortion and the distribution of the enhancement RE1 of the first harmonic were depicted in Figure 7.10. The detailed distribution of the relative enhancement RE1 for each of the NLFs and recording conditions is shown in Figure A.15.

The question of the component at zero frequency (the "offset") is taken up in Figure A.16. Any NLF other than odd produces a considerable offset in the distorted signal; it often exceeds the first harmonic by more than 10 dB. To quantitatively investigate this issue, the parameter RA0 (relative amplitude of offset) was defined in (7.19), and its behavior for selected even NLFs was

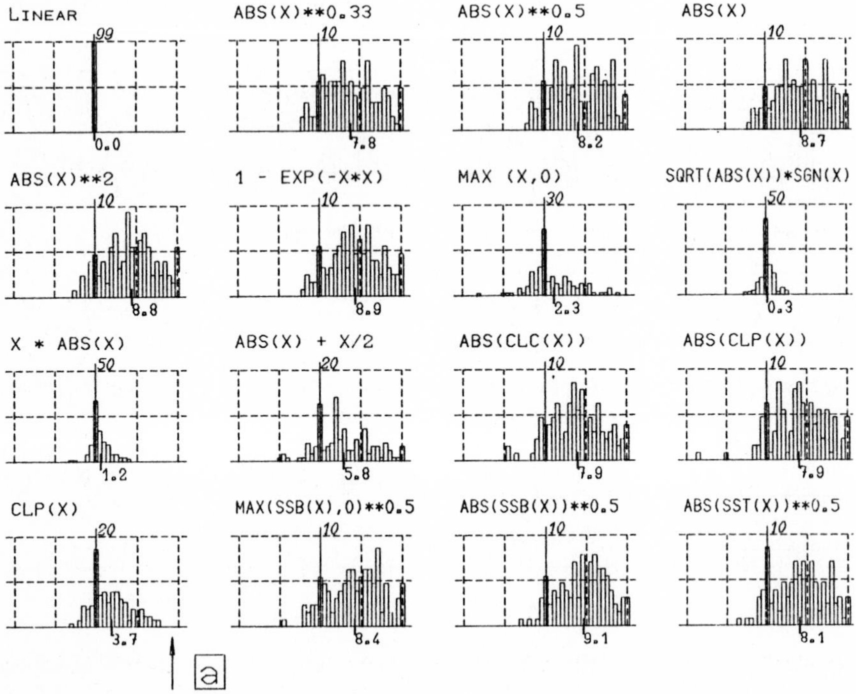

Fig.A.15a-c. Enhancement RE1 of the first harmonic by nonlinear distortion - detailed view. (a) Signal SiLU (undistorted, low fundamental frequency), speakers 1 and 2 (male); (b) (on opposite page) signal SiLL (band limited, low fundamental frequency), speakers 1 and 2 (male); (c) (on opposite page) signal SiH (high fundamental frequency): speaker 3 (female). (Abscissa) Enhancement (-20 to 20 dB); (ordinate) percentage of occurrence as indicated. The number below each figure indicates the average (averaged over the decibel scale)

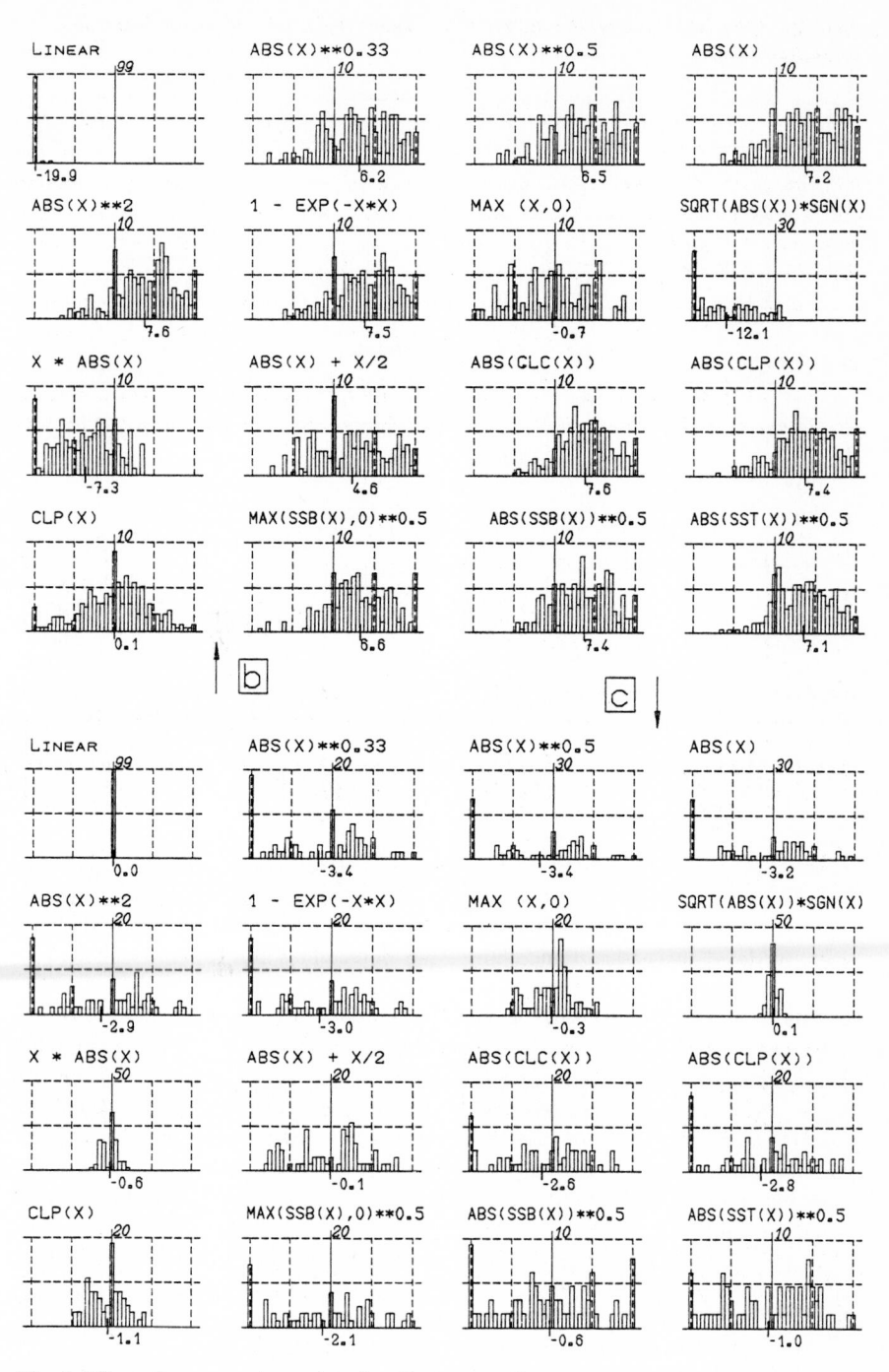

Fig. A.15b,c. See opposite page for figure caption

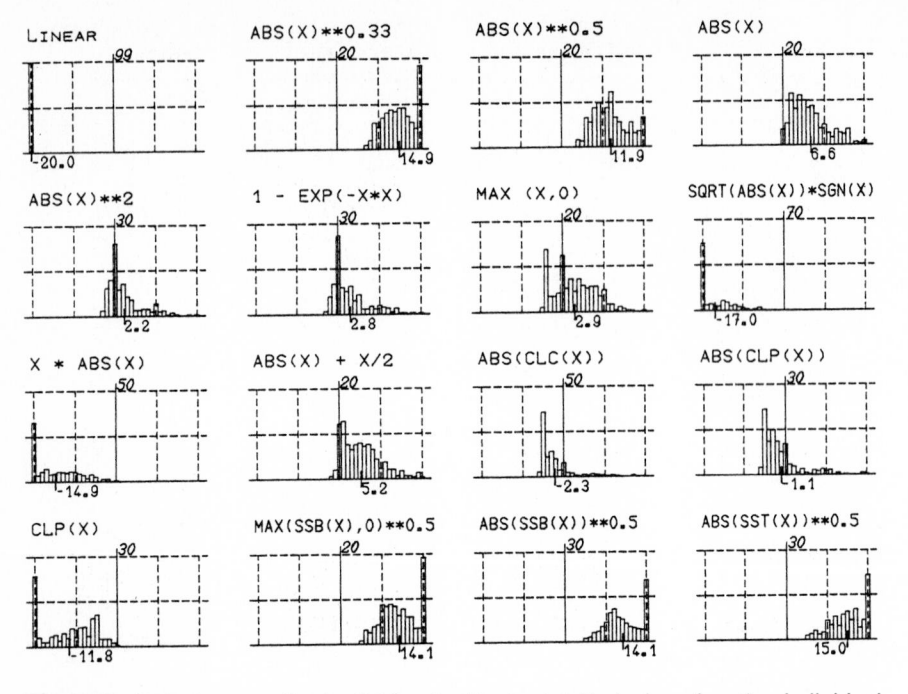

Fig.A.16. Relative amplitude RA0 of offset: detailed view for the individual nonlinear functions. (Abscissa) parameter RA0 (-20 to +20 dB), (ordinate) percentage of occurrence as indicated. (Number below each figure) Average (averaged over the decibel scale). The figure was compiled from the utterances of all three speakers. No band-limited signals were taken into account

shown in Figure 7.14. Figure A.16 presents a detailed view of the parameter RA0 for all the applied NLFs. Since the offset is time variant in the practical application, its removal by high-pass filtering gets more and more difficult when its level increases. Hence this issue determines the lower bound for the exponent a in case an even NLF of the form $y = |x|^a$ is applied.

The performance of RA1 and RE1 for the four investigated classes of NLFs (odd, even, mixed, and SSB modulation), plotted over RA1 of the undistorted signal, was displayed in Fig.7.11 as the main result of the investigation. Figure A.17 supplies some additional data on this issue. The figure shows the relative amplitude RA1 of the first harmonic with respect to the index of that harmonic which is actually the strongest. A dangerous situation is to be expected when a low-amplitude first harmonic coincides with a higher harmonic of low order (second or third) but high amplitude. The figure shows that this constellation does occur; it happens especially for band-limited signals in connection with odd NLFs, and for even NLFs in connection with high values of F_0.

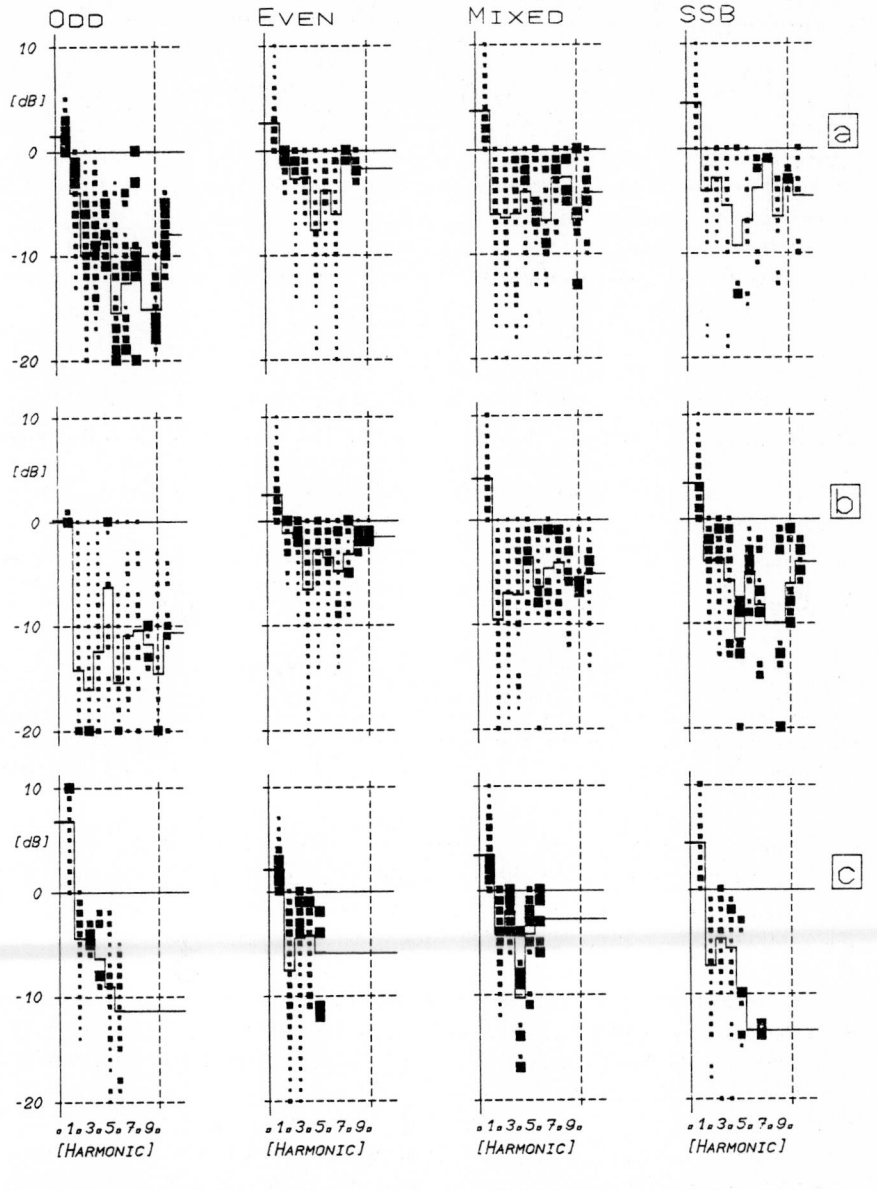

Fig.A.17a–c. Distribution of the relative amplitude RA1 of the first harmonic, depending on the index of the strongest harmonic. (**a**) Signal SiLU (undistorted, low fundamental frequency), speakers 1 and 2 (male); (**b**) signal SiLL (band limited, low fundamental frequency), speakers 1 and 2 (male); (**c**) signal SiH (high fundamental frequency): speaker 3 (female). (Number below each figure) Average (averaged on the decibel scale)

A.4 Relative Amplitude RASM of Spurious Maximum and Autocorrelation Threshold (Figs.A.18-20)

In the primary investigation the parameters RA1 and RE1 were preferred since they give a good measure of the behavior of NLFs independent of external circumstances such as the adjustment of an adaptive filter. For nonzero threshold analysis, however, the parameter RASM (relative amplitude of spurious maximum) is of primary importance. This parameter was defined in (7.23) as the ratio of the amplitude of the second-largest peak in the period (the spurious maximum) with respect to the amplitude of the principal peak which determines the period boundary. To be regarded as spurious in the sense of this definition, a peak must have the same polarity as the principal peak and be separated from the latter by at least one sample of opposite sign. The parameter RASM was displayed for all the applied NLFs in Fig.7.18; there a double comb filter was used to cancel the first formant.

Figure A.18 shows the influence of adaptive filtering on the performance of threshold analysis. This figure has to be interpreted with some caution. It displays differences of the parameter RASM between a single comb filter and the original signal (after removal of the higher formants by low-pass filtering), a double comb filter and the original, and the two comb filters, each time

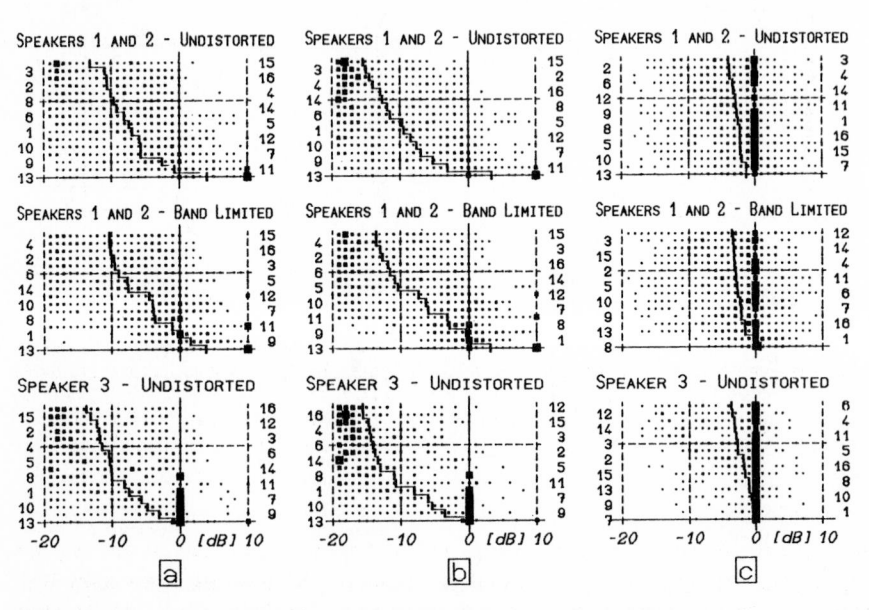

Fig.A.18a–c. Influence of adaptive filtering on nonzero threshold analysis. The figure displays the difference (in dB) of the parameter RASM (filtered signal) for the following situations: (a) single comb filter versus unprocessed signal, (b) double comb filter versus unprocessed signal, and (c) double comb filter versus single comb filter. Abscissa: difference RASM (-20 to 10 dB); (ordinate) type of NLF (marked left, see specifications in Table 7.4). (———) Average (averaged over decibel scale, marked right). For more details, see text

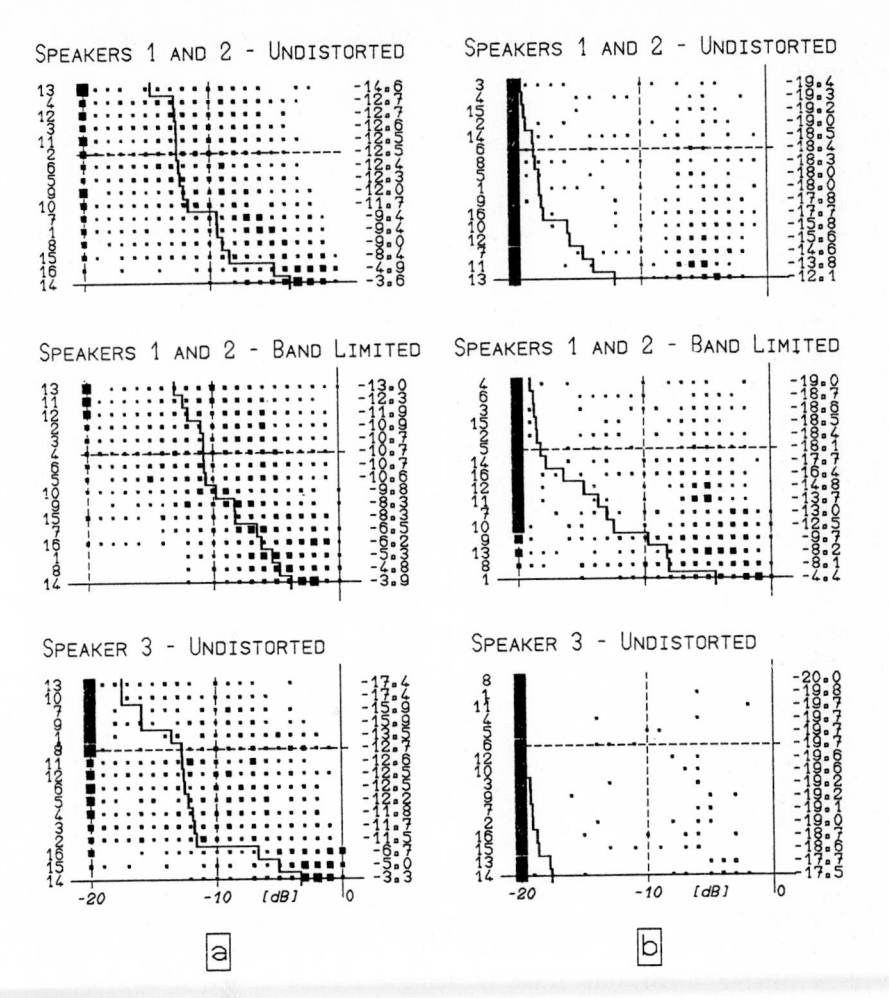

Fig.A.19a,b. Relative Amplitude RASM of spurious maximum for autocorrelation analysis. (**a**) Direct autocorrelation (AC) analysis, (**b**) AC analysis after adaptive filtering. (Abscissa) RASM (-20 to 0 dB); (ordinate) type of NLF (marked on the left-hand side of each display, see specifications in Table 7.4). (————) Average (averaged over decibel scale, marked on the right-hand side of each display)

with the respective NLF as parameter of the figure. Except for center clipping (which again proves the incompatibility of linear and nonlinear spectrum flattening techniques) the application of the filter greatly improves the performance of the parameter. The large number of zero dB indications in the comparison between the single and the double comb filter is due to those signals where the single comb filter has already caused the spurious maximum to disappear.

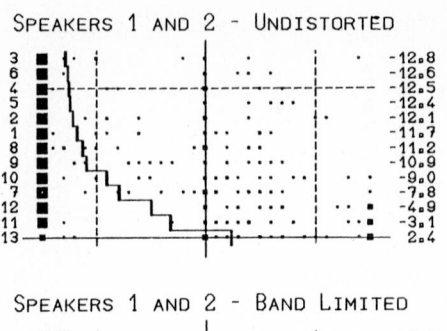

SPEAKERS 1 AND 2 - UNDISTORTED

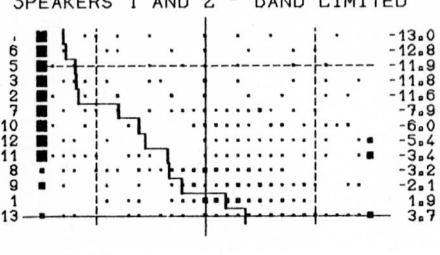

SPEAKERS 1 AND 2 - BAND LIMITED

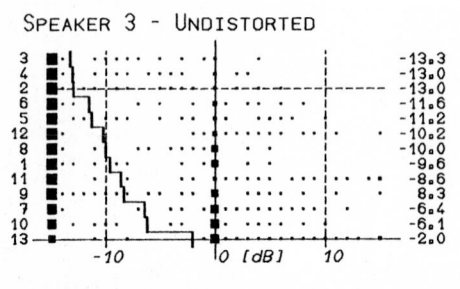

SPEAKER 3 - UNDISTORTED

Fig.A.20. Influence of adaptive filtering on autocorrelation analysis. The figure displays the difference (in dB) of the parameter RASM for autocorrelation (AC) analysis after adaptive filtering with respect to direct AC analysis. (Abscissa) Difference RASM (-20 to 10 dB), (ordinate) type of NLF (marked on the left-hand side of each display, see specifications in Table 7.4). (———) Average (averaged over decibel scale, marked on the right-hand side of each display)

The same parameter RASM is displayed in Fig.A.19 for the autocorrelation function of the unfiltered signal ("direct" AC) and of the filtered signal ("filtered" AC). For the original signal, center clipping as NLF performs best. With a filtered signal (the filter is the one applied in the linear PDA of Sect.7.1) center clipping spoils the performance whereas even nonlinear functions do a good job. Figure A.20 shows the performance of NLFs in autocorrelation analysis. The figure displays the difference (in dB) of RASM for AC analysis after filtering and direct AC analysis. A negative value of the difference indicates that (for a given type of signal and NLF) the AC analysis after filtering (which roughly corresponds to the SIFT PDA by Markel, 1972) behaves better than direct AC analysis. There is a distinct preference for center clipping in direct AC analysis and a slight preference in favor of full-wave rectification for the filtered AC analysis.

A.5 Processing Sequence, Preemphasis, Phase, Band Limitation (Figs.A.21-24)

Figure A.21 investigates the influence of the processing sequence on the parameters. The alternatives are 1) nonlinear distortion of the unprocessed signal with linear distortion thereafter, and 2) nonlinear processing after the first step of low-pass filtering (in the investigation the prefilter used in Sect.7.1 was taken as the low-pass filter). The figure shows the difference (for the parameters RA1, RASM for threshold analysis of signal, and RASM for autocorrelation analysis) between alternative s 1 and 2. The results suggest consistently but with different relevance that alternative 1, i.e., non-

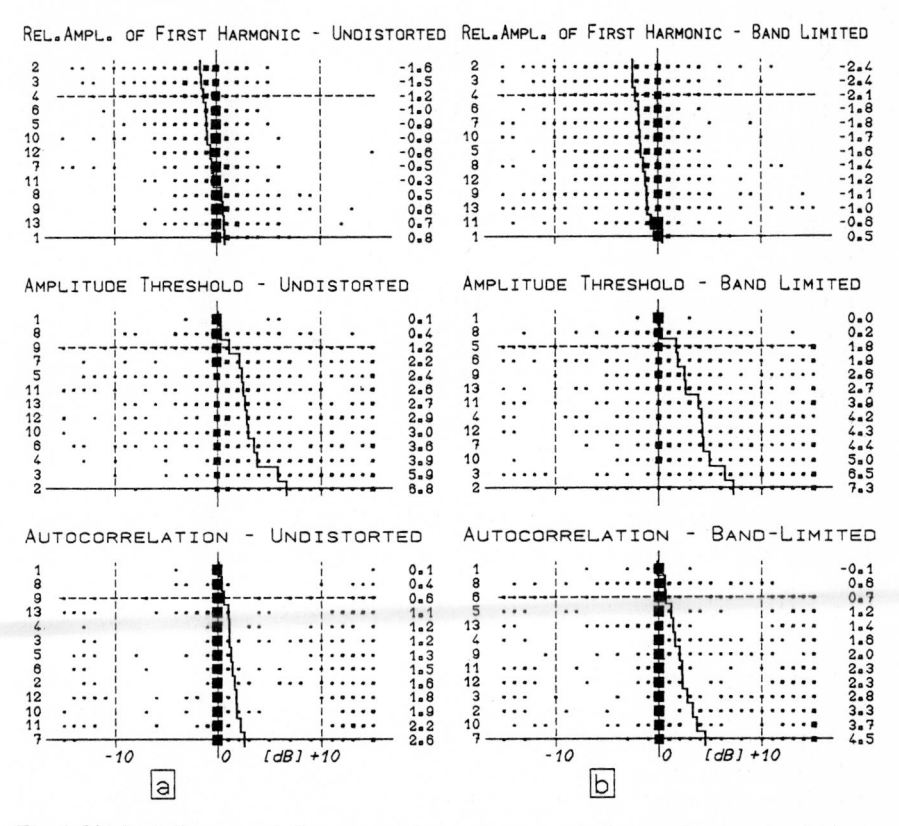

Fig.A.21a,b. Influence of the processing sequence on the performance of algorithms: (a) without band limitation, (b) with band limitation. The figure displays the difference of the indicated parameters between the two processing sequences: 1) nonlinear distortion before low-pass filtering, and 2) nonlinear distortion after low-pass filtering. The data by the female speaker were not taken into account. (Abscissa) Difference (-20 to 10 dB) of indicated parameter (upper part: RA1, middle and lower part: RASM); (ordinate) type of NLF (marked on the left-hand side of each display, see specifications in Table 7.4). (———) Average (averaged over decibel scale, marked on the right-hand side of each display)

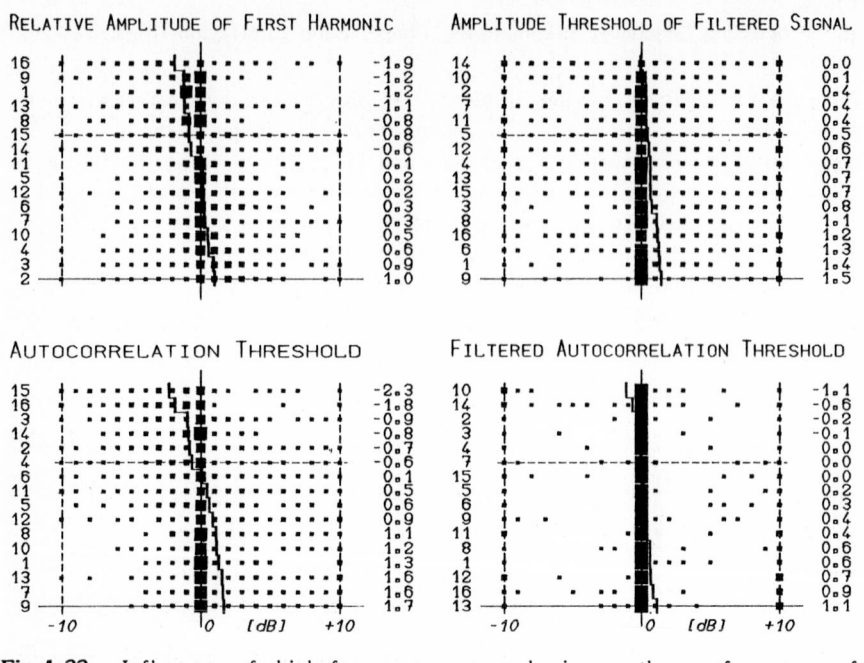

Fig.A.22. Influence of high-frequency preemphasis on the performance of algorithms. The figure displays the difference of the indicated parameters for the signal with preemphasis compared to the signal without preemphasis. No band limited signals were taken into account. (Abscissa) Difference as indicated (-20 to 20 dB); (ordinate) type of NLF (marked on the left-hand side of each display, see specifications in Table 7.4). (——) Average (averaged over decibel scale, marked on the right-hand side of each display)

linear processing first, yields a better performance. This means that the higher partials significantly contribute to the enhancement of the first harmonic by nonlinear distortion. An additional preemphasis of the higher frequencies (Fig.A.22; 3 dB at 1 kHz and 9 dB at 3 kHz), however, shows little effect.

Figure A.23 shows the influence of low-frequency phase distortions individually for each NLF. (The phase distortions applied in this experiment were specified in Fig.7.7.) There is a considerable variance in connection with phase conditions. The figure displays the difference (in dB) of RA1 for the phase distorted signal with respect to the undistorted signal. The figure evaluates the signals from all three speakers, but does not include band-limited signals. On the average, phase distortions do not bias the performance of a specific NLF as a whole; in the individual case, however, they may cause a drastic change of behavior.

The influence of low-frequency band limitation (to 220 Hz, for the male speakers only) is shown in Fig.A.24: 1) for the parameter RA1; and for the parameter RASM in the cases 2) threshold analysis of filtered signal,

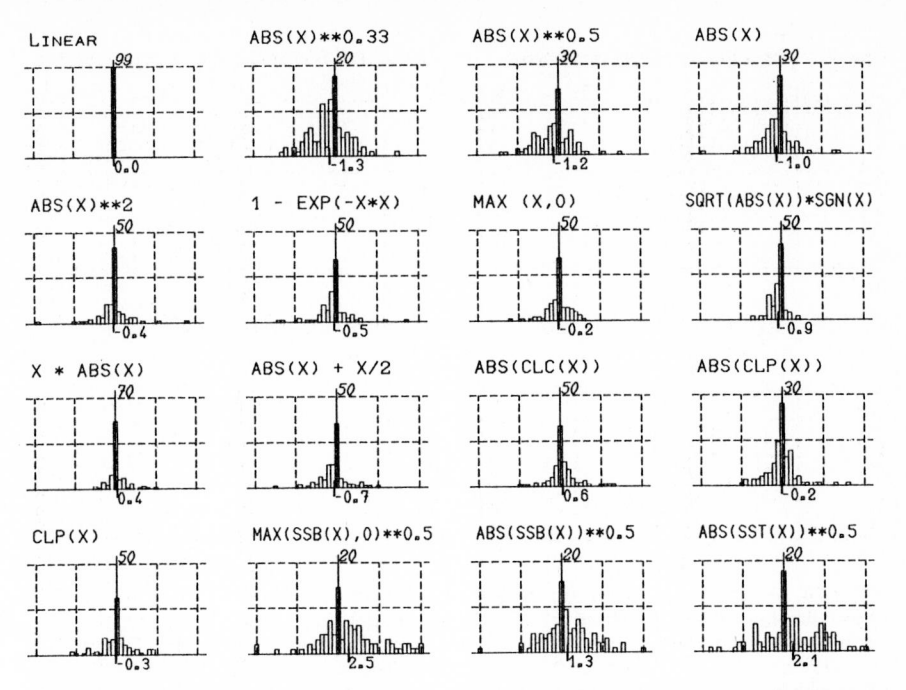

Fig.A.23. Influence of low-frequency phase distortions on the enhancement RE1 of the first harmonic: detailed view. The figure displays the difference (in dB) of RA1 for the low-frequency phase-distorted signal versus the undistorted signal. No band-limited signals were taken into account. The data were compiled from all three speakers. (Abscissa) Difference (-20 to +20 dB); (ordinate) percentage of occurrence as indicated. The number below each figure indicates the average (averaged over the decibel scale)

3) autocorrelation analysis, and 4) filtered AC analysis. Direct AC analysis proves least sensitive. For all the parameters the odd NLFs show the worst performance.

A.6 Optimal Performance of Nonlinear Functions (Figs.A.25-27)

In Figs.A.25-27 the optimal performance of nonlinear functions is investigated. The figures display the optimal choice out of the 16 NLFs for the parameters RA1, RS1(6), and RS1(12), the latter defined in (6.3,4) and important for the use of a ZXABE. According to McKinney's condition [1965, Eq.(6.1)], a ZXABE becomes applicable when the parameter RS1 takes on a value above 0 dB. In the score of preference a NLF is selected (or better: marked applicable) if 1) the function gives the optimal value of the parameter for all NLFs, or if 2) the value of the parameter is within 1 dB from the optimum, or if 3) the parameter takes on a value above 0 dB. According to Fig.A.25, there is practically always a NLF which can render the first

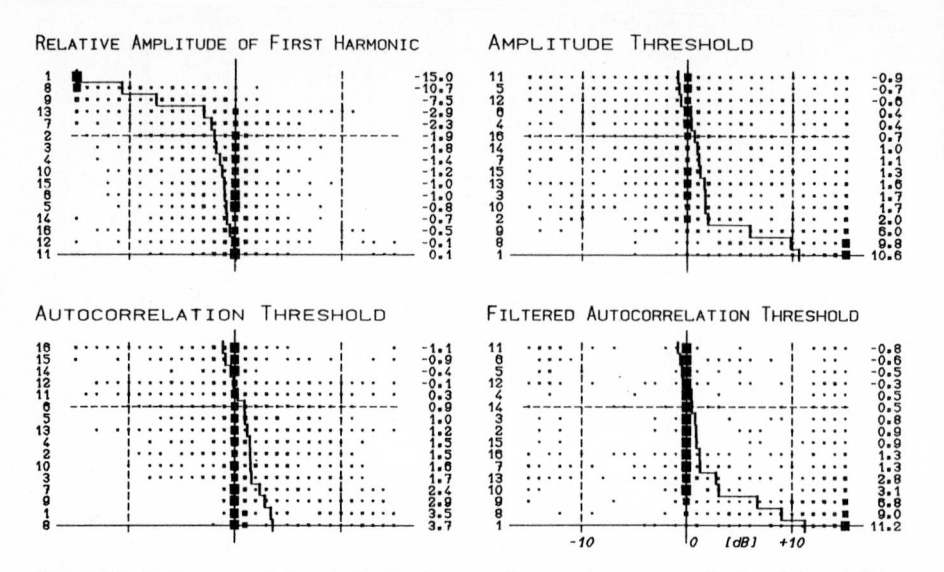

Fig.A.24. Influence of band limitation on the performance of algorithms. The figure displays the difference of the indicated parameters for the band-limited signal with respect to the (linearly) undistorted signal. (Abscissa) Difference of parameter as indicated (-15 to 15 dB); (ordinate) type of NLF (marked on the left-hand side of each display, see specifications in Table 7.4). (————) Average (averaged over decibel scale, marked on the right-hand side of each display)

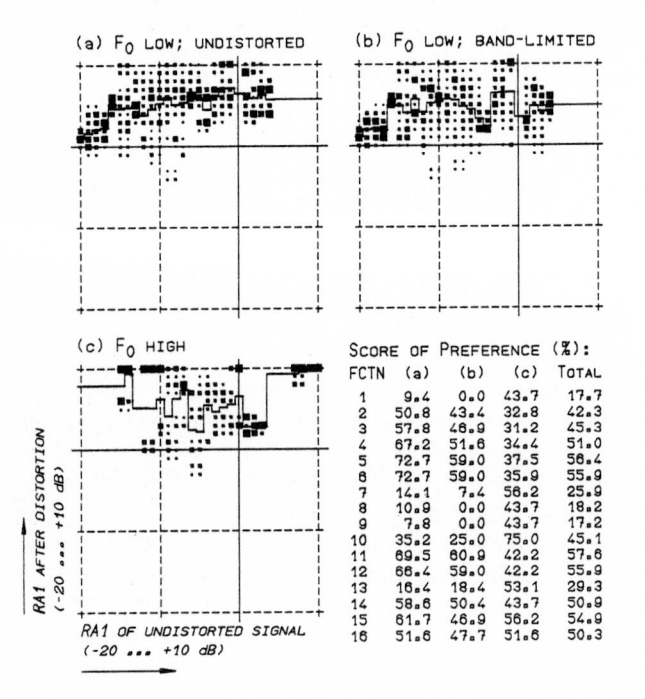

SCORE OF PREFERENCE (%):

FCTN	(a)	(b)	(c)	TOTAL
1	9.4	0.0	43.7	17.7
2	50.8	43.4	32.8	42.3
3	57.8	46.9	31.2	45.3
4	67.2	51.6	34.4	51.0
5	72.7	59.0	37.5	56.4
6	72.7	59.0	35.9	55.9
7	14.1	7.4	56.2	25.9
8	10.9	0.0	43.7	18.2
9	7.8	0.0	43.7	17.2
10	35.2	25.0	75.0	45.1
11	69.5	60.9	42.2	57.6
12	66.4	59.0	42.2	55.9
13	16.4	18.4	53.1	29.3
14	58.6	50.4	43.7	50.9
15	61.7	46.9	56.2	54.9
16	51.6	47.7	51.6	50.3

Fig.A.25a–c. Relative amplitude RA1 of first harmonic: optimal choice out of the 16 nonlinear functions applied. For details of the selection procedure, see text

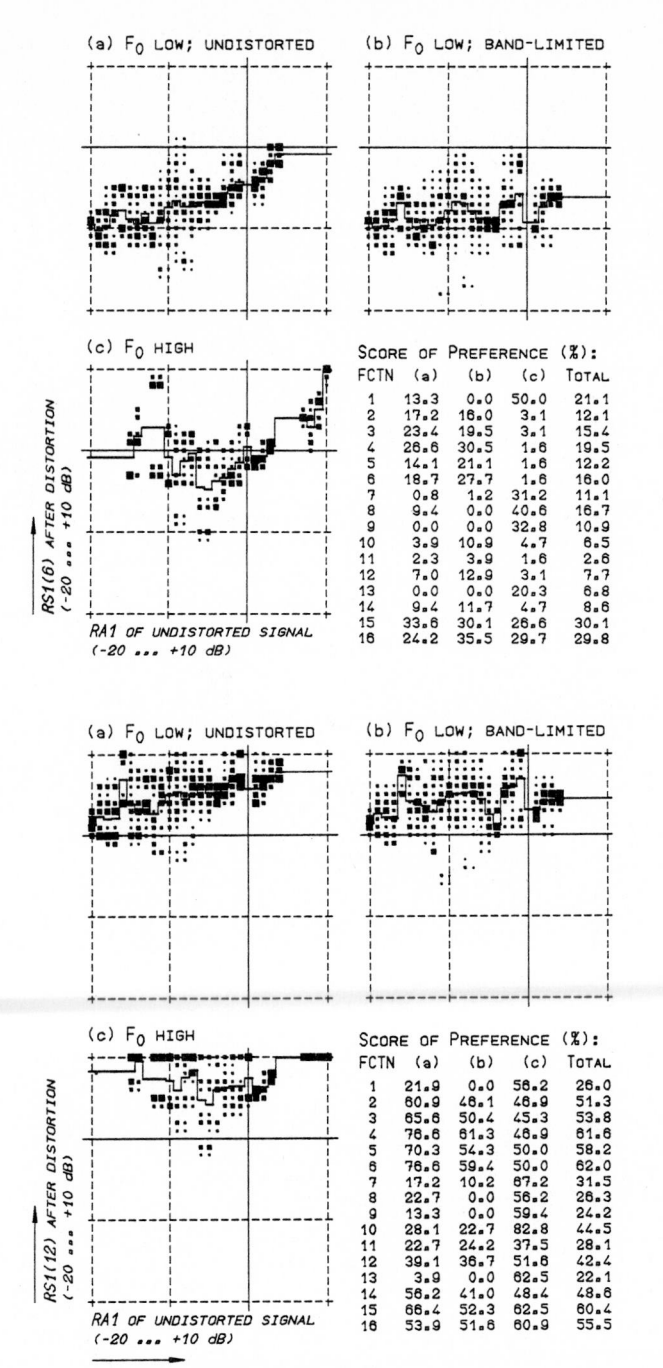

(a) F_0 LOW; UNDISTORTED

(b) F_0 LOW; BAND-LIMITED

(c) F_0 HIGH

RS1(6) AFTER DISTORTION (-20 ... +10 dB)

RA1 OF UNDISTORTED SIGNAL (-20 ... +10 dB)

SCORE OF PREFERENCE (%):

FCTN	(a)	(b)	(c)	TOTAL
1	13.3	0.0	50.0	21.1
2	17.2	16.0	3.1	12.1
3	23.4	19.5	3.1	15.4
4	28.6	30.5	1.6	19.5
5	14.1	21.1	1.6	12.2
6	18.7	27.7	1.6	16.0
7	0.8	1.2	31.2	11.1
8	9.4	0.0	40.6	16.7
9	0.0	0.0	32.8	10.9
10	3.9	10.9	4.7	6.5
11	2.3	3.9	1.6	2.6
12	7.0	12.9	3.1	7.7
13	0.0	0.0	20.3	6.8
14	9.4	11.7	4.7	8.6
15	33.6	30.1	26.6	30.1
16	24.2	35.5	29.7	29.8

Fig.A.26a–c. Relative amplitude RS1(6) of first harmonic as defined in (6.3,4): optimal choice out of the 16 nonlinear functions applied. For details of the selection procedure, see text

(a) F_0 LOW; UNDISTORTED

(b) F_0 LOW; BAND-LIMITED

(c) F_0 HIGH

RS1(12) AFTER DISTORTION (-20 ... +10 dB)

RA1 OF UNDISTORTED SIGNAL (-20 ... +10 dB)

SCORE OF PREFERENCE (%):

FCTN	(a)	(b)	(c)	TOTAL
1	21.9	0.0	58.2	26.0
2	60.9	48.1	46.9	51.3
3	65.6	50.4	45.3	53.8
4	76.6	61.3	46.9	61.6
5	70.3	54.3	50.0	58.2
6	76.6	59.4	50.0	62.0
7	17.2	10.2	87.2	31.5
8	22.7	0.0	56.2	26.3
9	13.3	0.0	59.4	24.2
10	28.1	22.7	82.8	44.5
11	22.7	24.2	37.5	28.1
12	39.1	36.7	51.6	42.4
13	3.9	0.0	62.5	22.1
14	56.2	41.0	48.4	48.6
15	66.4	52.3	62.5	60.4
16	53.9	51.6	60.9	55.5

Fig.A.27a–c. Relative amplitude RS1(12) of first harmonic as defined in (6.3,4): optimal choice out of the 16 nonlinear functions applied. For details of the selection procedure, see text

harmonic to be the strongest. If a ZXABE is to be applied, at least a 12-dB-per-octave higher harmonic attenuation is necessary, even under optimal conditions (Figs.A.26,27). If only those three NLFs are applied which were selected for the PDA in Chap.7, the value of RA1 is in 75% of the cases less than 1 dB off the optimum of all the 16 NLFs. In the rest of the cases, the difference is not much larger. The optimum value of RA1 for the three NLFs was already displayed in Chap.7 (Figs.7.12,13). For the parameter RASM of the filtered signal, the corresponding figure is rather uninformative; the score of preference is similar to that for the parameter RA1.

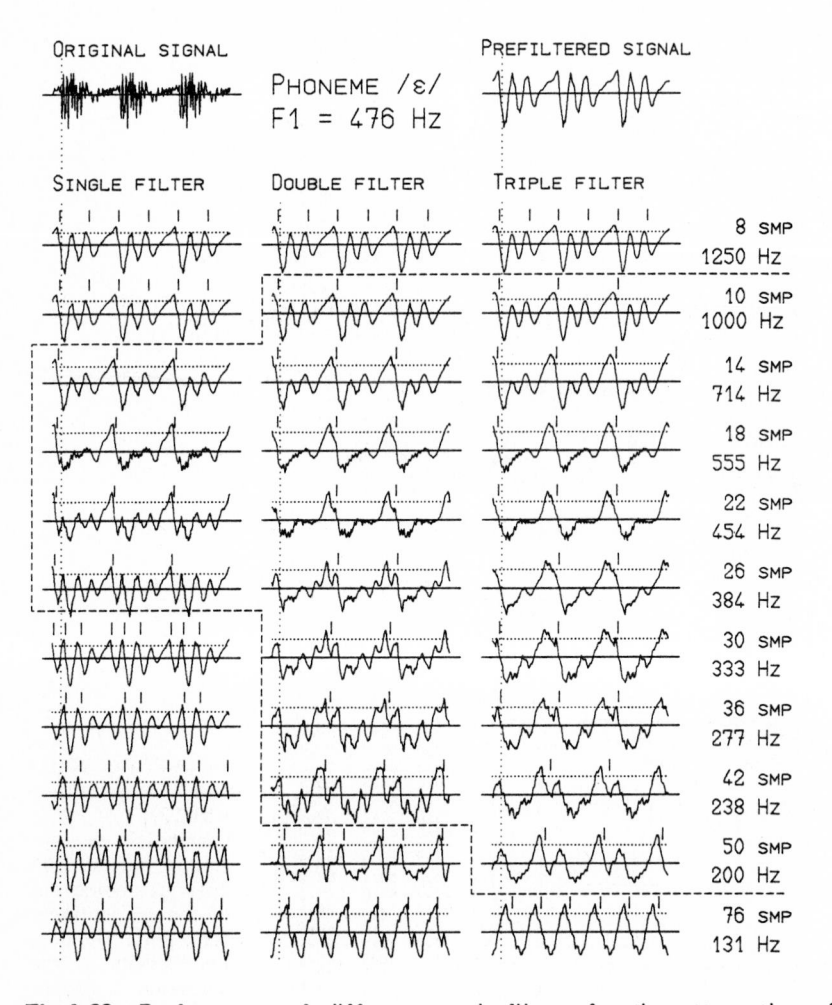

Fig.A.28. Performance of different comb filters for the attenuation of F1. (Example) phoneme /ɛ/, male speaker, no nonlinear distortions. Only one formant below 1200 Hz. (-----) Range for which the filter gives correct results; (smp) samples

A.7 Performance of the Comb Filters (Figs.A.28–30)

Figures A.28-30 display three samples for the tuning of the comb filter (in addition to the ones shown in Figs.7.4,5,18). In all these figures the pulse width of the filter and the position of the first zero is indicated. The original signal is not band limited. In the case of the double and triple comb filters, the indicated numbers apply to the filter with the lowest zero; the other components are calculated from them. The threshold was set to 65% of the maximum amplitude. A spurious peak was disregarded when it occurred

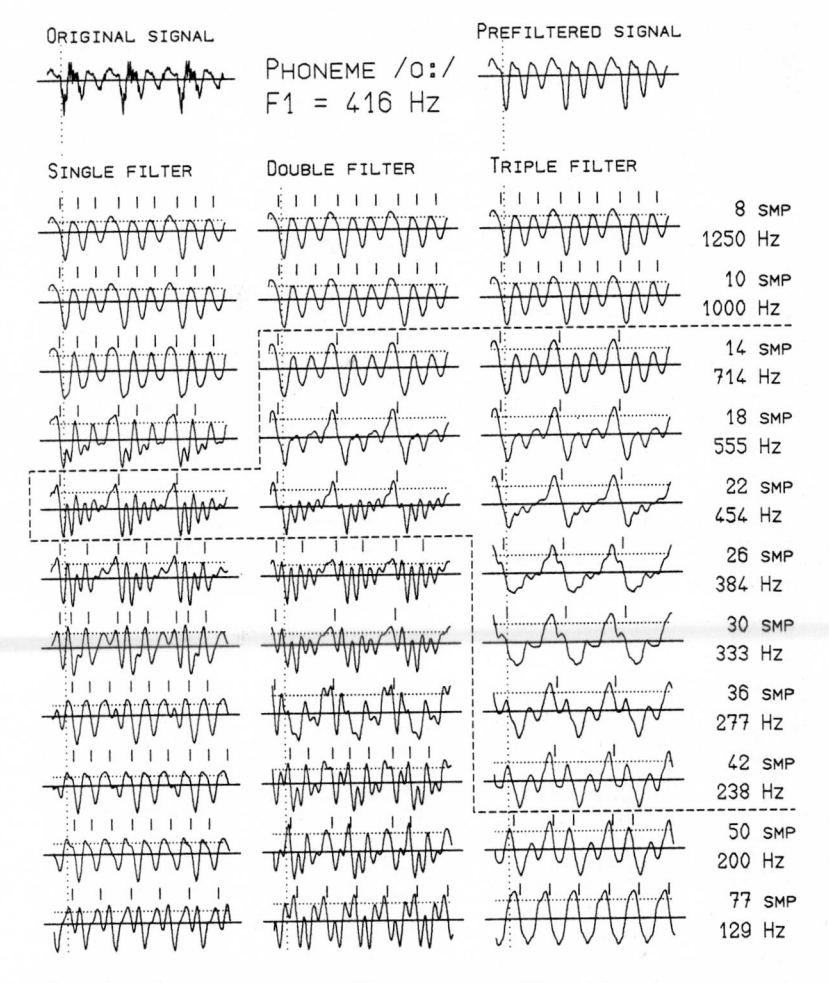

Fig.A.29. Performance of different comb filters for the attenuation of F1. (Example) phoneme /o:/, male speaker, no nonlinear distortions. Two formants below 1200 Hz. (-----) Range for which the filter gives correct results; (smp) samples

less than 1 ms apart from the principal peak. All the signals were normalized before plotting. The dashed line indicates the tuning range of the filter within which the performance of the system remains correct. In the bottom line the zero of the filter coincides with the fundamental frequency.

Things change drastically when the second formant is situated below 1000 Hz (Fig.A.29), where the double and especially the triple comb filter work much better. In case of previous nonlinear processing, however, the triple comb filter proves unnecessary; after all, the double filter has such a large tuning range that the triple filter hardly brings any additional advantage (Fig.A.30; the displayed example is rather bad).

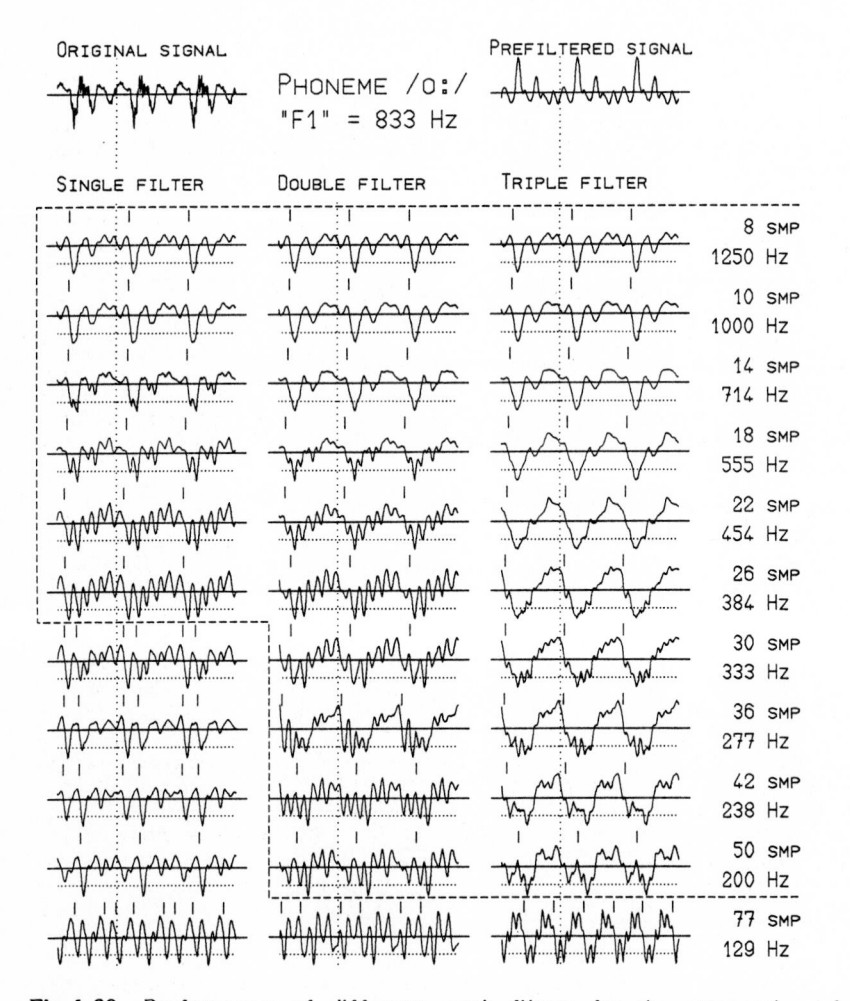

Fig.A.30. Performance of different comb filters for the attenuation of F1. Same signal as in Fig.A.29, but nonlinearly distorted by an even function. (-----) Range for which the filter gives correct results; (smp) samples

Appendix B. Original Text of the Quotations in Foreign Languages Throughout This Book

Человек легко определяет частоту ОТ [основного тона] как при прослушивании речи, так и при визуальном изучении осциллограмм и спектрограмм речи. Однако построить устройство, автоматически определяющее частоту ОТ с малой ошибкой и малой задержкой во времени даже про относительно низком уровне помех, довольно трудно. (Baronin and Kushtuev, 1974; quoted in Sect.9.4.1)

Отношение $H_L(F_0)/H_L(2F_0)$ должно быть достаточно велико во всем диапазоне возможных значений частоты ОТ. Чем больше $H_L(F_0)/H_L(2F_0)$, тем меньше вероятность срабатывания от второй гармоники. При больших значениях $H_L(F_0)/H_L(2F_0)$ амплитуда первой гармоники на выходе фильтра $H_L(F_0)$ при высокой частоте ОТ может оказаться соизмеримой с уровнем низкочастотных шумов, которые усиливаются фильтром, хотя полезный сигнал в указанных условиях сильно ослабляется. Схема ВОТ [выделения основного тона] начинает давать сбои на высоких голосах. Если уменьшить отношение $H_L(F_0)/H_L(2F_0)$, то процент сбоев на низких голосах вследствие срабатываний от высших гармоник увеличится. (Baronin and Kushtuev, 1974, pp.116-117; quoted in Sect.6.1.2)

Если считать, что средняя частота ОТ составляет 200 Гц, то относительная ошибка измерения схемой грубого выделения ОТ составит 17.5%. Казалось бы, что построить выделитель ОТ со столь низкими требованиями к точности не представляет больших трудностей. Однако следует иметь в виду, что значение 17.5% относится к максимально допустимому значению ошибки. Если бы ошибка была распределена по нормальному закону, то допустимое среднее значение ошибки сократилось бы до 6–8%. Но, как показывают исследования, распределение вероятностей величины ΔF обычно отличается от нормального.

Может, например, оказаться, что схема грубого выделения ОТ со средним значением ошибки 6% в течение 94% времени дает практически правильные замеры частоты ОТ, а в течение 6% времени – удвоенное значение частоты ОТ, причем ошибки появляются группами. В этом случае в течение 6% времини ошибочной работы схемы грубого выделения ОТ следящий фильтр оказывается настроенным на вторую гармонику частоты ОТ и вся система выделения ОТ работает неправильно, т.е. точность работы всей системы выделения ОТ близка к точности работы схемы грубого выделения ОТ. Таким образом, даже при сравительно высокой средней точности работы схемы грубого выделения ОТ большие ошибки (например, переходы на высшие гармоники у фильтровых выделителей ОТ, ложные срабатывания у схем амплитудной селекции), появляющиеся группами, существенно снижают эффективность системы выделения ОТ на перестраиваемом узкополосном фильтре разомкнутого типа, т.е. такой, в которой сигнал управления для перестройки полосового фильтра получают от устройства грубого определения частоты ОТ. (Baronin and Kushtuev, 1974, pp.150-151; quoted in Sect.6.1.4)

В настоящем разделе будет рассматриваться многоканальная схема ВОТ еще более общего вида. Дело в том, что ... способы выбора результата измерения в многоканальной схеме не позволяют наилучшим образом обработать полученные данные, так как:

1) Для правильной работы систем ... необходимо, чтобы в каждый момент времени не выходах нескольких элементарных выделителей получались результаты, близкие к правильным. Если это условие не выполняется, то измеренное в момент t значение частоты ОТ может сильно отличаться от истинного значения. В таких системах недостаточно учитывается сравнительно медленное изменение функции $F_0(t)$, поэтому значения частоты ОТ на небольшом интервале времени ... можно с большой точностью интерполировать по значениям частоты ОТ на соседних интервалах. Принципиально возможно построить многоканальную систему ВОТ, работающую правильно даже в те моменты, когда во всех каналах частота ОТ измеряется ошибочно.

2) ... Очевидно, что и точность измерения в различних условиях (например, про высокой и низкой частотах ОТ) у различных элементарных ВОТ не однакова. ... В системах ... все результаты измерений обрабатываются одинаково.

3) Применяемая в указанных системах операция выбора не является оптимальной обработкой, поскольку выбор связан с отбрасыванием части информации. Информация о частоте ОТ содержится и на выходе элементарного выделителя, дающего измерение, например, измеряющего частоту второй гармоники.

4) В системах ... никак не учтены особенности слухового восприятия ошибок выделения ОТ. (Baronin, 1974a, p.171; quoted in Sect.6.4.1)

При оптимальных значениях параметров ($\Delta d = 0.1$ мс, число членов в сумме, аппроксимирующей сдвиговую функтию, $k = 70$) точность выделения ОТ сдвиговым методом приближается к точности выделения автокорреляционным методом, однако сдвиговый метод более удобен как для аналогового, так и для цифрового моделирования. Кроме того, время усреднения сигнала ОТ при использовании сдвигового метода оказывается в 2-3 раза меньше (отсутствуют сглаживающие фильтры, поэтому этим методом можно измерять даже отдельные значения периодов ОТ. (Baronin, 1974a, p.200; quoted in Sect.8.3.1)

1) Большое разнообразие разработанных и применяемых на практике методов выделения основного тона свидетельствует об отсутствии одного такого метода, который обладал бы абсолютными преимуществами перед всеми остальными. В различных условиях целесообразно применять различные методы.

2) При выделении ОТ из высококачественного речевого сигнала (динамический микрофон, отсутствие подавления низкочастотных составляющих) и при работе с одним диктором хорошие результаты дают прострейшие схемы без адаптации, построенные по принципу выделения составляющей частоты ОТ с помощью фильтров с постоянными параметрами или по принципу амплитудной селекции. Качество звучания синтезированной речи при выделении ОТ схемами с амплитудной селекцией оказывается несколько выше из-за меньшей инерционности этих схем, однако при действии помех лучшие результаты получаются с фильтровыми выделителями.

3) При работе с разными дикторами заметными преимуществами обладают адаптивные системы выделения основного тона. Если в речевом сигнале сохранены низкочастотные составляющие, приемлемое для многих случаев качество выделения может быть получено при использовании сравнительно простых схем амплитудной селекции с параметрами, автоматически перестраиваемыми при изменении ОТ и логических схем коррекции сбоев. Дальнейшее улучшение качества выделения достигается при переходе к многоканальным анализаторам, работающим во временной (Голд, 1962) или в час-

тотной области (Баронин и Куштуев, 1969; Миллер, 1970); при малом уровне
шумов хорошие результаты дают также кепстральные анализаторы. При большом
уровние шумов предпочтительны корреляционные анализаторы (Вайс и Харрис,
1966) и многоканальные анализаторы, работающие в частотной области.

4) При работе от угольного микрофона могут быть использованы простые
фильтровые методы. Отсутствующую составляющую частоты ОТ можно восста-
навливать нелинейным преобразованием анализируемого сигнала. Однако
качество выделения ОТ при использовании а ких простых методов оказывается весьма низким. Для получения лучших результатов необходимо использо-
вать оптимальные или близкие к ним методы обработки (схемы с фазированием
гармоник ОТ, корреляционные схемы с предварительной адаптивной филь-
трацией речевого сигнала и др.). При принятии решения о частоте ОТ
необходимо учитывать предысторию процесса. Весьма желательно учитывать
также и будущий по отношению к моменту замера ход анализируемого процесса. Возникающие при этом задержки выдачи замеров ОТ на время, пример-
но равное нескольким десяткам миллисекунд, во многих случаях вполне
допустимы. (Baronin and Kushtuev, 1974, pp.207-208; quoted in Sect.9.4.4)

Soit dans un plan un point M décrivant une trajectoire quelconque autour
d'une origine O. Le mouvement de ce point est susceptible d'être décrit par
ses composantes x et y ou par sa distance à l'origine, E, et par l'angle polaire
φ. Dans le cas d'une trajectoire circulaire et d'un mouvement uniforme
$(x = A \cos 2\pi F_0 t, \ y = A \sin 2\pi F_0 t)$ on a:

$$E = A \quad \text{et} \quad \phi = \arctan y/x = 2\pi F_0 t$$

et la dérivée de φ par rapport au temps s'écrit:

$$\phi' = \frac{xy' - yx'}{E^2} = 2\pi F_0 \ . \tag{6.30}$$

La quantité $\phi'/2\pi$ est la fréquence de rotation du point M autour de l'origine.
L'idée sur laquelle repose la méthode de calcul est que le nombre F_0 qui la
mesure aurait été obtenu en considérant indifféremment x ou y et en effectuant
une analyse harmonique de l'une de ces fonctions. Autrement dit, dans le cas
du mouvement circulaire uniforme (et en toute rigueur seulement dans ce cas)
on peut identifier fréquence de rotation au sens de la cinématique du point
dans le plan (dérivée par rapport au temps de la phase du signal) et fréquence
au sens de l'analyse harmonique de l'une quelconque de ses composantes.
(Berthomier, 1979; quoted in Sect.6.3.2)

En général la détection des pics est satisfaisante. La correction des pics
aberrants n'intervenant que dans les cas suivants: fricative voisée faible, fin
de voyelle nasale et surtout début de voyelle précédée de la plosive /p/ ...
(Dours et al., 1977, p.69; quoted in Sect.6.3.1)

Un courant électrique alternatif de fréquence élevée: de l'ordre de 200000
oscillations par seconde, mais d'intensité très faible, est admis à parcourir le
larynx, de part en part, à travers la peau du cou. Deux électrodes métalliques
longitudinales, humidifiées avec une pâte au kaolin, sont appliquées systéma-
tiquement contre la saillie du cartilage thyroide. Une partie des lignes du
courant emprunte la voie de la glotte, en sorte que leur passage est facilité
d'autant plus que celle-ci est fermée. L'amplitude du courant électrique sera
donc modulée par l'accolement des cordes vocales. Il y aura proportionnalité
entre l'amplitude du courant modulé et le degré d'accolement glottique. La
résistance aù courant diminue en effet en raison de la surface selon laquelle
se joignent les lèvres des cordes vocales. L'inscription de la modulation du
courant après détection, traduira donc fidélement l'image de la cinétique
glottique. (Fabre, 1957, pp.67-68; quoted in Sect.5.2.3)

Das ... für das Verfahren ausschlaggebende Glied besteht aus einer Thyratron-röhre, welche über einen Kondensator und einen Widerstand in bekannter Weise Kippschwingungen erzeugt, deren Frequenz durch die Größe der Schaltelemente gegeben ist. Man wählt die Frequenz so, daß sie tief gegen die Sprachfrequen-zen, also bei 20-50 Hz liegt.

Durch einen Kathodenstrahloszillographen wird die Kippschwingung sichtbar gemacht und erscheint als Linie bestimmter Länge. Diese kann bei zeitlicher Auseinanderziehung als Dreieckskurve oder dergleichen festgelegt werden. Gibt man auf das Steuergitter des Thyratrons eine Wechselspannung höherer Fre-quenz als die Kippschwingung, so zündet das Thyratron dann, wenn das Gitter eine bestimmte Aufladung erhalten hat, die weitgehend unabhängig von den Spitzen dieser Steuerspannung sein kann. Die Zündungen folgen umso schneller aufeinander, je höher die Frequenz der an das Steuergitter gelegten Spannung ist, d.h. aber, daß die vorher eingestellte Aufladung der Kippschwingung eher unterbrochen wird und auf dem Leuchtschirm der Kathodenstrahlröhre eine mit zunehmender Frequenz wachsende Verkürzung des Leuchtstriches zu sehen ist. Durch diese Anordnung können auch die Sprachgrundschwingungen benutzt wer-den, um eine konstante Kippschwingung zu steuern. Naturgemäß arbeitet dieses Verfahren völlig trägheitslos, so daß damit das Problem grundsätzlich gelöst ist, auch ganz schnell verlaufende Vorgänge mit genügender Sicherheit aufzu-zeichnen. (Grützmacher and Lottermoser, 1937; p.243, quoted in Sect.6.6.1)

In een proefopstelling mit een vektorimpedantiemeter met de electroglotto-graafelektroden op de hals, waarbij von elektrodenpasta gebruikt wordt gemaakt, blijkt, dat bij toenemende oscillatorfrequentie de impedantie over de hals von de proefperson afneemt en dat de fasehoek naar nul gaat. In het model ... betekent dit dat er voornamelijk sprake is van een serieschakeling tussen een capaciteit en een weerstand. Bij open glottis betreft het de serieschakeling tussen C_e en R_s omdat we mogen aannemen dat de capaciteit tussen de elektroden en het halsweefsel groot is ten opzichte van de capaci-teit van het halsweefsel zelf, terwijl R_v zeer groot is. De capaciteiten C_t en C_s zijn klein ten, opzichte van R_t en R_s en kunnen verwaarloosd worden. Aangezien de fase van het complex C_e en R_s klein is, zullen we bij een meetfrequentie van 300 kHz of hoger de capaciteit C_e eveneens verwaarlozen. De glottiscapaciteit C_v wordt kortgesloten bij de stembandsluiting. Vlak vóór de sluiting van de stembanden is deze capaciteit maximaal. Dit zou inhouden dat, mocht deze capaciteit van enige betekenis zijn, het elektroglottogram vóór zou moeten lopen op het stroboscopisch beeld. Zodoende kan C_v eveneens verwaarloosd worden. Door deze reductie komt men op Fig.5.15b uit, een model dat alleen uitgaat van parallelle, zuiver ohmse weerstanden. Het model voldoet aan de volgende formule uit de electriciteitsleer:

$$1 / R_L = 1 / (2R_t + R_v) + 1 / R_s . \qquad (5.1)$$

We kunnen nun R_s bepalen door bij geopende stembanden door middel van een decadebank in de brugstelling van de elektroglottograaf, die gevonden wordt bij preparaat of proefperson, de zogenaamde "open impedantie" te meten. R_L is dan gelijk aan R_s.

R_s is voor het preparaat 130 Ω en voor de proefperson 30 Ω. R_s is bij de proefperson kleiner dan bij het preparaat. Dit komt voort uit het feit dat er bij de proefperson een veel dikkere subcutislaag aanwezig is, waarvan de weerstand omgekerd evenredig is met de doorsnede. Het praktisch naakte preparaat heeft minder weefsel waarlangs een parallelle stroomweg mogolijk is, zodat de weerstand hiervan hoger is. Tevens is bij een open glottis de weerstand R_v oneindig groot en bij gesloten glottis nul. De stembanden werken dus als een schakelaar. Nu is het bekend dat de impedantie bij stembandsluiting met 10% daalt bij het preparaat en met 1% bij de proefperson tot respectieve-lijk 117 Ω en 29.7 Ω (de "gesloten impedantie"). Wanneer men deze gegevens invult in (5.1) komt men tot het resultaat: $2R_t = 1170\,\Omega$ voor het preparaat en $2970\,\Omega$ voor de proefperson. (Lecluse, 1977, pp.151-152; quoted in Sect.5.2.3)

Alors que plus d'une centaine d'analyseurs de mélodie différents (par les principes et les résultats) ont été conçus dans le monde, il n'y a guère que deux méthodes, l'analyse oscillographique et l'analyse spectrographique, qui aient été utilisées par les phonéticiens - sans doute parce qu'au fond ce sont les plus sûres (mais aussi les plus lentes).
En effet le processus d'analyse introduit souvent, surtout lorsqu'il utilise des filtres, des délais, déphasages et distortions diverses dont on peut difficilement connaître les lois des variation, d'autant plus qu'il est difficile de traiter mathématiquement l'onde de la parole.
Le point le plus délicat, plus encore que l'analyse elle-même, qu'elle soit spectrale (transformée de Fourier) ou périodique (autocorrélation) semble être l'extraction de la fréquence ou de la période fondamentale. La réduction du taux d'information apportée par l'analyse pose souvent des problèmes difficiles particulièrement lorsque le signal comprend un niveau de bruit relativement important, et lorsqu'il manque de périodicité (transitions rapides de formants, changement brutal de hauteur mélodique).
Si a priori la stratégie de décision qui s'applique à l'ensemble de l'information contenue dans le spectre, et qui comporte le calcul du plus grand commun diviseur des fréquences harmoniques ou du plus petit commun multiple des périodes des harmoniques, semble résister à la présence de bruit, même important, sur une partie du spectre, des erreurs peuvent encore se produire, même pour des procédés très élaborés comme le "cepstre". Ceci est vraisemblablement dû au fait que la structure formantique (enveloppe spectrale) et la fondamentale (fine structure spectrale) sont inséparables et interactives. Sans doute faudrait-il élaborer une tactique de reconnaissance du type "pattern recognition" appliquée aux résultats de l'analyse spectrale ou périodique, et qui pourrait tenir compte de l'avenir du signal analysé.
En plus de l'extraction il reste le problème de la mesure et de la précision de la mesure de la fondamentale, le but à atteindre étant la meilleure approximation de la fréquence laryngienne. Ceci se pose surtout dans les méthodes les plus utilisées: l'analyse du tracé oscillographique "manuelle" qui demande des vitesses de défilement énormes (et coûteuses) pour obtenir une précision acceptable, et l'examen des spectrogrammes dont la précision est limitée par l'emploi de filtres.
Les solutions choisies doivent évidemment correspondre aux buts poursuivis, et à la fonction choisie. Si l'analyse spectrographique peut être parfois suffisante pour certaines recherches, des systèmes plus précis et travaillant en temps réel sont indispensables pour des applications telles que le vocoder et sont souhaitables lorsqu'un grand nombre de données doivent être traités. (Léon and Martin, 1969, p.132-133; quoted in Sect.9.4.2)

Cette solution est particulièrement appropriée à nos buts ultérieurs: ... la visualisation de la courbe mélodique en temps réel sur écran d'oscilloscope, et la reconnaissance automatique des patrons intonatifs, qui se rattachent toutes deux à ce qu'on pourrait appeler la perception "visuelle" de la courbe d'intonation. (Léon and Martin, 1969, p.170; quoted in Sect.9.4.2)

La méthode ne permet pas d'obtenir de résultats pour certains phonèmes prononcés par une voix aiguë lorsque la fréquence du fondamental se confond avec celle du premier formant. Dans ce cas, le signal filtré a une allure sinusoidale, sa période étant celle du fondamental.

L'algorithme ne détecte alors *aucune* impulsion (par opposition au bruit où il en détecte de façon irrégulière).

Dans ce cas, la période du fondamental est le temps qui sépare deux passages du signal d'une valeur positive ou nulle à une valeur négative. (Le Roux, 1975, p.8; quoted in Sect.6.3.1)

Le principe de la méthode du peigne réside dans la recherche de valeurs du spectre situées à des fréquences harmoniques et dont la somme soit maximale pour un intervalle fréquentiel donné. L'intercorrélation spectre-peigne revient au calcul de la somme des composantes spectrales correspondant à une structure harmonique donnée. ... La fondamentale correspondant à la structure harmonique donnant la somme la plus grande est alors retenue comme fréquence fondamentale du signal, du moins que cette somme est suffisament différente des valeurs obtenues pour d'autres structures dans le même spectre (cas qui correspond aux segments non voisés). (Martin, 1981, p.224; quoted in Sect.8.5.2)

On peut faire en sorte que seule la fréquence F_0 corresponde au maximum de la fonction d'intercorrélation en ponderant différemment les différentes raies du spectre, de manière à pénaliser les harmoniques élevées dans le calcul de $A_C(p)$. Les mêmes pics harmoniques du spectre correspondent à des dents du peigne d'ordre différent pour $p = F_0/2$, $F_0/3$, etc., on obtient ce résultat en conférant aux dents du peigne une amplitude décroissante, par exemple selon la loi

$$C(k) = k^{-1/s} \quad \text{avec } s \text{ entier} > 1. \tag{8.31}$$

Cette amplitude décroissante des dents du peigne est préférable à une normalisation de la fonction d'intercorrélation par le nombre de dents présentes dans la bande de fréquence utilisée. Par cette dernière méthode en effet, on obtient des maxima égaux pour $p = F_0$ and $p = 2F_0$ si le spectre ne présente que deux pics harmoniques et que le premier corresponde à la fondamentale. (Martin, 1981, p.225; quoted in Sect.8.5.2)

En fait, le spectre d'amplitude n'est pas un spectre de raie et il présente des valeurs non nulles entre ses différents sommets. Malgré une amplitude décroissante des dents du peigne, la valeur de $A_C(p)$ risque donc d'être supérieure, pour des valeurs de p sous harmoniques, à celle obtenue pour $p = F_0$, puisque des dents intermédiaires du peigne correspondent alors aux creux non nuls entre les pics harmoniques.

Pour être assuré d'obtenir un maximum de $A_C(p)$ pour $p = F_0$, il y a donc lieu d'éliminer l'effet des valeurs spectrales non nulles entre les maxima du spectre, en réalisant un *prélèvement fréquentiel* de chacun des sommets selon une fonction analogue aux fenêtres utilisées pour le prélèvement temporel du signal (Hamming ...). Ce prélèvement est préférable à une discrétisation réduisant le spectre à ses seuls maxima. En effet, l'imprécision du calcul numérique et les éventuelles distorsions de phase peuvent rendre les fréquences des pics du spectre non harmoniques exacts de la fréquence fondamentale. Il n'y aurait donc plus dans ce cas correspondance exacte des dents du peigne avec chacune des harmoniques. Au contraire, le prélèvement fréquentiel permet d'interpréter le sommet de la fonction d'intercorrélation comme résultat du meilleur ajustement d'un spectre donné avec une contrainte d'harmonicité de fréquence fondamentale. (Martin, 1981, p.226; quoted in Sect.8.5.2)

On voit donc là apparaître l'intérêt de la méthode par rapport au cepstre, pour lequel la mesure de la fondamentale dépend de la détermination d'une périodicité dans le spectre. La fonction d'intercorrélation présentera un maximum pour F_0 même si cette fréquence est la seule composante d'un spectre dépourvu de toute harmonique, ou si seules deux harmoniques consécutives sont présentes, alors que ce résultat n'est pas garanti dans le cas du cepstre. (Martin, 1981, p.226; quoted in Sect.8.5.2)

Das Gesamtverfahren bei der Umsetzung von Stimmkurven in Tonhöhenkurven zerfällt also, wenn diese höheren Anforderungen an Genauigkeit genügen soll, in folgende Momente: 1) Mikroskopische Ausmessung der Periodenlängen; 2) Feststellung der logarithmischen Werte dieser Periodenlängen; 3) Abtragen der einzelnen Periodenlängen längs der Abscissenachse; 4) Abtragen der logarithmischen Werte der Periodenlängen als Ordinaten zu den jeweiligen Abscissenwerten; 5) Verbindung der so erhaltenen Ordinatenpunkte zu einer Kurve; 6) Berechnung der den Tönen der Tonleiter entsprechenden Ordinatenwerte aufgrund der Geschwindigkeit der Zylindermantelfläche bei der Aufnahme der Stimmkurve. (Meyer, 1911, p.7-8; quoted in Sect.5.1.1)

Bei der Untersuchung stimmkranker Patienten spielt die eingehende Prüfung der verschiedenen stimmlichen Leistungen eine wichtige Rolle. ... Dies geschieht im allgemeinen so, daß die Grundtonhöhe des Stimmklanges mit Hilfe von Vergleichstönen - sei es von Klavier oder Stimmgabel - bestimmt und notiert wird. Abgesehen davon, daß diese Untersuchungsmethodik geschulte Prüfer mit gutem musikalischem Gehör erfordert, beinhaltet sie ausserdem manche Fehlerquellen, wodurch die Bestimmung der Tonhöhenwerte nur annähernd gelingen kann. Für Diagnostik und Therapie einer Stimmerkrankung mag das durchaus genügen, für eine exakte Dokumentation reicht es jedoch nicht aus und macht vergleichende Untersuchungen nur mit Vorbehalt möglich. (Schultz-Coulon and Arndt, 1972, p.241; quoted in Sect.9.4.3)

1) Ein Sinuston ruft eine Tonhöhe hervor, welche direkt mit dem Ort der maximalen Erregung des Corti'schen Organs ... zusammenhängt; sie wird als *Spektraltonhöhe* bezeichnet.
2) Die Spektraltonhöhen, welche zu den Teiltönen eines Klanges gehören, werden bei entsprechend gelenkter Aufmerksamkeit der Versuchsperson einzeln wahrgenommen, sofern der gegenseitige Frequenzabstand der Teiltöne eine gewisse Grenze nicht unterschreitet. Handelt es sich um einen *harmonischen* Klang, so sind die Spektraltonhöhen der ersten 6 bis 8 Harmonischen einzeln wahrnehmbar.
3) Außer den unter 1) und 2) genannten Tonhöhen ruft ein Klang im allgemeinen eine dominante "Gesamt-Tonhöhe" hervor, welche bei harmonischen Klängen der Grundfrequenz entspricht. ... Diese Tonhöhenempfindung stellt ein von der Spektraltonhöhe völlig verschiedenes Wahrnehmungsattribut dar; sie wird *virtuelle Tonhöhe* genannt.
4) Als für die virtuelle Tonhöhe eines Klanges maßgebende Klangmerkmale haben sich diejenigen Teiltöne erwiesen, welche unter den bei 2) genannten Bedingungen getrennte Spektraltonhöhen hervorrufen. (Terhardt, 1972a, p.63; quoted in Sect.4.2.1)

Pendant la première phase "décollement des cordes vocales" la lumière croît lentement, tandis que l'impédance croît rapidement. Ces variations traduisent l'augmentation lente d'une raie de lumière, mais aussi le décollement rapide des cordes vocales.
Dans la seconde phase "ouverture", la lumière croît rapidement et l'impédance parvient à saturation.
Au cours de la troisième phase "fermeture", la lumière décroit rapidement et l'impédance décroit lentement, marquant le début de l'accollement des cordes vocales. Enfin, pendant la quatrième phase "accollement", la lumière devient nulle, l'impédance décroit brusquement, et une impulsion acoustique de forte amplitude naît de la fermeture rapide des cordes vocales. (Vallancien et al., 1971, pp.375-76; quoted in Sect.5.2.3)

Bibliography

This bibliography is an exhaustive list of references on the problem of pitch determination and related fields. It goes somewhat beyond the scope of the text of this book and is thus more than a collection of the items referenced in the text. The field of pitch determination of the speech signal is covered as completely as the references available; the fields of instrumental pitch determination and voicing determination are partly covered. Selected references have been added on related fields, such as signal processing, larynx anatomy, phonetic and linguistic aspects of prosody, and application of pitch determination algorithms in speech communication.

How To Use The Bibliography

Nonstandard Abbreviations. A large number of entries are taken from conference proceedings or from reports published by universities or other scientific institutions. In these entries, for reason of space, the full reference is not repeated every time. Instead, a (nonstandard) abbreviation is given in boldface under which this reference is listed. Nonstandard reference abbreviations are themselves listed within the bibliography in the same alphabetic sequence as normal entries. For instance, if the user wants to know what **ICASSP-77** means, he may look for this item among the authors whose names begin with an "I."

Classification. This is a *classified* bibliography, i.e., the entries are labeled according to the particular field they deal with. For reasons of space, descriptors cannot be used; a subject index of the bibliography cannot be given either. The classification of the bibliography is thus displayed in the form of "section numbers" which correspond to the section numbers used in this book. Since the bibliography covers a wider field than the text of this book, some of the section numbers in the bibliography include more topics than the corresponding section in the text. In the following, these section numbers are listed together with the pertinent topics. Entries in the bibliography, which are labeled with a number not contained in the following list treat only subjects covered in the corresponding section of the book.

1.1   Tutorial surveys on speech processing
1.2   Surveys on speech analysis, speech processing, and application areas of pitch detection, that may serve as an introduction to the specific domain
1.3   Bibliographies and articles with a substantial number of references
2.2   Signal representation; sampling theorems, quantization, analog-to-digital and digital-to-analog conversion
2.3   Digital filters
2.4   Short-term analysis; discrete and fast Fourier transforms; computation of spectra
2.6   Surveys of pitch determination algorithms

Abberton E. (1972a): "Visual feedback and intonation learning." **ICPhS-7**, 813-819 [5.1;
  9.4.3]
Abberton E. (1972b): "Some laryngographic data for Korean stops." J. Int. Phonetics
  Assoc. **2**, 67-77 [3.4; 5.2.3; 9.4.2]
Abberton E. (1974a): "Listener identification of speakers from larynx frequency." **Phon.**
  London [5.2.3; 9.4.4]
Abberton E. (1974b): "Listener identification of speakers from larynx frequency." **ICA-8**,
  273 [9.4.4]
Abberton E., Fourcin A.J. (1969): "Fundamental frequency display and the acquisition of
  intonation skill." In *Proceedings, 2nd Congress of Applied Linguistics, Cambridge,*
  September 1969 (Cambridge UK) [5.1; 9.4.3]
Abberton E., Parker A. (1976): "Further work with laryngographic displays." **Phon. London,**
  64-73 [5.2.3; 9.4.3]
Abberton E., Parker A., Fourcin A.J. (1978): "Speech improvement in deaf adults using
  laryngograph displays." **Phon. London,** 33-60 [5.2.3; 9.4.3]
Abramson A.S., Lisker L., Cooper F.S. (1965): "Laryngeal activity in stop consonants."
  Speech Research 4 [3.1; 5.2.5]
Abry C., Boe L.J. (1977): "Traits, indices et proprietés du voisement: Vers une phono-
  logie des unités infra-phonémiques. Implications pour la détection automatique et
  perceptive." **Journées d'Etude sur la Parole,** 267-274 [5.3]
Abry C., Boe L.J. (1979): "Le trait de voisement, ses indices et leurs proprietes.
  Implications pour les detections automatique et perceptive d'une analyse phonologique
  au-dela du trait." In *Recherche sur la prosodie du français* (Univ. de Langues et
  Lettres, Grenoble), 57-94 [5.3]
Abry C., Boe L.J., Zurcher J.F. (1975): "La détection du voisement par les propriétés
  physiques résultant de l'excitation périodique du conduit vocal: Comparaison statistique
  de trois procédés." **Journées d'Etude sur la Parole 6,** 228-245 [5.3]
Abry C., Boe L.J., Zurcher J.F. (1979): "La détection du voisement par les propriétés
  physiques resultant de l'excitation périodique du conduit vocal: Comparaison statistique
  de trois procédés." In *Recherche sur la prosodie du français* (Univ. de Langues et
  Lettres, Grenoble), 95-115 [5.3]
Abry C., Cartier M., Serniclaès W. (1975): "Analyse et détection du voisement. Table
  ronde." **Journées d'Etude sur la Parole 6,** 194-204 [5.3]
Achilles D. (1971): "Über die diskrete Fourier-Transformation und ihre Anwendungen auf
  lineare zeitinvariante Systeme;" Habilitationsschrift, Universität Erlangen-Nürnberg
  [2.4; 8.5.4]
Achilles D. (1978): *Die Fouriertransformation in der Signalverarbeitung* (Springer, Berlin)
  [2.4; 8.5.4]
Acoustics Lab. (ed.) (1955): "Speech compression research: status report" (Acoustics Lab.,
  MIT, Cambridge MA; DDC-AD-59091) [6.1.1]
Acoustics Lab. (ed.) (1957): "Speech compression research. Final report" (Acoustics Lab.,
  MIT, Cambridge MA; DDC-AD-117049; AF-19(604)-626) [6.1.1; 6.2.1]
Adams C.M. (1969): "A survey of Australian English intonation." Phonetica **20**, 81-130
  [9.4.2]
Admiraal D.J.H. (1970): "A pitch follower for speech signals." **IPO Annual Progr. Rept.**
  5, 197-205 [6.2.1; 6.6.1]
Adolph A.R. (1957a): "Studies of pitch periodicity." **MIT-RLE Progr. Rept. 45**, 127-131
  [6.2.1]
Adolph A.R. (1957b): "Speech pitch-information extractor." **MIT-RLE Progr. Rept. 46**,
  121-130 [6.1.1]
Adoul J.P.: see (Daaboul and Adoul, 1977)
Adrien M., Martin Ph. (eds.) (1981): "Symposium Prosodie - Prosody symposium. Toronto,
  May 1981" (Experim. Phonetics Lab., Univ. of Toronto, Toronto, Ont.) [9.4.2]
Advani A.M. (1964): "Pick-up of throat-wall vibrations for the synthesis of speech."
  MIT-RLE Progr. Rept. 74, 191-192 [5.2.2]
Agnello J.G. (1975): "Measurements and analysis of visible speech." In *Measurement*
  procedures in speech, hearing, and language; ed. by S.Singh (University Park Press,
  Baltimore), 379-397 [5.1]
Aho G.: see (Tierney et al., 1964)
Ahrens R.E. (1979): "Syntax und Intonation in der gesprochenen deutschen Standard-
  sprache." In *Miszellen V;* ed. by J.P. Köster. Hambuger Phonetische Beiträge, Bd.24
  (Buske, Hamburg), 7-27 [9.4.2]

Aigrain P.R., Lair J.B. (1953): "Speech communication system;" United States Patent No. 2,640,880. Issued June 2, 1953; filed July 24, 1948

Ainsworth W.A. (1973): "A system for converting English text into speech." IEEE Trans. AU-21, 288-290 [9.4.2; 9.4.4]

Ainsworth W.A. (1974): "Influence of fundamental frequency on perceived vowel boundaries in English." Proc. 1974 Speech Commun. Sem. 3, 123-129 [3.4; 9.4.2]

Ainsworth W.A. (1978): "Adaptation of pitch change judgments by repetition." In Spring conference, Cambridge University, April 5-7, 1978 (Institute of Acoustics, Edinburgh), 24 [4.3]

Akinfiev N.N., Kharova S.S., Sobakin A.N. (1973): "Детектирование сигнала основного тона из озвученных звуков. В кн.: Труды АРСО-7. Н.Н.Акинфиев, С.С.Карова, А.Н.Собакин. (Detection of the fundamental tone from voiced sounds. In Russian.)." ARSO-7, 53-55 [6.3.1; 6.5.2]

Akinfiev N.N., Sobakin A.N. (1974): "Преобразование речевых сигналов для целей выделения основного тона. В кн.: Труды АРСО-8 Львов. Н.Н.Акинфиев и А.Н.Собакин. (Preprocessing of speech signals for pitch determination. In Russian.)." ARSO-8 [6.1.3; 7.1]

Allen J. (1968): "A study of the specification of prosodic features of speech from a grammatical analysis of printed text;" Dissertation, MIT, Cambridge MA [9.4.2]

Allen J. (1973): "Reading machines for the blind - Technical problems and the methods adopted for their solution." IEEE Trans. AU-21, 259-264 [9.4.2; 9.4.4]

Allen J. (1976): "Synthesis of speech from unrestricted text." Proc. IEEE 64, 433-442 [9.4.2; 9.4.4]

Allen J., O'Shaughnessy D. (1976): "A comprehensive model for fundamental frequency generation." ICASSP-76, 701-704 [9.4.2]

Allen J.B., Curtis T.H. (1973): "Automatic extraction of glottal pulses by linear estimation." J. Acoust. Soc. Am. 54 (A), 36 (Paper G2; 86th Meet. ASA) [6.5.1]

Allen J.B., Rabiner L.R. (1977): "A unified approach to short-time Fourier analysis and synthesis." Proc. IEEE 65, 1558-1564 [2.4; 8.4.1]

Almeida A. (1978): Nasalitätsdetektion und Vokalerkennung. Forum Phoneticum, Bd.17 (Buske, Hamburg) [9.4.2]

Almeida A., Fleischmann G., Heike G., Thurmann E. (1977): "Short-time statistics of the fundamental tone in verbal utterances under psychic stress." Phon. Köln 8, 67-77 [4.1; 9.4.3]

Altes R.A., Evans W.E., Johnson C.S. (1975): "Cetacean echolocation signals and a new model for the human glottal pulse." J. Acoust. Soc. Am. 57, 1221-1222 [3.3]

Altmann H. (1979): "Intonation: Bibliographische Hinweise" (Institut für Deutsche Philologie, Universität München) [1.3]

Ambikairajah E., Carey M.J. (1980): "Time-domain periodogram analysis (paper presented at the 1980 L'Aquila Meeting on Digital Signal Processing)" [1.3; 9.4.2]

Ambikairajah E., Carey M.J., Tattersall G. (1980): "A method of estimating the pitch period of voiced speech." Electron. Lett. 16, 464-466 [6.2.5; 8.6.3]

Ambler S.: see (Haggard et al., 1970)

Amos W.A.: see (Kingsbury and Amos, 1980)

Ananthapadmanabha T.V., Fant G. (1981): "Glottal flow calculations." J. Acoust. Soc. Am. 70 (A), S76 (Paper JJ3, 102nd Meet. ASA) [3.3]

Ananthapadmanabha T.V., Fant G. (1982a): "Calculation of the true glottal flow and its components." STL-QPSR (1), 1-30 [3.1; 3.3; 6.5.1]

Ananthapadmanabha T.V., Fant G. (1982b): "Calculation of the true glottal flow and its components." Speech Commun. 1, 167-184 [3.1; 3.3; 6.5.1]

Ananthapadmanabha T.V., Yegnanarayana B. (1975): "Epoch extraction of voiced speech." IEEE Trans. ASSP-23, 562-569 [6.3.2]

Ananthapadmanabha T.V., Yegnanarayana B. (1977): "Zero phase inverse filtering for extraction of source characteristics." ICASSP-77, 336-339 [6.3.2; 6.5.1]

Ananthapadmanabha T.V., Yegnanarayana B. (1978): "Epoch extraction from linear prediction residual." ICASSP-78, 8-11 [6.3.2]

Ananthapadmanabha T.V., Yegnanarayana B. (1979): "Epoch extraction from linear prediction residual for identification of closed glottis interval." IEEE Trans. ASSP-27, 309-319 [6.3.2; 6.5.2]

Anderson D.C.: see (Dolansky et al., 1969)

Anderson D.J., Deller J.R.jr., Stone R.E. (1976): "Computer analysis of time jitter in vowel sounds." ICASSP-76, 340-342 [4.3]

Anderson F. (1960): "An experimental pitch indicator for training deaf scholars." J. Acoust. Soc. Am. 32, 1065-1074 [6.2.1; 9.4.3]

Anderson W. (1969): "Specification of a digital vocoder system." J. Acoust. Soc. Am. **45** (A), 308 (Paper L1, 76th Meet. ASA) [6.2.1; 6.4.2]

André P., Filleau M., Hamelin F. (1976): "Glottométrie en temps réel - applications." **Journées d'Etude sur la Parole 7**, 235-246 [5.2.3]

André P., Lhote F. (1977): "Sur la commande par la glotte d'une prothese des membres superieurs." **Journées d'Etude sur la Parole 8**, 203-210 [9.4.3; 9.4.4]

Angelo E.J.jr. (1981): "A tutorial introduction to digital filtering." Bell Syst. Tech. J. **60**, 1499-1545 [2.2]

Ann. Bull. RILP --> Annual Bulletin, Research Institute of Logopedics and Phoniatrics (University of Tokyo, Tokyo)

Ann. Rept. Eng. Res. Inst. --> Annual Report of the Engineering Research Institute (Faculty of Engineering, Univ. of Tokyo)

Anon. (1963): "Apparatus for deriving pitch signals from a speech wave;" British Patent No. 918,941. Filed 20 Feb., 1963

Antoniadis Z., Strube H.W. (1980): "Messungen zum "intrinsic pitch" deutscher Vokale." In Fortschritte der Akustik, DAGA'80 München (VDE-Verlag, Berlin), 707-710 [3.1; 9.4.2]

Antoniadis Z., Strube H.W. (1981): "Untersuchungen zum "intrinsic pitch" deutscher Vokale." Phonetica **38**, 277-290 [4.1; 9.4.2]

ARIPUC --> Annual Report, Institute of Phonetics, University of Copenhagen

Appaix A.: see (Tardy-Renucci and Appaix, 1978)

Arjmand M.M., Doddington G.R. (1983): "Pitch-congruent baseband speech coding." **ICASSP-83** [9.4.5]

Arkhangelskiy S.V., Loktev N.A., Mil'chenko V.I. (1982): "Метод определения частоты основного тона. В кн.: АРСО-12, Киев. С.В. Архангелский, Н.А. Локтев, В.И. Мильченко. (Method for pitch frequency detection. In Russian)." **ARSO-12**, 57-60 [6.1.4; 6.4.1]

Arndt H.J.: see (Schultz-Coulon and Arndt, 1972)

Arnold G.E.: see (Luchsinger and Arnold, 1970)

Arnott G.: see (Lhote et al., 1969)

ARSO --> Автоматическое распознавание. слуховых образов (АРСО). (Automatic recognition of auditory signals and patterns. In Russian) (Moscow)

Artèmov V.A. (1968): Устройство для определения основного тона и громкости речи. А.С.#226310-СССР. В.А.Артемов и др. Опубл. в бюлл."Открытия, Изобретения, Промышленные образцы. Товарные знаки". В.А.Артемов. (Device for measuring the pitch and level of speech. In Russian.). Otkrytia, Izobretenia **28**, 115

Artèmov V.A. (1978): "Intonation und Prosodie." Phonetica **35**, 301-339 [1.2; 9.4.2]

Asano H. (1968): "Application of the ultrasonic pulse method to the larynx (in Japanese)." Jpn. J. Otology **71**, 895-916 [5.2.4]

ASA*50 --> Wolf J.J., Klatt D.H. (eds.): Speech communication papers. Presented at the 97th Meeting of the Acoustical Society of America, Cambridge MA, June 12-16, 1979 (Acoustical Society of America, New York NY 1979)

Aschkenasy E., Weiss M.R., Parsons T.W. (1974): "Determining vocal pitch from fine-resolution spectrograms." **IEEE Symp. Speech Recogn.**, 288-289 [8.5.1]

Askenfelt A. (1973a): "Bestämning av differenslimen för freqvenser i basomradet (Determination of difference limen for low frequencies; in Swedish);" Thesis Work, Royal Institute of Technology, Stockholm [4.2.3; 4.3]

Askenfelt A. (1973b): "Determination of difference limen at low frequencies." **STL-QPSR** (2-3), 37-40 [4.2.3; 4.3]

Askenfelt A. (1976): "Automatic notation of played music (Status report)." **STL-QPSR** (1), 2-11 [9.4.2]

Askenfelt A. (1979): "Automatic notation of played music: The Visa project." Fontes Artis Musicae **26**, 109-120 [9.4.2]

Askenfelt A., Gauffin J., Kitzing P., Sundberg J. (1977): "Electroglottograph and contact microphone for measuring vocal pitch." **STL-QPSR** (4), 13-21 [5.2.2; 5.2.3; 5.2.6]

Askenfelt A., Gauffin J., Sundberg J., Kitzing P. (1980): "A comparison of contact microphone and electroglottograph for the measurement of vocal fundamental frequency." J. Speech Hear. Res. **23**, 258-273 [5.2.2; 5.2.3; 5.2.6]

Askenfelt A., Hammarberg B. (1980): "Speech waveform perturbation analysis." **STL-QPSR** (4), 40-49 [4.3]

Askenfelt A., Sjölin A. (1980): "Voice analysis in depressed patients: Rate of change of fundamental frequency related to mental state." **STL-QPSR** (2), 71-84 [4.3; 9.4.3]

Atal B.S. (1968): "Automatic speaker recognition based on pitch contours;" Dissertation, Polytechnic Inst. of Brooklyn, Brooklyn NY [9.4.4]

Atal B.S. (1972): "Automatic speaker recognition based on pitch contours." J. Acoust. Soc. Am. **52**, 1687-1697 [9.4.4]

Atal B.S. (1974a): "Influence of pitch on formant frequencies and bandwiths obtained by linear prediction analysis." J. Acoust. Soc. Am. **55** (A), S81 (Paper NN2; 87th Meet. ASA) [3.3; 3.4; 6.3.1; 9.4.4]

Atal B.S. (1974b): "Effectiveness of linear prediction characteristics of the speech wave for automatic speaker identification and verification." J. Acoust. Soc. Am. **55**, 1304-1312 [6.3.1; 9.4.4]

Atal B.S. (1975): "Pitch determination by 41-pole linear prediction and Newton transformation. [Unpublished note and personal communication; briefly described in (Rabiner et al., 1976)]" (Bell Labs, Murray Hill NJ) [8.3.3; 9.3.1]

Atal B.S. (1976): "Automatic recognition of speakers from their voices." Proc. IEEE **64**, 460-475 [9.4.4]

Atal B.S., David N. (1978): "On finding the optimum excitation for LPC speech synthesis." J. Acoust. Soc. Am. **63** (A), S79 (Paper GG7, 95th Meet. ASA) [3.3]

Atal B.S., David N. (1979): "On synthesizing natural-sounding speech by linear prediction." **ICASSP-79**, 44-47 [3.3; 9.4.4]

Atal B.S., Hanauer S.L. (1971): "Speech analysis and synthesis by linear prediction of the speech wave." J. Acoust. Soc. Am. **50**, 637-655 [2.4; 6.3.1]

Atal B.S., Rabiner L.R. (1975): "Voiced-unvoiced decision without pitch detection." J. Acoust. Soc. Am. **58** (A), S62 (Paper FF7; 90th Meet. ASA) [5.3]

Atal B.S., Rabiner L.R. (1976): "A pattern recognition approach to voiced-unvoiced-silence classification with applications to speech recognition." IEEE Trans. ASSP-**24**, 201-212 [5.3]

Atal B.S., Remde J.R. (1982): "A new model of LPC excitation for producing natural-sounding speech at low bit rates." **ICASSP-82**, 614-617 [9.4.4; 9.4.5]

Atal B.S., Schroeder M.R. (1978): "Linear prediction analysis of speech based on a pole-zero representation." J. Acoust. Soc. Am. **64**, 1310-1318 [2.4; 3.4]

Atal B.S., Stover V. (1975): "Voice-excited predictive coding system for low-bit-rate transmission of speech." J. Acoust. Soc. Am. **57** (A), S35 (Paper R10; 89th Meet. ASA) [3.3; 9.4.5]

Atkinson J.E. (1973a): "Aspects of intonation in speech: Implications from an experimental study of fundamental frequency;" Dissertation, Univ. of Connecticut, Storrs CT [9.4.2]

Atkinson J.E. (1973b): "Aspects of intonation in speech: implications from an experimental study of fundamental frequency." **Speech Research 34** (A), 211-213 [9.4.2]

Atkinson J.E. (1973c): Physiological factors controlling F_0: Results of a correlation analysis. J. Acoust. Soc. Am. **53** (A), 74 (Paper CC4; 85th Meet. ASA) [3.1]

Atkinson J.E. (1973d): "Intrinsic F_0 in vowels: Physiological correlates." J. Acoust. Soc. Am. **53** (A), 346 (Paper DD8; 84th Meet. ASA) [3.1; 9.4.2]

Atkinson J.E. (1976): "Inter- and intraspeaker variability in fundamental voice frequency." J. Acoust. Soc. Am. **60**, 440-445 [4.1]

Atkinson J.E. (1978): "Correlation analysis of the physiological factors controlling fundamental voice frequency." J. Acoust. Soc. Am. **63**, 211-222 [3.1; 4.2]

Auloge J.Y. (1979): "Modélisation continue de la source vocale." **Larynx et parole,** 87-102 [3.3]

Autesserre D., Di Cristo A. (1972): "Recherches sur l'intonation du francais: Traits significatifs et non significatifs." **ICPhS-7**, 842-859 [9.4.2]

Autesserre D., Rossi M., Sarrat P., Giraud G., Visquis R., Demange R., Chevrot L. (1979): "Exploration radiologique de l'oropharynx, de l'hypopharynx et du larynx en phonation." **Larynx et parole,** 45-74 [3.1; 5.2.1]

Auth W., Lacroix A. (1971): "Dreidimensionale Darstellung von sprachgrundfrequenz-synchron berechneten Sprachspektrogrammen." Nachrichtentech. Z. **24**, 502-507 [9.4.5]

Babic H., Temes G.C. (1976): "Optimum low-order windows for discrete Fourier transform systems." IEEE Trans. ASSP-**24**, 512-517 [2.4; 8.1]

Bach R.E.: see (Chang et al., 1959)

Bachem A. (1955): "Absolute pitch." J. Acoust. Soc. Am. **27**, 1180-1185 [4.2.1; 4.3]

Badoyannis G.M. (1957): "Theoretical and experimental research in communication theory and application (7 reports)" (Rutgers University, New Brunswick NH; Contract DA-36-039-sc-64704) [6.2.1]

Baer T. (1973a): "Measurement of Vibration Patterns." J. Acoust. Soc. Am. **53** (A), 73 (Paper CC1; 85th Meet. ASA) [3.1; 5.2]

Baer T. (1973b): "Measurement of vibration patterns of excised larynxes." **MIT-RLE Progr. Rept. 110**, 169-175 [3.1; 3.2; 5.2]

Baer T. (1974): "Vibration patterns of excised larynxes." J. Acoust. Soc. Am. **55** (A), S79 (Paper MM3; 87th Meet. ASA) [3.1; 3.2; 5.2]

Baer T. (1975): "Investigation of phonation using excised larynxes;" Dissertation, MIT, Cambridge MA [3.1; 3.2; 5.2]

Baer T. (1978): "Effect of single-motor-unit firings on fundamental frequency of phonation." J. Acoust. Soc. Am. **64** (A), S90 (Paper I17, 96th Meet. ASA) [3.1]

Baer T., Gay T., Niimi S. (1975): "Control of fundamental frequency, intensity, and register of phonation." J. Acoust. Soc. Am. **58** (A), S12-13 (Paper F10; 90th Meet. ASA) [3.1; 3.2]

Baer T., Gay T., Niimi S. (1976): "Control of fundamental frequency, intensity, and register of phonation." **Speech Research 45**, 175-185 [3.1; 3.2]

Baer T., Löfqvist A., McGarr N.S. (1981): "Glottal area variations: a comparison between transillumination and high-speed filming." J. Acoust. Soc. Am. **70** (A), S12 (Paper F3, 102nd Meet. ASA) [5.2.1; 5.2.5; 5.2.6]

Baker J.M. (1974): "A new time domain analysis of fricatives and stop consonants." **IEEE Symp. Speech Recogn.**, 134-141 [5.3]

Baker J.M. (1976): "Time-domain analysis and segmentation of connected speech." **Proc. 1974 Speech Commun. Sem. 3**, 369-383 [5.3]

Ball E.: see (Hayes et al., 1981)

Bancroft J.C. (1979a): "A periodicity measure for voiced speech." **ASA*50**, 241-244 [8.2.4]

Bancroft J.C. (1979b): "A periodicity measure for voiced speech." J. Acoust. Soc. Am. **65** (A), S67 (Paper Y6; 97th Meet. ASA)

Bancroft J.C. (1980): "Separation of voiced and unvoiced speech, and silence, using energy and periodicity." J. Acoust. Soc. Am. **68** (A), S70 (Paper LL3, 100th Meet. ASA) [5.3]

Bannister L.H.: see (Fourcin et al., 1978)

Barabell A.J., Crochiere R.E. (1979): "Sub-band coder design incorporating quadrature filters and pitch prediction." **ICASSP-79**, 530-533 [9.4.5]

Barik H.C. (1972): "Some innovations in a computer approach to the analysis of speech patterns." Lang. Speech **15**, 196-207 [5.3]

Barlow (1874): "On the pneumatic action which accompanies the articulation of sounds by the human voice, as exhibited by a recording instrument." Proc. R. Soc. London **22**, 277 [5.1.1]

Barnes L.A.: see (Early et al., 1962)

Barnett J.: see (Newell et al., 1971)

Barney H.L. (1958): "Transmission and Reconstruction of Artificial Speech;" United States Patent No. 2,819,341. Issued Jan. 7, 1958; filed Sept. 30, 1954 [6.4.1]

Barnwell T.P., Bush A.M. (1974): "Gapped ADPCM for speech digitization." In *Proceedings of the National Electronics Conference* (National Engineering Consortium, Oakbrook IL), Vol.29, 422-427

Baronin S.P. (1961): "Автокоррелязионный метод выделения основного тона речи." В кн.: Сборн трудов Гос. НИИ Мин. связи СССР. С.П.Баронин. (The autocorrelation method for pitch determination of speech signals. In Russian). Sborn Trudov **3**, 93-102 [8.2.3]

Baronin S.P. (1964): "Исследование двух схем выделения основного тона речи фильтровым методом." В кн.: Сборн трудов Гос. НИИ Мин. связи. С.П.Баронин. (Investigation of two proposals for pitch determination of speech signals by means of filtering. In Russian). Sborn Trudov **3**, 48-60 [6.1.3]

Baronin S.P. (1965): "О построении многоканальных схем выделения основного тона речи." В кн.: Сборн трудов Гос. НИИ Мин. связи СССР. С.П.Баронин. (Development of a multichannel system for pitch determination of speech signals. In Russian). Sborn Trudov **3**, 17-24 [6.4.1]

Baronin S.P. (1966): "Статистические методы анализа речевых сигналов." В журн.: Электросвязь. С.П.Баронин. (Statistical methods in speech analysis. In Russian). Elektrosvyaz **5**, 50-56 [6.6.3]

Baronin S.P. (1974a): "Выделение основного тона голоса." В кн.: Вокодерная телефония. Под редакцией А.А. Пирогова. С.П.Баронин. (Determination of the fundamental tone of the voice. In Russian). In *Vokoder'naya telefoniya;* ed. by A.A.Pirogov (Svyaz, Moscow), 113-209 [1.2; 1.3; 2.6; 6.1.1; 6.1.4; 6.6.2; 6.6.3; 8.2.3; 8.5.1; 8.6.1; 9.1.1]

Baronin S.P. (1974b): "Автоматическое разделение звонких и глухих звуков речи." В кн.: Вокодерная телефония. Под редакцией А.А. Пирогова. С.П.Баронин. (Automatic discrimination of voiced and voiceless sounds. In Russian). In *Vokoder'naya telefoniya;* ed. by A.A.Pirogov (Izdatel'stvo "Svyaz", Moscow), 209-259 [5.3.1; 5.3.2; 5.3.3]

Baronin S.P., Kushtuev A.I. (1970a): "Устройство для измерения частоты основного тона речевых сигналов." А.С.#280561-СССР. Опубл. в бюлл. "Открытия, Изобретения, Промышленные образцы, Товарные знаки. С.П.Баронин и А.И.Куштуев. (Device for pitch determination of speech signals. In Russian.). Otkrytia, Izobretenia 28, 45

Baronin S.P., Kushtuev A.I. (1970b): "Многоканальный измеритель частоты основного тона речевых сигналов." А.С.#287346-СССР. Опубл. в бюлл. "Открытия, Изобретения, Промышленные образцы. Товарные знаки". С.П.Баронин и А.И.Куштуев. (Multichannel measuring device for pitch determination of speech signals. In Russian). Otkrytia, Izobretenia 35, 119 [6.4.1; 6.6.3]

Baronin S.P., Kushtuev A.I. (1971): "О построении схемы адаптации анализаторов частоты основного тона речи." В кн.: Тезиси докладов ВИИ Всесоюз.Акуст.Конф. С.П.Баронин и А.И.Куштуев. (How to design an adaptive pitch determination algorithm. In Russian.). Proc. of the 7th All-Union Acoustical Conference, Leningrad, 18 [6.1.4; 6.6.3]

Baronin S.P., Kushtuev A.I. (1973): "Следящий выделитель основного тона речи." А.С.#403090-СССР. Опубл. в бюлл. "Открытия, Изобретения, Промышленные образцы, Товарные знаки. С.П.Баронин и А.И.Куштуев. (Device for adaptive pitch tracking of speech signals. In Russian.). Otkrytia, Izobretenia 42, 158 [6.1.4]

Baronin S.P., Kushtuev A.I. (1974): see (Baronin, 1974a). [These two entries reference the same article, which was partly written by Baronin and Kushtuev, partly by Baronin alone.]

Barry W.J. (1981): "Prosodic functions revisited again!." Phonetica 38, 320-340 [9.4.2]

Bascom B.: see (Pike et al., 1959)

Basmaian J.V. (1967): *Muscles alive: their functions revealed by electromyography* (Williams and Wilkins, Baltimore MD) [5.2.1]

Bass S.D.: see (Dolanský et al., 1969)

Bastet L., Dolmazon J.M., Guérin B., Shuplyakov V. (1977): "Détection de mélodie par un modèle fonctionnel du système auditif périphérique." **Journées d'Etude sur la Parole 8,** 71-78 [6.1.1]

Batalov V.S.: see (Mataušek and Batalov, 1980)

Bate E.M., Fallside F., Gulian E., Hinds P., Keiller C. (1982): "A speech training aid for the dea with display of voicing, frication, and silence." **ICASSP-82,** 743-746 [5.3; 9.4.3]

Bates R.H.T.: see (Tucker and Bates, 1978)

Battmer R.D.: see (Schultz-Coulon et al., 1979)

Baudry M. (1978): "Etude du signal vocal dans sa représentation amplitude-temps: Algorithmes de segmentation et de reconnaissance de la parole;" Dissertation, Université Pierre et Marie Curie (Paris 6), Paris [5.3; 6.2.5]

Baudry M., Dupeyrat B. (1976): "Analyse du signal vocal - Utilisation des extrema du signal et de leurs amplitudes - détection du fondamental et recherche des formants." **Journées d'Etude sur la Parole 7,** 247-257 [6.2.5]

Baudry M., Dupeyrat B. (1978): "Utilisation de méthodes syntaxiques pour la détection automatique des traits phonétiques en reconnaissance de la parole." **Journées d'Etude sur la Parole 9,** 313-320 [6.2.5]

Baudry M., Dupeyrat B. (1979): "Speech segmentation and recognition using syntactic methods on the direct signal." **ICASSP-79,** 101-104 [6.2.5]

Baudry M., Dupeyrat B. (1980): "Speech analysis using syntactic methods and a pitch-synchronous formant detector of the direct signal." In *Spoken language generation and understanding.* Proceedings of the NATO Advanced Study Institute held at Bonas, France, June 23 - July 5, 1979; ed. by J.C. Simon (Reidel, Dordrecht), 279-292 [6.2.5]

Baudry M., Dupeyrat B., De Cosnac B. (1977): "Analyse du signal vocal en temps réel sur mini-ordinateur." **ICA-9,** 450 (Paper I44) [6.2.5]

Beach J.L., Kelsey C.A. (1969): "Ultrasonic Doppler monitoring of vocal-fold velocity and displacement." J. Acoust. Soc. Am. 46, 1045-1047 [5.2.4]

Beauchamp J.W. (1968): "A simple method for phase compensation in tape copying with identical tape recorders." J. Audio Eng. Soc. 16, 112-113 [9.2.1]

Beavers G.S.: see (Gupta et al., 1973)

Becker R.W., Poza F. (1973): "Acoustic phonetic research in speech understanding." IEEE Trans. ASSP-23, 416-426 [5.3]

Beckett R. (1969): "Pitch perturbations as a function of vocal constrictions." Folia Phoniatr. **21**, 416-425 [4.3]

Beckmann G. (1956): "Elektroakustische Untersuchungen zum primären Larynxton." Arch. Ohren-Nasen-Kehlkopfheilkunde **169**, 196-201

Beekmans R., Carré R., Jospa P. (1974): "Influence de la fréquence fondamentale sur l'espace perceptif des voyelles." **Journées d'Etude sur la Parole 5**, 146-154 [4.2; 9.4.2]

Bekesy G. von (1963): "Three experiments concerned with pitch perception." J. Acoust. Soc. Am. **35**, 602-606 [4.2]

Belar H.: see (Olson et al., 1967)

Belforte G., De Mori R., Ferraris F. (1979): "A contribution to the automatic processing of electrocardiograms using syntactic methods." IEEE Trans. BME-**26**, 125-136 [6.5.2]

Bell C.G., Fujisaki H., Heinz J.M., Stevens K.N., House A.S. (1961): "Reduction of speech spectra by analysis-by-synthesis techniques." J. Acoust. Soc. Am. **33**, 1725-1736 [9.4.5]

Bellanger M. (1980): *Traitement numérique du signal* (Masson, Paris) [2.3; 8.5.4]

Bellet G., Contini M., Boe L.J. (1982): "Normalisation temporelle et fréquentielle de F_0 intra- et interindividuelle." **Phon. Grenoble 11**, 265-283 [6.6.1; 9.4.2; 9.4.4]

Beltrame D., Castellino P. (1981): "Criteria and experiments for choosing a pitch period estimator (in Italian)" (Centro Studi e Laboratort Telecomunicazioni, Turin, Italy; CSELT Internal Rept.) [9.3.1]

Benade A.H. (1976): *Fundamentals of musical acoustics* (Oxford Univ. Press, New York) [1.2; 9.4.2]

Benedetto M.D. di: see (Bruno et al., 1981)

Benedetto M.G. di: see (Bruno et al., 1981)

Benguerel A.P., Hirose H., Sawashima M., Ushijima T. (1978): "Laryngeal control in French stop production: a fiberscopic, acoustic, and electromyographic study." Folia Phoniatr. **30**, 175-198 [3.1; 3.4; 5.2.1]

Bennett G. (1981): "Singing synthesis in electronic music." In *Research aspects on singing*. Papers given at a seminar organized by the Committee for the Acoustics of Music (Royal Swedish Academy of Music, Stockholm; Report #33), 34-50 [3.3; 9.4.2; 9.4.4]

Bennett S.: see (Weinberg and Bennett, 1972)

Bennett W.R. (1953): "The correlatograph." Bell Syst. Tech. J. **32**, 1173-1185 [8.2.3]

Bérard E.: see (Fonagy and Bérard, 1972)

Berg J. van den (1954a): "On the origin of the multiplication of the vocal period during speech." Folia Phoniatr. **6**, 228-232 [3.1; 4.3]

Berg J. van den (1954b): "Über die Kopplung bei der Stimmbildung." Z. Phonetik **8**, 281-293 [3.1; 3.4]

Berg J. van den (1955a): "Direct and indirect determination of the mean subglottic pressure." Folia Phoniatr. **8**, 1-24 [5.2]

Berg J. van den (1955b): "Calculations on a model of the vocal tract for vowel /i/ (meat) and on the larynx." J. Acoust. Soc. Am. **27**, 332-338 [3.4]

Berg J. van den (1957): "Subglottic pressure and vibrations of the vocal folds." Folia Phoniatr. **9**, 65-71 [3.1]

Berg J. van den (1958): "Myoelastic-aerodynamic theory of voice production." J. Speech Hear. Res. **1**, 227-244 [3.1]

Berg J. van den (1960): "Vocal ligaments versus registers." Folia Phoniatr. **1**, 19-34 [3.1; 3.2]

Berg J. van den (1962): "Modern research in experimental phoniatrics." Folia Phoniatr. **14**, 81-149 [1.2; 1.3; 9.4.3]

Berg J. van den (1968a): "Mechanisms of the larynx and the laryngeal vibrations." In *Manual of Phonetics*; ed. by B.Malmberg (North Holland Publishing Company, Amsterdam), 278-308 [3.1; 3.2]

Berg J. van den (1968b): "Register problems." Ann. NY Acad. Sci. **155**, 129-134 [3.2]

Berg J. van den, Minne J.M. van der (1962): "On the primary glottal sound source." Acta Physiol. Pharmacol. Neerl. **11**, [3.1]

Berg J. van den, Zantema J.T., Doornenbal P. (1957): "On the air resistance and the Bernoulli effect of the human larynx." J. Acoust. Soc. Am. **29**, 626-631 [3.1]

Bergeron L.E., Goldberg A.J., Kwon S.Y., Miller M. (1980): "A robust, adaptive transform coder for 9.6 kb/s speech transmission." **ICASSP-80**, 344-347 [9.4.5]

Bernhard R. (1980): "Solid state looks to VLSI." IEEE Spectrum **17**, 44-49 [9.1.5]

Berouti M.G. (1977): "Estimation of glottal volume velocity by the linear prediction inverse filter;" Dissertation, University of Florida, Gainesville FL [6.3.1; 6.5.1]

Berouti M.G., Childers D.G., Paige A. (1977): "Glottal area versus glottal volume-velocity." **ICASSP-77**, 33-36 [3.1; 3.3; 5.2.1]

Berouti M.G., Makhoul J. (1979): "An adaptive-transform baseband coder." **ASA*50,** 377-380 [9.4.5]

Berthomier C. (1979): "Calcul analogique de la frequence du fondamental." **ICPhS-9** (1) (A), 260 [6.3.2; 6.6.2]

Bertrand J.: see (Jackson and Bertrand, 1976)

Bessey A.C.: see (Siegel and Bessey, 1982)

Bethge W. (1952): "Phonometrische Untersuchungen zur Sprachmelodie." Z. Phonetik **6,** 229-247 [5.1; 9.4.2]

Bethge W. (1953): "Über abgehörte und gemessene Lautmelodie." Z. Phonetik **7,** 339-346 [4.3; 5.1]

Bethge W. (1957): "Geschätzte und gemessene Melodiewinkel." Phonetica **1,** 203-206 [4.3; 5.1; 6.6.1]

Bezooijen R. van (1981): "Characteristics of vocal expressions of emotion: pitch level and pitch range." **Phon. Nijmegen 5,** 1-18 [4.1; 9.4.3]

Bially T., Anderson W. (1970): "A digital channel vocoder." IEEE Trans. COM-**18,** 435-443 [9.4.4]

Biddulph R.: see (Shower and Biddulph, 1931)

Bierwisch M. (1966): "Regeln für die Intonation deutscher Sätze." Studia Grammatica **7,** 99-201 [9.4.2]

Bjuggren G. (1960): "Device for laryngeal phase-determinable flash photography." Folia Phoniatr. **12,** 36-41 [6.2.5]

Black J.W. (1961): "Relationships among fundamental frequency, vocal sound pressure and rate of speaking." Lang. Speech **4,** 96-99 [4.1; 9.4.2]

Black J.W. (1967): "The magnitude of pitch inflection." **ICPhS-6,** 177-181 [4.3]

Blackman R.B., Tukey J.W. (1958): *The measurement of power spectra* (Dover Publ, New York NY) [2.4]

Blasdell R.: see (Majewski and Blasdell, 1969)

Bleakley D. (1973): "The effect of fundamental frequency variations on the perception of stress in German." Phonetica **28,** 42-59 [4.3; 9.4.2]

Blis I.H. (1967): "What causes the voiced-voiceless distinction?." **ICPhS-6,** 841-844 [5.3]

Blokhina L.P. (1974): "On modelling frequency features in a phrase." **Proc. 1974 Speech Commun. Sem. 2,** 197-204 [9.4.2]

Blom J.G. (1972): "Pitch extraction by means of digital computers." **Phonetica Pragensia III,** 55-57 [6.2.1]

Bluestein L.I., Rader C.M. (1965): "Comparison of vector and autocorrelation pitch detectors." J. Acoust. Soc. Am. **37** (C), 751-752 [6.3.2; 8.2.4]

Bock D.E.: see (Flanagan et al., 1976)

Boddie J.R. (1981): "Digital Signel Processor. Overview: the device, support facilities, and applications." Bell Syst. Tech. J. **60,** 1431-1440 [9.1.5]

Boddie J.R., Daryanani G.T., Eldumiati I.I., Gadenz R.N., Thompson J.S., Walters S.M. (1981): "Digital Signal Processor: architecture and performance." Bell Syst. Tech. J. **60,** 1449-1462 [9.1.5]

Boe L.J. (1973a): "Etude acoustique du couplage larynx-conduit vocal (fréquence laryngienne des productions vocaliques)." Revue d'Acoustique (27), 235-244 [3.3; 3.4]

Boe L.J. (1973b): "Les faits prosodiques et la fréquence laryngienne, approche théorique et experimentale." **Bull. Audiophonologie 2,** 3-24 [4.1; 9.4.2]

Boe L.J. (1977): "La frequence fondamentale: études physio-acoustiques et instrumentales 1: Variations de la périodicité; régulation de la vibration des cordes vocales (Compte-rendu du rapporteur)." **Journées d'Etude sur la Parole 8,** 3-18 [3.2; 4.1]

Boe L.J., Bellet G. (1979): "Mesure et statistique de la fréquence laryngienne: appareillage et programmes." **Phon. Grenoble 8,** 23-34 [4.1]

Boe L.J., Contini M., Rakotofiringa H. (1975): "Etude statistique de la fréquence laryngienne - Application à l'analyse et à la synthèse des faits prosodiques du francais." Phonetica **32,** 1-23 [1.2; 4.1; 9.4.2]

Boe L.J., Descout R., Guérin B. (eds.) (1979): *Larynx et Parole*. Actes du Séminaire organisé à l'Institut de Phonétique de Grenoble, les 8 et 9 Février 1979, sous le patronnage du G.A.L.F (Institut de Phonétique, Université des Langues et Lettres, Grenoble) [3.1]

Boe L.J., Guérin B. (1979a): "Etude du fonctionnement d'un modele des cordes vocales a deux masses; determination des parametres de commande et de leurs influences respectives." In *Recherche sur la prosodie du français* (Univ. de Langues et Lettres, Grenoble), 7-56 [3.3]

Boe L.J., Guérin B. (1979b): "Caracteristiques de forme de l'onde de debit des cordes vocales: productions vocaliques." **ICPh-Miami,** 109-118 [3.1; 3.3]

Boe L.J., Larreur D. (1974): "Les characteristiques intrinsèques de la frequence laryngienne: production, réalisation, et perception." **Journées d'Etude sur la Parole 5,** 19-28 [4.1]

Boe L.J., Rakotofiringa H. (1971): "Exigences, réalisation et limite d'un appareillage destiné à l'étude de l'intensité et de la hauteur d'un signal acoustique." Revue d'Acoustique (14), 104-113 [4.3]

Boe L.J., Rakotofiringa H. (1972): "Une méthode systematique de la mesure de la fréquence laryngienne, de l'intensité et de la durée de la parole." **Phon. Grenoble 1,** 1-9 [9.4.2]

Boe L.J., Rakotofiringa H. (1975): "A statistical analysis of laryngeal frequency - its relationship to intensity level and duration." Lang. Speech **18,** 1-14 [4.1]

Boeke J.D. (1891): "Mikroskopische Phonogrammstudien." Pflügers Archiv für die gesamte Physiologie **50** [5.1.1; 9.4.2]

Boer E. de (1956): "On the 'Residue' in Hearing;" Dissertation, University of Amsterdam [4.2.1]

Boer E. de (1976): "On the residue and auditory pitch perception." In *Handbook of sensory physiology;* ed. by W.D. Keidel and W.D. Neff (Springer, Berlin), 479-583 [4.2.1]

Boer E. de (1977): "Pitch theories unified." In *Psychophysics and physiology of hearing;* ed. by E.F. Evans, J.P. Wilson (Academic Press, London), 323-334 [4.2.1]

Bogert B.P. (1953): "On the bandwidth of vowel formants." J. Acoust. Soc. Am. **25,** 791-792 [3.4]

Bogert B.P., Healy M.J.R., Tukey J.W. (1963): "The quefrency alanysis [sic!] of time series for echoes: cepstrum, pseudo-autocovariance, cross-cepstrum, and saphe cracking." In *Proceedings of the Symposium on Time Series Analysis;* ed. by M. Rosenblatt (Wiley, New York), 209-243 [8.4.1]

Bogert B.P., Ossanna J.F. (1966): "Computer experimentation on echo detection, using the cepstrum and pseudoautocovariance." J. Acoust. Soc. Am. **39** (A), 1258 (Paper 6G3; 71st Meet. ASA) [8.4.1]

Bolanowsky S.J.jr.: see (Rothenberg et al., 1977)

Bolinger D.L. (1955): "Intonation as stress-carrier." Litera **2,** 35-40 [9.4.2]

Bolinger D.L. (1961): "Intonation: Levels Versus Configurations." Word **7,** 199-210 [9.4.2]

Bolinger D.L. (ed.) (1972): *Intonation: selected readings* (Penguin Books, Harmondsworth) [1.2; 5.1; 9.4.2]

Boll S.F. (1974): "Selected methods for improving synthesis speech quality using LPC: System description, coefficient smoothing, and STREAK (simplified technique for recursively estimating autocorrelation k-parameters)" (Comp. Sci. Dept., Univ. of Utah, Salt Lake City UT; UTEC-CSc-74-151) [6.3.1; 8.2.4]

Bomas V.V.: see (Genina and Bomas, 1970)

Booth J.R., Childers D.G. (1979): "Automated analysis of ultra high-speed laryngeal films." IEEE Trans. BME-**26,** 185-192 [5.2.1]

Boothroyd A. (1972a): "Some experiments on the control of voice in the profoundly deaf using a pitch extractor and storage oscilloscope display." **ICSCP-72,** 234-237 (Paper G2) [9.4.3]

Boothroyd A. (1972b): "Control of voice pitch by the deaf - am experiment using a visible pitch device." Audiology **11,** 343-353 [6.1.2; 9.4.3]

Boothroyd A. (1973): "Some experiments on the control of voice in the profoundly deaf using a pitch extractor and storage oscilloscope display." IEEE Trans. AU-**21,** 274-278 [6.1.2; 9.4.3]

Bordone-Sacerdote C., Righini G.U. (1965): "The glottal wave detected as a high frequency modulation." **ICA-5** (Paper A-34) [5.2.3]

Bordone-Sacerdote C., Sacerdote G.G. (1965): "Investigations of the movement of the glottis by ultrasounds." **ICA-5** (Paper A-42) [5.2.4]

Bordone-Sacerdote C., Sacerdote G.G. (1969): "Some spectral properties of individual voices." Acustica **21,** 199-210 [3.2; 3.3]

Bordone-Sacerdote C., Sacerdote G.G. (1970): "Measurement of the behaviour of the vocal cords (preliminary note)." Acustica **23,** 46-47 [3.2; 3.3]

Bordone-Sacerdote C., Sacerdote G.G. (1976): "Distribution of pauses as a characteristic of individual voices." Acustica **34,** 245-247 [5.3]

Borel-Maisonny S. (1973): "Mélodie de la voix." **Bull. Audiophonologie 2,** 107-146 [9.4.2]

Boring E.G. (1940): "The size of the differential limen for pitch." Am. J. Psychol. **53,** 450-455 [4.2.3]

Bosscha G.J., Sluyter R.J. (1982): "DFT-Vocoder using harmonic-sieve pitch extraction."
ICASSP-82, 1952-1955 [8.5.2; 9.4.5]

Bot K.L.J. de (1980a): "Intonation teaching and pitch control." Phon. Nijmegen 4, 53-63
[3.1; 9.4.3]

Bot K.L.J. de (1980b): "Relative reliability in judging intonation." Phon. Nijmegen 4,
64-78 [4.3]

Bot K.L.J. de (1981a): "Feedback modality and learning behaviour in intonation teaching."
Phon. Nijmegen 5, 19-23 [5.1; 9.4.3]

Bot K.L.J. de (1981b): "Visual feedback of English intonation: an experimental approach."
Phon. Nijmegen 5, 24-40 [5.1; 9.4.3]

Boulogne M. (1972): "Détecteur de mélodie à autocorrelation;" Dissertation, Ecole Nat.
Sup. Electr. Radioélectr, Grenoble [8.2.3]

Boulogne M. (1979): "Détection du voisement et de la fréquence fondamentale d'un signal
de parole;" Thèse d'Etat, Inst. Nat. de Polytechnique, Grenoble [5.3.1; 6.3.2; 8.2.4]

Boulogne M., Carré R., Charras J.P. (1973): "La fréquence fondamentale, les formants,
éléments d'identification des locuteurs." Revue d'Acoustique (23), 343-350 [4.1; 9.4.2]

Boves L. (1979a): "Instrumentation pour la comparaison des résultats du filtrage inverse
avec des donnés physiologiques du comportement des cordes vocales." Larynx et
parole, 301-312 [5.2.3; 5.2.5; 5.2.6; 6.5.1]

Boves L. (1979b): "A set-up for the comparison of glottal inverse filtering with physiological registrations." Phon. Nijmegen 3, 103-123 [5.2.1; 5.2.6; 6.5.1]

Boves L., Cranen B. (1982a): "Comparison of reconstructed glottal-flow waveforms with
physiological registrations." In Fortschritte der Akustik, FASE/DAGA'82 Göttingen
(VDE-Verlag, Berlin), 979-982 [5.2.3; 5.2.5; 5.2.6; 6.5.1]

Boves L., Cranen B. (1982b): "Evaluation of glottal inverse filtering by means of physiological registrations." ICASSP-82, 1988-1991 [5.2.3; 5.2.5; 5.2.6; 6.5.1; 9.3.1]

Boves L., Have B.L. ten (1980): "Instrumental verification of a perceptually based system
for the transcription of the intonation of utterances in Dutch; first results." Phon.
Nijmegen 4, 1-30 [9.4.2]

Boves L., Rietveld T. (1978): "Stylization of pitch contours." Phon. Nijmegen 2, 34-40
[9.4.2]

Bowler N.W. (1964): "A fundamental frequency analysis of harsh vocal quality." Speech
Monogr. 31, 128-134 [6.5.3; 9.4.3]

Boxhall T.R.: see (Hales et al., 1956)

Brachman M.L.: see (Rothenberg et al., 1977)

Bradley C.B. (1916): "On plotting the inflections of the voice." University of California
Pulications in American Archeology and Ethnology 12, 195-218 [5.1; 5.1.1]

Bradshaw F.T.: see (Pillay et al., 1975)

Brady P.T. (1968): "A statistical analysis of on-off-patterns in 16 conversations." Bell
Syst. Tech. J. 47, 73-91 [5.3; 9.4.2]

Brady P.T. (1970): "Fixed-scale mechanism of absolute pitch." J. Acoust. Soc. Am. 48,
883-887 [4.2.1]

Brady P.T. (1972): "Objective measures of peak clipping and threshold crossings in
continuous speech." Bell Syst. Tech. J. 51, 933-946 [6.1.2]

Braida L.D.: see (Frazier et al., 1975)

Bralley R.C.: see (Bull et al., 1979)

Braun H.J. (1975): "Digitale inverse Filterung von Sprachsignalen mit einem Zeitreihen-
analysator" (FTZ der Deutschen Bundespost, Darmstadt; N-I 630 722 300) [6.5.1]

Braun H.J. (1977): "Effects of pitch-synchronous PARCOR analysis-synthesis on the articulation score of CV-elements" (FTZ der Deutschen Bundespost, Darmstadt) [9.4.5]

Brehm H.: see (Wolf and Brehm, 1973)

Brend R.M. (ed.) (1975): "Studies in tone and intonation." Bibliotheca Phonetica 11,
[9.4.2]

Bridle J.S.: see (Hunt et al., 1978)

Briess B., Fant G. (1962): "Studies of voice pathology by means of inverse filtering."
STL-QPSR (1) [6.3.1; 6.5.1; 9.4.3]

Brigham E.O., Morrow R.E. (1967): "The fast Fourier Transform." IEEE Spectrum 4, 63-70
[2.4]

Brings A. (1968): "Beziehungen zwischen Tonhöhenverlauf und Intensitätsverlauf beim
Sprechen. Eine statistische Untersuchung am deutschen Sprachmaterial;" Dissertation,
Univ. Bonn [3.4; 4.1; 9.4.2]

Brink G. van den (1970): "2 experiments on pitch perception: Displacing of harmonic AM signals and pitch of inharmonic AM signals." J. Acoust. Soc. Am. **48**, 1355-1365 [4.2.1; 4.2.3]

Bristow G.J. (1980): "Speech training with colour graphics;" Dissertation, Univ. of Cambridge [5.1; 9.4.3]

Bristow G.J., Fallside F. (1982): "An autocorrelation pitch detector with error correction." **ICASSP-82**, 184-187 [6.6.3; 8.2.4; 9.4.3]

Brocaar M.P.: see (Lecluse et al., 1975)

Brodersen R.W.: see (Dalrymple et al., 1979)

Brokx J.P.L., Nooteboom S.G., Cohen A. (1979): "Pitch differences and the intelligibility of speech masked by speech." **IPO Annual Progr. Rept. 14**, 55-60 [4.3]

Browman C.P.: see (Coker et al., 1973)

Brown B.L., Strong W.J., Rencher A.C. (1974): "Fifty-four voices from two: The effects of simultaneous manipulations of rate, mean fundamental frequency, and variance of fundamental frequency on ratings of personality from speech." J. Acoust. Soc. Am. **55**, 313-318 [9.4.2]

Bruce G. (1982): "Developing the Swedish intonation model." **Phon. Lund 22**, 51-116 [9.4.2]

Bruno G., Di Benedetto M.D., Di Benedetto M.G., Gilio A. (1981): "A Bayesian approach for voiced/unvoiced classification of segments for a speech signal." **F.A.S.E.-81**, 181-184 [5.3.3]

Bruno G., Di Benedetto M.G., Gilio A. (1982): "A probabilistic pitch estimation method with combination of several techniques." In *Fortschritte der Akustik, FASE/DAGA'82 Göttingen* (VDE-Verlag, Berlin), 975-978 [6.6.3]

Brunt M.: see (Takefuta et al., 1972)

Bryzgunova E.A. (1963): "Практическая фонетика и интонация Русского языка." Е.А.Брызгунова. Москва: Университет. (*Manual of Russian Phonetics and Intonation.* In Russian) (University of Moscow) [9.4.2]

Buddenhagen R.G.: see (Westin et al., 1966)

Buhr R.D. (1978): "More on doubling .." J. Acoust. Soc. Am. **63** (A), S34 (Paper L12, 95th Meet. ASA) [3.3; 4.3]

Buhr R.D., Keating P.A. (1977): "Spectrographic effects of register shifts in speech production." J. Acoust. Soc. Am. **62** (A), S25 (Paper K1; 94th Meet. ASA) [3.2; 5.1.2]

Buisson L., Mercier G. (1975): "Utilisation de l'information prosodique en segmentation de la parole continue." **Recherches/Acoustique** 2, 115-121 [9.4.2; 9.4.4]

Bukhtilov L.D., Legtyarev N.P., Lobanov B.M. (1970): "Об автоматическом выделении из речевого сигнала акустического признака 'тон - не тон'." В кн.: Труды акустического инситутата. Л.Д.Бухтилов, Н.П.Легтярев, Б.М.Лобанов. (Automatic voiced-unvoiced discrimination of speech signals. In Russian.). Trudy Akusticheskogo Instituta **12**, 17-20 [5.3]

Bull G.L., Johns M.E., McDonald W.E., Bralley R.C. (1979): "The effects of teflon injection on laryngeal dynamics." **ICASSP-79**, 475-478 [3.1; 9.4.3]

Bull. Audiophonologie --> *Bulletin d'Audiophonologie* (Association Franc-Comtoise d'Audiophonologie, Besançon)

Bullock T.H. (ed.) (1977): *Recognition of complex acoustic signals;* ed. by T.H.Bullock (Abakon-Verlag, Berlin; Life Sciences Report No. 5)

Buning J., Jurgens J.E., Van Schooneveld C.H. (1961): *The sentence intonation of contemporary standard Russian as a linguistic structure* (Mouton and Company, 's-Gravenhage) [9.4.2]

Burgstahler P.: see (Straka and Burgstahler, 1963)

Burk K.W.: see (Saxman and Burk, 1967)

Bush A.M.: see (Barnwell and Bush, 1974)

Buss D.D.: see (Hewes et al., 1979)

Caelen G. (1979): "Structures prosodiques de la phrase enonciative simple et étendue;" Dissertation, Université des Lettres, Toulouse [9.4.2]

Caelen J., Cazenave P. (1977): "Mesure du fondamental par filtrage variable." **Journées d'Etude sur la Parole 8**, 53-61 [6.1.3]

Calcaterra C., see (Koike et al., 1977)

Callow M., see (Haggard et al., 1970)

Calzia-Panconcelli G. (1912): "Über Sprachmelodie und den heutigen Stand der Forschungen auf diesem Gebiete." Die neueren Sprachen **115** [5.1; 9.4.2]

Campanella S.J. (1958): "A survey of speech bandwidth compression techniques." IRE Trans. AU-7, 104-116 [9.4.5]

Campanella S.J. (1966): "Baseband excitation of the formant vocoder." J. Acoust. Soc. Am. **39** (A), 1239 (Paper 4C4; 71st Meet. ASA) [9.4.5]

Canas M.A., see (Pillay et al., 1975)

Cannon T.M., see (Stockham et al., 1975)

Cao Y.S., see (Lin M.C. et al., 1979)

Cappellini V., Constantinides A.G., Emiliani P. (1978): *Digital filters and their applications* (Academic Press, London) [2.3]

Caprio J.R., see (Wise et al., 1976)

Carayannis G., Gueguen C. (1976): "The factorial linear modelling: A Karhunen-Loève approach to speech analysis." **ICASSP-76**, 489-492

Carcaud M. (1972): "Simulation de la source vocale." **Recherches/Acoustique 1**, 92-104 [3.3]

Carcaud M. (1973): "Contribution à la synthèse de la parole - Simulation de la source vocale;" Dissertation, Université des Sciences, Orsay (France) [3.3; 9.4.4]

Carcaud M., Courbon J.L., Génin J., Lucas J.P. (1976): "A hardware vocal source simulator." **ICASSP-76**, 51-54 [3.3; 9.4.5]

Cardozo B.L. (1970): "The perception of jittered pulse trains." In *Frequency analysis and periodicity detection in hearing;* ed. by R.Plomp and G.F.Smoorenburg (Sijthoff, Leiden), 339-349 [4.2.1; 4.3]

Cardozo B.L., Ritsma R.J. (1965): "Short-time characteristics of periodicity pitch." **ICA-5** (Paper B-37) [4.2.1]

Cardozo B.L., Ritsma R.J. (1968): "On the perception of imperfect periodicity." IEEE Trans. AU-**16**, 159-164 [4.2.1; 4.3]

Carey B.J. (1971): "A method for automatic time-domain analysis of human speech" (Computer Research Lab., Univ. of California, Santa Barbara CA; Rept. CRL-19)

Carey B.J., Howard J.A. (1972): "A method for speech analysis by a wavefunction representation." **ICSCP-72**, 372-375 (Paper J6) [5.3]

Carey M.J., see (Goodman and Carey, 1977)

Carlborg B., see (Kitzing et al., 1982)

Carlson R., Granström B. (1980): "Speech research in China. Impressions from a visit." **STL-QPSR** (4), 1-13 [1.2; 9.4.2]

Carlson R., Granström B., Lindblom B.E.F., Rapp K. (1972): "Some timing and fundamental frequency characteristics of Swedish sentences: Data, rules, and a perceptual evaluation." **STL-QPSR** (4), 11-19 [9.4.4]

Carr P.B., Trill D. (1964): "Long-term larynx-excitation spectra." J. Acoust. Soc. Am. **36**, 2033-2040 [3.3]

Carré R., Descout R., Wajskop M. (eds.) (1977): *Modèles articulatoires et Phonétique - Articulatory Modeling and Phonetics*. Proceedings of a Symposium held at Grenoble July 10-12, 1977; ed. by R.Carré, R.Descout, M.Wajskop (G.A.L.F, F-22301 Lannion) [3.4]

Carré R., Guérin B. (1980): "Glottal impedance of a two-mass model of the vocal cords." J. Acoust. Soc. Am. **67** (A), S95 (Paper NN16; 99th Meet. ASA) [3.3]

Carré R., Lancia R., Paille J., Gsell R. (1963): "Etude et réalisation d'un détecteur de mélodie pour l'analyse de la parole." Onde électrique, 556-562 [6.1.2; 6.6.1]

Cartier M. (1977): La fréquence fondamentale: études physio-acoustiques et instrumentales II: Détection de F_0 et du voisement (Compte-rendu du rapporteur). **Journées d'Etude sur la Parole 8**, 19-32 [2.6; 5.3; 9.4.2; 9.4.4]

Carton F. (1973): "Intonation et pédagogie." **Bull. Audiophonologie 2**, 25-39 [9.4.3]

Castelaz P.F., Niederjohn R.L. (1976): "A comparative study of the use of zero-crossing analysis methods for vowel recognition." **ICASSP-76**, 170-173 [5.3; 9.4.4]

Castellino P.: see (Beltrame and Castellino, 1981)

Catford J.C. (1964): "Phonation types: the classification of some laryngeal components of speech production." In *In honour of Daniel Jones;* ed. by D. Abercrombie et al (Longmans, London), 26-37 [3.1; 4.1]

Catford J.C. (1968): "The articulatory possibilites of man." In *Manual of Phonetics;* ed. by B.Malmberg (North Holland Publishing Company, Amsterdam), 309-333 [3.1; 4.1]

Caudel E.: see (McDonough K. et al., 1982)

Cazenave P.: see (Caelen and Cazenave, 1977)

Cederlund C., Krokstad A., Kringlebotn M. (1960): "Voice Source Studies." **STL-QPSR** (1), 1-2 [3.3; 6.5.1]

Ceplitis L.K. (1974): "Анализ речевой интонации. Рига: Академия Наук Латвийской ССР, Инст. языка и Литературы. Л.К. Цеплитис. (*Intonation analysis of speech*. In Russian)" (Latvian Academy of Sciences, Riga) [1.2; 9.4.2]

Chafcouloff M.: see (Rossi and Chafcouloff, 1972)

Chamzas Ch.: see (Papoulis and Chamzas, 1979)

Chandra S., Lin W.C. (1974): "Experimental comparison between stationary and nonstationary formulations of LP applied to voiced speech analysis." IEEE Trans. ASSP-**22**, 403-415 [6.3.1]

Chang S.H. (1957): "Visual message presentation. Final report" (Northeastern Univ. Electronics Lab, Boston MA; DDC-AD-110283) [6.2.1; 9.4.3]

Chang S.H., Bach R.E., Howard C.R., Sukys R. (1959): "Speech analysis. Final report for Contract AF-19(604)3465" (Northeastern Univ. Electronics Lab, Boston MA) [6.2.1]

Chang S.H., Pihl G.E., Wiren J. (1951): "The intervalgram as a visual representation of speech sounds." J. Acoust. Soc. Am. **23**, 675-679 [5.1]

Charras J.P.: see (Boulogne et al., 1973)

Cheng M.J., Rabiner L.R., Rosenberg A.E., McGonegal C.A. (1975): "Comparative performance study of several pitch detection algorithms." J. Acoust. Soc. Am. **58** (A), S61-62 (Paper FF6; 90th Meet. ASA) [9.3.1]

Cherry E.C., Phillips V.L. (1961): "Some possible uses of single sideband signals in formant tracking systems." J. Acoust. Soc. Am. **33**, 1067-1077 [7.2]

Cherry L.: see (Flanagan and Cherry, 1969)

Chevrie-Muller C. (1976): "Physiologie du larynx au cours de la phonation: L'historique et données récentes." Revue d'Acoustique (37), 113-125 [3.1]

Chevrie-Muller C., Decante P. (1973): "Etude de la fréquence fondamentale en pathologie." **Bull. Audiophonologie 3**, 147-194 [9.4.3]

Chevrie-Muller C., Decante P. (1977): "Programme pour le traitement automatique des donnés obtenues par extraction du fondamental de la parole (application à la pathologie de la prosodie en psychiatrie)." **Journées d'Etude sur la Parole 8**, 5-12 [9.4.3]

Chevrot L.: see (Autesserre et al., 1979)

Chiba S.: see (Sakoe and Chiba, 1978)

Chiba T., Kajiyama M. (1941): The vowel - its nature and structure (Phonetic Society of Japan, Tokyo) [1.1; 3.4]

Childers D.G. (1977): "Laryngeal pathology detection." Critical Review of Biomedical Engineering **2**, 375-425 [5.2.1; 9.4.3]

Childers D.G. (ed.) (1978): Modern spectrum analysis (IEEE Press, New York NY) [2.4; 8.1; 8.4.1]

Childers D.G., Mott J.S., Moore G.P. (1980): "Automatic parameterization of vocal cord motion from ultra high speed films." ICASSP-80, 65-68 [5.2.1]

Childers D.G., Paige A., Moore G.P. (1972): "Laryngeal vibration patterns, machine-aided measurements from high-speed film." Arch. Otolaryng. **102**, 407-410 [5.2.1]

Childers D.G., Paige A., Moore G.P., Nadal-Suris M. (1976): "Automation of the measurement of laryngal vibration patterns from high speed film." ICASSP-76, 63-66 [5.2.1]

Childers D.G., Skinner D.P., Kemerait R.C. (1977): "The cepstrum: a guide to processing." Proc. IEEE **65**, 1428-1443 [8.4.1]

Chistovich L.A.: see (Liang and Chistovich, 1961)

Chlumsky J. (1913): "L'appareil de M.Meyer pour mesurer la hauteur de la parole." Revue de Phonétique **3**, 84-89 [5.1.1]

Chlumsky J. (1929): "Quelques observations sur l'accent d'intensité, la mélodie et la quantité articulatoire et acoustique." Revue de Phonétique **6**, 1-34 [5.1.1; 9.4.2]

Chollet G.F., Kahane J.C. (1979): "Laryngeal patterns of consonant productions in sentences observed with an impedance glottograph." **ICPh-Miami**, 119-128 [5.2.3; 9.4.2]

Choppy C. (1977): "Introduction de la prosodie dans la synthese vocale automatique;" Dissertation, Univ. Pierre et Marie Curie (Paris 6), Paris [9.4.2; 9.4.4]

Choppy C., Liénard J.S. (1977): "Prosodie automatique pour la synthèse par diphonèmes." **Journées d'Etude sur la Parole 8**, 211-218 [9.4.2; 9.4.4]

Christiansen H.M. (1981): "Verfahren zur Ermittlung der Grundfrequenz eines wenigstens zeitweise periodischen Signals;" Deutsche Patentschrift Nr. 2062589. Angemeldet am 18.12.1970; bekanntgemacht am 3.7.1980 [8.3.2]

Christiansen H.M., Schweizer L., Sethy A., Hoffenreich F. (1966): "New correlation vocoder." J. Acoust. Soc. Am. **40**, 614-620 [8.2.3; 9.4.5]

Christopher D.: see (Flanagan et al., 1976)

Christov Ph. (1965): "Experiments for changing the envelope form of vowels." **ICA-5** (Paper A-11) [6.2.1]

Chuang C.K., Hiki Sh., Sorre T., Nimura T. (1971): "The acoustical features and perceptual cues of the four tones of standard colloquial Chinese." **ICA-7** (3), 297-300 (Paper 25C13) [4.2; 9.4.2]

Chuang C.K., Wang W.S. (1978): "Psychophysical pitch biases related to vowel quality, intensity differences, and sequential order." J. Acoust. Soc. Am. **64**, 1004-1014 [4.1; 9.4.2]

Claasen T.A.C.M.: see (Sluyter et al., 1982)

Clark J.A.: see (Lawrence et al., 1956)

Clarke G.P.: see (Fourcin et al., 1978)

Clements M.A., Braida L.D., Durlach N.I. (1983): "Speech processing for artificial tactile speech displays." **ICASSP-83** [9.4.3]

Clemoes P.: *Liturgical influence on punctuation in late old English and early Middle English manuscripts* (Dept.of Anglo-Saxon, University of Cambridge, Cambridge UK) [5.1.1; 9.4.2]

Cohen A., Collier R., Hart J. 't (1982): "Declination: construct or intrinsic feature of speech pitch?." Phonetica **39**, 254-274 [9.4.2]

Cohen A., Nooteboom S.G. (eds.) (1975): *Structures and processes in speech perception.* Proceedings of the Symposium on Dynamic Aspects of Speech Perception, Eindhoven, August 4-6, 1975; ed. by A.Cohen and S.G.Nooteboom (Springer, Berlin)

Cohen J.R. (1982a): "A theory of the neural coding of pitch" (Univ. of Connecticut, Storrs CT; Dissertation) [4.2.1]

Cohen J.R. (1982b): "A pitch measurement algorithm for speech." **ICASSP-82**, 176-179 [4.2.1; 8.5.2]

Cohn D.L., Melsa J.L. (1975): "The residual encoder - an improved ADPCM system for speech digitization." IEEE Trans. COM-**23**, 935-941 [9.4.5]

Cohn D.L., Melsa J.L. (1976): "A pitch compensating quantizer." **ICASSP-76**, 258-261 [9.4.5]

Cohn D.L., Melsa J.L. (1979): "A new configuration for speech digitization at 9600 bits per second." **ICASSP-79**, 550-553 [8.3.1; 9.4.5]

Coker C.H., Umeda N., Browman C.P. (1973): "Automatic synthesis from ordinary English text." IEEE Trans. AU-**21**, 293-298 [9.4.4]

Coleman H.O. (1914): "Intonation and emphasis." Miscellania phonetica [9.4.2]

Coleman R.F., Mabis J.H., Hinson J.K. (1977): "Fundamental frequency - sound pressure level profiles of adult male and female voices." J. Speech Hear. Res. **20**, 197-204 [4.1]

Coleman R.F., Mott J.B. (1978): "Fundamental frequency and sound pressure level profiles of young female singers." Folia Phoniatr. **30**, 94-102 [3.2; 4.1]

Coleman R.O. (1963): "Decay characteristics of vocal fry." Folia Phoniatr. **15**, 256-263 [3.2]

Coleman R.O. (1973): "Speaker identification in the absence of inter-subject differences in glottal source characteristics." J. Acoust. Soc. Am. **53**, 1741-1743 [9.4.4]

Coleman R.O., Wendahl R.W. (1968): "On the validity of laryngeal photosensor monitoring." J. Acoust. Soc. Am. **44**, 1733-1735 [5.2.5]

Collier R. (1970): "The optimum position of prominence-lending pitch rises." **IPO Annual Progr. Rept. 5**, 52-85 [9.4.2]

Collier R. (1972): "From pitch to intonation;" Dissertation, Catholic University, Leuven [9.4.2]

Collier R. (1975): "Physiological correlates of intonation patterns." J. Acoust. Soc. Am. **58**, 249-255 [3.1]

Collier R., Hart J.'t (1971): "A grammar of pitch movements in Dutch intonation." **IPO Annual Progr. Rept. 6**, 17-20 [9.4.2]

Collier R., Hart J.'t (1972): "Perceptual experiments on Dutch intonation." **ICPhS-7**, 880-884 [9.4.2]

Collier R., Hart J.'t (1975): "The role of intonation in speech perception." In *Structures and processes in speech perception.* Proceedings of the Symposium on Dynamic Aspects of Speech Perception, Eindhoven, August 4-6, 1975; ed. by A.Cohen and S.G.Nooteboom (Springer, Berlin), 107-123 [4.2; 9.4.2]

Collier R., Hart J.'t (1978): *Cursus nederlandse intonatie* (Course of Dutch intonation. in Dutch) (Institute of Perception Research, Eindhoven; IPO manuscript # 333) [9.4.2]

Colton R.H., Estill J. (1976): "Observations of glottal waveforms in selected phonatory tasks." J. Acoust. Soc. Am. **59** (A), S15 (Paper G4; 91st Meet. ASA) [6.5.1]

Colton R.H., Hollien H. (1972): "Phonational range in the modal and falsetto registers." J. Speech Hear. Res. **15**, 708-713 [3.2; 4.1]

Comer D.J. (1968): "The use of waveform asymmetry to identify voiced sounds." IEEE Trans. AU-**16**, 500-506 [5.3]

Conradi G. (1894): *Der harmonische Analysator* (Zürich) [5.1]

Constantinides A.G.: see (Cappellini et al., 1978)

Contini M., Bellet G., Boe L.J. (1981): Normalisation temporelle de F_0: étude intralocu-
teur sur une phrase énonciative en Français. In *Symposium Prosodie - Prosody sym-
posium*. Toronto, May 1981 (Experim. Phonetics Lab., Univ. of Toronto, Toronto,
Ont.), 85-103 [9.4.2]
Contini M., Boe L.J. (1979): "Etude quantitative de l'intonation en français." In *Recherche
sur la prosodie du français* (Univ. de Langues et Lettres, Grenoble), 117-129 [9.4.2]
Cooley J.W., Tukey J.W. (1965): "An algorithm for the machine calculation of complex
Fourier series." Math. Computation **19**, 287-301 [2.4]
Cooper F.S., Rand T.C., Music R.S., Mattingly I.G. (1970): "A voice for the laboratory
computer." **Speech Research 24**, 91-94 [9.4.4]
Cooper M. (1974): "Spectrographic analysis of fundamental frequency and hoarseness
before and after vocal rehabilitation." J. Speech Hear. Disorders **39**, 286-297 [5.1.2;
9.4.3]
Cooper W.E., Sorensen J.M. (1978): "Declination of fundamental frequency (F_0) in speech
production." J. Acoust. Soc. Am. **63** (A), S67 (Paper AA3; 95th Meet. ASA) [3.1]
Corlew M.M. (1968): "A developmental study of intonation recognition." J. Speech Hear.
Res. **11**, 825-832 [9.4.2; 9.4.4]
Cosnac B. de: see (Baudry et al., 1977)
Coulter D.C. (1973): "Some baseband speech processing experiments." J. Acoust. Soc.
Am. **53** (A), 321 (Paper Q7; 84th Meet. ASA) [9.4.5]
Coulter D.C., Early D.M., Irons R.E. (1961): "A digitalized speech compression system"
(Naval Research Lab, Falls Church VA; ASD-TR-61-494, DDC-AD-266959)
Coulter D.C., Ludlow C., Gentges F. (1979): "Determination of lower limits of pitch
variation in normal larynges." J. Acoust. Soc. Am. **66** (A), S12 (Paper E11; 98th
Meet. ASA) [3.3; 4.3]
Courbon J.L. (1972): "Régulation automatique du niveau - étude et réalisation de deux
compresseurs de dynamique." **Recherches/Acoustique 1**, 47-60 [6.2.1]
Courtois J.: see (Sovak et al., 1971)
Covert G.D. (1982): "A 32-point monolithic FFT processor chip." **ICASSP-82**, 1081-1084
[2.4; 9.1.5]
Cowan J.M. (1936): "Pitch and intensity characteristics of stage speech." Arch. Speech,
1-90 [4.1]
Cowan J.M. (1939): "A technique for the measurement of intonation." Arch. vergl.
Phonetik **3**, 223-234
Cowan J.M. (1962): "Graphical representation of perceived pitch." **ICPhS-4**, 567-570 [5.1]
Cowan J.M., Fairbanks F., Lewis D. (1940): "Pitch and frequency modulation." J. Exper.
Psychol. **27**, 23-36
Cox B.V., Timothy L.K. (1979): "Rank-order speech classification algorithm." **ICASSP-79**,
759-763 [5.3]
Cox B.V., Timothy L.K. (1980): "Nonparametric rank-order statistics applied to robust
voiced-unvoiced-silence classification." IEEE Trans. ASSP-**28**, 550-561 [5.3.3]
Cox R.V., Crochiere R.E. (1980): "Multiple-user variable-rate coding for TASI and packet
transmission systems." IEEE Trans. COM-**28**, 334-344 [5.3; 9.4.4]
Cox R.V., Crochiere R.E. (1982): "A single chip speech periodicity detector." **ICASSP-82**,
525-528 [8.2.4; 9.1.5]
Cox R.V., Robinson D.M. (1980): "Some notes on phase in speech signals." **ICASSP-80**,
150-153 [9.2.1]
Coyne A.E. (1938): "More about the voice pitch indicator." Volta Review **40**, 549-552
[5.2.2]
Crane G.W.: see (Gault and Crane, 1928)
Cranen B.: see (Boves and Cranen, 1982)
Cremer L. (1947): "Sichtbare Sprache." AEÜ **1**, 152 [5.1.2]
Cretchley R.R. (1954): "Further investigations into fundamental extractors for speech
sounds" (Canadian Defense Ressearch Board, Telecommun. Establ. Electronics Labs,
Ottawa; Rept. EM 5037-66) [2.6; 9.1.1]
Creutzfeldt O., Scheich H., Schreiner Ch. (eds.) (1979): *Hearing Mechanisms and Speech*.
EBBS-Workshop Göttingen, April 26-28, 1979. Experimental Brain Research Suppl. 2
(Springer, Berlin) [4.2]
Crighton R.G., Fallside F. (1974): "The development of a deaf speech training aid using
linear prediction." **Proc. 1974 Speech Commun. Sem. 4**, 35-40 [9.4.3]
Cristo A. di: see Di Cristo A.
Crochiere R.E. (1979): "A novel approach for implementing pitch prediction in sub-band
coding." **ICASSP-79**, 526-529 [9.4.5]

Crochiere R.E., Flanagan J.L. (1983): "Current perspectives in digital speech." IEEE Commun. Magazine, 32-40 [9.2.1; 7.4.4; 9.4.5]

Crochiere R.E., Tribolet J.M. (1979): "Frequency domain techniques for speech coding." J. Acoust. Soc. Am. **66**, 1642-1646 [5.3; 9.4.5]

Crowell D.H.: see (Tenold et al., 1974)

Crowither W.R., Rader C.M. (1966): "Efficient coding of vocoder pitch periods, using variable-length coding." J. Acoust. Soc. Am. **40** (A), 1242 [9.4.4]

Crystal D. (1969): *Prosodic systems and intonation in English* (University Press, Cambridge UK) [1.3; 5.1; 9.4.2]

Crystal D. (1981): *Clinical linguistics.* Disorders of human communication, vol.3 (Springer, Wien) [9.4.2; 9.4.3]

Crystal T.H. (1966): "Further experiments using a larynx model." J. Acoust. Soc. Am. **39** (A), 1218 (Paper 1F4; 71st Meet. ASA) [3.3]

Crystal T.H., Jackson T.L. (1970): "Extracting and processing vocal pitch for laryngeal disorder detection." J. Acoust. Soc. Am. **48** (A), 118 (Paper DD3, 79th Meet. ASA) [6.2.1; 9.4.3]

Culler G., Greenwood E., Harrison D. (1982): "A high-performance VLSI CMOS arithmetic processor chip." **ICASSP-82**, 1053-1056 [2.3; 9.1.5]

Cullis A.D.S.: see (Hales et al., 1956)

Current K.W., Mow D.A., Youssef-Digaleh S. (1979): "A high data rate, low power all-digital correlation circuit design." **ICASSP-79**, 859-862 [8.2.1]

Curry E.T. (1940): "The pitch characteristics of the adolescent male voice." Speech Monogr. **7**, 48-62 [4.1]

Curry R. (1937): "The mechanism of pitch change in the voice." J. Psychol. **91** [3.1]

Curtis J.F.: see (Hollien and Curtis, 1960)

Curtis J.H. (1973): "Inter-speaker versus intra-speaker variability of glottal pulse shapes." J. Acoust. Soc. Am. **54** (A), 102 (Paper LL10; 86th Meet. ASA) [3.3; 9.4.2; 9.4.4]

Czajka S.: see (Jassem et al., 1973)

Czekajewski J.A., Woyczinski W.A. (1968): "A real-time method for fast determination of the fundamental frequency (In Polish.)." Zastowany matematyki **9**, 253-259 [6.2.3]

Czermak J. (1861): "De l'application de la photographie à la laryngoscopie et a la rhinoscopie." Comptes Rendus des Séances de l'Académie des Sciences, Paris **53**, 966-968 [5.1.1; 5.2.1]

Daaboul F., Adoul J.P. (1977): "Parametric segmentation of speech into voiced-unvoiced-silence intervals." **ICASSP-77**, 327-331 [5.3]

Daggett N.L. (1966): "A computer for vocoder pitch extraction;" MIT Lincoln Labs, Cambridge MA, TN 1966-3; AF-19(628)-5167 [6.2.4; 9.4.5]

Daggett N.L., Gold B. (1964): "Compact pitch computer." J. Acoust. Soc. Am. **36**, 1030 (Paper N8; 67th Meet. ASA) [6.2.4]

Dal Degan N. (1982): "Vocoder quality: an automatic procedure to measure the performance of pitch extractors." In *Proceedings, Globecom'82* [9.3.1; 9.4.4]

Dalrymple M., Senderowicz D., Brodersen R.W. (1979): "Pitch extraction using MOS-LSI circuitry." **ICASSP-79**, 768-772 [6.2.4; 9.1.5]

Damsté P.H., Hollien H., Moore G.P., Murry T. (1968): "An X-ray study of vocal fold length." Folia Phoniatr. **20**, 349-359 [3.1; 5.2.1]

Daniel T.H.: see (Tenold et al., 1974)

Danielsson B. (1963): *John Hart's work on English orthography and pronunciation* (Almqvist and Wiksell, Stockholm) [5.1.1; 9.4.2]

Danielsson K., Gustrin K. (1961): "Grundtonsanalysator (Pitch analyzer; in Swedish)" (National Swedish Defense Div.3, Res.Inst, Stockholm; Report A483) [6.1.1; 6.4.2]

Dankberg M.D., Wong D.Y. (1979): "Development of a 4.8-9.6 kbps RELP vocoder." **ICASSP-79**, 554-557 [9.4.5]

Dara-Abrams B.: see (O'Malley M.H. et al., 1973)

Darby J.K. (ed.) (1981a): *Speech evaluation in psychiatry* (Grune and Stratton, New York) [9.4.3]

Darby J.K. (ed.) (1981b): *Speech evaluation in medicine* (Grune and Stratton, New York) [9.4.3]

Darwin C.J. (1975): "On the dynamic use of prosody in speech perception." In *Structures and processes in speech perception*. Proceedings of the Symposium on Dynamic Aspects of Speech Perception, Eindhoven, August 4-6, 1975; ed. by A.Cohen and S.G.Nooteboom (Springer, Berlin), 178-194 [4.2.1; 9.4.2]

Daryanani G.T.: see (Boddie et al., 1981)

David E.E.jr. (1956): "Signal theory in speech transmission." IRE Trans. CT-3, 232-244 [6.1.1; 6.1.2]

David E.E.jr. (1958): "Artificial auditory recognition in telephony." IBM J. Res. Development **2**, 294-309 [6.1.2]

David E.E.jr., McDonald H.S. (1956): "Note on pitch-synchronous processing of speech." J. Acoust. Soc. Am. **28**, 1261-1266 [9.4.5]

David E.E.jr., Pierce J.R. (1963): "Pitch synchronous auto-correlation vocoder;" United States Patent No. 3,109,070. October 29, 1963; Application August 9, 1960 [8.2.3; 9.4.5]

David E.E.jr., Schodder G.R. (1961): "Pitch discrimination of complex sounds." **ICA-3**, 106-109 [4.2.3]

David E.E.jr., Schroeder M.R., Logan B.F., Prestigiacomo A.J. (1962a): "New applications of voice excitation to vocoders." **Proc. 1962 Speech Commun. Sem.** [9.4.5]

David E.E.jr., Schroeder M.R., Logan B.F., Prestigiacomo A.J. (1962b): "Voice-excited vocoders for practical speech bandwidth reduction." IRE Trans. IT-6, 101-105 [9.4.5]

David E.E.jr., Schroeder M.R., Sondhi M.M. (1968): "Apparatus for determining the periodicity and aperiodicity of a complex wave;" United States Patent No.3,405,237. Issued Oct. 8, 1968; filed June 1, 1965 [8.2.4]

David N.: see (Atal and David, 1978)

Davis C.R.: see (Maitra and Davis, 1979)

Davis H. (1936): "Psychophysiological acoustics: pitch and loudness." J. Acoust. Soc. Am. **8**, 1-13 [4.2.1]

Davis H., Silverman S.R., McAuliffe D.R. (1951): "Some observations on pitch and frequency." J. Acoust. Soc. Am. **23**, 40-42 [4.2]

Davis R.K.: see (Oppenheim et al., 1983)

Davis S.B. (1975): "Preliminary results using inverse filtering of speech for automatic evaluation of laryngeal pathology." J. Acoust. Soc. Am. **58** (A), S111 (Paper BBB4; 90th Meet. ASA) [6.5; 9.4.3]

Davis S.B. (1976a): "Computer evaluation of laryngeal pathology based on inverse filtering of speech" (Speech Communication Research Lab, Santa Barbara CA; SCRL Monograph # 13) [6.5.1; 9.4.3]

Davis S.B. (1976b): "Determination of glottal area based on digital image processing of high-speed motion pictures of the vocal folds." J. Acoust. Soc. Am. **60** (A), S65 (Paper CC12; 92nd Meet. ASA) [3.1; 5.2.1]

Davis S.B. (1978): "Acoustic characteristics of normal and pathologic voices." **Speech Research 54**, 133-164 [6.5.1; 9.4.3]

Davis S.B. (1981): "Acoustic characteristics of laryngeal pathology." In *Speech evaluation in medicine;* ed. by J.K. Darby (Grune and Stratton, New York)

Dawe P.G.M., Deutsch M.A. (1955): "An audio frequency meter for graphing frequency veriations in the human voice." Electron. Eng. **27**, 2-6 [5.1; 6.1.4]

De Boer E.: see Boer E. de

De Bot K.L.J.: see Bot K.L.J. de

De Cosnac B.: see (Baudry et al., 1977)

De Mori R., Laface P., Makhonine V.A., Mezzalama M. (1977): "A syntactic procedure for the recognition of glottal pulses in continuous speech." Pattern Recognition **9**, 181-189 [6.3.2; 6.4.1]

De Mori R., Serra A. (1972): "Digital speech data transmission using pitch synchronous analysis and extremal coding." J. Audio Eng. Soc. **20**, 562-567 [9.4.5]

De Rooij J.J. (1976): "Perception of prosodic boundaries." **IPO Annual Progr. Rept. 11**, 20-24 [4.2.3; 9.4.2]

De Rooij J.J. (1979): "Speech punctuation, an acoustic and perceptual study of some aspects of speech prosody in Dutch;" Dissertation, University of Utrecht [9.4.2]

Deal R.E., Emanuel F.W. (1978): "Some waveform and spectral features of vowel roughness." J. Speech Hear. Res. **21**, 250-264 [3.4; 4.3]

Decante P.: see (Chevrie-Muller and Decante, 1973)

Degryse D.: see (Guérin et al., 1977)

Deinse J.B. van (1981): "Registers." Folia Phoniatr. **33**, 37-50 [3.2]

Dejonckere P.H. (1981): "Comparison of two methods of photoglottography in relation to electroglottography." Folia Phoniatr. **33**, 338-347 [5.2.3; 5.2.5; 5.2.6]

Delattre P. (1963): "Comparing the prosodic features of English, German, Spanish, and French." Int. J. Am. Linguistics **1**, 193-210 [9.4.2]

Delattre P. (1966): "Les dix intonations de base du français." French Review (Baltimore) **40**, 1-14 [9.4.2]

Delgutte B. (1978): Technique for the perceptual investigation of F_0 contours with application to French. J. Acoust. Soc. Am. **64**, 1319-1332 [4.2; 4.3; 9.4.2]

Deller J.R.jr. (1979): "The use of the digital inverse filter for the acoustic diagnosis and assessment of laryngeal dysfunction;" Dissertation, Univ. of Michigan, Ann Arbor MI [6.5.1; 9.4.3]

Deller J.R.jr. (1981): "Some notes on glottal inverse filtering." IEEE Trans. ASSP-**30**, 917-919 [6.3.1; 6.5.1]

Deller J.R.jr. (1982): "Evaluation of laryngeal dysfunction based on features of an accurate estimate of the glottal waveform." **ICASSP-82**, 759-762 [6.5.1; 9.4.3]

Deller J.R.jr., Anderson D.J. (1979a): "Toward the use of the inverse filter as a diagnostic aid." **ASA*50**, 253-256 [6.3.1; 6.5.1; 9.4.3]

Deller J.R.jr., Anderson D.J. (1979b): "Toward the use of the inverse filter as a diagnostic aid." J. Acoust. Soc. Am. **65** (A), S67 (Paper Y9; 97th Meet. ASA) [6.5.1; 9.4.3]

Deller J.R.jr., Anderson D.J. (1980): "Automatic classification of laryngeal dysfunction unsing the roots of the digital inverse filter." IEEE Trans. BME-**27**, 714-721 [6.5.1; 9.4.3]

Delos M.: see (Guérin et al., 1976)

Demange R.: see (Autesserre et al., 1979)

Demenko G.: see (Jassem and Demenko, 1981)

Dempsey M.E. (1955): "Design and evaluation of a fundamental frequency recorder for complex sounds" (Purdue University, West Lafayette IN) [6.1.1; 6.1.3]

Dempsey M.E., Draegert G.L., Siskind R.P., Steer M.D. (1950): "The Purdue pitch meter - a direct reading fundamental frequency analyzer." J. Speech Hear. Disorders **15**, 135-141 [4.1; 6.1.1; 6.1.3]

Dempsey M.E., Siskind R.P., Hanley T.D., Steer M.D. (1953): "A fundamental frequency recorder for complex sounds" (Purdue Univ., Lafayette IN; DDC-AD-45504) [6.1.1; 6.1.3]

Demytenko N., English K.S. (1977): "Echo cancellation on time-variant circuits." Proc. IEEE **65**, 444-453 [9.4.5]

Denes P.B. (1962): "Computer processing of acoustic and linguistic information in automatic speech recognition" (DDC-AD 293449, Dept. of Phonetics, Univ. of London) [9.4.4]

Denes P.B. (1970): "On-line computers for speech research." IEEE Trans. AU-**18**, 418-425 [1.2; 9.1.5; 9.4.4]

Denes P.B., Milton-Williams J. (1962): "Further studies in intonation." Lang. Speech **5**, 1-14 [9.4.2]

Denes P.B., Pinson E.N. (1973): *The speech chain. The physics and biology of spoken language* (Anchor Books, Garden City NY) [1.2]

Derby J.H. (1978): "Analysis and representation of composite signals by cepstral inverse filtering." **ICASSP-78**, 214-217 [6.5.1; 8.4.1]

Dersch W.C. (1965): "Sound analyzing system;" United States Patent No.3,198,884. issued 1965, filed 29 August 1960 [5.3.1]

Deutsch M.A.: see (Dawe and Deutsch, 1955)

Dew D.: see (Hollien et al., 1971)

Di Benedetto M.D. (1982): "The design and construction of digital speech processing systems to serve as an aid to the hard of hearing (In French);" Dissertation, Univ. Pierre et Marie Curie (Paris 6), Paris [9.4.3]

Di Benedetto M.G.: see (Bruno et al., 1981)

Di Cristo A. (1975): *Soixante et dix ans de recherches en prosodie* (Editions de l'Université de Provence, Aix-en-Provence) [1.2; 1.3; 9.4.2]

Di Cristo A. (1977): "L'intonation: aspects linguistiques et reconnaissance de la parole; méthodes et modèles d'analyse dans les recherches sur l'intonation (Compte-rendu du rapporteur)." **Journées d'Etude sur la Parole 8**, 41-60 [9.4.2; 9.4.4]

Di Cristo A. (1978a): "De la microprosodie à l'intonosyntaxe;" Thèse d'Etat, Université de Provence, Aix-en-Provence [3.1; 9.4.2]

Di Cristo A. (1978b): "Méthodes et mdèles d'analyse dans les recherches sur l'intonation." **Journées d'Etude sur la Parole 9**

Di Cristo A. (1980): "Variabilité acoustique et intégration perceptive des cibles." **Journées d'Etude sur la Parole 11**, 91-113 [4.3; 9.4.2]

Di Cristo A., Chafcouloff M. (1977): "Les faits microprosodiques du français: voyelles, consonnes, coarticulation." **Journées d'Etude sur la Parole 8**, 147-158 [3.1; 9.4.2]

Di Toro M.J., Graham W., Dwork B.M. (1955): "Fundamental pitch detector system;" United States Patent No. 2,699,464. Issued Jan. 11, 1955; filed May 22, 1952 [6.4.1]

Dibbern U. (1972): "Grundfrequenzmessung bei der menschlichen Sprache." In *Akustik und Schwingungstechnik.* Plenarvorträge und Kurzreferate der Gemeinschaftstagung Stuttgart 1972 (VDE-Verlag, Berlin), 345-348 [6.4.1]

Dibbern U. (1976): "Multiplex vocoder for voice response." **ICASSP-76,** 709-712 [6.4.1; 9.4.5]

Dickinson B.: see (Morf et al., 1977)

Divenyi P.L. (1979): "Is pitch a learned attribute of sounds? Two points in support of Terhardt's pitch theory." J. Acoust. Soc. Am. **66** (C), 1210-1213 [4.2.1]

Dobrogowska K.: see (Jassem and Dobrogowska, 1980)

Dodd L.R. (1926): "The adaption of the phonolescope as a precision pitch indicator, and an application to vocal tones." J. Opt. Soc. Am. **12,** 119 [5.1.1]

Doddington G.R., Schalk T.B. (1981): "Speech recognition: Turning theory to practice." IEEE Spectrum **9,** 26-32 [9.1.5]

Dodson H.: see (Fourcin et al., 1978)

Doetsch W. (1982): *Anleitung zum praktischen Gebrauch der Laplace-Transformation und der z-Transfomation* (Oldenbourg, München) [2.3; 8.5.4]

Doherty E.T.: see (Hollien et al., 1973)

Dolanský L.O. (1954): "Instantaneous Pitch-Period Indicator." J. Acoust. Soc. Am. **26** (A), 953 [6.2.1]

Dolanský L.O. (1955): "An instantaneous pitch-period indicator." J. Acoust. Soc. Am. **27,** 67-72 [4.1; 6.2.1]

Dolanský L.O. (1960): "Choice of base signals in speech signal analysis." IRE Trans. AU-**8,** 221-227

Dolanský L.O., Bass S.D., Savage C.J., Pronovost W.L. (1971): "Demonstration of the Instantaneous Pitch Period Indicator in classrooms of deaf children" [U.S. Department of Health, Education, and Welfare, Washington DC; Final report, Grant No. OEG-0-9 312139-4094(032)] [1.2; 6.2.1; 9.4.3]

Dolanský L.O., Ferullo R.J., O'Donnell M.C. (1966): "Pitch extraction and display for the deaf." J. Acoust. Soc. Am. **39** (A), 1257 (Paper 6C7; 71st Meet. ASA) [9.4.3]

Dolanský L.O., Manley H.J. (1960): "Speech analysis (A survey report)" (Applied Research Laboratory, Waltham, Mass; Sylvania Proj. 72-401) [1.2; 6.2.1; 9.4.3]

Dolanský L.O., Phillips N.D., Bass S.P., Pronovost W.L., Anderson D.C. (1971): "An intonation display system for the deaf." Acustica **25,** 189-202 [1.2; 6.2.1; 9.4.3]

Dolanský L.O., Pronovost W.L., Anderson D.C., Bass S.D., Phillips N.D. (1969): "Teaching of intonation patterns to the deaf using the instantaneous pitch period indicator" (Northeastern University, Boston MA; VRA Grant 2360-S) [6.2.1; 9.4.3]

Dolanský L.O., Tjernlund P. (1967): "On certain irregularities of voiced speech waveorms." **STL-QPSR** (2-3), 58-65 [3.3; 4.3]

Dolanský L.O., Tjernlund P. (1968): "On certain irregularities of voiced speech waveforms." IEEE Trans. AU-**16,** 51-56 [3.3; 4.3]

Dolmazon J.M.: see (Bastet et al., 1977)

Douek E.E., Fourcin A.J., Moore B.C.J., Clarke G.P. (1977): "A new approach to the cochlear implant." Proc. R. Soc. Medicine (London) **70** [4.2; 9.4.3]

Douglas B.P. (1979): "A robust vocoder with pitch-adaptive spectral envelope estimation and an integrated maximum-likelihood pitch estimator." J. Acoust. Soc. Am. **65** (A), S100 (Paper MM13; 97th Meet. ASA) [8.6.1; 9.4.5]

Douovan R. (1967): "Variables of laryngeal tone." Folia Phoniatr. **19,** 281-296

Dours D., Facca R., Perennou G. (1977): "Automate de segmentation et detection du fondamental." **Journées d'Etude sur la Parole 8,** 61-70 [6.2.5]

Dove W.P., Oppenheim A.V. (1980): "Event location using recursive least squares signal processing." **ICASSP-80,** 848-850 [6.3.2]

Draegert G.L.: see (Dempsey et al., 1950)

Drago P.G., Molinari A.M., Vagliani F.C. (1978): "Digital dynamic speech detectors." IEEE Trans. COM-**26,** 140-144 [5.3]

Drajic D.: see (Lukatela et al., 1976)

Dreyfus-Graf J.A. (1977): "Analyses des rédondances de systemes symboliques et degrés de voisement." **Journées d'Etude sur la Parole 8,** [2.1; 3.5; 5.3]

Drommel R. (1974): "Ein Überblick über die bisherigen Arbeiten zur Sprechpause." Phonetica **30,** 221-238 [5.3]

Droogleever-Fortuyn J., Ritsma R.J. (1979): "Virtual pitch and our own voice." In *Hearing Mechanisms and Speech.* EBBS-Workshop Göttingen, April 26-28, 1979; ed. by O.Creutzfeld et al. Experimental Brain Research Suppl. 2 (Springer, Berlin), 268-280 [4.2.1]

Drucker H. (1968): "Speech processing in a high ambient noise environment." IEEE Trans. AU-16, 165-168 [5.3]

Dubnowski J.J., Schafer R.W. (1974): "Digital hardware for pitch detection." J. Acoust. Soc. Am. 56 (A), S16 (Paper H12; 88th Meet. ASA) [8.2.4]

Dubnowski J.J., Schafer R.W., Rabiner L.R. (1976): "Real-time digital hardware pitch detector." IEEE Trans. ASSP-24, 2-8 [8.2.4; 9.3.1]

Dubois C.: see (Luchsinger and Dubois, 1963)

Dubus F.: see (Galand et al., 1976)

Duda R.O., Hart P.E. (1973): Pattern classification and scene analysis (Wiley, New York) [5.3.3]

Dudgeon D.E. (1970): "Two-mass model of the vocal cords." J. Acoust. Soc. Am. 48 (A), 118 (Paper DD2, 79th Meet. ASA) [3.3]

Dudley H.W. (1937): "Signaling system;" United States Patent No. 2,098,956. Issued Nov. 16, 1937; filed Dec. 2, 1936 [4.1; 6.1.2]

Dudley H.W. (1939a): "Signal transmission;" United States Patent No. 2,151,091. Issued March 21, 1939; filed Oct. 30, 1935 [6.1.2; 9.4.4]

Dudley H.W. (1939b): "Remaking speech." J. Acoust. Soc. Am. 11, 169-177 [5.3; 6.1.1; 9.4.4]

Dudley H.W. (1939c): "The vocoder." Bell Labs. Record 18, 122-126 [6.1]

Dudley H.W. (1940a): "The carrier nature of speech." Bell Syst. Tech. J. 19, 495-513 [1.1]

Dudley H.W. (1940b): "The carrier nature of speech." Reprinted in Acoustic Phonetics - A course of basic readings; ed. by D.B.Fry (Cambridge University Press, Cambridge), 21-30 [1.1]

Dudley H.W. (1941): "Production of artificial speech;" United States Patents No. 2,243,526 and 2,243,527. Issued May 27, 1941; filed March 16, 1940 [6.1.1]

Dudley H.W. (1964): "State of the art in speech compression and digitization" (ITT Communic. Systems, Paramus NY; Report # 64-TR-471) [1.2; 9.4.5]

Duffy R.J.: see (Lerman and Duffy, 1970)

Duifhuis H. (1969): "Limitations of frequency analysis in hearing." IPO Annual Progr. Rept. 4, 12-17 [4.2.1; 4.3]

Duifhuis H. (1971): "Audibility of high harmonics in a periodic pulse II: time effect." J. Acoust. Soc. Am. 49, 1155-1162 [4.2.1; 4.2.3]

Duifhuis H., Willems L.F., Sluyter R.J. (1978): "Measuring pitch in speech." IPO Annual Progr. Rept. 13, 24-30 [4.2.1; 8.5.2]

Duifhuis H., Willems L.F., Sluyter R.J. (1979a): "Pitch in speech: A hearing theory approach." ASA*50, 245-248 [8.5.2]

Duifhuis H., Willems L.F., Sluyter R.J. (1979b): "An outline of pitch analysis in speech: a hearing theory approach." In Hearing Mechanisms and Speech. EBBS-Workshop Göttingen, April 26-28, 1979; ed. by O.Creutzfeld et al. Experimental Brain Research Suppl. 2 (Springer, Berlin), 254-259 [4.2.1; 4.4; 8.5.2; 8.5.4]

Duifhuis H., Willems L.F., Sluyter R.J. (1979c): "Pitch in speech: A hearing theory approach." J. Acoust. Soc. Am. 65 (A), S67 (Paper Y7; 97th Meet. ASA) [4.2.1; 8.5.2]

Duifhuis H., Willems L.F., Sluyter R.J. (1982): "Measurement of pitch in speech: an implementation of Goldstein's theory of pitch perception." J. Acoust. Soc. Am. 71, 1568-1580 [4.2.1; 8.5.2]

Duker S. (ed.) (1974): Time-compressed speech. An anthology and bibliography in three volumes (Scarecrow Press, Metuchen NJ) [1.3]

Dukiewicz K. (1965): "Pitch extraction from speech signal." Bull. de la Soc. des Amis des Sciences et des Lettres de Poznań 19, 19-26 [6.4.1]

Dumaman J.A.: see (Tierney et al., 1964)

Dunn C. (1962): "Pitch characteristics of adult males" (Communication Sciences Laboratory, Gainesville FL; NB-04398-01) [4.1]

Dunn H.K. (1961): "Methods of measuring formant bandwidths." J. Acoust. Soc. Am. 33, 1737-1746 [3.4]

Dupeyrat B. (1973): "Reconnaissance de la parole. Méthode des passages par zéro du signal. Reconnaissance automatique de voyelles isolées;" Dissertation, Univ. Pierre et Marie Curie (Paris 6), Paris [5.3; 6.2.5]

Durlach N.I.: see (Clements et al., 1983)

Dwork B.M.: see (Di Toro et al., 1955)

Dziurnikowski A. (1975): "Microphonemes as fundamental segments of speech wave. Primary segmentation. Automatic searching for microphonemes." In *Advance Papers of the 4th International Joint Conference on Artificial Intelligence,* Tbilisi, USSR, 3.-8.9.1975 (Artif. Intell. Lab. MIT, Cambridge MA), Vol.2, 476-482 [6.1.1]

Dziurnikowski A. (1980): "Automatic determination of the frequency behaviour of the larynx tone by the method of linear prediction." Arch. Acoustics **5**, 31-46 [6.3.1]

Early D.M. (1962): "Design of a digitalized speech compression system" (Melpar, Falls Church, Virginia; DDC-AD 271616) [9.4.5]

Early D.M., Campanella S.J., Coulter D.C., Barnes L.A. (1962): "Investigation of vocal quality and articulation improvements of a speech bandwidth compression system" (Melpar, Fall Church, Virginia; DDC-AD-291104) [9.4.5]

Ebihara S.: see (Kasuya et al., 1983)

Eck T. van (1981): "Quantitative determination of periodicities from short time series." IEEE Trans. ASSP-**29**, 329-331 [8.6.1; 8.6.3]

Edie J., Sebestyen G. (1962): "Voice identification general criteria" (Litton Systems, Waltham, Mass; DDC-AD-278565) [9.4.5]

Edmondson G.D. (1974): "A time-domain waveform display/analyzer for speech research" (Naval Postgraduate School, Monterey CA; DDC-AD-783783/4GA) [5.1]

Edson J.O., Feldman C.B.H. (1959): "Derivation of vocoder pitch signals;" United States Patent No. 2,906,955. Issued Sept. 29, 1959; filed Feb. 17, 1956 [6.1.3]

Egan J.P.: see (Stevens S.S. et al., 1947)

Eggert B. (1908): "Untersuchungen über Sprachmelodie." Z. Psychologie **49**, 218-237 [5.1.1; 9.4.2]

Eichler I.: see (Schlatter and Eichler, 1979)

El Mallawany I. (1975a): "Détermination de l'intervalle de fermeture de la glotte." **Recherches/Acoustique 2**, 171-177 [6.5.2]

El Mallawany I. (1975b): "Analyse/synthese par prédiction linéaire." **Recherches/Acoustique 2**, 179-188 [6.3.1; 9.4.4]

El Mallawany I. (1976a): "L'analyse sur l'intervalle de fermeture de la glotte." **Recherches/Acoustique 3**, 149-165 [3.4; 6.5.2; 9.4.4]

El Mallawany I. (1976b): "Approches à la détection de et à l'analyse sur l'intervalle de fermeture de la glotte." **Journées d'Etude sur la Parole 7**, 319-331 [6.5.1; 9.4.4]

El Mallawany I. (1977a): "Analysis over the closed glottis interval." **ICA-9** (Paper I65) [6.5.2]

El Mallawany I. (1977b): "Area function over the closed glottis interval." In *Modèles articulatoires et Phonétique - Articulatory Modeling and Phonetics.* Proceedings of a Symposium held at Grenoble July 10-12, 1977; ed. by R.Carré, R.Descout, M.Wajskop (G.A.L.F, F-22301 Lannion), 65-76 [3.3; 6.5.2]

El Mallawany I. (1977c): "Evaluation comparative de mélographes numériques." **Recherches/Acoustique 4**, 15-34 [6.2.1; 9.3]

El Mallawany I., Zurcher J.F. (1977): "Détecteur numérique de mélodie." **Journées d'Etude sur la Parole 8**, 87 [6.2.1]

Eldumiati I.I.: see (Boddie et al., 1981)

Elias P. (1955): "Predictive coding." IRE Trans. IT-**1**, 16-33 [6.3.1]

Elman J.L. (1980): "The effects of pitch-distorted feedback on vocal productions." J. Acoust. Soc. Am. **68** (A), S50 (Paper BB6, 100th Meet. ASA) [3.4]

Els T.J.M. van: see (Kolk et al., 1977)

Emanuel F.W.: see (Deal and Emanuel, 1978)

Emerard F. (1977): "Synthèse par diphones et traitement de la prosodie" (CNET, F-22301 Lannion) [9.4.2; 9.4.4]

Emerard F. (1979): "Les diphones et le traitement de la prosodie dans la synthese de la parole." In *Recherche sur la prosodie du français* (Univ. de Langues et Lettres, Grenoble), 131-168 [9.4.2; 9.4.4]

Emerard F., Larreur D. (1976): "Synthèse par diphones et traitement automatique de la prosodie." **Journées d'Etude sur la Parole 7**, 179-192 [9.4.2; 9.4.4]

Emiliani P.: see (Cappelini et al., 1978)

Endres W.K. (1967): "Probleme der Analyse und Synthese von Sprache durch Automaten." In *Jahrbuch des elektrischen Fernmeldewesens* (Heidecker, Bad Windsheim), Vol.18, 132-191

Engebretson A.M.: see (Monsen and Engebretson, 1977)

English K.S.: see (Demytenko and English, 1977)

Enriquez E.: see (Gerber and Enriquez, 1962)

Eras H. (1967): "Die Entwicklung eines Tonhöhenschreibers für phonetische Forschungen." **ICPhS-6**, 297-303 [6.2.1; 6.6.1; 9.4.2]

Erb H.J. (1972a): "Untersuchungen an einer Anordnung zur Erkennung der Grundfrequenz in harmonischen Zeitsignalen;" Bericht Nr. 45, Inst. f. Übertragungstechnik, TH Darmstadt [6.1.1; 6.1.3]

Erb H.J. (1972b): "Untersuchung und Vergleich von Verfahren zur Erkennung der Sprachgrundfrequenz;" Bericht Nr. 41, Inst. f. Übertragungstechnik, TH Darmstadt [6.1.1; 9.3]

Erb H.J. (1973a): "Eine Anordnung zur Erkennung von Pausen in Sprachsignalen und zur Unterscheidung von stimmhaften und stimmlosen Lauten;" Bericht Nr. 51, Inst. f. Übertragungstechnik, TH Darmstadt [5.3]

Erb H.J. (1973b): "Verringerung des Messfehlers eines Sprachgrundfrequenzanalysators nach dem Prinzip einer Integratorkette;" Bericht Nr. 54, Fachgebiet Übertragungstechnik, TH Darmstadt [6.1.1; 9.3.1]

Erb H.J. (1974a): "Bestimmung der Sprachgrundfrequenz aus fliessender Sprache in Echtzeit." ICA-8, 251 [6.1.1]

Erb H.J. (1974b): "Ein Verfahren zur Bestimmung der Sprachgrundfrequenz in Echtzeit." Frequenz 28, 23-28 [4.1; 6.1.1]

Erickson D.: see (Ojamaa et al., 1970)

Erikson Y. (1973): "Preliminary evidence of syllable locked temporal control of F_0." STL-QPSR (2-3), 23-30 [3.1; 9.4.2]

Espesser R. (1981): "Détection de voisement par classification automatique." In Symposium Prosodie – Prosody symposium. Toronto, May 1981 (Experim. Phonetics Lab., Univ. of Toronto, Toronto, Ont.), 104-112 [5.3]

Essen O. von (1956): Grundzüge der hochdeutschen Satzintonation (Hein, Düsseldorf-Ratingen) [9.4.2]

Essen O. von (1961): Mathematische Analyse periodischer Vorgänge (Buske, Hamburg)

Essen O. von (ed.) (1972): Sprechmelodie als Ausdrucksgestaltung. Hamburger Phonetische Beiträge, Bd.1 (Buske, Hamburg) [9.4.2]

Esser J. (1978): "Contrastive intonation of German and English." Phonetica 35, 41-55 [9.4.2]

Essigmann M.W., Dolanský L.O., Wiren J. (1954): "Visual-message presentation: covering special equipment for speech-analysis purposes" (Northeastern University, Boston; DDC-AD-30493) [5.1]

Esteban D.J., Galand C., Mauduit D., Menez J. (1978): "9.6/7.2 kbps voice excited predictive coder (VEPC)." ICASSP-78, 307-311 [9.4.5]

Esteban D.J., Galand C., Mauduit D., Menez J. (1979): "A 4800 bps voice excited predictive coder (VEPC) based on improved baseband/sub-band filters." ICASSP-79, 975-980 [9.4.5]

Esteban D.J., Mauduit D., Maurel O. (1979): "Real time signal processing software for multiplierless microprocessors." ICASSP-79, 985-989 [8.5.4]

Estill J.: see (Colton and Estill, 1976)

EUSIPCO-80 --> Signal processing: theory and applications. Proceedings of the First Eoropean Signal Processing Conference, Lausanne, September 1980; ed. by M. Kunt and F. de Coulon (North-Holland Publishing Company, Amsterdam 1980)

Evans W.E.: see (Altes et al., 1975)

Ewan W.G. (1977): "Can the intrinsic F_0 differences between vowels be explained by source/tract coupling?" Speech Research 51, 197-199 [3.1; 3.4]

Ewan W.G. (1979a): "Laryngeal behavior in speech" (Phonology Laboratory, University of California, Berkeley CA; Report No. 3) [3.1; 3.2]

Ewan W.G. (1979b): "Can intrinsic vowel F_0 be explained by source/tract coupling?" J. Acoust. Soc. Am. 66, 358-362 [3.1; 3.4]

Ewanowski S.J.: see (Minifie et al., 1972)

Ewing: see (Jenkin and Ewing, 1878)

Exner M.L. (1954): "Untersuchung unperiodischer Zeitvorgänge mit der Autokorrelations- und Fourieranalyse." Acustica 4, 365-379 [6.5.3]

F.A.S.E.-81 --> Ferrero F. (ed.): Proceedings of the Fourth F.A.S.E. Symposium on Acoustics and Speech, Venice, April 21-24, 1981; ed. by F.Ferrero (Edizioni Scientifiche Associate, Roma 1981)

Faaborg-Andersen K., Yanagihara N., Leden H. von (1967): "Vocal pitch and intensity regulation - A comparative study of electrical activity in the cricothyroid muscle and the airflow rate." Arch. Otolaryng. 85, 448-454 [3.1]

Fabre P. (1957): "Un procédé électrique percutané d'inscription de l'accolement glottique au cours de la phonation: Glottographie de haute fréquence: premiers resultats." Bull. Acad. Nat. Méd. (Paris), 66-69 [5.2.3]

Fabre P. (1958): "Etude comparée des glottogrammes et des phonogrammes de la voix humaine." Ann. Oto-Laryngol. (Paris) **75**, 768-775 [5.2.3]

Fabre P. (1959): "La glottographie en haute fréquence, particularités de l'appareillage." Comptes Rendus des Séances de la Société de Biologie **153**, 1361-1364 [5.2.3]

Fabre P. (1961): "Glottographie respiratoire." Comptes Rendus des Séances de l'Académie des Sciences, Paris **6**, 1386-1388 [5.2.3]

Fabre P., Husson R., Roelens R. (1959): "Etude experimentale du comportement phonatoire des cordes vocales par la glottographie électrique." Comptes Rendus des Séances de l'Académie des Sciences, Paris **248**, 3041-3042 [3.3; 5.2.3]

Facca R.: see (Dours et al., 1977)

Fairbanks G. (1940): "Recent experimental investigations of vocal pitch in speech." J. Acoust. Soc. Am. **11**, 457-466 [1.2; 3.3; 4.1; 5.2.1]

Fairbanks G. (1960): *Voice and articulation drillbook* (Harper & Row, New York NY) [3.0; 9.4.2]

Fairbanks G., Pronovost W.L. (1939): "An experimental study of the pitch characteristics of the voice during the expression of emotion." Speech Monogr. **6**, 87-104 [9.4.2]

Fallside F.: see (Crighton and Fallside, 1974; Bristow and Fallside, 1982; Bate et al., 1982)

Falter J.W., Otten K.W. (1967): "Cybernetics and speech communication: a survey of Russian literature." IEEE Trans. AU-**15**, 27-36

Fant G. (1957): "Modern instruments and methods for acoustic studies of speech" (Speech Transmission Lab., KTH, Stockholm; Rept.No.8; DDC-AD-146813) [6.6.1]

Fant G. (1958): "Modern instruments and methods for acoustic studies of speech." In *Proceedings of the 8th International Congress of Linguists* (University Press, Oslo), 282-358 [5.1; 5.2; 9.4.2]

Fant G. (1959): "Acoustic Analysis and Synthesis of Speech with Applications to Swedish." Ericsson Techniques [3.4]

Fant G. (1960): *Acoustic theory of speech production.* With calculations based on X-ray studies of Russian articulations (Mouton and Company, 's-Gravenhage) [1.1; 3.3; 3.4]

Fant G. (1961): "A new antiresonance circuit for inverse filtering." **STL-QPSR** (4), 1-6 [6.5.1]

Fant G. (ed.) (1962a): *Proceedings of the Speech Communication Seminar, Stockholm 1962* (Speech Transmission Laboratory, Royal Institute of Technology, Stockholm)

Fant G. (1962b): "Speech analysis and synthesis: Technical (Final) Report" (Speech Transmission Lab, Stockholm; Report No.26, June 4th, 1962; DDC-AD-298309) [6.1.2; 6.4.2; 9.4.4]

Fant G. (1968): "Analysis and synthesis of speech processes." In *Manual of Phonetics;* ed. by B.Malmberg (North Holland Publishing Company, Amsterdam), 173-278 [1.2; 3.3; 3.4]

Fant G. (1970): *Acoustic theory of speech production.* With calculations based on X-ray studies of Russian articulations; 2nd ed. (Mouton and Company, 's-Gravenhage) [1.1; 1.2; 3.4]

Fant G. (ed.) (1974): *Proceedings of the 1974 Speech Communication Seminar, Stockholm, Sweden, August 1-3, 1974;* 4 volumes (Almqvist and Wiksell, Stockholm)

Fant G. (1979a): "Voice source analysis - a progress report." **STL-QPSR** (3-4), 31-53 [3.1; 3.3; 6.5.1]

Fant G. (1979b): "Glottal source and excitation analysis." **STL-QPSR** (1), 85-107 [3.3; 9.4.4]

Fant G. (1979c): "Temporal time structures of formant damping and excitation." J. Acoust. Soc. Am. **65** (A), S31 (Paper L1; 97th Meet. ASA) [3.3; 3.4]

Fant G. (1980a): "Voice source dynamics." **STL-QPSR** (2-3) [3.3]

Fant G. (1980b): "Voice source dynamics." **ICA-10** (Paper A1-9.2) [3.3]

Fant G. (1981a): "The source filter concept in voice production." **STL-QPSR** (1), 21-37 [3.3; 9.4.4]

Fant G. (1981b): "The source filter concept in voice production." **F.A.S.E.-81** (2), 39-56 [3.3; 9.4.4]

Fant G., Ishizaka K., Lindqvist J., Sundberg J. (1972): "Subglottal formants." **STL-QPSR** (1), 1-12 [3.3; 3.4]

Fant G., Liljencrants J. (1979): "Perception of vowels with truncated intra-period decay envelopes." **STL-QPSR** (1), 79-84 [3.4; 4.2]

Fant G., Ondráčková J., Lindqvist J., Sonesson B. (1966): "Electrical glottography." **STL-QPSR** (4), 15-21 [5.2.3]

Fant G., Risberg A. (1962): "Speech analysis techniques [several reports]" (Speech Transmission Lab, Stockholm; DDC-AD-250559/254069/256730/269293/291799) [6.1.2; 6.1.3]

Fant G., Scully C. (eds.) (1977): "The larynx and language. Proceedings of a discussion seminar at the 8th Intern.Congress of Phonetic Sciences, Leeds 1975." Phonetica **34**, 259 [3.1]

Fant G., Sonesson B. (1962): "Indirect studies of glottal cycles by synchronous inverse filtering and photo-electrical glottography." **STL-QPSR** (4) [5.2.5; 6.5.1]

Fant G., Tatham M.A.A. (eds.) (1975): *Auditory analysis and perception of speech* (Academic Press, London) [4.2]

Faris W.R., Timothy L.K. (1974): "Linear predictive coding with zeros and glottal wave." In *Proceedings of the National Electronics Conference* (National Engineering Consortium, Oakbrook IL), Vol.29, 409-411 [3.3; 6.3.1; 6.5]

Farnsworth D.W. (1940): "High speed motion pictures of the human vocal cords." Bell Labs. Record **18**, 203-208 [3.1; 5.2.1]

Fastl H. (1976): "Temporal masking effects: I. Broad band noise masker." Acustica **35**, 287-302 [4.2; 8.5.2]

Fastl H. (1977): "Temporal masking effects: II. Critical band noise masker." Acustica **36**, 317-331 [4.2; 8.5.2]

Fastl H. (1979): "Temporal masking effects: III. Pure tone masker." Acustica **43**, 282-294 [4.2; 8.5.2]

Faulhaber J.: see (Vallancien and Faulhaber, 1967)

Faure G. (1962): "L'intonation et l'identification des mots dans la chaine parlée." **ICPhS-4**, 598-609 [9.4.2]

Faure G. (1973): "Tendances et perspectives de la recherche intonologique." **Bull. Audiophonologie 3**, 5-29 [9.4.2]

Faure M.A. (1979): "Physiologie laryngée dans le chant et de la parole." **Bull. Audiophonologie 9**, 1-93 [3.1; 3.3]

Fears R.R. (1978): "Speech prosodics and speech understanding systems - a review." In *Spring conference, Cambridge University, April 5-7, 1978* (Institute of Acoustics, Edinburgh), 21 [2.6; 9.4.4]

Fedders B. (1977): "Ein schnelles Periodendauermeßgerät mit logarithmischer Frequenzanzeige für Sprachuntersuchungen." Biomediz. Technik **22**, 36-38 [6.4.1]

Fedders B., Schultz-Coulon H.J. (1975): "Ein Grundtonanalysator für die klinische und experimentelle Phoniatrie." Acustica **33**, 17-24 [6.4.1; 9.4.3]

Fedorinchik S.M.: see (Lyudovik and Fedorinchik, 1980)

Feldman C.B.H. (1957): "Speech transmission system;" United States Patent No. 2,817,711. Issued Dec. 24, 1957 [9.4.5]

Feldman C.B.H., Norwine A.C. (1958): "Derivation of vocoder pitch signals;" United States Patent No. 2,859,405. Issued Nov. 4, 1958; filed Feb. 17, 1956 [6.1.3]

Feldtkeller R.: see (Zwicker and Feldtkeller, 1967)

Fellbaum K. (1979): "Verfahren der digitalen Sprachsignalübertragung." Nachrichtentech. Z. **32**, 603-607 [9.4.4; 9.4.5]

Fellbaum K. (1982): "Bericht über Vorhaben zur vergleichenden Untersuchung von GFB-Algorithmen beim Rundgespräch der DFG am 21.1.1982 in Frankfurt" [private communication] [9.3.2]

Fenton A.K.: see (Sommers et al., 1961)

Ferber L.A. (1972): "Three parameter speech display." **ICSCP-72**, 252-254 (Paper G6) [5.1]

Fernald A., Simon Th. (1977): "Analyse von Grundfrequenz und Sprachsegmentlänge bei der Kommunikation von Müttern und Neugeborenen." **Phon. München 7**, 21-39 [6.1.1; 9.4.2; 9.4.3]

Ferraris F.: see (Belforte et al., 1979)

Ferrein A. (1741): "De la formation de la voix de l'homme." Histoire de l'Académie Royale des Sciences de Paris, 402-432 [3.1; 5.2.1]

Ferrero F., Vagges K., Zovato S. (1981): "Laryngeal analog synthesis of diplophonic voice." **F.A.S.E.-81**, 105-108 [9.4.3; 9.4.4]

Ferullo R.J.: see (Dolanský et al., 1966)

Fette B., Gibson R., Greenwood E. (1980): "Windowing functions for the average magnitude difference function pitch extractor." **ICASSP-80**, 49-52 [2.4; 8.3.1]

Filip M. (1962): "Unpublished notes on pitch determination by envelope modeling [referenced in (Filip, 1969, 1972)]" [6.2.1]

Filip M. (1967): "Some aspects of high-accuracy analog fundamental frequency recording." **ICPhS-6**, 319-322 [6.2.1]

Filip M. (1969): "Envelope periodicity detection." J. Acoust. Soc. Am. **45**, 719-732 [1.2; 4.2.2; 6.2.1; 9.4.2]

Filip M. (1970): "Auditory and electronic envelope periodicity detection." J. Acoust. Soc. Am. **47**, 654-657 [6.2.1]

Filip M. (1971): "Pitch period extraction in monophonic music and similar signals." **ICA-7** (3), 649-652 (Paper 21S5) [6.2.1; 9.4.2]

Filip M. (1972): "A survey of pitch-frequency recording methods." **Phonetica Pragensia III,** 81-90 [1.3; 6.2.1; 9.1.1; 9.4.2]

Filleau M., Lhote F. (1973): "Mesure logarithmique de fréquence instantanée par un dispositif digital." Revue d'Acoustique (17), 231-234

Fink B.R. (1975): *The human larynx - a functional study* (Raven Press, New York NY) [3.1]

Fischer-Jørgensen E. (1958): "What can the new techniques of acoustic phonetics contribute to linguists?." In *Proceedings of the 8th International Congress of Linguists* (University Press, Oslo), 433-499 [9.4.2]

Fischer-Jørgensen E., Frøkjaer-Jensen B., Rischel J. (1966): "Preliminary experiments with the Fabre glottograph." **ARIPUC 1,** 22-30 [5.2.3; 9.4.2]

Fisher J., Fourcin A.J., King A., Parker A., Timms C., Wright R. (1978): "Speech production by normal and hearing impaired children." In *Spring conference, Cambridge University, April 5-7, 1978* (Institute of Acoustics, Edinburgh), 35 [9.4.3]

Fisher W.M., Monsen R.B., Engebretson A.M. (1975): "Variations in the normal male glottal wave." J. Acoust. Soc. Am. **58** (A), S41 (Paper V10; 90th Meet. ASA) [3.3; 4.3]

Flanagan J.L. (1955a): "A speech analyzer for a formant-coding compression system" (Acoustics Lab., MIT, Cambridge MA; DDC-AD-80771) [6.1.1]

Flanagan J.L. (1955b): "A difference limen for vowel formant frequency." J. Acoust. Soc. Am. **27**, 613-617 [4.3]

Flanagan J.L. (1955c): "Difference limen for the intensity of a vowel sound." J. Acoust. Soc. Am. **27**, 1223-1225 [4.3]

Flanagan J.L. (1958a): "Some properties of the glottal sound source." J. Speech Hear. Res. **1**, 99-111 [3.1]

Flanagan J.L. (1958b): "Some properties of the glottal sound source." Reprinted in *Acoustic Phonetics - A course of basic readings*; ed. by D.B.Fry (Cambridge University Press, Cambridge), 31-51 [3.1]

Flanagan J.L. (1959a): "A resonance-vocoder and baseband system for speech transmission." **ICA-3,** 211-213 [9.4.5]

Flanagan J.L. (1959b): "Estimates of interglottal pressure during phonation." J. Speech Hear. Res. **2**, 168-172 [3.1]

Flanagan J.L. (1960): "A resonance-vocoder baseband complement - a hybrid system for speech transmission." IRE Trans. AU-**8**, 95-102 [9.4.5]

Flanagan J.L. (1961): "Audibility of periodic pulses and a model for the threshold." J. Acoust. Soc. Am. **33**, 1540-1549 [4.3]

Flanagan J.L. (1962a): "Perceptual criteria in speech processing." **Proc. 1962 Speech Commun. Sem.** (Paper D2) [4.2]

Flanagan J.L. (1962b): "Some influences of the glottal wave upon vowel quality." **ICPhS-4,** 34-49 [3.3; 3.4; 9.4.2]

Flanagan J.L. (1965): *Speech analysis, synthesis, and perception* (Springer, Berlin) [1.1; 3.0; 4.2.1]

Flanagan J.L. (1969): "Use of an interactive laboratory computer to study an acoustic oscillator model of the vocal cords." IEEE Trans. AU-**17**, 2-6 [3.3]

Flanagan J.L. (1971a): "Focal points in speech communication research." **ICA-7** (1), 57-74 (Paper 21G3) [1.2]

Flanagan J.L. (1971b): "Focal points in speech communication research." IEEE Trans. COM-**19**, 1006-1015 [1.2]

Flanagan J.L. (1972a): *Speech analysis, synthesis, and perception* (Springer, Berlin) [1.1; 3.0; 4.2.1]

Flanagan J.L. (1972b): "Voices of men and machines." J. Acoust. Soc. Am. **51**, 1375-1387 [1.2; 3.3; 9.4.5]

Flanagan J.L. (1976): "Computers that talk and listen: Man-machine communication by voice." Proc. IEEE **64**, 405-415 [1.2; 9.4.4]

Flanagan J.L. (1978): "Computational models of speech sound generation." J. Acoust. Soc. Am. **64** (A), S40 (Paper P1; 96th Meet. ASA) [3.3]

Flanagan J.L., Guttman N. (1960a): "On the pitch of periodic pulses." J. Acoust. Soc. Am. **32**, 1308-1319 [4.2.3]

Flanagan J.L., Guttman N. (1960b): "Pitch of periodic pulses without fundamental frequency." J. Acoust. Soc. Am. **32**, 1319-1328 [4.2.3]

Flanagan J.L., Ishizaka K. (1975): "Automatic generation of voiceless excitation in a vocal-cord/vocal-tract speech synthesizer." J. Acoust. Soc. Am. **57** (A), S50 (Paper X15; 89th Meet. ASA) [3.5; 5.3; 9.4.4]

Flanagan J.L., Ishizaka K. (1976): "Automatic generation of voiceless excitation in a vocal cord-vocal tract speech synthesizer." IEEE Trans. ASSP-**24**, 163-169 [5.3; 9.4.4]

Flanagan J.L., Ishizaka K. (1978): "Computer model to characterize the air volume displaced by the vibrating vocal cords." J. Acoust. Soc. Am. **63**, 1559-1565 [3.3]

Flanagan J.L., Ishizaka K., Shipley K.L. (1975): "Synthesis of speech from a dynamic model of the vocal chords and vocal tract." Bell Syst. Tech. J. **54**, 485-506 [3.3; 9.4.4]

Flanagan J.L., Landgraf L.L. (1968a): "Self-oscillating source for vocal tract synthesizers." IEEE Trans. AU-**16**, 57-64 [3.3; 9.4.4]

Flanagan J.L., Landgraf L.L. (1968b): "Excitation of vocal tract synthesizers." ICA-**6**, 179-182 [3.3; 9.4.4]

Flanagan J.L., Meinhart D.I.S. (1964): "Source-System Interaction in the Vocal Tract." J. Acoust. Soc. Am. **36**, 2001-2002 (Paper J5; 68th Meet. ASA) [3.3; 3.4; 9.4.4]

Flanagan J.L., Rabiner L.R., Christopher D., Bock D.E. (1976): "Digital analysis of laryngeal control in speech production." J. Acoust. Soc. Am. **60**, 446-455 [3.1]

Flanagan J.L., Saslow M.G. (1958): "Pitch discrimination for synthetic vowels." J. Acoust. Soc. Am. **30**, 435-442 [4.2.3]

Flanagan J.L., Schroeder M.R., Atal B.S., Crochiere R.E., Jayant N.S., Tribolet J.M. (1979): "Speech coding." IEEE Trans. COM-**27**, 710-736 [1.2; 2.2; 5.3; 9.4.4]

Fleischmann G.: see (Almeida et al., 1977)

Fletcher H. (1924): "Physical criterion for determining the pitch of a musical tone." Physical Review **23**, 427-437

Fletcher H. (1931): "Some physical characteristics of speech and music." J. Acoust. Soc. Am. **3**, S1-25 [4.2; 9.4.2]

Fletcher H. (1953): *Speech and hearing in communication* (D. Van Nostrand Company Inc., New York) [4.2]

Fletcher H., Galt R.H. (1950): "The perception of speech and its relation to telephony." J. Acoust. Soc. Am. **22**, 89-151 [4.2]

Fletcher S.G.: see (Stilwell et al., 1981)

Fletcher W.W. (1950): "A study of internal laryngeal activity in relation to vocal intensity;" Dissertation, Northwestern Univ., Evanston IL [3.1]

Floyd W. (1964): "Voice identification techniques" (Litton Syst, Waltham MA; DDC-AD-606634) [4.1; 6.4.2; 9.4.4]

Fónagy I. (1978): "A new method of investigating the perception of prosodic features." Lang. Speech **21**, 34-49 [4.2; 5.2.3]

Fónagy I. (1979): "Artistic vocal communication at the prosodic level." **ICPh-Miami**, 245-260 [9.4.2]

Fónagy I. (1981): "Emotions, voice, and music." In *Research aspects on singing*. Papers given at a seminar organized by the Committee for the Acoustics of Music (Royal Swedish Academy of Music, Stockholm; Report #33), 51-79 [3.2; 9.4.2; 9.4.3]

Fonagy I., Bérard E. (1972): "Il est huit heures": contribution à l'analyse semantique de la vive voix." Phonetica **26**, 157-192 [9.4.2]

Fónagy I., Magdics K. (1963): "Emotional patterns in intonation and music." Z. Phonetik **16**, 293-326 [9.4.2]

Forgie J.: see (Newell et al., 1971)

Foster S.H.: see (Maitra et al., 1980)

Fourcin A.J. (1968): "Speech source inference." IEEE Trans. AU-**16**, 65-67 [3.1]

Fourcin A.J. (1970): "Central pitch and auditory lateralization." In *Frequency analysis and periodicity detection in hearing*; ed. by R.Plomp and G.F.Smoorenburg (Sijthoff, Leiden), 319-327 [4.2]

Fourcin A.J. (1972): "Perceptual mechanisms at the first level of speech processing." **ICPhS-7**, 48-62 [4.2]

Fourcin A.J. (1974a): "Laryngographic examinations of vocal fold vibrations." In *Ventilatory and phonatory control*; ed. by F. Wyke (Oxford University Press, London), 315-333 [3.3; 5.2.3]

Fourcin A.J. (1974a): "Laryngographic examinations of vocal fold vibrations." In *Ventilatory and phonatory control*; ed. by F. Wyke (Oxford University Press, London), 315-333 [3.3; 5.2.3]

Fourcin A.J. (1974b): "Regularity of vocal fold vibration." **ICA-8**, 225 [4.3]

Fourcin A.J. (1977): "Speech processing by man and machine. Group report." In *Recognition of complex acoustic signals*; ed. by T.H.Bullock (Abakon-Verlag, Berlin; Life Sciences Report No. 5) [1.2; 3.3; 4.2; 5.1; 5.2]

Fourcin A.J. (1978a): "Acoustic patterns and speech acquisition." In *The development of communication*; ed. by N. Waterson and C. Snow (Wiley, London), 47-72 [9.4.3; 9.4.4]

Fourcin A.J. (1978b): "Perception and production of speech patterns by hearing-impaired children." **Phon. London**, 173-204 [9.4.3]

Fourcin A.J., Abberton E. (1971): "First applications of a new laryngograph." Medical and Biological Illustration **21**, 172-182 [3.3; 5.2.3]

Fourcin A.J., Abberton E. (1976): "The laryngograph and the voiscope in speech therapy." In *Proceedings of the 16th International Congress on Logopedics and Phoniatrics, Interlaken, 1974*; ed. by E.Loebell (S.Karger AG, Basel), 116-122 [5.3.3; 9.4.3]

Fourcin A.J., Abberton E. (1977): "Laryngograph studies of vocal-fold vibration." Phonetica **34**, 313-315 [5.2.3]

Fourcin A.J., Norgate M. (1965): "Measurement of the transglottal impedance." **Phon. London**, 34 [5.2.3]

Fourcin A.J., Rosen S.M., Moore B.C.J., Douek E.E., Clarke G.P., Dodson H., Bannister L.H. (1978): "External electrical stimulation of the cochlea: Clinical, psychophysical, speech-perceptual and histological findings." **Phon. London**, 1-32 [4.2; 9.4.3]

Fourcin A.J., Rosen S.M., Moore B.C.J., Douek E.E., Clarke G.P., Dodson H., Bannister L.H. (1979): "External electrical stimulation of the cochlea: Clinical, psychophysical, speech-perceptual and histological findings." Brit. J. Audiology **13**, 85-107 [4.2; 9.4.3; 9.4.4]

Fourcin A.J., West J.E. (1968): "Larynx movement detector." **Phon. London** [5.2.3]

Fourcin A.J., West J.E. (1972): "Le laryngographe." Revue d'Acoustique [5.2.3]

Fransen L.J.: see (Kang et al., 1981)

Fransson F.: see (Mörner et al., 1964)

Frazier R.H., Samsam S., Braida L.D., Oppenheim A.V. (1975): "Enhancement of speech by adaptive filtering." **MIT-RLE Progr. Rept. 116**, 177-181 [6.3.1]

French N.R.: see (Steinberg and French, 1946)

French T.R. (1883): "On photographing the larynx." Arch. Laryngology (New York) **4**, 235-243 [5.1.1; 5.2.1]

Freudberg R.: see (Ross et al., 1974)

Freystedt E. (1935): "Das Tonfrequenzspektrometer." Z. Tech. Physik **16**, 533-534 [5.1.2]

Frichon M.: see (Zurcher et al., 1975)

Friedlander B., Maitra S. (1981): "Speech deconvolution by recursive ARMA lattice filters." **ICASSP-81**, 343-346 [6.3.1; 9.4.5]

Friedman D.H. (1977a): "Determining speech pitch by modified least-squares periodic approximation." **ICASSP-77**, 315-318 [8.6.1]

Friedman D.H. (1977b): "Pseudo-maximum-likelihood speech pitch extraction." IEEE Trans. ASSP-**25**, 213-221 [8.6.1]

Friedman D.H. (1978): "Multidimensional pseudo-maximum-likelihood pitch estimation." IEEE Trans. ASSP-**26**, 185-196 [4.1; 6.3.2; 8.6.2]

Friedman D.H. (1979): "Multichannel zero-crossing-interval pitch estimation." **ICASSP-79**, 764-767 [6.4.1; 8.3.3]

Friedrich G.: see (Ungeheuer et al., 1965)

Fritzell B. (1981): "Singing and the health of voice." In *Research aspects on singing*. Papers given at a seminar organized by the Committee for the Acoustics of Music (Royal Swedish Academy of Music, Stockholm; Report #33), 97-108 [9.4.2; 9.4.3]

Fritzell B., Hammarberg B., Wedin L. (1977): "Clinical applications of acoustic voice analysis 1 - background and perceptual factors." **STL-QPSR** (2-3), 31-38 [9.4.3]

Frøkjaer-Jensen B. (1967): "A photo-electric glottograph." **ARIPUC 2**, 5-19 [5.2.5]

Frøkjaer-Jensen B. (1968): "Comparison between a Fabre Glottograph and a photo-electric glottograph." **ARIPUC 3**, 9-16 [5.2.3; 5.2.5; 5.2.6]

Frøkjaer-Jensen B. (1970): "Manual for photo-electric glottograph" (F-J Electronics, Holte, Denmark) [5.2.5]

Frøkjaer-Jensen B. (1980a): "Application notes, Pitch Computer" (F-J Electronics A/S, Gentofte, Denmark) [6.4.2]

Frøkjaer-Jensen B. (1980b): "Application note, Electro-Glottograph" (F-J Electronics A/S, Gentofte, Denmark) [5.2.3]

Frøkjaer-Jensen B. (1980c): "Application note, fundamental frequency meter" (F-J Electronics A/S, Gentofte, Denmark) [6.1.1]

Frøkjaer-Jensen B., Thorvaldsen P. (1968): "Construction of a Fabre glottograph." ARIPUC 3, 1-9 [5.2.3]

Fromkin V.A. (1978): *Tone, a linguistic survey* (Academic Press, New York NY) [9.4.2]

Fromkin V.A., Ladefoged P. (1966): "Electromyography in speech research." Phonetica 15, 219-242 [5.2.1]

Fry D.B. (1968): "Prosodic phenomena." In *Manual of Phonetics;* ed. by B.Malmberg (North Holland Publishing Company, Amsterdam), 365-411 [5.1; 9.4.2]

Fry D.B. (ed.) (1976): *Acoustic Phonetics – A course of basic readings* (Cambridge University Press, Cambridge)

Fry D.B. (1979): *The physics of speech* (Cambridge University Press, Cambridge) [1.2]

Fry D.B., Manen L. (1957): "Basis for the acoustical study of singing." J. Acoust. Soc. Am. 29, 690-692 [5.1.2; 9.4.2]

Frykberg S.D.: see (Tucker et al., 1977)

Fuchi K., Itahashi S. (1973): "Estimation of fundamental frequency by direct linear prediction." PIPS Rept., 27-29 [6.3.1]

Fuchi K., Itahashi S. (1974a): "Reconstruction of a vocoder base band from linear prediction coefficients." PIPS Rept., 23-25 [6.3.1; 9.4.5]

Fuchi K., Itahashi S. (1974b): "Improvements of the direct linear prediction method for fundamental frequency estimation." PIPS Rept., 73-75 [6.3.1]

Fuchi K., Itahashi S. (1976): "Direct linear prediction for fundamental frequency analysis." ICASSP-76, 318-321 [6.3.1]

Fucks W. (1964): "Über mathematische Musikanalyse." Nachrichtentech. Z. 17, 41-47 [9.4.2]

Fürstenberg F. (1929): "Zwei einfache Tonhöhenmeßapparate." Vox [5.1]

Fujimura O. (1968): "An approximation to voice aperiodicity." IEEE Trans. AU-16, 68-72 [4.3; 6.5.3; 9.4.3]

Fujimura O. (1972): "Acoustics of speech." Ann. Bull. RILP 6, 149-199 [1.2]

Fujimura O. (1976): "Stereo-fiberscope." In *Dynamic aspects of speech production;* ed. by M. Sawashima and F.S. Cooper (Univ. of Tokyo Press), 133-138 [3.1; 5.2.1]

Fujimura O. (1977): "Control of the larynx in speech." Phonetica 34, 280-288 [3.1]

Fujimura O. (1979a): "Physiological functions of the larynx in phonetic control." ICPh-Miami, 129-163 [3.1]

Fujimura O. (1979b): "Modern methods of investigation in speech production." ICPhS-9 (1), 161-166 [3.1; 5.2.1]

Fujimura O. (1980): "Body-cover theory and its phonetic implication." Vocal Fold Physiology Conf. Kurume, Japan [3.1]

Fujimura O., Baer T., Niimi S. (1979): "A stereo-fiberscope with a magnetic interlens bridge for laryngeal observation." J. Acoust. Soc. Am. 65, 478-480 [5.2.1]

Fujisaki H. (1959): "Automatic extraction of fundamental period in speech sounds;" Master's Thesis, MIT, Cambridge MA [8.2]

Fujisaki H. (1960a): "Theoretical studies on pitch extension and formant tracking" (Speech Transmission Laboratory, KTH, Stockholm; Internal Report) [6.1.2; 7.2]

Fujisaki H. (1960b): "Automatic extraction of fundamental period of speech by auto-correlation analysis and peak detection." J. Acoust. Soc. Am. 32 (A), 1518 [8.2.3]

Fujisaki H. (1963): "Studies on the extraction of pitch and formant of speech sounds." Ann. Rept. Eng. Res. Inst. 11, 7-11 [3.4; 8.2]

Fujisaki H. (1981a): "Dynamic characteristics of voice fundamental frequency in speech and singing." STL-QPSR (1), 1-35 [3.1; 3.2; 3.3; 4.3; 9.4.5]

Fujisaki H. (1981b): "Dynamic characteristics of voice fundamental frequency in speech and singing – acoustical analysis and physiological interpretations." F.A.S.E.-81 (2), 57-69 [3.1; 3.3; 4.3]

Fujisaki H., Hirose H. (1982): "Analysis and synthesis of fundamental frequency contours of English sentences." J. Acoust. Soc. Am. 71 (A), S7 [9.4.2]

Fujisaki H., Hirose K., Ohta K. (1979): "Acoustic features of the fundamental frequency contours of declarative sentences in Japanese." Ann. Bull. RILP 13, 163-173 [9.4.2]

Fujisaki H., Kawashima T. (1968): "The role of pitch and higher formants in the perception of vowels." IEEE Trans. AU-16, 73-77 [3.4; 9.4.2]

Fujisaki H., Nagashima S. (1969): "A model for synthesis of pitch contours of connected speech." Ann. Rept. Eng. Res. Inst. 28, 53-60 [9.4.2]

Fujisaki H., Nagashima S. (1970): "A new type of baseband vocoder and its computer simulation." Ann. Rept. Eng. Res. Inst. 29, 187-194 [9.4.5]

Fujisaki H., Sudo H. (1971): "Synthesis by rule of prosodic features of connected Japanese." **ICA-7** (3), 133-136 (Paper 23C2) [9.4.2]

Fujisaki H., Sudo H. (1972): "A generative model for the prosody of connected speech in Japanese." **ICSCP-72**, 140-143 (Paper D4) [9.4.2]

Fujisaki H., Tanabe Y. (1972): "A time-domain technique for pitch extraction of speech." **Ann. Rept. Eng. Res. Inst. 31**, 259-266 [8.2.4]

Fujisaki H., Tatsumi M., Niguchi N. (1980): "Analysis of pitch control in singing." **Ann. Bull. RILP 14**, 101-112 [3.1; 3.3; 9.4.2]

Fukuda H.: see (Saito et al., 1978; Ono et al., 1978)

Funada T. (1979): "A new method for extraction of power spectrum peaks of acoustic signals with pitch." Trans. of the IECE of Japan **62**, 382-388 [8.5]

Furst M.: see (Goldstein et al., 1978)

G-AE Subcommittee on Measurement Concepts (1967): "What is the fast Fourier Transform?" IEEE Trans. AU-**15**, 45-55 [2.4; 8.1; 8.4; 8.5]

Gabor D. (1947): "New possibilities in speech transmission." J. IEE (London) **94**, 111 [8.2.1]

Gadenz R.N.: see (Boddie et al., 1981)

Galand C. (1974): "Etude et réalisation d'un détecteur de mélodie en temps réel;" Dissertation, Université de Nice [5.2.3; 8.2.4]

Galand C. (1976): "Nonlinear autocorrelation function (NLAF) real-time pitch extractor." J. Acoust. Soc. Am. **59** (A), S98 (Paper RR13; 91st Meet. ASA) [8.2.4]

Galand C., Esteban D.J., Dubus F. (1976): "Détection de la mélodie par autocorrélation non linéaire." **Journées d'Etude sur la Parole 7**, 333-345 [8.2.4]

Gall V. (1978): "Fotokymographische Befunde bei funktionellen Dysphonien, Kehlkopflähmungen und Stimmlippentumoren." Folia Phoniatr. **30**, 28-35 [3.1; 5.1.1; 9.4.3]

Galt R.H.: see (Fletcher and Galt, 1950)

Galunov V.I., Tampel' I.B. (1981): "Operating mechanism of the voice sound source." Soviet Physics - Acoustics **27**, 177-184 [3.2; 3.3]

Gandour J.T. (1975): "The features of the larynx: N-ary or binary?." Phonetica **32**, 241-253 [3.1]

Gandour J.T., Harshman R. (1978): "Cross-language differences in tone perception." Lang. Speech **21**, 1-33 [4.2; 9.4.2]

Gandour J.T., Maddieson I. (1976): "Measuring larynx movement in standard Thai using the cricothyrometer." Phonetica **33**, 241-267 [5.2.1]

García Gómez R., Tribolet J.M. (1982): "Speech analysis and modelling using a sequential ARMA estimation technique." **ICASSP-82**, 1585-1588 [6.3.1; 6.3.2]

Garcia M. (1855): *Physiological observations on the human voice* (London) [3.1; 3.2]

Garnceva A.A. (1968): К вопросу методах и приемах экспериментального исследования речевой интонации. А.А.Гарнцева. (On the question of methods and means of experimental investigation of speech intonation. In Russian). U.Z.-MGPI-VL 282, 7-14

Gauffin J. (1977): "Mechanisms of larynx tube constriction." Phonetica **34**, 307-309 [3.1]

Gauffin J. (1979): "Filtrage inverse à l'aide du masque de Rothenberg." **Larynx et parole**, 277-296 [5.2.2; 6.5.1]

Gauffin J., Sundberg J. (1977): "Clinical applications of acoustic voice analysis II - acoustic analysis, results, and discussion." **STL-QPSR** (2-3), 39-43 [3.1; 5.2; 5.2.6]

Gauffin J., Sundberg J. (1980): "Data on the glottal voice source behavior in vowel production." **STL-QPSR** (2-3), 61 [3.1]

Gault R.H., Crane G.W. (1928): "Tactual patterns for certain vowel qualities instrumentally communicated from a speaker to a subject's fingers." J. Gen. Psychol. **1**, 353-359 [9.4.3]

Gautheron B. (1973): "La rôle des variations de la fréquence fondamentale et de l'intensité dans la reconnaissance auditive ou tactile de la parole." **Bull. Audiophonologie 3**, 59-93 [9.4.3]

Gay T., Hirose H., Strome M., Sawashima M. (1971): "Electromyography of the intrinsic laryngeal muscles during phonation." **Speech Research 25**, 97-106 [3.1; 5.2.1]

Geçkinli N.C. (1976): "Some novel, optimal spectral windows, speech segmentation, recognition, and pitch extraction algorithms;" Dissertation, Middle East Technical Univ., Ankara [2.4; 6.2.3; 8.1]

Geçkinli N.C., Yavuz D. (1977): "Algorithm for pitch extraction using zero-crossing interval sequence." IEEE Trans. ASSP-**25**, 559-564 [6.2.3]

Geçkinli N.C., Yavuz D. (1978): "Some novel windows and a concise tutorial comparison of window families." IEEE Trans. ASSP-**26**, 501-507 [2.4]

Geçkinli N.C., Yavuz D. (1983): "Discrete Fourier transformation and its application to power spectra estimation" (Elsevier, Amsterdam) [2.4; 8.1]

Génin J. (1977): "La melodie: Etudes perceptives et synthèse de la parole; rôle des traits prosodiques dans la transmission et la synthèse de la parole" (Compte rendu du rapporteur). **Journées d'Etude sur la Parole 8**, 33-40 [9.4.2; 9.4.4]

Génin J., Carcaud M. (1975): "Simulation de la source vocale." **Recherches/Acoustique 2**, 129-140 [3.3]

Genina B.G., Bomas V.V. (1970): "Статистическая обработка акустических характеристик при изучении интонации." В журн.: Уч.Зап.МГПИ-ВЛ. Б.Г.Генина и В.В.Бомас. (Statistical treatment of acoustic characteristics in the study of intonation. In Russian). **U.Z.-MGPI-VL 57**, 371-384 [9.4.2]

Gentges F.: see (Coulter et al., 1979)

Gerber S.E., Enriquez E. (1962): "Pitch coding by delta modulation." **Proc. 1962 Speech Commun. Sem.** [9.4.4]

Gerson A., Goldstein J.L. (1978): "Evidence for a general template in central optimal processing for pitch of complex tones." J. Acoust. Soc. Am. **63**, 498-510 [4.2.1]

Gertheiss G., Fellbaum K., Mangold H., Schunk H., Stall D. (1974): "Sprachsignalverarbeitung" [Bundesministerium für Forschung u. Technologie (BMFT), Bonn; Bericht Nr. T74-44] [6.2.1]

Gibbs K.M. (ed.) (1963): "Bionics and related research; a report bibliography prepared by ASTIA" (Armed Services Technical Information Agency, Arlington, Virginia; DDC-AD-294150) [1.3]

Gibson R.: see (Fette et al., 1980)

Gikis I.I. (1968a): "К вопросу векторного способа выделения основного тона речевого сигнала." В кн.: Техническая кибернетика. Материалы XVIII республ. научно-технической конференции. И.И.Гикис. (On the vector method of pitch determination of speech signals. In Russian.) (Kaunas), 57-59 [6.3.2]

Gikis I.I. (1968b): "Способ выделения основного тона речевого сигнала." А.С.# 222036-СССР. Опубл. в бюлл. "Открытия, Изобретения, Промышленные образцы, Товарные знаки. И.И.Гикис. (Method for pitch determination of speech signals. In Russian.). Otkrytia, Izobretenia **22**, 108 [6.3.2; 8.3.3]

Gikis I.I. (1969): "Об одном алгоритме векторного метода выделения основного тона речевого сигнала." В кн.: Труды АРСО-4, Киев 1968. И.И.Гикис. (Algorithmic realization of the vector method for pitch determination of speech signals. In Russian.). **ARSO-4**, 213-217 [6.3.2]

Gikis I.I. (1970a): "Исследование векторных методов автоматического выделения основного тона речевого сигнала." Диссертация. И.И.Гикис. (Investigation of the vector method for pitch determination of speech signals. In Russian.); Dissertation; Vilnius [6.3.2]

Gikis I.I. (1970b): "Обзор методов выделения основного тоно речевого сигнала." И.И.Гикис. В кн.: Звуки Речи Интонация. (Survey of methods for extraction of the fundamental frequency of the speech signal. In Russian) (Vilnius, USSR), 214-236 [2.6]

Gilio A.: see (Bruno et al., 1981, 1982)

Gill J.S. (1959a): "Automatic extraction of the excitation function of speech with particular reference to the use of correlation methods." **ICA-3**, 217-220 [8.2.4]

Gill J.S. (1959b): "A study of the requirements for excitation control in synthetic speech." **ICA-3**, 221-224 [9.4.4]

Gill J.S. (1961): "Estimation of vocal excitation during speech with particular reference to the requirements of analysis-synthesis telephony;" Dissertation, Univ. of London [6.5.1]

Gill J.S. (1962a): "Estimation of larynx pulse timing during speech." **ICA-4** (Paper G33)

Gill J.S. (1962b): "Estimation of larynx-pulse timing during speech." **Proc. 1962 Speech Commun. Sem.** [3.3]

Gill J.S. (1962c): "Recent research on methods for automatic estimation of vocal excitation." **ICPhS-4**, 167-172 [3.3; 4.3]

Gillmann R.A. (1975): "A fast frequency domain pitch algorithm." J. Acoust. Soc. Am. **58** (A), S62 (Paper FF8; 90th Meet. ASA) [8.2.4]

Giordano T.A., Rothman H.B., Hollien H. (1973): "Helium speech unscramblers - A critical review of the state of the art." IEEE Trans. AU-**21**, 436-444 [9.4.4]

Girard G.T.: see (Hollien et al., 1977)

Giraud G.: see (Autesserre et al., 1979)

Glass H.C. (1960): "Technical note for speech bandwidth reduction equipment" (Scope Inc., Fairfax, Virginia; DDC-AD-245925) [6.1.1]

Goff K.W. (1955): "An analog electronic correlator for acoustic measurements." J. Acoust. Soc. Am. **27**, 223-236 [8.2.2]

Gohbara M., Suminaga Y., Miyaguchi H., Ozawa S. (1978): "Real-time pitch detection by two mini-computers." J. Acoust. Soc. Am. **64** (A), S91 (Paper II11; 96th Meet. ASA) [6.2.5]

Gold B. (1961): "Pitch extraction on TX-2 computer" (MIT Lincoln Lab, Cambridge MA; Quarterly Progress Report March 1961; DDC-AD-256371), 21-23 [6.2.4; 6.6.3]

Gold B. (1962a): "Description of a computer program for pitch detection." **ICA-4** (Paper G34) [6.2.4; 6.6.3]

Gold B. (1962b): "Computer program for pitch extraction." J. Acoust. Soc. Am. **34**, 916-921 [6.2.4; 6.6.3; 9.3.1]

Gold B. (1964a): "Note on buzz-hiss detection." J. Acoust. Soc. Am. **36**, 1030 (Paper N7; 67th Meet. ASA) [5.3]

Gold B. (1964b): "Note on buzz-hiss detection." J. Acoust. Soc. Am. **36**, 1659-1661 [5.3]

Gold B. (1964c): "Experiment with speechlike phase in a spectrally flattened pitch-excited channel vocoder." J. Acoust. Soc. Am. **36**, 1892-1894 [6.2.4; 9.4.5]

Gold B. (1965): "Techniques for speech bandwidth compression, using combinations of channel vocoders and formant vocoders." J. Acoust. Soc. Am. **38**, 2-10 [9.4.5]

Gold B. (1976): "Some comments on linear predictive coding." **Proc. 1974 Speech Commun. Sem. 1**, 10-21 [6.3.1]

Gold B. (1977): "Digital Speech Networks." Proc. IEEE **65**, 1636-1658 [1.2; 3.5; 5.3; 6.3.1; 9.4.4; 9.4.5]

Gold B., Lebow I.L., McHugh P.G., Rader C.M. (1971): "The FDP, a fast programmable signal processor." IEEE Trans. C-**20**, 33-38 [9.1.5]

Gold B., Rabiner L.R. (1969): "Parallel processing techniques for estimating pitch periods of speech in the time domain." J. Acoust. Soc. Am. **46**, 442-448 [4.1; 6.2.4; 7.2; 9.3.1]

Gold B., Rader C.M. (1967): "The channel vocoder." IEEE Trans. AU-**15**, 148-161 [9.4.5]

Gold B., Rader C.M. (1969): *Digital processing of signals* (McGraw-Hill, New York) [2.2; 2.3]

Gold B., Tierney J. (1963): "Pitch-induced spectral distortion in channel vocoders." J. Acoust. Soc. Am. **35** (C), 730-731 [9.4.4; 9.4.5]

Gold B., Tierney J. (1965): "Digitized voice-excited vocoder for telephone-quality inputs, using bandpass sampling of the baseband signal." J. Acoust. Soc. Am. **37** (C), 753-754 [2.1; 9.4.5]

Goldberg A.J. (1967): "Vocoded speech in the absence of the laryngeal fundamental;" Thesis Work, MIT, Cambridge MA [6.2.4; 9.4.4]

Goldberg A.J. (1979): "Practical implementations of speech waveform coders for the present day and for the mid 1980s." J. Acoust. Soc. Am. **66**, 1653-1657 [9.4.4; 9.4.5]

Goldberg A.J., Shaffer H.L. (1975): "A real-time adaptive predictive coder using small computers." IEEE Trans. COM-**23**, 1443-1451 [9.4.5]

Golden R.M. (1963): "Digital computer simulation of a sampled-data voice-excited vocoder." J. Acoust. Soc. Am. **35**, 1358-1366 [2.3; 9.4.5]

Golden R.M., MacLean D.J., Prestigiacomo A.J. (1964): "Frequency multiplex system for a 10-spectrum-channel voice-excited vocoder." J. Acoust. Soc. Am. **36**, 1022 (Paper I3; 67th Meet. ASA) [9.4.5]

Goldstein J.L. (1973): "An optimum processor theory for the central formation of the pitch of complex tones." J. Acoust. Soc. Am. **54**, 1496-1516 [4.2.1]

Goldstein J.L., Gerson A., Srulowicz P., Furst M. (1978): "Verification of the optimal probabilistic basis of aural processing in pitch of complex tones." J. Acoust. Soc. Am. **63**, 586 [4.2.1]

Goldstein L.D. (1958): "Functional speech disorders and personality. 1) A survey of the literature; 2) methodological and thepretical considerations." J. Speech Hear. Res. **1**, 359-382 [1.2; 9.4.3]

Goldstein M.H., Stark R.E., Yeni-Konishian G.H., Grant D.G. (1976): "Tactile stimulation as an aid for the deaf." **ICASSP-76**, 598-601 [9.4.3]

Gonzalez R.C.: see (Tou and Gonzalaz, 1974)

Goodman A.C. (1952): "Imitation of intonation patterns" (University of Michigan, Ann Arbor MI) [4.3; 9.4.2]

Goodman D.J., Carey M.J. (1977): "Nine digital filters for decimation and interpolation." IEEE Trans. ASSP-**25**, 121-126 [2.3; 8.5.4]

Gould W.J. (1976): "Newer aspects of high-speed photography of the vocal folds." In *Dynamic aspects of speech production;* ed. by M. Sawashima and F.S. Cooper (Univ. of Tokyo Press), 139-144 [3.1; 5.2.1]

Graham W.: see (Di Toro et al., 1955)

Grange M.F.: see (Lhote et al., 1975)

Granström B.: see (Carlson et al., 1972; Rothenberg et al., 1974; Carlson and Granström, 1980)

Grant D.G.: see (Goldstein M.H. et al., 1976)

Gray A.H., Markel J.D. (1974): "A spectral flatness measure for studying the autocorrelation method of linear prediction of speech analysis." IEEE Trans. ASSP-**22**, 207-217 [6.3.1]

Gray A.H., Wong D.Y. (1980): "The Burg algorithm for LPC speech analysis/synthesis." IEEE Trans. ASSP-**28**, 609-615 [6.3.1]

Green C.: see (Newell et al., 1971)

Green D.G. (1967): "A transistor instantaneous frequency meter." Medical and Biological Engineering **5**, 387-390 [6.1.1]

Green H.C.: see (Kopp and Green, 1946)

Greenwood E.: see (Fette et al., 1980)

Grigoryan A.A., Lazaryan L.M., Shatverov E.A. (1976): "Метод метранения слежения высшими гармониками в системах ФАПЧ при выделений основного тона речи." В кн.: Труды АРСО-9 (Минск). А.А.Григориан и др. Pitch determination from higher harmonics using the system "FAPCh" [a phase-locked loop system]. In Russian). **ARSO-9** [6.1.4]

Gruenz O.O., Schott L.O. (1949): "Extraction and portrayal of pitch of speech sounds." J. Acoust. Soc. Am. **21**, 487-495 [6.2.1; 9.4.3]

Grützmacher M. (1927): "Eine neue Methode der Klanganalyse." Elektr. Nachrichtentech. **4**, 533 [6.1]

Grützmacher M. (1965): "Demonstration eines Tonhöhenschreibers." Phonetica **13**, 3-17 [6.1.2; 6.6.1; 9.4.2]

Grützmacher M., Lottermoser W. (1937): "Über ein Verfahren zur trägheitsfreien Aufzeichnung von Melodiekurven." Akustische Z. **2**, 242-248 [4.1; 6.1.2; 6.6.1; 6.6.2]

Grützmacher M., Lottermoser W. (1938a): "Ein neuer Tonhöhenschreiber." In *Proceedings of the Third International Congress on Phonetic Sciences, Ghent 1938*. 105-109 [6.1.2; 6.6.1]

Grützmacher M., Lottermoser W. (1938b): "Die Verwendung des Tonhöhenschreibers bei mathematischen, phonetischen und musikalischen Aufgaben." Akustische Z. **3**, 183-196 [6.6.1; 9.4.2]

Grützmacher M., Lottermoser W. (1940): "Die Aufzeichnung kleiner Tonhöhenschwankungen." Akustische Z. **5**, 1-6 [9.4.2]

Gruszka-Koscielak H.: see (Jassem et al., 1973)

Gsell R.: see (Carré et al., 1963)

GTE Sylvania Inc. (1972): "Narrow band autocorrelation study" (Final report) (GTE Sylvania Inc., Needham MA; DAAB07-71-C-0207) [8.2.3]

Gubrynowicz R., Kacprowski J., Mikiel W., Zarnecki P. (1981): "Detection and evaluation of laryngeal pathology based on pitch period measurements in continuous speech." F.A.S.E.-**81**, 131-135 [4.1; 9.4.3]

Gubrynowicz R., Mikiel W., Zarnecki P. (1980a): "Acoustical analysis for evaluation of laryngeal dysfunction in the case of vocal chords paralysis." In *Speech analysis and synthesis*; ed. by W. Jassem (Inst. of Fundamental Research, Polish Academy of Sciences, Warsaw), Vol.5 [9.4.3]

Gubrynowicz R., Mikiel W., Zarnecki P. (1980b): "An acoustic method for the evaluation of the state of the larynx source in cases involving paathological changes in the vocal folds." Arch. Acoustics **5**, 3-30 [4.3; 9.4.3]

Gueguen C., Scharf L.L. (1980): "Exact maximum-likelihood identification of ARMA models: a signal processing perspective." EUSIPCO-**80**, 759-769 [6.3.1]

Guérin B. (1978a): "Contribution aux recherches sur la production de la parole. Etude du fonctionnement de la source vocale. Simulation d'un modele;" Thèse d'Etat, Univ. scientifique et médicale, Grenoble [3.1; 3.3]

Guérin B. (1978b): "Modelisation du larynx, étude des pertes dans la glotte: leur effet sur la bande passante des voyelles." **Phon. Grenoble 7**, 67-85 [3.1; 3.3; 3.4]

Guérin B., Boe L.J. (1977a): "A two-mass model of the vocal cords: determination of control parameters and their respective consequences." ICASSP-**77**, 583-586 [3.3]

Guérin B., Boe L.J. (1977b): "La régulation de la vibration des cordes vocales: simulation à l'aide d'un modèle à deux masses." **Journées d'Etude sur la Parole 8**, 37-42 [3.3]

Guérin B., Boe L.J. (1979): "Simulation du larynx par un modèle à deux masses. Influence du couplage acoustique avec le conduit vocal sue quelques caractéristiques intrinsèques du français." **Larynx et parole**, 103-144 [3.3]

Guérin B., Boe L.J. (1980): Etude d'influence du couplage acoustique source-conduit vocale sur F_0 des voyelles orales - conséquences pour l'étude des caractéristiques intrinsèques. Phonetica **37**, 161-196 [3.3; 9.4.2]

Guerin B., Boe L.J., Lancia R. (1979): "L'influence du couplage acoustique larynx - conduit vocal sur la fréquence fondamentale des voyelles: une simulation." **ICPhS-9** (1) (A), 264 [3.1; 3.3; 3.4]

Guérin B., Degryse D., Boe L.J. (1977): "Influences acoustiques des paramètres de commande d'un modèle de source couplé avec le conduit vocal." In *Modèles articulatoires et Phonétique - Articulatory Modeling and Phonetics.* Proceedings of a Symposium held at Grenoble July 10-12, 1977; ed. by R.Carré, R.Descout, M.Wajskop (G.A.L.F, F-22301 Lannion), 263-278 [1.2; 3.1; 3.4; 9.4.2]

Guérin B., Delos M., Mrayati M. (1976): "Comportement d'un modèle de la source vocale chargé par l'impedance d'entrée du conduit vocal." **Journées d'Etude sur la Parole 7**, 193-204 [3.1; 3.4]

Guérin B., Mrayati M., Carré R. (1976): "A voice source taking account of coupling with the supraglottal cavities." **ICASSP-76**, 47-50 [3.1; 3.4]

Guichard J.: see (Lafon et al., 1973)

Guisez D.: see (Vallancien et al., 1971)

Gulian E.: see (Bate et al., 1982)

Gulut Y.K.: see (Holden and Gulut, 1976)

Gupta V., Wilson T.A., Beavers G.S. (1973): "A model for vocal cord excitation." J. Acoust. Soc. Am. **54**, 1607-1617 [3.3]

Gustrin K.: see (Danielsson and Gustrin, 1961)

Guttman N., Flanagan J.L. (1964): "Pitch of high-pass filtered pulse trains." J. Acoust. Soc. Am. **36**, 757-765 [4.2.3]

Guttman N., Pruzansky S. (1962): "Lower limits of pitch and musical pitch." J. Speech Hear. Res. **5**, 207-214 [4.1; 4.2.1]

Haas C.: see (Sovak et al., 1971)

Haavel R. (1976a): "Studying melody by means of speech signal periodicity variation." **Phon. Tallinn 6**, 25-36 [6.1.3; 9.4.2; 9.4.4]

Haavel R. (1976b): "Temporal characteristics of the pitch contour I/II." Acustica **34**, 147-157 [9.4.2]

Habib M., Robinson D.M., Sincoskie W.D. (1978): "Real-zeros in pitch detection." **ICASSP-78**, 31-34

Hadding K., Studdert-Kennedy M. (1973): "Are you asking me, telling me, or asking to yourself?." **Speech Research 33**, 1-12 [9.4.2]

Hadding-Koch K. (1956): "Recent work on intonation." Studia Linguistica **10**, 77-96 [4.1; 9.4.2]

Hadding-Koch K. (1961): *Acoustico-phonetic studies in the intonation of southern Swedish.* Research Reports, Institute of Phonetics, University of Lund (Gleerup, Lund) [5.1.2; 9.4.2]

Hadding-Koch K., Studdert-Kennedy M. (1964): "An experimental study of some intonation contours." Phonetica **11**, 175-185 [5.1.2; 9.4.2]

Hadley G. (1964): *Nonlinear and dynamic programming* (Addison-Wesley, Reading MA) [6.6.3; 8.3.3]

Haggard N., Ambler S., Callow M. (1970): "Pitch as a voicing cue." J. Acoust. Soc. Am. **47**, 613-617 [5.3]

Hala B., Honty L. (1931): "La cinématographie des cordes vocales à l'aide du stroboscope et de la grande vitesse." Oto-Laryngol. Slavica **3** [5.2.1]

Halbedl G.: see (Seidner et al., 1972; Wendler et al., 1973)

Hales A.C., Boxhall T.R., Cullis A.D.S. (1956): "A summary of vocoder research over the period 1947 to 1955" (Post Office Eng. Dollis Hill, London; Res.Rept.#13961; DDC-AD-119258) [9.4.5]

Hall G.T. (1978): "Computer modeling of voice signals with adjustable pitch and formant frequencies" (Naval Postgraduate School, Monterey CA; DDC-AD-A067556/1GA) [9.4.4]

Hall J.L.: see (Schroeder et al., 1979)

Halle M. (1959): *The sound patterns of Russian* - A linguistic and acoustical investigation (Mouton & Co., s'Gravenhage) [9.4.2]

Halle M., Stevens K.N. (1967): "On the mechanism of glottal vibration for vowels and consonants." **MIT-RLE Progr. Rept. 85**, 267-270 [3.1; 3.3]

Halle M., Stevens K.N. (1971): "A note on laryngeal features." **MIT-RLE Progr. Rept.**, 198-213 [3.1; 9.4.2]

Halsey R.J., Swaffield J. (1948): "Analysis-synthesis telephony, with special reference to the vocoder." J. IEE (London), 391-412 [4.1; 6.1.1; 6.4.1; 9.3; 9.4.5]

Hamelin F.: see (André et al., 1976)

Hamlet S.L. (1971): "Location of scope discontinuities in glottal pulse shapes during vocal fry." J. Acoust. Soc. Am. **50** (C), 1561-1562 [3.2; 5.2.4]

Hamlet S.L. (1972): "Vocal fold articulatory activity during whispered sibilants." Arch. Otolaryng. **95**, 211-213 [3.5; 5.2.4]

Hamlet S.L. (1973): "Vocal compensation: an ultrasonic study of vocal-fold vibration in normal and nasal vowels." Cleft Palate J. **10**, 267-285 [5.2.4]

Hamlet S.L. (1978): "Evaluation of a rectangular ultrasonic transducer for laryngeal use." J. Acoust. Soc. Am. **64** (A), S91 (Paper II13; 96th Meet. ASA) [5.2.4]

Hamlet S.L. (1980): "Ultrasonic measurements of larynx height and vocal-fold vibratory patterns." J. Acoust. Soc. Am. **68**, 121-126 [5.2.4]

Hamlet S.L., Palmer J.M. (1974): "Investigation of laryngeal trills using the transmission of ultrasound through the larynx." Folia Phoniatr. **26**, 362-377 [5.2.4]

Hamlet S.L., Reid J.M. (1972): "Transmission of ultrasound through the larynx as a measure of determining vocal fold activity." IEEE Trans. BME-**19**, 34-37 [3.3; 5.2.4]

Hamm R.A. (1977): "Fast pitch detection" [paper presented at the 58th Convention, New York 1977] (Audio Eng. Society, New York; Preprint No.1265 E-4) [6.4.1]

Hammarberg B., Fritzell B., Gauffin J., Sundberg J., Wedin L. (1980): "Perceptual and acoustic correlates of abnormal voice qualities." Acta Otolaryngol. (Stockholm) **90**, 441-451 [3.3; 4.3; 9.4.3]

Hammarström G. (1975): *Linguistic units and items* (Springer, Berlin) [2.1]

Hanauer S.L.: see (Atal and Hanauer, 1971)

Handzel L., Jarosz A. (1967): "Über die Hörbarkeit der Sprachmelodie." **ICPhS-6**, 421-423 [4.2.1; 4.3]

Hanley T.D.: see (Dempsey et al., 1953)

Hanne J.R. (1965): "Formant Analysis" (Commun. Sci. Lab., Univ. of Michigan, Ann Arbor MI; Report No. 12) [3.4; 6.2.1]

Hanson R.J. (1975): "Fundamental frequency dynamics in VCV sequences." **ICPhS-8** (A) [3.1]

Hanson R.J. (1978): "A two-state model of F_0 control." J. Acoust. Soc. Am. **64** (A), S43-44 (Paper P14; 96th Meet. ASA) [3.3; 9.4.4]

Hanson R.J., Atal B.S., Kohut J. (1982): "Real-time determination of multiphase excitation for LPC speech synthesis." J. Acoust. Soc. Am. **72** (A), S80 (Paper RR12, 104th Meet. ASA) [9.4.4]

Hardcastle W.J. (1974): "Instrumental investigations of lingual activity during speech: A survey." Phonetica **29**, 129-157 [5.2]

Hardcastle W.J. (1976): *Physiology of speech production* (Academic Press, London) [3.1; 3.4]

Harden R.J. (1975): "Comparison of glottal area changes as measured from ultra-high speed photographs and photoelectric glottographs." J. Speech Hear. Res. **18**, 728-738 [5.2.1; 5.2.5; 5.2.6]

Harris C.M., Waite W.M. (1963): "Response of spectrum analyzers of the bank-of-filters type to signals generated by vowel sounds." J. Acoust. Soc. Am. **35**, 1972-1977 [8.5]

Harris C.M., Waite W.M. (1964): "Display of sound spectrographs in real time." J. Acoust. Soc. Am. **36**, 729 [5.1.2]

Harris C.M., Weiss M.R. (1963): "Pitch extraction by computer processing of high-resolution Fourier analysis data." J. Acoust. Soc. Am. **35**, 339-343 [8.5.3]

Harris C.M., Weiss M.R. (1964): "Effects of speaking condition on pitch." J. Acoust. Soc. Am. **36**, 933-936 [3.2; 9.4.2]

Harris F.J. (1978): "On the use of windows for harmonic analysis with the discrete Fourier transform." Proc. IEEE **66**, 51-83 [2.4]

Harris J.D. (1952a): "Remarks on the determination of a differential threshold by the so-called ABX technique." J. Acoust. Soc. Am. **24**, 417 [4.3]

Harris J.D. (1952b): "Pitch discrimination." J. Acoust. Soc. Am. **24**, 750-755 [4.3]

Harris J.D. (1960): "Scaling of pitch intervals." J. Acoust. Soc. Am. **32**, 1575-1581 [4.3]

Harris K.S.: see (Lieberman et al., 1970)

Harris M.S., Quinn A.M.S. (1979): "Difference limens for fundamental frequency contours in sentence level stimuli." J. Acoust. Soc. Am. **66** (A), S50 (Paper Y3; 98th Meet. ASA) [4.3]

Harrison D.: see (Culler et al., 1982)

Harshman R.: see (Gandour and Harshman, 1978)

Hart J. (1569): *An orthographie, conteyning the due order and reason, how to write or paint th'image of mannes voice, most like to the life of nature.* (Reprinted 1955 by Almqvist and Wiksell, Stockholm; ed. by B. Daniesson) (Serres, London) [5.1; 9.4.2]

Hart J. 't (1971): "A speaker grammar for Dutch intonation" (Institute for Perception Research, Eindhoven; Report No. 227) [9.4.2]

Hart J. 't (1973): "Intonation by rule: A perceptual quest." J. Phonetics **1**, 309-328 [4.2; 9.4.4]

Hart J. 't (1974): "Discriminability of the size of pitch movements in speech." **IPO Annual Progr. Rept. 9**, 56-63 [4.2.3]

Hart J. 't (1976): "Psychoacoustic backgrounds of pitch contour stylization." **IPO Annual Progr. Rept. 11**, 11-19 [9.4.2]

Hart J. 't (1977): "Vers une base psychophonetique de la stylisation intonative." **Journées d'Etude sur la Parole 8**, 167-174 [9.4.2]

Hart J. 't (1981): "Differential sensitivity to pitch distance, particularly in speech." J. Acoust. Soc. Am. **69**, 811-822 [4.2.3]

Hart J. 't, Collier R. (1975): "Integrating different levels of intonation analysis." J. Phonetics **3**, 235-255 [9.4.2]

Hart J. 't, Collier R. (1978): "A course in Dutch intonation." **IPO Annual Progr. Rept. 13**, 31-35 [9.4.2]

Hart P.E.: see (Duda and Hart, 1973)

Hartlieb K. (1967): "Stimm- und Sprachheilkunde aus biokybernetischer Sicht." Folia Phoniatr. **19**, 368-387 [3.1; 3.2; 9.4.3]

Hartwell W.T., Prezas D.P. (1981): "A pulse driving function generator for LPC synthesis of voiced segments of speech." IEEE Trans. ASSP-**29**, 1113-1116 [3.3; 9.4.5]

Harvey N., Howell P. (1980): "Isotonic vocalis contraction as a means of producing rapid decreases in F_0." J. Speech Hear. Res. **23**, 576-592 [3.1; 4.3]

Hasegawa A.: see (Stilwell et al., 1981)

Hashimoto S.: see (Sugimoto and Hashimoto, 1962)

Hasquin-Delaval J.: see (Lebrun and Hasquin-Delaval, 1971)

Hast M.H. (1975): "Experimental physiology of the larynx." In *Measurement procedures in speech, hearing, and language*; ed. by S.Singh (University Park Press, Baltimore), 345-362 [3.1]

Hast M.H., Holtzmark E.B. (1969): "The larynx, organ of voice by Julius Casserius. Translated by M.H.Hast and E.B.Holtzmark." Acta Otolaryngol. (Stockholm), 1-36 [3.1; 5.1.1]

Haton J.P. (ed.) (1982): *Automatic speech analysis and recognition.* Proceedings of the NATO Advanced Study Institute held at Bonas, France, June 29 - July 10, 1981 (Reidel, Dordrecht)

Haton M.C., Haton J.P. (1975): "Essai de caractérisation des voix d'enfants sourds par analyse polynomiale de la mélodie." **Journées d'Etude sur la Parole 5**, 317-324 [9.4.3]

Haton M.C., Haton J.P. (1977a): "Computer-aided speech analysis and reeducation for deaf children." **ICA-9** (Paper I23) [9.4.3]

Haton M.C., Haton J.P. (1977b): "Analyse et reeducation des parametres prosodiques chez l'enfant sourd." **Journées d'Etude sur la Parole 8**, 363-370 [9.4.3]

Haton M.C., Haton J.P. (1978): "Une aide à la communication pour les sourds: le système SIRENE." **Bull. Audiophonologie 8**, 85-92 [9.4.3]

Haton M.C., Haton J.P. (1979): "Sirene, a system for speech training of deaf people." **ICASSP-79**, 482-485 [9.4.3]

Hatori K.: see (Miyanaga et al., 1982)

Have B.L. ten (1978): "Some problems concerning the measurement of intonation contours." **Phon. Nijmegen 2**, 29-33 [9.4.2]

Hayasagi K.: see (Kaneko et al., 1974)

Hayashibe H.: see (Uyeno et al., 1979)

Hayden E., Koike Y. (1972): "A data processing scheme for frame-by-frame film analysis." Folia Phoniatr. **24**, 169-186 [5.2.1]

Hayes P., Ball E., Reddy D.R. (1981): "Breaking the man-machine communication barrier." IEEE Computer **18**, 19-30 [5.1; 9.2]

Healy M.J.R.: see (Bogert et al., 1963)

Hedelin P. (1981): "A tone-oriented voice-excited vocoder." **ICASSP-81**, 205-208 [9.4.5]

Hedelin P., Hult G. (1981): "QD - an algorithm for nonlinear inverse filtering." **ICASSP-81**, 366-369 [6.3.1; 6.3.2]

Hegedüs L. (1958): "Tonhöhen- und Aktionsstromuntersuchungen bei der Bildung stimmhafter Konsonanten im Ungarischen." Z. Phonetik **11**, 1-22 [9.4.2]

Heike G. (1977): "Computer generations of a parameter controlled articulatory model as an aid for the deaf." **ICA-9,** 423 (Paper I17) [9.4.3]

Heinz J.M.: see (Bell et al., 1961)

Helfrich H.: see (Tolkmitt et al., 1981)

Hellwarth G.A. (1962): "An automatic speech formant tracking filter" (Commun. Sci. Lab., Univ. of Michigan, Ann Arbor MI; DDC-AD-282147)

Hellwarth G.A., Jones G.D. (1968): "Automatic conditioning of speech signals." IEEE Trans. AU-**16,** 169-179

Helmholtz H.J. (1870): *Die Lehre von den Tonempfindungen als physiologische Grundlage für die Theorie der Musik* (Vieweg, Braunschweig) [4.2; 9.4.2]

Helms H.D.: see (Rabiner et al., 1972)

Helstrom C.W. (1960): *Statistical theory of signal detection* (Pergamon Press, New York NY) [2.0; 8.6]

Henderson J.A. (1977): "The larynx and language: a missing dimension?" Phonetica **34,** 256-263 [3.1; 9.4.2]

Henke W.L. (1974a): "Signals from external accelerometers during phonation: Attributes and their internal physical correlates" (MIT Res.Lab.Electronics, Cambridge MA; DDC-AD-A007503/6GA) [5.2.2]

Henke W.L. (1974b): "Display and interpretation of glottal activity as transduced by external accelerometers." J. Acoust. Soc. Am. **55** (A) [5.2.2]

Henke W.L. (1974c): "Signals from external accelerometers during phonation: attributes and their internal physical correlates." **MIT-RLE Progr. Rept. 114,** 224-231 [5.2.2]

Hensen (1887): "Über die Schrift von Schallbewegungen." Z. Biologie **23,** 291 [5.1.1]

Hequet B.: see (Lhote et al., 1969)

Herbst L. (1967): "Über Faktoren, welche die Perzeption der individuellen Hauptsprechtonhöhe fördern oder hemmen." **ICPhS-6,** 439-443 [4.1; 9.4.2]

Hermann (1889): "Phonophotographische Untersuchungen." Pflügers Archiv für die gesamte Physiologie **45,** 582 [5.1.1]

Hertz C., Lindström B., Sonesson B. (1970): "Ultrasonic recording of the vibrating vocal folds." Acta Otolaryngol. (Stockholm) **69,** 223-230 [5.2.4]

Hess W.J. (1972a): "Pitch synchronous, non-harmonic analysis of speech signals." **ICSCP-72,** 66-69 (Paper B5) [9.4.5]

Hess W.J. (1972b): "Digitale grundfrequenzsynchrone Analyse von Sprachsignalen als Teil eines automatischen Spracherkennungssystems;" Dissertation, TU München [5.3.2; 7.1]

Hess W.J. (1973): "Bestimmung der Grundfrequenz von Sprachsignalen im Zeitbereich mit Hilfe eines digitalen Filters." In *Signalverarbeitung (Signal Processing).* Vorträge der gleichnamigen Fachtagung, Erlangen 1973; ed. by H.W.Schüssler (Universität Erlangen-Nürnberg), 386-393 [7.1]

Hess W.J. (1974a): "On-line, digital pitch period extractor for speech signals." In *Proceedings, 1974 International Zurich Seminar on Digital Communications. Zurich, Switzerland, 12-15 March 1974* (Inst.f.Fernmeldetechnik, ETH, Zürich) [7.1]

Hess W.J. (1974b): "A pitch-synchronous digital feature extraction system for phonemic recognition of speech." **IEEE Symp. Speech Recogn.,** 112-123 (Paper T8) [6.3.1; 7.1; 9.4.4; 9.4.5]

Hess W.J. (1974c): "On-line digital pitch period extractor for speech signals." In *Proceedings of the Summer School on Circuit Theory, Prague, 2-6 September 1974;* Vol.2 (Short Contributions) (Inst.of Radio Engineering, Czechoslovak Academy of Sciences, Prague), 413-419 [7.1]

Hess W.J. (1974d): "A pitch-synchronous digital feature extraction system for phonemic recognition of speech." **Proc. 1974 Speech Commun. Sem. 3,** 201-213 [6.3.1; 7.1; 9.4.4; 9.4.5]

Hess W.J. (1974e): "Time-domain pitch extraction of speech signals using a digital filter." J. Acoust. Soc. Am. **55** (A), S21 (Paper J8; 87th Meet. ASA) [7.1]

Hess W.J. (1976a): "A pitch-synchronous digital feature extraction system for phonemic recognition of speech." IEEE Trans. ASSP-**24,** 14-25 [6.3.1; 7.1; 9.4.4; 9.4.5]

Hess W.J. (1976b): "An algorithm for digital time-domain pitch period determination of speech signals and its application to detect F_0 dynamics in VCV utterances." **ICASSP-76,** 322-325 [6.1.2; 7.5]

Hess W.J. (1977): "Détermination de la période fondamentale du signal vocal avec un filtre digital." **Journées d'Etude sur la Parole 8,** 79-86 [7.1; 7.5]

Hess W.J. (1979a): "Three-channel time-domain pitch detector using different nonlinear digital filters." **ASA*50,** 237-240 [7.2; 7.5]

Hess W.J. (1979b): "Pitch determination of speech signals by nonlinear digital filtering." **ICPhS-9** (1) (A), 267 [7.5]

Hess W.J. (1979c): "Pitch determination of speech signals - a survey." In *Hearing Mechanisms and Speech.* EBBS-Workshop Göttingen, April 26-28, 1979; ed. by O.Creutzfeld et al. Experimental Brain Research Suppl. 2 (Springer, Berlin), 260-267 [2.6]

Hess W.J. (1979d): "Time-domain pitch period extraction of speech signals using three nonlinear digital filters." **ICASSP-79**, 773-776 [4.1; 7.5]

Hess W.J. (1979e): "Three-channel time-domain pitch detector using different nonlinear digital filters." J. Acoust. Soc. Am. **65** (A), S67 (Paper Y5; 97th Meet. ASA) [7.5]

Hess W.J. (1980a): "Pitch determination of speech signals - a survey." In *Spoken language generation and understanding.* Proceedings of the NATO Advanced Study Institute held at Bonas, France, June 23 - July 5, 1979; ed. by J.C. Simon (Reidel, Dordrecht) [2.6]

Hess W.J. (1980b): "Pitch determination - an example for the application of signal processing methods in the speech domain." **EUSIPCO-80**, 625-634

Hess W.J. (1980c): "Grundfrequenzbestimmung von Sprachsignalen im Zeitbereich mit Hilfe eines nichtlinearen digitalen Filters." Frequenz **34**, 152-156 [7.5]

Hess W.J. (1981): "Algorithmes et méthodes pour la détermination du fondamental." **Journées d'Etude sur la Parole** 12, 202-220 [2.6; 9.1.1]

Hess W.J. (1982a): "Algorithms and devices for pitch determination of speech signals." In *Automatic speech analysis and recognition.* Proceedings of the NATO Advanced Study Institute held at Bonas, France, June 29 - July 10, 1981; ed. by J.P. Haton (Reidel, Dordrecht), 49-69

Hess W.J. (1982b): "Algorithms and devices for pitch determination of speech signals." Phonetica **39**, 219-240 [1.2; 9.1.1]

Hettwer G., Fellbaum K. (1981): "Ein modifiziertes Sprachgrundfrequenzanalyseverfahren für lineare Prädiktionsvocoder." In *Fortschritte der Akustik, DAGA'81 Berlin* (VDE-Verlag, Berlin) [8.3.1]

Hettwer G., Richter W., Fellbaum K. (1980): "Gesichtspunkte zur Auswahl sowie Realisierung eines Sprachgrundfrequenzanalyseverfahrens." In *Fortschritte der Akustik, DAGA'80 München* (VDE-Verlag, Berlin), 663-666 [8.3.1; 9.3; 9.4.4]

Heute U., Vary P. (1981): "A digital filter bank with polyphase network and FFT hardware: Measurements and applications." Signal Processing **3**, 307-319 [2.3; 2.4; 8.1; 8.5.4; 8.6.2]

Hewes C.R., Brodersen R.W., Buss D.D. (1979): "Applications of CCD and switched capacitor filter technology." Proc. IEEE **67**, 1403-1415 [2.3]

Hewitt T.L.: see (Wood and Hewitt, 1963)

Hickman C.N. (1934): "An acoustic spectrometer." J. Acoust. Soc. Am. **6**, 108-111 [5.1.2]

Hicks J.W.jr.: see (Hollien and Hicks, 1976)

Hieronymus J.L. (1974): "Pitch synchronous acoustic segmentation." **IEEE Symp. Speech Recogn.**, 131-133 [9.4.5]

Hieronymus J.L. (1975): "Vocal tract excitation during voicing and its relation to optimal short-time frequency analysis of speech." J. Acoust. Soc. Am. **58** (A), S97 (Paper V V7; 90th Meet. ASA) [3.3; 6.5.2; 9.4.4]

Higgins A.L., Viswanathan V.R., Russell W.H. (1979): "New high-frequency regeneration (HFR) techniques for voice-excited speech coders." J. Acoust. Soc. Am. **66** (A), S22-23 (Paper K1; 98th Meet. ASA) [9.4.5]

Hiki Sh., Imaizumi S., Hirano M., Matsushita H., Kakita Y. (1976): "Acoustical analysis for voice disorders." **ICASSP-76**, 613-616 [9.4.3]

Hill D.R., Reid N.A. (1977): "An experiment on the perception of intonational features." Int. J. Man-Machine Studies **9**, 337-348 [4.2; 4.3; 9.4.2]

Hill L.A. (1965): *Stress and intonation step by step* (University Press, Oxford UK) [9.4.2]

Hill R.: see (Yost et al., 1978)

Hill T. (1966): "The technique of prosodic analysis." In *In Memory of J.R.Firth;* ed. by C.E.Basell et al. (Longmans, London), 198-225 [9.4.2]

Hill W.A.: see (Saunders et al., 1976)

Hiller S.: see (Laver et al., 1982)

Hinds P.: see (Bate et al., 1982)

Hinson J.K.: see (Coleman et al., 1977)

Hiramatsu K. (1960): "A new pitch extractor." J. Acoust. Soc. Am. **32** (A), 279-281 [8.2.3]

Hirano M. (1972): "Regulatory mechanism of voice in singing." **ICSCP-72**, 416-419 (Paper K8) [3.1; 3.2]

Hirano M. (1974): "Morphological structure of the vocal cord as a vibrator and its vibrations." Folia Phoniatr. **26**, 89-94 [3.1]

Hirano M. (1976): "Structure and vibratory behavior of the vocal folds." In *Dynamic aspects of speech production;* ed. by M. Sawashima and F.S. Cooper (Univ. of Tokyo Press), 13-27 [3.1]

Hirano M. (1978): "Physiological aspects of the glottal sound generator." J. Acoust. Soc. Am. **64** (A), S41 (Paper P3; 96th Meet. ASA) [3.1]

Hirano M. (1981a): "The laryngeal examination." In *Speech evaluation in medicine;* ed. by J.K. Darby (Grune and Stratton, New York) [3.1; 5.2.1]

Hirano M. (1981b): *Clinical examination of voice* (Springer, Berlin) [3.1; 9.4.3]

Hirano M., Mihashi S., Kasuya T., Kurita S. (1979): "Structure of the vocal fold as a sound generator." **ICPhS-9** (1) (A), 189 [3.1]

Hirano M., Ohala J., Vennard W. (1969): "The function of laryngeal muscles in regulating fundamental frequency and intensity of phonation." J. Speech Hear. Res. **12**, 616-628 [3.1]

Hirano M., Vennard W., Ohala J. (1970): "Regulation of registers, pitch, and intensity of voice. An electromyographic investigation of intrinsic laryngeal muscles." Folia Phoniatr. **22**, 1-20 [3.1; 5.2.1]

Hirdes F.P. (1975): "Toonhoogtebepaling mit natuurlijke spraak ten behoeve va een sprekerverificatie system (Pitch extraction of natural speech for a speaker verification system; in Dutch)" (Afstudeerverlag Werkgroep Signaalverwerking, Delft) [2.6]

Hirose H. (1977): "Laryngeal adjustments in consonant production." Phonetica **34**, 289-294 [3.1; 3.3; 3.4]

Hirose H., Gay T. (1971): "The activity of the intrinsic laryngeal muscles in voicing control: an electromyographic study." **Speech Research 28**, 115-143 [3.1; 5.2.1]

Hirschberg J., Szepe Gy., Vass-Kovács E. (eds.) (1971): *Papers in interdisciplinary speech research.* Proceedings of the Speech Symposium Szeged 1971 (Akademiai Kiadó, Budapest)

Hirsh I.J.: see (Nabĕlek and Hirsh, 1969)

Hirst D.J., Di Cristo A., Nishinuma Y. (1979): "Etude comparative de la F_0 spécifique des voyelles." **ICPhS-9** (1) (A), 381 [3.1; 9.4.2]

Hirt C.C.: see (Rubin and Hirt, 1960)

Hixon T.J., Klatt D.H., Mead J. (1971): "Influence of forced transglottal pressure changes on vocal fundamental frequency." J. Acoust. Soc. Am. **49**, 105 [3.1; 3.2]

Hlavac S. (1972): "Ein Tonhöhenmesser nach glottographischem Prinzip." **Phonetica Pragensia III**, 111-116 [5.2.3]

Höffe W.L. (1960): "Über Beziehungen von Sprachmelodie und Lautstärke." Phonetica **5**, 129-159 [4.1]

Hoff M.E.jr. (1980): "IC technology: Trends and impact on digital signal processing." **ICASSP-80**, 1-6 [9.1.5]

Hoffenreich F.: see (Christiansen et al., 1966)

Hoffmann K., Auth W., Erb H.J., Kühlwetter J., Lacroix A., Reiss R. (1974): "Ein adaptives Sprachübertragungssystem." Nachrichtentech. Z. **27**, 260-265 [6.1.1; 9.4.5]

Holden A.D.C., Gulut Y.K. (1976): "A new method for accurate analysis of voiced speech." **ICASSP-76**, 458-461

Holler J.R. (1951): "Cycle by cycle recorder of fundamental pitch;" Thesis, Purdue Univ., West Lafayette IN [6.1.1]

Hollien H. (1960a): "Some laryngeal correlates of vocal pitch." J. Speech Hear. Res. **3**, 52-58 [3.1]

Hollien H. (1960b): "Vocal pitch variation related to changes in vocal fold length." J. Speech Hear. Res. **3**, 150-156 [3.1; 3.3]

Hollien H. (1962): "Vocal fold thickness and fundamental frequency of phonation." J. Speech Hear. Res. **5**, 237-243 [3.1; 3.3]

Hollien H. (1963): "Fundamental frequency indicator." Am. Speech Hear. Assoc (Paper M10; 39th Meet., ASHA) [6.4.1]

Hollien H. (1964): "Stroboscopic laminagraphy of the vocal folds." In *Proceedings of the Fifth International Congress on Phonetic Sciences, Münster 1964* (Karger, Basel), 362-364 [5.2.1]

Hollien H. (1972): "Three major vocal registers: a proposal." **ICPhS-7**, 320-331 [3.1; 4.1]

Hollien H. (1974): "On vocal registers." J. Phonetics **2**, 125-143 [3.2]

Hollien H. (1981): "Analog instrumentation for acoustic speech analysis." In *Speech evaluation in psychiatry;* ed. by J.K. Darby (Grune and Stratton, New York)

Hollien H., Brown W.S., Hollien K. (1971): "Vocal fold length associates with modal, folsetto and varying vocal intensity phonations." Folia Phoniatr. **23**, 66-78 [3.1; 3.2]

Hollien H., Curtis J.F. (1960): "A laminographic study of vocal pitch." J. Speech Hear. Res. **3**, 361-371 [5.2.1]

Hollien H., Curtis J.F. (1962): "Elevation and titting of vocal folds as a function of vocal pitch." Folia Phoniatr. **14**, 23-36 [3.1]

Hollien H., Dew D., Philips P. (1971): "Phonational frequency ranges of adults." J. Speech Hear. Res. **14**, 755-760 [4.1]

Hollien H., Girard G.T., Coleman R.F. (1977): "Vocal fold vibration patterns of pulse register phonation." Folia Phoniatr. **29**, 200-205 [3.2; 3.3]

Hollien H., Hicks J.W.jr. (1976): "Voice fundamental frequency in saturation diving." J. Acoust. Soc. Am. **60** (A), S46 (Paper T17; 92nd Meet.ASA) [3.1; 9.4.3]

Hollien H., Hollien P., Majewski W. (1974): "Analysis of fundamental frequency as a speaker identification technique." Conv. Assoc. Phonetic Sci. [9.4.4]

Hollien H., Jackson B. (1973): "Normative data on the speaking fundamental frequency characteristics of young and adult males." J. Phonetics **1**, 117-120 [4.1]

Hollien H., Michel J.F. (1968): "Vocal fry as a phonational register." J. Speech Hear. Res. **11**, 600-604 [3.2]

Hollien H., Michel J.F., Doherty E.T. (1973): "A method for analyzing vocal jitter in sustained phonation." J. Phonetics **1**, 85-91 [4.3]

Hollien H., Moore G.P. (1960): "Measurements of the vocal folds during changes in pitch." J. Speech Hear. Res. **3**, 157-165 [3.1; 4.3]

Hollien H., Moore G.P., Wendahl R.W., Michel J.F. (1966): "On the nature of vocal fry." J. Speech Hear. Res. **9**, 245-247 [3.2; 4.3]

Hollien H., Shipp T. (1972): "Speaking fundamental frequency and chronologic age in males." J. Speech Hear. Res. **15**, 155-159 [4.1]

Hollien H., Wendahl R.W. (1968): "Perceptual study of vocal fry." J. Acoust. Soc. Am. **43**, 506-509 [3.2; 4.2; 4.3]

Hollien P.A., Brown W.S.jr., Hollien H. (1978): "Speaking fundamental frequency (SFF) characteristics of children." J. Acoust. Soc. Am. **64** (A), S90 (Paper II9; 96th Meet. ASA) [4.1]

Holmer N.G., Kitzing P., Lindström K. (1973): "Echo-glottography, ultrasonic recording of vocal fold vibrations in preparation of human larynges." Acta Otolaryngol. (Stockholm) **75**, 454-463 [5.2.4]

Holmer N.G., Rundqvist H.E. (1975): "Ultrasonic registration of the fundamental frequency of a voice during normal speech." J. Acoust. Soc. Am. **58**, 1073-1077 [5.2.4]

Holmes J.N. (1962a): "The effect of simulating natural larynx behaviour on the quality of synthetic speech." **Proc. 1962 Speech Commun. Sem.** (Paper F6) [3.3; 9.4.4]

Holmes J.N. (1962b): "An investigation of the volume velocity waveform at the larynx during speech by means of an inverse filter." **Proc. 1962 Speech Commun. Sem.** (Paper B4) [3.3; 6.5.1]

Holmes J.N. (1962c): "An investigation of the volume velocity waveform at the larynx during speech by means of an inverse filter." ICA-4 (Paper G13) [3.3; 6.5.1]

Holmes J.N. (1973): "The influence of glottal waveform on the naturalness of speech from a parallel formant synthesizer." IEEE Trans. AU-**21**, 298-305 [3.3; 4.2.3; 9.4.4]

Holmes J.N. (1975): "Low-frequency phase distortion of speech recordings." J. Acoust. Soc. Am. **58**, 747-749 [6.2.1; 9.2.1]

Holmes J.N. (1976): "Formant excitation before and after glottal closure." ICASSP-76, 39-42 [3.3; 3.4; 4.2.3; 6.5.2; 9.4.4]

Holmes J.N. (1977): "The acoustic consequences of vocal-cord action." Phonetica **34**, 316-317 [3.3]

Holmes J.N., Judd M.W., Walesby D.H. (1978): "A high-quality all digital sound spectrograph developed for speech analysis." ICASSP-78, 43-46 [5.1.2]

Holmgren G.L. (1966): "Speaker recognition, speech characteristics, speech evaluation, and modification of speech signal - A selected bibliography." IEEE Trans. AU-**14**, 32-39 [1.3]

Holtzmark E.B.: see (Hast and Holtzmark, 1969)

Hombert J.M. (1976): "Difficulty of producing different F_0 in speech." J. Acoust. Soc. Am. **60** (A), S44-45 (Paper T9; 92nd Meet. ASA) [3.1; 9.4.2]

Hombert J.M. (1977): "Difficulty of producing different F0 in speech." **Phon. Los Angeles 34**, 12-19 [3.1; 9.4.2]

Honty L.: see (Hala and Honty, 1931)

Hoptner N.: see (Lacroix and Hoptner, 1977)

Horii Y. (1973): "Skewness characteristics of vocal period and fundamental frequency distributions." **Speech Research 34**, 57-61 [4.1; 4.3]

Horii Y. (1975): "Some statistical characteristics of voice fundamental frequency." J. Speech Hear. Res. **18**, 192-201 [4.1]

Horii Y. (1979a): "Fundamental frequency perturbation observed in sustained phonation." J. Speech Hear. Res. **22**, 5-19 [4.3; 9.4.3]

Horii Y. (1979b): "Jitter and shimmer as physical correlates of roughness in sustained phonation reexamination." J. Acoust. Soc. Am. **66** (A), S65 (Paper EE12; 98th Meet. ASA) [4.3]

Horii Y. (1980): "Vocal shimmer in sustained phonation." J. Speech Hear. Res. **23**, 202-209 [4.3]

Horvitch G.M. (1951): "Means for indicating sound, pitch or voice inflection." J. Acoust. Soc. Am. **23** (A), 258

Houde R.A.: see (Larkin et al., 1975; Steward et al., 1976)

House A.S. (1959): "Note on optimal vocal frequency." J. Speech Hear. Res. **2**, 55-60 [4.1]

Houtsma A.J.M., Goldstein J.L. (1972): "The central origin of the pitch of complex tomes: evidence from musical interval recognition." J. Acoust. Soc. Am. **51**, 520-529 [4.2.1]

Howard C.R. (1964): "Real-time measurement of pitch perturbations." J. Acoust. Soc. Am. **36**, 1048 (Paper X7; 67th Meet. ASA) [4.3]

Howard C.R. (1965): "Pitch perturbation detection." IEEE Trans. AU-**13**, 9-14 [4.3]

Howard J.A.: see (Carey and Howard, 1972)

Howarth R.J.: see (Tucker et al., 1977)

Howell P.: see (Harvey and Howell, 1980)

Howie J.M. (1974): "On the domain of tone in Mandarin." Phonetica **30**, 129-148 [9.4.2]

Hudziak F.: see (McLaughlin et al., 1980)

Huggins A.W.F. (1982): "Is periodicity detection central?." J. Acoust. Soc. Am. **71**, 963-967 [4.2]

Huggins W.H. (1954): "A note on autocorrelation analysis of speech sounds." J. Acoust. Soc. Am. **26**, 790-792 [8.2.1]

Hughes G.W., House A.S., Li K.P. (1968): "Research on word spotting" (Purdue Research Foundation, West Lafayette IN; AFCRL-69-0240) [9.4.4]

Hull A.W. (1929): "Hot-cathode thyratrons - part II: operation." Gen. Electr. Review **32**, 390-399 [6.1.1]

Hult G.: see (Hedelin and Hult, 1981)

Hunt F.V. (1935): "A direct-reading frequency meter suitable for high speed recording." Review Scientific Instrum. **6**, 43-46 [6.1.2]

Hunt M.J., Bridle J.S., Holmes J.N. (1978): "Interactive digital inverse filtering and its relation to linear prediction methods." **ICASSP-78**, 15-19 [6.3.1; 6.5.1]

Husson R. (1959): "Der gegenwärtige Stand der physiologischen Phonetik." Phonetica **4**, 1-32 [3.1]

Husson R. (1960a): *La voix chantée* (Gauthier-Villars, Paris) [3.1; 3.2; 5.2.3]

Husson R. (1960b): "Le fonctionnement du larynx comparativement dans la parole et dans le chant." Acta Linguistica **10**, 19-54 [3.1; 3.2; 9.4.2]

Husson R. (1962): *Physiologie de la phonation* (Masson, Paris) [3.1; 3.2; 5.2.3]

Husson R. (1967): "Sur la physiologie de la phonation et du langage oral." **Ann. Bull. RILP 1**, 20-30 [3.1]

Huttar G.L. (1968): "Relations between prosodic variables and emotions in normal American English utterances." J. Speech Hear. Res. **11**, 481-487 [9.4.2; 9.4.3]

Hutters B. (1976): "Problems in the use of the photo-electric glottograph." **ARIPUC 10**, 275-302 [5.2.5; 5.2.6]

Hyde S.R. (1972): "Automatic speech recognition. A critical survey and discussion of the literature." In *Human communication. A unified view;* ed. by E.E. David jr. and P.B.Denes (McGraw-Hill, New York), 339-438 [3.1; 9.4.4]

ICA-10 --> *Proceedings of the Tenth International Congress on Acoustics, Sydney, Australia, 9-16 July 1980* (Australian Acoustical Society, Sydney 1980)

ICA-3 --> *Proceedings of the Third International Congress on Acoustics, Stuttgart 1959;* ed. by L.Cremer (Elsevier, Amsterdam 1962)

ICA-4 --> *Congress report, 4th International Congress on Acoustics, Copenhagen 1962;* ed. by A.K. Nielsen (Harlang and Toksvig, Kopenhagen 1962; 2 volumes)

ICA-5 --> *Proceedings of the 5th International Congress on Acoustics, Liège 1965*; ed. by D.E.Commins (Imprimerie G.Thoné, Liège 1966)

ICA-6 --> *Proceedings of the 6th International Congress on Acoustics, Tokyo 1968* (American Elsevier)

ICA-7 --> *Proceedings of the 7th International Congress on Acoustics, Budapest 1971* (Akademiai Kiadó, Budapest 1971)

ICA-8 --> *Proceedings of the 8th International Congress on Acoustics, London 1974.* Vol.1 - Contributed Papers; vol.2 - Invited Papers; ed. by the Inst. of Acoustics (Goldcrest Press, Trowbridge Wiltsh. 1974)

ICA-9 --> *Proceedings of the 9th International Congress on Acoustics, Madrid 1977* (Sociedad Espanola de Acustica, Madrid 1977)

ICASSP-76 --> *Proceedings, 1976 IEEE International Conference on Acoustics, Speech, and Signal Processing, Philadelphia PA, April 12-14, 1976* (Inst. of Elec. and Electronics Engineers, New York 1976)

ICASSP-77 --> *Proceedings, 1977 IEEE International Conference on Acoustics, Speech, and Signal Processing. Hartford CT, May 9-11, 1977* (Inst. of Elec. and Electronics Engineers, New York 1977; IEEE Catalog #77CH1197-3 ASSP)

ICASSP-78 --> *Proceedings, 1978 IEEE International Conference on Acoustics, Speech, and Signal Processing, Tulsa OK, April 10-12, 1978* (Inst. of Elec. and Electronics Engineers, New York 1978; IEEE Catalog #78CH1285-6 ASSP)

ICASSP-79 --> *Proceedings, 1979 IEEE International Conference on Acoustics, Speech, and Signal Processing, Washington DC, April 2-4, 1979* (Institute of Electrical and Electronics Engineers, New York, IEEE Catalog #79CH1379-7 ASSP)

ICASSP-80 --> *Proceedings, 1980 IEEE International Conference on Acoustics, Speech, and Signal Processing, Denver CO, April 9-11, 1980* (Inst. of Elec. and Electronics Engineers, New York 1980)

ICASSP-81 --> *Proceedings, 1981 IEEE International Conference on Acoustics, Speech, and Signal Processing, Atlanta GA, April 14-16, 1981* (Inst. of Elec. and Electronics Engineers, New York 1981)

ICASSP-82 --> *Proceedings, 1982 IEEE International Conference on Acoustics, Speech, and Signal Processing, Paris, May 2-5, 1982* (Inst. of Elec. and Electronics Engineers, New York 1982; IEEE Catalog #82CH1746-7 ASSP)

ICASSP-83 --> *Proceedings, 1983 IEEE International Conference on Acoustics, Speech, and Signal Processing, Boston, MA, April 14-16, 1983* (Inst. of Elec. and Electronics Engineers, New York 1983)

Ichikawa T.: see (Ono et al., 1978)

ICPh-Miami --> *Current issues in the phonetic sciences.* Proc. of the Intern.Conf. on Phonetics, Miami FL, December 1977; ed. by H. and P.Hollien (John Benjamins B.V, Amsterdam 1979)

ICPhS-4 --> *Proceedings of the 4th International Congress on Phonetic Sciences, Helsinki 1961* (Mouton, Den Haag 1962)

ICPhS-6 --> *Proceedings of the 6th International Congress on Phonetic Sciences, Prague, 7.-13.Sept.1967*; ed. by B. Hala et al. (Hueber, München 1967)

ICPhS-7 --> *Actes du 7ème Congrès International des Sciences Phonétiques, Montréal 1971*; ed. by A. Rigault et M. Charbonneau (Mouton, Den Haag 1972)

ICPhS-8 --> *Abstracts of papers, 8th International Congress of Phonetic Sciences, Leeds, August 1975* (Dept. of Phonetics, University of Leeds, Leeds 1975)

ICPhS-9 --> *Proceedings of the Ninth International Congress of Phonetic Sciences, Copenhagen, August 6-11, 1979*; ed. by E. Fischer-Jörgensen (University, Institute of Phonetics, Copenhagen 1979)

ICSCP-72 --> *Conference Record, 1972 International Conference on Speech Communication and Processing, Newton MA, April 24 to 26, 1972*; ed. by IEEE/AFCRL (AFCRL, Bedford MA 1972; IEEE Cat. 72CHO596-7AE)

IEEE Symp. Speech Recogn. --> *IEEE Symposium on Speech Recognition, Carnegie-Mellon University, Pittsburgh PA, April 1974: Contributed papers*; ed. by L.Erman (Inst.of Elec. and Electronics Eng. (IEEE), New York, Pittsburgh 1974; IEEE Cat.No. 74CH0878-9AE)

Ideno K.: see (Kumagai and Ideno, 1960)

Idler H.: see (Timme et al., 1973)

Iles M. (1972): "Speaker identification as a function of fundamental frequency and resonant frequencies;" Dissertation, Univ. of Florida, Gainesville FL [9.4.4]

Imagawa H., Kiritani S., Saito S. (1980): "Random perturbations on the pitch pattern and the naturalness of synthetic speech." **Ann. Bull. RILP 14**, 201-208 [4.3; 9.4.4]

Imagawa H., Kiritani S., Saito S. (1982): Study on the pitch perception of sentences, based on the F_0 contour generation model. **Ann. Bull. RILP 16**, 63-72 [3.3; 4.3; 9.4.2]

Imai K.: see (Uyeno et al., 1979)

Imai S.: see (Nguyen and Imai, 1977)

Imaizumi S.: see (Hiki et al., 1976)

Indefrey H., Hess W.J. (1983): "Vergleichende Untersuchungen von Grundfrequenzbestimmungsalgorithmen mit Hilfe des Ausgangssignals eines Laryngographen." In 6. *DFG-Kolloquium Signalverarbeitung, Göttingen 1983;* ed. by V. Neuhoff and H.D.Zimmer (Göttingen), 59-62 [9.2.3]

Ingebretsen R.B.: see (Stockham et al., 1975)

Inomata S. (1960): "A new method of pitch extraction using a digital computer." J. Acoust. Soc. Jpn. **16**, 283-285 [8.2.3]

Inomata S. (1963): "Speech recognition and generation by a digital computer" (Electrotechnical Laboratory, Tokyo; Report #645) [8.2.3]

Institut de Phonétique de Grenoble (ed.) (1979): *Recherche sur la prosodie du français* (Univ. de Langues et Lettres, Grenoble) [3.1; 9.4.2]

IPO Annual Progr. Rept. --> *IPO Annual Progress Report;* ed. by The Institute for Perception Research (IPO, Eindhoven, The Netherlands)

Irons R.E.: see (Coulter et al., 1961)

Irwin M.J. (1979): "Periodicity estimation in the presence of noise." In *Proceedings of the 1979 Autumn Conference* (Institute of Acoustics, Edinburgh), 55-58 [8.6.1]

Isacenko A.V., Schädlich H.J. (1966): "Untersuchungen über die deutsche Satzintonation." Studia Grammatica **7**, 7-67 [9.4.2]

Isacenko A.V., Schädlich H.J. (1970): *A model of standard German intonation* (Mouton, Den Haag) [4.2.3; 9.4.2]

Ishida H.: see (Miyazaki et al., 1973)

Ishigami H.: see (Kurematsu et al., 1979)

Ishizaka K., Flanagan J.L. (1972): "Synthesis of voiced sounds from a two-mass model of the vocal cords." Bell Syst. Tech. J. **51**, 1233-1268 [3.1; 3.3; 9.4.4]

Ishizaka K., Isshiki N. (1975): "Computer simulation of pathologic vocal-cord vibration." J. Acoust. Soc. Am. **58** (A), S111 (Paper BBB2; 90th Meet. ASA) [3.3; 9.4.3]

Ishizaka K., Isshiki N. (1976): "Computer simulation of pathological vocal-cord vibration." J. Acoust. Soc. Am. **60**, 1193-1198 [3.3; 9.4.3]

Ishizaka K., Matsudaira M. (1968a): "Fluid mechanical considerations of vocal cord vibrations" (Speech Communication Research Labs, Santa Barbara CA; SCRL Monograph #8) [3.1]

Ishizaka K., Matsudaira M. (1968b): "Analysis of the vibration of the vocal cords." J. Acoust. Soc. Jpn. **24**, 311-312 [3.1]

Ishizaka K., Matsudaira M., Kaneko T. (1976): "Input acoustic-impedance measurement of the subglottal system." J. Acoust. Soc. Am. **60**, 190-197 [3.1; 5.2.1]

Isogai Y.: see (Saito et al., 1978)

Isomichi Y. (1968): "The optimum filter for frequency analysis of speech." ICA-6 (Paper C-5-17)

Isshiki N. (1959): "Regulatory mechanism of the pitch and volume of voice." Oto-Rhino-Laryngol. Clinic **52**, 1065-1094 [3.1]

Isshiki N., Yanagihara N., Morimoto N. (1966): "Approach to the objective diagnosis of hoarseness." Folia Phoniatr. **18**, 393-400 [4.3; 9.4.3]

Itahashi S. (1977): "Description of fundamental frequency trajectory with continuous lines." **PIPS Rept.,** 1-4 [9.4.2]

Itahashi S. (1978): "Description of speech data patterns by several functions with applications to formant and fundamental frequency trajectories." **STL-QPSR** (2-3), 1-22 [9.4.2]

Itakura F. (1975): "Minimum prediction residual principle applied to speech recognition." IEEE Trans. ASSP-**23**, 67-72 [5.3.3]

Itakura F., Saito S. (1968): "Analysis synthesis telephony based on the maximum likelihood method." **ICA-6** (Paper C-5-5) [6.3.1; 8.4.1]

Itakura F., Saito S. (1971): "Digital filtering techniques for speech analysis and synthesis." **ICA-7** (3), 261-264 (Paper 25C1) [6.3.1; 8.4.1]

Ito M.R., Donaldson R.W. (1971): "Zero crossings measurements for analysis and recognition of speech sounds." IEEE Trans. AU-**19**, 235-242 [5.3]

Iwata Sh. (1972): "Periodicities of pitch perturbations in normal and pathologic larynges." Laryngoscope (St.Louis) **82**, 87-96 [4.3; 9.4.3]

Iwata Sh., Leden H. von (1970): "Pitch perturbation in normal and pathologic voices." Folia Phoniatr. **22**, 413-424 [4.3; 9.4.3]

Jackson B.: see (Hollien and Jackson, 1973)

Jackson K.: see (Schafer et al., 1976)

Jackson L.B., Bertrand J. (1976): "An adaptive inverse digital filter for formant analysis of speech." **ICASSP-76**, 84-87 [6.5.1]

Jackson T.L.: see (Crystal and Jackson, 1970)

Jaeger R.C. (1982): "Tutorial: Analog data acquisition technology; part I - digital-to-analog conversion." IEEE Micro, 20-37 [2.2; 9.2.1]

Jain J.R.: see (Maitra and Jain, 1982)

Jancosek E.: see (Takefuta et al., 1972)

Janota P., Liljencrants J. (1969): "The effect of fundamental frequency changes on the perception of stress by Czech listeners." **STL-QPSR** (4), 32-38 [4.3; 9.4.2]

Jarosz A.: see (Handzel and Jarosz, 1967)

Jassem W. (1959): "A note on plotting pitch curves." Le Maître phonétique **111**, 2-3 [5.1]

Jassem W. (1972): "Pitch and compass of the speaking voice." J. Int. Phonetics Assoc. **1**, 59-68 [4.1]

Jassem W., Demenko G. (1981): "Normalization and mathematical description of linguistically relevant pitch curves." **F.A.S.E.-81**, 145-148 [9.4.2; 9.4.4]

Jassem W., Dobrogowska K. (1980): "Speaker-independent intonation curves." In *The melody of Language;* ed. by L.R. Waugh and C. van Schooneveld (Univ.Park Press, Baltimore), 135-148 [9.4.2]

Jassem W., Steefen-Batog M., Czajka S. (1973): "Statistical characteristics of short-term average F_0 distributions as personal voice features." In *Speech analysis and synthesis;* ed. by W. Jassem (Inst. of Fundamental Research, Polish Academy of Sciences, Warsaw), Vol.3, 209-225 [4.1]

Jassem W., Steffen-Batog M., Gruszka-Koscielak H. (1973): Statistical distribution of short-term F_0 values as a personal voice characteristic. In *Speech analysis and synthesis;* ed. by W. Jassem (Inst. of Fundamental Research, Polish Academy of Sciences, Warsaw), Vol.2, 197-208 [4.1]

Javkin H.R., Maddieson I. (1982): "An inverse-filtering study of Burmese creaky voice." J. Acoust. Soc. Am. **72** (A), S102 (Paper BBB10, 104th Meet. ASA) [6.5.1; 9.4.2]

Jayant N.S. (1973): "Delta modulation of pitch, formant and amplitude signals for the synthesis of voiced speech." IEEE Trans. AU-**21**, 135-140 [9.4.5]

Jayant N.S. (1976): "Average and median-based smoothing techniques for improving digital speech quality in the presence of transmission errors." IEEE Trans. COM-**24**, 1043-1045 [6.6.4]

Jayant N.S. (1977): "Pitch-adaptive DPCM coding of speech with two-bit quantization and fixed spectrum prediction." Bell Syst. Tech. J. **56**, 439-454 [9.4.5]

Jeel V. (1975): "An investigation of the fundamental frequency of vowels after various Danish consonants, in particular stop consonants." **ARIPUC 9**, 191-211 [3.1; 9.4.2]

Jenkin, Ewing (1878): "The phonograph and vowel theories." Nature **18**, 167 [5.1.1]

Jenkins R.A. (1960): "Amplitude-time envelope as a cue to pitch." J. Acoust. Soc. Am. **32** (A), 1506 (Paper J10; 60th Meet. ASA) [6.2.1]

Jenkins R.A. (1961): "Perception of pitch, timbre, and loudness." J. Acoust. Soc. Am. **33**, 1550-1557 [4.2; 6.2.1]

Jentzsch H., Sasama R., Unger E. (1978): "Elektroglottographische Untersuchungen zur Problematik des Stimmeinsatzes bei zusammenhängendem Sprechen." Folia Phoniatr. **30**, 59-66 [3.4; 5.2.3; 9.4.2]

Jentzsch H., Unger E., Sasama R. (1981): "Elektroglottographische Verlaufskontrollen bei Patienten mit funktionellen Stimmstörungen." Folia Phoniatr. **33**, 234-241 [5.2.3; 9.4.3]

Jerri A.J. (1977): "The Shannon sampling theorem - its various extensions and applications." Proc. IEEE **65**, 1575-1596 [2.2]

Jesorsky P. (1976): "Untersuchung der Effektivität verschiedener Analyseverfahren." **Phon. München 5**, 209-221 [5.3; 9.4.4]

Jewett W.M. (1963): "Pitch variation in vocoded voice" (US Naval Res.Lab, Washington DC; NRL-MR-1669) [4.3; 9.4.4; 9.4.5]

Joffe L. (1977): "A hardware pitch exercising game for deaf children." **ICA-9**, 424 (Paper I18) [9.4.3]

Johns M.E.: see (Bull et al., 1979)

Johnson C.S.: see (Altes et al., 1975)

Jones D. (1909): *Intonation curves* (Teubner, Leipzig) [5.1.1; 9.4.2]
Jones E.T.: see (Stead and Jones, 1961)
Jones G.D.: see (Hellwarth and Jones, 1968)
Jones R.H.: see (Tenold et al., 1974)
Joos M. (1948): "Acoustic phonetics." Lang. Monogr. **24** [5.1.2; 9.4.2]
Journées d'Etude sur la Parole --> *Journées d'Etudes sur la Parole;* ed. by Groupe de la communication parlée du G.A.L.F. (Groupement des Acousticiens de Langue Française, F-22301 Lannion)
Jospa P.: see (Beekmans et al., 1974)
Judd M.W.: see (Holmes et al., 1978)
Junien-Lavillauroy C. (1979): "Examen radiologique du larynx." **Larynx et parole,** 27-44 [5.2.1]
Jurgens J.E.: see (Buning et al., 1961)
Jury E.I. (1973): *Theory and application of the z-transform method* (R.E.Krieger Publ. Comp, New York NY) [2.2]
Justice J.H. (1979): "Analytic signal processing in music computation." IEEE Trans. ASSP-**27,** 670-684 [2.3; 6.3.2]
Kacprowski J. (1979): "Objective acoustical methods in phoniatric diagnostics of speech organ disorders." Arch. Acoustics **4,** 287-292 [5.2; 9.4.3]
Kacprowski J., Mikiel W., Gubrynowicz R. (1977): "Acoustic methods in phoniatric diagnostics of larynx and vocal tract pathology." **ICA-9,** 467 (Paper I61) [4.1; 9.4.3]
Kahane J.C.: see (Chollet and Kahane, 1979)
Kailath T.L. (ed.) (1977): *Linear least-squares estimation.* Benchmark papers in electrical engineering and computer science, vol.17 (Dowden, Hutchinson, and Ross, Stroudsburg PA) [8.6]
Kaiser J.F.: see (Rabiner et al., 1972)
Kaiser L. (1936): "Biological and statistical research concerning the speech of 216 Dutch students." Arch. Néerl. Phonétique Exp., 1-76 [4.1]
Kajiyama M.: see (Chiba and Kajiyama, 1941)
Kakita Y., Hiki Sh. (1974): "A study of laryngeal control for voice pitch based on anatomical model." **Proc. 1974 Speech Commun. Sem. 2,** 45-54 [3.1; 3.3]
Kakita Y., Hiki Sh. (1976a): "A study on laryngeal control for pitch change by use of anatomical structure model." **ICASSP-76,** 43-46 [3.3; 4.3]
Kakita Y., Hiki Sh. (1976b): "Investigation of laryngeal control in speech by use of thyrometer." J. Acoust. Soc. Am. **59,** 669-674 [6.2.5]
Kalfaian M.V. (1954): "Phonetic printer of spoken words;" United States Patent No. 2,673,893. Issued March 30, 1954; filed Feb. 24, 1953 [6.2.1]
Kalfaian M.V. (1958): "Fundamental frequency selector from speech sound waves;" United States Patent No. 2,872,517. Issued Feb. 3, 1959
Kalfaian M.V. (1960): "Fundamental frequency extractor from speech waves;" United States Patent No. 2,957,134. Issued Oct. 18, 1960
Kalikow D.N.: see (Stevens et al., 1975; Nickerson et al., 1976)
Kallenbach W. (1951): "Eine Weiterentwicklung des Tonhöhenschreibers mit Anwendungen bei phonetischen Untersuchungen." Akustische Beihefte **1,** 37-42 [6.1.2; 6.6.1; 9.4.2]
Kallenbach W. (1962): "Die Untersuchung der Sprache mit dem Tonhöhenschreiber." Frequenz **16,** 37-42 [6.1.2; 9.4.2]
Kallenbach W., Schröder H.J. (1961): "Zur Technik der Tonbandaufnahme bei Sprachunter-suchungen." Phonetica **7,** 95-108 [9.2.1]
Kaneko T., Kobayashi N., Asano H., Hayasagi K., Kitamura T. (1974): "Ultrasonoglotto-graphie." Ann. Oto-Laryngol. (Paris) **91,** 403-410 [5.2.4]
Kang G.S., Fransen L.J., Kline E.L. (1981): "Mediumband speech processor with baseband residual spectrum encoding." **ICASSP-81,** 820-823 [9.4.5]
Karplus W.J. (1981): "Architectural and software issues in the design and application of peripheral array processors." IEEE Computer (9), 11-17 [9.1.5]
Karwoski R.J. (1980a): "A four-cycle butterfly arithmetic architecture" (TRW, El Segundo CA) [2.3; 9.1.5]
Karwoski R.J. (1980b): "Second-order recursive digital filter design with the TRW multi-plier-accumulators" (TRW, El Segundo CA) [2.3; 9.1.5]
Karwoski R.J. (1980c): "Architecture development for a general-purpose digital filter" (TRW, El Segundo CA) [2.3; 9.1.5]
Karwoski R.J. (1980d): "An introduction to digital spectrum analysis including a high-speed FFT processor design" (TRW, El Segundo CA) [2.3; 9.1.5]

Kasuya H. (1980): "An improved autocorrelation pitch detector." J. Acoust. Soc. Jpn. (E) 1, 263-264 [8.2.4]

Kasuya H., Kobayashi Y., Ebihara S. (1983): "Characteristics of pitch period and amplitude perturbations in pathologic voice." ICASSP-83 [4.3; 9.4.3]

Katagiri S.: see (Ooyama et al., 1978)

Kato K.: see (Saito et al., 1958)

Kato Y.: see (Nagata and Kato, 1960)

Katterfeldt H. (1981): "A DFT-based residual-excited predictive coder (RELP) for 4.8 and 9.6 kb/s." ICASSP-81, 824-827 [9.4.5]

Katterfeldt H. (1983): "Implementation of a robust RELP speech coder." ICASSP-83 [6.3.1; 9.4.5]

Kaunisto M.: see (Vuorenowski et al., 1970)

Kawashima T.: see (Fujisaki and Kawashima, 1968)

Kay Elemetrics Corp. (ed.) (1980a): "Application notes: prosodic and stress features; acoustic analysis, Visi-Pitch" (Kay Elemetrics Corp., Pine Brook NJ) [5.1.2; 6.1.1; 9.4.2; 9.4.3]

Kay Elemetrics Corp. (ed.) (1980b): "References on the use of sound spectrography in speech disorders" (Kay Elemetrics Corp., Pine Brook NJ) [1.3; 5.1.2; 9.4.3]

Keating P.A., Buhr R.D. (1978): "Fundamental frequency in the speech of infants and children." J. Acoust. Soc. Am. 63, 567-571 [4.1]

Kelly D.H., Sansone F.E. (1976): "Clinical estimation of fundamental frequency: the 3M Plastiform Magnetic Tape Viewer." J. Acoust. Soc. Am. 59 (A), S15 (Paper G5; 91th Meet. ASA) [5.1; 9.4.3]

Kelly J.M., Kennedy R. (1966): "Experimental cepstrum pitch detector for use in a 2400 bit/sec channel vocoder." J. Acoust. Soc. Am. 41 (A), 1241 (Paper H3, 72nd Meet. ASA) [8.4.2; 9.4.5]

Kelman A.W. (1981): "Vibratory pattern of the vocal folds." Folia Phoniatr. 33, 73-99 [3.0; 5.2.3]

Kelsey C.A., Hixon T.J., Minifie F.D. (1969): "Ultrasonic measurement of lateral pharyngeal wall displacement." IEEE Trans. BME-16, 143-147 [5.2.4]

Kelsey C.A., Minifie F.D., Hixon T.J. (1969): "Applications of ultrasound in speech research." J. Speech Hear. Res. 12, 564-575 [5.2.4]

Kemerait R.C.: see (Childers et al., 1977)

Kennedy J.G.: see (Watkin wt al., 1978)

Kennedy R.: see (Kelly and Kennedy, 1966)

Kennedy W.K.: see (Tucker et al., 1977)

Kharova S.S.: see (Akinfiev et al., 1973)

Kiang N.Y.S., Moxon E.C. (1972): "Physiological considerations in artificial stimulation of the inner ear." Ann. Otol. Rhinol. Laryngol. 81, 714-730 [4.2.1; 9.4.3]

Kido K.: see (Ooyama et al., 1978)

Kim B. (1968): "F_0 variations according to consonantal environments;" Dissertation, Phonology Lab., Univ. of California, Berkeley CA [3.1; 9.4.2]

Kincaid T.G., Scudder H.J.III (1967): "Estimation of periodic signals in noise." J. Acoust. Soc. Am. 42 (A), 1166 [8.6.1]

King A.: see (Fisher J. et al., 1978)

Kingsbury N.G., Amos W.A. (1980): "A robust channel vocoder for adverse environments." ICASSP-80, 19-22 [6.2.4; 9.4.5]

Kinoshita T.: see (Tatsumi et al., 1973)

Kiritani S.: see (Imagawa et al., 1982)

Kitahara S.: see (Saito et al., 1978)

Kitajima K.: see (Tanabe et al., 1975)

Kitamura T., Kaneko T., Asano H. (1964): "Ultrasonic diagnosis of the laryngeal diseases (in Japanese)." Jpn. Med. Ultrason. 2, 14-15 [5.2.4]

Kitamura T., Kaneko T., Asano H., Miura S. (1967): "Ultrasonoglottography: a preliminary report (in Japanese)." Jpn. Med. Ultrason. 5, 40-41 [5.2.4]

Kitayama S.: see (Kurematsu et al., 1979)

Kittel G.: see (Moser and Kittel, 1977)

Kitzing P. (1977): "Methode zur kombinierten photo- und elektroglottographischen Registrierung der Stimmlippenschwingungen." Folia Phoniatr. 29, 249-260 [5.2.3; 5.2.5; 5.2.6]

Kitzing P. (1979): "Glottografisk freqvensindikering (Glottographic frequency indication. In Swedish);" Dissertation, University of Lund [5.2.3]

Kitzing P. (1982): "Photo- and electroglottographical recording of the laryngeal vibratory pattern during different registers." Folia Phoniatr. **34,** 234-241 [3.2; 5.2.3; 5.2.5; 5.2.6]

Kitzing P., Carlborg B., Löfqvist A. (1982): "Aerodynamic and glottographic studies of the laryngeal vibratory cycle." Folia Phoniatr. **34,** 216-224 [5.2.1; 5.2.3; 5.2.6]

Kitzing P., Löfqvist A. (1979): "Evaluation of voice therapy by means of photoglottography." Folia Phoniatr. **31,** 103-109 [5.2.5; 9.4.3]

Kitzing P., Rundqvist H.E., Tayo E. (1975): "Fundamental frequency of the voice in continuous speech. Preliminary report on a device for determining mean and distribution of frequencies." **Phon. Lund 12,** 85-92 [4.1; 4.3; 9.4.3]

Kitzing P., Sonesson B. (1974): "A photoglottographical study of the female vocal folds during phonation." Folia Phoniatr. **26,** 138-149 [3.0; 5.2.5]

Kjellin O. (1975): "Progress report on an acoustical study of pitch in Tibetan." **Ann. Bull. RILP 9,** 137-150 [8.2.4; 9.4.2]

Kjellin O. (1976): "Progress report on an acoustical study of pitch in Tibetan. - How to explain the "tones" in Tibetan." **Ann. Bull. RILP 10,** 137-166 [9.4.2]

Klatt D.H. (1973): "Discrimation of fundamental frequency contours in synthetic speech; implications for models of pitch perception." J. Acoust. Soc. Am. **53,** 8-16 [4.2; 4.3; 9.4.2; 9.4.4]

Klatt D.H. (1977): "Review of the ARPA speech understanding project." J. Acoust. Soc. Am. **62,** 1345-1366

Kleinschmidt K. (1957): "The effect of quasi-periodicity on the naturalness of synthetic vowels." **MIT-RLE Progr. Rept.,** 10-11 [9.4.4; 9.4.5]

Kline E.L.: see (Kang et al., 1981)

Kloker D.R. (1976): "A technique for the automatic location and description of pitch contours." **ICASSP-76,** 55-58 [6.6.1; 9.4.2]

Kloker K.: see (McLaughlin et al., 1980)

Kloster-Jensen M. (1957): "Die Erleichterung der instrumentellen Worttonforschung durch Tonhöhenschreiber." Studia Linguistica **9,** 70-77 [2.6; 9.4.2]

Knipper A.V., Orlov I.A. (1976): Синхронное с импульсами основного тона измерение частоты формант." В кн.: Труды АРСО-9 (Минск). А.В.Книппер и И.А.Орлов. (Pitch Ⴈ synchronous formant frequency measurement. In Russian). **ARSO-9** [9.4.5]

Knorr S.G. (1979): "Reliable voiced/unvoiced decision." IEEE Trans. ASSP-**27,** 263-267 [5.3]

Knudson R.: see (Lieberman et al., 1969)

Kobayashi N.: see (Kaneko et al., 1974)

Kobayashi R. (1960): "On a pitch period indicator combined with zero-crossing interval indication." J. Acoust. Soc. Jpn. **16** (A), 286-288 [6.1.1]

Kobayashi R. (1961): "A new pitch and intensity indicator for speech analysis." Study of Sounds **9,** 137 [6.1.1]

Kobayashi Y.: see (Kasuya et al., 1983)

Kock W.E. (1937): "A new interpretation of the results of experiments on the differential pitch sensitivity of the ear." J. Acoust. Soc. Am. **9,** 129-134 [4.2.3]

Kodzasov S.V., Otryasenkov M. (1968): Измерение основной частоты при исследовании интонации. С.В.Кодзасов и М.Отрясенков. В кн.: Семантические и фонологические проблемы прикладной лингвистики. Москва: Государственный Университет. (Measurement of fundamental frequency in the study of intonation. In Russian.)" (Moscow), 165-181

König E. (1957): "Effect of time on pitch discrimation thresholds under several psychophysical procedures; comparison with intensity discrimination thresholds." J. Acoust. Soc. Am. **29,** 606-612 [4.2.3; 4.3]

Koenig W., Dunn H.K., Lacy L.Y. (1946): "The sound spectrograph." J. Acoust. Soc. Am. **18,** 19-49 [5.1.2]

Köster J.P. (1971): "Rôle de l'intonation dans l'information émotionnelle de phrases allemandes synthétisées." **ICPhS-7,** 914-921 [9.4.3; 9.4.4]

Köster J.P. (1972): "Extreme Intonationsmodulationen und ihre Registrierung." **Phonetica Pragensia III,** 143-147 [4.3; 9.4.2]

Köster J.P. (1973): *Historische Entwicklung von Syntheseapparaten zur Erzeugung statischer und vokalartiger Signale nebst Untersuchungen zur Synthese deutscher Vokale.* Hamburger Phonetische Beiträge, Bd.4 (Buske, Hamburg) [1.2; 3.2; 9.4.4]

Köster J.P. (1974): "Künstliche Intonationsmuster im Urteil deutscher und nicht-deutscher Schüler." In *Miszellen II*; ed. by J.P. Köster. Hamburger Phonetische Beiträge, Bd.13 (Buske, Hamburg), 21-39 [9.4.3]

Köster J.P. (1976): "Abriss historischer Ansätze der Sprachsynthese." In *Speech analysis and synthesis;* ed. by W. Jassem (Inst. of Fundamental Research, Polish Academy of Sciences, Warsaw), Vol.4, 1-104 [5.1.1]

Köster J.P., Smith S.L. (1970): "Zur Interpretation elektrischer und photoelektrischer Glottogramme." Folia Phoniatr. **22** [5.2.3]

Kohut R.: see (Hanson et al., 1982)

Koike Y., Hirano M. (1973): "Glottal-area time function and subglottal pressure variation." J. Acoust. Soc. Am. **54**, 1618-1627 [3.1; 3.3]

Koike Y., Takahashi H., Calcaterra T. (1977): "Acoustic measures for detecting laryngeal pathology." Acta Otolaryngol. (Stockholm) **84**, 105-117 [9.4.3]

Kojima H., Gould W.J., Lambiase A. (1979): "Computer analysis of hoarseness." **ASA*50**, 249-252 [4.3; 8.5.2; 9.4.3]

Kokawa N.: see (Saito et al., 1978)

Kolb H.-J. (1982): "Prozessorkonzepte zur digitalen Signalverarbeitung." Elektronik **21**, 107-114 [9.1.5]

Kolk H.H.T., Els T.J.M. van, Rietveld A.C.M. (1977): "Die Visualisierung von Intonationsmustern im Fremdsprachenunterricht." In *Phonetische Grundlagen der Ausspracheschulung;* ed. by H.P.Kelz. Forum phoneticum 4 (Buske, Hamburg), 199-214 [9.4.3]

Konvalinka I.S., Mataušek M.R. (1979): "Simultaneous estimation of poles and zeros in speech analysis and ITIF - Iterative Inverse Filtering Algorithm." IEEE Trans. ASSP-**27**, 485-492 [6.3.1; 6.5.1; 9.4.2]

Kopec G.E., Oppenheim A.V., Tribolet J.M. (1977): "Speech analysis by homomorphic prediction." IEEE Trans. ASSP-**25**, 40-49 [8.4.1]

Kopp G.A., Green H.C. (1946): "Basic phonetic principles of visible speech." J. Acoust. Soc. Am. **18**, 74-89 [5.1.2]

Kopp G.A., Kopp H.G. (1963): "Visible speech for the deaf" (Speech and Hearing Clinic, Wayne State Univ.) [9.4.3]

Kopp H.G.: see (Kopp and Kopp, 1963)

Koshikawa T., Sugimoto T. (1960): "Time variation of fundamental frequencies of speech sounds." J. Acoust. Soc. Jpn. **16**, 289-291 [4.1]

Koshikawa T., Sugimoto T. (1962): "The information rate in the pitch signal in speech." **Proc. 1962 Speech Commun. Sem.** [4.1; 5.2.2; 9.4.5]

Kosiel U. (1968): "Relations between vowel spectra and fundamental frequency in Polish." In *Speech analysis and synthesis;* ed. by W. Jassem (Inst. of Fundamental Research, Polish Academy of Sciences, Warsaw), Vol.1, 69-92 [3.4; 9.4.2]

Kosiel U. (1973): "Correlations between fundamental frequency and formant frequencies in polish vowels." In *Speech analysis and synthesis;* ed. by W. Jassem (Inst. of Fundamental Research, Polish Academy of Sciences, Warsaw), Vol.3, 117-120 [3.4; 9.4.2]

Kotmans H.J.: see (Sluyter et al., 1980)

Kotten K. (1981): "Einige Verfahren zur Analyse des Grundtonverlaufs gesprochener Sprache." **Phon. Köln 11**, 8-11 [2.6; 9.1.1]

Krasner N.F., Regier W.L., Wong D.L. (1973): "Theory and design of pipelined FFT processors." In *Signalverarbeitung (Signal Processing).* Vorträge der gleichnamigen Fachtagung, Erlangen 1973; ed. by H.W.Schüssler (Universität Erlangen-Nürnberg), 319-340 [2.4; 8.1]

Krause M. (1982): "Bericht über Vorhaben zur vergleichenden Untersuchung von GFB-Algorithmen beim Rundgespräch der DFG am 21.1.1982 in Frankfurt" [private communication] [9.3.2]

Kretschmar J. (1978): "Untersuchungen zum Tonhöhenverlauf deutscher Sätze für die Sprachsynthese." In *Fortschritte der Akustik, DAGA'78 Bochum* (VDE-Verlag, Berlin), 455-458 [9.4.2; 9.4.4]

Kretschmar J. (1980a): "Automatic synthesis of the pitch contour of German sentences." **ICA-10** (Paper A1-12.3) [9.4.2]

Kretschmar J. (1980b): "Untersuchungen zur Natürlichkeit sythetisierter Satzmelodien." In *Fortschritte der Akustik, DAGA'80 München* (VDE-Verlag, Berlin), 699-702 [9.4.2; 9.4.4]

Kriger L.V. (1958): "Pitch circuit of AFCRC scanvocoder" (AFCRC, Cambridge MA; ERD-CRRS-TM-58-126) [6.1]

Kriger L.V. (1959): "Pitch extraction for speech synthesis with special techniques for use in digitized bandwidth compression systems" (US Air Force, Bedford MA; AFCRC-TR-59-116; DDC-AD-214449) [6.1.2; 7.2; 7.4; 9.4.4]

Kriger L.V. (1963): "Accurate log and inverse-log function generation circuits for use in digital vocoder channels" (AFCRL, Cambridge MA; AFCRL Report #63186) [7.2]

Kringlebotn M.: see (Cederlund et al., 1960)

Krishnamurthy A.K., Childers D.G. (1981): "Vocal fold vibratory patterns: comparison of film and inverse filtering." ICASSP-81, 133-136 [5.2.1; 5.2.6; 6.5.1]

Krishnamurthy A.K., Naik J.M., Childers D.G. (1982): "Laryngeal function: electroglotto-graph, ultra-high-speed films, and speech inverse filtering." J. Acoust. Soc. Am. 72 (A), S65 (Paper LL8, 104th Meet. ASA) [3.1; 5.2.1; 5.2.3; 5.2.6; 6.5.1]

Kriz S. (1975): "A 16-bit A-D-A conversion system for high fidelity audio research." IEEE Trans. ASSP-23, 146-149 [9.2.1]

Krokstad A.: see (Cederlund et al., 1960)

Krones R. (1968): "Pitch meters." Internal Memo, Univ. of California, 72-88 [2.6]

Krüger F. (1906): Die Messung der Sprachmelodie (Bonn) [5.1.1; 9.4.2]

Kryukova R.V. (1968): "Исследование основного тона речевых сигналов." В кн.: Труды Воронежцкого инженерно-строительного института. Р.В.Крюкова. (Pitch determination of speech signals. In Russian) (Institute of Civil Engineering, Voronets), Vol.14, 98-103

Kubina J.G., O'Shaughnessy D. (1981): "Quality limitations in residual-excited LP speech coding." J. Acoust. Soc. Am. 70 (A), S41 (Paper R9, 102nd Meet. ASA) [9.4.5]

Kubzdela H. (1973): "Automatyczna ekstrakcja czestotliwości tonu podstawowego oraz pierwszych trzech formantów sygnalu mowy. (Automatic extraction of the fundamental frequency and the first three formants of speech. In Polish)." Reports of the Inst. of Fund. Technol. Res., Polish Acad. of Sci. 51, 3-72 [6.1.3]

Kubzdela H. (1976): "An analogue fundamental frequency extractor." In Speech analysis and synthesis; ed. by W. Jassem (Inst. of Fundamental Research, Polish Academy of Sciences, Warsaw), Vol.4, 269-279 [6.1.2; 6.6.1; 9.4.2]

Kühlwetter J. (1971): "Ein Verfahren zur Bestimmung der Sprachgrundfrequenz mit einem Rechner" (Bericht Nr.24, Inst. f. Übertragungstechnik, TH Darmstadt) [6.1.1]

Kühlwetter J. (1972): "Sprachgrundfrequenzbestimmung mit dem Rechner." Angew. Infor-matik 14, 326-330 [6.1.1]

Kühlwetter J. (1973a): "Grundperiodensynchrone analoge Bestimmung der Frequenzen und Amplituden der Koeffizienten des Spektrums von Sprachsignalen in Echtzeit" (Bericht Nr.49, Fachgebiet Übertragungstechnik, TH Darmstadt) [9.4.5]

Kühlwetter J. (1973b): "Ein Verfahren zur Echtzeitbestimmung der Maxima und Minima grundperiodensynchroner Amplitudenspektren aus fliessender Sprache." Nachrichtentech. Z. 26, 503-507 [5.1.1; 9.4.5]

Kühlwetter J. (1973c): "Grundperiodensynchrone Bestimmung des Amplitudenspektrums von Signalen mit veränderlicher Grundfrequenz." Frequenz 27, 21-23 [9.4.5]

Kühlwetter J. (1981): "Bestimmung der Grundfrequenz von Sprachsignalen in Echtzeit." Elektronik 11, 79-86 [6.1.1]

Kumagai K., Ideno K. (1960): "The improvement in Halsey's pitch extraction method." J. Acoust. Soc. Jpn. 16 (A), 274-276 [6.1.1]

Kumar A.: see (Ojamaa et al., 1970)

Kunt M. (1980): Traitement numérique des signaux. Traité d'Electricité, vol.20 (Editions Georgi, St.Saphorin) [2.0]

Kurematsu A., Ishigami H., Kitayama S., Yato F., Tamura J. (1979): "A linear predictive vocoder with new pitch extraction and exciting source." ICASSP-79, 69-72 [6.4.2; 8.2.4; 9.4.5]

Kurita S.: see (Hirano et al., 1979)

Kusch H. (1967): "Das Segment, ein Baustein der Sprache." Nachrichtentech. Z. 20, 495-502 [6.1.1; 9.4.5]

Kusch H. (19/1): "Ein neues Verfahren zur Verbesserung der Sprache in Heliumatmo-sphäre." Acustica 25, 42-46 [6.1.1; 9.4.5]

Kushtuev A.I. (1968): "Об одной статистической задаче анализа речи." В кн.: Труды НИИР. А.И. Куштуев. (On a statistical task in speech analysis. In Russian). Trudy NIIR 2, 83-89 [8.5.1]

Kushtuev A.I. (1969a): "Исследование многоканальной схемы выделения основного тона речи." В журн.: "Труды НИИР". А.И.Куштуев. (Implementation of a multi-channel method for pitch determination of speech signals. In Russian.). Trudy NIIR 2, 85-90 [8.5.1]

Kushtuev A.I. (1969b): "Адаптивный многоканальный анализатор основного тона речи." В кн.: Труды АРСО-4, Киев 1968. А.И.Куштуев. (Adaptive multichannel device for pitch determination. In Russian.). ARSO-4, 208-212 [8.5.1]

Kushtuev A.I. (1976a): "Выделение основного тона речи с сеткой пробных частот." В кн.: Труды АРСО-9 (Минск). А.И.Куштуев. (Pitch determination using a set of trial frequencies. In Russian). **ARSO-9**

Kushtuev A.I. (1976b): "Адаптивная коррекция ошибок выделения основного тона речи." В кн.: Труды АРСО-9 (Минск). А.И.Куштуев. (Adaptive correction algorithm for pitch determination of speech signals. In Russian). **ARSO-9** [6.6.3]

Kvavik K.H., Olsen C.L. (1974): "Theories and methods in Spanish intonational studies." Phonetica **30**, 65-100 [9.4.2]

Kwon S.Y.: see (Bergeron et al., 1980)

La Riviere C. (1975): "Contributions of fundamental frequency and formant frequencies to speaker identification." Phonetica **31**, 185-197 [9.4.4]

Lacroix A. (1976): "Pitch detection with the aid of digital filters." In *14.Conf. on Acoustics, October 5-9, 1976, Tatranska Lomnica* (The Czechoslovak Scientific and Technical Society, Prague), 93-96 [6.1.1]

Lacroix A. (1977): "Estimation of pitch contours with the aid of digital filters." **ICA-9** (1), 520 [6.1.1]

Lacroix A. (1980): *Digitale Filter. Eine Einführung in zeitdiskrete Signale und Systeme* (Oldenbourg, München) [2.3]

Lacroix A., Hoptner N. (1977): "Accurate pitch estimation using digital filters." **ICASSP-77**, 319-322 [6.1.1]

Lacy L.Y.: see (Koenig et al., 1946)

Ladefoged P. (1971): *Preliminaries to linguistic phonetics* (Univ. of Chicago Press, Chicago IL) [3.1]

Ladefoged P. (1973): "The features of the larynx." J. Phonetics **1**, 73-84 [3.1]

Längle D. (1976): "Digital encoding of variable-length vectors with application to pitch extraction and pitch-synchronous speech analysis and synthesis." **ICASSP-76**, 254-257 [6.3.2; 9.4.5]

Längle D. (1977): "Sprachübertragung mit grundfrequenzsynchroner Blockquantisierung;" Dissertation, Technische Universität München [6.3.2; 9.4.5]

Laface P.: see (De Mori et al., 1977)

Lafon J.C., Guichard J. (1973): "Etude de la mélodie de la voix." **Bull. Audiophonologie** **3**, 91-104 [9.4.2]

Laguaita J.K., Waltrop W.F. (1964): "Acoustic analysis of fundamental frequency of voices before and after therapy." Folia Phoniatr. **16**, 183-192 [5.1.2; 9.4.3]

Lair J.B.: see (Aigrain and Lair, 1953)

Lamb M.R., Bates R.H.T. (1978): "Computerized aural training: An interactive system designed to help both teachers and students." J. Computer-Based Instruction **5**, 30-37 [4.2; 4.3; 9.4.2]

Lambiase A.: see (Tanabe et al., 1975)

Lancia R.: see (Carré et al., 1963; Paille and Lancia, 1969)

Landercy A. (1972): "Contribution à l'étude de la source vocale et de ses variations à court terme;" Dissertation, Université Libre de Bruxelles, Bruxelles [3.1]

Landercy A. (1977): "Variations à court terme de la fonction de source vocale." **Journées d'Etude sur la Parole 8**, 29-36 [4.3]

Landgraf L.L.: see (Flanagan and Landgraf, 1968)

Large J.W. (1973): "Acoustic study of register equalization in singing." Folia Phoniatr. **25**, 39-61 [3.2]

Large J.W. (1976): "Preliminary analysis of vocal chant." J. Acoust. Soc. Am. **60** (A), S43 (Paper S12; 92nd Meet. ASA) [3.2]

Large J.W. (1979): "Studies of the Garcian model for vocal registration." J. Acoust. Soc. Am. **66** (A), S56 (Paper BB6; 98th Meet. ASA)

Large J.W., Iwata Sh., Leden H. von (1972): "The male head register versus falsetto in singing." Folia Phoniatr. **24**, 19-29 [3.2]

Larkin W.D., Houde R.A., Stewart L.C. (1975): "Use of the visual speech training aid (VSTA) in speech training for the deaf." J. Acoust. Soc. Am. **57** (A), S18 (Paper I1; 89th Meet. ASA) [9.4.3]

Larreur D., Emerard F. (1976): "Speech synthesis by dyads and automatic intonation processing." **ICASSP-76**, 694-697 [9.4.2; 9.4.4]

Larreur D., Emerard F. (1977): "Analyse de la structure intonative de la phrase en francais." **Journées d'Etude sur la Parole 8**, 227-236 [9.4.2]

Larsson B. (1977): "Pitch tracking in music signals." **STL-QPSR** (4), 1-8 [6.1.1; 6.3.2; 9.2; 9.4.2]

Larynx et Parole -- Boe L.J., Descout R., Guérin B. (eds.): *Larynx et Parole.*Actes du Séminaire organisé à l'Institut de Phonétique de Grenoble, les 8 et 9 Février 1979, sous le patronnage du G.A.L.F (Institut de Phonétique, Université des Langues et Lettres, Grenoble 1979) [3.1]

Lashbrook W.: see (Tosi et al., 1972)

Lasse L.T. (1937): "The effect of pitch and intensity on the quality of vowels of speech." Arch. Speech **2**, 41-60 [3.4; 9.4.2]

Lauter J.L.: see (Vemula et al., 1979)

Laver J. (1979): *Voice quality: a classified research bibliography* (John Benjamins B.V, Amsterdam) [1.3; 3.1; 9.4.3]

Laver J., Hanson R.J. (1981): "Describing the normal voice." In *Speech evaluation in psychiatry;* ed. by J.K. Darby (Grune and Stratton, New York) [3.1; 3.2]

Laver J., Hiller S., Hanson R.J. (1982): "Comparative performance of pitch detection algorithms on dysphonic voices." **ICASSP-82**, 192-195 [9.3.1; 9.4.3]

Lawrence W., Stead L.G., Jones E.T., McLean D.J., Clark J.A. (1956): *Methods and purposes of speech synthesis.* A Report of the Colloquium held at the S.R.D.E. in Christchurch, September 1955 (Signals Research and Development Establishment, Christchurch (Hants., England); S.R.D.E.Rept.No.1100; DDC-AD-98950) [6.1.2]

Lay T.: see (Timme et al., 1973)

Lazaryan L.M.: see (Grigoryan et al., 1976)

Le Corre C.: see (Vivès et al., 1977)

Le Roux J. (1975): "Une méthode synchrone d'extraction en temps réel du fondamental." **Journées d'Etude sur la Parole 6**, 1-11 [6.2.5; 6.3.1]

Lea W.A. (1972): "An approach to syntactic recognition without phonemics." **ICSCP-72**, 198-201 (Paper F2) [9.4.2; 9.4.4]

Lea W.A. (1973a): "An approach to syntactic recognition without phonemics." IEEE Trans. AU-**21**, 249-258 [9.4.2; 9.4.4]

Lea W.A. (1973b): "Influences of phonetic sequences and stress on fundamental frequency contours of isolated words." J. Acoust. Soc. Am. **53** (A), 346 (Paper DD9; 84th Meet. ASA) [9.4.2]

Lea W.A. (1982): "Selecting, designing, and using practical speech recognizers." In *Automatic speech analysis and recognition.* Proceedings of the NATO Advanced Study Institute held at Bonas, France, June 29 - July 10, 1981; ed. by J.P. Haton (Reidel, Dordrecht) [9.4.2]

Lea W.A., Medress M.F., Skinner T.F. (1974): "A prosodically guided speech understanding strategy." **IEEE Symp. Speech Recogn.**, 38-44 (Paper M7) [9.4.4]

Lea W.A., Medress M.F., Skinner T.F. (1975): "A prosodically guided speech understanding strategy." IEEE Trans. ASSP-**23**, 30-38 [9.4.4]

Lebow I.L.: see (Gold et al., 1971)

Lebrun Y., Hasquin-Deleval J. (1971): "Variations in vocal wave duration." J. Laryngology **85**, 43-56 [4.3]

Leclu̧se F.L.E. (1977): "Elektroglottografie" (Electroglottography; in Dutch. English, French, German abstr.); Dissertation, Univ. of Rotterdam [3.1; 5.2.1; 5.2.3; 5.2.6]

Leclu̧se F.L.E., Brocaar M.P., Verschure J. (1975): "The electroglottography and its relation to glottal activity." Folia Phoniatr. **27**, 215-224 [5.2.3; 5.2.6]

Leden H. von, Moore G.P., Timcke R.H. (1960): "Laryngeal vibrations: measurement of the glottic wave, part 3: The pathologic larynx." Arch. Otolaryng. **71**, 16-35 [9.4.3]

Lee D.T.L., Morf M. (1980): "A novel innovations based time-domain pitch detector." **ICASSP-80**, 40-44 [6.3.1; 6.3.2]

Lee F.F.: see (Willemain and Lee, 1972)

Lee H.C.: see (McKinnon and Lee, 1976)

Lee H.H.: see (Un and Lee, 1980)

Lee J.R.: see (Un and Lee, 1982)

Lee Y.W. (1960): *Statistical theory of communication* (Wiley, New York NY) [2.2]

Lees H. (1980): "Computerized aural training - a personal view." Austr. J. Music Educ. **27**, 33-35 [9.4.2; 9.4.3]

Leeuwaarden A. van: see (Sluyter et al., 1980)

Legtyarev A.: see (Bukhtilov et al., 1970)

Lehiste I. (1970): *Suprasegmentals* (MIT Press, Cambridge MA) [1.2; 3.0; 9.4.2]

Lehiste I. (1976): "Influence of fundamental frequency pattern on the perception of duration." J. Phonetics **4**, 113-117 [9.4.2]

Lehiste I., Peterson G.E. (1961): "Some basic considerations in the analysis of intonation." J. Acoust. Soc. Am. **33**, 419-425 [9.4.2]

Lehiste I., Peterson G.E. (1961): "Some basic considerations in the analysis of intonation."
    Reprinted in *Acoustic Phonetics – A course of basic readings;* ed. by D.B.Fry (Cam-
    bridge University Press, Cambridge), 378-393 [1.2; 9.4.2]
Leigh A.: see (McDonough et al., 1982)
Léon P. (1972): "Où en sont les études sur l'intonation." **ICPhS-7,** 113-156 [1.2; 1.3;
    9.4.2]
Léon P. (1973): "Problèmes de l'étude intonative." **Bull. Audiophonologie 3,** 31-42 [9.4.2]
Léon P., Faure G., Rigault A. (eds.) (1970): *Prosodic feature analysis.* Studia Phonetica,
    vol.3 (Didier, Montréal) [9.4.2]
Léon P., Martin Ph. (1969): *Prolégomènes à l'étude des structures intonatives.* Studia
    Phonetica, vol.2 (Didier, Montréal) [1.2; 1.3; 2.6; 4.1; 4.3; 5.1; 6.4.1; 6.6.1; 6.6.2;
    8.2.2; 8.4.2; 9.1; 9.4.2]
Léon P., Martin Ph. (1971): "Linguistique appliquee et enseignement de l'intonation."
    Etudes de Linguistique Appliquée **3,** 36-45 [9.4.3]
Léon P., Martin Ph. (1972): "Machines and measurements." In *Intonation: selected read-
    ings;* ed. by D.L.Bolinger (Penguin Books, Harmondsworth), 30-47 [6.4.1]
Leonard R.J., Ringel R.L. (1979): "Vowel shadowing under conditions of normal and
    altered laryngeal sensation." J. Speech Hear. Res. **22,** 794-817 [4.3]
Lepeshkin V.A., Pak S.P., Rodionov I.E. (1980):"Простэй многоканальный выделитель
    основного тона." В кн.: Труды АРСО-11, Ереван. В.А.Лепешкин, С.П.Пак, И.Е.Род-
    ионов. (Simple multichannel pitch detector. In Russian). **ARSO-11,** 62-63 [6.2.1; 6.4.1]
Lerman J.W., Duffy R.J. (1970): "Recognition of falsetto voice quality." Folia Phoniatr.
    22, 21-27 [3.2]
Lerner R.M. (1956): "Pitch-synchronous chopping of speech." **MIT-RLE Progr. Rept.,**
    99-103 [9.4.5]
Lerner R.M. (1959): "A method of speech compression" (Group Rept.36-41,Lincoln Lab.,
    MIT, Lexington MA; DDC-AD-244920) [6.2.1; 6.3.2]
Lesmo L., Mezzalama M., Torasso P. (1978): "A text-to-speech translation system for
    Italian." Int. J. Man-Machine Studies **10,** 569-591 [9.4.2]
Levin H., Lord W. (1975): "Speech pitch frequency as an emotional state indicator." IEEE
    Trans. SMC-**5,** 259-273 [4.1; 9.4.3]
Levinson S.E.: see (Tufts et al., 1976)
Levitt H. (1972a): "Speech processing aids for the deaf: An overview." **ICSCP-72,** 230-233
    (Paper G1) [9.4.3]
Levitt H. (1972b): "Acoustic analysis of deaf speech using digital processing techniques."
    IEEE Trans. AU-**20,** 35-41 [9.4.3]
Levitt H. (1973): "Speech processing aids for the deaf: An overview." IEEE Trans. AU-**21**
    [9.4.3]
Levitt H., Rabiner L.R. (1971): "Analysis of fundamental frequency contours in speech."
    J. Acoust. Soc. Am. **49,** 569-582 [9.4.2; 9.4.4]
Lewis D.: see (Cowan et al., 1940)
Lhote E. (1970): "La méthode glottospectrographique et la simulation de la parole;"
    Dissertation, Univ. des Langues et Lettres, Strasbourg [5.1.2; 5.2.3]
Lhote E. (1972): "Apport de la glottospectrographie a l'étude des tons." **ICPhS-7,** 944
    [5.1.2; 5.2.3]
Lhote E. (1973): "Contribution à l'étude de la fonction linguistique du larynx." Phonetica
    28, 26-41 [5.1.2; 5.2.3]
Lhote E. (1979): "Glottographie, glottographes et glottogramme." **Larynx et parole,**
    259-277 [5.2.3; 5.2.6]
Lhote E. (1981a): "Quelques aspects du comportement linguistique dans la perception de
    hauteur propre à la parole." In *Miszellen VII;* ed. by J.P. Köster. Hamburger Phone-
    tische Beiträge, Bd.35 (Buske, Hamburg), 63-73 [4.2; 9.4.2]
Lhote E. (1981b): *La parole et la voix. Analyse et synthèse de faits de langue au niveau
    du larynx.* Hamburger Phonetische Beiträge, Bd.37 (Buske, Hamburg) [3.1]
Lhote E., Arnott G., Hequet B., Milbled G. (1969): "Glotto-spectrographie et phonétique."
    Lille Médical **14,** 264-266 [5.1.2; 5.2.3]
Lhote E., Arnott G., Milbled G. (1969): "La glotto-spectrographie. Nouvelle méthode
    d'activité phonatoire." Comptes Rendus des Séances de la Société de Biologie **163,**
    [5.1.2; 5.2.3]
Lhote E., Filleau M., Grange M.F. (1975): "Reconnaissance de patrons intonatifs." **Journées
    d'Etude sur la Parole 6,** 28-37 [5.2.3; 9.4.2]

Lhote E., Milbled G., Arnott G. (1969): "Simulation de la fonction de transfert du fractus vocal à partir de la glotto-spectrographie." Comptes rendus des Séances de la Société de Biologie **163**, [5.1.2; 5.2.3]

Lhote F.: see (Filleau and Lhote, 1973; André and Lhote, 1977)

Li K.P.: see (Hughes et al., 1968)

Liang G., Chistovich L.A. (1961): "Frequency difference limens as a function of tonal duration." Soviet Physics - Acoustics **6**, 75-80 [4.2.3]

Licklider J.C.R. (1946): "Effects of amplitude distortion upon the intelligibility of speech." J. Acoust. Soc. Am. **18**, 429-434 [4.3; 6.1.2; 8.2.4]

Licklider J.C.R. (1960): "Audio pitch control;" United States Patent No. 2,943,152. Issued June 28, 1960; filed Nov. 7, 1957 [6.2.1]

Licklider J.C.R., Pollack I. (1948): "Effects of differentiation, integration, and infinite peak clipping upon the intelligibility of speech." J. Acoust. Soc. Am. **20**, 42-51 [4.3; 8.2.4]

Lidikh A.K.: see (Zorin and Lidikh, 1968)

Lieberman P. (1961): "Perturbations in vocal pitch." J. Acoust. Soc. Am. **33**, 597-603 [4.3; 6.5.3]

Lieberman P. (1962a): "Pitch perturbations of normal and pathologic larynxes." **ICA-4** (Paper G37) [4.3; 9.4.3]

Lieberman P. (1962b): "Pitch perturbations of normal and pathologic larynxes." **Proc. 1962 Speech Commun. Sem.** (Paper C6) [4.3; 9.4.3]

Lieberman P. (1963): "Some acoustic measures of the fundamental periodicity of normal and pathologic larynges." J. Acoust. Soc. Am. **35**, 344-353 [4.3; 6.5.3; 9.4.3]

Lieberman P. (1965): "On the acoustic basis of the perception of intonation by linguists." Word **21**, 40-54 [4.3; 9.4.2]

Lieberman P. (1967a): *Intonation, perception, and language.* Research Monograph #38 (MIT Press, Cambridge MA) [4.3; 9.4.2]

Lieberman P. (1967b): "Intonation and the syntactic processing of speech." In *Models for the perception of speech and visual form.* Symposium Nov. 11-14, 1964, AFCRL; ed. by W.Wathen-Dunn (MIT Press, Cambridge MA), 314-320 [5.1]

Lieberman P. (1968): "Determination of the rate of change of fundamental frequency with respect to subglottal air pressure during sustained phonation." **Speech Research 13**, 153-175 [3.1; 3.3; 4.3]

Lieberman P., Knudson R., Mead J. (1969): "Determination of the rate of change of fundamental frequency with respect to glottal air pressure during sustained phonation." J. Acoust. Soc. Am. **45**, 1537-1543 [3.1; 4.3]

Lieberman P., Michaels S.B. (1962a): "On the discrimination of missing pitch pulses." **Proc. 1962 Speech Commun. Sem.** (Paper C7) [4.3]

Lieberman P., Michaels S.B. (1962b): "Some aspects of fundamental frequency and envelope amplitude as related to the emotional content of speech." J. Acoust. Soc. Am. **34**, 922-927 [4.1; 9.4.3]

Lieberman P., Michaels S.B., Soron H.I. (1964): "Emotional effects of fundamental frequency transformations." J. Acoust. Soc. Am. **36** (A) (Paper X8; 67th Meet. ASA) [9.4.3]

Lieberman P., Sawashima M., Harris K.S., Gay T. (1970): "The articulation implementation of the breath-group and prominence: crico-thyroid muscular activity in intonation." Language **46**, 312-327 [3.1; 3.3; 9.4.2]

Lieberman P., Tseng C.Y. (1980): "On the fall of the declination theory: breath-group versus "declination" as the base form for intonation." J. Acoust. Soc. Am. **67** (A), S63 (Paper AA1; 99th Meet. ASA) [4.3; 9.4.2]

Liedtke C.E. (1971): "Rechnergesteuerte Spracherzeugung" (Heinrich-Hertz-Institut, Berlin; Technischer Bericht Nr.137) [6.2.2; 8.4.2]

Liénard J.S. (1977): "L'intonation: prosodie et reconnaissance automatique de la parole (Compte-rendu du rapporteur)." **Journées d'Etude sur la Parole 8**, 75-85 [9.4.2]

Liénard J.S., Manceron F. (1981): "Analyse impulsionnelle de la parole: expériences préliminaires." **Journées d'Etude sur la Parole 12**, 263-272 [6.2.1; 6.4.1]

Liiv G., Remmel M. (1974): "Estimate of the distinctive parameters in the domain of timing, fundamental frequency and intensity, with implications." **Proc. 1974 Speech Commun. Sem. 2**, 179-185 [4.3; 9.4.2]

Liljencrants J. (1962): "A few experiments of voiced-voiceless identification and time segmentation of speech." **STL-QPSR** (3), 16-24 [5.3]

Liljencrants J. (1965): "A filter bank speech spectrum analyzer." **ICA-5** (Paper A27) [3.4; 5.3]

Lim J.S., Oppenheim A.V. (1979): "Enhancement and bandwidth compression of noisy speech." Proc. IEEE **67**, 1586-1604 [5.3; 9.4.5]

Lin L.H.: see (Lin M.C. et al., 1979)

Lin M.C., Lin L.H., Xia G.R., Cao Y.S. (1979): "A study of tone-sandhi in standard Chinese with computer." **ICPhS-9** (1) (A), 388 [9.4.2]

Lin M.T. (1964): "Pitch indicator and pitch characteristics of standard colloquial Chinese tones" (Linguistic Institute, Peking, China), 23-44 [9.4.2]

Lin W.C.: see (Chandra and Lin, 1974; Pillay et al., 1975)

Lindblom B.E.F., Öhman S.E.G. (eds.) (1979): *Frontiers of speech communication research* (Academic Press, London)

Lindqvist J. (1963): "Inverse filtering equipment." **STL-QPSR** (1), 13 [6.5.1]

Lindqvist J. (1964): "Inverse Filtering. Instrumentation and techniques." **STL-QPSR** (4), 1-4 [6.5.1]

Lindqvist J. (1965a): "Studies of the voice source by means of inverse filtering technique." **ICA-5** (Paper A35) [3.3; 6.5.1]

Lindqvist J. (1965b): "Studies of the voice source by means of inverse filtering." **STL-QPSR** (2), 8-13 [3.3; 6.5.1]

Lindqvist J. (1969): "Laryngeal mechanisms in speech." **STL-QPSR** (2), 26-32 [3.1; 3.3; 6.5.1]

Lindqvist J. (1970): "The voice source studies by means of inverse filtering." **STL-QPSR** (1), 3-9 [3.3; 6.5.1]

Lindqvist J. (1972a): "A descriptive model of laryngeal articulation in speech." **STL-QPSR** (2-3), 1-9 [3.1]

Lindqvist J. (1972b): "Laryngeal articulation studied on Swedish subjects." **STL-QPSR** (2-3), 10-27 [3.1]

Lindqvist J. (1972c): "Laryngeal articulation in Swedish." **ICPhS-7**, 362-366 [3.1]

Lindqvist J., Sundberg J. (1971): "Octaves and pitch." **STL-QPSR** (1), 51-67 [4.2]

Lindsey W.C., Simon M.K. (1973): *Telecommunication systems engineering* (Prentice-Hall, Englewood Cliffs NJ) [6.1.4]

Lindström B.: see (Hertz et al., 1970)

Linggard R., Millar W. (1982): "Pitch detection by harmonic histograms." Speech Commun. **1**, 113-124 [8.5.2]

Linke E. (1973): "A study of pitch characteristics of female voices and their relationship to vocal effectiveness." Folia Phoniatr. **25**, 173-185 [4.1]

Lippus U., Remmel M. (1976): "Some contributions to the study of Estonian word intonation." **Phon. Tallinn 6**, 37-66 [9.4.2]

Lisker L., Abramson A.S. (1970): "The voicing dimension: Some experiments in comparative phonetics." **ICPhS-6**, 563-567 [5.3]

Lisker L., Abramson A.S. (1970): "The voicing dimension: Some experiments in comparative phonetics." Reprinted in *Acoustic Phonetics – A course of basic readings;* ed. by D.B.Fry (Cambridge University Press, Cambridge), 348-352 [5.3]

Lisker L., Abramson A.S. (1971): "Distinctive features and laryngeal control." **Speech Research 27**, 133-151 [3.1; 3.5]

Lisker L., Abramson A.S., Cooper F.S., Schvey M.H. (1966): "Transillumination of the larynx in running speech." J. Acoust. Soc. Am. **39** (A), 1218 (Paper 1F3; 71st Meet. ASA) [5.2.5]

Lisker L., Abramson A.S., Cooper F.S., Schvey M.H. (1969): "Transillumination of the larynx in running speech." J. Acoust. Soc. Am. **45**, 1544-1546 [5.2.5]

Lisker L., Sawashima M., Abramson A.S., Cooper F.S. (1970): "Cinegraphic observations of the larynx during voiced and voiceless stops." J. Acoust. Soc. Am. **48** (A), 119 (Paper DD5, 79th Meet. ASA) [5.2.5]

Liu B.: see (Peled and Liu, 1976)

Lobanov B.M. (1970): "Automatic discrimination of noisy and quasi periodic speech sounds by the phase plane method." [Original (in Russian) in "Akusticheskiy Zhurnal 16 (1970), pp.425-428]. Soviet Physics - Acoustics **16**, 353-356 [5.3]

Lochbaum C.C. (1960): "Segmentation of speech into voiced sounds, unvoiced sounds, and silence." J. Acoust. Soc. Am. **32**, 914 [5.3]

Lochbaum C.C., David E.E.jr., Mathews M.V. (1961): "Decision functions for voiced-unvoiced-silence detection." J. Acoust. Soc. Am. **33** (A), 852 [5.3]

Loebell E. (1968): "Über den klinischen Wert der Elektroglottographie." Arch. Ohren-Nasen-Kehlkopfheilkunde **191**, 760-764 [5.2.3]

Loebner H. (1982): "Zur Funktion und Beurteilung von Kompandersystemen." Frequenz **36**, 222-227 [6.2.1]

Löfqvist A. (1975a): "Some phonetic correlates of emphatic stress in Swedish." **Phon. Lund 10**, 105-117 [9.4.2]

Löfqvist A. (1975b): "Intrinsic and extrinsic F_0 variations in Swedish tonal accents." Phonetica **31**, 228-247 [3.1; 9.4.2]

Löfqvist A., Baer T., Yoshioka H. (1980): "Scaling of glottal opening." J. Acoust. Soc. Am. **67** (A), S53 (Paper V15; 99th Meet. ASA) [3.1; 3.3; 5.2.1]

Logan B.F.: see (David et al., 1962)

Lokerson D.C. (1979): "A conceptually unique speech training system." **ICASSP-79**, 479-482 [9.4.3]

Loktev N.A.: see (Arkhangelskiy et al., 1982)

Loo C. (1972): "An experimental study of the effect of filtering and compression on speech probability distribution." **ICSCP-72**, 364-367 (Paper J4)

Loonen A.R.: see (Willems and Loonen, 1967)

Lopes Cardozo B.: see (Ritsma and Lopes Cardozo, 1962)

Lord W.: see (Levin and Lord, 1975)

Lottermoser W. (1950): "Melodiekurven europäischer Rundfunksprecher." Z. Phonetik **4**, 369-381 [4.1; 9.4.2]

Louie T. (1981): "Array processors: a selected bibliography." IEEE Computer (9), 53-57 [1.3; 9.1.5]

Lubker J.: see (Risberg and Lubker, 1978)

Lublinskaya V. (1969):"Воспроисдение простых контуров изменения частоты основного тона звуков." В.Лублинская. Ленинград: Институт Физиологии Имени И.В.Павлова, Ак.Наук СССР. (Reproduction of simple contours of pitch change. In Russian) (Leningrad, USSR), Vol.8 [4.3; 9.4.2]

Lucas J.P.: see (Carcaud et al., 1974)

Luchsinger R. (1958): "Über die Bedeutung der synchronen Registrierung der Sprachmelodie und des dynamischen Akzentes für die Sprachpathologiebeschreibung eines Sprachspektrometers." Folia Phoniatr. **10**, 84-96 [5.2.1; 5.2.6; 9.4.3]

Luchsinger R. (1975): "Zeitdehneraufnahmen der Stimmlippen beim offenen und gedeckten Singen." Folia Phoniatr. **27**, 88-92 [3.2; 3.3]

Luchsinger R., Arnold G.E. (eds.) (1970): *Handbuch der Stimm- und Sprachheilkunde;* ed. by R.Luchsinger and G.E.Arnold (Springer, Wien) [3.0; 9.4.3]

Luchsinger R., Arnold G.E. (1975): *Voice-speech-language clinical communicology: its physiology and pathology* (Wadsworth Pub. Co., Belmont CA) [3.1; 9.4.3]

Luchsinger R., Dubois C. (1963): "Ein Vergleich der Sprachmelodie- und Lautstärkekurve bei Normalen, Gehirnkranken und Stotterern." Folia Phoniatr. **15**, 21-41 [4.1; 9.4.3]

Luchsinger R., Pfister K. (1961): "Die Messung der Stimmlippenverlängerung bein Steigern der Tonhöhe." Folia Phoniatr. **13**, 1-12 [3.1]

Ludlow C., Coulter D.C., Cardano C. (1979): "Application of pitch perturbation measures to the assessment of hoarseness in Parkinson's disease." J. Acoust. Soc. Am. **66** (A), S64-65 (Paper EE10; 98th Meet. ASA) [4.3; 9.4.3]

Lukatela G. (1973): "Pitch determination by adaptive autocorrelation method." **Speech Research 33**, 185-193 [8.2.4]

Lukatela G. (1975): "A Hilbert transformation preprocessing of telephone quality speech signal at pitch extractor input." In *Proc. of the First FASE Congress, Paris, October 1975* (Paris), 17-26 [6.3.2]

Lukatela G., Tomic T. (1974): "Adaptive autocorrelation techniques for vocal pitch detection." **ICA-8**, 244 [8.2.4]

Lukatela G., Tomić T., Drajic D. (1976): "Efficiency and accuracy in vocal pitch determination by some methods using digital computer." **Proc. 1974 Speech Commun. Sem. 1**, 139-147 [8.2.4; 9.3.1]

Lunday A.M. (1967): "The vocal quality and pitch of voices suspected of laryngeal pathology;" Dissertation, Ohio State University, Columbus OH [3.3; 4.1; 9.4.3]

Lyberg B. (1979): "Final lengthening - partly a consequence of restrictions on the speed of fundamental frequency change?" J. Phonetics **7**, 187-196 [4.3; 9.4.2]

Lyon J.G.: see (Wängler and Lyon, 1974)

Lyons M.C., Towers B. (eds.) (1962): *Galen on anatomical procedures: the later books.* Translated by W.L.H.Duckworth (University Press, Cambridge UK) [3.1; 5.1.1; 5.2.1]

Lytle E.G.: see (Melby et al., 1975)

Lyudovik E.K. (1976):"Определение периода основного тона с помощью динамического программирования." В кн.: Труды АРСО-9, Минск. Е.К.Людовик. (Pitch period determination by dynamic programming. In Russian.). **ARSO-9** [6.6.3; 8.3.3]

Lyudovik E.K., Fedorinchik S.M. (1980): "Алгоритм для определения мгновенной длины периода основного тона." В кн.: Труды АРСО-11, Ереван. Е.К. Людовик и С.М. Федоринчик. (Algorithm for detecting the momentary length of the fundamental period. In Russian). **ARSO-11**, 64-67 [8.3.3]

Maack A. von (1957a): "Verzerrungsfreie Melodiewinkel aus der Tonhöhenkurve." Phonetica 1, 206-215 [4.3; 5.1; 6.6.1; 9.4.2]

Maack A. von (1957b): "Melodiewinkel und Einsatztonhöhe." Phonetica 1, 216-229 [9.4.2]

Mabis J.H.: see (Coleman et al., 1977)

MacKinnon D.A., Lee H.C. (1976): "Real time recognition of unvoiced fricatives in continuous speech to aid the deaf." **ICASSP-76**, 586-589 [3.5; 5.3; 9.4.4]

MacLean D.L.: see (Golden et al., 1964)

Maddieson I., Gandour J.T. (1974): "An annotated bibliography on tone." **Phon. Los Angeles 28** [1.3; 9.4.2]

Maeda Sh. (1968): "Voice aperiodicity in terms of pulsed shape and intervals." **ICA-6**, 43-46 [3.3; 4.3]

Maeda Sh. (1974a): "Characterization of fundamental-frequency contours of speech." J. Acoust. Soc. Am. 56 (A), S33 (Paper P12; 88th Meet. ASA) [9.4.2]

Maeda Sh. (1974b): "A characterization of fundamental frequency contours of speech." **MIT-RLE Progr. Rept. 114**, 193-212 [9.4.2]

Maeda Sh. (1975): "Electromyographic study of intonational contours." **MIT-RLE Progr. Rept. 115**, 261-269 [5.2.1; 9.4.2]

Maeda Sh. (1977): "Sur les corrélatifs physiologiques de la fréquence du fondamental de la parole." **Journées d'Etude sur la Parole 8**, 43-49 [9.4.2]

Maeda Sh. (1979): "On the F_0 control mechanisms of the larynx." **Larynx et parole**, 243-258 [3.1]

Maeda Sh., Fujimura O. (1969): "Factors of the glottal wave that contribute to the naturalness of speech." **Ann. Bull. RILP 3**, 46-51 [3.3]

Maezono K. (1963): "Autocorrelation-type pitch extractor." J. of the IECE of Japan 46, 1083-1091 [8.2.3]

Magar S.: see (McDonough et al., 1982)

Magdics K. (1963): "Research on intonation during the past ten years." Acta Linguistica Hafniensia (Copenhagen) 13, 133-165 [9.4.2]

Magdics K. (1964a): "First findings in the comparative study of intonation of Hungarian dialects, part 1." Phonetica 11, 19-38 [9.4.2]

Magdics K. (1964b): "First findings in the comparative study of intonation of Hungarian dialects, part 2." Phonetica 11, 101-115 [9.4.2]

Magill D.T., Un C.K. (1974a): "Speech digitization by LPC estimation techniques" (Stanford Research Inst, Menlo Park CA; DDC-AD-785738/6GA) [6.3.1; 9.4.5]

Magill D.T., Un C.K. (1974b): "Residual excited linear predictive coder." J. Acoust. Soc. Am. 55 (A), S81 (Paper NN3; 87th Meet. ASA) [9.4.5]

Mahshie J.: see (Rothenberg and Mahshie, 1977)

Maidment J. (1978): "Linear approximations to fundamental frequency." **Phon. London**, 61-70

Maissis A.H. (1972): "Méthode d'extraction du pitch." **Journées d'Etude sur la Parole 3**, 309-315 [6.2.5]

Maissis A.H. (1973a): "Le traitement de l'information acoustique, étape fondamentale de la reconnaissance de la parole;" Thèse d'Etat, Université Pierre et Marie Curie (Paris 6), Paris [9.4.4]

Maissis A.H. (1973b): "Une méthode d'extraction du fondamental." L'Onde Electrique 53 [6.2.5]

Maitra S., Davis C.R. (1979): "A speech digitizer at 2400 bits/s." IEEE Trans. ASSP-27, 729-733 [8.2.4; 9.4.5]

Maitra S., Davis C.R. (1980): "Improvements in the classical model for better speech quality." **ICASSP-80**, 23-27 [3.3; 9.4.5]

Maitra S., Foster S.H., Davis C.R. (1980): "A maximum peakiness criterion for deconvolving speech waveforms." **ICASSP-80**, 154-157 [6.3.1]

Maitra S., Jain J.R. (1982): "Event detection: a generalized concept for pitch identification (Presented at ICASSP-82)" (Vocom Inc., 625 Ellis St, Mountain View CA) [6.3.2]

Majewski W., Blasdell R. (1969): "Influence of fundamental frequency cues on the perception of some synthetic intonation contours." J. Acoust. Soc. Am. 45, 450-457 [4.2; 9.4.4]

Majewski W., Hollien H., Zalewski J. (1972): "Speaking fundamental frequency of Polish adult males." Phonetica 25, 119-125 [4.1]

Majewski W., Myślecki W., Zalewski J. (1979): "Optimal intonation contours for Polish speech synthesis." **ICPhS-9** (1) (A), 354 [9.4.2; 9.4.4]

Makhonin V.A.: see (De Mori et al., 1977)

Makhoul J. (1975a): "Spectral linear prediction: Properties and applications." IEEE Trans. ASSP-**23**, 283-296

Makhoul J. (1975b): "Linear prediction: a tutorial review." Proc. IEEE **63**, 561-580 [6.3.1; 6.5.1; 9.4.4; 9.4.5]

Makhoul J., Berouti M.G. (1979a): "High-frequency regeneration in speech coding systems." **ICASSP-79**, 428-431 [9.4.5]

Makhoul J., Berouti M.G. (1979b): "Predictive and residual encoding of speech." J. Acoust. Soc. Am. **66**, 1633-1641 [9.4.5]

Makhoul J., Viswanathan V.R., Schwartz R., Huggins A.W.F. (1978): "A mixed-source model for speech compression and synthesis." **ICASSP-78**, 163-166 [4.3; 5.3; 9.4.5]

Makhoul J., Wolf J.J. (1972): "Linear prediction and the spectral analysis of speech" (Bolt Beranek and Newman Inc., Cambridge MA; Report No. 2304) [2.3; 6.3.1; 8.4.1]

Maksym J.N. (1972a): "Iterative adjustment of predictive quantizers;" Dissertation, Carleton University, Ottawa [6.3.1]

Maksym J.N. (1972b): "Real time pitch extraction by adaptive prediction of the speech waveform." **ICSCP-72**, 70-73 (Paper B6) [6.3.1]

Maksym J.N. (1973): "Real time pitch extraction by adaptive prediction of the speech waveform." IEEE Trans. AU-**21**, 149-150 [6.3.1]

Malah D. (1979): "Time-domain algorithms for harmonic bandwidth reduction and time scaling of speech signals." IEEE Trans. ASSP-**27**, 121-133 [9.4.5]

Malah D. (1980): "Combined time-domain harmonic compression and CVSD for 7.2 kbit/s transmission of speech signals." **ICASSP-80**, 504-507 [9.4.5]

Malah D., Crochiere R.E., Cox R.V. (1981): "Performance of transform and subband coding system combined with harmonic scaling of speech." IEEE Trans. ASSP-**29**, 273-283 [9.4.5]

Malécot A., Peebles K. (1965): "An optical device for recording glottal adduction/abduction during normal speech." Z. Phonetik **18**, 545-550 [5.2.5]

Malmberg B. (1962): "Analyse instrumentale et structurale des faits d'accent." **ICPhS-4**, 456-475 [9.4.2]

Malmberg B. (ed.) (1968): *Manual of Phonetics* (North Holland Publishing Company, Amsterdam)

Manceron F.: see (Liénard and Manceron, 1981)

Manen L.: see (Fry and Manen, 1957)

Mangold H.: see (Gertheiss et al., 1974)

Manley H.J. (1963): "Analysis-synthesis of connected speech in terms of orthogonalized exponentially damped sinusoids." J. Acoust. Soc. Am. **35**, 464-474

Manolson A. (1972): "Comparative study of intonation patterns in normal hearing and hearing impaired infants." **ICPhS-7**, 962-965 [9.4.3]

Markel J.D. (1971a): "Formant trajectory estimation from a linear least-squares inverse filter formulation" (Speech Communications Res. Lab, Santa Barbara CA; SCRL Monograph No. 7) [3.4; 6.5.1]

Markel J.D. (1971b): "FFT Pruning." IEEE Trans. AU-**19**, 305-311 [8.5.4]

Markel J.D. (1972a): "Application of a digital inverse filter for automatic formant and F_0 analysis." **ICSCP-72**, 81-84 (Paper B9) [6.3.1; 8.2.5]

Markel J.D. (1972b): "The SIFT algorithm for fundamental frequency estimation." IEEE Trans. AU-**20**, 367-377 [8.2.5; 9.3.1]

Markel J.D. (1973): "Application of a digital inverse filter for automatic formant and F0 analysis." IEEE Trans. AU-**21**, 154-160 [6.3.1; 8.2.5]

Markel J.D., Gray A.H. (1975): "An optimal linear prediction synthesizer structure for array processor implementation." In *Speech Recognition*. Invited papers of the 1974 IEEE Symposium; ed. by D.R.Reddy (Academic Press, New York), 231-240 [2.3; 6.3.1]

Markel J.D., Gray A.H. (1976): *Linear prediction of speech*. Communications and Cybernetics, vol.12 (Springer, Berlin) [1.1; 6.3.1; 8.2.5]

Markel J.D., Koike Y. (1973): "Early acoustic detection of laryngeal pathology." J. Acoust. Soc. Am. **54** (A), 38 (Paper G16; 86th Meet. ASA) [9.4.3]

Markel J.D., Wong D.Y. (1976): "Considerations in the estimation of glottal volume velocity waveforms." J. Acoust. Soc. Am. **59** (Paper RR6, 91th Meet.ASA) [6.5.1]

Marko H. (1980): *Methoden der Systemtheorie* (Springer, Berlin) [2.2]

Martin J.A.M. (1981): *Voice, speech, and language in the child: development and disorder*. Disorders of human communication, vol.4 (Springer, Wien) [9.4.3]

Martin Ph. (1971): "Linguistique appliquée et enseignement de l'intonation." Etudes de Linguistique Appliquée **3**, 36-45 [9.4.2; 9.4.3]

Martin Ph. (1972a): "Reconnaissance automatique de patrons intonatifs." **Phonetica Pragensia III**, 77-81 [9.4.2; 9.4.4]

Martin Ph. (1972b): "L'analyseur de mélodie du laboratoire de phonétique expérimentale de l'Université de Toronto." **ICPhS-7**, 1272-1274 [5.1; 6.4.1; 9.4.2]

Martin Ph. (1973): "Réalisation d'un analyseur de mélodie fonctionnant en temps réel en vue de la recherche phonétique et l'enseignement;" Dissertation, Université Libre de Bruxelles [6.4.1; 9.4.2; 9.4.3]

Martin Ph. (1975): "Intonation et reconnaissance automatique de la structure syntaxique." **Journées d'Etude sur la Parole 6**, 51-62 [9.4.2]

Martin Ph. (1977a): "Résumé d'une théorie de l'intonation de l'italien." **Phon. Grenoble 6**, 57-87 [9.4.2]

Martin Ph. (1977b): "L'intonation: méthodes et modèles d'étude des rapports intonation-syntaxe (Compte-rendu du rapporteur)." **Journées d'Etude sur la Parole 8**, 61-74 [9.4.2]

Martin Ph. (1977c): "Un analyseur-visualiseur de mélodie à microprocesseur." **Journées d'Etude sur la Parole 8**, 95-102

Martin Ph. (1977d): "Problèmes de neutralisation des marques prosodiques. Application à la reconnaissance automatique." **Journées d'Etude sur la Parole 8**, 305-314 [9.4.2; 9.4.4]

Martin Ph. (1979): "Perception des séquences de contours prosodiques de phrases synthétisées." **Journées d'Etude sur la Parole 9**, 21-29 [9.4.2; 9.4.4]

Martin Ph. (1981): "Détection de F_0 par intercorrélation avec une function peigne." **Journées d'Etude sur la Parole 12**, 221-232 [4.1; 8.1; 8.5.2; 8.5.4]

Martin Ph. (1982a): "Comparison of pitch detection by cepstrum and spectral comb analysis." **ICASSP-82**, 180-183 [8.4.2; 8.5.1; 9.3.1]

Martin Ph. (1982b): "Phonetic realization of prosodic contours in French." Speech Commun. **1**, 283-294 [9.4.2]

Martin T.B. (1976): "Practical application of voice input to machines." Proc. IEEE **64**, 487-501

Martin T.B. (1977): "One way to talk to computers." IEEE Spectrum **14**, 35-39

Mártony J. (1962): "Studies of the voice source." **STL-QPSR** (2), 9 [3.3]

Mártony J. (1965): "Studies of the voice source." **STL-QPSR** (1), 4-9 [3.1; 6.5.1]

Mártony J. (1968): "On the correction of the voice pitch level for severely hard of hearing subjects." Am. Ann. Deaf **113** [9.4.3]

Mártony J. (1971): "Pitch range in deaf children." In *Papers in interdisciplinary speech research.* Proceedings of the Speech Symposium Szeged 1971; ed. by J.Hirschberg, Gy.Szepe, E.Vass-Kovács (Akademiai Kiadó, Budapest), 171-173 [9.4.3]

Martynov V.S. (1958): "Pitch determination by amplitude selection [referenced by Baronin and Kushtuev, 1974]" [6.2.1]

Martynov V.S. (1965): "Статистические параметры основного тона речи." Кандидатская диссертация. ЛЭИС,Ленинград. В.С. Мартынов. (Statistic parameters of the fundamental frequency of speech signals. In Russian.) (Leningrad) [4.1]

Martynov V.S. (1966): "Способ выделения сигнала тон-шум." А.С.№ 177939-СССР. Опубл. в бюлл. "Открытия, Изобретения, Промышленные образцы, Товарные знаки. Б.С.Мартынов. (Method for buzz-hiss detection. In Russian.). Otkrytia, Izobretenia **2**, 43 [5.3]

Martynov V.S. (1970): "Устройство для выделения основного тона речевых звуков." А.С.№ 120660-СССР. Опубл. в бюлл. "Открытия, Изобретения, Промышленные образцы, Товарные знаки. В.С.Мартынов. (Device for pitch determination of speech sounds. In Russian.). Otkrytia, Izobretenia **12**, 43

Mataušek M.R., Batalov V.S. (1980): "A new approach to the determination of the glottal waveform." IEEE Trans. ASSP-**28**, 616-622 [6.5.1]

Mathews M.V., Miller J.E., David E.E.jr. (1961a): "Pitch synchronous analysis of voiced sounds." J. Acoust. Soc. Am. **33**, 179-186 [9.4.5]

Mathews M.V., Miller J.E., David E.E.jr. (1961b): "An accurate estimate of the glottal waveshape." J. Acoust. Soc. Am. **33** (A), 843 [6.5.1]

Mathews M.V., Pierce J.R. (1980): "Harmony and nonharmonic partials." J. Acoust. Soc. Am. **68**, 1252-1257 [4.2; 8.5.2]

Matsudaira M.: see (Ishizaka and Matsudaira, 1968)

Matsushita H. (1975): "The vibratory mode of the vocal folds in the excised larynx." Folia Phoniatr. **27**, 7-18 [3.1; 3.3]

Mattingly I.G. (1966): "Synthesis rule of prosodic features." Lang. Speech **1**, 1-13 [9.4.2]

Mattingly I.G. (1968): "Synthesis by rule of general American English." **Speech Research,** [9.4.2; 9.4.4]

Matuszkina O. (1978): "Variations of the fundamental frequency in Polish voiced consonants." Arch. Acoustics **3**, 105-119 [3.1; 4.1; 9.4.2]

Mauduit D.: see (Esteban et al., 1978)

Maurel O.: see (Esteban et al., 1979)

McAulay R.J. (1978): "Maximum likelihood pitch estimation using state variable techniques." **ICASSP-78,** 12-14 [8.6.1]

McAulay R.J. (1979): "Design of a robust maximum-likelihood pitch estimator for speech in additive noise" (MIT Lincoln Labs, Lexington MA; DDC-AD-A077159/2) [8.6.1; 9.4.4]

McAuliffe D.R.: see (Davis H. et al., 1951)

McCandless S.S. (1974): "An algorithm for automatic formant extraction using linear prediction spectra." IEEE Trans. ASSP-**22**, 135-141 [6.4.1]

McClellan J.H.: see (Parks and McClellan, 1972)

McCutcheon M.J.: see (Stilwell et al., 1981)

McDonald H.S.: see (David and McDonald, 1956)

McDonald W.E.: see (Bull et al., 1979)

McDonough K., Caudel E., Magar S., Leigh A. (1982): "Microcomputer with 32-bit arithmetic does high-precision number crunching." Electronics (Feb. 24), 105-110 [9.1.5]

McGarr N.S.: see (Baer et al., 1981)

McGlone R. (1967): "Air flow during vocal fry." J. Speech Hear. Res. **10**, 299-304 [3.2]

McGlone R. (1969): "Laryngeal dynamics associated with voice frequency changes." J. Acoust. Soc. Am. **46**, 1033-1036 [3.1]

McGlone R. (1977): "Acoustic analysis of voice disguise related to voice identification." In *Proceedings of the 1977 International Carnahan Conference on Crime Counter-measures* (Univ. of Kentucky, Lexington KY), 31-35 [9.4.4]

McGlone R., Hollien H. (1963): "Vocal pitch characteristics of aged women." J. Speech Hear. Res. **6**, 164-170 [4.1]

McGlone R., Shipp T. (1969): "Identification of the "Shift" between vocal registers." J. Acoust. Soc. Am. **46**, 1033-1036 [3.2]

McGonegal C.A., Rabiner L.R., Rosenberg A.E. (1975): "A semiautomatic pitch detector (SAPD)." IEEE Trans. ASSP-**23**, 570-574 [6.2.5; 8.2.4; 8.4; 9.2.2]

McGonegal C.A., Rabiner L.R., Rosenberg A.E. (1976): "Perceptual evaluation of several pitch detection algorithms." J. Acoust. Soc. Am. **60** (A), S107 (Paper RR5; 92nd Meet. ASA) [4.2; 9.3.1]

McGonegal C.A., Rabiner L.R., Rosenberg A.E. (1977): "A subjective evaluation of pitch detection methods using LPC synthesized speech." IEEE Trans. ASSP-**25**, 221-229 [4.2; 9.3.1]

McHugh P.G.: see (Gold et al., 1971)

McKinney N.P. (1965): "Laryngeal frequency analysis for linguistic research" (Commun. Sci. Lab., Univ. of Michigan, Ann Arbor MI; Res. Rept. #14) [1.2; 1.3; 2.1; 2.6; 4.3; 5.1; 5.2.2; 6.1; 6.1.1; 6.6.2; 6.6.3; 7.2; 7.4; 9.4.2]

McKinney N.P., Carlson R. (1976): "A hybrid multiple-channel pitch-frequency analysis system." **Phon. Los Angeles 33**, 203-215 [6.4.2]

McLaughlin M., Hudziak F., Gerson I., Kloker K. (1980): "High-performance processor for real-time speech applications." **ICASSP-80,** 859-863 [9.1.5]

McLean D.J.: see (Lawrence et al., 1956)

McPherson D.F.: see (Tenold et al., 1974)

Mead J.: see (Lieberman et al., 1969; Hixon et al., 1971)

Medress M.F.: see (Lea et al., 1974)

Medvedev V.I. (1959): "Физиологический анализ колебаний голосовых цвясок." В журн.: "Проблемы физиологической акустики". В.И.Медведев. (Physiological analysis of voice oscillations. In Russian.). Problemy fiziologicheskoy akustiki **4**, 208-215

Meerbergen J.L. van: see (Zuidweg et al., 1982)

Megna J.J. (1964): "Pitch-period reiterating speech-compression system." J. Acoust. Soc. Am. **36**, 1030 (Paper N6; 67th Meet. ASA) [9.4.5]

Meinhart D.I.S.: see (Flanagan and Meinhart, 1964)

Meisel W.S. (1972): *Computer-oriented approaches to pattern recognition* (Academic Press, New York) [5.3.3]

Melby A.K., Millet R., Strong W.J., Lytle E.G. (1975): "Modifying fundamental frequency envelopes." J. Acoust. Soc. Am. **58** (A), S120 (Paper EEE8; 90th Meet. ASA) [9.4.5]

Melka A. (1970): "Messungen der Klangeinsatzdauer bei Musikinstrumenten." Acustica **23**, 108-117 [9.4.2]

Melas J.L.: see (Cohn and Melsa, 1975)

Menez J.: see (Esteban et al., 1978)

Mensch B. (1964): "Analyse par exploration ultrasonique du mouvement des cordes vocales isolées." Comptes Rendus des Séances de la Société de Biologie **158**, 2295-2296 [5.2.4]

Meo A.R., Righini G.U. (1971a): "A new technique for analyzing speech by computer." ICA-7 (1), 317-320 (Paper 19E15) [8.3.1]

Meo A.R., Righini G.U. (1971b): "A new technique for analyzing speech by computer." Acustica **25**, 261-268 [5.3; 6.4.1]

Mercier G. (1976): "Analyse phonétique." **Recherches/Acoustique 3**, 177-190 [5.3]

Mermelstein P. (1971): "Computer generated spectrogram displays for on-line speech research." IEEE Trans. AU-**19**, 44-47 [5.1.2]

Metfessel M. (1926): "Technique for objective studies of the vocal art." Psychol. Monogr. **36**, 1-40 [5.1.1]

Metfessel M. (1928): "A photographic method of measuring pitch." Science **68**, 430-432 [5.1.1]

Metfessel M. (1929): "The strobophotograph: A device for measuring pitch." J. Gen. Psychol. **2**, 135-138 [5.1.1; 5.2.1]

Mettas O. (1963): *Etude sur les facteurs ecto-sémantiques de l'intonation en français* (Travaux de Linguistique et de Litérature, Strasbourg) [9.4.2]

Mettas O. (1971): *Les techniques de la phonétique instrumentale et l'intonation* (Presses Universitaires de Bruxelles) [5.1; 9.4.2]

Meulen M.L. van der: see (Zuidweg et al., 1982)

Meyer E.A. (1911): "Ein neues Verfahren zur graphischen Bestimmung des musikalischen Akzents." Medizinisch-pädagogische Monatsschrift für die gesamte Sprachheilkunde [5.1.1]

Meyer E.A. (1937): *Die Intonation im Schwedischen* (Fritzes Bokförlag, Stockholm) [5.1.1; 9.4.2]

Meyer E.A., Schneider C. (1913): "Theorie des Tonhöhen-Messapparates." Vox **25**, 152-163 [5.1.1]

Meyer M.F. (1960): "Temporal irregularity of excitations: how much is accepted by the brain for reporting pitch?" J. Acoust. Soc. Am. **32**, 391-393 [4.3]

Meyer-Eppler W. (1948): "Tonhöhenschreiber." Z. Phonetik **2**, 16-38 [2.6; 9.3]

Meyer-Eppler W. (1950): "Reserved speech and repetition systems as means of phonetic research." J. Acoust. Soc. Am. **22**, 804-806 [9.2.1; 9.4.2]

Meyer-Eppler W., Endres W. (1967): "Automatische Spracherkennung und synthetische Sprache." In *Taschenbuch der Nachrichtenübertragung;* ed. by K. Steinbuch (Springer, Berlin), 787-814 [2.1]

Meyers W.I.: see (Sommers et al., 1961)

Mezzalama M. (1979): "Influence of the position of the analysis frame in LPC pitch-synchronous analysis." Signal Processing **1**, 191-204 [9.4.5]

Mezzalama M., Rusconi E. (1976): "Intonation in speech synthesis: a preliminary study for the Italian language." **Proc. 1974 Speech Commun. Sem. 2**, 317-327 [9.4.2]

Mezzalama M., Rusconi E., Torasso P. (1976): "Automatic generation of pitch contour for speech synthesis." ICASSP-76, 82-87 [9.4.2]

Michaels S.B., Lieberman P. (1962): "Discrimination of missing pitch pulses." J. Acoust. Soc. Am. **34** (A), 729 [4.2.3]

Michaels S.B., Soron H.I., Strong W.J., Lieberman P. (1965): "Analysis-synthesis of glottal excitation." J. Acoust. Soc. Am. **37** (A), 1186 (Paper I3; 71st Meet. ASA) [3.3; 6.5.1; 9.4.4]

Michaels S.B., Strong W.J. (1965): "Analysis-Synthesis of Glottal Excitation." J. Acoust. Soc. Am. **38** (A), 935 (Paper P8; 71st Meet. ASA) [6.5.1; 9.4.4]

Michel C. van (1966): "Mouvements glottiques phonatoires sans émission sonore. Etude électroglottographique." Folia Phoniatr. **18**, 1-8 [3.3; 3.5; 5.2.3]

Michel C. van, Pfister K., Luchsinger R. (1970): "Electroglottographie et cinématographie laryngée ultra-rapide." Folia Phoniatr. **22**, 81-91 [5.2.1; 5.2.3; 5.2.6]

Michel C. van, Raskin L. (1969): "L'électroglottomètre Mark 4, son principe, ses possibilités." Folia Phoniatr. **21**, 145-157 [5.2.3]

Michel J.F. (1968): "Fundamental frequency investigation of vocal fry and harshness." J. Speech Hear. Res. **11**, 590-594 [3.2; 4.3]

Michel J.F., Hollien H. (1968): "Perceptual differentiation of vocal fry and harshness." J. Speech Hear. Res. **11**, 439-443 [3.2; 4.2.1; 4.3]

Michel J.F., Hollien H., Monroe P. (1966): "Speaking fundamental frequency characteristics of 15-16 and 17 year old girls." Lang. Speech **9**, 46-51 [4.1]

Mihashi S.: see (Hirano et al., 1979)

Mikheev Yu.V. (1971): "Statistical distribution of the periods of the fundamental tone of Russian speech [Original (in Russian) in Akusticheskiy Zhurnal 16 (1970), pp.558-562]." Soviet Physics - Acoustics **16**, 424-472 [4.1]

Miki N.: see (Miyanaga Y., 1982)

Mikiel W.: see (Niedźwiecki and Mikiel, 1976; Kacprowski et al., 1977; Gubrynowicz et al., 1980)

Milbled G.: see (Lhote et al., 1969)

Mil'chenko V.I.: see (Arkhangelskij et al., 1982)

Mildenberger O. (1977): "Eine Methode zur Messung von Korrelationsfunktionen bei periodisch stationären Zufallssignalen." Nachrichtentech. Z. **30**, 717-720 [8.2.1]

Milios E.: see (Protonotarios et al., 1980)

Millar B.F.: see (Rozsypal and Millar, 1979)

Millar J.B., Underwood M.J. (1975): "Pitch synchronous gating in a time domain analysis of vowel waveforms." Acustica **32**, 314-320 [9.4.5]

Millar W., Linggard R. (1981): "Periodic pitch detection." In *Proceedings of the Spring Conference, 22-24 April 1981* (Inst. Acoustics, Edinburgh) [8.5.2]

Miller G.A.: see (Stevens S.S. et al., 1947)

Miller J.E. (1982): "Pitch extraction by adaptive window time averaging." J. Acoust. Soc. Am. **72** (A), S78 (Paper RR1, 104th Meet. ASA) [8.3.2]

Miller J.E., Fujimura O. (1979): "A graphic display for combined presentation of acoustic and articulatory information." **ASA*50**, 221-224 [5.1]

Miller J.E., Mathews M.V. (1963): "Investigation of the glottal waveshape by automatic inverse filtering." J. Acoust. Soc. Am. **35**, 1876 (Paper B3, 66th Meet. ASA) [3.3; 6.5.1]

Miller J.E., Mathews M.V., David E.E.jr. (1959): "Pitch synchronous spectral analysis of vowel sounds." J. Acoust. Soc. Am. **31** (A) [9.4.5]

Miller M.: see (Bergeron et al., 1980)

Miller N.J. (1974): "Pitch detection by data reduction." **IEEE Symp. Speech Recogn.**, 122-128 (Paper T9) [6.2.3]

Miller N.J. (1975): "Pitch detection by data reduction." IEEE Trans. ASSP-**23**, 72-79 [6.2.3; 7.2; 9.3.1]

Miller R.L. (1951): "Determination of pitch frequency of complex waves." J. Acoust. Soc. Am. **25** (A), 1035 [6.4.1]

Miller R.L. (1953a): "Determination of pitch frequency of complex waves;" United States Patent No. 2,627,541. Issued Feb. 3, 1953; filed June 20, 1951 [8.5.1]

Miller R.L. (1953b): "Determination of the pitch frequency of complex waves;" United States Patent No. 2,927,969. March 8, 1960 [6.1.1]

Miller R.L. (1959): "Nature of the vocal cord wave." J. Acoust. Soc. Am. **31**, 667-677 [3.3; 6.5.1]

Miller R.L. (1968): "Pitch determination by measurement of harmonics." J. Acoust. Soc. Am. **44** (A), 390 [8.5.1]

Miller R.L. (1970): "Performance characteristics of an experimental harmonic identification pitch extraction system (HIPEX)." J. Acoust. Soc. Am. **47**, 1593-1601 [8.5.1]

Miller R.L., Weibel E.S. (1956): "Measurements of the fundamental period of speech using a delay line." J. Acoust. Soc. Am. **28** (A) (Paper DD10; 51th Meet. ASA) [8.2.2; 8.3]

Millet R.: see (Melby et al., 1975)

Milton-Williams J.: see (Denes and Milton-Williams, 1962)

Mimpen A.M.: see (Plomp and Mimpen, 1968)

Minifie F.D., Kelsey C.A., Hixon T.J. (1969): "Measurement of vocal fold motion using an ultrasonic Doppler velocity monitor." J. Acoust. Soc. Am. **43**, 1165-1169 [5.2.4]

Minifie F.D., Kim B.W., Ewanowski S.J. (1972): "The nature of the glottal tone: a clinical case report." J. Speech Hear. Res. **15**, 142-147 [3.3; 4.1; 5.1.2; 9.4.3]

Minne J.M. van der: see (van den Berg and van der Minne, 1962)

MIT-RLE Progr. Rept. --> *MIT-RLE (Quarterly) Progress Report* (MIT Research Lab. of Electronics, Cambridge MA)

Mitchell D. (1961): "Analyzing system for determining the fundamental frequency of a complex wave;" United States Patent No. 2,561,478. Issued July 24, 1961

Miura T.: see (Kitamura et al., 1967)

Miyaguchi H.: see (Gohbara et al., 1978)

Miyanaga Y., Miki N., Nagai N., Hatori K. (1982): "A speech analysis algorithm which eliminates the influence of pitch using the model reference adaptive system." IEEE Trans. ASSP-**30**, 88-96 [6.5.1; 9.4.5]

Miyazaki S., Sekimoto S., Ishida H., Sawashima M. (1973): "A computerized method of frame-by-frame film analysis for fiberscopic measurement of the glottis." **Ann. Bull. RILP 7**, 35-38 [5.2.1]

Möller A.: see (Risberg et al., 1960)

Mörner M., Fransson F., Fant G. (1964): "Voice register terminology and standard pitch." **STL-QPSR** (4) [3.2; 4.1]

Molinari A.M.: see (Drago et al., 1978)

Molitor R.D.: see (Rothenberg and Molitor, 1979)

Moll K.L. (1960): "Cinefluorographic techniques in speech research." J. Speech Hear. Res. **3**, 227-241 [5.2.1]

Mondshein L.F.: see (Weinstein et al., 1975)

Monroe P.: see (Michel et al., 1966)

Monsen R.B., Engebretson A.M. (1977): "Study of variations in the male and female glottal wave." J. Acoust. Soc. Am. **62**, 981-993 [3.3; 4.1; 6.5.1]

Monsen R.B., Engebretson A.M., Fisher W.M. (1974): "Some characteristics of the glottal sound source of deaf children." J. Acoust. Soc. Am. **56** (A), S61 (Paper DD4; 88th Meet. ASA) [3.3; 9.4.3]

Monsen R.B., Engebretson A.M., Vemula N.R. (1978): "Indirect assessment of the contribution of subglottal air pressure and vocal-fold tension to changes of fundamental frequency in English." J. Acoust. Soc. Am. **64**, 65-80 [3.3; 5.2]

Monsen R.B., Engebretson A.M., Vemula N.R. (1979): "Some effects of deafness in the generation of voice." J. Acoust. Soc. Am. **66**, 1680-1690 [3.3; 9.4.3]

Montgomery A.A. (1967): "Laryngeal frequency under selected phonatory conditions;" Dissertation, Purdue Univ., Lafayette IN [4.1]

Montgomery H.C.: see (Munson and Montgomery, 1950)

Moore B.C.J.: see (Fourcin et al., 1978)

Moore G.P. (1937): "Vocal fold movement during vocalization." Speech Monogr. **4**, 44-55 [3.3]

Moore G.P., Leden H. von (1956): *Larynx and voice: function of the normal larynx* (Gould Foundation Laryngeal Research Labs, Chicago) [3.1]

Moore G.P., Leden H. von (1958): "Dynamic observation of the vibratory pattern in the normal larynx." Folia Phoniatr. **10**, 205-238 [3.3; 5.2]

Moore G.P., White F.D., Leden H. von (1962a): "Ultra-high-speed photography in laryngeal physiology." J. Speech Hear. Disorders **27**, 165-171 [5.2.1]

Moore G.P., White F.D., Leden H. von (1962b): "Ultra high-speed photography in laryngeal physiology." J. Speech Hear. Disorders **27**, 265 [5.2.5]

Moorer J.A. (1974): "The optimum comb method of pitch period analysis of continuous digitized speech." IEEE Trans. ASSP-**22**, 330-338 [4.1; 8.3.1]

Moorer J.A. (1975): "On the segmentation and analysis of continuous musical sound by digital computer;" Dissertation, Stanford Univ., San Francisco CA [9.4.2]

Moorer J.A. (1977): "Signal processing aspects of computer music: a survey." Proc. IEEE **65**, 1108-1137 [2.2; 2.4; 9.4.2]

Morf M., Dickinson B., Kailath T.L., Vieira A. (1977): "Efficient solution of covariance equations for linear prediction." IEEE Trans. ASSP-**25** [6.3.1]

Morf M., Lee D.T.L. (1978): "Fast algorithms for speech modeling" (Inf. Syst. Lab., Stanford Univ., Stanford CA; DDC-AD-A070050/0GA) [2.4; 6.3.1; 9.1.5]

Morgan D.R. (1975): "A note on real time linear prediction of speech waveforms." IEEE Trans. ASSP-**23** (C), 386-387 [6.3.1]

Mori R. de: see De Mori R.

Morikawa H., Fujisaki H. (1977): "A new method for speech analysis based on pole-zero model." **ICA-9**, 448 (Paper I42) [6.5.1]

Morikawa H., Fujisaki H. (1982): "Adaptive analysis of speech based on a pole-zero representation." IEEE Trans. ASSP-**30**, 77-88 [6.5.1; 9.4.5]

Morimoto N.: see (Isshiki et al., 1966)

Morris L.R. (1975a): "A simple real-time error waveform extraction using analog adaptive prediction of the speech waveform." Unpubl. Memorandum; cited by Markel and Gray (1976) [6.3.1]

Morris L.R. (1975b): "Fast speed spectrogram reduction and display on minicomputer/graphics processors." IEEE Trans. ASSP-**23**, 297-300 [5.1.2]

Morrison R.L. (1978): "Microcomputers invade the linear world." IEEE Spectrum **15**, 38-41 [9.1.5]

Morrow R.E.: see (Brigham and Morrow, 1967)

Moser M., Kittel G. (1977): "Rechnergestützte Tonhöhenbestimmung." Folia Phoniatr. **29**, 119-126 [4.3; 6.2.5; 8.3.1; 9.4.3]

Mott J.S.: see (Childers et al., 1980)

Mourikis K. (1979): "Conception d'un vocodeur à excitation vocale à 9600 bits/s;" Dissertation, Ecole Nationale Supérieure de Télécommunication, Paris [9.4.5]

Mow D.A.: see (Current et al., 1979)

Moxon E.C.: see (Kiang and Moxon, 1972)

Mrayati M.: see (Guérin et al., 1976)

Müller J. (1837): "Von der Stimme und Sprache." In *Handbuch der Physiologia des Menschen* (J.Holscher, Koblenz), 133-245 [3.1; 5.2.1]

Muller F.A. (1953): "A new principle for a pitch finder." **MIT-RLE Progr. Rept.**, 46

Munson J.: see (Newell et al., 1971)

Munson W.A., Montgomery H.C. (1950): "A speech analyzer and synthesizer." J. Acoust. Soc. Am. **22** (A), 678 [6.1.1]

Muravyev V.E. (1959): "Гармоническая система кодирования речи." В кн.: Сборн трудов Гос. НИИ Мин. свяси СССР. В.Е.Муравьев. (Speech encoding by a harmonic system. In Russian.). Sborn Trudov 1 [9.4.5]

Murray T. (1979): "Vocal jitter in singers' voice." J. Acoust. Soc. Am. **66** (A), S55 (Paper BB3; 98th Meet. ASA) [4.3]

Murry T. (1971): "Subglottal pressure and airflow measures during vocal fry phonation." J. Speech Hear. Res. **14**, 544-551 [3.2]

Murry T. (1975): "Some acoustic features of hoarseness." J. Acoust. Soc. Am. **58** (A), S111 (Paper BBB1; 90th Meet. ASA) [9.4.3]

Murry T. (1978): "Speaking fundamental frequency characteristics associated with voice pathologies." J. Speech Hear. Disorders **43**, 374-379 [4.1; 9.4.3]

Music R.S.: see (Cooper et al., 1970)

Myers W. (1981): "Computer graphics: reaching the user." IEEE Computer **18**, 7-17 [5.1]

Mysak E.D. (1959): "Pitch and duration characteristics of older males." J. Speech Hear. Res. **2**, 46-54 [4.1]

Myślecki W.: see (Majewski et al., 1979)

Nábělek I., Hirsh I.J. (1969): "On the discrimination of frequency transitions." J. Acoust. Soc. Am. **45**, 1510-1519 [4.3]

Nadal-Suris M.: see (Childers et al., 1976)

Nagai N.: see (Miyanaga et al., 1982)

Nagashima S.: see (Fujisaki and Nagashima, 1969, 1970)

Nagata K., Kato Y. (1960): "Some results of an experimental pitch detector." J. Acoust. Soc. Jpn. **16**, 277-279 [6.4.1]

Nagle H.T., Nelson V.P. (1981): "Digital filter implementation on 16-bit microcomputers." IEEE Micro **2**, 23-41 [2.3, 9.1.5]

Naik J.M.: see (Krishnamurthy et al., 1982; Yea et al., 1983)

Nakatsui M., Mermelstein P., Stevenson D.C. (1980): "Intelligibility evaluation of a 4.8 kb/s residual-excited linear prediction coder." J. Acoust. Soc. Am. **68** (A), S88 (Paper TT14, 100th Meet. ASA) [9.4.5]

Nakatsui M., Stevenson D.C., Mermelstein P. (1981): "Subjective evaluation of a 4.8-kbits/s residual-excited linear-prediction coder." IEEE Trans. COM-**29**, 1389-1393 [9.4.5]

Nakatsui M., Suzuki J. (1970): "Method of observation of glottal-source wave using digital inverse filtering in time domain." J. Acoust. Soc. Am. **47** (C), 664-665 [6.5.1]

Nash E.: see (Tosi et al., 1972)

Nash R. (1970): "John likes Mary more than Bill. An experiment in disambiguation using synthesized intonation contours." Phonetica **22**, 170-188 [9.4.4]

Nash R. (1973a): *Turkish intonation. An instrumental study* (Mouton, Den Haag) [9.4.2]

Nash R. (1973b): "Soviet work on intonation, 1968-70 with additional bibliography including 1971." Linguistics **113**, 63-104 [1.3]

Naval Research Office (ed.) (1958): "An annotated bibliography and critical review of voice communications" (Naval Res. Office, Washington DC; ACR-26; DDC-AD-152263) [1.3]

Navarro-Tomas T. (1944): *Manual de entonacion espanola* (Manual of Spanish intonation. In Spanish) (New York) [9.4.2]

Negus V.E. (1929): *The mechanism of the larynx* (C.V.Mosby, St.Louis MI) [3.1]

Negus V.E. (1949): *The comparative anatomy and physiology of the larynx* (William Heinemann Medical Books Ltd., London) [3.1]

Nelson V.P.: see (Nagle and Nelson, 1981)

Neuburg E.P. (1978a): "Improvement of voicing decisions by use of context." **ICASSP-78**, 5-7 [5.3]

Neuburg E.P. (1978b): "Simple pitch-dependent algorithm for high-quality speech rate changing." J. Acoust. Soc. Am. **63** (C), 624-625 [9.4.5]

Neuburg E.P. (1979): "Automatic thresholding for voicing detection algorithms." ICASSP-79, 756-759 [5.3]

Neweklowsky G. (1975): "Spezifische Dauer und spezifische Tonhöhe der Vokale." Phonetica **32**, 38-60 [9.4.2]

Newell A., Barnett J., Forgie J., Green C., Klatt D., Klatt D.H., Munson J., Reddy D.R., Woods W. (1971): *Speech understanding systems: Final report of a study group* (Dept. of Computer Science, Carnegie-Mellon Univ., Pittsburgh PA)

Newman E.B.: see (Stevens S.S. et al., 1937)

Ney H. (1981): "Bestimmung der Zeitverläufe von Intensität und Grundperiode der Sprache für die automatische Spracherkennung." Frequenz **35**, 265-270 [9.4.4]

Ney H. (1982): "A time warping approach to fundamental period estimation." IEEE Trans. SMC-**12**, 383-388 [8.3.3]

Nguyen L.P., Imai S. (1977): "Vocal pitch detection using generalized distance function associated with a voiced-unvoiced decision logic." Bull. P.M.E. (T.I.T.) **39**, 11-21 [5.3; 8.3.2]

Nichol D.G., Bogner R.E. (1978): "Quasi-periodic instability in a linear prediction analysis of voiced speech." IEEE Trans. ASSP-**26**, 210-217 [6.3.1; 9.4.5]

Nickerson R.S., Kalikow D.N., Stevens K.N. (1976): "Computer-aided speech training for the deaf." J. Speech Hear. Disorders **41**, 120-132 [5.1; 9.4.3]

Nickerson R.S., Stevens K.N. (1973): "Teaching speech to the deaf - can a computer help?." IEEE Trans. AU-**21**, 445-455 [9.4.3]

Nicol J.: see (Tosi et al., 1972)

Niederjohn R.L. (1975): "A mathematical formulation and comparison of zero crossings analysis techniques which have been applied to automatic speech recognition." IEEE Trans. ASSP-**23**, 373-379 [5.3]

Niedźwiecki A., Mikiel W. (1976): "Digital measurement of the fundamental frequency in a harmonic speech signal." In *Speech analysis and synthesis;* ed. by W. Jassem (Inst. of Fundamental Research, Polish Academy of Sciences, Warsaw), Vol.4, 293-300 [6.1.1]

Nielsen O. (1969): "A new method in stroboscopy." Tech. Review Brüel & Kjaer (Copenhagen) (1), 26-30 [5.2.1]

Niemann H. (1974): *Methoden der Mustererkennung* (Akademische Verlagsgesellschaft, Frankfurt/M) [5.3.3]

Niemann H. (1981): *Pattern analysis* (Springer, Berlin) [5.3.3]

Niguchi N.: see (Fujisaki et al., 1980)

Niimi S., Baer T., Fujimura O. (1976): "Stereo fiberscopic investigation of the larynx." J. Acoust. Soc. Am. **60** (A), S64 (Paper CC9; 92nd Meet. ASA) [5.2.1]

Nilsson N.J. (1965): *Learning machines* (McGraw-Hill, New York) [5.3.3]

Nimura T.: see (Chuang et al., 1971; Tsumara et al., 1973)

Nishinuma Y.: see (Hirst et al., 1979; Rossi et al., 1981)

Noll A.M. (1964): "Short time spectrum and "cepstrum" techniques for vocal pitch detection." J. Acoust. Soc. Am. **36**, 296-302 [8.4]

Noll A.M. (1966): "Computer-simulated cepstrum pitch detection for vocoder excitation." J. Acoust. Soc. Am. **40** (A), 1241 (Paper 1H2; 72nd Meet. ASA) [8.4; 9.4.5]

Noll A.M. (1967): "Cepstrum pitch determination." J. Acoust. Soc. Am. **41**, 293-309 [4.1; 8.4; 9.2.2; 9.3.1]

Noll A.M. (1968a): "Computer graphics in acoustic research." IEEE Trans. AU-**16**, 213-220 [5.1; 8.4]

Noll A.M. (1968b): "Clipstrum Pitch Determination." J. Acoust. Soc. Am. **44**, 1585-1592 [8.4; 8.4.2]

Noll A.M. (1970): "Pitch determination of human speech by the harmonic product spectrum, the harmonic sum spectrum, and a maximum likelihood estimate." In *Symposium on Computer Processing in Communication;* ed. by the Microwave Institute (Univ. of Brooklyn Press, New York NY), Vol.19, 779-797 [8.4; 8.5.1; 8.5.4; 8.6.1]

Noll A.M. (1971): "Pitch detector;" United States Patent No. 3,555,191. Issued Jan. 12, 1971 [8.4.2]

Noll A.M. (1973): "The cepstrum and some close relatives." In *Signal Processing;* ed. by J.W.R. Griffiths, P.L. Stocklin, and C. van Schooneveld (Academic Press, London), 11-22 [8.4.1; 8.4.2]

Noll A.M., Schroeder M.R. (1964): "Short-time "cepstrum" pitch detection." J. Acoust. Soc. Am. **36**, 1030 (Paper N5; 67th Meet. ASA) [8.4.2]

Noll A.M., Schroeder M.R. (1971): "Real time cepstrum analyzer;" United States Patent No.3,566,035. Issued Feb. 23, 1971; filed July 17, 1969 [8.4.2]

Nooteboom S.G., Brokx J.P.L., De Rooij J.J. (1976): "Contributions of prosody to speech perception." **IPO Annual Progr. Rept.** 11, 34-54 [9.4.2]

Norgate M.: see (Fourcin and Norgate, 1965)

Norwine A.C.: see (Feldman and Norwine, 1958)

Nussbaumer H.J. (1981): *Fast Fourier transform* (Springer, Berlin) [2.4; 8.5.4]

Obata J., Kobayashi R. (1937): "Apparatus for direct recording the pitch and intensity of sound." J. Acoust. Soc. Am. 9, 156-161 [6.1.2; 6.6.2]

Obata J., Kobayashi R. (1938): "A direct reading pitch recorder and its applications to music and speech." J. Acoust. Soc. Am. 10, 147-149 [6.1.2; 6.6.2]

Obata J., Kobayashi R. (1940): "Further applications of our direct-reading pitch and intensity recorder." J. Acoust. Soc. Am. 12, 188-192 [6.1.2; 6.6.2]

O'Donnell M.C.: see (Dolanský et al., 1966)

Ögren J.E.: see (Wedin and Ögren, 1982)

Öhman S.E.G. (1966): "Computer program for pitch measurements." STL-QPSR (1), 11 [5.1.1; 9.2.1]

Öhman S.E.G. (1967): "Word and sentence intonation - a quantitative model." STL-QPSR (2-3), 20-54 [9.4.2]

Öhman S.E.G., Lindqvist J. (1968): "Analysis by synthesis of prosodic pitch contours." Z. Phonetik 21, 164-170 [9.4.2]

Oertel M. (1878): "Über eine neue laryngostroboskopische Untersuchungsmethode." Zentralblatt der Medizinischen Wissenschaften 16, 81-82 [5.1.1; 5.2.1]

Oetinger R. (1959): "Die Grenzen der Hörbarkeit von Frequenz- und Tonzahländerungen bei Tonimpulsen." Acustica 9, 430-434 [4.2.3]

Ohala J.J. (1966): "A new photo-electric glottograph." Phon. Los Angeles 4, 40-52 [5.2.3; 5.2.5]

Ohala J.J. (1973): "Explanations for the intrinsic pitch of vowels." Monthly Internal Memorandum, Phonology Lab., Univ. of Calif., Berkeley CA, 9-26 [3.1; 9.4.2]

Ohala J.J. (1977): "Speculations on pitch regulation." Phonetica 34, 310-312 [3.1]

Ohala J.J., Ewan W.G. (1973): "Speed of pitch change." J. Acoust. Soc. Am. 53 (A), 345 (Paper DD6; 84th Meet. ASA) [4.3]

Ohala J.J., Hirano M. (1967a): "Studies of pitch change in speech." Phon. Los Angeles 7, 80-90 [4.3]

Ohala J.J., Hirano M. (1967b): "An experimental investigation of pitch change in speech." J. Acoust. Soc. Am. 42 (A), 1208-1209 [4.2.3; 4.3]

Ohm G.S. (1843): "Über die Definition des Tones, nebst daran geknüpfter Theorie der Sirene und ähnlicher tonbildender Vorrichtungen." Ann. Phys. Chem. 59, 516-565 [4.2.1]

Ohta K.: see (Fujisaki et al., 1979)

Ojamaa K., Erickson D., Kumar A. (1970): "Coarticulation effects on pitch." J. Acoust. Soc. Am. 48 (A), 84 (Paper I6, 79th Meet. ASA) [9.4.2]

Olbrecht D.H.: see (Westin et al., 1966)

Olive J.P. (1973): "Pitch rules for the synthesis of simple declarative English sentences." J. Acoust. Soc. Am. 54 (A), 115 (Paper RR5; 86th Meet. ASA) [9.4.2; 9.4.4]

Olive J.P. (1975): "Fundamental frequency rules for the synthesis of simple declarative English sentences." J. Acoust. Soc. Am. 57, 476-482 [9.4.2; 9.4.4]

Olsen C.L.: see (Kvavik and Olsen, 1974)

Olson H.F., Belar H., Rogers E.S. (1967): "Speech Processing Techniques." IEEE Trans. AU-15, 120-126 [5.3]

O'Malley M.H., Kloker D.R., Dara-Abrams B. (1973): "Recovering parentheses from spoken algebraic expressions." IEEE Trans. AU-21, 217-220 [9.4.2; 9.4.4]

O'Malley M.H., Peterson G.E. (1966): "An experimental method for prosodic analysis." Phonetica 15, 1-13 [9.4.2]

Ondráčková J. (1968): "Glottographical research in Czech sound groups." Phon. München 21, 145-152 [5.2.3; 9.4.2]

Ondráčková J. (1972): "Vocal chords activity: its dynamics and role in speech production." ICPhS-7, 29-47 [3.1]

Ondráčková J., Lindqvist J., Fant G. (1967): "Electrical glottography." ICPhS-6, 709-712 [5.2.3]

O'Neal J.B.jr. (1980): "Waveform encoding of voiceband data signals." Proc. IEEE 68, 232-247 [9.4.5]

Ono H., Saito S., Fukuda H., Ichikawa T. (1978): "A voice analysis system using a digital processor during phonosurgery and its clinical application." J. Acoust. Soc. Am. 64 (A), S91 (Paper II12; 96th Meet. ASA) [5.2.1; 9.4.3]

Ooyama G., Katagiri S., Kido K. (1978): "A new method of cepstrum analysis by using comb lifter." **ICASSP-78**, 19-22 [8.4.1]

Oppenheim A.V. (1969): "Speech analysis-synthesis system based on homomorphic filtering." J. Acoust. Soc. Am. **45**, 458-465 [2.4; 8.4; 9.4.4]

Oppenheim A.V. (1970): "Speech spectrogram using the fast Fourier transform." IEEE Spectrum **7**, 57-62 [2.4; 5.1.2]

Oppenheim A.V., Davis R.K., Dove W.P. (1983): "Knowledge-based pitch detection." **ICASSP-83**

Oppenheim A.V., Kopec G.E., Tribolet J.M. (1976): "Signal analysis by homomorphic prediction." IEEE Trans. ASSP-**24**, 327-332 [2.4; 8.4; 9.4.4]

Oppenheim A.V., Schafer R.W. (1975): *Digital signal processing* (Prentice Hall, Englewood Cliffs NJ) [2.0; 8.1; 8.2.1]

Oppenheim A.V., Schafer R.W., Stockham T.G. (1968): "Non-linear filtering of multiplied and convolved signals." Proc. IEEE **56**, 1264-1291 [8.4.1]

Orlov I.A.: see (Knipper and Orlov, 1976)

O'Shaughnessy D. (1978): "Stress perception in words varied in fundamental frequency." J. Acoust. Soc. Am. **64** (A), S20 (Paper J21; 96th Meet. ASA) [9.4.2]

O'Shaughnessy D. (1979): "Linguistic features in fundamental frequency patterns." J. Phonetics **7**, 119-145 [9.4.2]

O'Shaughnessy D. (1982): "Coding of fundamental frequency at low rates." J. Acoust. Soc. Am. **72** (A), S80 (Paper RR11, 104th Meet. ASA) [9.4.4]

Ossanna J.F.: see (Bogert and Ossanna, 1966)

Otryasenkov M.: see (Kodzasov and Otryasenkov, 1968)

Otten K.W.: see (Falter and Otten, 1967)

Oyer H.: see (Tosi et al., 1972)

Ozawa S.: see (Gohbara et al., 1978)

Paige A.: see (Childers et al., 1972, 1976; Berouti et al., 1977)

Paillé J. (1969): "Source vocale pour synthétisateurs à formants." Revue d'Acoustique (6), 111-114 [3.3; 9.4.4]

Paille J., Lancia R. (1969): "Synthese parametrique de la parole source vocale pour synthesiteur a formants." Teleinformatique [9.4.4]

Paley B.: see (Vallancien et al., 1971)

Paliwal K.K., Rao P.V.S. (1979): "Windowing in linear prediction analysis of voiced speech." J. Acoust. Soc. Am. **66** (A), S63 (Paper EE2; 98th Meet. ASA) [2.4]

Paliwal K.K., Rao P.V.S. (1983): "A synthesis-based method for pitch extraction." Speech Commun. **2** [8.5.2]

Palmer E.P.: see (Strong and Palmer, 1975)

Palmer J.M.: see (Hamlet and Palmer, 1974)

Panconcelli-Calzia G. (1961): *3000 Jahre Stimmforschung* (Marburg) [9.4.2]

Papoulis A. (1965): "Probability, random variables, and stochastic processes" (McGraw-Hill, London) [6.6.3; 8.6]

Papoulis A., Chamzas Ch. (1979): "Detection of hidden periodicities by adaptive extrapolation." IEEE Trans. ASSP-**27**, 492-500

Parker A. (1974): "Voice and intonation training using a laryngographic display." **Phon. London**, 1-38 [5.2.3; 9.4.3]

Parker F.: see (Walsh and Parker, 1980)

Parks T.W., McClellan J.H. (1972): "Chebychev approximation for nonrecursive digital filters with linear phase." IEEE Trans. CT-**19**, 189-194 [2.3]

Parmentier C.E., Trevino S.N. (1932): "A technique for the analysis of pitch in connected discourse." Arch. Néerl. Phonétique Exp. **7**, 1-29 [5.1.1]

Parsons T.W. (1976): "Separation of speech from interfering speech by means of harmonic selection." J. Acoust. Soc. Am. **60**, 911-918 [8.5.2; 9.4.5]

Partridge G.R. (1965): "New method for the rapid determination of speech frequencies." J. Acoust. Soc. Am. **38** (A), 911 (Paper B5; 71st Meet. ASA)

Pasternak L.: see (Vallancien et al., 1971)

Patterson R.D., Wightman F.L. (1976): "Residue pitch as a function of component spacing." J. Acoust. Soc. Am. **59**, 1450-1459 [4.2.1]

Paul D.B. (1979): "A robust vocoder with pitch-adaptive spectral envelope estimation and an integrated maximum-likelihood pitch estimator." **ICASSP-79**, 64-68 [8.6.1; 9.4.5]

Paul P. (1978): "Large-scale vector/array processors." In *Advances in digital image processing;* ed. by P. Stucki (Plenum Press, New York), 277-300 [8.1; 9.1.5]

Paulsen K. (1966): "The primary tone of the human larynx." Arch. Ohren-Nasen-Kehlkopfheilkunde **167**, 807-815 [3.1; 3.3]

Paulus E., Längle D. (1973): "Die Anwendung der Karhunen-Loève-Entwicklung für die digitale Sprachanalyse und -synthese." In *Signalverarbeitung (Signal Processing)*. Vorträge der gleichnamigen Fachtagung, Erlangen 1973; ed. by H.W.Schüssler (Universität Erlangen-Nürnberg), 362-369 [6.3.2]

Peck C.W. (1969): "An acoustic investigation of the intonation of American English" (Univ. of Michigan, Ann Arbor MI; Natural Language Studies, No. 1) [9.4.2]

Peckels P., Riso V. (1965): "On the use of an accelerometer as a pitch extractor." **ICA-5** (Paper A25) [5.2.2]

Peckham J.B. (1979): "A device for tracking the fundamental frequency of speech and its application in the assessment of strain in pilots and air traffic controllers" (Aircr. Establ, Farnsborough, England; Rept. #N80-22573/3) [9.4.4]

Pedersen M.F. (1977): "Electroglottography compared with stroboscopy in normal persons." Folia Phoniatr. **29**, 191-199 [5.2.1; 5.2.3; 5.2.6; 9.4.3]

Pedrey C.: see (Tosi et al., 1972)

Peebles K.: see (Malécot and Peebles, 1965)

Peled A., Liu B. (1976): *Digital signal processing. Theory, design, and implementation* (Wiley, New York) [2.3]

Perennou G.: see (Dours et al., 1977)

Perez-Falcon T.: see (Yost et al., 1978)

Perkell J.S. (1969): *Physiology of speech production: results and implications of a quantitative cineradiographic study* (MIT Press, Cambridge MA) [3.4; 5.2.1]

Perkins H.W., Yanagihara N. (1968): "Parameters on voice production I: Some mechanisms for the regulation of pitch." J. Speech Hear. Res. **11**, 246-267 [3.1]

Petersen N.R. (1978): "Intrinsic fundamental frequency of Danish vowels." J. Phonetics **6**, 177-190 [3.1; 9.4.2]

Peterson E. (1951): "Frequency Detection and Speech Formants." J. Acoust. Soc. Am. **23**, 668-674 [6.1.2]

Peterson E. (1952a): "Analyzer for determining fundamental frequency of a complex wave;" United States Patent No. 2,593,695. Issued April 22, 1952; filed May 10, 1948 [6.1.2; 6.1.3]

Peterson E. (1952b): "Wave analyzer for determining fundamental frequency of a complex wave;" United States Patent No. 2,593,694. Issued April 22, 1952; filed March 26, 1948 [6.1.2]

Peterson G.E. (1952): "The information-bearing elements of speech." J. Acoust. Soc. Am. **24**, 629-637 [1.2; 9.4.2]

Peterson G.E. (1954): "Design of visible speech devices." J. Acoust. Soc. Am. **26**, 406-413 [5.1.2; 9.4.2]

Peterson G.E., Barney H.L. (1952): "Control methods used in a study of the vowels." J. Acoust. Soc. Am. **24**, 175-184 [1.2; 3.4]

Peterson G.E., Peterson G.G. (1968): "Fundamental frequency detector utilizing plural filters and gates;" United States Patent No.3,364,425. Issued Jan. 16, 1968 [6.4.1]

Peterson G.E., Shoup J.E. (1966a): "A physiological theory of phonetics." J. Speech Hear. Res. **9**, 5-67 [1.2; 2.1]

Peterson G.E., Shoup J.E. (1966b): "The elements of an acoustic phonetic theory." J. Speech Hear. Res. **9**, 68-99 [1.2; 2.1]

Peterson G.E., Shoup J.E. (1966c): "Glossary of terms from the physiological and acoustic phonetic theories." J. Speech Hear. Res. **9**, 100-120 [1.2; 2.1]

Peterson G.G.: see (Peterson and Peterson, 1968)

Petrov M. (1957): "Audio frequency meter;" United States Patent No. 2,779,920. Issued Jan. 29, 2957 [6.4.1]

Pétursson M. (1976): "Aspiration et activité glottale." Phonetica **33**, 169-198 [3.1; 3.5]

Pétursson M. (1977): "Glottographische Untersuchungen der Anlautgruppen hl-, hr-, hn- im modernen Isländischen." In *Miszellen IV;* ed. by J.P. Köster. Hamburger Phonetische Beiträge, Bd.22 (Buske, Hamburg), 113-125 [5.2.3; 9.4.2]

Petzinger J.: see (Steiglitz et al., 1975)

Pfister K.: see (Luchsinger and Pfister, 1961; van Michel et al., 1970)

Philipp J.E. (1978): "Automatische Transskription mehrstimmiger Musik mit dem Prony-Algorithmus." In *1. DFG-Kolloqium Signalverarbeitung, Göttingen. Oktober 1978;* ed. by Z.D.Zimmer and V.Neuhoff (Göttingen), 28 [6.3.1; 9.4.2]

Philips P.: see (Hollien et al., 1971)

Phillips N.D. (1968): "Detection of the voice fundamental for visual speech detection." J. Acoust. Soc. Am. **44** (A), 366 [6.2.1]

Phillips V.L.: see (Cherry and Phillips, 1961)

Phon. Aix-en-Provence --> *Travaux de l'Institut de Phonétique d'Aix* (Aix-en-Provence)

Phon. Grenoble --> *Bulletin de l'Institut de Phonétique de Grenoble* (Grenoble)

Phon. Köln --> *IPK-Berichte.* Reports of the Institute of Phonetics, University of Cologne; ed. by G.Heike (Inst. f. Phonetik, Universität Köln)

Phon. London --> *Speech and Hearing - Work in Progress* (Department of Phonetics and Linguistics, University College, London)

Phon. Los Angeles --> *Working Papers in Phonetics;* ed. by the University of California at Los Angeles - Department of Phonetics (Los Angeles CA)

Phon. Lund --> *Working papers* (Phonetics Lab., Dept. of General Linguistics, Lund Univ., Lund)

Phon. München --> *Institut für Phonetik und Sprachliche Kommunikation der Universität München - Forschungsberichte;* ed. by H.G.Tillmann and H.Günther (Inst. f. Phonetik und sprachliche Kommunikation, Universität München)

Phon. Nijmegen --> *Proceedings, Institute of Phonetics* (Catholic University, Nijmegen)

Phon. Tallinn --> *Estonian papers in phonetics* (Acad. of Sci. of the Estonian SSR, Tallinn)

Phonetica Pragensia III --> Romportl M., Janota P. (eds.): *Phonetica Pragensia III - Symposium on Intonology, Prague, October 1970* (Universita Karlova, Praha 1972) [9.4.2]

Pierce J.R. (1969): "Whither speech recognition?." J. Acoust. Soc. Am. **46** (C), 1049-1051 [9.1.5]

Pierrat T. (1975): "Etude comparative des méthodes appliquées à la parole, de la détection de la periodicité ou de la non periodicité d'un signal;" Dissertation, Université Laval, Quebec [2.6; 5.3; 9.3]

Pierrehumbert J. (1979a): "Intonation synthesis based on metrical grids." J. Acoust. Soc. Am. **65** (A), S131 (Paper YY5; 97th Meet. ASA) [9.4.2; 9.4.4]

Pierrehumbert J. (1979b): "The perception of fundamental frequency declination." J. Acoust. Soc. Am. **66**, 363-369 [4.3; 9.4.2]

Pierrehumbert J. (1981): "Synthesizing intonation." J. Acoust. Soc. Am. **70**, 985-995 [9.4.2]

Pihl G.E.: see (Chang et al., 1951)

Pike K.L. (1945): *The intonation of American English* (University of Michigan Press, Ann Arbor) [9.4.2]

Pike K.L. (1948): *Tone languages: A technique for determining the number and type of pitch contrasts in a language, with studies in tonemic substitution and fusion* (Univ. Of Michigan, Ann Arbor MI) [1.2; 9.4.2]

Pike K.L., Barrett R.P., Bascom B. (1959): "Instrumental collaboration on a Tepehuan (Uto-Aztecan) pitch problem." Phonetica **3**, 1-22 [5.1.2; 9.4.2]

Pilch H. (1977): "Intonation in discourse analysis. With material from Finnish, English, Alemannic German." Phonetica **34**, 1-81 [9.4.2]

Pillay S.K., Canas M.A., Bradshaw F.T., Lin W.C. (1975a): "Voicing detection by Kalman filtering." J. Acoust. Soc. Am. **58** (A), S61 (Paper FF3; 90th Meet. ASA) [5.3.3]

Pillay S.K., Canas M.A., Bradshaw F.T., Lin W.C. (1975b): "Pitch detection using slope overload noise in an adaptive delta modulation process." J. Acoust. Soc. Am. **58** (A), S61 (Paper FF2; 90th Meet. ASA) [6.3.1]

Pinson E.N. (1963): "Pitch synchronous time domain estimation of formant frequencies and bandwidths." J. Acoust. Soc. Am. **35**, 1264-1273 [3.4; 7.1.5; 9.4.4; 9.4.5]

PIPS Rept. --> *Progress report on speech research;* ed. by Speech Processing Section, Information Science Division, Electrotechn.Lab., Ministry of Trade and Industry (2-6-1 Nagata-cho, Chiyoda-ku, Tokyo)

Pirogov A.A. (1959): "Гармоническая система сжатия спектров речи." В журн.: Електросвяз. А.А.Пирогов. (Harmonic system for spectral compression of speech. In Russian). Elektrosvyaz **3**, 8-17 [9.4.5]

Pirogov A.A. (1960): "Устройство для автоматического определеня частоты основного тона." А.С.№ 129739-СССР. Опубл. в бюлл. "Открытия, Изобретения, Промышленные образцы, Товарные знаки. А.А.Пирогов.(Device for automatic pitch determination. In Russian.). Otkrytia, Izobretenia **13**, 38

Pirogov A.A. (1963): "Синтетическая телефония." А.А.Пирогов. (*Synthetic telephony.* In Russian)" (Svyazizdat, Moscow) [6.6.2]

Pirogov A.A. (ed.) (1974): "Вокодерьная телефония. Методы и проблемы." Под редакцией А.А.Пирогова. (*Vocoder telephony - methods and problems.* In Russian.) (Svyaz, Moscow) [1.1; 1.2; 9.4.4; 9.4.5]

Pisani R.U. (1971): "A melody detector." Acustica **25**, 179-182 [6.1.2; 6.4.1]

Pisani R.U. (1975): "Representation with the sonagaph of different speech characteristics." Acustica **32**, 96-99 [5.1.2; 9.4.2]

Pisani R.U., Sacerdote G.G. (1969): "Experimental determination of musical intervals." Acustica 21, 26-29 [5.1.2; 9.4.2]

Plant G.R.G. (1959): "Frequency indicators;" United States Patent No. 2,915,589. Issued Dec. 1, 1959; filed Feb. 13, 1958 [6.1.1]

Plomp R. (1964): "The ear as a frequency analyzer." J. Acoust. Soc. Am. 36, 1628-1636 [4.2]

Plomp R. (1967): "Pitch of complex tones." J. Acoust. Soc. Am. 41, 1526-1532 [4.2]

Plomp R. (1976): Aspects of tone sensation (Academic Press, London) [1.2; 4.2; 4.2.1; 9.1]

Plomp R., Mimpen A.M. (1968): "The ear as a frequency analyzer II.." J. Acoust. Soc. Am. 43, 764-767 [4.2]

Plomp R., Smoorenburg G.F. (eds.) (1970): Frequency analysis and periodicity detection in hearing (Sijthoff, Leiden)

Pohlink M.: see (Tscheschner and Pohlink)

Pollack I. (1970): "Jitter detection for repeated auditory pulse patterns." In Frequency analysis and periodicity detection in hearing; ed. by R.Plomp and G.F.Smoorenburg (Sijthoff, Leiden), 329-338 [4.2.1; 4.3]

Polykowskij A.M. (1959): "Матричний гармонический анализатор." В кн.: Сборн трудов Гос. НИИ Мин. связи. А.М. Полыкоский.   (Matrix harmonic analyzer. In Russian.). Sborn Trudov 1, 51-53

Popov V.A. (1966): Анализ интонационной характеристики речи как показателя эмоционального состояния человека в условиях космического полёта. В.А.Попов. В журн.: Вышей Нервной Деятельности. (Analysis of intonational features as characteristics of the emotional state during space orbit, In Russian.)." Vyshej Nervnoj Deyatel'nosti 16, 974-983 [3.1; 9.4.3]

Popper A.N.: see (Tenold et al., 1974)

Porter H.C. (1963): "Extraction of pitch from the trachea" (AFCRL, Cambridge MA; Res.Note No.63-24; DDC-AD-401920) [5.2.2]

Potter R.K., Kopp G.A., Kopp K.G. (1966): Visible speech (Dover Pub.Inc., New York) [5.1.2]

Poza F.: see (Becker and Poza, 1973)

Preis D. (1981): "Phase equalization for magnetic recording." ICASSP-81, 790-795 [9.2.1]

Prestigiacomo A.J. (1962): "Apparatus for deriving pitch information from a speech wave;" United States Patent No. 3,020,344. Issued Feb. 6, 1962; filed Dec. 27, 1960

Prestigiacomo A.J. (1965): "Anordnung zum Ableiten der Tonhöhen-Information eines Sprachsignals;" Deutsche Patentschrift Nr. 1186912. Bekanntgemacht am 7.10.1965

Prezas D.P.: see (Hartwell and Prezas, 1981)

Priestley T.M. (1976): "A note on the glottal stop." Phonetica 33, 268-274 [3.1; 9.4.2]

Prisse d'Avennes A. (1972): "Essai d'étude statistique de la détermination de niveaux intonatifs de la phrase énonciative francaise." Phon. Grenoble 1, 67-75 [4.1]

Proc. 1962 Speech Commun. Sem. --> Proceedings of the Speech Communication Seminar, Stockholm 1962; ed. by G.Fant (Speech Transmission Laboratory, Royal Institute of Technology, Stockholm 1962)

Proc. 1974 Speech Commun. Sem. --> Proceedings of the 1974 Speech Communication Seminar, Stockholm, Sweden, August 1-3, 1974. 4 volumes; ed. by G. Fant (Almqvist and Wiksell, Stockholm 1974)

Proctor D.F. (1980): Breathing, speech, and song (Springer, Wien) [3.1]

Pronovost W.L. (1942): "An experimental study of methods for determining natural and habitual pitch." Speech Monogr. 9, 111-123 [4.1; 9.4.2]

Protonotarios E.N., Carayannis G., Milios E. (1980): "Time series monitoring using fast recursive schemes." EUSIPCO-80, 635-640 [6.3.1]

Pruzansky S.: see (Guttman and Pruzansky, 1962)

Quinn A.M.S.: see (Harris and Quinn, 1979)

Rabiner L.R. (1977a): "On the use of autocorrelation analysis for pitch detection." IEEE Trans. ASSP-25, 24-33 [8.2.1; 8.2.4]

Rabiner L.R. (1977b): "Evaluation of a statistical approach to voiced-unvoiced-silence analysis for telephone quality speech." Bell Syst. Tech. J. 56, 455-482 [5.3]

Rabiner L.R., Atal B.S., Sambur M.R. (1976): "The LPC error signal-analysis of its variation with the position of the analysis frame." J. Acoust. Soc. Am. 60 (A), S108 (Paper RR12; 92nd Meet. ASA) [5.3; 9.4.5]

Rabiner L.R., Atal B.S., Sambur M.R. (1977): "LPC prediction error - analysis of its variation with the position of the analysis frame." IEEE Trans. ASSP-25, 434-442 [6.3.1; 9.4.5]

Rabiner L.R., Cheng M.J., Rosenberg A.E., McGonegal C.A. (1976a): "A comparative
    study of several pitch detection algorithms." IEEE Trans. ASSP-**24**, 399-413 [3.2; 4.1;
    8.5.4; 9.3; 9.4.4]
Rabiner L.R., Cheng M.J., Rosenberg A.E., McGonegal C.A. (1976b): "A comparative
    study of several pitch detection algorithms." ICASSP-**76**, 340-343 [3.2; 4.1; 9.3; 9.4.4]
Rabiner L.R., Cooley J.W., Helms H.D., Jackson L.B., Kaiser J.F., Rader C.M., Schafer
    R.W., Steiglitz K., Weinstein C.J. (1972): "Terminology in digital signal processing."
    IEEE Trans. AU-**20**, 322-337 [2.1; 2.2; 2.3]
Rabiner L.R., Gold B. (1975): *Theory and application of digital signal processing* (Prentice
    Hall, Englewood Cliffs NJ) [2.0]
Rabiner L.R., Sambur M.R. (1974): "Algorithm for determining the endpoints of isolated
    utterances." J. Acoust. Soc. Am. **56** (A), S31 (Paper P2; 88th Meet. ASA) [5.3]
Rabiner L.R., Sambur M.R. (1975): "An algorithm determining the endpoints of isolated
    utterances." Bell Syst. Tech. J. **54**, 297-315 [5.3]
Rabiner L.R., Sambur M.R. (1977a): "Application of an LPC distance measure to the
    voiced-unvoiced-silence detection problem." IEEE Trans. ASSP-**25**, 338-343 [5.3.3]
Rabiner L.R., Sambur M.R. (1977b): "Voiced-unvoiced-silence detection using the Itakura
    LPC distance measure." ICASSP-**77**, 323-326 [5.3.3]
Rabiner L.R., Sambur M.R., Schmidt C.E. (1975): "Applications of nonlinear smoothing
    algorithm to speech processing." IEEE Trans. ASSP-**23**, 552-557 [6.6.4; 9.3.1]
Rabiner L.R., Schafer R.W. (1976): "Digital techniques for computer voice response:
    Implementations and applications." Proc. IEEE **64**, 416-433 [2.2; 9.4.5]
Rabiner L.R., Schafer R.W. (1978): *Digital processing of speech signals* (Prentice Hall,
    Englewood Cliffs NJ; Prentice-Hall Signal Proc. Series, ed. by A.V.Oppenheim) [1.1;
    2.2; 2.6]
Rabiner L.R., Schafer R.W., Rader C.M. (1969a): "The chirp-z-transformation algorithm."
    IEEE Trans. AU-**17**, 86-92 [2.4; 8.1]
Rabiner L.R., Schafer R.W., Rader C.M. (1969b): "The chirp-z transform algorithm and
    its application." Bell Syst. Tech. J. **48**, 1249-1292 [2.4; 8.1]
Rabiner L.R., Schmidt C.E., Atal B.S. (1977): "Evaluation of a statistical approach to
    voiced-unvoiced-silence analysis for telephone-quality speech." Bell Syst. Tech. J. **56**,
    455-481 [5.3.3]
Rader C.M. (1964): "Vector pitch detection." J. Acoust. Soc. Am. **36** (C), 1463 [6.3.2]
Raisbeck G. (1959): "Voice pitch determination;" United States Patent No. 2,908,761.
    Issued October 1959, filed 1954 [8.2.2]
Rajappan K.P.: see (Sridhar and Rajappan, 1979)
Rajcan E. (1971): "An experimental study of the pitch dependence on the initial phase of
    short tone pulses." ICA-**7** (3), 365-368 (Paper 20H4) [9.2.1]
Rakotoforinga H.: see (Boe and Rakotoforinga, 1971, 1972, 1975)
Ramamoorthy V. (1980): "Voiced/unvoiced detection based on a composite-Gaussian source
    model of speech." ICASSP-**80**, 57-60 [5.3.3]
Rameau J.P. (1750): *Démonstration du principe de l'harmonie* [Reprinted 1930; English
    translation 1971: *Treatise on harmony*, Dover, New York] (Paris) [4.2; 8.5.2; 9.4.2]
Ramsey J.L. (1964): "Logical techniques for glottal source measurements" (AFCRL,
    Cambridge MA; AFCRL-64-661)
Rand T.C.: see (Cooper et al., 1970)
Rao P.V.S.: see (Sreenivas and Rao, 1979, 1980, 1981; Paliwal and Rao, 1979, 1983)
Rao R.: see (Tufts et al., 1976)
Rapp K.: see (Carlson et al., 1972)
Rappaport W. (1956): "Über die Messung der Tonhöhenverteilung in der deutschen Sprache;"
    Dissertation, Technische Hochschule Aachen [4.1]
Rappaport W. (1958): "Messungen der Tonhöhenverteilung der deutschen Sprache." Acustica
    8, 220-225 [4.1]
Raskin L.: see (van Michel and Raskin, 1969)
Recherches/Acoustique --> *Recherches/Acoustique;* ed. by CNET (Centre National d'Etudes
    des Telecommunications) (CNET, F-22301 Lannion)
Reddy D.R. (1966): "An approach to computer speech recognition by direct analysis of
    the speech wave" (Stanford University, Berkeley CA; Tech. Rept. CS-49) [5.3.2; 6.2.2;
    6.6.3]
Reddy D.R. (1967): "Pitch period determination of speech sounds." Commun. ACM **10**,
    343-348 [5.1; 5.3.2; 6.2.2; 6.6.3; 7.2]
Reed C.G. (1980): "Voice therapy: a need for research." J. Speech Hear. Disorders **45**,
    157-189 [9.4.3]

Rees M.J. (1958a): "Some variables affecting perceived harshness." J. Speech Hear. Res. 1, 155-168 [3.3; 4.3]

Rees M.J. (1958b): "Harshness and glottal attack." J. Speech Hear. Res. 1, 344-349 [3.3; 4.3]

Regier W.L.: see (Krasner et al., 1973)

Rehman I.: see (Snidecor et al., 1959)

Reid J.M.: see (Hamlet and Reid, 1972)

Reid N.A.: see (Hill and Reid, 1977)

Reiss R.: see (Hoffmann et al., 1974)

Remde J.R.: see (Atal and Remde, 1982)

Remmel M., Ruutel I., Sarv J., Sule R. (1975): "Automatic notation of one-voiced song" (Inst. of Language and Literature, Acad. of Science of the Estonian SSR, Tallinn, USSR; Preprint KKT-4) [9.4.2]

Rencher A.C.: see (Brown et al., 1974)

Riccitelli J.M. (1965): "Tone analysis: a practical approach." Bible Translator 16, 54-73

Rice D.L. (1971): "Glottal pulse waveform effects in line analog speech synthesis." **Phon. Los Angeles 19**, 48-54 [3.3; 9.4.4]

Richter W., Fellbaum K. (1980): "Zum Einsatz von Mikroprozessoren und LSI-Bausteine für die aufwandsgünstige Realisierung eines LPC-Vocoders." In *Fortschritte der Akustik, DAGA'80 München* (VDE-Verlag, Berlin), 695-698 [9.4.5]

Riedel P. (1963): "Tonhöhe und Intensität beim Sprechen;" Dissertation, RWTH Aachen [4.1; 9.4.2]

Riesz R.R. (1939): "Wave translation;" United States Patent No. 2,183,248. Issued Dec. 12, 1939; filed Sep. 11, 1936 [6.1.1; 6.1.2]

Riesz R.R. (1950): "Frequency control for synthesizing systems;" United States Patent No. 2,522,539. Issued Sep. 19, 1950; filed July 2, 1948 [6.1.1]

Riesz R.R. (1952): "Apparatus for determining pitch frequency in a complex wave;" United States Patent No. 2,593,698. Issued April 22, 1952; filed May 10, 1948 [6.1.1; 6.1.3]

Riesz R.R. (1965): "Frequency measuring system;" United States Patent No.3,197,560. Issued 1965; filed Dec. 29, 1944 [6.1.2]

Riesz R.R., Schott L.O. (1946): "Visible speech cathode-ray translator." J. Acoust. Soc. Am. 18, 50-61 [6.1.3; 9.4.2]

Rietveld A.C.M., Boves L. (1979): "Automatic detection of prominence in the Durch language." **Phon. Nijmegen 3**, 72-78 [6.2.1; 9.4.2]

Rietveld T. (1978): "Pitch and timbre; pitch and duration; pitch judgments by untrained listeners." **Phon. Nijmegen 2**, 41-50 [4.3]

Rigault A.: see (Léon et al., 1970)

Righini G.U. (1963): "A pitch extractor of the voice." Acustica 13 (C), 315-316 [6.4.1; 6.6.2]

Righini G.U. (1968): "Extraction of pitch by means of a phase locked tracking filter." Ist. Elettr. Nazion. Galileo Ferraris (Turin) 45, 1101 [6.1.4]

Ringel R.L., Isshiki N. (1964): "Intra-oral voice recordings: an aid to laryngeal photography." Folia Phoniatr. 16, 19-28 [5.2.1]

Risberg A. (1960a): "Voice fundamental frequency tracking." **STL-QPSR** (2) [4.1; 6.1.2; 6.4.2]

Risberg A. (1960b): "Voice fundamental frequency tracking." **STL-QPSR** (1), 3-6 [6.1.2]

Risberg A. (1961a): "Fundamental frequency tracking." **STL-QPSR** (3), 1-2 [6.1.2; 6.4.2]

Risberg A. (1961b): "Statistical studies of fundamental frequency range and rate of change." **STL-QPSR** (4), 7-8 [3.4; 4.1]

Risberg A. (1962): "Fundamental frequency tracking." **ICPhS-4**, 227-231 [6.1.2; 6.4.2]

Risberg A. (1964a): "Recoding speech for the deaf and hard of hearing. A status report." **STL-QPSR** (4), 15-20 [9.4.3]

Risberg A. (1964b): "A bibliography on recoding speech for the deaf and hard of hearing." **STL-QPSR** (4), 21-30 [1.3; 9.4.3]

Risberg A. (1968): "Visual aids for speech correction." Am. Ann. Deaf 113 [9.4.3]

Risberg A. (1969): "A critical review of work on speech analyzing hearing aids." IEEE Trans. AU-17, 290-297 [4.2; 9.4.3]

Risberg A., Lubker J. (1978): "Prosody and speech reading." **STL-QPSR** (1), 1-16 [9.4.3]

Risberg A., Möller A., Fujisaki H. (1960): "Voice fundamental frequency tracking." **STL-QPSR** (1), 3-5 [6.1.2; 7.2; 9.3]

Rischel J.: see (Fischer-Jørgensen et al., 1966)

Riso V.: see (Peckels and Riso, 1965)

Ritsma R.J. (1962a): "A model of human pitch extraction based on additive correlation." **ICA-4** (Paper H58) [4.2.1; 8.2.2]

Ritsma R.J. (1962b): "Existence region of the tonal residue." J. Acoust. Soc. Am. **34**, 1224-1229 [4.2.1]

Ritsma R.J. (1970): "Periodicity detection." In *Frequency analysis and periodicity detection in hearing;* ed. by R.Plomp and G.F.Smoorenburg (Sijthoff, Leiden), 250-263 [4.2.1]

Ritsma R.J., Lopes Cardozo B. (1962): "The perception of pitch." Philips Tech. Tijdschrift, 341-346 [4.2.1]

Roach P.J., Hardcastle W.J. (1979): "Instrumental measurement of phonation types: a laryngographic contribution." **ICPh-Miami**, 201-207 [4.1; 5.2.3]

Robinson D.M.: see (Habib et al., 1978)

Rodet X. (1977): "Analyse du signal vocal dans sa représentation amplitude-temps; synthèse de la parole par règles;" Thèse d'Etat, Univ. Pierre et Marie Curie (Paris 6), F-91190 Gif-sur-Yvette [6.2.5; 9.4.2; 9.4.4]

Rodionov I.E.: see (Lepeshkin et al., 1980)

Roelens R.: see (Fabre et al., 1959)

Rogers E.S.: see (Olson et al., 1967)

Rokhtla M. (1969): "Малоинерционный интонометр основного тона." В кн.: Труды АРСО-4, Киев. М.Рохтла. (Quasi-instantaneous fundamental frequency measuring device. In Russian.). **ARSO-4**, 218-222 [6.2.1]

Romportl M., Janota P. (eds.) (1972): *Phonetica Pragensia III – Symposium on Intonology, Prague, October 1970* (Universita Karlova, Praha) [9.4.2]

Rooij J.J. de: see De Rooij J.J.

Rosen S.M.: see (Fourcin et al., 1978)

Rosenberg A.E. (1966): "Pitch discrimination of jittered impulse trains." J. Acoust. Soc. Am. **39**, 920-928 [4.2.3]

Rosenberg A.E. (1968): "Effect of pitch averaging on the quality of natural vowels." J. Acoust. Soc. Am. **44**, 1592-1595 [3.4; 4.3; 9.4.4]

Rosenberg A.E. (1971): "Effect of glottal pulse shape on the quality of natural vowels." J. Acoust. Soc. Am. **49**, 583-590 [3.3; 9.4.2]

Rosenberg A.E. (1976a): "Evaluation of an automatic speaker verification system over telephone lines." Bell Syst. Tech. J. **55**, 723-744 [9.4.4]

Rosenberg A.E. (1976b): "Automatic speaker verification - a review." Proc. IEEE **64**, 475-487 [9.4.4]

Rosenberg A.E., Schafer R.W., Rabiner L.R. (1971): "Effects of smoothing and quantizing the parameters of formant-coded voiced speech." J. Acoust. Soc. Am. **50**, 1532-1538 [6.6.4; 9.4.5]

Rosenblith W.A., Stevens K.N. (1953): "On the DL for frequency." J. Acoust. Soc. Am. **25**, 980-985 [4.2.3]

Rosenthal L.H., Schafer R.W., Rabiner L.R. (1974): "An algorithm for locating the beginning and end of an utterance using ADPCM coded speech." Bell Syst. Tech. J. **53**, 1127-1136 [5.3]

Ross M.J., Shaffer H.L., Cohen A., Freudberg R., Manley H.J. (1974): "Average magnitude difference function pitch extractor." IEEE Trans. ASSP-22, 353-361 [4.1; 8.3.1; 9.3.1]

Rossi M. (1967): "Sur la hierarchie des paramètres de l'accent." **ICPhS-6**, 779-786 [9.4.2]

Rossi M. (1971): "Le seuil de glissando ou seuil de perception des variations tonales pour les sons de la parole." Phonetica **23**, 1-33 [4.3]

Rossi M. (1972): "Le seuil de glissando ou seuil de perception des variations tonales pour les sons de la parole." **Phonetica Pragensia III**, 229-238 [4.3]

Rossi M. (1973): "L'intonation prédicative en français dans les phrases transformées par permutation." Linguistics **103**, 64-94 [9.4.2]

Rossi M. (1978): "La perception des glissandos descendants dans les contours prosodiques." Phonetica **35**, 11-40 [4.3]

Rossi M. (1979a): "Interactions between intensity glides and frequency glissandos." **ICPh-Miami**, 1117-1130 [4.3]

Rossi M. (1979b): "Interaction of intensity glides and frequency glissandos." Lang. Speech **22** [4.3]

Rossi M. (1981): "Prosodical aspects of speech production." **F.A.S.E.-81** (2), 125-158 [9.4.2]

Rossi M., Autesserre D. (1981): "Movements of the hyoid and the larynx and the intrinsic frequency of vowels." J. Phonetics **9**, 233-249 [3.1]

Rossi M., Chafcouloff M. (1972a): "Recherches sur le seuil différentiel de fréquence fondamentale dans la parole." **Phon. Aix-en-Provence 1**, 177-184 [4.3]

Rossi M., Chafcouloff M. (1972b): "Les niveaux intonatifs." **Phon. Aix-en-Provence 1**, 167-178 [4.2.3; 9.4.2]

Rossi M., Di Cristo A. (1977): "Propositions pour un modèle d'analyse de l'intonation." **Journées d'Etude sur la Parole 8**, 323-329 [9.4.2]

Rossi M., Di Cristo A. (1980): "Un modèle de détection automatique des fromtières intonatives et syntaxiques." **Journées d'Etude sur la Parole 11**, 211-232 [9.4.2; 9.4.4]

Rossi M., Di Cristo A., Hirst D.J., Martin Ph., Nishinuma Y. (1981): *L'intonation de l'acoustique à la sémantique* (Klincksieck, Paris) [9.4.2]

Rossum N. van, Boves L. (1978): "An analog pitch-period extractor." **Phon. Nijmegen 2**, 1-17 [6.2.1]

Rossum N. van, Cranen B., Bot K.L.J. de (1982): "The visualization of pitch fluctuations as an aid in language teaching." **Phon. Nijmegen 6**, 36-52 [5.1; 9.4.3]

Rostron A.B., Welbourn C.P. (1976): "A computer assisted system for the extraction and visual display of pitch." Behav. Res. Methods Instrum. **8**, 456-459 [5.1]

Rothenberg M. (1970): "New inverse-filtering technique for deriving the glottal air flow waveform during voicing." J. Acoust. Soc. Am. **48** (A), 130 (Paper II2, 79th Meet. ASA) [5.2.2]

Rothenberg M. (1972): "The glottal volume velocity waveform during loose and tight voiced glottal adjustment." **ICPhS-7**, 380-388 [3.3; 5.2.2]

Rothenberg M. (1973a): "Nonlinear inverse filtering." J. Acoust. Soc. Am. **53** (A), 294 (Paper B4; 84th Meet. ASA) [6.5]

Rothenberg M. (1973b): "A new inverse-filtering technique for deriving the glottal air flow waveform during voicing." J. Acoust. Soc. Am. **53**, 1632-1645 [5.2.2; 6.5]

Rothenberg M. (1974): "Glottal noise during speech." **STL-QPSR** (2-3), 1-10 [3.3; 3.5]

Rothenberg M. (1977): "Measurement of airflow in speech." J. Speech Hear. Res. **20**, 155-176 [3.1; 5.2.2]

Rothenberg M. (1981): "The voice source in singing." In *Research aspects on singing*. Papers given at a seminar organized by the Committee for the Acoustics of Music (Royal Swedish Academy of Music, Stockholm; Report #33), 34-50 [3.1; 3.2; 3.3; 6.5.1]

Rothenberg M., Carlson R., Granström B., Lindqvist-Gauffin J. (1974): "A three parameter voice source for speech synthesis." **Proc. 1974 Speech Commun. Sem. 2**, 235-243 [3.3; 9.4.4]

Rothenberg M., Mahshie J. (1977): "Induced transglottal pressure variations during voicing." J. Acoust. Soc. Am. **62** (A), S14 (Paper F2; 94th Meet. ASA) [3.3; 3.4; 5.2.1]

Rothenberg M., Molitor R.D. (1979): "Encoding voice fundamental frequency into vibrotactile frequency." J. Acoust. Soc. Am. **66**, 1029-1038 [9.4.3]

Rothenberg M., Verrillo R.T., Zahorian S.A., Brachman M.L., Bolanowsky S.J.jr. (1977): "Vibrotactile frequency for encoding a speech parameter." J. Acoust. Soc. Am. **62**, 1003-1012 [9.4.3]

Rothenberg M., Zahorian S.A. (1977): "Nonlinear inverse filtering technique for estimating the glottal-area waveform." J. Acoust. Soc. Am. **61**, 1063-1071 [3.3; 5.2.2; 6.5]

Rothman H.B.: see (Giordano et al., 1973)

Rousselot P. (1908): *Principes de phonétique expérimentale* (Didier, Paris) [5.1.1]

Rozsypal A.J., Millar B.F. (1979): "Percepton of jitter and shimmer in synthetic vowels." J. Phonetics **7**, 343-355 [4.3]

Rubin M.J. (1963): "Experimental studies on vocal pitch and intensity in phonation." Laryngoscope (St.Louis) **73**, 973-1015 [3.1; 4.1]

Rubin M.J., Hirt C.C. (1960): "The falsetto: a high-speed cinematographic study." Laryngoscope (St.Louis) **70**, 1305-1325 [5.2.1]

Rundqvist H.E. (1975): "Fundamental frequency spectrum analyzing of human speech;" Dissertation, Inst. of Technology, Lund [5.2.4; 8.5]

Rupprath R. (1971): *Der Periodizitätsanalysator*. IPK-Forschungsberichte, Bd.39 (Buske, Hamburg) [6.2.1]

Rusconi E.: see (Mezzalama et al., 1976)

Ruske G., Schotola T. (1978): "An approach to speech recognition using syllabic decision units." **ICASSP-78**, 722-725 [5.3.1]

Russell G.O. (1934): "First preliminary X-ray consonant study." J. Acoust. Soc. Am. **5**, 247-251 [5.2.1]

Russell W.H.: see (Viswanathan et al., 1979, 1980; Higgins et al., 1979)

Ruth W. (1955): *Untersuchungen über die Sichtbarmachung der Tonhöhenbewegung beim Sprechen und Singen* (Akademie der Wissenschaften, Wien) [5.1; 9.4.2]

Ruutel I.: see (Remmel et al., 1975)

Rydbeck N., Tjernlund P., Uddenfeldt J. (1980): "A 4.8 kbps voice excited DFT vocoder with time encoded baseband." **ICASSP-80**, 138-141 [9.4.5]

Saal R. (1978): *Handbuch zum Filterentwurf - Handbook of filter design* (AEG-Telefunken, Frankfurt) [2.3]

Sacerdote G.G., Schindler O. (1971): "Phonetic structure of voiced/unvoiced utterances." Acustica **25**, 1-9 [5.3]

Saito S., Fukuda H., Kitahara S., Kokawa N. (1978): "Stroboscopic observation of vocal fold vibration with fiberoptics." Folia Phoniatr. **30**, 241-244 [5.2.1]

Saito S., Fukuda H., Ono H., Isogai Y. (1978): "Observation of vocal fold vibration by x-ray stroboscopy." J. Acoust. Soc. Am. **64** (A), S90-91 (Paper II10; 96th Meet. ASA) [5.2.1]

Saito S., Kato K., Teranishi N. (1958): "Statistical properties of the fundamental frequencies of Japanese speech voices." J. Acoust. Soc. Jpn. **14**, 111-116 [4.1]

Sakoe H., Chiba S. (1971): "A dynamic-programming approach to continuous-speech recognition." **ICA-7** (3), 65-68 (Paper 20C13) [6.6.3; 9.4.4]

Sakoe H., Chiba S. (1978): "Dynamic programming algorithm optimization for spoken word recognition." IEEE Trans. ASSP-26, 43-49

Salles J.L. (1980): "Essai d'un enseignement de l'intonation aux jeunes déficients auditifs profonds par visualisation graphique" (Centre de Traitement de l'Ouie et de la Parole, F-35301 Fougères) [5.1; 9.4.3]

Sambur M.R.: see (Rabiner and Sambur, 1974, 1975, 1977)

Samsam S.: see (Frazier et al., 1975)

Sánchez Gonzalez F.J. (1977a): "Application of dissimilarity and a periodicity function to fundamental frequency measure of speech and voiced/unvoiced decision." **ICA-9**, 523 (Paper I117) [8.3.2]

Sánchez Gonzalez F.J. (1977b): "Dissimilarity and aperiodic functions. Temporal processing of quasiperiodic signals." **ICA-9**, 859 (Paper R13) [8.3.2]

Sánchez Gonzalez F.J. (1979): "Tratamento de señales pseudoperiodicas y aplicacion a la estimacion de los rasgos prosodicos de la palabra." (Treatment of pseudo-periodic signals. Application to analysis of prosody in speech. In Spanish); Dissertation, Universidad Politecnica di Madrid [8.3.2; 9.4.2]

Sansone F.E.: see (Kelly and Sansone, 1976)

Santamarina C. (1975): "Synthèse par règles du signal vocal dans sa répresentation amplitude-temps. Etude de la melodie. Programme de transcription phonétique;" Dissertation, Univ. de Paris, Orsay [9.4.2; 9.4.4]

Sapozhkov M.A. (1963): (The speech signal in cybernetics and communication. In Russian) (Svyazizdat, Moscow) [1.1]

Sarma V.V.S., Venugopal V. (1978): "Studies on pattern recognition approach to voiced-unvoiced-silence classification." **ICASSP-78**, 1-4 [5.3.3]

Sarma V.V.S., Yegnanarayana B. (1976): "Cascade realization of digital inverse filter for extracting speaker dependent features." **ICASSP-76**, 723-726 [6.3.1; 6.5.1]

Sarrat P.: see (Autesserre et al., 1979)

Sarv J.: see (Remmel et al., 1975)

Sasama R.: see (Jentzsch et al., 1978, 1981)

Saslow M.G.: see (Flanagan and Saslow, 1958)

Saunders F.A., Hill W.A., Simpson C.A. (1976): "Speech perception via the tactile mode: A progress report." **ICASSP-76**, 594-597 [5.1; 9.4.3]

Savaga C.J.: see (Dolanský et al., 1971)

Sawada M.: see (Tanabe et al., 1979)

Sawashima M. (1970): "Laryngeal research in experimental phonetics." **Speech Research 23**, 69-115 [3.1; 5.2.1]

Sawashima M. (1974): "Laryngeal research in experimental phonetics." In *Current Trends in Linguistics*, Vol.12, 303-348 [3.1; 9.4.2]

Sawashima M. (1976): "Fiberoptic Observation of the larynx and other speech organs." In *Dynamic aspects of speech production;* ed. by M. Sawashima and F.S. Cooper (Univ. of Tokyo Press), 31-47 [3.1; 5.2.1]

Sawashima M., Abramson A.S., Cooper F.S., Lisker L. (1970): "Observing laryngeal adjustments during running speech by use of a fiberoptics system." Phonetica **22**, 193-201 [3.3; 5.2.1]

Sawashima M., Cooper F.S. (eds.) (1977): *Dynamic aspects of speech production* (Univ. of Tokyo Press)

Sawashima M., Hirose H. (1968): "A new laryngoscopic technique by use of fiberoptics." J. Acoust. Soc. Am. **43**, 168-169 [3.1; 5.2.1]

Sawashima M., Hirose H. (1980): "Laryngeal gestures in speech production." **Ann. Bull. RILP 14** [3.1]

Sawashima M., Hirose H., Yoshioka H. (1978): "Laryngeal muscle activities in relation to glottal shape in voiceless sound production." J. Acoust. Soc. Am. **64** (A), S89 (Paper II4; 96th Meet. ASA) [3.1; 3.5]

Sawashima M., Miyazaki S. (1974): "Stereo-fiberscopic measurement of the larynx: a preliminary experiment by use of ordinary laryngeal fiberscopes." **Ann. Bull. RILP 8**, 7-12 [5.2.1]

Sawashima M., Ushijima T. (1971): "Use of the fiberscope in speech research." **Ann. Bull. RILP 5**, 25-34 [5.2.1]

Saxman J.H., Burk K.W. (1967): "Speaking fundamental frequency characteristics of middle aged females." Folia Phoniatr. **19**, 167-172 [4.1]

Schaaf G.: see (Wendler et al., 1973)

Schäfer-Vincent K. (1982): "Significant points: pitch period detection as a problem of segmentation." Phonetica **39**, 241-254 [6.2.5]

Schafer R.W. (1972): "A survey of digital speech processing techniques." IEEE Trans. AU-**20**, 28-35 [1.2]

Schafer R.W., Jackson K., Dubnowski J.J., Rabiner L.R. (1976): "Detecting the presence of speech using ADPCM coding." IEEE Trans. COM-**24**, 563-567 [5.3]

Schafer R.W., Rabiner L.R. (1970): "System for automatic formant analysis of voiced speech." J. Acoust. Soc. Am. **47**, 634-648 [5.3]

Schafer R.W., Rabiner L.R. (1975): "Digital representations of speech signals." Proc. IEEE **63**, 662-677 [2.2]

Schafer R.W., Rabiner L.R. (1976): "Parametric representations of speech." In *Speech Recognition.* Invited papers of the 1974 IEEE Symposium; ed. by D.R.Reddy (Academic Press, New York), 99-150 [2.2]

Schalk T.B.: see (Doddington and Schalk, 1981)

Scharf L.L.: see (Gueguen and Scharf, 1980)

Scheich H.: see (Creutzfeldt et al., 1979)

Scherer K.R. (1977): "Effect of stress on fundamental frequency of the voice." J. Acoust. Soc. Am. **62** (A), S25-26 (Paper K2; 94th Meet. ASA) [1.2; 9.4.4]

Scherer K.R. (1981a): "Vocal indicators of stress." In *Speech evaluation in psychiatry;* ed. by J.K. Darby (Grune and Stratton, New York) [9.4.3]

Scherer K.R. (1981b): "Speech and emotional states." In *Speech evaluation in psychiatry;* ed. by J.K. Darby (Grune and Stratton, New York) [9.4.3]

Scherer R.C., Curtis J.F., Titze I.R. (1980): "Pressure-flow relationships within static models of the larynx." J. Acoust. Soc. Am. **68** (A), S101 (Paper 3, 100th Meet. ASA) [3.1; 3.3]

Scherer R.C., Titze I.R., Curtis J.F. (1983): "Pressure-flow relationships in two models of the larynx having rectangularr glottal shapes." J. Acoust. Soc. Am. **72**, 668-676 [3.1; 3.3]

Schief R. (1963): "Koinzidenz-Filter für Tonhöhenuntersuchung." Kybernetik **2**, 8-15 [4.2.1]

Schindler O.: see (Sacerdote and Schindler, 1971)

Schirm L. IV (1979): "Get to know the FFT and take advantage of speedy LSI building blocks." Electron. Design (April 26) [8.5.4; 9.1.5]

Schirm L. IV (1980): "Multiplier-accumulator application notes" (TRW, El Segundo CA) [2.3; 9.1.5]

Schlatter M., Eichler J. (1979): "An introduction to the Gaussian least squares approximation and its application in signal processing and system modeling." Signal Processing **1**, 211-225 [8.6]

Schmidt C.E.: see (Rabiner et al., 1975, 1977)

Schmidt K.O. (1942): "Verfahren zur besseren Ausnutzung des Übertragungsweges;" Deutsche Patentschrift Nr. 722607. Bekanntgenacht am 28.5.1942 [6.1.3; 9.4.5]

Schneebeli (1878): "Expériences avec le phonautographe." Arch. Sci. Phys. Nat. Genève **64**, 79 [5.1.1]

Schneider C.: see (Meyer and Schneider, 1913)

Schodder G.R.: see (David and Schodder, 1961)

Schönhärl E. (1960): *Die Stroboskopie in der praktischen Laryngologie* (G.Thieme, Stuttgart) [3.1; 5.2.1; 9.4.3]

Schooneveld C.H. van: see (Buning et al., 1961)

Schotola Th.: see (Ruske and Schotola, 1978)

Schouten J.F. (1938): "The perception of subjective tones." Proc. Kon. Ned. Akad. Wetensch. **41**, 1086-1093 [4.2.1]

Schouten J.F. (1940): "The residue and the mechanism of hearing." Proc. Kon. Ned. Akad. Wetensch. **43**, 991-999 [4.2.1]

Schouten J.F., Ritsma R.J., Cardozo B.L. (1962): "Pitch of the residue." J. Acoust. Soc. Am. **34**, 1418-1424 [4.2.1]

Schrack G.F. (1983): "Gaining access to the literature of computer graphics." IEEE Trans. E-**26**, 7-10 [5.1]

Schreiber L. (1962): "Eine Apparatur zur ohrgemässen Tonhöhenanzeige." **ICA-4** (Paper N53)

Schreiner Ch.: see (Creutzfeldt et al., 1979)

Schröder H.J.: see (Kallenbach and Schröder, 1961)

Schröder M.C. (1976): "Implementations of pitch detection and extraction techniques with regard to their usefulness in linguistic research." **Phon. Köln 5**, 1-83 [2.6; 5.1; 5.2]

Schroeder M.R. (1959): "Recent progress in speech coding at Bell Telephone Laboratories." **ICA-3**, 201-210 [9.4.5]

Schroeder M.R. (1961): "Einrichtung zur Übertragung von Sprache unter Frequenzband-pressung nach Art eines Vocoders;" Deutsche Patentschrift Nr.1114851. bekanntgemacht am 12.10.1961, angemeldet am 26.10.1959 [9.4.5]

Schroeder M.R. (1962): "Band compression system;" United States Patent No. 3,030,450. Issued April 17, 1962; filed Nov. 17, 1958 [9.4.5]

Schroeder M.R. (1963): "Vocoder excitation generator;" United States Patent No. 3,102,928. Issued Sep. 3, 1963; filed Dec. 23, 1960 [3.3; 9.4.5]

Schroeder M.R. (1966): "Vocoders: Analysis and synthesis of speech. A review of 30 years of applied speech research." Proc. IEEE **54**, 720-734 [1.2; 5.3; 9.4.4; 9.4.5]

Schroeder M.R. (1967): "Cepstrum analysis of speech signals;" United States Patent No. 3,321,582. Issued May 1967, filed Aug. 1963 [8.4]

Schroeder M.R. (1968): "Period histogram and product spectrum: new methods for funda-mental-frequency measurement." J. Acoust. Soc. Am. **43**, 829-834 [4.4; 8.5.1]

Schroeder M.R. (1969): "Computers in acoustics: Symbiosis of an old science and a new tool." J. Acoust. Soc. Am. **45**, 1077-1088 [1.2; 2.0]

Schroeder M.R. (1970a): "Fundamental frequency detector;" United States Patent No.3,496,465. Issued Feb. 17, 1970, filed May 19, 1967 [8.5.1]

Schroeder M.R. (1970b): "Speech-noise discriminator;" United States Patent No.3,507,999. Issued April 21, 1970; filed Dec. 20, 1967 [5.3]

Schroeder M.R. (1970c): "Interpolation of data with continuous speech signals;" United States Patent No.3,492,429. Issued January 27, 1970; filed June 1, 1967 [9.4.5]

Schroeder M.R. (1970d): "Parameter estimation in speech: a lesson in unorthodoxy." Proc. IEEE **58**, 707-712 [1.2; 2.0; 8.2.3; 9.4.4]

Schroeder M.R. (1981a): "Speech coding and the human ear." **F.A.S.E.-81** (2), 159-176 [4.2; 9.4.4]

Schroeder M.R. (1981b): "Direct (nonrecursive) relations between cepstrum and predictor coefficients." IEEE Trans. ASSP-**29**, 297-301 [6.3.1; 8.4.1]

Schroeder M.R., Atal B.S. (1962): "Generalized short-time power spectra and autocorrela-tion functions." J. Acoust. Soc. Am. **34**, 1679-1683 [8.2.1]

Schroeder M.R., Atal B.S., Hall J.L. (1979a): "Objective measure of certain speech signal degradations based on masking properties of human auditory perception." In *Frontiers of speech communication research;* ed. by B. Lindblom and S. Öhman (Academic Press, London), 217-229 [4.2]

Schroeder M.R., Atal B.S., Hall J.L. (1979b): "Optimizing digital speech coders by exploit-ing masking properties of the human ear." J. Acoust. Soc. Am. **66**, 1647-1652 [4.2]

Schroeder M.R., David E.E.jr. (1960): "A vocoder for transmitting 10 kc/s speech over a 3.5 kc/s channel." Acustica **10**, 35-43 [9.4.5]

Schubert K. (1951): "Über die Prüfung des Tonhöhenunterscheidungsvermögens." Arch. Ohren-Nasen-Kehlkopfheilkunde **159**, 339-354 [4.2.3; 4.3]

Schubiger M. (1958): *English intonation, its form and function* (M.Niemeyer, Tübingen) [9.4.2]

Schüssler H.W. (1973): *Digitale Systeme zur Signalverarbeitung* (Springer, Berlin) [2.3]

Schulman R.J. (1973): "Time domain pitch extraction algorithm." J. Acoust. Soc. Am. **54** (A), 36 (Paper G4; 86th Meet. ASA)

Schultz-Coulon H.J. (1975a): "Grundtonanalyse - ein Beitrag zur objektiven Beurteilung der Sprechstimme." HNO **23**, 218-225 [2.6; 9.4.3]

Schultz-Coulon H.J. (1975b): "Bestimmung und Beurteilung der individuellen mittleren Sprechstimmlage." Folia Phoniatr. **27**, 375-386 [4.1; 9.4.4]

Schultz-Coulon H.J., Arndt H.J. (1972): "Tonhöhenschreibung in der phoniatrischen Praxis." Folia Phoniatr. **24**, 241-258 [2.6; 5.1; 9.3; 9.4.4]

Schultz-Coulon H.J., Battmer R.D. (1981): "Die quantitative Bewertung des Sängervibratos." Folia Phoniatr. **33**, 1-14 [3.2; 4.3; 9.4.2; 9.4.3]

Schultz-Coulon H.J., Battmer R.D., Fedders B. (1979): "Measurement of fundamental-frequency variations." Folia Phoniatr. **31**, 56-69 [9.4.3]

Schunk H. (1971): "Ein Sprechgerät für Heliumatmosphäre." Techn. Mitteilungen AEG-Telefunken **61**, 378-381 [6.2.1; 9.4.5]

Schunk H. (1972): "Entzerrung von Heliumsprache." In *Akustik und Schwingungstechnik*. Plenarvorträge und Kurzreferate der Gemeinschaftstagung Stuttgart 1972 (VDE-Verlag, Berlin), 341-344 [6.2.1; 6.3.2; 9.4.4]

Schvey M.H.: see (Cooper et al., 1966; Lisker et al., 1969)

Schwartz M.R.: see (Makhoul et al., 1978; Viswanathan et al., 1982)

Schweizer L. (1970): "Quasi-random pulse excitation for synthesis of unvoiced sounds." Acustica **22**, 107-110 [3.5; 9.4.4]

Scott (1861): "Inscription automatique des sons de l'air au moyen d'une oreille artificielle." Annales du Conservatoire des Arts et Métiers, Paris [5.1.1]

Scott B.L. (1978): "Development of a real-time voice pitch extractor." J. Acoust. Soc. Am. **63** (A), S79 (Paper GG8; 95th Meet. ASA)

Scott R.J., Gerber S.E. (1972): "Pitch synchronous time compression of speech." **ICSCP-72**, 62-65 (Paper B4) [9.4.5]

Scripture E.W. (1902): *The elements of experimental phonetics*. Reprinted 1973 by AMS Press, New York NY (C.Scribner's Sons, New York) [5.1.1]

Scripture E.W. (1903): "A record of the melody of the Lord's prayer." Die neueren Sprachen **10** [5.1.1]

Scripture E.W. (1906): "Researches in experimental phonetics. The studies of pitch curves" (Carnegie Institute of Washington, Washington DC; Publ.No.44) [5.1.1]

Scripture E.W. (1935): "Registration of Speech Sounds." J. Acoust. Soc. Am. **7**, 139-141 [5.1.1]

Scudder H.J.: see (Kincaid and Scudder, 1967)

Scully C.: see (Fant and Scully, 1977)

Seashore C.E. (1914): "The tonoscope." Psychol. Monogr. **16**, 1-12 [5.1.1]

Secrest B.G., Doddington G.R. (1982): "Postprocessing techniques for voice pitch trackers." **ICASSP-82**, 172-175 [6.6.3]

Secrest B.G., Doddington G.R. (1983): "An integrated pitch tracking algorithm for speech systems." **ICASSP-83**

Sedlacek K., Sychra A. (1963): "Die Melodie als Faktor emotionellen Ausdrucks." Folia Phoniatr. **15**, 89-98 [9.4.3]

Seebeck A. (1841): "Beobachtungen über einige Bedingungen der Entstehung von Tönen." Ann. Phys. Chem. **53**, 417-436 [4.2.1]

Seebeck A. (1843): "Über die Definition des Tones." Ann. Phys. Chem. **60**, 449-481 [4.2.1]

Seewann M.: see (Terhardt et al., 1982)

Séguy J. (1953): "Un combiné magnétophone-électrokymographe en vue de l'analyse tonométrique." Orbis **2**, 518-520 [5.1.1]

Seidner W., Wendler J., Halbedl G. (1972): "Mikrostroboskopie." Folia Phoniatr. **24**, 81-85 [5.2.1]

Sekimoto S.: see (Miyazaki et al., 1973)

Semmelink A. (1974): "The use of a real-time spectrum analyzer in speech analysis." **ICA-8**, 233 [9.4.3]

Senderowicz D.: see (Dalrymple et al., 1979)

Seneff S. (1976): "Real-time harmonic pitch detector." J. Acoust. Soc. Am. **60** (A), S107 (Paper RR6; 92nd Meet. ASA) [8.1; 8.5.3]

Seneff S. (1977): "Real time harmonic pitch detector" (M.I.T. Lincoln Labs, Lexington MA; DDC-AD A038542/7GA) [8.1; 8.5.3]

Seneff S. (1978): "Real-time harmonic pitch detector." IEEE Trans. ASSP-**26**, 358-364 [8.1; 8.5.3; 8.5.4]

Seneff S. (1982): "System to independently modify excitation and/or spectrum of speech waveform without explicit pitch extraction." IEEE Trans. ASSP-**30**, 566-578 [9.4.4; 9.4.5]

Senmoto S. (1972): "Adaptive decomposition of a composite signal of identical unknown wavelets in noise." IEEE Trans. SMC-**2**, 59-66 [8.6]

Serniclaès W.: see (Abry et al., 1975)

Serra A.: see (De Mori and Serra, 1972)

Sethy A.: see (Christiansen et al., 1966)

Sette W.J.: see (Wolf et al., 1935)

Seymour J. (1972a): "Acoustic analysis of singing voices I: Sound samples and room responses." Acustica **27**, 203-209 [3.2; 9.4.2]

Seymour J. (1972b): "Acoustic analysis of singing voices II: Frequency and amplitude vibrato analyses." Acustica **27**, 209-217 [3.2; 9.4.2]

Seymour J. (1972c): "Acoustic analysis of singing voices III: Spectral components, formants, and glottal source." Acustica **27**, 218-227 [3.2; 9.4.2]

Seymour J. (1976): "Formants and source spectrum of partially trained singing voices." Acustica **35**, 253-256 [3.2; 9.4.2]

Sferrino V.: see (Tierney et al., 1964)

Shaffer H.L. (1964): "Information rate necessary to transmit pitch-period durations for connected speech." J. Acoust. Soc. Am. **36**, 1895-1900 [4.1; 9.4.5]

Shaffer H.L., Ross M.J., Cohen A. (1973): "AMDF pitch extractor." J. Acoust. Soc. Am. **53** (A), 95 (Paper NN9; 85th Meet. ASA) [8.3.1]

Shankland R.S. (1941): "The analysis of pulses by means of the harmonic analyzer." J. Acoust. Soc. Am. **12**, 383-386

Shannon C.E. (1948): "A mathemathical theory of communication." Bell Syst. Tech. J. **27**, 623-656 [2.2]

Shapiro L.G. (1978a): "Sampling theory in digital processing." Electron. Eng. **50**, 45-53 [2.2]

Shapiro L.G. (1978b): "The design of digital filters." Electron. Eng. **50**, 51-69 [2.3]

Shatverov E.A.: see (Grigoryan et al., 1976)

Sheshradi S., Waldron M.B. (1979): "A pattern recognition approach to compare natural and synthesized speech." **ICASSP-79**, 777-780 [5.3.3]

Shigenaga M. (1962): "Extraction of principal characteristics of speech by short-time autocorrelation functions." J. of the IECE of Japan **45**, 1076-1085 [2.4; 8.2.2]

Shipley K.L.: see (Flanagan et al., 1975)

Shipp T. (1975): "Vertical laryngeal position during continuous and discrete vocal frequency change." J. Speech Hear. Res. **18**, 707-718 [3.1; 3.3]

Shoup J.E.: see (Peterson and Shoup, 1966)

Shower E.G., Biddulph R. (1931): "Differential pitch sensitivity of the ear." J. Acoust. Soc. Am. **3**, 275-287 [4.2.3]

Shuplyakov V., Murrak T., Liljencrants J. (1968): "Phase dependent pitch sensation." **STL-QPSR** (4), 7-14 [4.2.1]

Siegel L.J. (1977): "A pattern classification algorithm for the voiced-unvoiced decision in speech analysis;" Dissertation, Univ. of Princeton, Princeton NJ [5.3.3]

Siegel L.J. (1979a): "Features for the identification of mixed excitation in speech analysis." **ICASSP-79**, 752-755 [5.3.3]

Siegel L.J. (1979b): "A procedure for using pattern classification techniques to obtain a voiced/unvoiced classifier." IEEE Trans. ASSP-**27**, 83-89 [5.3.3]

Siegel L.J., Bessey A.C. (1980): "A decision tree procedure for voiced/unvoiced/mixed excitation classification of speech." **ICASSP-80**, 53-56 [5.3]

Siegel L.J., Bessey A.C. (1982): "Voiced/unvoiced/mixed excitation classification of speech." IEEE Trans. ASSP-**30**, 451-461 [5.3.3]

Siegel L.J., Steiglitz K. (1976): "A pattern classification algorithm for the voiced/unvoiced decision." **ICASSP-76**, 326-329 [5.3]

Signal Technology Inc. (ed.) (1980): "Technical Manual, Interactive Laboratory System (ILS)" (Signal Technology Inc., Santa Barbara CA) [8.2.5; 9.4.4]

Silverman S.R.: see (Davis et al., 1951)

Simon C. (1927): "The variability of consecutive wave lengths in vocal and instrumental sounds." Psychol. Monogr. **36**, 41-83 [4.3]

Simon J.C. (ed.) (1980): *Spoken language generation and understanding.* Proceedings of the NATO Advanced Study Institute held at Bonas, France, June 23 - July 5, 1979 (Reidel, Dordrecht) [1.2]

Simon J.C. (1982): "Complexity of algorithms and pattern recognition." In *Automatic speech analysis and recognition.* Proceedings of the NATO Advanced Study Institute held at Bonas, France, June 29 - July 10, 1981; ed. by J.P. Haton (Reidel, Dordrecht), 309-330 [8.5.4; 9.1.5]

Simon M.K.: see (Lindsey and Simon, 1973)

Simon Th. (1977a): "Die Verwendung eines Computers bei einem Einführungskurs über Experimentalphonetik." **Phon. München 7**, 101-110 [9.4.2]

Simon Th. (1977b): "Digitale Grundfrequenzanalyse in Echtzeit." **Phon. München 7**, 125-132 [6.1.1]

Simpson C.A.: see (Saunders et al., 1976)

Sincoskie W.D.: see (Habib et al., 1978)

Singh S. (ed.) (1975): *Measurement procedures in speech, hearing, and language* (University Park Press, Baltimore)

Singhal S., Atal B.S. (1982): "On reducing pitch sensitivity of LPC parameters." J. Acoust. Soc. Am. **72** (A), S78 (Paper RR2, 104th Meet. ASA) [6.3.1; 9.4.4]

Singleton H.E. (1951): "Harmonic amplitude selector for signalling systems;" United States Patent No. 2,553,610. Issued May 22, 1951; filed Jan. 21, 1950 [8.2.3]

Siskind R.P.: see (Dempsey et al., 1950, 1953)

Six W.: see (Vermeulen and Six, 1949)

Sjölin A.: see (Askenfelt and Sjölin, 1980)

Skinner B.F. (1935): "A calibrated recording and analysis of the pitch, force, and quality of vocal tones expressing happiness and sadness,and a determination of the pitch and force of the subjective concepts of ordinary, soft, and loud tones." Speech Monogr. **2**, 81-137 [3.2; 9.4.3]

Skinner D.P.: see (Childers et al., 1977)

Skinner T.F. (1974): "Unpublished memoranda on F_0 analysis: 1) Autocorrelation method for determining F_0; 2) F_0 determination - absolute addition vs. multiplication; 3) F_0 determination - analysis of an octave error; 4) F_0 analysis - conversion from hertz to tones; 5) F_0 computation" (Sperry Univac, Def.Syst.Div. St.Paul MN; Univac Intercommunication) [8.5.4]

Slawson A.W. (1968): "Vowel quality and musical timbre as function of spectrum envelope and fundamental frequency." J. Acoust. Soc. Am. **43**, 87-101 [9.4.2]

Slepokurova N.A. (1972): "Effect of fundamental tone pitch on the position of the phoneme boundary between vowels (Original in Russian, translated from Akusticheskii Zhurnal 18, 426-432)." Soviet Physics - Acoustics **18**, 356-361 [3.3; 9.4.2]

Slis I.H. (1967): "What causes the voiced-voiceless distinction?." **ICPhS-6**, 841-843 [5.3]

Slis I.H., Damsté P.H. (1967): "Transillumination of the glottis during voiced and voiceless consonants." **IPO Annual Progr. Rept. 2**, 103-109 [5.2.5]

Sluyter R.J., Kotmans H.J., Claasen T.A.C.M. (1982): "Improvements of the harmonic-sieve pitch extraction scheme and an appropriate method for voiced-unvoiced detection." **ICASSP-82**, 188-191 [5.3; 8.5.2]

Sluyter R.J., Kotmans H.J., Leeuwaarden A. van (1980): "A novel method for pitch extraction from speech and a hardware model applicable to vocoder systems." **ICASSP-80**, 45-48 [8.1; 8.5.2; 8.5.4]

Smiarowski R.A. (1977): "On measuring the DL for pitch memory with the transformed up-down psychophysical method." J. Acoust. Soc. Am. **61** (A), S49 (Paper AA3; 93rd Meet. ASA) [4.2.3]

Smith A.M. (1981): "Feature extraction for laryngeal evaluation." **ICASSP-81**, 137-140 [5.2.3; 6.3.1; 6.5.1; 9.4.3]

Smith C.P. (1951): "A phoneme detector." J. Acoust. Soc. Am. **23**, 446-451 [6.3.2]

Smith C.P. (1954): "Device for extracting the excitation function from speech signals;" United States Patent No. 2,691,137. Issued Oct. 5, 1954; filed June 27, 1952; reissued 1956 [6.3.2]

Smith C.P. (1957): "Speech data reduction: Voice communications by means of binary signals at rates under 1000 bits/sec" (AFCRC, Bedford MA; DDC-AD-117290) [6.3.2]

Smith S.C.: see (Stilwell et al., 1981)

Smith S.L. (1954): "Remarks on the physiology of the vibrations of the vocal cords." Folia Phoniatr. **6**, 166-178 [3.1]

Smith S.L. (1957): "Chest register versus head register in the membrane cushion model of the vocal chords." Folia Phoniatr. **9**, 32-36 [3.2]

Smith S.L. (1959): "On pitch variation." Folia Phoniatr. **11**, 173-177 [4.3]

Smith S.L. (1968): "A valve device for difficult cases of fundamental recording." Folia Phoniatr. **20**, 202-206 [5.2.2]

Smith S.L. (1977): "Electroglottography." In *Proceedings, 17th International Congress of Logopedics and Phoniatrics (IALP), Copenhagen, August 1977* (Copenhagen) [5.2.3]

Smith S.L. (1981): "Research on the principle of electroglottography." Folia Phoniatr. **33**, 105-114 [5.2.3]

Smith S.L. (1982): "Untersuchungen zum Prinzip der Elektroglottographie." In *Miszellen VIII*; ed. by J.P.Köster. Hamburger Phonetische Beiträge, Bd.40 (Buske, Hamburg) [5.2.3]

Smoorenburg G.F.: see (Plomp and Smoorenburg, 1970)

Snidecor J.C., Rehman I., Washburn D.D. (1959): "Speech pickup by contact microphone at head and neck positions." J. Speech Hear. Res. **2**, 277-281 [5.2.2]

Snovalshchikov N.A. (1976): Сравнительное исследование методов выделения основного тона речи, использующих фазирование частотных компонент. В кн.: АРСО-9 (Минск). Н.А.Сновальщиков. (Comparative investigation of pitch determination methods. In Russian). **ARSO-9** [9.3.1]

Snow T.B., Hughes G.W. (1965): "Fundamental frequency estimation by harmonic identification." J. Acoust. Soc. Am. **45**, 316 [8.5.1]

Snow T.B., Montgomery A.A. (1973): "Observations on a multiple harmonic F_0-tracking program." J. Acoust. Soc. Am. **53** (A), 345 (Paper DD7; 84th Meet. ASA) [8.5.1]

Sobakin A.N. (1972): "Digital computer determination of the formant parameters of the vocal tract from a speech signal." (Original in Russian; translated from Akusticheskii Zhurnal 18, 106-114). Soviet Physics - Acoustics **18**, 84-90 [6.5.2]

Sobakin A.N. (1976): "Адаптивныи метод выделения основного тона речи." В кн.: Труды АРСО-9 (Минск). А.Н.Собакин. (Adaptive method of pitch determination. In Russian). **ARSO-9**

Sobakin A.N. (1978): "Метод выделения основного тона речи, использующей адаптивную комб-фильтрацию (АКФ)." В кн.: Труды АРСО-10, Тбилиси. А.Н.Собакин. (Pitch determination algorithm using an adaptive comb filter. In Russian).. **ARSO-10** [8.3.2]

Sobolev V.N. (1968): "Экспериментальное исследование коррелятсионного метода выделения основного тона речи." В журн.: Акустический Журнал. В.Н. Соболев. (Experimental investigation of the correlation method of fundamental frequency extraction. In Russian). Akusticheskii Zhurnal **14**, 441-448 [8.2.4]

Sobolev V.N. (1972): "Сравительное адаптивного алгоритма выделения мелодии речи." В кн.: Труды учебных институтов связи. В.Н. Соболев. (Investigation of an adaptive algorithm for determination of the speech melody. In Russian). Trudy Uchebnykh Institutov Svyazi **59**, 3-9

Sobolev V.N. (1977): "Discrimination of vocal pitch by an error correcting time-spectral method." Soviet Physics - Acoustics **23**, 69-73

Sobolev V.N. (1982): "Выделение основного тона методом гребенчатой филтрации.-" В кн.: Труды АРСО-12, Киев. В.Н. Соболев. (Pitch determination by means of a comb filter. In Russian). **ARSO-12**, 141-143 [8.3.3; 8.6.3]

Sobolev V.N., Baronin S.P. (1968):"Исследование сдвигового метода выделения основного тона речи." В журн.: Электросвяз. В.Н.Соболев и С.П.Баронин. (Investigation of the shift method for pitch determination. In Russian.). Elektrosvyaz **12**, 30-36 [8.3.1]

Somin N.V. (1973): "Сравнение нескольких спектральных методов выделения основного тона в условых шума и ограниченной полосы частот." В кн.: Труды АРСО-7. Н.В.Сомин. (Comparison of several spectral pitch determination methods for noisy and band-limited signals. In Russian). **ARSO-7**, 56-59 [8.5.2; 9.3.1]

Sommers R.K., Meyers W.I., Fenton A.K. (1961): "Pitch discrimination and articulation." J. Speech Hear. Res. **4**, 55-60

Sondhi M.M. (1964): "Equivalence of "vector" and autocorrelation pitch detectors." J. Acoust. Soc. Am. **36** (C), 1964 [6.3.2; 8.2.3]

Sondhi M.M. (1968): "New methods of pitch extraction." IEEE Trans. AU-**16**, 262-266 [6.2.4; 8.2.4; 9.2]

Sondhi M.M. (1973): "Measurement of the glottal pulses." J. Acoust. Soc. Am. **54** (A), 36 (Paper G1; 86th Meet. ASA) [5.2.2]

Sondhi M.M. (1975a): "Apparatus for determining the glottal waveform;" United States Patent No. 3,925,616. Issued Dec. 9, 1975; filed April 30, 1974 [5.2.2]

Sondhi M.M. (1975b): "Measurement of the glottal waveform." J. Acoust. Soc. Am. **57**, 228-232 [3.3; 5.2.2]

Sone T.: see (Tsumara et al., 1973)

Sonesson B. (1960): "On the anatomy and vibratory pattern of the human vocal folds, with special reference to a photo-electrical method for studying the vibratory movements." Acta Otolaryngol. Suppl. #156 (Stockholm) [3.0; 5.2.5]

Sonesson B. (1968): "The functional anatomy of the speech organs." In *Manual of Phonetics;* ed. by B.Malmberg (North Holland Publishing Company, Amsterdam), 45-75 [3.1]

Song K.H., Un C.K. (1980): "On pole-zero modeling of speech." **ICASSP-80**, 162-165 [6.3.1; 9.4.5]

Song K.H., Un C.K. (1982): "Pole-zero modeling of noisy speech and its application to vocoding." **ICASSP-82**, 1581-1584 [9.4.5]

Sonninen A., Damsté P.H. (1972): "An international terminology in the field of logopedics and phoniatrics." Folia Phoniatr. **23**, 1-32 [2.1; 9.4.3]

Sorensen D., Horii Y., Leonard R.J. (1980): "Effects of laryngeal topical anaesthesia on voice fundamental frequency perturbation." J. Speech Hear. Res. **23**, 274-284 [3.1; 4.3]

Sorokin V.N. (1981): "The vocal sound source as a distributed-parameter system." Soviet Physics - Acoustics **27**, 239-242 [3.2; 3.3]

Soron H.I. (1967): "High-speed photography in speech research." J. Speech Hear. Res. **10**, 768-776 [5.2.1]

Sorre T.: see (Chuang et al., 1971)

Sovak M., Courtois J., Haas C., Smith S.L. (1971): "Observations on the mechanism of phonation investigated by ultraspeed cinefluorography." Folia Phoniatr. **23**, 277-287 [5.2.1]

Specker P. (1982): "A powerful postprocessing algorithm for time-domain pitch trackers." J. Acoust. Soc. Am. **72** (A), S32 (Paper R14, 104th Meet. ASA) [6.6.3]

Speech Research --> Haskins Labs. (ed.): *Status Report on Speech Research* (Haskins Labs, New Haven CT)

Spens K.E. (1975): "Pitch information displayed on a vibrator matrix as a speech reading aid. Some preliminary results." **STL-QPSR** (2-3), 34-39 [5.1; 9.4.3]

Spens K.E. (1980): "Tactile speech communication aids for the deaf: a comparison." **STL-QPSR** (4), 23-39 [9.4.3]

Sreenivas T.V. (1981): "Pitch estimation of aperiodic and noisy speech signals;" Dissertation, Department of Electrical Engineering, Indian Institute of Technology, Bombay [4.3; 8.5.2; 8.5.4; 9.1; 9.3]

Sreenivas T.V., Rao P.V.S. (1979): "Pitch extraction from corrupted harmonics of the power spectrum." J. Acoust. Soc. Am. **65**, 223-228 [4.3; 8.5.2]

Sreenivas T.V., Rao P.V.S. (1980a): "Problem of fundamental frequency analysis in speech" (Speech and Digital Systems Group, Tata Institute of Fundamental Research, Bombay) [8.5.2; 9.1; 9.3]

Sreenivas T.V., Rao P.V.S. (1980b): "High-resolution narrow-band spectra by FFT pruning." IEEE Trans. ASSP-**28**, 254-257 [8.5.4]

Sreenivas T.V., Rao P.V.S. (1981): "Functional demarcation of pitch." Signal Processing **3**, 277-284 [9.1]

Sridhar C.S., Rajappan K.P. (1979): "A simple pitch extractor." Z. Elektr. Inf. Energietech. **9**, 91-93

Srulowicz P.: see (Goldstein J.L. et al., 1978)

Stall D.: see (Gertheiss et al., 1974)

Standke R.: see (Tolkmitt et al., 1981)

Stanley D.: see (Wolf et al., 1935)

Stark R.E.: see (Goldstein M.H. et al., 1976)

Stead L.G., Jones E.T. (1961): "The S.R.D.E. speech bandwidth compression project" (Signals Research and Development Establishment, Christchurch, Hants (England); S.R.D.E.Report No.1123; DDC-AD-259275) [6.1.1]

Stead L.G., Weston R.C. (1962): "Sampling and quantizing the parameters of a formant-tracking vocoder system." **Proc. 1962 Speech Commun. Sem.** (Paper G9)

Steele R.: see (Xydeas and Steele, 1976)

Steer M.D. (1960): "Modern instrumentation for diagnosis, therapy, and research." Folia Phoniatr. **12**, 196-204 [9.4.3]

Steffen-Batog M.: see (Jassem et al., 1973)

Steffen-Batogowa M. (1966): "Versuch einer strukturellen Analyse der polnischen Aussage-melodie." Z. Phonetik **19**, 397-440 [9.4.2]

Steiglitz K. (1977): "On the simultaneous estimation of poles and zeros in speech analysis." IEEE Trans. ASSP-**27**, 229-234 [6.5.1]

Steiglitz K., Winham G., Petzinger J. (1975): "Pitch extraction by trigonometric curve fitting." IEEE Trans. ASSP-**23** (C), 321-323 [8.6.1]

Steinberg J.C. (1934): "Application of sound measuring instruments to the study of phonetic problems." J. Acoust. Soc. Am. **6**, 16-24

Steinberg J.C. (1953): "Speech analyzing and synthesizing communication system;" United States Patent No. 2,635,146. Issued April 14, 1953; filed Dec. 15, 1949 [6.1.1]

Steinberg J.C., French N.R. (1946): "The portrayal of visible speech." J. Acoust. Soc. Am. **18**, 4-18 [5.1]

Steingrube A., Schroeder M.R. (1977): "Modification of pitch and duration of speech signals." **ICA-9**, 521 (Paper I115) [6.3.1; 9.4.4]

Sterne T.A., Zimmerman H.J. (1939): "A thyratron indicator for teaching the deaf." J. Sci. Instrum. **16**, 334-336 [9.4.3]

Stevens K.N. (1950): "Autocorrelation analysis of speech sounds." J. Acoust. Soc. Am. **22**, 769-771 [8.2.1; 8.2.2]

Stevens K.N. (1952): "Frequency discrimination for damped waves." J. Acoust. Soc. Am. **24**, 76-79 [4.2.3]

Stevens K.N. (1972): "Sources of inter- and intra-speaker variability in the acoustic properties of speech sounds." **ICPhS-7**, 206-232 [9.4.4]

Stevens K.N. (1975): "Modes of conversion of airflow to sound, and their utilization in speech." **ICPhS-8** [3.1]

Stevens K.N. (1977): "Physics of laryngeal behavior and larynx modes." Phonetica **34**, 264-279 [3.1]

Stevens K.N., Atkinson J.E. (1973): "Theory of vocal-cord vibration and its relation to laryngeal features." J. Acoust. Soc. Am. **54** (A), 23 (Paper A1, 86th Meet. ASA) [3.1]

Stevens K.N., Kalikow D.N., Willemain T.R. (1975): "A miniature accelerometer for detecting glottal waveforms and nasalization." J. Speech Hear. Res. **18**, 594-699 [5.2.2; 9.4.3]

Stevens K.N., Klatt D.H. (1974): "Role of formant transitions in the voiced-voiceless distinction for stops." J. Acoust. Soc. Am. **55**, 653-659 [5.3]

Stevens S.S., Egan J.P., Miller G.A. (1947): "Methods of measuring speech spectra." J. Acoust. Soc. Am. **19**, 771-780 [5.1.2]

Stevens S.S., Volkman J. (1940): "The relation of pitch to frequency - a revised scale." Am. J. Psychol. **53**, 329-338 [4.2]

Stevens S.S., Volkman J., Newman E.B. (1937): "A scale for the measurement of the psychological magnitude pitch." J. Acoust. Soc. Am. **8**, 185-190 [4.2]

Stevenson D.C.: see (Nakatsui et al., 1980)

Stewart L.C., Larkin W.D., Houde R.A. (1976): "A real time sound spectrograph with implications for speech training for the deaf." **ICASSP-76**, 590-593 [5.1.2; 9.4.3]

Stilwell D.J., McCutcheon M.J., Hasegawa A., Fletcher S.G., Smith S.C. (1981): "A real-time voice fundamental frequency acquisition and processing system." J. Acoust. Soc. Am. **70** (A), S40 (Paper R3, 102nd Meet. ASA) [5.2.2]

Stirling W.C., Turner J.M. (1983): "Joint estimation of excitation and vocal tract response." **ICASSP-83**

STL-QPSR --> *Quarterly progress and status report;* ed. by Speech Transmission Laboratory (Royal Institute of Technology, Stockholm)

Stock D. (1966): *Zum Problem der Periodizitätsdetektion im Sprachschall.* IPK-Forschungsberichte, Bd.66/3 (Buske, Hamburg) [9.3.1]

Stock D. (1971): *Untersuchungen zur Stimmhaftigkeit hochdeutscher Phonemrealisationen.* IPK-Forschungsberichte, Bd.28 (Buske, Hamburg) [5.3]

Stock D. (1972): "Ein Computerprogramm zur Untersuchung von Einzelperioden phonetischer Signale." **ICPhS-7**, 613-617 [9.4.5]

Stock D., Rupprath R. (1971): "Die Stimmhaft-Stimmlos-Distinktion im DAWID-II." **ICA-7** (3), 57-60 (Paper 20C11) [5.3]

Stock D., Tillmann H.G. (1974): "Instrumentalphonetische Lautabgrenzung: zwei Verfahren zur automatischen Segmentation akustischer Sprachsignale." In *Kommunikationsforschung und Phonetik* (Hamburg), 247-260 [5.3]

Stock E., Zacharias C. (1971): *Deutsche Satzintonation* (Leipzig) [9.4.2]

Stockham T.G., Cannon T.M., Ingebretsen R.B. (1975): "Blind deconvolution through digital signal processing." Proc. IEEE **63**, 678-692 [2.4; 8.4.1]

Stockhoff K.C.H. (1959): "A twelve-channel transistorized vocoder." In *Proceedings of the seminar on speech compression and processing* (Electronics Research Directorate, Bedford, Mass; DDC-AD-233717,233720) [6.1.2; 9.4.4]

Stoll G.: see (Terhardt et al., 1982)

Stone R.B., White G.M. (1963): "Digital correlator detects voice fundamental." Electronics **36**, 26-30 [8.3.3]

Stone R.E.: see (Anderson et al., 1976)

Stout B. (1938): "Waveform and spectrum of the glottal voice source." J. Acoust. Soc. Am. **10**, 137-146 [3.1; 3.3]

Stover V.: see (Atal and Stover, 1975)

Straka G., Burgstahler P. (1963): "Etude du rythme à l'aide de l'oscillographe cathodique combiné avec le sonomètre." Trav. Ling. Litt., 125-141 [5.1.1]

Stratton W.D. (1974): "Information feedback for the deaf through a tactile display." Volta Review **76**, 26-35 [9.4.3]

Striglioni L. (1963): "Contribution à l'étude de la physiologie laryngée en voix parlée. Apport de l'électro-glottographie;" Dissertation, Université de Provence, Aix-en-Provence [3.3; 5.2.3]

Strome M.: see (Gay et al., 1971)

Strong W.J., Palmer E.P. (1975): "Computer-based sound spectrograph system." J. Acoust. Soc. Am. **58**, 899-904 [5.1.2]

Strube H.W. (1974a): "Bestimmung der Querschnittsfunktion des menschlichen Stimmkanals aus dem Sprachsignal;" Dissertation, Universität Göttingen [3.4; 6.5.2; 9.4.4]

Strube H.W. (1974b): "Determination of the instant of glottal closure from the speech wave." J. Acoust. Soc. Am. **56**, 1625-1629 [6.5.2]

Strube H.W. (1980): "Comments on "Least-squares glottal inverse filtering from the acoustic speech waveform" [Wong et al., 1979]." IEEE Trans. ASSP-**28**, 343 [6.5.1]

Stubbs H.L.: see (Wiren and Stubbs, 1956)

Studdert-Kennedy M.: see (Hadding-Koch and Studdert-Kennedy, 1964)

Sudo H.: see (Fujisaki and Sudo, 1971)

Sugimoto T. (1963): "Pitch sensation and fundamental frequency extraction of speech sounds" (Emmanuel Collection, Boston MA; Report No.ETJ-632; DDC-AD-456752)

Sugimoto T. (1972): "On the pitch sensation and the fundamental frequency estimation." IEEE Trans. AU-**21**, 140-148

Sugimoto T., Hashimoto S. (1962): "The voice fundamental pitch and formant tracking computer program by short term autocorrelation function." **Proc. 1962 Speech Commun. Sem.** [8.2.3]

Sugimoto T., Hashimoto S. (1963): "Voice fundamental pitch and formant-tracking computer program by short-term autocorrelation function." J. Acoust. Soc. Am. **35** (A), 1114 [8.2.3]

Sugimoto T., Hiki Sh. (1962a): "On the extraction of the pitch signal using the body wall vibration at the throat of the talker." ICA-**4** (Paper G26) [5.2.2]

Sugimoto T., Hiki Sh. (1962b): "On the extraction of the pitch signal using the body wall vibration at the throat of the talker." **Proc. 1962 Speech Commun. Sem.** [5.2.2]

Sukys R.: see (Chang et al., 1959)

Sule R.: see (Remmel et al., 1975)

Suminaga Y.: see (Gohbara et al., 1978)

Sundberg J. (1971): "Voice source properties of bass singers." ICA-**7** (3), 593-596 (Paper 20S5) [3.2]

Sundberg J. (1972): "Pitch of synthetic sung vowels." STL-QPSR (1), 34-44 [3.2; 9.4.4]

Sundberg J. (1973a): "Data on maximum speed of pitch changes." STL-QPSR (4), 39-47 [4.3]

Sundberg J. (1973b): "The source spectrum in professional singing." Folia Phoniatr. **25**, 71-90 [3.1; 3.2; 3.4]

Sundberg J. (1975): "Vibrato and vowel identification." STL-QPSR (2), 49-60 [3.2]

Sundberg J. (1979): "Maximum speed of pitch changes in singers and untrained subjects." J. Phonetics **7**, 71-79 [3.2; 4.3]

Sundberg J. (1981a): "The voice as a sound generator." In *Research aspects on singing*. Papers given at a seminar organized by the Committee for the Acoustics of Music (Royal Swedish Academy of Music, Stockholm; Report #33), 6-14 [3.1; 3.3]

Sundberg J. (1981b): "To perceive one's own voice and another person's voice." In *Research aspects on singing*. Papers given at a seminar organized by the Committee for the Acoustics of Music (Royal Swedish Academy of Music, Stockholm; Report #33), 80-97 [4.2]

Sundberg J., Askenfelt A. (1981): "Larynx height and voice source. A relationship?." STL-QPSR (2-3), 23-36 [3.1]

Sundberg J., Gauffin J. (1978): "Waveform and spectrum of the glottal voice source." STL-QPSR (2-3), 35-50 [3.1; 3.2; 3.3; 5.2; 6.5]

Sundberg J., Gauffin J. (1979): "Waveform and spectrum of the glottal voice source." In *Frontiers of speech communication research;* ed. by B. Lindblom and S. Öhman (Academic Press, London), 301-320 [3.1; 3.3]

Sundberg J., Tjernlund P. (1970): "A computer program for the notation of played music." STL-QPSR (2-3), 46-49 [9.4.2]

Sundberg J., Tjernlund P. (1971): "Real time notation of performed melodies by means of a computer." ICA-**7** (3), 653-656 (Paper 21S6) [9.4.2]

Sung W.Y., Un C.K. (1980): "A high-speed pitch extraction method based on peak detection and AMDF." J. Korea Inst. Electron. Eng. **17**, 38-44 [8.3.1]

Suzuki G. (1960): "Pitch extractor." J. Acoust. Soc. Jpn. **16**, 281-283

Suzuki J., Nakatsui M., Tanaka R. (1977): "Translation of helium speech by the method of segmentation, partial-rejection, and expansion." J. Radio Res. Labs. (Tokyo) **24**, 1-16 [9.4.5]

Suzuki J., Takasugi T. (1971): "Analysis of voice sound by reference to glottal waveform." ICA-7 (3), 265-268 (Paper 25C2) [6.5.1]

Suzuki J., Tanaka R. (1980): "LPC vocoder without pitch extractor." **ICA-10** (Paper A1-3.3) [9.4.4]

Swaffield J. (1959): "Some progress with vocoder-type systems." In *Proceedings of the seminar on speech compression and processing* (Electronics Research Directorate, Bedford, Mass; DDC-AD-233717,233720) [9.3.1; 9.4.5]

Sychra A.: see (Sedlacek and Sychra, 1963)

Sylvania Electronics Systems (1962): "Optimum speech signals mapping techniques (final report)" (Sylvania Electr.Syst. Rept.F428-1, Waltham MA; DDC-AD-273443) [5.1]

Szepe Gy.: see (Hirschberg et al., 1971)

Takahashi H.: see (Koike et al., 1977)

Takasugi T. (1971): ""Analysis-by-synthesis" method utilizing spectral features of voice source, and measurement of glottal waveform parameters." J. Radio Res. Labs. (Tokyo) **18**, 209-220 [6.5.1; 9.4.4; 9.4.5]

Takasugi T., Nakatsui M., Suzuki J. (1970): "Observations of glottal waveform from speech waves." J. Acoust. Soc. Jpn. **26**, 141-149 [6.5.1]

Takasugi T., Suzuki J. (1968): "Speculation of glottal waveform from speech wave." J. Radio Res. Labs. (Tokyo) **15**, 279-293 [6.5.1]

Takefuta Y. (1966): "A study of relative efficiency of acoustic parameters in the intonational signal of American English;" Dissertation, Ohio State Univ., Columbus OH [9.4.2]

Takefuta Y. (1968): "Automatic detection of intonational signals in American English." **ICA-6,** 39-42 (Paper B22)

Takefuta Y. (1975): "Method of acoustic analysis of intonation." In *Measurement procedures in speech, hearing, and language;* ed. by S.Singh (University Park Press, Baltimore), 363-378

Takefuta Y. (1976): "Perception, recognition, and analysis of American intonation by human listeners and by a computer." Study of Sounds **17**, 89-109 [9.4.2]

Takefuta Y. (1978): "Defining intonation patterns by a computer program." Study of Sounds **18**, 291-305 [9.4.2]

Takefuta Y. (1979): "Some experiments in the digital extraction of American intonation patterns." **ICPhS-9** (1) (A), 402

Takefuta Y., Jancosek E., Brunt M. (1972): "A statistical analysis of melody curves in the intonation of American English." **ICPhS-7**, 1035-1039 [4.1]

Talkin D.T.: see (Titze and Talkin, 1977)

Tampel' I.B.: see (Galunov and Tampel', 1981)

Tamura J.: see (Kurematsu et al., 1979)

Tanabe M., Isshiki N., Sawada M. (1979): "Damping ratio of the vocal cords." Folia Phoniatr. **31**, 27-34 [3.1; 3.3]

Tanabe M., Kitajima K., Gould W.J., Lambiase A. (1975): "Analysis of high-speed motion pictures of the vocal folds." Folia Phoniatr. **27**, 77-87 [5.2.1]

Tanabe Y.: see (Fujisaki and Tanabe, 1972)

Tardy-Renucci M., Appaix A. (1978): "Etude sur l'analyse sonagraphique des cris de nouveau-nés." Folia Phoniatr. **30**, 1-12 [5.1.2]

Tarnóczy T. (1951): "The opening-time and opening-quotient of the vocal cords during phonation." J. Acoust. Soc. Am. **23**, 42-44 [3.3]

Tatham M.A.A.: see (Fant and Tatham, 1975)

Tatsumi M., Sawashima M., Ushijima K., Kinoshita T. (1973): "Fiberoptic observation of the larynx in singing." **Ann. Bull. RILP 7**, 11-18 [3.2; 5.2.1]

Tattersall G.: see (Ambikairajah et al., 1980)

Tayo E.: see (Kitzing et al., 1975)

Teil D. (1979): "Comparaison de plusieurs algorithmes de marquage prosodique." **Journées d'Etude sur la Parole 10**, 266-272 [9.4.2]

Temes G.C.: see (Babic and Temes, 1976)

Ten Have B.L.: see Have B.L. ten

Tenold J., Crowell D.H., Jones R.H., Daniel T.H., McPherson D.F., Popper A.N. (1974): "Cepstral and stationarity analysis of full-term and premature infants' cries." J. Acoust. Soc. Am. **56**, 975-980 [8.4; 9.4.3]

Teranishi N.: see (Saito et al., 1958)

Terhardt E. (1970): "Prequency analysis and periodicity detection in the sensations of roughness and periodicity pitch." In *Frequency analysis and periodicity detection in hearing;* ed. by R.Plomp and G.F.Smoorenburg (Sijthoff, Leiden), 278-290 [4.2.1]

Terhardt E. (1972a): "Tonhöhenwahrnehmung und harmonisches Empfinden." In *Akustik und Schwingungstechnik.* Plenarvorträge und Kurzreferate der Gemeinschaftstagung Stuttgart 1972 (VDE-Verlag, Berlin) [4.2.1]

Terhardt E. (1972b): "Zur Tonhöhenwahrnehmung von Klängen. I. Psychoakustische Grundlagen; II. Ein Funktionsschema.." Acustica **26**, 173-199 [4.2.1; 8.5.2]

Terhardt E. (1974): "Pitch, consonance, and harmony." J. Acoust. Soc. Am. **55**, 1061-1069 [2.1; 4.2.1; 8.5.2]

Terhardt E. (1979a): "On the perception of spectral information in speech." In *Hearing Mechanisms and Speech.* EBBS-Workshop Göttingen, April 26-28, 1979; ed. by O.Creutzfeld et al. Experimental Brain Research Suppl. 2 (Springer, Berlin), 281-291 [4.2.1; 8.5.2]

Terhardt E. (1979b): "Calculating virtual pitch." Hearing Res. **1**, 155-182 [2.1; 4.2.1; 4.2.2; 4.4; 8.5.2; 9.1]

Terhardt E. (1980): "Sprachgrundfrequenzextraktion nach Prinzipien der Tonhöhenwahrnehmung." In *Fortschritte der Akustik, DAGA'80 München* (VDE-Verlag, Berlin), 667-670 [8.5.2]

Terhardt E., Stoll G., Seewann M. (1982a): "Pitch of complex signals according to virtual-pitch theory: tests, examples, and predictions." J. Acoust. Soc. Am. **71**, 671-678 [4.2.1; 8.5.2]

Terhardt E., Stoll G., Seewann M. (1982b): "Algorithm for extraction of pitch and pitch salience from complex tonal signals." J. Acoust. Soc. Am. **71**, 679-688 [4.2.1; 8.5.2]

Teston B. (1972a): "Description d'une unité d'analyse des paramètres prosodiques." **ICPhS-7**, 1040-1046 [6.1.1]

Teston B. (1972b): "Description d'un système de détection de l'intensité et de la mélodie de la voix." **Phon. Aix-en-Provence 1**, 129-145 [6.1.1]

Teston B. (1979): "La détection instrumentale des paramètres prosodiques." In *Hommage à Georges Faure* (Didier, Paris)

Teston B., Rossi M. (1977): "Un système de détection automatique du fondamental et de l'intensité." **Journées d'Etude sur la Parole 8**, 111-118 [5.3]

't Hart J.: see Hart J. 't

Theis D.J. (1981): "Array processor architecture: Guest Editor's introduction." IEEE Computer (9), 8-9 [9.1.5]

Thompson J.S.: see (Boddie et al., 1981)

Thorsen N.G. (1977): "On the interpretation of raw fundamental frequency tracings." **Journées d'Etude sur la Parole 8**, 175-182 [3.1; 9.4.2]

Thorsen N.G. (1978a): "An acoustical investigation of Danish intonation." J. Phonetics **6**, 151-176 [9.4.2]

Thorsen N.G. (1978b): "On the identification of selected Danish intonation contours." **ARIPUC 12**, 17-73 [9.4.2]

Thorsen N.G. (1980): "Neutral stress, emphatic stress, and sentence intonation in advanced standard copenhagen Danish." **ARIPUC 14**, 121-203 [9.4.2]

Thorvaldsen P.: see (Frøkjaer-Jensen and Thorvaldsen, 1968)

Thurman W.L. (1958): "Frequency-intensity relationships and optimum pitch level." J. Speech Hear. Res. **1**, 117-123 [4.1]

Thurmann E.: see (Almeida et al., 1977)

Tierney J. (1964): "Some work on channel-vocoder excitation." J. Acoust. Soc. Am. **36**, 1022-1023 (Paper I7; 67th Meet. ASA) [9.4.5]

Tierney J., Gold B., Sferrino V., Dumaman J.A., Aho G. (1964): "Channel vocoder with digital pitch extractor." J. Acoust. Soc. Am. **36**, 1901-1905 [6.2.4; 9.4.5]

Tietze G.: see (Unger et al., 1981)

Tiffin J.C. (1934): "Applications of pitch and intensity measurement of connected speech." J. Acoust. Soc. Am. **6**, 225-234 [5.1.1; 5.1.2]

Tillmann H.G. (1978): "Bestimmung der Stimmperiode in der Zeitfunktion des digitalen Sprachschallsignals." **Phon. München 9**, 207-213 [6.2.4]

Tillmann H.G. (1981): "Digitale Signalverarbeitung und Experimentalphonetik: Grundsätzliche Bemerkungen und zwei Beispiele." **Phon. Köln 11**, 43-51 [9.2; 9.4.2]

Timcke R.H.   (1956):   "Die Synchron-Stroboskopie von menschlichen Stimmlippen bzw. ähnlichen Schallquellen und Messung der Öffnungszeit." Z. Laryngologie, Rhinologie, Otologie **35**, 331-335 [5.2.1]

Timcke R.H., Leden H. von, Moore G.P. (1959): "Laryngeal vibrations: Measurements of the glottic wave, part 1: The normal vibratory cycle." Arch. Otolaryng. **69**, 438-444 [3.1; 3.2]

Timcke R.H., Leden P. von, Moore G.P. (1959): "Laryngeal vibrations: measurement of the glottic wave, part 2: physiological variations." Arch. Otolaryng. **69**, 438-444 [3.1; 3.2]

Timme M. (1973): "Ein Verfahren zur Verarbeitung gestörter Telefongespräche." In *Fortschritte der Akustik, DAGA'73 Aachen* (VDI-Verlag, Düsseldorf) [5.3; 8.4]

Timme M., Idler H., Lay T. (1973): "Auswertung von Echtzeit-Cepstra zur schnellen Detektion stimmhafter Laute." Nachrichtentech. Z. **26**, 312-316 [5.3; 8.4]

Timms C.: see (Fisher J. et al., 1978)

Titze I.R. (1973a): "The human vocal chords: A mathematical model (Part 1)." Phonetica **28**, 129-170 [3.1]

Titze I.R. (1973b): "Simulation of vocal fry." J. Acoust. Soc. Am. **54** (A), 23 (Paper A3; 86th Meet. ASA) [3.1; 3.2]

Titze I.R. (1974): "The human vocal chords: A mathematical model (Part 2)." Phonetica **29**, 1-21 [3.1]

Titze I.R. (1975): "Normal modes in vocal cord tissues." J. Acoust. Soc. Am. **57**, 736-744 [3.1]

Titze I.R. (1976): "On the mechanics of vowel-fold vibration." J. Acoust. Soc. Am. **60**, 1366-1380 [3.1; 3.3]

Titze I.R. (1979a): "Physical and physiological dimensions of intrinsic voice quality." **ICPh-Miami**, 217-223 [3.1; 3.2]

Titze I.R. (1979b): "Comments on the myoelastic-aerodynamic theory of phonation." **ICPhS-9** (1) (A), 278 [3.1]

Titze I.R. (1979c): "A physiological interpretation of vocal registers." J. Acoust. Soc. Am. **66** (A), S56 (Paper BB5; 98th Meet. ASA) [3.2]

Titze I.R. (1980a): "Comments on the myoelastic-aerodynamic theory of phonation." J. Speech Hear. Res. **23**, 495-510 [3.1]

Titze I.R. (1980b): "Parametrization of the glottal source function and glottographic waveforms." J. Acoust. Soc. Am. **68** (A), S71 (Paper LL10, 100th Meet. ASA) [3.3]

Titze I.R. (1982): "Mathematical models of glottal area and vocal-fold contact area." J. Acoust. Soc. Am. **72** (A), S48 (Paper CC1, 104th Meet. ASA) [3.3; 5.2.3; 5.2.5]

Titze I.R., Strong W. (1975): "Normal modes in vocal cord tissues." J. Acoust. Soc. Am. **57**, 736-744 [3.1]

Titze I.R., Strong W.J. (1973): "Computer modeling of vocal cord pathologies and voice registers." J. Acoust. Soc. Am. **53** (A), 344-345 (Paper DD1; 84th Meet. ASA) [3.1; 3.2; 3.3; 9.4.3]

Titze I.R., Talkin D.T. (1977): "Simulation and classification of glottal waveforms." J. Acoust. Soc. Am. **61** (A), S32 (Paper P10; 93rd Meet. ASA) [3.1; 3.3; 6.5]

Titze I.R., Talkin D.T. (1978): "Estimation of effective mass and stiffness of the vocal folds from distributed models." J. Acoust. Soc. Am. **64** (A), S41 (Paper P4; 96th Meet. ASA) [3.1]

Tjernlund P. (1964a): "Grundtonsindikator för dövundervisning (Pitch determination instrument for the education of the deaf. In Swedish);" Dissertation, Speech Transmission Laboratory, KTH, Stockholm [5.2.2]

Tjernlund P. (1964b): "A pitch extractor with larynx pick-up." **STL-QPSR** (3), 32-34 [5.2.2]

Toffler J.E., Wade F.B. (1964): "Pitch extractor, using clippers." J. Acoust. Soc. Am. **36** (A) (Paper N4; 67th Meet. ASA) [6.1.2]

Tolkmitt F., Helfrich H., Standke R., Scherer K.R. (1981): "Vocal indicators of psychiatric treatment effects in depressives and schizophrenics." J. Commun. Disorders [9.4.3]

Tomic T.: see (Lukatela and Tomic, 1974)

Torasso P.: see (Mezzalama et al., 1976; Lesmo et al., 1978)

Toro M.J. di: see Di Toro M.J.

Tosi O., Oyer H., Lashbrook W., Pedrey C., Nicol J., Nash E. (1972): "Experiment on voice identification." J. Acoust. Soc. Am. **51**, 2030-2043

Tostmann R.: see (Wendler and Tostmann, 1981)

Tou J.T., Gonzalez R.C. (1974): *Pattern recognition principles* (Addison-Wesley, London) [5.3.3]

Tove P.A., Czekajewski J. (1974): "Pulse period meter with short response time, applied to cardiotachometry." Electron. Eng., 290-295 [6.5.3]

Towers B.: see (Lyons and Towers, 1962)

Trevino S.N.: see (Parmentier and Trevino, 1932)

Tribolet J.M.: see (Oppenheim et al., 1976; Kopec et al., 1977; Flanagan et al., 1979; Crochiere and Tribolet, 1979; García Gómez and Tribolet, 1982)

Trill D.: see (Carr and Trill, 1964)

TRW (ed.) (1980): "Digital signal processing - LSI, your design keys for the 1980's" (TRW, El Segundo CA) [2.3; 2.4; 8.5.4; 9.1.5]

Tscheschner W. (1964): "Ein Beitrag zur Messung der Stimmbandgrundfrequenz" (III. Akustische Konferenz, Budapest) [6.2.1]

Tscheschner W., Pohlink M.: "Schaltungsanordnung zur Ermittlung der Stimmbandgrundfrequenz;" Patentschrift der DDR Nr. 49355, Berlin (DDR)

Tseng C.Y.: see (Lieberman and Tseng, 1980)

Tsumara T., Sone T., Nimura T. (1973): "Auditory detection of frequency transition." J. Acoust. Soc. Am. **53**, 17-25 [4.3]

Tucker W.H. (1977): "Interactive computer based music systems;" Dissertation, Univ. of Canterbury, Christchurch [6.2.4; 9.4.2]

Tucker W.H., Bates R.H.T. (1977): "Efficient pitch estimation for speech and music." Electron. Lett. **13**, 357-358 [6.2.4]

Tucker W.H., Bates R.H.T. (1978): "A pitch estimation algorithm for speech and music." IEEE Trans. ASSP-**26**, 597-604 [4.1; 6.2.4; 7.2; 9.4.2]

Tucker W.H., Bates R.H.T., Frykberg S.D., Howarth R.J., Kennedy W.K., Lamb M.R., Vaughan R.G. (1977): "An interactive aid for musicians." Int. J. Man-Machine Studies **9**, 635-651 [6.2.4; 9.4.2]

Tufts D.W., Levinson S.E., Rao R. (1976): "Measuring pitch and formant frequencies for a speech understanding system." ICASSP-76, 314-317

Tukey J.W. (1974): "Nonlinear (nonsuperposable) methods for smoothing data." IEEE EASCON Record, 673 [6.6.4]

Turner J.M.: see (Stirling and Turner, 1983)

U.Z.-MGPI-VL --> "Ученые Записки, Московский Государственний Педагогический Институт Имени В.И. Ленина." (*Proceedings of the Moscow State Pedagogic Institute.* In Russian) (Moscow)

Uddenfeldt J.: see (Rydbeck et al., 1980)

Umeda N. (1980): "F$_0$ declination is situation-dependent." J. Acoust. Soc. Am. **68** (A), S70 (Paper LL5, 100th Meet. ASA) [9.4.2]

Un C.K., Lee H.H. (1980): "Voiced-unvoiced-silence discrimination of speech by delta modulation." IEEE Trans. ASSP-**28**, 398-407 [5.3.1]

Un C.K., Lee J.R. (1982): "On spectral flattening techniques in residual-excited linear prediction vocoding." ICASSP-82, 216-219 [9.4.5]

Un C.K., Magill D.T. (1975): "The residual-excited linear prediction vocoder with transmission rate below 9.6 kbit/s." IEEE Trans. COM-**23**, 1466-1473 [9.4.5]

Un C.K., Sung W.Y. (1980): "A 4800 bps LPC vocoder with improved excitation." ICASSP-80, 142-145 [9.4.5]

Un C.K., Yang S.C. (1977): "A pitch extraction algorithm based on LPC inverse filtering and AMDF." IEEE Trans. ASSP-**25**, 565-572 [8.3.1]

Underwood M.J.: see (Millar and Underwood, 1975)

Ungeheuer G. (1962): *Elemente einer akustischen Theorie der Vokalartikulation* (Springer, Berlin) [1.1; 3.4]

Ungeheuer G. (1963): "Zur Periodizitätsanalyse phonetischer Signale (Gegenwärtiger Stand der Entwicklung von Tonhöhenschreibern)." Phonetica **10**, 174-186 [2.6; 9.3]

Ungeheuer G. (1967): "Periodizität und Tonhöhe." In *Colloquium amicorum* - Joseph Schmid-Görg zum 70. Geburtstag (Bonn), 394-409 [9.1.1]

Ungeheuer G., Rupprath R., Friedrich G. (1965): "Zur Entwicklung eines Verbundsystems von Periodizitätsanalysator (Tonhöhenschreiber) und Intensimeter." ICA-5 (Paper J11) [5.3.2; 6.2.1]

Unger E., Unger H., Tietze G. (1981): "Stimmuntersuchungen mittels der elektroglottographischen Einzelkurven. Eine kritische Wertung." Folia Phoniatr. **33**, 168-180 [5.2.3; 9.4.3]

Unger H.: see (Unger E. et al., 1981)

Ushijima K.: see (Sawashima and Ushijima, 1971; Tatsumi et al., 1973)

Uyeno T., Hayashibe H., Imai K. (1979): "On pitch contours of declarative complex sentences in Japanese." **Ann. Bull. RILP 13**, 175-188 [9.4.2]

Vagges K.: see (Ferrero et al., 1981)

Vaissière J. (1974): "On French prosody." **MIT-RLE Progr. Rept. 114**, 212-223 [9.4.2]

Vaissière J. (1975a): "Caractérisation des variations de la fréquence du fondamental dans les phrases françaises." **Journées d'Etude sur la Parole 6**, 39-50 [9.4.2]

Vaissière J. (1975b): "Further note on French prosody." **MIT-RLE Progr. Rept. 115**, 251-261 [9.4.2]

Vaissière J. (1976a): "Premiers essais de segmentation automatique de la parole continue en mots à partir des variations du fondamental dans la phrase." **Recherches/Acoustique 3**, 191-201 [9.4.2]

Vaissière J. (1976b): "Une procédure de segmentation automatique de la parole en mots prosodiques, en français." **Journées d'Etude sur la Parole 7**, 103-114 [1.2; 9.4.2]

Vallancien B. (1973): "Essai d'interprétation des courbes de micromélodie ou mélodie articulatoire." **Bull. Audiophonologie 3**, 41-57 [3.3; 9.4.2]

Vallancien B., Faulhaber J. (1967): "What to think of glottography." Folia Phoniatr. **19**, 39-44 [5.2.3]

Vallancien B., Gautheron B., Pasternak L., Guisez D., Paley B. (1971): "Comparaison des signaux microphoniques, diaphanographiques et glottographiques, avec application au laryngographe." Folia Phoniatr. **23**, 371-380 [5.2.3; 5.2.6]

Van den Berg J.W.: see Berg J.W. van den

Van der Meulen M.L.: see (Zuidweg et al., 1982)

Van der Minne J.M.: see (van den Berg and van der Minne, 1962)

Van Eck T.: see Eck T. van

Van Els T.J.M.: see (Kolk et al., 1977)

Van Leeuwaarden A.: see (Sluyter et al., 1980)

Van Meerbergen J.L.: see (Zuidweg et al., 1982)

Van Michel C.: see Michel C. van

Van Rossum N.: see Rossum N. van

Van Schooneveld C.H.: see (Buning et al., 1961)

Vardanian R. (1964): "Teaching English intonation through oscilloscope displays." Language Learning **14**, 109-118 [5.1; 9.4.3]

Vary P., Heute U. (1980): "A short-time spectrum analyzer with polyphase-network and DFT." Signal Processing **2**, 55-65 [2.4]

Vass-Kovács E.: see (Hirschberg et al., 1971)

Vaughan R.G.: see (Tucker et al., 1977)

Velichko V.G. (1965): "О статистике приращений соседних периодов основного тона речи. В журн.. Труды учебных институтов связи. В.Г.Величко. Statistical investigation of the differences between adjacent pitch periods. In Russian.). Trudy Uchebnykh Institutov Svyazi (Novosibirsk) **25**, 183-190 [4.3; 9.4.4]

Velichko V.G., Zagoruyko N.G. (1970): "Automatic recognition of 200 words." Int. J. Man-Machine Studies **2**, 223-234

Vemula N.R., Engebretson A.M., Monsen R.B., Lauter J.L. (1979): "A speech microscope." ASA*50, [5.1; 9.2.2]

Vende K. (1972): "Intrinsic pitch of Estonian vowels - measurement and perception." **Phon. Tallinn 1**, 44-109 [4.1; 5.1.1; 9.4.2]

Vende K. (1977): "On people's ability to analyze pitch movement." **Phon. Tallinn 7**, 99-102 [4.2.3]

Vennard W.: see (Hirano et al., 1969, 1970)

Venugopal V.: see (Sarma and Venugopal, 1978)

Vermeulen R., Six W. (1949): "Device for artificially generating speech sounds by electrical means;" United States Patent No. 2,458,227. Issued Jan. 4, 1949; filed April 18, 1946 [6.2.1]

Verrillo R.T.: see (Rothenberg et al., 1977)

Verschure J.: see (Leclusc et al., 1975)

Vesa L.: see (Vuorenowski et al., 1970)

Vezza A. (1964): "Mathematical analysis of glottal waveforms." J. Acoust. Soc. Am. **36**, 1048 (Paper X6; 67th Meet. ASA) [6.5.1]

Vieira A.: see (Morf et al., 1977)

Visquis R.: see (Autesserre et al., 1979)

Viswanathan V.R., Higgins A.L., Russell W.H. (1982): "Design of a robust baseband LPC coder for speech transmission over 9.6 kbit/s noisy channels." IEEE Trans. ASSP-**30**, 664-673 [6.3.1; 9.4.5]

Viswanathan V.R., Higgins A.L., Russell W.H., Makhoul J. (1980): "Baseband LPC coders for speech transmission over 9.6 kb/s noisy channels." **ICASSP-80,** 348-351 [9.4.5]

Viswanathan V.R., Makhoul J., Huggins A.W.F. (1977): "New source model for narrow-band vocoder." J. Acoust. Soc. Am. **62** (A), S62 (Paper BB5; 94th Meet. ASA) [9.4.5]

Viswanathan V.R., Makhoul J., Schwartz R. (1982): "Medium and low bit rate speech transmission." In *Automatic speech analysis and recognition.* Proceedings of the NATO Advanced Study Institute held at Bonas, France, June 29 - July 10, 1981; ed. by J.P. Haton (Reidel, Dordrecht), 21-48 [9.4.5]

Viswanathan V.R., Russell W.H., Higgins A.L., Berouti M.G., Makhoul J. (1980): "Speech-quality optimization of 16 kb/s adaptive predictive coders." **ICASSP-80,** 520-525 [9.4.5]

Viswanathan V.R., Russell W.H., Makhoul J. (1979): "Voice-excited LPC coders for 9,6 kbps speech transmission." **ICASSP-79,** 558-562 [9.4.5]

Vivès R., Le Corre C., Mercier G., Vaissière J. (1977): "Utilisation, pour la reconnaissance de la parole continue, de marqueurs prosodiques extraits de la frequence du fondamental." **Journées d'Etude sur la Parole 8,** 353-362 [9.4.4]

Vogel R.P.: see (Weiss et al., 1966)

Vogelsänger G.T. (1954): "Experimentelle Prüfung der Stimmleistung beim Singen." Folia Phoniatr. **6,** 193-227 [3.2; 9.4.2]

Volkman J.: see (Stevens S.S. et al., 1937)

Voloshenko Y.Y. (1968): "О регистрации частоты колебаний голосовых связок." В журн.: "Вопросы радиоэлектроники". Ю.Я. Волошенко. (Registration of the frequency of the voice excitation signal. In Russian.). Voprosy Radioelektroniki **7,** 30-37

Von Essen O.: see Essen O. von

Von Leden H.: see Leden H. von

Vuorenowski V., Kaunisto M., Tjernlund P., Vesa L. (1970): "Cry detector: a clinical apparatus for surveillance of pitch and intensity of a newborn infant." Acta Paediatr. Scand. **59,** 103-104 [9.4.3]

Vyssotsky V.A., David E.E.jr., Mathews M.V. (1961): "Recognition of voicing, voice pitch and formant frequencies with a digital computer." **ICA-3,** 229-231 [5.3]

Wade F.B. (1964): "Computer evaluation of pitch extractors." J. Acoust. Soc. Am. **36** (A), 2001 [9.3.1]

Wadia A.B. (1980): "Error correction scheme for telephone line transmission of RELP vocoder." **ICASSP-80,** 360-363 [9.4.5]

Wängler H.-H., Lyon J.G. (1974): "Instructing the deaf in intonation with an instantaneous visual feedback system: case presentations." In *Miszellen II;* ed. by J.P. Köster. Hamburger Phonetische Beiträge, Bd.13 (Buske, Hamburg), 223-243 [5.1; 9.4.3]

Waibel A. (1982): "Very large vocabulary recognition (VLVR): using prosodic and spectral filters." J. Acoust. Soc. Am. **72** (A), S32 (Paper R15, 104th Meet. ASA) [9.4.4]

Waite W.M.: see (Harris and Waite, 1963, 1964)

Wakita H. (1982): "Linear prediction of speech and its application to speech processing." In *Automatic speech analysis and recognition.* Proceedings of the NATO Advanced Study Institute held at Bonas, France, June 29 - July 10, 1981; ed. by J.P. Haton (Reidel, Dordrecht), 1-20 [6.3.1]

Wakita H., Fant G. (1978): "Toward a better vocal tract model." **STL-QPSR** (1), 9-29 [3.3; 3.4]

Waldron M.B.: see (Sheshradi and Waldron, 1979)

Walesby D.H.: see (Holmes et al., 1978)

Wallace J. (1963): "Comparative evaluation of pitch-signal indicators." J. Acoust. Soc. Am. **35** (A), 790 [9.3.1]

Walliser K. (1969): "Über ein Funktionsschema für die Bildung der Periodentonhöhe aus dem Schallreiz." Kybernetik **6,** 65-72 [4.2.1; 8.5.2]

Walsh T., Parker F. (1980): "Vocal fry: A cue for voicing in postvocalic stops." J. Acoust. Soc. Am. **67** (A), S51 (Paper V5; 99th Meet. ASA) [3.2; 3.5; 5.3]

Walters S.M.: see (Boddie et al., 1981)

Waltrop W.F.: see (Laguaita and Waltrop, 1964)

Wang J.Sh. (1980): "A method of pitch extraction by LPC analysis with predetermining voiced sounds (In Chinese; English abstr.)." Acta Acustica (Beijing) **4,** 286-290 [6.3.1]

Wang W.S.Y. (1972): "The many uses of F_0." In *Papers in linguistics and phonetics to the memory of P. Delattre* (Mouton, The Hague), 487-501 [1.3; 9.4.2]

Wang Z., Hess W.J. (1981): "Preparatory experiments for the automatic recognition of the four tones in spoken Chinese." **F.A.S.E.-81** [9.4.4]

Waser Sh. (1978): "High-speed monolithic multipliers for real-time digital signal processing." IEEE Computer 11, 19-29 [9.1.5]

Waser Sh. (1980): "Survey of VLSI for digital signal processing." ICASSP-80, 376-379
[9.1.5]

Washburn D.D.: see (Snidecor et al., 1959)

Watkin K.L., Minifie F.D., Kennedy J.G. (1978): "An ultrasonic-EMB transducer for
biodynamic research." J. Speech Hear. Res. 21, 174-182 [5.2.4]

Watkins H.E. (1979): "Description of a hybrid 7.2 kbps vocoder." ICASSP-79, 546-549
[9.4.5]

Wechsler E. (1976): "Laryngographic study of voice disorders." Phon. London, 12-29
[5.2.3; 9.4.3]

Wedin L.: see (Fritzell et al., 1977)

Wedin S., Ögren J.E. (1982): "Analysis of the fundamental frequency of the human voice
and its frequency distribution before and after a voice training program." Folia
Phoniatr. 34, 143-149 [4.1; 8.5.3]

Wegel R.L. (1930a): "Theory of vibration of the larynx." J. Acoust. Soc. Am. 1, 1-21
[3.1]

Wegel R.L. (1930b): "Theory of vibrations of the larynx." Bell Syst. Tech. J. 9, 207-227
[3.1]

Weibel E.S.: see (Miller and Weibel, 1956)

Weinberg B., Bennett S. (1972): "A comparison of the fundamental frequency characteristics of esophagal speech measured on a wave-by-wave and averaging basis." J. Speech
Hear. Res. 15, 351-355 [4.1; 6.5.3; 9.4.3]

Weinstein C.J. (1975): "A linear-prediction vocoder with voice excitation." IEEE EASCON
Record, 30 [9.4.5]

Weinstein C.J. (1978): "Fractional speech loss and talker activity model for TASI and
packet-switched speech." IEEE Trans. C-26 (C), 1253-1257 [5.3; 9.4.4; 9.4.5]

Weinstein C.J., McCandless S.S., Mondshein L.F., Zue V.W. (1975): "A system for acous-
tic-phonetic analysis of continuous speech." IEEE Trans. ASSP-23, 54-67 [5.3]

Weiss M.R., Harris C.M. (1962): "High-resolution analysis of speech by computer techniques." Proc. 1962 Speech Commun. Sem. (Paper C12) [8.4.2]

Weiss M.R., Harris C.M. (1963): "Computer technique for high speed extraction of speech
parameters." J. Acoust. Soc. Am. 35, 207-214 [8.4.2]

Weiss M.R., Vogel R.P., Harris C.M. (1966): "Implementation of a pitch-extractor of the
double spectrum analysis type." J. Acoust. Soc. Am. 40, 657-662 [8.4.2]

Welbourn C.P.: see (Rostron and Welbourn, 1976)

Welch J.R. (1980): "Automatic speech recognition - putting it to work in industry." IEEE
Computer, 65-73

Wendahl R.W. (1963): "Laryngeal analog synthesis of harsh voice quality." Folia Phoniatr.
15, 241-250 [9.4.3]

Wendahl R.W. (1966): "Laryngeal analog synthesis of jitter and shimmer auditory parameters of harshness." Folia Phoniatr. 18, 98-108 [4.3]

Wendahl R.W., Moore G.P., Hollien H. (1963): "Comments on vocal fry." Folia Phoniatr.
15, 241-250 [3.2]

Wendler J. (1970): "Zur auditiven Steuerung der Sprechintonation." ICPhS-6, 1009-1013
[4.2; 9.4.2]

Wendler J. (1975): "Practical hints for the application of laryngostroboscopy" (VEB
Transformatoren- und Röntgenwerk "Hermann Matern", Dresden) [5.2.1]

Wendler J., Seidner W., Halbedl G., Schaaf G. (1973): "Tele-Mikrostroboskopie." Folia
Phoniatr. 25, 281-287 [5.2.1]

Wendler J., Tostmann R. (1981): "Stroboglottometric measures in phoniatric voice diagnostics." F.A.S.E.-81, 113 [5.2.1]

Werner P.A. (1954): "Circuit for indicating the vocal frequency." MIT-RLE Progr. Rept.
[6.2.1]

West J.E.: see (Fourcin and West, 1968)

Westin K., Buddenhagen R.G., Olbrecht D.H. (1966): "An experimental analysis of the
relative importance of pitch, quantity, and intensity as cues to phonemic distinctions
in southern Swedish." Lang. Speech 9, 114-126 [3.4; 9.4.2]

Weston R.C.: see (Stead and Weston, 1962)

Wethlo F. (1950): "Simplified phonetic measurement of pitch." Z. Phonetik 4, 123-126

White F.D.: see (Moore et al., 1962)

White G.M.: see (Stone and White, 1963)

Wichern P.U.N. (1978): "On the design of efficient median smoothers." **Phon. Nijmegen 2,** 18-28 [6.6.4]

Wichern P.U.N., Boves L. (1980): Visual feedback of F_0 curves as an aid in learning intonation-contours. **Phon. Nijmegen 4,** 31-52 [5.1; 6.2.1; 9.4.3]

Wightman F.L. (1971): "A model of perceived pitch based on weighted autocorrelation." **ICA-7** (3), 361-364 (Paper 20H3) [4.2.1]

Wightman F.L. (1973a): "Pitch and stimulus fine structure." J. Acoust. Soc. Am. **54,** 397-406 [4.2.1]

Wightman F.L. (1973b): "The pattern-transformation model of pitch." J. Acoust. Soc. Am. **54,** 407-416 [4.2.1]

Willemain T.R., Lee F.F. (1972): "Tactile pitch displays for the deaf." IEEE Trans. AU-**20,** 9-16 [9.4.3]

Willems L.F. (1966): "The intonator." **IPO Annual Progr. Rept. 1,** 123-125 [3.3; 9.4.4]

Willems L.F. (1968): "Variations in larynx periodicity for monotonous CV utterances." **IPO Annual Progr. Rept. 3,** 63-68 [4.3; 5.2.2]

Willems L.F. (1982): "Programs which implement the DWS [Duifhuis-Willems-Sluyter] pitch detector" (Inst. for Perception Research, Eindhoven; IPO Rept. #394) [8.5.2]

Willems L.F., Loonen A.R. (1967): "Intonation contour generator." **IPO Annual Progr. Rept. 2,** 197-200 [3.3; 9.4.4]

Williams C.E., Stevens K.N. (1972): "Emotions and speech: some acoustical correlates." J. Acoust. Soc. Am. **52,** 1238-1250 [9.4.3]

Williams F.A. (1982): "An expandable single-IC digital filter/correlator." **ICASSP-82,** 1077-1080 [2.3; 9.1.5]

Wilson T.A.: see (Gupta et al., 1973)

Winckel F. (1953): "Elektroakustische Untersuchungen an der menschlichen Stimme." Folia Phoniatr. **5,** 232-252 [3.2]

Winckel F. (1954): "Neuentwicklung Lichtblitz-Stroboskop für Stimmuntersuchung." Wegweiser f. d. ärztl. Praxis, 210-212 [5.2.1]

Winckel F. (1962): "Development of a pitch extraction device for studies in synthetic speech." **Proc. 1962 Speech Commun. Sem.** [6.2.1]

Winckel F. (1963a): "Kybernetische Funktionen bei der Stimmgebung und beim Sprechen." Phonetica **9,** 108-126 [3.2]

Winckel F. (1963b): "Tonhöhenextraktor zur Messung und Steuerung von Stimme und Sprache." Arch. Ohren-Nasen-Kehlkopfheilkunde **182,** 651-655 [6.2.1]

Winckel F. (1964): "Tonhöhenextraktor für Sprache mit Gleichstromanzeige." Phonetica **11,** 248-256 [6.2.1]

Winckel F. (1967): "Darstellung des Sprechverhaltens als statistische Tonhöhen- und Formantverteilung mittels Langzeitanalyse." **ICPhS-6,** 1031-1035 [4.1]

Winckel F. (1968): "Acoustical foundations of phonetics." In *Manual of Phonetics;* ed. by B.Malmberg (North Holland Publishing Company, Amsterdam), 17-44 [1.2; 3.4]

Winckel F. (1970): "Akustik der Stimmbildung." In *Handbuch der Stimm- und Sprachheilkunde;* ed. by R.Luchsinger and G.E.Arnold (Springer, Wien), 43-113 [3.2]

Winckel F. (1973): "Die Regelmechanismen der menschlichen Stimme." In *Fortschritte der Akustik, DAGA'73 Aachen* (VDI-Verlag, Düsseldorf), 481-484 [3.1]

Winckel F. (1974): "Acoustical cues in the voice for detecting laryngeal diseases." In *Ventilatory and phonatory control;* ed. by F. Wyke (Oxford University Press, London), 248-259 [9.4.3]

Winckel F., Krause M. (1965): "Ermittlung von Spracheigenschaften aus statistischen Verteilungen von Amplitude und Tonhöhe." **ICA-5** (Paper A41) [4.1]

Wingfield A. (1975): "The intonation-syntax interaction: Prosodic features in perceptual processing of sentences." In *Structures and processes in speech perception.* Proceedings of the Symposium on Dynamic Aspects of Speech Perception, Eindhoven, August 4-6, 1975; ed. by A.Cohen and S.G.Nooteboom (Springer, Berlin), 146-160 [9.4.2]

Winham G.: see (Steiglitz et al., 1975)

Winograd S. (1980): "Signal processing and complexity of computation - extended summary." **ICASSP-80,** 94-101 [2.4; 8.5.4; 9.1.5]

Wiren J., Stubbs H.L. (1956): "Electronic binary selection system for phoneme classification." J. Acoust. Soc. Am. **28,** 1082-1091 [9.4.4]

Wise J.D., Caprio J.R., Parks T.W. (1976): "Maximum likelihood pitch estimation." IEEE Trans. ASSP-**24,** 418-423 [8.6.1]

Witting C. (1962): "A method of evaluating listeners' transcriptions of intonation on the basis of instrumental data." Lang. Speech **5,** 138-150 [9.4.2]

Wolf D. (1977): "Analytische Beschreibung von Sprachsignalen." AEÜ **31,** 392-398 [6. .2]

Wolf D., Brehm H. (1973): "Zweidimensionale Verteilungsdichte der Amplituden von Sprachsignalen." In *Signalverarbeitung (Signal Processing)*. Vorträge der gleichnamigen Fachtagung, Erlangen 1973; ed. by H.W.Schüssler (Universität Erlangen/Nürnberg), 378-385

Wolf H.E. (1981): "Control of prosodic parameters for a formant synthesizer based on diphone concatenation." ICASSP-81, 106-109 [9.4.2; 9.4.4]

Wolf J.J. (1981): "Acoustic analysis and computer systems." In *Speech evaluation in psychiatry*; ed. by J.K. Darby (Grune and Stratton, New York) [1.2]

Wolf S.K., Stanley D., Sette W.J. (1935): "Quantitative studies on the singing voice." J. Acoust. Soc. Am. **6**, 255-266 [3.2; 5.1.2]

Wong D.L.: see (Krasner et al., 1973)

Wong D.Y., Markel J.D. (1978): "An excitation function for LPC synthesis which retains the human glottal phase characteristics." ICASSP-78, 171-174 [6.5.1; 9.2.1; 9.4.4]

Wong D.Y., Markel J.D., Gray A.H. (1977): "Considerations in glottal inverse filtering from the acoustic speech waveform" (Speech Communication Research Lab, Santa Barbara CA) [6.3.1; 6.5.1]

Wong D.Y., Markel J.D., Gray A.H. (1979): "Least squares glottal inverse filtering from the acoustic speech waveform." IEEE Trans. ASSP-**27**, 350-355 [6.3.1; 6.5.1]

Wood C.A. (1977): "Source/vocal tract influences on speaker discrimination." ICA-9 [6.5.1]

Wood D.E.: "Fundamental pitch frequency signal extraction system for complex signals;" United States Patent No. 3,549,806. Issued Dec. 22, 1970

Wood D.E. (1964): "New display format and a flexible-time integrator for spectral-analysis instrumentation." J. Acoust. Soc. Am. **36**, 639-643 [5.1.2]

Wood D.E. (1965): "Experiments in pitch detection." J. Acoust. Soc. Am. **37** (A), 1211 (Paper Y3; 69th Meet. ASA)

Wood D.E., Hewitt T.L. (1963): "New instrumentation for making spectrographic pictures of speech." J. Acoust. Soc. Am. **35**, 1274-1278 [5.1.2]

Woyczinski W.A.: see (Czekajewski and Woyczinski, 1968)

Wright R.: see (Fisher J. et al., 1978)

Wurm S.A. (1967): "Pitch and intensity recording devices for the study of Australian languages." Z. Phonetik **20**, 251-257

Wyke B.F. (ed.) (1974a): *Ventilatory and phonatory control* (Oxford University Press, London)

Wyke B.F. (1974b): "Laryngeal neuromuscular control systems in singing." Folia Phoniatr. **26**, 295-306 [3.1; 3.2]

Xia G.R.: see (Lin M.C. et al., 1979)

Xydeas C.S., Steele R. (1976): "Pitch synchronous 1st-order linear D.P.C.M. system." Electron. Lett. **12**, 93-95 [9.4.5]

Yaggi L.A. (1962): "Full duplex digital vocoder. Scientific Report No.1" (Texas Instruments, Dallas TX; SP14-A62; DDC-AD-282986) [4.1; 6.3.2; 6.4.1; 9.4.5]

Yaggi L.A. (1963): "Full duplex digital vocoder, final report No.2" (Texas Instruments Inc., Dallas TX; Rept.No. SP16-A63) [6.3.2; 6.4.1; 9.4.5]

Yanagihara N. (1967): "Significance of harmonic changes and noise components in hoarseness." J. Speech Hear. Res. **10**, 531-541 [4.2; 8.5.2; 9.4.3]

Yang S.C.: see (Un and Yang, 1977)

Yasuhiro T. (1980): "Fundamental frequency estimation by peak-picking of squared speech wave (in Japanese; English abstr.)." J. Acoust. Soc. Jpn. **36**, 487-495 [6.1.3; 6.2.5]

Yasuo T. (1962): "System for pitch determination of speech signals (in Japanese);" Japanese patent No. 37/9929. Issued Aug. 1, 1962; filed Dec. 5; 1959 [6.4.1]

Yato F., Kitayama S., Kurematsu A. (1981): "Real-time pitch extraction method by adaptive searching procedure (In Japanese, Engl. abstract)." Trans. of the IECE of Japan **64**, 39 [8.2.4]

Yavuz D., Geçkinli N.C. (1978): "A new algorithm for pitch extraction and voiced/unvoiced segmentation of speech using only zero crossing interval sequences (ZCIS) and voiced phenomene recognition with ZCIS." In *Pattern recognition and signal processing*; ed. by C.H.Chen (Sijthoff & Noordhoff, Alphen a.d.Rijn, The Netherlands), 309-321 [5.3; 6.2.3]

Yea J.J., Krishnamurthy A.K., Naik J.M., Moore G.P., Childers D.G. (1983): "Glottal sensing for speech analysis and synthesis." ICASSP-83

Yegnanarayana B. (1976): "Effect of noise and distortion in speech on parametric extraction." ICASSP-76, 336-339 [4.3]

Yegnanarayana B. (1979a): "Speech analysis by pole-zero decomposition." **ASA*50**, 233-236 [6.3.1; 6.5.1]

Yegnanarayana B. (1979b): "Comparison of glottal pulse shapes derived from linear prediction residual and pole-zero model residual of speech." J. Acoust. Soc. Am. **66** (A), S64 (Paper EE5; 98th Meet. ASA) [6.5.1]

Yegnanarayana B. (1980): "Pole-zero decomposition: A new technique for design of digital filters." **ICASSP-80**, 279-282 [2.3; 6.3.1]

Yegnanarayana B. (1981): "Speech analysis by pole-zero decomposition of short-time spectra." Signal Processing **3**, 5-17 [3.4; 6.5.1; 9.4.4]

Yegnanarayana B., Ananthapadmanabha T.V. (1979): "On improving the reliability of cepstral pitch estimation" (Carnegie-Mellon University, Pittsburgh PA; Technical report) [8.4.2]

Yeni-Konishian G.H.: see (Goldstein M.H. et al., 1976)

Yoshioka H.: see (Sawashima et al., 1978; Löfqvist et al., 1980)

Yost W.A., Hill R., Perez-Falcon T. (1978): "Pitch and pitch discrimination of broad-band signals with rippled power spectra." J. Acoust. Soc. Am. **63**, 1166-1173 [4.3]

Youngberg J.E. (1979): "Rate/pitch modification using the constant-Q transform." **ICASSP-79**, 748-751 [2.4; 9.4.5]

Youssef-Digaleh S.: see (Current et al., 1979)

Yuan B.C. (1960): "К вопросу о частотной компресии и качестве - передачи речи в системе 'анализ-синтез'." Кандидатская диссертация, Ленинград. Юан Бао-Цзун. "(On the problem of band compression in speech transmission by means of an analysis-synthesis system. In Russian) (Leningrad) [6.1.4; 9.4.4]

Yuan B.C. (1961): "Об одном методе выделения основного тона." В кн.: Сборн трудов ЛИИЖТ. Юан Бао-Цзун. (A method of pitch determination. In Russian). Sborn Trudov **179**, 48-54 [6.1.4]

Zacharias C.: see (Stock and Zacharias, 1971)

Zagoruyko N.G.: see (Velichko and Zaguruyko, 1970)

Zahorian S.A.: see (Rothenberg and Zahorian, 1977)

Zalewski J., Myślecki W. (1979): "Selection of glottal excitation parameters optimizing the naturalness of synthetic speech." **ICPhS-9** (1) (A), 358 [3.3; 9.4.4]

Zantema J.T.: see (van den Berg et al., 1957)

Zarnecki P.: see (Gubrynowicz et al., 1980, 1981)

Zayac A.A., Lobanov B.M. (1968): Автоматическое измерение частоты основного тона в интонографических исследованиях. В кн.: Методы экспериментального анализа речи. А.А.Заяц и Б.М.Лобанов. (Automatic fundamental frequency determination in intonational studies. In Russian.) (Minsk), 90-91

Zemlin W. (1959): "A comparison of high-speed cinematography and a transillumination-photoconductive technique in the study of the glottis during voice production;" Dissertation, Univ. of Minnesota [5.2]

Zhinkin N.I. (1958): "Механизмы речи." Н.И.Жинкин. (The mechanisms of speech. In Russian.) (Russian Academy of Sciences, Moscow) [1.2]

Zhinkin N.I. (1968): "Глотографический метод выделения основного тона речи." В кн.: Исследивания по рецевой информации. Н.И.Жинкин. (Glottographic method for pitch determination of speech signals. In Russian.) (Univ. of Moscow), 54-65 [5.2.3]

Zimmerman H.J.: see (Sterne and Zimmerman, 1939)

Zorin V.M., Lidikh A.K. (1968): "К вопросу о выделении основного тона речи." В кн.: 6. Вцесоюзн. Акуст. Конференция, Москва. В.М.Сорин и А.К.Лидих. (On the question of pitch determination. In Russian.) (Moscow), 312-315

Zovato S.: see (Ferrero et al., 1981)

Zue V.W.: see (Weinstein et al., 1975)

Zuidweg P., Meerbergen J.L. van, Meulen M.L. van der (1982): "Custom LSI chip-set for speech analysis." **ICASSP-82**, 521-524 [8.5.2; 9.1.5]

Zurcher J.F. (1972): "Dispositif de détection et de mesure du fondamental de la parole humaine." **Recherches/Acoustique 1**, 7-16 [6.2.1]

Zurcher J.F. (1977): "La mesure du fondamental par la détection de crêtes. Techniques employés, résultats." **Journées d'Etude sur la Parole 8**, 119-126 [6.2.1]

Zurcher J.F., Cartier M., Boe L.J. (1975): "Détection et mesure du fondamental." **Journées d'Etude sur la Parole 6**, 12-21 [6.2.1]

Zurcher J.F., Cartier M., Frichon M. (1975): "Vocodeur à bande de base." **Recherches/Acoustique 2**, 67-71 [9.4.5]

Zwicker E., Feldtkeller R. (1967): Das Ohr als Nachrichtenempfänger (Hirzel, Stuttgart) [4.2; 4.2.1]

Zwicker E., Hess W., Terhardt E. (1967): "Erkennung gesprochener Zahlworte mit Funk-
tionsmodell und Rechenanlage." Kybernetik 3, 267-272 [1.2; 4.4; 9.1]
Zwicker E., Terhardt E. (1980): "Analytical expressions for critical-band rate and critical
bandwidth as a function of frequency." J. Acoust. Soc. Am. 68, 1523-1525 [1.2; 9.4.2]
Zwirner E. (1929a): "Tonhöhenmessung mit Hilfe eines neuen automatischen Frequenz-
schreibers." Vox 16, 1-3
Zwirner E. (1929b): "Über Tonhöhenmessung und einen neuen Frequenzschreiber." Journal
für Psychologie und Neurologie 40, 99-107 [5.1.1]
Zwirner E. (1952): "Probleme der Sprachmelodie." Z. Phonetik 6, 1 [9.4.2]
Zwirner E. (1962): "Deutsches Spracharchiv 1932-1962" (Münster, Westf.) [1.3]
Zwirner E., Zwirner K. (1937): "Über Hören und Messen der Sprachmelodie." Arch. vergl.
Phonetik 1, 35-47 [5.1.1]

List of Abbreviations

Abbr.	Page	Meaning
A-D	15	Analog-to-digital
AC	351	Autocorrelation
ACF	351	Autocorrelation function
ADC	15	Analog-to-digital converter
ADS	384	Autodissimilarity
AMDF	373	Average magnitude difference function
ASDF	384	Average squared difference function
CCD	407	Charge-coupled device
DAC	15	Digital-to-analog converter
DC	232	Direct current
DF	18	Digital filter
DFT	29	Discrete Fourier transform
DL	78	Difference limen
DSP	8	Digital signal processing
EC	202	Excursion cycle
F_0	10	Fundamental frequency (frequency-domain representation of pitch)
F1	56	First formant
F2	56	Second formant
FFT	30	Fast Fourier transform
FIR	436	Finite impulse response
FS	305	Filtered signal
GWDA	254	Glottal-waveform determination algorithm
H	18	Transfer function
HF	117	High frequency
HIPEX	412	Harmonic identification pitch extraction system (Miller, 1970)
HLCF	431	Highest likely common fundamental
IDFT	400	Inverse discrete Fourier transform
IGC	255	Instant of glottal closure
IGCDA	255	Algorithm which determines the instant of glottal closure
LPC	222	Linear predictive coding (linear prediction)
LSI	490	Large scale integration
MDPML	454	Multidimensional pseudo-maximum-likelihood (Friedman, 1978)
ML	446	Maximum likelihood

Abbr.	Page	Meaning
MML	460	Modified maximum likelihood
NLF	171	Nonlinear function
P_s	39	Subglottal pressure
PARCOR	229	Partial correlation (in linear prediction)
PDA	11	Pitch determination algorithm
PDD	11	Pitch determination device
PDI	11	Pitch determination instrument
PDS	506	Pitch determination system (combination of PDI and PDA)
PFS	303	Prefiltered signal
PLL	179	Phase-locked loop
PMC	335	Preliminary marker candidate
PML	450	Pseude-maximum-likelihood (Friedman, 1977)
PR	144	Pattern recognition
RA0	326	Relative amplitude of offset (component at zero frequency)
RA1	314	Relative amplitude of first harmonic with respect to the strongest higher harmonic
RAM	361	Random access memory
RASM	333	Relative amplitude of largest spurious peak (maximum)
RC	184	Resistor-capacitor (circuit or time constant)
RE1	317	Relative enhancement of first harmonic
ROM	491	Read only memory
RS1	157	Relative amplitude of first harmonic with respect to the sum of the amplitudes of the higher harmonics
SIFT	369	Simplified inverse filter transformation
SNR	434	Signal-to-noise ratio
SPL	70	Sound pressure level
SSB	168	Single sideband
SSBM	169	Single sideband modulation
STA	35	Short-term analysis
T	9	Sampling interval (of signal)
T_0	10	Fundamental period (time-domain or lag-domain representation of pitch)
TABE	36	Threshold analysis basic extractor
TPF	304	Two-pulse filter
UPDA	311	Universal pitch determination algorithm
V-UV	133	Voiced-unvoiced
VCO	177	Voltage-controlled oscillator
VDA	12	Voicing determination algorithm
VDD	133	Voicing determination device
VLSI	490	Very large scale integration
W	30	Transformation matrix of short-term transform
ZX	208	Zero crossing(s)
ZXABE	36	Zero-crossings analysis basic extractor

Author and Subject Index

This index is a combined author and subject index. Page numbers given in boldface refer to basic definitions or detailed discussions of PDAs. The bibliography is not referenced here.

Volume 10
Digital Pattern Recognition

Editor: **K.-s. Fu**
2nd corrected and updated edition. 1980. 59 figures,
7 tables. XI, 234 pages. ISBN 3-540-10207-8

Contents: *K.-s. Fu:* Introduction. – *T. M. Cover,
T. J. Wagner:* Topics in Statistical Pattern Recognition. – *E. Diday, J. C. Simon:* Clustering Analysis. –
K.-s. Fu: Syntactic (Linguistic) Pattern Recognition. –
A. Rosenfeld, J. S. Weszka: Picture Recognition. –
J. J. Wolf: Speech Recognition and Understanding. –
K.-s. Fu, A. Rosenfeld, J. J. Wolf: Recent Developments in Digital Pattern Recognition. – Subject
Index.

Volume 14
Syntactic Pattern Recognition, Application

Editor: **K.-s Fu**
1977. 135 figures, 19 tables. XI, 270 pages
ISBN 3-540-07841-X

Contents: *K.-s. Fu:* Introduction to Syntactic Pattern
Recognition. – *S. L. Horowitz:* Peak Recognition in
Waveforms. – *J. E. Albus:* Electrocardiogram
Interpretation Using a Stochastic Finite State Model.
– *R. DeMori:* Syntactic Recognition of Speech Patterns. – *W. W. Stallings:* Chinese Character Recognition. – *H.-Y. F. Feng, T. Pavlidis:* Shape Discrimination. – *R. H. Anderson:* Two-Dimensional Mathematical Notation. – *K.-s. Fu, B. Moayer:* Fingerprint Classification. – *J. M. Brayer, P. H. Swain:* Modeling of
Earth Resources Satellite Data. – *T. Vámos:* Industrial Objects and Machine Parts Recognition.

Volume 15: **P. Kümmel**
Formalization of Natural Languages

1979. 62 figures, 5 tables. X, 223 pages
ISBN 3-540-08271-9

Contents: Historical Survey on Formalization Efforts
of Natural Languages. – Formalizing Stimuli by
Understanding Brain Functions in Living Organisms. – Analyses of Natural Language Morphology.
– Syntheses and Formalization of Natural Language
Morphology. – Analyses of Natural Language's Syntax. – Syntheses and Formalization of Natural Language Syntax. – Analyses of Natural Language Content. – Syntheses and Formalization of Natural Language Content. – Application of Natural Language
Formalizations.

Springer-Verlag
Berlin
Heidelberg
New York
Tokyo